T0189278

Communications in Computer and Information Science 610

Commenced Publication in 2007
Founding and Former Series Editors:
Alfredo Cuzzocrea, Dominik Ślęzak, and Xiaokang Yang

More information about this series at http://www.springer.com/series/7899

Joao Paulo Carvalho · Marie-Jeanne Lesot
Uzay Kaymak · Susana Vieira
Bernadette Bouchon-Meunier
Ronald R. Yager (Eds.)

Information Processing and Management of Uncertainty in Knowledge-Based Systems

16th International Conference, IPMU 2016
Eindhoven, The Netherlands, June 20–24, 2016
Proceedings, Part I

Editors
Joao Paulo Carvalho
INESC-ID, Instituto Superior Técnico
Universidade de Lisboa
Lisboa
Portugal

Susana Vieira
IDMEC, Instituto Superior Técnico
Universidade de Lisboa
Lisboa
Portugal

Marie-Jeanne Lesot
LIP6
Université Pierre et Marie Curie
Paris
France

Bernadette Bouchon-Meunier
LIP6
Université Pierre et Marie Curie, CNRS
Paris
France

Uzay Kaymak
School of Industrial Engineering
Eindhoven University of Technology
Eindhoven
The Netherlands

Ronald R. Yager
Machine Intelligence Institute
Iona College
New Rochelle, NY
USA

ISSN 1865-0929 ISSN 1865-0937 (electronic)
Communications in Computer and Information Science
ISBN 978-3-319-40595-7 ISBN 978-3-319-40596-4 (eBook)
DOI 10.1007/978-3-319-40596-4

Library of Congress Control Number: 2016941088

Printed on acid-free paper

This Springer imprint is published by Springer Nature
The registered company is Springer International Publishing AG Switzerland

Preface

These are the proceedings of the 16th International Conference on Information Processing and Management of Uncertainty in Knowledge-Based Systems, IPMU 2016. The conference was held during June 20–24, 2016, in Eindhoven, The Netherlands: one of the vibrant hi-tech hot spots of Europe. The IPMU conference is organized every two years with the aim of bringing together scientists working on methods for the management of uncertainty and aggregation of information in intelligent systems.

Since 1986, the IPMU conference has been providing a forum for the exchange of ideas between theoreticians and practitioners working in these areas and related fields. In addition to the many contributed scientific papers, the conference has in the past attracted prominent plenary speakers, including the Nobel Prize winners Kenneth Arrow, Daniel Kahneman, and Ilya Prigogine. Another important feature of the conference is the presentation of the Kampé de Fériet Award for outstanding contributions to the field of uncertainty and management of uncertainty. Past winners of this prestigious award were Lotfi A. Zadeh (1992), Ilya Prigogine (1994), Toshiro Terano (1996), Kenneth Arrow (1998), Richard Jeffrey (2000), Arthur Dempster (2002), Janos Aczel (2004), Daniel Kahneman (2006), Enric Trillas (2008), James Bezdek (2010), Michio Sugeno (2012), and Vladimir N. Vapnik (2014). This year, the recipient was Joseph Y. Halpern from Cornell University, USA.

IPMU 2016 had a rich scientific program. Four invited overview talks (tutorials) were given on the first day, identifying the challenges and discussing the various methods in the field of information processing and the management of uncertainty. Further, the program consisted of five invited plenary talks, 13 special sessions, 127 contributed papers that were authored by researchers from 34 different countries, industry round tables, and discussion panels. The plenary presentations were given by the following distinguished researchers: Chris Dyer (Carnegie Mellon University, USA), Joseph Y. Halpern (Cornell University, USA), Katharina Morik (Technische Universität Dortmund, Germany), Peter P. Wakker (Erasmus University Rotterdam, The Netherlands), and Ronald R. Yager (Iona College, USA). All contributed papers underwent the same review process and were judged by at least two reviewers; 90 % of the papers were reviewed by three or more referees, and some papers by as many as five referees. Furthermore, all papers were scrutinized by the program chairs, meaning that each paper was studied by three to six independent researchers. The review process also respected the usual conflict-of-interest standards, so that all papers received blinded, independent evaluations.

Organizing a conference like IPMU 2016 is not possible without the assistance, dedication, and support of many people and institutions. We want to thank our industry sponsors, the institutional sponsors, and the material sponsors. Our sponsor chair, Paul Grefen, did an excellent job in attracting the interest and support from industry for the success of IPMU 2016. We are also particularly grateful to the organizers of sessions on dedicated topics that took place during the conference—these special sessions have always been a characteristic element of the IPMU conference. Special thanks go to Joao Sousa, who helped evaluate and select the special session proposals. The help of the members of the international Program Committee as well as multiple reviewers was essential in safeguarding the scientific quality of the conference. The local Organizing Committee is very grateful for the efforts of multiple student volunteers who provided practical support during the conference.

Finally, we gratefully acknowledge the technical support of several organizations and institutions, notably the IEEE Computational Intelligence Society, the European Society for Fuzzy Logic and Technology (EUSFLAT), and the Netherlands Research School for Information and Knowledge Systems (SIKS). Last, but not least, our greatest gratitude goes to the authors who submitted their work and presented it at the conference!

April 2016

Rui J. Almeida
Joao Paulo Carvalho
Marie-Jeanne Lesot
Anna M. Wilbik
Bernadette Bouchon-Meunier
Uzay Kaymak
Susana Vieira
Ronald R. Yager

Organization

General Chair

Uzay Kaymak	Technische Universiteit Eindhoven, The Netherlands

Program Chairs

Joao Paulo Carvalho	INESC-ID, Instituto Superior Técnico, Universidade de Lisboa, Portugal
Marie-Jeanne Lesot	LIP6 - Université Pierre et Marie Curie, Paris, France

Executive Directors

Bernadette Bouchon-Meunier	LIP6 - Université Pierre et Marie Curie, CNRS, Paris, France
Ronald R. Yager	Iona College, USA

Finance Chair

Anna Wilbik	Technische Universiteit Eindhoven, The Netherlands

Publicity Chair

Rui Jorge de Almeida	Technische Universiteit Eindhoven, The Netherlands

Special Session Chair

João M.C. Sousa	IDMEC, Instituto Superior Técnico, Universidade de Lisboa, Portugal

Publication Chair

Susana Vieira	IDMEC, Instituto Superior Técnico, Universidade de Lisboa, Portugal

Sponsor Chair

Paul W.P.J. Grefen	Technische Universiteit Eindhoven, The Netherlands

Local Organizing Committee

Uzay Kaymak	Technische Universiteit Eindhoven, The Netherlands
Rui Jorge de Almeida	Technische Universiteit Eindhoven, The Netherlands
Anna Wilbik	Technische Universiteit Eindhoven, The Netherlands
Annemarie van der Aa	Technische Universiteit Eindhoven, The Netherlands
Caro Fuchs	Technische Universiteit Eindhoven, The Netherlands

International Advisory Board

Giulianella Coletti, Italy
Miguel Delgado, Spain
Mario Fedrizzi, Italy
Laurent Foulloy, France
Salvatore Greco, Italy
Julio Gutierrez-Rios, Spain
Eyke Hüllermeier, Germany
Anne Laurent, France
Luis Magdalena, Spain
Christophe Marsala, France
Benedetto Matarazzo, Italy
Manuel Ojeda-Aciego, Spain
Maria Rifqi, France
Lorenza Saitta, Italy
Olivier Strauss, France
Enric Trillas, Spain
Llorenç Valverde, Spain
José Luis Verdegay, Spain
Maria-Amparo Vila, Spain
Lotfi A. Zadeh, USA

Program Committee

Rui Jorge de Almeida
Michał Baczyński
Nalan Bastürk
Gleb Beliakov
Radim Belohlavek
Salem Benferhat
Jim Bezdek
Isabelle Bloch
Ulrich Bodenhofer
Piero Bonissone
Humberto Bustince
Guoqing Chen
Esma Nur Cinicioglu
Carlos Coello Coello
Oscar Cordon
Ana Colubi
Didier Coquin
Inés Couso
Keeley Crockett
Fabio Cuzzolin
Bernard De Baets
Guy De Tré
Sébastien Destercke
Marcin Detyniecki
Antonio Di Nola
Remco Dijkman
Didier Dubois
Fabrizio Durante
Francesc Esteva
Janos Fodor
David Fogel
Sylvie Galichet
Patrick Gallinari
Maria Angeles Gil
Lluis Godo
Fernando Gomide
Gil González Rodríguez
Michel Grabisch
Steve Grossberg
Przemysław Grzegorzewski
Lawrence Hall
Francisco Herrera
Enrique Herrera-Viedma
Ludmilla Himmelspach
Kaoru Hirota
Janusz Kacprzyk
Uzay Kaymak
Cengiz Kahraman
Abraham Kandel
James Keller
Frank Klawonn
Erich Peter Klement
Laszlo Koczy
Vladik Kreinovich
Tomas Kroupa
Rudolf Kruse
Christophe Labreuche
Jérome Lang
Henrik Larsen
Mark Last
Weldon Lodwick
Edwin Lughofer

Jean-Luc Marichal
Trevor Martin
Sebastian Massanet
Mylène Masson
Silvia Massruhá
Gilles Mauris
Gaspar Mayor
Jerry Mendel
Radko Mesiar
Ralf Mikut
Enrique Miranda
Javier Montero
Jacky Montmain
Serafín Moral
Zbigniew Nahorski
Yusuke Nojima
Vilem Novak
Hannu Nurmi
Nikhil Pal
Endre Pap
Simon Parsons
Gabriella Pasi
Witold Pedrycz
Irina Perfilieva
Fred Petry
Vincenzo Piuri
Olivier Pivert
Henri Prade
Anca Ralescu
Dan Ralescu
Mohammed Ramdani
Marek Reformat
Adrien Revault d'Allonnes
Beloslav Riecan
Maria Dolores Ruiz
Thomas Runkler
Enrique Ruspini
Daniel Sanchez
Mika Sato-Ilic
Glen Shafer
Roman Słowinski
Grégory Smits
Joao Sousa
Pilar Sobrevilla
Martin Stepnicka
Umberto Straccia
Michio Sugeno
Eulalia Szmidt
Kay-Chen Tan
Bruno Teheux
Settimo Termini
Konstantin Todorov
Vicenc Torra
I. Burhan Turksen
Bülent Tütmez
Linda van der Gaag
Herman van Dijk
Barbara Vantaggi
Michel Verleysen
Thomas Vetterlein
Susana Vieira
Anna Wilbik
Sławomir Zadrożny
Hans-Jürgen Zimmermann
Jacek Zurada

Additional Reviewers

Cristina Alcalde
Alessandro Antonussi
Louis Aslett
Edurne Barrenechea
Maciej Bartoszuk
Fernando Batista
Benjamin Bedregal
Libor Behounek
Gleb Beliakov
Sarra Ben Hariz
Chiheb-Eddine Ben N'Cir
Magdalena Bendova
Mara José Benítez Caballero
Hanen Borchani
Felix Bou
Reda Boukezzoula
Denis Bouyssou
Yamine Bouzembrak
Christian Braune
Alberto Bugarìn
Ana Burusco
Inma Cabrera
Marta Cardin
Anna Cena
Alireza Chakeri
Mouna Chebbah
Petr Cintula
Maria Eugenia Cornejo Piñero
Miguel Couceiro
Maria Jose Del Jesus
Cyril De Runz
Denisa Diaconescu
Graaliz Dimuro
Alexander Dockhorn
Christoph Doell
Paweł Drygaś
Zied Elouedi
Javier Fernandez

Tommaso Flaminio
María Angeles Galán García
Marek Gagolewski
Juan Gomez Romero
Petra Hodakova
Aoi Honda
Petr Hurtik
Johan Jacquemin
Simon James
Balasubramaniam Jayaram
Ilyes Jenhani
Chee Kau Lim
Frank Klawonn
David Lobo
Nicolas Madrid
Enrico Marchioni
Nicolas Marin
Arnaud Martin
Maria J. Martin-Bautista
Brice Mayag
Jesús Medina
Thuraya Mellah
Jos M. Merigó
Pedro Miranda
Gildas Morvan
Mirko Navara
Tatiane Nogueira Rios
Michael Oberguggenberger
Pere Pardo
Barbara Pękala
Davide Petturiti
Marc Pirlot
Badran Raddaoui
Eloisa Ramírez Poussa
Alejandro Ramos
Jordi Recasens
Ricardo Ribeiro
Juan Vicente Riera
Ricardo Oscar Rodriguez
Jonas Rogger
Christoph Roschger
Daniel Ruiz-Aguilera
Lorenza Saitta
Ahmed Samet
Laura Schnüriger
Karima Sedki
Jose-Maria Serrano
Prakash Shenoy
Andrzej Skowron
Alexander Sostak
Jana Spirkova
Eiichiro Takahagi
Joan Torrens
Gracian Trivino
Esko Turunen
Lev Utkin
Lionel Valet
Llorenc Valverde
Francisco J. Valverde Albacete
Jan Van den Berg
Amanda Vidal Wandelmer
Zdenek Wagner
Gero Walter

Special Session Organizers

Mohamed Anis Bach Tobji	University of Mannouba, Tunisia
Humberto Bustince	Public University of Navarra, Spain
Mouna Chebbah	University of Jendouba, Tunisia
Marek Gagolewski	Polish Academy of Sciences, Warsaw, Poland
Lluis Godo	Artificial Intelligence Research Institute IIIA, Spain
Maria J. Martin-Bautista	University of Granada, Spain
Daniel Sánchez	University of Granada, Spain
Bruno Teheux	University of Luxembourg, Luxembourg
Michał Baczynski	University of Silesia, Poland

Gleb Beliakov	School of Information Technology, Melbourne, Australia
Boutheina Ben Yaghlane	University of Carthage, Tunisia
Didier Coquin	LISTIC - Polytech Annecy-Chambéry, France
Bernard De Baets	Ghent University, Belgium
Sébastien Destercke	Université de Technologie de Compiègne, France
Juan Gómez Romero	University of Granada, Spain
Michel Grabisch	University Paris Sorbonne, Paris, France
Christophe Labreuche	Thales Research and Technology, Palaiseau, France
Sebastia Massanet	University of the Balearic Islands, Spain
Jesús Medina	University of Cádiz, Spain
Javier Montero	Universidad Complutense de Madrid, Spain
Manuel Ojeda Aciego	University of Málaga, Spain
Irina Perfilieva	University of Ostrava, Czech Republic
M. Dolores Ruiz	University of Granada, Spain
Gero Walter	Eindhoven University of Technology, The Netherlands

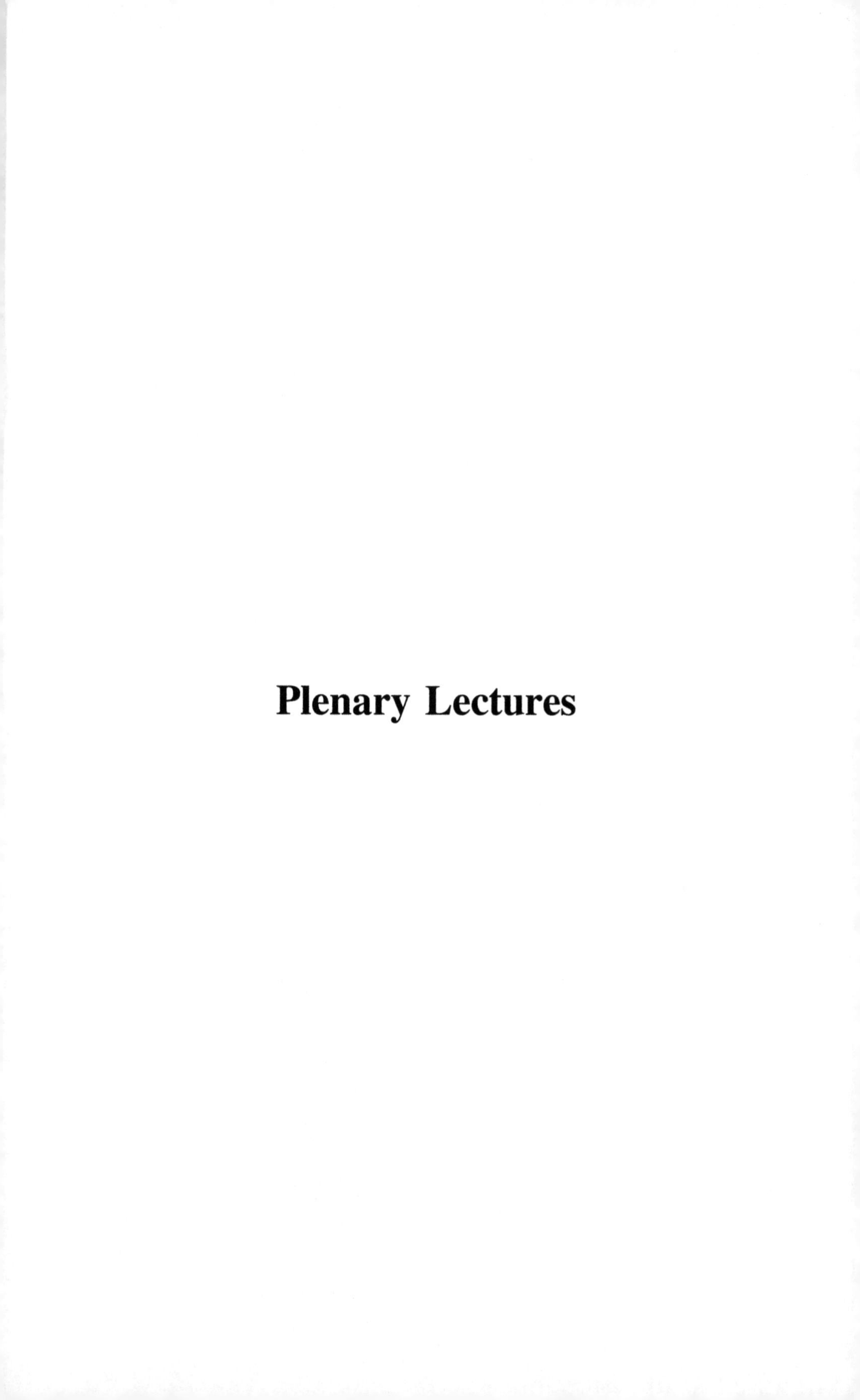

Plenary Lectures

Actual Causality: A Survey

Joseph Y. Halpern

Cornell University, Computer Science Department, 414 Gates Hall, Ithaca, NY 14853, USA
halpern@cs.cornell.edu

Abstract. What does it mean that an event C "actually caused" event E? The problem of defining actual causation goes beyond mere philosophical speculation. For example, in many legal arguments, it is precisely what needs to be established in order to determine responsibility. (What exactly was the actual cause of the car accident or the medical problem?) The philosophy literature has been struggling with the problem of defining causality since the days of Hume, in the 1700s. Many of the definitions have been couched in terms of counterfactuals. (C is a cause of E if, had C not happened, then E would not have happened.) In 2001, Judea Pearl and I introduced a new definition of actual cause, using Pearl's notion of structural equations to model counterfactuals. The definition has been revised twice since then, extended to deal with notions like "responsibility" and "blame", and applied in databases and program verification. I survey the last 15 years of work here, including joint work with Judea Pearl, Hana Chockler, and Chris Hitchcock. The talk will be completely self-contained.

Biography Joseph Halpern received a B.Sc. in mathematics from the University of Toronto in 1975 and a Ph.D. in mathematics from Harvard in 1981. In between, he spent two years as the head of the Mathematics Department at Bawku Secondary School, in Ghana. After a year as a visiting scientist at MIT, he joined the IBM Almaden Research Center in 1982, where he remained until 1996, also serving as a consulting professor at Stanford. In 1996, he joined the CS Department at Cornell, and was department chair 2010-14.

Halpern's major research interests are in reasoning about knowledge and uncertainty, security, distributed computation, decision theory, and game theory. He is a Fellow of AAAI, AAAS (American Association for the Advancement of Science), the American Academy of Arts and Sciences, ACM, IEEE, and SEAT (Society for the Advancement of Economic Theory). Among other awards, he received the ACM SIGART Autonomous Agents Research Award in 2011, the Dijkstra Prize in 2009, the ACM/AAAI Newell Award in 2008, the Godel Prize in 1997, was a Guggenheim Fellow in 2001-02, and a Fulbright Fellow in 2001-02 and 2009-10. Two of his papers have won best-paper prizes at IJCAI (1985 and 1991), and another two received best-paper awards at the Knowledge Representation and Reasoning Conference (2006 and 2012). He was editor-in-chief of the Journal of the ACM (1997-2003) and has been program chair of a number of conferences.

The Present State of the Art of Modeling Uncertainty in Decision Theory, Resulting from an Interaction between Mathematical Economists and Empirical Psychologists

Peter P. Wakker

Econometric Institute, Erasmus University, Rotterdam, The Netherlands
wakker@ese.eur.nl

Abstract. In decision theory, more than in other fields of IPMU, the modeling of uncertainty is driven by empirical findings about human behavior. Decision theorists are strict in the requirement that for every mathematical detail the empirical meaning must be exactly specified. For example, taking the lower bound of possible probabilities of an event, while accepted uncritically in most information management theories, is meaningless to a decision theorist until it has been specified whether the event in question yields good or bad outcomes.

This lecture describes how the current state of the art in uncertainty-decision theory could only come about from interactions between empirically oriented psychologists and mathematically oriented economists. At several stages in history, the next step forward could be made only by empirical intuitions from psychologists. Following up on that, the next step forward could be made only by theoretical inputs from economists with advanced technical skills. Modern views on the proper modeling of uncertainty attitudes could only arise from the merger of ideas from all the fields mentioned. It, for instance, led to a measure of information-insensitivity that is more refined than just taking supremums or infimums of uncertainty measures.

Biography Peter Wakker is a professor of decisions under uncertainty at Erasmus School of Economics of the Erasmus University Rotterdam. He works in behavioral economics, primarily on the differences between normative and descriptive decisions, and on decisions under risk and uncertainty. Wakker has published in leading journals in economics, business, medicine, psychology, statistics, and mathematics. He was nominated the best-publishing Dutch economist in the years 1994, 1998, 2003, and 2007, and was ranked 90th in the world in the ISI's most cited scientists in economics and business in 2003. He received a Frank P. Ramsey Medal in 2013 and the Medical Decision Making Career Achievement Award in 2007. Wakker regularly gives advices on insurance in the media. Wakker is director, jointly with Professor Han Bleichrodt, of the research group Behavioral Economics.

Decision Making with Multi-criteria

Ronald R. Yager

Machine Intelligence Institute, Iona College
yager@panix.com

Abstract. The construction of multi-criteria decision functions is strongly dependent upon the use of aggregation operators. Here if $D(x) = Agg(C_1(x), C_2(x), \ldots, C_n(x))$ represents the satisfaction of alternative x to the collection of criteria a central problem becomes the formulation of the decision function D. The structure of the function Agg must be a reflection of the decision makers perceived relationship between the different criteria. We must provide some approaches that can used to help in the construction of these decision functions. One approach is to allow the decision-maker to express their perceived relationship between the criteria in a linguistic like manner and then try to model this relationship using fuzzy logic formalisms. Another approach is the use of set measures for the representation of the relationship between criteria. Once having a formal representation of the decision function D we must evaluate it for each alternative. In many real world environments the values of the $C_j(x)$ can only be provided with some uncertainty. Among the different types of imprecise valuations are intervals, probability distributions, D-S belief structures, fuzzy sets, intuitionistic, Pythagorean and generalized orthopair fuzzy sets as well ordinallinguistic valuations. Finally we must choose among these alternatives based their values for $D(x)$. In the case of uncertainty in the $C_j(x)$ the value of $D(x)$ also manifests uncertainty. Choosing requires that we provide an ordering of these uncertain values. In our talk we shall discuss various topics from the above.

Biography Ronald R. Yager is Director of the Machine Intelligence Institute and Professor of Information Systems at Iona College. He is editor and chief of the International Journal of Intelligent Systems. He has published over 500 papers and edited over 30 books in areas related to fuzzy sets, human behavioral modeling, decision-making under uncertainty and the fusion of information. He is among the worlds top 1 % most highly cited researchers with over 45,000 citations in Google Scholar. He was the recipient of the IEEE Computational Intelligence Society Pioneer award in Fuzzy Systems. He received the special honorary medal of the 50-th Anniversary of the Polish Academy of Sciences. He received the Lifetime Outstanding Achievement Award from International the Fuzzy Systems Association. He recently received honorary doctorate degrees, honoris causa, from the Azerbaijan Technical University and the State University of Information Technologies, Sofia Bulgaria. Dr. Yager is a fellow of the IEEE, the New York Academy of Sciences and the Fuzzy Systems Association. He has served at the National Science Foundation as program director in the Information Sciences program. He was a NASA/Stanford visiting fellow and a research associate at the University of California, Berkeley. He has been a lecturer at NATO Advanced Study Institutes. He was a program director at the National Science Foundation. He is a visiting distinguished scientist at King Saud University,

Riyadh Saudi Arabia. He was an adjunct professor at Aalborg University in Denmark. He received his undergraduate degree from the City College of New York and his Ph. D. from the Polytechnic Institute New York University. He is the 2016 recipient of the IEEE Frank Rosenblatt Award the most prestigious honor given out by the IEEE Computational Intelligent Society.

Resource-Constrained Data Analysis and Exploration

Katharina Morik

Faculty for Computer Science, Artificial Intelligence Group,
TU Dortmund University, Dortmund, Germany
katharina.morik@tu-dortmund.de

Abstract. Computer science has always taken into account some resources needed for the execution of algorithms, namely runtime and memory space. Since the triumph of very large data centers, *energy* has become a resource of importance, additionally. In 2008, Google had its millionth server. Google's estimated yearly energy consumption is about 2024 watt hours (Wh). A search request consumes 0.3 Wh, asking and reading the result at a home computer consumes about the same, so that each query costs about 0.6 Wh[1].

Where data centers challenge resources at a global scale, the energy of cyber-physical systems and smartphones is restricted at the local device. The battery of a smartphone has a capacity of about 8 Wh. The user wants a long battery duration together with a high quality of service. Regarding machine learning, there are two ways, in which energy may be saved. On the one hand, a learning algorithm may learn from compiler logs[2] or from user behavior[3] how to enhance the heuristics of the system's software. On the other hand, the learning algorithm itself has to become energy-efficient. This can be achieved through approximations which reduce the operations that cost the most energy[4].

Cyber-physical systems populate diverse parts of our everyday life, they are the nodes of the Internet of Things and they produce big data. If we focus again on smartphones, each user generates about 60 GB of data per year. Learning a personal model of app usage could allow early warnings when to recharge the battery[5]. However, the analysis of such data is not easy: data may be missing, their incompleteness is not easy to recognize, and they may be wrong due to several reasons. Labels, which are needed for classifier training, are missing.

---

[1] E. Gelenbe, Y. Caseau (2015) The impact of information technology on energy consumption and carbon emissions, in: *Ubiquity*, June, 1–15

[2] P. Lokuciejewski, M. Stolpe, K. Morik, P. Marwedel (2010) Automatic Selection of Machine Learning Models for WCET-aware Compiler Heuristic Generation, in: 4th Workshop on Statistical and Machine Learning Approaches to ARchitecture and compilaTion (SMART)

[3] P. Fricke, F. Jungermann, K.Morik, N. Piatkowski, O. Spinczyk, M. Stolpe (2010) Towards Adjusting Mobile Devices to User's Behaviour, in: Intern. Workshop at ECML PKDD on Mining Ubiquitous and Social Environments

[4] N. Piatkowski, S. Lee, K. Morik (2016) Integer undirected graphical models for resource-constrained systems, in: *Neurocomputing*, 173(1), 9–23

[5] N. Piatkowski, S. Lee, K. Morik (2013) Spatio-Temporal Random Fields: Compressible Representation and Distributed Estimation, in: Machine Learning Journal 93(1), 115–139

Data exploration is an important, though often under-estimated first part of data analysis.

In the talk, several probabilistic graphical models will be presented together with their applications.

Biography Katharina Morik is full professor for computer science at the TU Dortmund University, Germany. She earned her Ph.D. (1981) at the University of Hamburg and her habilitation (1988) at the TU Berlin. Starting with natural language processing, her interest moved to machine learning ranging from inductive logic programming to statistical learning, then to the analysis of very large data collections, high-dimensional data, and resource awareness. She is a member of the National Academy of Science and Engineering and the North-Rhine- Westphalia Academy of Science and Art. She is the author of more than 200 papers in well acknowledged conferences and journals. Her latest results include spatio-temporal random fields and integer Markov random fields, both allowing for complex graphical models under resource constraints. Her interest in interdisciplinary research covers a large variety of fields. She successfully collaborated with linguists, engineers, physicians, and astrophysicists.

She was one of those starting the IEEE International Conference on Data Mining together with Xindong Wu, and was chairing the program of this conference in 2004. She was the program chair of the European Conference on Machine Learning (ECML) in 1989 and one of the program chairs of ECML PKDD 2008. She is in the editorial boards of the international journals Knowledge and Information Systems and Data Mining and Knowledge Discovery.

Her aim to share scientific results strongly supports open source developments. For instance, the first efficient implementation of the support vector machine, SVM_light, was developed at her lab by Thorsten Joachims. Also the leading data mining platform RapidMiner started out at her lab, which continues to contribute to it. Currently, the Java streams framework is developed, which abstracts processes on distributed data streams.

Since 2011, she is leading the collaborative research center SFB876 on data analysis under resource-constraints, an interdisciplinary center comprising 14 projects, 20 professors, and about 50 Ph.D. students or Postdocs.

Learning Representations of Complex Structures in Natural Language with Neural Networks

Chris Dyer

Machine Learning Department, Carnegie Mellon University, Pittsburgh, USA
cdyer@cs.cmu.edu

Abstract. Effective processing of natural language requires integrating information from a variety of sources: an individual word's meaning depends on the context it is used in; the proper interpretation of a sentence depends on understanding the discursive context it occurs in; and, reasoning about the truth of a linguistically encoded proposition requires drawing on world knowledge. However, if we take stock of what progress has been made in language processing applications to date, it is precisely those that depend on a narrow view of context rather than those that require significant integration of contextual information where we find the most success.

In this talk I argue that the challenge of developing next-generation models that are sensitive to broader contextual information can be helpfully cast as a representation learning problem. Given a basic representation of the input signal and relevant contextual information, a unified representation suitable for making predictions needs to be computed. I discuss work from my group on using neural networks to integrate basic representations of component linguistic elements and combining them recursively to obtain composite representations of complex objects. Our work has demonstrated that taking inspiration from the linguistic structures when designing architectures is more effective than task-agnostic architectures. Applications ranging from text categorization, to language modeling, to machine translation will be discussed.

Biography Chris Dyer is an assistant professor at Carnegie Mellon University. Dyer graduated from the Duke University in 2000, where he studied computer science. He went on to obtain a Ph.D. in linguistics in 2010 from University of Maryland under the supervision of Prof. Philip Resnik. Chris Dyer's research interests line in the intersection of statistical machine translation, unsupervised learning, computational morphology and phonology, large-scale data processing, probabilistic models of natural language processing, Bayesian techniques and machine learning. He is currently supported by grants from The National Science Foundation (Lexical Borrowing), DARPA (LORELEI), Google (A Hybrid Neural Phrase-Based Model for Machine Translation) and The Army Research Office (MT/NLP for Low-Resource Languages).

Learning Representations of Complex Structures in Natural Language with Neural Networks

Contents – Part I

Fuzzy Measures and Integrals

Decomposition Integral Based Generalizations of OWA Operators 3
Radko Mesiar and Andrea Stupňanová

Finding the Set of k-additive Dominating Measures Viewed as a Flow Problem 11
Pedro Miranda and Michel Grabisch

On Capacities Characterized by Two Weight Vectors 23
Christophe Labreuche

Computing Superdifferentials of Lovász Extension with Application to Coalitional Games 35
Lukáš Adam and Tomáš Kroupa

Conjoint Axiomatization of the Choquet Integral for Heterogeneous Product Sets 46
Mikhail Timonin

Aggregation of Choquet Integrals 58
Radko Mesiar, Ladislav Šipeky, and Alexandra Šipošová

Inclusion-Exclusion Integral and t-norm Based Data Analysis Model Construction 65
Aoi Honda and Yoshiaki Okazaki

Fuzzy Integral for Rule Aggregation in Fuzzy Inference Systems 78
Leary Tomlin, Derek T. Anderson, Christian Wagner, Timothy C. Havens, and James M. Keller

On a Fuzzy Integral as the Product-Sum Calculation Between a Set Function and a Fuzzy Measure 91
Eiichiro Takahagi

A 2-Additive Choquet Integral Model for French Hospitals Rankings in Weight Loss Surgery 101
Brice Mayag

Benchmarking over Distributive Lattices 117
Marta Cardin

Uncertainty Quantification with Imprecise Probability

Efficient Simulation Approaches for Reliability Analysis of Large Systems. . . . 129
Edoardo Patelli and Geng Feng

Bivariate *p*-boxes and Maxitive Functions 141
Ignacio Montes and Enrique Miranda

Sets of Priors Reflecting Prior-Data Conflict and Agreement 153
Gero Walter and Frank P.A. Coolen

On Imprecise Statistical Inference for Accelerated Life Testing. 165
Frank P.A. Coolen, Yi-Chao Yin, and Tahani Coolen-Maturi

The Mathematical Gnostics *(Advanced Data Analysis)*. 177
Pavel Kovanic

Textual Data Processing

The Role of Graduality for Referring Expression Generation in Visual Scenes 191
Albert Gatt, Nicolás Marín, François Portet, and Daniel Sánchez

Impact of the Shape of Membership Functions on the Truth Values of Linguistic Protoform Summaries 204
Akshay Jain, Tianqi Jiang, and James M. Keller

A Solution of the Multiaspect Text Categorization Problem by a Hybrid HMM and LDA Based Technique. 214
Sławomir Zadrożny, Janusz Kacprzyk, and Marek Gajewski

How Much Is "About"? Fuzzy Interpretation of Approximate Numerical Expressions 226
Sébastien Lefort, Marie-Jeanne Lesot, Elisabetta Zibetti, Charles Tijus, and Marcin Detyniecki

Towards a Non-oriented Approach for the Evaluation of Odor Quality 238
Massissilia Medjkoune, Sébastien Harispe, Jacky Montmain, Stéphane Cariou, Jean-Louis Fanlo, and Nicolas Fiorini

Belief Function Theory and Its Applications

Joint Feature Transformation and Selection Based on Dempster-Shafer Theory 253
Chunfeng Lian, Su Ruan, and Thierry Denœux

Recognition of Confusing Objects for NAO Robot 262
Thanh-Long Nguyen, Didier Coquin, and Reda Boukezzoula

Evidential Missing Link Prediction in Uncertain Social Networks 274
Sabrine Mallek, Imen Boukhris, Zied Elouedi, and Eric Lefevre

An Evidential Filter for Indoor Navigation of a Mobile Robot in Dynamic Environment 286
Quentin Labourey, Olivier Aycard, Denis Pellerin, Michèle Rombaut, and Catherine Garbay

A Solution for the Learning Problem in Evidential (Partially) Hidden Markov Models Based on Conditional Belief Functions and EM. 299
Emmanuel Ramasso

Graphical Models

Determination of Variables for a Bayesian Network and the Most Precious One 313
Esma Nur Cinicioglu and Taylan Yenilmez

Incremental Junction Tree Inference 326
Hamza Agli, Philippe Bonnard, Christophe Gonzales, and Pierre-Henri Wuillemin

Real Time Learning of Non-stationary Processes with Dynamic Bayesian Networks 338
Matthieu Hourbracq, Pierre-Henri Wuillemin, Christophe Gonzales, and Philippe Baumard

Fuzzy Implication Functions

About the Use of Admissible Order for Defining Implication Operators 353
Maria Jose Asiain, Humberto Bustince, Benjamin Bedregal, Zdenko Takáč, Michal Baczyński, Daniel Paternain, and Graçaliz Dimuro

Generalized Sugeno Integrals 363
Didier Dubois, Henri Prade, Agnès Rico, and Bruno Teheux

A New Look on Fuzzy Implication Functions: *FNI*-implications 375
Isabel Aguiló, Jaume Suñer, and Joan Torrens

On a Generalization of the Modus Ponens: *U*-conditionality 387
Margalida Mas, Miquel Monserrat, Daniel Ruiz-Aguilera, and Joan Torrens

A New Look on the Ordinal Sum of Fuzzy Implication Functions. 399
Sebastia Massanet, Juan Vicente Riera, and Joan Torrens

Distributivity of Implication Functions over Decomposable Uninorms Generated from Representable Uninorms in Interval-Valued Fuzzy Sets Theory 411
Michał Baczyński and Wanda Niemyska

On Functions Derived from Fuzzy Implications 423
Przemysław Grzegorzewski

Applications in Medicine and Bioinformatics

Non-commutative Quantales for Many-Valuedness in Applications 437
Patrik Eklund, Ulrich Höhle, and Jari Kortelainen

Evaluating Tests in Medical Diagnosis: Combining Machine Learning with Game-Theoretical Concepts. 450
Karlson Pfannschmidt, Eyke Hüllermeier, Susanne Held, and Reto Neiger

Fuzzy Modeling for Vitamin B12 Deficiency 462
Anna Wilbik, Saskia van Loon, Arjen-Kars Boer, Uzay Kaymak, and Volkher Scharnhorst

Real-World Applications

Using Geographic Information Systems and Smartphone-Based Vibration Data to Support Decision Making on Pavement Rehabilitation 475
Chun-Hsing Ho, Chieh-Ping Lai, and Anas Almonnieay

Automatic Synthesis of Fuzzy Inference Systems for Classification 486
Jorge Paredes, Ricardo Tanscheit, Marley Vellasco, and Adriano Koshiyama

A Proposal for Modelling Agrifood Chains as Multi Agent Systems 498
Madalina Croitoru, Patrice Buche, Brigitte Charnomordic, Jerome Fortin, Hazael Jones, Pascal Neveu, Danai Symeonidou, and Rallou Thomopoulos

Predictive Model Based on the Evidence Theory for Assessing Critical Micelle Concentration Property 510
Ahmed Samet, Théophile Gaudin, Huiling Lu, Anne Wadouachi, Gwladys Pourceau, Elisabeth Van Hecke, Isabelle Pezron, Karim El Kirat, and Tien-Tuan Dao

Fuzzy Methods in Data Mining and Knowledge Discovery

An Incremental Fuzzy Approach to Finding Event Sequences. 525
Trevor P. Martin and Ben Azvine

Scenario Query Based on Association Rules (SQAR) 537
Carlos Molina, Belen Prados-Suárez, and Daniel Sanchez

POSGRAMI: Possibilistic Frequent Subgraph Mining in a Single Large Graph 549
Mohamed Moussaoui, Montaceur Zaghdoud, and Jalel Akaichi

Mining Consumer Characteristics from Smart Metering Data through Fuzzy Modelling 562
Joaquim L. Viegas, Susana M. Vieira, and João M.C. Sousa

Soft Computing for Image Processing

Approximate Pattern Matching Algorithm 577
Petr Hurtik, Petra Hodáková, and Irina Perfilieva

Image Reconstruction by the Patch Based Inpainting 588
Pavel Vlašánek and Irina Perfilieva

Similarity Measures for Radial Data 599
Carlos Lopez-Molina, Cedric Marco-Detchart, Javier Fernandez, Juan Cerron, Mikel Galar, and Humberto Bustince

Application of a Mamdani-Type Fuzzy Rule-Based System to Segment Periventricular Cerebral Veins in Susceptibility-Weighted Images 612
Francesc Xavier Aymerich, Pilar Sobrevilla, Eduard Montseny, and Alex Rovira

On the Use of Lattice OWA Operators in Image Reduction and the Importance of the Orness Measure 624
Daniel Paternain, Gustavo Ochoa, Inmaculada Lizasoain, Edurne Barrenechea, Humberto Bustince, and Radko Mesiar

A Methodology for Hierarchical Image Segmentation Evaluation 635
J. Tinguaro Rodríguez, Carely Guada, Daniel Gómez, Javier Yáñez, and Javier Montero

Higher Degree F-transforms Based on B-splines of Two Variables 648
Martins Kokainis and Svetlana Asmuss

Gaussian Noise Reduction Using Fuzzy Morphological Amoebas 660
Manuel González-Hidalgo, Sebastia Massanet, Arnau Mir, and Daniel Ruiz-Aguilera

Clustering

Proximal Optimization for Fuzzy Subspace Clustering 675
Arthur Guillon, Marie-Jeanne Lesot, Christophe Marsala, and Nikhil R. Pal

Participatory Learning Fuzzy Clustering for Interval-Valued Data 687
Leandro Maciel, Rosangela Ballini, Fernando Gomide, and Ronald R. Yager

Fuzzy *c*-Means Clustering of Incomplete Data Using Dimension-Wise Fuzzy Variances of Clusters 699
Ludmila Himmelspach and Stefan Conrad

On a Generalized Objective Function for Possibilistic Fuzzy Clustering 711
József Mezei and Peter Sarlin

Seasonal Clustering of Residential Natural Gas Consumers. 723
Marta P. Fernandes, Joaquim L. Viegas, Susana M. Vieira, and João M.C. Sousa

Author Index 735

Contents – Part II

Fuzzy Logic, Formal Concept Analysis and Rough Sets

(Ir)relevant T-norm Joint Distributions in the Arithmetic of Fuzzy Quantities 3
Andrea Sgarro and Laura Franzoi

Knowledge Extraction from L-Fuzzy Hypercontexts 12
Cristina Alcalde and Ana Burusco

A Semantical Approach to Rough Sets and Dominance-Based Rough Sets . . . 23
Lynn D'eer, Chris Cornelis, and Yiyu Yao

Graded Generalized Hexagon in Fuzzy Natural Logic 36
Petra Murinová and Vilém Novák

On a Category of Extensional Fuzzy Rough Approximation L-valued Spaces 48
Aleksandrs Eļkins, Alexander Šostak, and Ingrīda Uļjane

The Syntax of Many-Valued Relations 61
Patrik Eklund

Reduct-Irreducible α-cut Concept Lattices: An Efficient Reduction Procedure to Multi-adjoint Concept Lattices 69
M. Eugenia Cornejo, Jesús Medina, and Eloísa Ramírez-Poussa

Towards Galois Connections over Positive Semifields 81
Francisco J. Valverde-Albacete and Carmen Peláez-Moreno

Graded and Many-Valued Modal Logics

From Kripke to Neighborhood Semantics for Modal Fuzzy Logics 95
Petr Cintula, Carles Noguera, and Jonas Rogger

Łukasiewicz Public Announcement Logic 108
Leonardo Cabrer, Umberto Rivieccio, and Ricardo Oscar Rodriguez

Possibilistic Semantics for a Modal *KD*45 Extension of Gödel Fuzzy Logic . . . 123
Félix Bou, Francesc Esteva, Lluís Godo, and Ricardo Oscar Rodriguez

A Calculus for Rational Łukasiewicz Logic and Related Systems 136
Paolo Baldi

Negation of Graded Beliefs 148
Bénédicte Legastelois, Marie-Jeanne Lesot, and Adrien Revault d'Allonnes

Comparing Some Substructural Strategies Dealing with Vagueness 161
Pablo Cobreros, Paul Egré, David Ripley, and Robert van Rooij

Imperfect Databases

An Incremental Algorithm for Repairing Training Sets with Missing Values . . . 175
Bas van Stein and Wojtek Kowalczyk

Analysis and Visualization of Missing Value Patterns 187
Bas van Stein, Wojtek Kowalczyk, and Thomas Bäck

Efficient Skyline Maintenance over Frequently Updated Evidential Databases 199
Sayda Elmi, Mohamed Anis Bach Tobji, Allel Hadjali, and Boutheina Ben Yaghlane

Multiple Criteria Decision Methods

Prediction Model with Interval Data -Toward Practical Applications-. 213
Michihiro Amagasa and Kiyoshi Nagata

β-Robustness Approach for Fuzzy Multi-objective Problems. 225
Oumayma Bahri, Nahla Ben Amor, and El-Ghazali Talbi

Construction of an Outranking Relation Based on Semantic Criteria with ELECTRE-III 238
Miriam Martínez-García, Aida Valls, and Antonio Moreno

Argumentation and Belief Revision

Argumentation Framework Based on Evidence Theory 253
Ahmed Samet, Badran Raddaoui, Tien-Tuan Dao, and Allel Hadjali

Constrained Value-Based Argumentation Framework. 265
Karima Sedki and Safa Yahi

Belief Revision and the EM Algorithm 279
Inés Couso and Didier Dubois

Causal Belief Inference in Multiply Connected Networks. 291
Oumaima Boussarsar, Imen Boukhris, and Zied Elouedi

Databases and Information Systems

Indexing Possibilistic Numerical Data: The Interval B^+-tree Approach. 305
Guy De Tré, Robin De Mol, and Antoon Bronselaer

Ordinal Assessment of Data Consistency Based on Regular Expressions 317
Antoon Bronselaer, Joachim Nielandt, Robin De Mol, and Guy De Tré

A Fuzzy Approach to the Characterization of Database Query Answers. 329
Aurélien Moreau, Olivier Pivert, and Grégory Smits

Making the Skyline Larger: A Fuzzy-Neighborhood-Based Approach 341
Djamal Belkasmi, Allel Hadjali, and Hamid Azzoune

Describing Rough Approximations by Indiscernibility Relations
in Information Tables with Incomplete Information 355
Michinori Nakata and Hiroshi Sakai

A Possibilistic Treatment of Data Quality Measurement. 367
Antoon Bronselaer and Guy De Tré

Computing Theoretically-Sound Upper Bounds to Expected Support
for Frequent Pattern Mining Problems over Uncertain Big Data 379
Alfredo Cuzzocrea and Carson K. Leung

In-Database Feature Selection Using Rough Set Theory. 393
Frank Beer and Ulrich Bühler

Computational Aspects of Data Aggregation and Complex Data Fusion

Linear Optimization for Ecological Indices Based on
Aggregation Functions. 411
Gleb Beliakov, Andrew Geschke, Simon James, and Dale Nimmo

A Qualitative Approach to Set Achievable Goals During the Design Phase
of Complex Systems 423
Diadie Sow, Abdelhak Imoussaten, Pierre Couturier,
and Jacky Montmain

Unbalanced OWA Operators for Atanassov Intuitionistic Fuzzy Sets 435
Laura De Miguel, Edurne Barrenechea, Miguel Pagola, Aranzazu Jurio,
Jose Sanz, Mikel Elkano, and Humberto Bustince

Fuzzy K-Minpen Clustering and K-nearest-minpen Classification
Procedures Incorporating Generic Distance-Based Penalty Minimizers 445
Anna Cena and Marek Gagolewski

Fuzzy Sets and Fuzzy Logic

Adjoint Fuzzy Partition and Generalized Sampling Theorem. 459
Irina Perfilieva, Michal Holčapek, and Vladik Kreinovich

How to Incorporate Excluding Features in Fuzzy Relational Compositions and What for 470
Nhung Cao and Martin Štěpnička

Towards Fuzzy Partial Set Theory. 482
Libor Běhounek and Martina Daňková

On Perception-based Logical Deduction with Fuzzy Inputs. 495
Antonín Dvořák and Martin Štěpnička

Graded Dominance and Cantor-Bernstein Equipollence of Fuzzy Sets 510
Michal Holčapek

Uninorms on Interval-Valued Fuzzy Sets 522
Martin Kalina and Pavol Král

Algorithm for Generating Finite Totally Ordered Monoids 532
Milan Petrík and Thomas Vetterlein

Decision Support

Constructing Preference Relations from Utilities and Vice Versa. 547
Thomas A. Runkler

A Characterization of the Performance of Ordering Methods in TTRP with Fuzzy Coefficients in the Capacity Constraints 559
Isis Torres-Pérez, Carlos Cruz, Alejandro Rosete-Suárez, and José Luis Verdegay

Preferences on Gambles Representable by a Choquet Expected Value with Respect to Conditional Belief and Plausibility Functions. 569
Letizia Caldari, Giulianella Coletti, Davide Petturiti, and Barbara Vantaggi

A New Vision of Zadeh's Z-numbers 581
Sebastia Massanet, Juan Vicente Riera, and Joan Torrens

Comparison Measures

Comparing Interval-Valued Estimations with Point-Valued Estimations 595
Hugo Saulnier, Olivier Strauss, and Ines Couso

On Different Ways to be (dis)similar to Elements in a Set. Boolean Analysis and Graded Extension 605
Henri Prade and Gilles Richard

Comparing System Reliabilities with Ill-Known Probabilities 619
Lanting Yu, Sébastien Destercke, Mohamed Sallak, and Walter Schon

Machine Learning

Visualization of Individual Ensemble Classifier Contributions. 633
Catarina Silva and Bernardete Ribeiro

Feature Selection from Partially Uncertain Data Within the Belief Function Framework 643
Asma Trabelsi, Zied Elouedi, and Eric Lefevre

On the Suitability of Type-1 Fuzzy Regression Tree Forests for Complex Datasets. 656
Fathi Gasir and Keeley Crockett

Social Data Processing

Dynamic Analysis of Participatory Learning in Linked Open Data: Certainty and Adaptation 667
Marek Z. Reformat, Ronald R. Yager, and Jesse Xi Chen

Online Fuzzy Community Detection by Using Nearest Hubs 678
Pascal Held and Rudolf Kruse

Creating Extended Gender Labelled Datasets of Twitter Users 690
Marco Vicente, Fernando Batista, and Joao Paulo Carvalho

Temporal Data Processing

Suppression of High Frequencies in Time Series Using Fuzzy Transform of Higher Degree 705
Michal Holčapek and Linh Nguyen

A Modular Fuzzy Expert System Architecture for Data and Event Streams Processing 717
Jean-Philippe Poli and Laurence Boudet

Estimation and Characterization of Activity Duration in Business Processes . . . 729
Rodrigo M.T. Gonçalves, Rui Jorge Almeida, João M.C. Sousa, and Remco M. Dijkman

Fuzzy Modeling Based on Mixed Fuzzy Clustering for Multivariate Time Series of Unequal Lengths 741
Cátia M. Salgado, Susana M. Vieira, and João M.C. Sousa

Time Varying Correlation Estimation Using Probabilistic Fuzzy Systems 752
Nalan Baştürk and Rui Jorge Almeida

Aggregation

Fitting Aggregation Functions to Data: Part I - Linearization and Regularization 767
Maciej Bartoszuk, Gleb Beliakov, Marek Gagolewski, and Simon James

Fitting Aggregation Functions to Data: Part II - Idempotization. 780
Maciej Bartoszuk, Gleb Beliakov, Marek Gagolewski, and Simon James

Mean Estimation Based on FWA Using Ranked Set Sampling with Single and Multiple Rankers 790
Bekir Cetintav, Gozde Ulutagay, Selma Gurler, and Neslihan Demirel

On the Sensitivity of the Weighted Relevance Aggregation Operator and Its Application to Fuzzy Signatures 798
István Á. Harmati and László T. Kóczy

Some Results on Extension of Lattice-Valued XOR, XOR-Implications and E-Implications 809
Eduardo Palmeira and Benjamín Bedregal

Fuzzy Block-Pulse Functions and Its Application to Solve Linear Fuzzy Fredholm Integral Equations of the Second Kind 821
Shokrollah Ziari and Reza Ezzati

Author Index 833

Fuzzy Measures and Integrals

Decomposition Integral Based Generalizations of OWA Operators

Radko Mesiar and Andrea Stupňanová([✉])

Faculty of Civil Engineering, Slovak University of Technology,
Radlinského 11, 81005 Bratislava, Slovak Republic
{radko.mesiar,andrea.stupnanova}@stuba.sk

Abstract. Based on the representation of OWA operators as Choquet integrals with respect to symmetric capacities, a new kind of OWA generalizations based on decomposition integrals is proposed and discussed. The symmetry of the underlying capacity is not sufficient to guarantee the symmetry of the resulting operator, and thus we deal with symmetric saturated decomposition systems only. All possible generalized OWA operators on $X = \{1, 2\}$ are introduced. Similarly, when considering the maximal decomposition system on $X = \{1, 2, 3\}$, all generalized OWA operators are shown, based on the ordinal structure of the normed weighting vector $\mathbf{w} = (w_1, w_2, w_3)$.

Keywords: Choquet integral · Decomposition integral · OWA operator · Pan-integral · Symmetric capacity

1 Introduction

OWA operators, as a special class of aggregation functions (i.e., increasing functions $A : [0,1]^n \to [0,1]$ satisfying two boundary conditions $A(0,\ldots,0) = 0$, $A(1,\ldots,1) = 1$, for more details see [5]) covering the standard min, max and arithmetic mean operators, were introduced by Yager [13] in 1988.

Definition 1. *Let* $\mathbf{w} = (w_1, \ldots, w_n) \in [0,1]^n$ *be a normed weighting vector, i.e.,* $\sum_{i=1}^{n} w_i = 1$. *A function* $\mathrm{OWA}_{\mathbf{w}} : [0,1]^n \to [0,1]$ *given by*

$$\mathrm{OWA}_{\mathbf{w}}(\mathbf{x}) = \sum_{i=1}^{n} w_i \, x_{\sigma(i)}, \tag{1}$$

where $\sigma : \{1,\ldots,n\} \to \{1,\ldots,n\}$ *is a permutation of* $\{1,\ldots n\}$ *such that* $x_{\sigma(1)} \geq \cdots \geq x_{\sigma(n)}$, *is called an OWA operator.*

Very soon they became an important tool in many domains, especially in decision problems. Rather early, several generalizations of OWA's appeared, such as GOWA [15] (Generalized OWA), IOWA [14,16] (Induced OWA), etc.

J.P. Carvalho et al. (Eds.): IPMU 2016, Part I, CCIS 610, pp. 3–10, 2016.
DOI: 10.1007/978-3-319-40596-4_1

Recently, a survey of OWA literature using a citation network analysis was published in [2], including 537 OWA related sources in supplementary document.

Note that OWA operators can be seen as Choquet integrals with respect to symmetric capacities [4] too. We recall that Choquet integral was introduced by Choquet in 1953 [1]. When dealing with a finite universe $X = \{1, \ldots, n\}$, functions $f : X \to [0,1]$ can be identified with vectors $\mathbf{x} \in [0,1]^n$, $x_i = f(i)$, $i = 1, \ldots, n$. A capacity (fuzzy measure) $m : 2^X \to [0,1]$ is a monotone set function constrained by the two boundary conditions, $m(\emptyset) = 0, m(X) = 1$. Since now, we will deal in this contribution with $X = \{1, \ldots, n\}$.

Definition 2 ([5]). *For a given vector $\mathbf{x} \in [0,1]^n$ and capacity m on X the corresponding Choquet integral is given by*

$$\mathbf{Ch}_m(\mathbf{x}) = \sum_{i=1}^{n} x_{\sigma(i)} \left(m(E_{\sigma,i}) - m(E_{\sigma,i-1}) \right), \tag{2}$$

where $\sigma : \{1, \ldots, n\} \to \{1, \ldots, n\}$ is a permutation such that $x_{\sigma(1)} \geq \cdots \geq x_{\sigma(n)}, E_{\sigma,0} = \emptyset$, and for $i = 1, \ldots, n$, $E_{\sigma,i} = \{\sigma(1), \ldots, \sigma(i)\}$.

As already observed in the case of formula (1), also in the case of formula (2) it may happen, that the permutation σ is not unique. This fact does not hurt the correctness of formula (2). As observed by Grabisch [4], formulae (1) and (2) may coincide for each $\mathbf{x} \in [0,1]^n$ if and only if $m(E_{\sigma,i})$ does not depend on the considered permutation σ. This means that only $\text{card}\, E_{\sigma,i} = i$ matters, i.e., $m(E) = m(\sigma(E))$ for any $E \in 2^X$ and permutation σ, $\sigma(E) = \{\sigma(i) |\ i \in E\}$. Such capacities are called symmetric. Now, it is enough to put $w_i = m(E_{\sigma,i}) - m(E_{\sigma,i-1})$ to see that

$$\text{OWA}_{\mathbf{w}} = \mathbf{Ch}_m \tag{3}$$

Vice-versa, for any normed weighting vector $\mathbf{w}$, it is enough to define a symmetric capacity $m : 2^X \to [0,1]$ by

$$m(E) = \sum_{i=1}^{\text{card}E} w_i \tag{4}$$

to see the representation (3).

Recall that OWA operators are symmetric, positively homogeneous, idempotent, comonotone additive [5] and piece-wise linear. To define them axiomatically, the comonotone additivity and symmetry are sufficient [4]. Any proper generalization of OWA operators should violate some of the above properties. We have discussed some of such generalizations in our recent paper [10].

Throughout this contribution, we will consider generalizations which are symmetric, positively homogeneous, idempotent and piece-wise linear, but not comonotone additive. From alternative approaches, recall the OWA generalizations based on p-symmetric capacities and Choquet integral [11], where the symmetry is replaced by p-symmetry, and the remaining above mentioned properties are preserved. Our approach deals with symmetric capacities and decomposition integrals recently introduced by Even and Lehrer [3].

2 Decomposition Integrals

Any non-empty set of non-empty subsets of X is called a collection. Any non-empty set $\mathcal{H}$ of collections is called a decomposition system.

Definition 3 ([3]). *Let a decomposition system $\mathcal{H}$ be fixed. For a capacity m on X the corresponding $\mathcal{H}$-decomposition integral $\mathbf{I}_{\mathcal{H},m}$ is given by*

$$\mathbf{I}_{\mathcal{H},m}(\mathbf{x}) = \max\left\{\sum_{i\in J} a_i\, m(A_i) \middle| \; (A_i)_{i\in J} \in \mathcal{H}, a_i \geq 0 \;\; \textit{for each } i \in J, \sum_{i\in J} a_i \mathbf{1}_{A_i} \leq \mathbf{x}\right\}. \tag{5}$$

Alternatively, we can write

$$\mathbf{I}_{\mathcal{H},m}(\mathbf{x}) = \max_{\zeta\in\mathcal{H}}\left\{\sum_{A_i\in\zeta} a_i\, m(A_i) \middle| \; a_i \geq 0 \;\text{ for each } i, \sum_{A_i\in\zeta} a_i \mathbf{1}_{A_i} \leq \mathbf{x}\right\} = \max_{\zeta\in\mathcal{H}} \mathbf{I}_{\zeta,m}.$$

For any collection ζ, the functional $\mathbf{I}_{\zeta,m}$ is positively homogeneous, monotone and piece-wise linear, and thus also the functional $\mathbf{I}_{\mathcal{H},m}$ is positively homogeneous, monotone and piece-wise linear. Obviously, it is symmetric whenever the capacity m is symmetric. In general, it need not be idempotent and neither an aggregation function. However, due to the positive homogeneity, the mapping

$$A_{\mathcal{H},m} = \frac{\mathbf{I}_{\mathcal{H},m}}{\mathbf{I}_{\mathcal{H},m}(\mathbf{1})} : [0,1]^n \to [0,1] \tag{6}$$

is an idempotent aggregation function whenever $\mathbf{I}_{\mathcal{H},m}(\mathbf{1}) > 0$.

Recall that due to [9] $\mathbf{I}_{\mathcal{H},m}(\mathbf{1}) = 1$ for each capacity m whenever the decomposition system $\mathcal{H}$ is complete (i.e., each non-empty subset E of X is contained in at least one collection from $\mathcal{H}$) and any of its collections is formed by logically independent subsets of X (i.e., their intersection is non-empty). We introduce now some decomposition systems and related decomposition integrals.

Example 1. Consider $X = \{1,\ldots,n\}$.

(1) Let $\mathcal{H}^{(i)} = \{\mathcal{B} \mid \mathcal{B}$ is a chain in X of length $i\}$, $i \in \{1,\ldots,n\}$. As shown in [9], these decomposition systems yield the only kind of decomposition integrals which are also universal integrals in the sense of Klement et al. [6]. Note that $\mathbf{I}_{\mathcal{H}^{(1)},m}$ is the Shilkret integral [12], while $\mathbf{I}_{\mathcal{H}^{(n)},m} = \mathbf{Ch}_m$ is the Choquet integral [1]. Note that $\mathbf{I}_{\mathcal{H}^{(i)},m}$ is an aggregation function for each capacity m and $i \in \{1,\ldots,n\}$.
(2) For $\emptyset \neq A \subseteq X$, let $\mathcal{H}_A = \{\{A\}\}$. Then

$$\mathbf{I}_{\mathcal{H}_A,m}(\mathbf{x}) = \min\{x_i \mid i \in A\} \cdot m(A),$$

and $\mathbf{I}_{\mathcal{H}_A,m}$ is an aggregation function only if $m(A) = 1$.
(3) Let $\mathcal{H}_{Pan} = \{\mathcal{B} \mid \mathcal{B}$ is a partition of $X\}$. Then $\mathcal{H}_{Pan}$ is the Pan-integral introduced in [17], and it is not an aggregation function, in general.

(4) Let $\mathcal{H}_L = \{\mathcal{B}|\ \emptyset \neq \mathcal{B} \subseteq 2^X - \{\emptyset\}\}$. This decomposition system is the greatest one and thus also $\mathbf{I}_{\mathcal{H}_L,m}$ is the greatest decomposition integral. It was introduced by Lehrer [7], where it is called a concave integral, and it is not an aggregation function, in general.
(5) Let $\mathcal{H}_S = \{\{\{1\},\{1,2\},\dots,\{1,2,\dots,n\}\},\{\{2\}\},\dots\}$ consists of one maximal chain in X and remaining singleton collections $\{A\}$, $A \notin \{\{1\},\{1,2\},\dots,\{1,2,\dots,n\}\}$. Clearly, $\mathcal{H}^{(1)} \subsetneqq \mathcal{H}_S \subsetneqq \mathcal{H}^{(n)}$ whenever $n > 1$, and thus the decomposition integral based on $\mathcal{H}_S$ is between the Shilkret and Choquet integral, i.e.,

$$\mathbf{I}_{\mathcal{H}^{(1)},m} \leq \mathbf{I}_{\mathcal{H}_S,m} \leq \mathbf{I}_{\mathcal{H}^{(n)},m}.$$

Note that $\mathbf{I}_{\mathcal{H}_S,m}$ is an aggregation function for any capacity m.
For $n = 2$, it holds

$$\mathbf{I}_{\mathcal{H}_S,m}(x,y) = \begin{cases} m(\{1\})\cdot x + (1-m(\{1\}))\cdot y & \text{if } x \geq y \\ \max(x, m(\{2\})\cdot y) & \text{otherwise} \end{cases}. \tag{7}$$

3 $\mathcal{H}$-OWA Operators

To guarantee the symmetry of decomposition integrals, the symmetry of a considered capacity m is not sufficient. For example, the formula (7) in the case of a symmetric capacity m determined by $m(1) = m(2) = a$ yields

$$\mathbf{I}_{\mathcal{H}_S,m}(x,y) = \begin{cases} a\cdot x + (1-a)\cdot y & \text{if } x \geq y \\ \max(x, a\cdot y) & \text{otherwise} \end{cases},$$

which is symmetric only if $a \in \{0,1\}$. This fact is due to non-symmetry of the decomposition system $\mathcal{H}_S$.

Definition 4. *Let $\sigma : X \to X$ be a permutation on X. For any non-empty subset E of X, denote $E_\sigma = \{\sigma(i)|\ i \in E\}$. For any collection $\mathcal{B} = \{E_1,\dots,E_k\}$, denote $\mathcal{B}_\sigma = \{(E_1)_\sigma,\dots,(E_k)_\sigma\}$. Similarly, for any decomposition system $\mathcal{H}$, denote $\mathcal{H}_\sigma = \{\mathcal{B}_\sigma|\ \mathcal{B} \in \mathcal{H}\}$. A decomposition system $\mathcal{H}$ is called symmetric if and only if $\mathcal{H} = \mathcal{H}_\sigma$ for any permutation σ on X.*

Observe that considering decomposition systems introduced in Example 1, symmetric are only the systems $\mathcal{H}^{(i)}, i = 1,\dots,n$, $\mathcal{H}_{Pan}$ and $\mathcal{H}_L$.

As already mentioned, there are decomposition system on $\mathcal{H}$ such that $\mathbf{I}_{\mathcal{H},m}(\mathbf{1}) = 0$ may happen for some capacity m.

Proposition 1. *Let $\mathcal{H}$ be a decomposition system on X. Then $\mathbf{I}_{\mathcal{H},m}(\mathbf{1}) > 0$ for any capacity m on X if and only if there is a collection $\mathcal{B} \in \mathcal{H}$ such that $X \in \mathcal{B}$.*

Proof. Note that if $m_1 \geq m_2$ then $\mathbf{I}_{\mathcal{H},m_1} \geq \mathbf{I}_{\mathcal{H},m_2}$ and thus it is enough to deal with the smallest capacity m_* on X, $m_*(E) = \begin{cases} 1 & \text{if } E = X \\ 0 & \text{otherwise} \end{cases}$. Clearly, if $X \notin \mathcal{B}$ for each $\mathcal{B} \in \mathcal{H}$, then $\mathbf{I}_{\mathcal{H},m_*}(\mathbf{1}) = 0$. On the other hand, if $X \in \mathcal{B}$ for some $\mathcal{B} \in \mathcal{H}$, then $\mathbf{I}_{\mathcal{H},m_*}(\mathbf{1}) = 1$. □

Definition 5. *Let $\mathcal{H}$ be a decomposition system on X such that $X \in \mathcal{B}$ for some $\mathcal{B} \in \mathcal{H}$. Them $\mathcal{H}$ is called a saturated decomposition system.*

Note that each complete decomposition system is saturated, but not vice versa. Systems $\mathcal{H}^{(i)}$, $i = 1, \ldots, n$, $\mathcal{H}_{Pan}$, $\mathcal{H}_L$ and $\mathcal{H}_S$ introduced in Example 1 are saturated, but not the systems $\mathcal{H}_A$ for $A \neq X$.

Definition 6. *Let $\mathcal{H}$ be a symmetric saturated decomposition system on X, and let m be a symmetric capacity on X. Then the functional $\mathcal{H} - \mathrm{OWA}_{\mathbf{w}} : [0,1]^n \to [0,1]$ given by*

$$\mathcal{H} - \mathrm{OWA}_{\mathbf{w}}(\mathbf{x}) = \frac{\mathbf{I}_{\mathcal{H},m}(\mathbf{x})}{\mathbf{I}_{\mathcal{H},m}(\mathbf{1})} \tag{8}$$

is called a decomposition OWA operator. Here $\mathbf{w} = (w_1, \ldots, w_n)$ with $w_1 = m(\{1\}), w_2 = m(\{1,2\}) - m(\{1\}), \ldots, w_n = m(\{1,\ldots,n\}) - m(\{1,\ldots,n-1\})$, i.e.,

$$m(E) = \sum_{i=1}^{\mathrm{card} E} w_i, \quad E \in 2^X.$$

Note that $\mathcal{H}^{(n)} - \mathrm{OWA}_{\mathbf{w}} = \mathrm{OWA}_{\mathbf{w}}$ is the standard OWA operator due to the fact that $\mathcal{H}^{(n)}$ generates just the Choquet integral. We summarize some properties of $\mathcal{H}$-OWA operators, which form a particular class of aggregation functions:

- symmetry
- positive homogeneity
- idempotency
- piece-wise linearity
- monotonicity in weights, i.e.,

$$\mathcal{H} - \mathrm{OWA}_{\mathbf{w}^{(1)}} \leq \mathcal{H} - \mathrm{OWA}_{\mathbf{w}^{(2)}}$$

  whenever
  $w_1^{(1)} \leq w_1^{(2)}, w_1^{(1)} + w_2^{(1)} \leq w_1^{(2)} + w_2^{(2)}, \ldots, w_1^{(1)} + \cdots + w_n^{(1)} \leq w_1^{(2)} + \cdots + w_n^{(2)}$.

Example 2. For $n = 2$, the only symmetric saturated decomposition systems are

$\mathcal{H}^{(1)} = \{\{\{1\}\}, \{\{2\}\}, \{\{1,2\}\}\}$,
$\mathcal{H}^{(2)} = \{\{\{1\}, \{1,2\}\}, \{\{2\}, \{1,2\}\}\}$,
$\mathcal{H}_{Pan} = \{\{\{1\}, \{2\}\}, \{\{1,2\}\}\}$,
$\mathcal{H}_L = \{\{\{1\}\}, \{\{2\}\}, \{\{1,2\}\}, \{\{1\}, \{1,2\}\}, \{\{2\}, \{1,2\}\}, \{\{1\}, \{2\}\}, \{\{1\}, \{2\}, \{1,2\}\}\}$,
$\mathcal{H}_X = \{\{\{1,2\}\}\}$
and
$\mathcal{H}_1 = \{\{\{1\}, \{1,2\}\}, \{\{2\}, \{1,2\}\}, \{\{1,2\}\}\}$,
$\mathcal{H}_2 = \{\{\{1\}\}, \{\{2\}\}, \{\{1\}, \{1,2\}\}, \{\{2\}, \{1,2\}\}\}$.

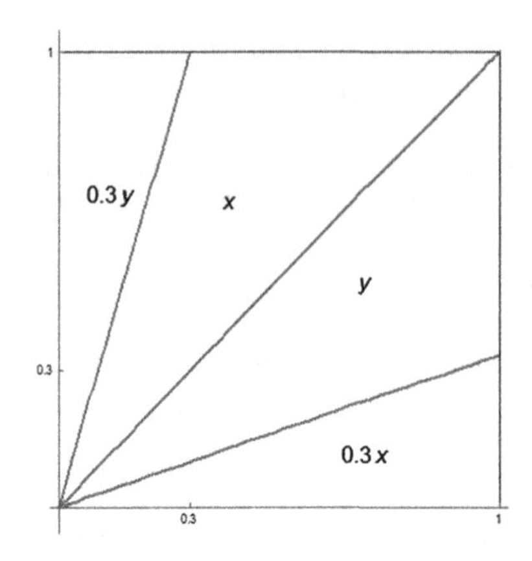

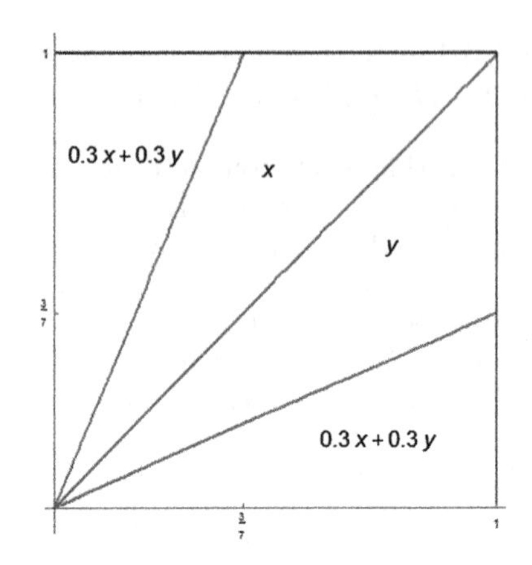

Fig. 1. $\mathcal{H}^{(1)} - \mathrm{OWA}_{(0.3,0.7)}$ on the left and $\mathcal{H}_{Pan} - \mathrm{OWA}_{(0.3,0.7)}$ on the right

The corresponding $\mathcal{H}$-OWA operators are then given by: (see also Fig. 1)

$$\mathcal{H}^{(1)} - \mathrm{OWA}_{(w_1,1-w_1)}(x,y) = \max(w_1x, w_1y, \min(x,y));$$

$$\begin{aligned}\mathcal{H}^{(2)} - \mathrm{OWA}_{(w_1,1-w_1)}(x,y) &= \mathrm{OWA}_{(w_1,1-w_1)}(x,y) = \\ &= w_1 \max(x,y) + (1-w_1)\min(x,y);\end{aligned}$$

$$\begin{aligned}\mathcal{H}_{Pan} - \mathrm{OWA}_{(w_1,1-w_1)}(x,y) &= \frac{\max(w_1x + w_1y, \min(x,y))}{\max(2w_1, 1)} = \\ &= \begin{cases} \frac{x+y}{2} & \text{if } w_1 \geq \frac{1}{2} \\ \max(w_1x + w_1y, \min(x,y)) & \text{otherwise} \end{cases};\end{aligned}$$

$$\mathcal{H}_L - \mathrm{OWA}_{(w_1,1-w_1)}(x,y) = \begin{cases} \frac{x+y}{2} & \text{if } w_1 \geq \frac{1}{2} \\ \mathrm{OWA}_{(w_1,1-w_1)}(x,y) & \text{otherwise} \end{cases};$$

$$\mathcal{H}_X - \mathrm{OWA}_{(w_1,1-w_1)}(x,y) = \min(x,y);$$

$$\mathcal{H}_1 - \mathrm{OWA}_{(w_1,1-w_1)}(x,y) = \mathrm{OWA}_{(w_1,1-w_1)}(x,y);$$

$$\mathcal{H}_2 - \mathrm{OWA}_{(w_1,1-w_1)}(x,y) = \mathrm{OWA}_{(w_1,1-w_1)}(x,y).$$

As we can see, in some cases the resulting $\mathcal{H}$-OWA operators coincide for different capacities, and in some cases they result into the standard OWA operators. We have the next general result.

Theorem 1. *Let $\mathcal{H}$ be a symmetric saturated decomposition system on X. Then:*

(i) if $\{\{1\},\ldots,\{n\}\} \in \mathcal{H}$ and $w_1 \geq w_i, i = 2,\ldots,n$, then

$$\mathcal{H} - \mathrm{OWA}_{\mathbf{w}}(\mathbf{x}) = \mathrm{AM}(\mathbf{x}),$$

where AM is the arithmetic mean;

(ii) if $\mathcal{H}^{(n)} \subseteq \mathcal{H}$, and for any $\mathcal{B} \in \mathcal{H}$ there is $\mathcal{D} \in \mathcal{H}^{(n)}$ such that $\mathcal{B} \subseteq \mathcal{D}$ (i.e., any $\mathcal{B} \in \mathcal{H}$ is a chain), then

$$\mathcal{H} - \mathrm{OWA}_{\mathbf{w}}(\mathbf{x}) = \mathrm{OWA}_{\mathbf{w}}(\mathbf{x}),$$

i.e., $\mathcal{H}$-OWA is the standard OWA operator.

Thought Theorem 1 shows some constraints of the $\mathcal{H}$-based generalization of OWA operators, there is still a rich variety of new kinds of aggregation functions obtained by this proposed generalization method. We illustrate this fact for $n = 3$ and $\mathcal{H} = \mathcal{H}_L$, i.e., when concave integral based generalization of ternary OWA operators is considered.

Example 3. Let a symmetric capacity m on $X = \{1, 2, 3\}$ be related to a weighting vector $\mathbf{w} = (w_1, w_2, w_3) \in [0, 1]^3, w_1 + w_2 + w_3 = 1$. Due to Theorem 1, if $w_1 \geq w_2$ and $w_1 \geq w_3$, $\mathcal{H}_L - \text{OWA}_\mathbf{w} = AM$.

On the other hand, if $w_1 \leq w_2 \leq w_3$, then $\mathcal{H}_L - \text{OWA}_\mathbf{w} = \text{OWA}_\mathbf{w}$. We have still 3 remaining cases considering the ordinal structure of weighting vector $\mathbf{w}$:

- if $w_3 \leq w_1 \leq w_2$ then

$$\mathcal{H}_L - \text{OWA}_\mathbf{w}(x_1, x_2, x_3) = \begin{cases} AM(x_1, x_2, x_3) & \text{if } x_{\sigma(1)} \leq x_{\sigma(2)} \leq x_{\sigma(3)} \\ \frac{2w_1}{3(1-w_3)} x_{\sigma(1)} + \frac{2w_2}{3(1-w_1)}(x_{\sigma(2)} + x_{\sigma(3)}) & \text{otherwise} \end{cases},$$

  where $\sigma : X \to X$ is a permutation such that $x_{\sigma(1)} \geq x_{\sigma(2)} \geq x_{\sigma(3)}$;
- if $w_1 \leq w_3 \leq w_2$ then

$$\begin{aligned} &\mathcal{H}_L - \text{OWA}_\mathbf{w}(x_1, x_2, x_3) \\ &= \min\left(1, \frac{2}{3(1 - w_3)}\right) \cdot \max\left(\text{OWA}_\mathbf{w}(x_1, x_2, x_3), A_\mathbf{w}(x_1, x_2, x_3)\right) \end{aligned}$$

  where

$$A_\mathbf{w}(x_1, x_2, x_3) = \begin{cases} \frac{1-w_3}{2}(x_1 + x_2 + x_3) & \text{if } x_{\sigma(1)} \leq x_{\sigma(2)} + x_{\sigma(3)}; \\ w_1 x_{\sigma(1)} + 2w_2(x_{\sigma(2)} + x_{\sigma(3)}) & \text{otherwise} \end{cases};$$

- if $w_2 \leq w_1 \leq w_3$ then

$$\begin{aligned} &\mathcal{H}_L - \text{OWA}_\mathbf{w}(x_1, x_2, x_3) = \min\left(1, \frac{1}{3w_1}\right) \\ &\cdot \max\left(\text{OWA}_\mathbf{w}(x_1, x_2, x_3), x_{\sigma(3)} + w_1(x_{\sigma(1)} + x_{\sigma(2)} - 2x_{\sigma(3)}), w_1(x_1 + x_2 + x_3)\right). \end{aligned}$$

4 Concluding Remarks

We have introduced a generalization of OWA operators based on a symmetric capacity m and a $\mathcal{H}$-decomposition integral, where $\mathcal{H}$ is a saturated symmetric decomposition system. The introduced $\mathcal{H}$-OWA operators are not only symmetric, idempotent and positively homogeneous aggregation functions, but they are also piece-wise linear and continuous. Note that there is a dual concept to decomposition integrals, namely the superdecomposition integrals introduced in [8]. Based on symmetric capacities m and $\mathcal{H}$-superdecomposition integrals, where again $\mathcal{H}$ should be symmetric and saturated, a further type of OWA generalizations could be introduced. Note that when considering $\mathcal{H}^{(n)}$, then also the superdecomposition integral yields the Choquet integral, and hence the standard OWA operators are included in this new intended family, too.

Acknowledgments. The support of the grants APVV-14-0013 and VEGA 1/0682/16 is kindly announced.

References

1. Choquet, G.: Theory of capacities. Ann. Inst. Fourier **5**, 131–295 (1953)
2. Emrouznejad, A., Marra, M.: Ordered weighted averaging operator: a citation-based literature survey. Int. J. Intell. Syst. **29**, 994–1014 (2014)
3. Even, Y., Lehrer, E.: Decomposition-integral: unifying Choquet and the concave integrals. Econ. Theory **56**(1), 1–26 (2014)
4. Grabisch, M.: Fuzzy integral in multicriteria decision making. Fuzzy Sets Syst. **69**(3), 279–298 (1995)
5. Grabisch, M., Marichal, J.L., Mesiar, R., Pap, E.: Aggregation Functions. Cambridge University Press, Cambridge (2009)
6. Klement, E.P., Mesiar, R., Pap, E.: A universal integral as common frame for Choquet and Sugeno integral. IEEE Trans. Fuzzy Syst. **18**, 178–187 (2010)
7. Lehrer, E.: A new integral for capacities. Econ. Theory **39**, 157–176 (2009)
8. Mesiar, R., Li, J., Pap, E.: Superdecomposition integrals. Fuzzy Sets and Syst. **259**, 3–11 (2015)
9. Mesiar, R., Stupňanová, A.: Decomposition integrals. Int. J. Approximate Reason. **54**, 1252–1259 (2013)
10. Mesiar, R., Stupňanová, A., Yager, R.R.: Generalizations of OWA operators. IEEE Trans. Fuzzy Syst. **23**(6), 2154–2162 (2015)
11. Miranda, P., Grabisch, M., Gil, P.: p-symmetric fuzzy measures. Int. J. Uncertainty Fuzziness Knowl.-Based Syst. **10**, 105–123 (2002)
12. Shilkret, N.: Maxitive measure and integration. Indag. Math. **33**, 109–116 (1971)
13. Yager, R.R.: On ordered weighted averaging aggregation operators in multicriteria decisionmaking. IEEE Trans. Syst. Man Cybern. **18**(1), 183–190 (1988)
14. Yager, R.R.: Induced aggregation operators. Fuzzy Sets Syst. **137**(1), 59–69 (2003)
15. Yager, R.R.: Generalized OWA aggregation operators. Fuzzy Optim. Decis. Making **3**(1), 93–107 (2004)
16. Yager, R.R., Filev, D.P.: Induced ordered weighted averaging operators. IEEE Trans. Syst. Man Cybern. Part B Cybern. **29**(2), 141–150 (1999)
17. Yang, Q.: The pan-integral on the fuzzy measure space. Fuzzy Math. **3**, 107–114 (1985)

Finding the Set of k-additive Dominating Measures Viewed as a Flow Problem

Pedro Miranda[1(✉)] and Michel Grabisch[2]

[1] Universidad Complutense de Madrid, Madrid, Spain
pmiranda@mat.ucm.es

[2] Paris School of Economics, Université Paris I-Panthéon-Sorbonne, Paris, France
Michel.Grabisch@univ-paris1.fr

Abstract. In this paper we deal with the problem of obtaining the set of k-additive measures dominating a fuzzy measure. This problem extends the problem of deriving the set of probabilities dominating a fuzzy measure, an important problem appearing in Decision Making and Game Theory. The solution proposed in the paper follows the line developed by Chateauneuf and Jaffray for dominating probabilities and continued by Miranda et al. for dominating k-additive belief functions. Here, we address the general case transforming the problem into a similar one such that the involved set functions have non-negative Möbius transform; this simplifies the problem and allows a result similar to the one developed for belief functions. Although the set obtained is very large, we show that the conditions cannot be sharpened. On the other hand, we also show that it is possible to define a more restrictive subset, providing a more natural extension of the result for probabilities, such that it is possible to derive any k-additive dominating measure from it.

Keywords: Fuzzy measures · Dominance · k-additivity

1 Introduction

Fuzzy measures, also called capacities, nonadditive measures, are widely used in the representation of uncertainty, decision making and cooperative game theory. A particular class of fuzzy measures which is of interest in this paper can be found in the Theory of Evidence developed by Dempster [4] and Shafer [22]. In this theory, uncertainty is represented by a pair of "lower probability" (or "degree of belief") and "upper probability" (or "degree of plausibility") assigned to every event. These upper and lower probabilities have been well studied [25,26]; they are not additive in general, and are called by Shafer belief and plausibility functions.

The problem of finding the set of probability measures dominating a given fuzzy measure appears in many situations, especially in decision theory and in cooperative game theory. In decision theory, it may happen that the available information is not sufficient to assign an exact probability to events, but it only

J.P. Carvalho et al. (Eds.): IPMU 2016, Part I, CCIS 610, pp. 11–22, 2016.
DOI: 10.1007/978-3-319-40596-4_2

allows an interval of compatible probability values. In this case, we obtain a set of possible probabilities, denoted by $\mathcal{P}$. If we consider $\mu := \inf_{P \in \mathcal{P}} P$, then μ is a fuzzy measure (but not necessarily a belief function [24]), called the "coherent lower probability". As a consequence, for any probability P' dominating μ, it follows that $E_{P'}(f) \geq \mathcal{C}_\mu(f)$, for any function f, where $\mathcal{C}_\mu$ denotes the Choquet integral [3]. In [2], Chateauneuf and Jaffray use this result and the fact that $\mu \leq P, \forall P \in \mathcal{P}$ to obtain an easy method to compute $\inf_{P \in \mathcal{P}} E_P(f)$. Note that this method is based on the knowledge of the set of all probability distributions dominating μ. The same can be applied for obtaining an upper bound.

In cooperative game theory, a TU-game is a set function μ vanishing on the empty set (it is not necessarily a fuzzy measure, however). One of the most important problems in this field is to obtain a sharing function for the game, that is, assuming that the grand coalition X is formed and the benefit $\mu(X)$ is obtained, we are looking for a rational and equitable way to divide $\mu(X)$ among all players. Any possible sharing function is called a solution of the game. Among the many concepts of solutions in the literature (see, e.g., [5]), one of the most popular is the *core* of the game [23], which is defined as the set of additive games dominating μ and coinciding with μ on the grand coalition X. The core is a bounded polyhedron, possibly empty, and much research has been devoted to its study (see a survey in [10]).

On the other hand, a natural extension of probabilities or additive measures is the concept of k-additive measure [7,8]. They constitute a mid-term between probability measures (which are too restrictive in many situations) and general fuzzy measures (whose complexity is too high to deal with in practice). Thus, a natural extension of the previous dominance problem is to look for the set of k-additive measures dominating a given fuzzy measure. There are some cases where this could be useful. First, suppose a situation that can be modelled via a k-additive measure (an axiomatic characterization to this situation can be found in [19]), but where our information is not enough to completely determine the measure. Then, we have to work with a set of compatible k-additive measures (let us call it $\mathcal{U}_k$). A second example is the identification of a capacity in a practical situation. It can be proved that the available information may not be sufficient to determine a single solution, but there exists a set of k-additive measures, all equally suitable [16]. Moreover, the set of all these measures is a convex set and consequently, the measure for an event $A \subseteq X$ lies in an interval of possible values (a deeper study about the uniqueness of the solution and the structure of the set of solutions can be found in [18]). As for probabilities, if $\mu = \inf_{\mu_k \in \mathcal{U}_k} \mu_k$, then μ is a fuzzy measure and $\mathcal{C}_{\mu'_k}(f) \geq \mathcal{C}_\mu(f)$, for any k-additive measure μ'_k dominating μ. Therefore, it seems interesting to find the set of all k-additive fuzzy measures dominating μ, thus extending the results in [2].

Another interest in finding the set of dominating k-additive measures can be found in game theory. As we have said above, the core of a game μ may be empty [1]. Considering instead the set of its k-additive dominating games, called the k-additive core, it is shown in [17] that the k-additive core is never empty, as soon as $k \geq 2$.

In this paper we deal with the problem of characterizing the set of all k-additive measures dominating a given fuzzy measure μ. Previous attempts in this direction appear in [9,20]. As it will become apparent below, we have to face many difficulties that do not arise in the case of probabilities, except in very restrictive situations. One of these situations is the case of k-additive belief functions dominating a belief function. We will use the results in this case to derive a general result for any fuzzy measure and any dominating k-additive measure.

The rest of the paper is organized as follows: in the next section, we explain the basic facts and results in order to fix notation and to be self-contained; then, we derive the results for characterizing the set of dominating k-additive measures. We end the paper with concluding remarks and open problems.

2 Basic Results

Consider a finite referential set of n elements, $X = \{1, ..., n\}$. The set of subsets of X is denoted by $\mathcal{P}(X)$ and we denote $\mathcal{P}^*(X) = \mathcal{P}(X)\backslash\{\emptyset\}$; the set of subsets whose cardinality is less or equal than k is denoted by $\mathcal{P}^k(X)$, or $\mathcal{P}^k_*(X)$ if the emptyset is not included. Subsets of X are denoted $A, B, ...$; we will sometimes write $i_1 \cdots i_k$ instead of $\{i_1, \dots, i_k\}$ in order to avoid a heavy notation; braces are usually omitted for singletons and subsets of two elements.

We define a **fuzzy measure** as a set function $\mu : \mathcal{P}(X) \to [0,1]$ satisfying the boundary conditions $\mu(\emptyset) = 0, \mu(X) = 1$ and monotonicity ($\mu(A) \leq \mu(B)$ whenever $A \subseteq B$). Fuzzy measures are denoted by μ, μ^* and so on, and the set of all fuzzy measures on X is denoted $\mathcal{FM}(X)$.

Given a set function μ (not necessarily a fuzzy measure), an equivalent representation of μ can be obtained via the **Möbius transform** [21], given by

$$m(A) := \sum_{B\subseteq A} (-1)^{|A\backslash B|}\mu(B), \forall A \subseteq X.$$

The Möbius transform is also widely used in the field of Game Theory, where it is known as *dividends* [13]. It is worth noting that $m(A)$ can attain negative values. The set of fuzzy measures μ such that the corresponding Möbius transform satisfies $m(A) \geq 0, \forall A \subseteq X$ is known as the set of **belief functions**, denoted $\mathcal{BEL}(X)$. Belief functions come from the Theory of Evidence developed by Dempster [4] and Shafer [22]. Given the Möbius transform, it is possible to recover the original fuzzy measure through the *Zeta transform*:

$$\mu(A) = \sum_{B\subseteq A} m(B).$$

Contrarily to fuzzy measures, for which it holds $0 \leq \mu(A) \leq 1, \forall A \subseteq X$, the upper and lower bounds for the Möbius transform are not trivial. These bounds are given in the next result.

Theorem 1. *[12] For any fuzzy measure μ, its Möbius transform satisfies for any $A \subseteq N$, $|A| > 1$:*

$$l_{|A|} := -\binom{|A|-1}{c'_{|A|}} \leqslant m(A) \leqslant \binom{|A|-1}{c_{|A|}} := u_{|A|},$$

with

$$c_{|A|} = 2\left\lfloor \frac{|A|}{4} \right\rfloor, \quad c'_{|A|} = 2\left\lfloor \frac{|A|-1}{4} \right\rfloor + 1,$$

and for $|A| = 1 < n$:

$$0 \leqslant m(A) \leqslant 1,$$

and $m(A) = 1$ if $|A| = n = 1$. These upper and lower bounds are attained by the fuzzy measures μ^*_A, μ_{A*}, respectively:

$$\mu^*_A(B) = \begin{cases} 1, & \text{if } |A| - l_{|A|} \leqslant |B \cap A| \\ 0, & \text{otherwise} \end{cases},$$

$$\mu_{A*}(B) = \begin{cases} 1, & \text{if } |A| - l'_{|A|} \leqslant |B \cap A| \\ 0, & \text{otherwise} \end{cases},$$

for any $B \subseteq N$.

We give in Table 1 the first values of the bounds.

Table 1. Lower and upper bounds for the Möbius transform of a fuzzy measure

$\|A\|$	1	2	3	4	5	6	7	8	9	10	11	12
u.b. of $m(A)$	1	1	1	3	6	10	15	35	70	126	210	462
l.b. of $m(A)$	1(0)	−1	−2	−3	−4	−10	−20	−35	−56	−126	−252	−462

Let us now introduce the concept of k-additivity. A problem appearing in the practical use of fuzzy measures is their complexity. Contrary to the case of probabilities, where just $n-1$ values suffice to completely determine a probability on a set of cardinality n, in order to determine a fuzzy measure $2^n - 2$ values are necessary. As a consequence, complexity grows exponentially with n. In an attempt to reduce this complexity, Grabisch has defined the concept of k-additive measure [7].

A fuzzy measure μ is said to be **k-additive** if its Möbius transform vanishes for any $A \subseteq X$ such that $|A| > k$ and there exists at least one subset A with exactly k elements such that $m(A) \neq 0$.

Thus, it can be seen that probabilities are just 1-additive measures (and also 1-additive belief functions). As a consequence, k-additive measures generalize probability measures and they fill the gap between probability measures and general fuzzy measures. For a k-additive measure, the number of coefficients is reduced to

$$\sum_{i=1}^{k} \binom{n}{i}.$$

More about k-additive measures can be found, e.g., in [8]. We define the set $\mathcal{FM}^k(X)$ (resp. $\mathcal{BEL}^k(X)$) as the set of fuzzy measures (resp. belief functions) μ whose corresponding Möbius transform m satisfies $m(A) = 0$ if $|A| > k$.

Finally, we say that a fuzzy measure μ^* **dominates** μ, and we denote it by $\mu^* \geq \mu$, if

$$\mu^*(A) \geq \mu(A), \, \forall A \subseteq X.$$

For general set functions, dominance is defined by

$$\mu^*(A) \geq \mu(A), \, \forall A \subseteq X, \mu^*(X) = \mu(X).$$

Given a fuzzy measure μ, we define the set $\mathcal{FM}^k_{\geq}(\mu)$ (or $\mathcal{BEL}^k_{\geq}(\mu)$ if we restrict to dominating k-additive belief functions) as the set of fuzzy measures (resp. belief functions) in $\mathcal{FM}^k(X)$ (resp. $\mathcal{BEL}^k(X)$) dominating μ.

3 Characterizing the Set of Dominating Fuzzy Measures

Consider a fuzzy measure μ and let us turn to the problem of obtaining the set $\mathcal{FM}^k_{\geq}(\mu)$. In [2], the following result is proved.

Theorem 2. *Let μ be a fuzzy measure on X, m its Möbius transform, and suppose $P \in \mathcal{FM}^1_{\geq}(\mu)$. Then, P can be put under the following form:*

$$P(\{i\}) = \sum_{B \ni i} \lambda(B, i) m(B), \, \forall i \in X.$$

The function $\lambda : \mathcal{P}_*(X) \times X \to [0, 1]$ is a weight function satisfying:

$$\sum_{i \in B} \lambda(B, i) = 1, \, \forall \emptyset \neq B \subseteq X.$$

$$\lambda(B, i) = 0 \text{ whenever } i \notin B.$$

Dempster has shown the same result in [4] and also Shapley in [23], but both of them only for belief functions.

If we restrict our attention to the case of the set of k-additive belief functions dominating a belief function, the following result appears in [20].

Theorem 3. *Let $\mu, m, \mu^* : \mathcal{P}(X) \to \mathbb{R}$, where μ is a fuzzy measure, m its Möbius inverse, and $\mu^* \in \mathcal{BEL}^k_{\geq}(\mu)$. Then, necessarily the Möbius transform m^* of μ^* can be put under the following form:*

$$m^*(A) = \sum_{B | A \cap B \neq \emptyset} \lambda(B, A) m(B), \, \forall A \in \mathcal{P}^k_*(X),$$

where function $\lambda : \mathcal{P}_*(X) \times \mathcal{P}^k_*(X) \to [0,1]$ *is such that*

$$\sum_{A|B\cap A\neq\emptyset} \lambda(B,A) = 1, \forall B \in \mathcal{P}_*(X). \tag{1}$$

$$\lambda(B,A) = 0, \; if \; A \cap B = \emptyset. \tag{2}$$

We have to keep in mind that Eqs. 1 and 2 lead to a non-empty intersection condition; from a mathematical point of view, another possibility (with better properties) of generalizing Theorem 2 could be a more restrictive inclusion condition, i.e. satisfying $\lambda(B,A) = 0$ whenever $A \not\subseteq B$. However, this condition fails to obtain all dominating k-additive belief functions, as it is shown in [20]. When dealing with general fuzzy measures, it happens that we have to permit functions λ attaining negative values [9]. Thus, we obtain a very wide set of functions, many of them failing to satisfy monotonicity or dominance.

In this paper, we are going to apply the result for belief functions to obtain a more handy result for the general case.

Theorem 4. *Let* $\mu, \mu^* : \mathcal{P}(X) \to \mathbb{R}$, *where* $\mu \in \mathcal{FM}^k(X)$ *and* $\mu^* \in \mathcal{FM}^k_{\geq}(\mu)$, *for* $k = 1, ..., n$, *and let us denote by* m *and* m^* *their respective Möbius transforms. Let us define:*

$$m_{aux}(A) = m(A) - l_{|A|}, \; m^*_{aux}(A) = m^*(A) - l_{|A|},$$

where l_i *denotes the lower bound for the Möbius transform of subsets of cardinality* $i, i = 1, ..., k$. *Then, necessarily* m^*_{aux} *can be put under the following form:*

$$m^*_{aux}(A) = \sum_{B|A\cap B\neq\emptyset} \lambda(B,A) m_{aux}(B), \forall A \in \mathcal{P}^k_*(X),$$

where function $\lambda : \mathcal{P}^k_*(X) \times \mathcal{P}^k_*(X) \to [0,1]$ *is such that*

$$\sum_{A|A\cap B\neq\emptyset} \lambda(B,A) = 1, \forall B \in \mathcal{P}^k_*(X). \tag{3}$$

$$\lambda(B,A) = 0, \; if \; A \cap B = \emptyset. \tag{4}$$

Indeed, in this result, function λ is a sharing function of $m_{aux}(B)$ among any subset A such that $A \cap B \neq \emptyset$. Thus, this problem can be turned into a transshipment problem in a flow network. Figure 1 shows the corresponding flow network for $k = 2$.

The proof of the result is based on Gale's Theorem for a transshipment network [6], where subset A offers $m(A) - l_{|A|}$ to be shared among subsets intersecting with A. However, the underlying idea of the result relies on the result for k-additive dominating belief functions. For belief functions, Theorem 3 is an extension of Theorem 2; on the other hand, this result cannot be applied for general $k-$additive dominating measures, as shown in [9]. The idea then is to transform μ and μ^* into other set functions μ_{aux} and μ^*_{aux} resp., having

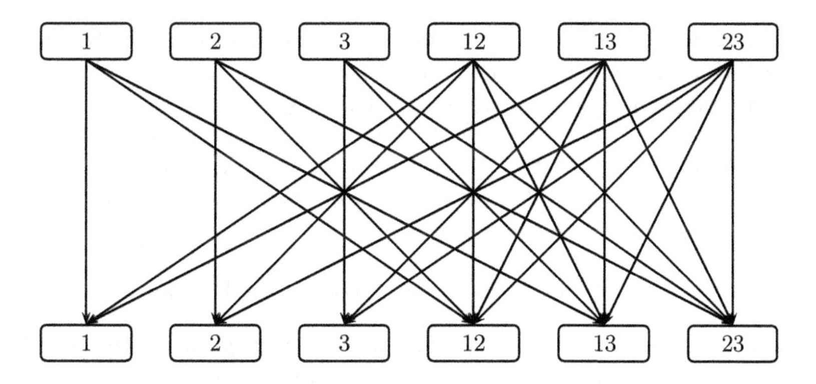

Fig. 1. Example of flow network for $k = 2$

properties similar to belief functions. More concretely, the corresponding Möbius transform is non-negative (but these set functions are not normalized!). In this sense, we can add other constraints instead of $l_{|A|}$, or even the same constant regardless the cardinality, but these are the more accurate [12].

Remark that the condition allowing positive weights for non-empty intersections in Eqs. 3 and 4 cannot be turned into an inclusion condition, as the next example shows.

Example 1. Consider $|X| = 3$ and the 2-additive fuzzy measure μ whose Möbius transform m, and whose corresponding m_{aux} are given in next table

A	$\{1\}$	$\{2\}$	$\{3\}$	$\{1,2\}$	$\{1,3\}$	$\{2,3\}$
μ	0.4	0.3	0.4	0.4	1	0.7
m	0.4	0.3	0.4	-0.3	0.2	0
m_{aux}	0.4	0.3	0.4	0.7	1.2	1

Now, consider μ^* the 2-additive measure, with m^*, m^*_{aux} given by

A	$\{1\}$	$\{2\}$	$\{3\}$	$\{1,2\}$	$\{1,3\}$	$\{2,3\}$
μ^*	0.4	0.3	0.5	0.5	1	0.7
m^*	0.4	0.3	0.5	-0.2	0.1	-0.1
m^*_{aux}	0.4	0.3	0.5	0.8	1.1	0.9

Then, $m^*_{aux}(12) > m_{aux}(12)$, while the only subset containing $\{1,2\}$ is $\{1,2\}$ itself.

The previous result can be extended when we are looking for k'-additive measures dominating k-additive measures when $k \neq k'$. For this, it suffices to notice that $\mathcal{FM}^{k'}(X) \subset \mathcal{FM}^k(X)$ if $k' < k$. Consequently, any measure in

$\mathcal{FM}^{k'}_{\geq}(\mu)$ can be derived from the previous theorem considering k. Similarly, if $k' > k$, any k'-additive measure dominating μ can be derived from the previous theorem just taking into account that $\mu \in \mathcal{FM}^{k'}(X)$.

Corollary 1. *Let* $\mu, \mu^* : \mathcal{P}(X) \rightarrow \mathbb{R}$, *where* $\mu \in \mathcal{FM}^k(X)$ *and* $\mu^* \in \mathcal{FM}^{k'}_{\geq}(\mu)$, *for* $k, k' = 1, ..., n$, *and let us denote by* m *and* m^* *their respective Möbius transforms. Assume* $k \geq k'$ *and let us define:*

$$m_{aux}(A) = m(A) - l_{|A|},\ m^*_{aux}(A) = m^*(A) - l_{|A|},$$

where l_i *denotes the lower bound for the Möbius transform of subsets of cardinality* $i, i = 1, ..., k$. *Then, necessarily* m^*_{aux} *can be put under the following form:*

$$m^*_{aux}(A) = \sum_{B | A \cap B \neq \emptyset} \lambda(B, A) m_{aux}(B),\ \forall A \in \mathcal{P}^{k'}_*(X),$$

where function $\lambda : \mathcal{P}^k_*(X) \times \mathcal{P}^{k'}_*(X) \rightarrow [0, 1]$ *is such that*

$$\sum_{A | A \cap B \neq \emptyset} \lambda(B, A) = 1,\ \forall B \in \mathcal{P}^k_*(X).$$

$$\lambda(B, A) = 0,\ if\ A \cap B = \emptyset.$$

Corollary 2. *Let* $\mu, \mu^* : \mathcal{P}(X) \rightarrow \mathbb{R}$, *where* $\mu \in \mathcal{FM}^k(X)$ *and* $\mu^* \in \mathcal{FM}^{k'}_{\geq}(\mu)$, *for* $k, k' = 1, ..., n$, *and let us denote by* m *and* m^* *their respective Möbius transforms. Assume* $k \leq k'$ *and let us define:*

$$m_{aux}(A) = m(A) - l_{|A|},\ m^*_{aux}(A) = m^*(A) - l_{|A|},$$

where l_i *denotes the lower bound for the Möbius transform of subsets of cardinality* $i, i = 1, ..., k'$. *Then, necessarily* m^*_{aux} *can be put under the following form:*

$$m^*_{aux}(A) = \sum_{B | A \cap B \neq \emptyset} \lambda(B, A) m_{aux}(B),\ \forall A \in \mathcal{P}^{k'}_*(X),$$

where function $\lambda : \mathcal{P}^k_*(X) \times \mathcal{P}^{k'}_*(X) \rightarrow [0, 1]$ *is such that*

$$\sum_{A | A \cap B \neq \emptyset} \lambda(B, A) = 1,\ \forall B \in \mathcal{P}^k_*(X).$$

$$\lambda(B, A) = 0,\ if\ A \cap B = \emptyset.$$

As we have seen in Example 1, a non-empty intersection condition is needed. However, it is possible to obtain all dominating k-additive dominating measures from set functions that can be derived via an inclusion condition. This is stated in next result.

Theorem 5. *Let* $\mu, m, \mu^*, m^* : \mathcal{P}(X) \to \mathbb{R}$, *where* $\mu \in \mathcal{FM}^k(X)$, $\mu^* \in \mathcal{FM}^k_{\geq}(\mu)$ *and* m, m^* *their corresponding Möbius inverses. Let us define*

$$m_{aux}(A) = m(A) - l_{|A|},\; m^*_{aux}(A) = m^*(A) - l_{|A|},$$

where l_i *denotes the lower bound for the Möbius transform of subsets of cardinality* $i, i = 1, ..., k$. *Then, there exists a set function (not necessarily a fuzzy measure)* μ' *dominating* μ *whose Möbius transform* m' *is such that the corresponding* m'_{aux} *can be written as*

$$m'_{aux}(B) = \sum_{A|B\subseteq A} \lambda'(A,B) m_{aux}(A), \forall B \in \mathcal{P}^k_*(X),$$

where $\lambda' : \mathcal{P}^k_*(X) \times \mathcal{P}^k_*(X) \to [0,1]$ *is such that*

$$\sum_{B|B\subseteq A} \lambda'(A,B) = 1, \forall A \in \mathcal{P}^k_*(X). \tag{5}$$

$$\lambda'(A,B) = 0 \; if \; B \not\subseteq A, \tag{6}$$

and m^*_{aux} *can be derived from* m'_{aux} *through*

$$m^*_{aux}(C) = \sum_{B|B\subseteq C} \lambda^*(B,C) m'_{aux}(B), \forall C \in \mathcal{P}^k_*(X),$$

where $\lambda^* : \mathcal{P}^k_*(X) \times \mathcal{P}^k_*(X) \to [0,1]$ *is such that*

$$\sum_{C|B\subseteq C} \lambda^*(B,C) = 1, \forall B \in \mathcal{P}^k_*(X).$$

$$\lambda^*(B,C) = 0 \; if \; B \not\subseteq C.$$

This result is explained in Fig. 2 for $|X| = 3$ and $k = 2$.

It is worthnoting the differences with a similar result appearing in [20]; in that result, applying for dominating k-additive belief functions, any set function obtained using Eqs. 5 and 6 is a k-additive dominating belief function. However, in this more general situation, we cannot ensure monotonicity, as next example shows.

Example 2. Consider $|X| = 3$ and let μ be the $\{0,1\}$-fuzzy measure such that $\mu(A) = 1$ if and only if $\{1,2\} \subseteq A$. Then, the corresponding m is given by $m(1,2) = 1$ and $m(A) = 0$ otherwise. Then, $\mu \in \mathcal{FM}^2(X)$ and $m_{aux}(1,2) = 2, m_{aux}(i,j) = 1$ for any other pair and $m_{aux}(i) = 0$ for any singleton. Now, if we define

$$\lambda(A,B) = \begin{cases} 1 \text{ if } A = B \\ 0 \text{ otherwise} \end{cases}, \text{ if } A \neq \{1,2\}, \; \lambda(\{1,2\},B) = \begin{cases} 1 \text{ if } B = \{1\} \\ 0 \text{ otherwise} \end{cases}$$

Then, $m'_{aux}(1) = 2, m'_{aux}(1,2) = 0, m'_{aux}(2) = 0$, whence it follows $m'(1) = 2, m'(1,2) = -1, m'(2) = 0$ and thus, $\mu'(1) = 2 > \mu'(1,2) = 1$, violating monotonicity.

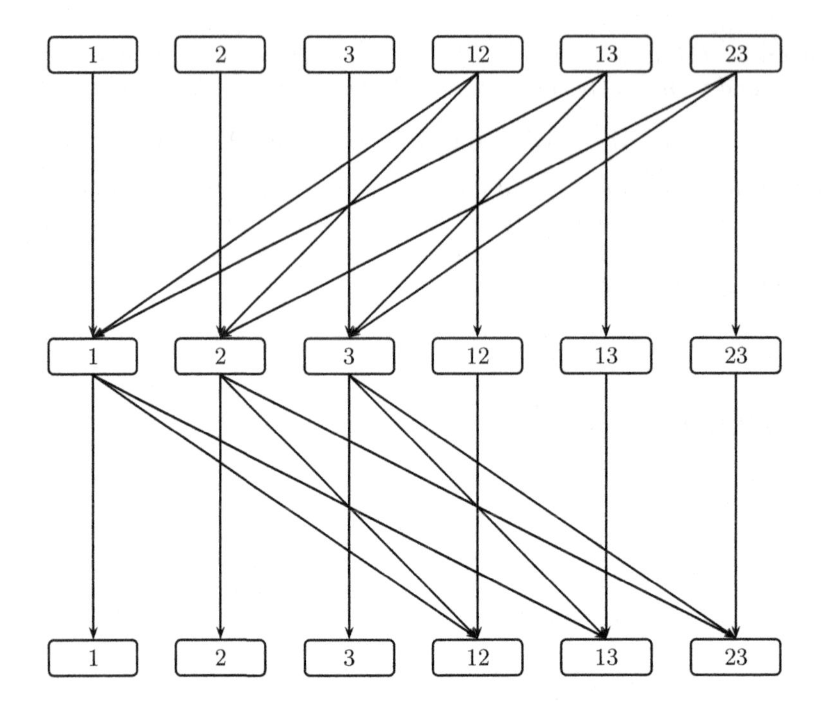

Fig. 2. Example of flow network for $k = 2$.

4 Conclusions and Open Problems

In this paper we have dealt with the problem of obtaining the set of k-additive dominating measures of a k-additive measure. For this, we have used a previous result valid for the special case of belief functions. The result follows the same philosophy of other results derived by Chateauneuf and Jaffray [2] and Miranda et al. [18]; along this line, we have obtained a superset of the set $\mathcal{FM}^k_{\geq}(\mu)$. A natural question is whether $\mathcal{FM}^k_{\geq}(\mu)$ is strictly contained into that set.

We have proved that in general, the non-empty intersection condition cannot be removed, but it seems interesting to search for special cases for which non-empty intersection can be turned into an inclusion condition because this condition seems easier to handle.

However, we feel that the most interesting open problem is to apply these results in a procedure for obtaining the set of vertices of $\mathcal{FM}^k_{\geq}(\mu)$. As it can be easily found, $\mathcal{FM}^k_{\geq}(\mu)$ is a polytope and thus, it is completely determined by its vertices. There are several results concerning the vertices of the core and the set of dominating probabilities, i.e. $\mathcal{FM}^1_{\geq}(\mu)$ [14,23]. For $k \geq 2$, several results have been obtained in [11]. The problem is particularly difficult for $2 < k < n$ because the set of vertices of $\mathcal{FM}^k(X)$ is not the set of $\{0, 1\}$-valued measures in $\mathcal{FM}^k(X)$ [15] and the general form of these vertices is not known. Even for a seemingly simple situation, the set $\mathcal{FM}^2_{\geq}(\mu)$, up to our knowledge, the set of vertices is not known for any μ. The results in the paper could shed light on these problems, as they provide bounds for these sets.

Acknowledgements. This research has been partially supported by Spanish Grant MTM2012-33740 and MTM-2015-67057.

References

1. Bondareva, O.: Some applications of linear programming to the theory of cooperative games. Problemy Kibernet **10**, 119–139 (1963)
2. Chateauneuf, A., Jaffray, J.-Y.: Some characterizations of lower probabilities and other monotone capacities through the use of Möbius inversion. Math. Soc. Sci. **17**, 263–283 (1989)
3. Choquet, G.: Theory of capacities. Annales de l'Institut Fourier **5**, 131–295 (1953)
4. Dempster, A.P.: Upper and lower probabilities induced by a multivalued mapping. Ann. Math. Statististics **38**, 325–339 (1967)
5. Driessen, T.: Cooperative Games. Kluwer Academic, Dordrecht (1988)
6. Gale, D.: The Theory of Linear Economic Models. Mc Graw Hill, New York (1960)
7. Grabisch, M.: k-order additive discrete fuzzy measures. In: Proceedings of 6th International Conference on Information Processing and Management of Uncertainty in Knowledge-Based Systems (IPMU), Granada, Spain, pp. 1345–1350 (1996)
8. Grabisch, M.: k-order additive discrete fuzzy measures and their representation. Fuzzy Sets Syst. **92**, 167–189 (1997)
9. Grabisch, M.: On lower and upper approximation of fuzzy measures by k-order additive measures. In: Bouchon-Meunier, B., Yager, R.R., Zadeh, L. (eds.) Information, Uncertainty, Fusion, pp. 105–118. Kluwer Sci. Publ. (2000). Selected papers from IPMU'98
10. Grabisch, M.: The core of games on ordered structures and graphs. Ann. Oper. Res. **204**, 33–64 (2013). doi:10.1007/s10479-012-1265-4
11. Grabisch, M., Miranda, P.: On the vertices of the k-additive core. Discrete Math. **308**, 5204–5217 (2008)
12. Grabisch, M., Miranda, P.: Exact bounds of the Möbius inverse of monotone set functions. Discrete Appl. Math. **186**, 7–12 (2015)
13. Harsanyi, J.C.: A simplified bargaining model for the n-person cooperative game. Int. Econ. Rev. **4**, 194–220 (1963)
14. Ichiishi, T.: Super-modularity: Applications to convex games and to the Greedy algorithm for LP. J. Econ. Theory **25**, 283–286 (1981)
15. Miranda, P., Combarro, E.F., Gil, P.: Extreme points of some families of non-additive measures. Eur. J. Oper. Res. **33**(10), 3046–3066 (2006)
16. Miranda, P., Grabisch, M.: Optimization issues for fuzzy measures. Int. J. Uncertainty, Fuzziness, Knowl.-Based Syst. **7**(6), 545–560 (1999). Selected papers from IPMU'98
17. Miranda, P., Grabisch, M.: k-balanced games and capacities. Eur. J. Oper. Res. **200**, 465–472 (2010)
18. Miranda, P., Grabisch, M., Gil, P.: Identification of non-additive measures from sample data. In: Kruse, R., Della Riccia, G., Dubois, D., Lenz, H.-J. (eds.) Planning based on Decision Theory. CISM Courses and Lectures, vol. 472, pp. 43–63. Springer, Vienna (2003)

19. Miranda, P., Grabisch, M., Gil, P.: Axiomatic structure of k-additive capacities. Mathe. Soc. Sci. **49**, 153–178 (2005)
20. Miranda, P., Grabisch, M., Gil, P.: Dominance of capacities by k-additive belief functions. Eur. J. Oper. Res. **175**, 912–930 (2006)
21. Rota, G.C.: On the foundations of combinatorial theory I. Theory of Möbius functions. Zeitschrift für Wahrscheinlichkeitstheorie und Verwandte Gebiete (2), 340–368 (1964)
22. Shafer, G.: A Mathematical Theory of Evidence. Princeton University Press, Princeton (1976)
23. Shapley, L.S.: Cores of convex games. Int. J. Game Theory **1**, 11–26 (1971)
24. Walley, P.: Coherent lower (and upper) probabilities. Technical report 22, U. of Warwick, Coventry, (UK) (1981)
25. Walley, P., Fine, T.L.: Towards a frequentist theory of upper and lower probability. Ann. Stat. **10**, 741–761 (1982)
26. Wolfenson, M., Fine, T.L.: Bayes-like decision making with upper and lower probabilities. J. Amer. Statis. Assoc. **77**, 80–88 (1982)

On Capacities Characterized by Two Weight Vectors

Christophe Labreuche(✉)

Thales Research & Technology, 1 avenue Augustin Fresnel,
91767 Palaiseau Cedex, France
christophe.labreuche@thalesgroup.com

Abstract. We are interested in aggregation function based on two weights vectors: the criteria weights p and the rank weights w. The main drawback of the existing proposals based on p and w (in particular the *Weighted OWA* (WOWA) and the *Semi-Uninorm OWA* (SUOWA) operators) is that their expression is rather complex and the contribution of the weights p and w in the aggregation is obscure as there is no clear interpretation of these weights. We propose a new approach to define aggregation functions based on the weights p and w. We consider the class of capacities (which subsumes the WOWA and SUOWA). We start by providing clear interpretations of these weights that are seen as constraints on the capacity. We consider thus the whole class of capacities fulfilling these constraints. A simulation shows that the WOWA and SUOWA almost never satisfy these constraints in a strict sense.

Keywords: Choquet integral · Ordered Weighted Average · Shapley value · Intolerance index

1 Introduction

Aggregation functions are widely used in many applications. The Weighted Sum (WS) and Ordered Weighted Average (OWA) – defined by Yager [1,2]– operators are very popular. They are both based on a vector on n components, if there are n criteria to aggregate. In the WS, the parameters p are criteria weights. In the OWA, the parameters w are rank weights. The main interest of these aggregation functions is their simplicity and interpretability. Thanks to that, these two weight vectors can be directly provided by a decision maker. The first one corresponds to the importance of criteria. This type of information is often asked to decision makers. The second one is related to the assessment of solutions having good marks on only one criterion, on only two criteria, etc., on all criteria [3]. A decision maker can also provide such an information.

It is then natural to combine criteria weights and rank weights, in order to take advantage of the flexibility of both WS and OWA – being thereby able to capture relative importance of criteria and interaction among them – while using only a very small amount of information from the decision maker. By contrast,

J.P. Carvalho et al. (Eds.): IPMU 2016, Part I, CCIS 610, pp. 23–34, 2016.
DOI: 10.1007/978-3-319-40596-4_3

the most general aggregation function able to represent the importance of criteria and interaction among them is the Choquet integral w.r.t. a capacity. The latter contains 2^n parameters, making it difficult to elicit in applications. The interest of having a generalization of WS and OWA based on only $2n$ parameters is thus very high.

Several aggregation functions based on criteria weights and rank weights have been proposed in the literature (see Sect. 2). The *Weighted OWA* (WOWA) operator [4], the *Hybrid Weighted Averaging* (HWA) operator [5], the *Semi-Uninorm OWA* (SUOWA) operator [6], and the Ordered Weighted Averaging Weighted Average (OWAWA) operator [7]. HWA has a simple expression but fails to fulfill basic important properties, such as idempotency. OWAWA is a simple linear combination of WS and OWA, and its interpretation is not so intuitive. On the other, WOWA and SUOWA operators have quite complex expressions, which are not intuitive for a decision maker. Moreover, the two weights vectors p and w are combined in a way that is difficult to understand, and one cannot readily see what is the contribution of p and w in the formula.

The aim of this paper is to propose a new approach to define aggregation functions based on two weight vectors p and w, in which these two vectors have a clear semantics in the aggregation function. The existing approaches first define an aggregation formula parametrized by w and p, and then analyze their properties. We proceed in the opposite way (see Sect. 3). We start with a class of very general aggregation functions able to capture both importance of criteria and interaction among them. We naturally consider the Choquet integral w.r.t. general capacities. Then we define clear semantics of p and w. More precisely, it takes the form of linear constraints on the capacity.

Going back to WOWA and SUOWA, we performed a numerical simulation to check whether these constraints are fulfilled, and if not how far they are to be fulfilled (see Sect. 4).

2 Background

The set of criteria is denoted by $N = \{1, \ldots, n\}$. We are interested in an aggregation function $H : \mathbb{R}^N \to \mathbb{R}$.

2.1 Background on the Choquet Integral

Capacities and Choquet Integral. A *capacity* (also called *fuzzy measure*) on $N = \{1, \ldots, n\}$ is a set function $\mu : 2^N \to [0, 1]$ such that [8,9]

- boundary conditions: $\mu(\emptyset) = 0$, $\mu(N) = 1$,
- monotonicity: $\forall A \subseteq B \subseteq N$, $\mu(A) \leq \mu(B)$.

Let $\mathcal{M}_N$ be the set of all capacities on N.

A capacity is said to be *additive* if $\mu(A \cap B) = \mu(A) + \mu(B)$ for every pair (A, B) of disjoint coalitions. A capacity is said to be *symmetric* if $\mu(A)$ depends only on the cardinality of A.

The *Choquet integral* of $a = (a_1, \ldots, a_n) \in \mathbb{R}^n$ defined w.r.t. a capacity μ has the following expression [8]:

$$C_\mu(a_1, \ldots, a_n) = \sum_{i=1}^{n} a_{\sigma(i)} \times [\mu(\{\sigma(i), \cdots, \sigma(n)\}) - \mu(\{\sigma(i+1), \cdots, \sigma(n)\})], \tag{1}$$

where σ is a permutation on N such that $a_{\sigma(1)} \leq a_{\sigma(2)} \leq \cdots \leq a_{\sigma(n)}$, and $\mu(\{\sigma(i+1), \cdots, \sigma(n)\}) = 0$ when $i = n$. The Choquet integral has been proved to be able to model both the importance of criteria and the interaction between criteria.

The weighted sum is given by

$$\text{WS}_p(a) = \sum_{i \in N} p_i \, a_i$$

where $p = (p_1, \ldots, p_n)$ is the weight of criteria. It corresponds to a Choquet integral with respect to an additive capacity: $\mu_{\text{WS}_p}(S) = \sum_{i \in S} p_i$.

The *Ordered Weighted Average* (OWA) is defined by for any $a \in \mathbb{R}^N$ [1,2]

$$\text{OWA}_w(a) = \sum_{j=1}^{n} w_j \, a_{\sigma(n-j+1)}$$

where w_j are the weights allotted to the jth largest value of vector a. It corresponds to a Choquet integral with respect to a symmetric capacity: $\mu_{\text{OWA}_w}(S) = \sum_{j=1}^{|S|} w_j$.

Shapley Value. It is not easy to interpret a capacity in a synthetic way, as it contains 2^n parameters. The most helpful importance index helping to interpret a capacity is the concept of mean importance, defined as the Shapley value of the capacity [10]:

$$\phi_i(\mu) = \sum_{S \subseteq N \setminus \{i\}} \frac{|S|!(n - |S| - 1)!}{n!} (\mu(S \cup \{i\}) - \mu(S)).$$

Index of Tolerance and Intolerance. In decision under uncertainty, the use of belief functions (infinitely monotone capacities) allows to represent risk aversion. This is a general behavior of the decision maker towards risk or uncertainty, that is not specific to a given state of nature. The counterpart of risk aversion in MCDA is the concept of *intolerance*. Roughly speaking, a decision maker is intolerant if one needs to be good on most of criteria to have a good overall evaluation [11]. Marichal introduced tolerance and intolerance indices to formalize this idea [12].

Vetos and favors are situations of extreme intolerance and tolerance respectively. Criterion i is said to be a veto (resp. a favor) if $C_\mu(a) = 0$ (resp. $C_\mu(a) = 1$)

whenever $a_i = 0$ (resp. $a_i = 1$) [13]. In other words, a bad evaluation on a veto criterion cannot be compensated by the other criteria. The concept of a veto can be generalized to subsets of cardinality k: A Choquet integral C_μ (or equivalently its underlying capacity μ) is *at most k-intolerant* if $C_\mu(a) = 0$ whenever $a_{\sigma(k)} = 0$ [11]. It is *k-intolerant* if, in addition, $C_\mu(a) \neq 0$ for some $a_{\sigma(k-1)} = 0$. One can prove that C_μ is at most k-intolerant if and only if $\mu(A) = 0$, $\forall A \subseteq N$ with $|A| \leq n-k$ [11]. The index of k-intolerance is the mean value of $C_\mu(a)$ over all a such that $a_{\sigma(k)} = 0$ [12]:

$$\text{intol}_k(\mu) = \frac{n-k+1}{(n-k)\binom{n}{k}} \sum_{\substack{K \subseteq N \\ |K|=k}} E(C_\mu(0_K, Z_{-K}))$$

where E denotes expectation, assuming that the random inputs $Z_1, \ldots, Z_n$ are independent and uniformly distributed. This gives:

$$\text{intol}_k(\mu) = 1 - \frac{1}{n-k} \sum_{t=0}^{n-k} \frac{1}{\binom{n}{t}} \sum_{\substack{T \subseteq N \\ |T|=t}} \mu(T).$$

Likewise, a Choquet integral C_μ is *at most k-tolerant* if $C_\mu(a) = 1$ whenever $a_{\sigma(n-k+1)} = 1$ [11]. The index of k-tolerance is the mean value of $C_\mu(a)$ over all a such that $a_{\sigma(n-k+1)} = 1$ [12]:

$$\text{tol}_k(\mu) = \frac{n-k+1}{(n-k)\binom{n}{k}} \sum_{\substack{K \subseteq N \\ |K|=k}} E(C_\mu(1_K, Z_{-k})) - \frac{1}{n-k},$$

This gives:

$$\text{tol}_k(\mu) = \frac{1}{n-k} \sum_{t=k}^{n} \frac{1}{\binom{n}{t}} \sum_{\substack{T \subseteq N \\ |T|=t}} \mu(T) - \frac{1}{n-k}.$$

2.2 Existing Proposals Based on Two Weight Vectors

We describe in this section the existing proposals based on two weight vectors: the criteria weights p and the rank weights w.

Wished Properties. We start by giving some important properties that the aggregation function $H : \mathbb{R}^N \to \mathbb{R}$ should satisfy. The following properties are taken from [14]:

- Continuity: H is continuous
- Idempotency: $H(\alpha, \ldots, \alpha) = \alpha$ for every α
- Monotonicity: H is monotone

- Compensation: $\min_{i \in N} a_i \leq H(a) \leq \max_{i \in N} a_i$. Note that this property follows directly from Idempotency and Monotonicity

Consider now an aggregation function $H_{p,w}$ based on the two weights p and w. The following property is wished:

- Generalization of WS_p and OWA_w: One says that $H_{p,w}$ generalizes both the WS and OWA, if $H_{p,\eta} = \text{WS}_p$ and $H_{\eta,w} = \text{OWA}_w$, where η denotes the uniform vector $(\frac{1}{n}, \ldots, \frac{1}{n})$ [6].

WOWA. Torra introduced an aggregation function based on the criteria weights p and the rank weights w. It is called *Weighted OWA* (WOWA) operator [4]. It depends also on a quantifier $Q : [0,1] \to [0,1]$, which is a non-decreasing function such that $Q(0) = 0$ and $Q(1) = 1$. The quantifier Q is defined from rank weights w in the following way [2,15]

$$Q\left(\frac{k}{n}\right) - Q\left(\frac{k-1}{n}\right) = w_k, \qquad \text{for } k = 1, \ldots, n.$$

The WOWA has the following expression:

$$\text{WOWA}^Q_{p,w}(a) = \sum_{j=1}^{n} q_j \, a_{\tau(n-j+1)}$$

where

$$q_j = Q\left(\sum_{k=1}^{j} p_{\tau(n-k+1)}\right) - Q\left(\sum_{k=1}^{j-1} p_{\tau(n-k+1)}\right).$$

A WOWA are a particular case of a Choquet integral for the following capacity

$$\mu_{\text{WOWA}^Q_{p,w}}(S) = Q\left(\sum_{i \in S} p_i\right).$$

SUOWA. Another aggregation function constructed from the two vectors p and w has recently been proposed by Llamazares [6,16]. Its definition is based a semi-uninorm and is thus called *Semi-Uninorm OWA* operator (SUOWA in short). A *semi-uninorm* is a mapping $U : [0,1]^2 \to [0,1]$ if it is monotonic and possesses a neutral element $e \in [0,1]$ (such that $U(e,x) = U(x,e) = x$ for every x).

Let us define a set function v by

$$v(S) = |S| \, U\left(\frac{\mu_{\text{WS}_p}(S)}{|S|}, \frac{\mu_{\text{OWA}_w}(S)}{|S|}\right) \tag{2}$$

The factor $\frac{1}{|S|}$ comes from the fact that when $p = \eta$ (resp. $w = \eta$), $\frac{\mu_{\text{WS}_p}(S)}{|S|}$ (resp. $\frac{\mu_{\text{OWA}_w}(S)}{|S|}$) is independent of S and is always equal to $\frac{1}{n}$. Hence property "generalization of WS_p and OWA_w" is satisfied, if the neutral element is $e = \frac{1}{n}$.

The drawback with this expression is that it may be non-monotone. Hence Llamazares considers the monotonic cover of v:

$$\mu_{\text{SUOWA}^U_{p,w}}(S) = \max_{T \subseteq S} v(T).$$

The monotonic cover can be computed recursively [16]

$$\mu_{\text{SUOWA}^U_{p,w}}(S) = \max\left(v(S), \max_{i \in S} \mu_{\text{SUOWA}^U_{p,w}}(S \setminus \{i\}) \right).$$

Then the SUOWA is the Choquet integral with respect to $\mu_{\text{SUOWA}^U_{p,w}}$.

Set-function v is bounded by 1 if and only if $U \in \widetilde{\mathcal{U}}^{\frac{1}{n}}$, where $\widetilde{\mathcal{U}}^{\frac{1}{n}}$ is the set of semi-uninorms with neutral element $e = \frac{1}{n}$ such that $U(\frac{1}{k}, \frac{1}{k}) \leq \frac{1}{k}$ for all $k \in \mathbb{N}$ [16, Proposition 5]. In the context of MCDA, we will generally consider idempotent semi-uninorms. They are necessarily elements in $\widetilde{\mathcal{U}}^{\frac{1}{n}}$. Let us give some interesting examples of semi-uninorms with neutral element $e = \frac{1}{n}$ [17]:

$$U_{\min}(x,y) = \begin{cases} \max(x,y) & \text{if } (x,y) \in [\frac{1}{n}, 1]^2, \\ \min(x,y) & \text{otherwise} \end{cases}$$

$$U_{\max}(x,y) = \begin{cases} \min(x,y) & \text{if } (x,y) \in [0, \frac{1}{n}]^2, \\ \max(x,y) & \text{otherwise} \end{cases}$$

$$U_{AM}(x,y) = \begin{cases} \min(x,y) & \text{if } (x,y) \in [0, \frac{1}{n}]^2, \\ \max(x,y) & \text{if } (x,y) \in [\frac{1}{n}, 1]^2 \setminus \{(\frac{1}{n}, \frac{1}{n})\} \\ \frac{x+y}{2} & \text{otherwise} \end{cases}$$

$$U_P(x,y) = n \min\left(x, y, \frac{1}{n}\right) \max\left(x, y, \frac{1}{n}\right) = \begin{cases} \max(x,y) & \text{if } (x,y) \in [\frac{1}{n}, 1]^2, \\ \min(x,y) & \text{if } (x,y) \in [0, \frac{1}{n}]^2, \\ n\, x\, y & \text{otherwise} \end{cases}$$

$$U_{T_M}(x,y) = \min\left(x, y, \frac{1}{n}\right) + \max\left(x, y, \frac{1}{n}\right) - \frac{1}{n} = \begin{cases} \max(x,y) & \text{if } (x,y) \in [\frac{1}{n}, 1]^2, \\ \min(x,y) & \text{if } (x,y) \in [0, \frac{1}{n}]^2, \\ x + y - \frac{1}{n} & \text{otherwise} \end{cases}$$

$$U_{T_L}(x,y) = \begin{cases} \max(x,y) & \text{if } (x,y) \in [\frac{1}{n}, 1]^2, \\ \max(x + y - \frac{1}{n}, 0) & \text{otherwise} \end{cases}$$

All previous semi-uninorms are idempotent except U_{T_L}, as they subsumes to the min or max operator on the diagonal. Functions $U_{\min}$ and $U_{\max}$ are the smallest and largest idempotent semi-uninorms respectively. Note that U_{AM} is basically the average mean between $U_{\min}$ and $U_{\max}$. Semi uninorm U_P is a continuous idempotent semi-uninorm defined from generating the logarithm function. Finally, T_M and T_L are defined from the largest and the smallest (i.e. the Lukasiewicz t-norm) quasi-copula.

Concerning U_{T_L}, if $\min_{i \in N} p_i + \min_{j \in N} w_j \geq \frac{1}{n}$, and for every $1 \leq k \leq n$, either $\frac{1}{k} \sum_{1 \leq j \leq k} w_j \leq \frac{1}{n}$ or for all $S \subseteq N$ with $|S| = k$, $\frac{1}{k} \sum_{i \in S} p_i \leq \frac{1}{n}$, then

$$\mu_{\text{SUOWA}^{U_{T_L}}_{p,w}}(S) = \mu_{\text{WS}_p}(S) + \mu_{\text{OWA}_w}(S) - \frac{|S|}{n}$$

HWA. The hybrid weighted averaging (HWA) operator is an aggregation function $H_p^w : \mathbb{R}^n \to \mathbb{R}$ defined by [5]

$$\mathrm{HWA}_p^w(a) = \sum_{i=1}^{n} w_i \times n p_{\sigma(i)} a_{\sigma(i)}$$

where σ is a permutation of $\{1, \ldots, n\}$ such that $p_{\sigma(1)} a_{\sigma(1)} \geq \cdots \geq p_{\sigma(n)} a_{\sigma(n)}$. This formula can be seen as the OWA operator applied to vector $(np_1 a_1, \ldots, np_n a_n)$.

HWA fails to satisfy Idempotency and Compensation properties, which make it not suitable for applications in MCDA.

OWAWA. The Ordered Weighted Averaging Weighted Average (OWAWA) operator is defined by [7]

$$\mathrm{OWAWA}_p^W(a) = \alpha\, \mathrm{WS}_p(a) + (1-\alpha)\, \mathrm{OWA}_w(a)$$

where $\alpha \in [0, 1]$ is a parameter. This expression is too simple and fails to combine the weights w and p in a subtle way. In particular, it does not satisfy the "Generalization of WS_p and OWA_w" property [6].

3 Definition of a Semantics on the Two Weight Vectors *p* and *w*

We are given two weight vectors: the criteria weights $p = (p_1, \ldots, p_n)$, and the rank weights $w = (w_1, \ldots, w_n)$. As we have seen, a major problem with the existing proposals combining these two weight vectors p and w is their interpretation.

The existing approaches first define an aggregation formula parametrized by w and p, and then analyze their properties. We proceed in the opposite way. We start by providing some properties that we would like to have for an aggregation function based on the two weight vectors p and w. These properties will be considered as constraints on the capacity μ. One obtains thus a set of admissible capacities.

The aim of this section is to provide a clear interpretation of p and w, and derive constraints on the capacity from these interpretations. Intuitively, the two vectors p and w are orthogonal in their interpretations, as p relates the importance of criteria and w relates to the type of interaction among criteria.

3.1 Constraints on *p*

Let us first identify the constraints on μ derived from the criteria weight vector p. The weighted sum WS_p is the simplest aggregation function based on p. In the weighted sum, p_i has a clear interpretation. It is simply the importance of criterion i. We observe that $p_i = \mathrm{WS}_p(1_i, 0_{-i}) - \mathrm{WS}_p(0, \ldots, 0)$.

When the DM provides as inputs the importance p_i of criterion i, the mean importance of criteria i should be precisely p_i. The Shapley value has been advocated to represent the mean importance of the criteria for a capacity. Then the Shapley values of μ shall be equal to p:

$$\forall i \in N \qquad \phi_i(\mu) = p_i. \tag{3}$$

3.2 Constraints on w

The rank weight vector w that we wish to use, is taken from the OWA operator OWA_w. In an OWA operator, all criteria are symmetric and have thus the same importance. Hence w describes only the way criteria are interacting together.

We have used the Shapley value to interpret p. The $*$-intolerant indices depict interaction among criteria, and are not dependent on a particular pair or subset of criteria – unlike the interaction indices. Hence the k-intolerant index might be suitable to interpret the weights w. In order to check this, let compute the k-intolerant index for the OWA operator OWA_w:

$$\begin{aligned}\text{intol}_k(\mu_{\text{OWA}_w}) &= 1 - \frac{1}{n-k}\sum_{t=1}^{n-k}\frac{1}{\binom{n}{t}}\sum_{\substack{T\subseteq N\\|T|=t}}\mu_{\text{OWA}_w}(T)\\ &= 1 - \frac{1}{n-k}\sum_{t=1}^{n-k}\sum_{j=1}^{t} w_j = 1 - \frac{1}{n-k}\sum_{t=1}^{n-k} t\, w_t.\end{aligned}$$

There is a clear relation between the k-intolerant index and the value of the $n-k$ first values in the weight vector w. However, this relation is not so trivial, and it will not be convenient as a constraint.

An essential property of the weights w is the following [3]: $\text{OWA}_w(1_S, 0_{N\setminus S}) = w_1 + w_2 + \cdots + w_s$ for any $S \subseteq N$ with $|S| = s$. We note that $\text{OWA}_w(1_S, 0_{N\setminus S})$ is equal to the OWA capacity μ_{OWA_w} at S. In order to generalize this property to a non-symmetric capacity, we just need to replace $\text{OWA}_w(1_S, 0_{N\setminus S})$ by the average value of $\mu(S)$ over all subsets of cardinality s. This corresponds to the term $\frac{1}{\binom{n}{t}}\sum_{\substack{T\subseteq N\\|T|=t}}\mu(T) =: A_t(\mu)$ in the expression of intol_k. There is a simple linear relation between the $*$-intolerant indices and the A_* indices:

$$\text{intol}_1(\mu) = 1 - \frac{1}{n-1}\sum_{t=1}^{n-1} A_t(\mu)$$

$$\cdots$$

$$\text{intol}_k(\mu) = 1 - \frac{1}{n-k}\sum_{t=1}^{n-k} A_t(\mu)$$

$$\cdots$$

$$\text{intol}_{n-1}(\mu) = 1 - A_1(\mu).$$

Enforcing constraints on the $*$-intolerant indices or on the A_* indices is completely equivalent. Due to the simple relations $A_t(\mu_{\text{OWA}_w}) = \sum_{j=1}^{t} w_j$, we choose to interpret the weights w from the A indices. Note that $A_t(\mu)$ is the average value of an alternative being very good on t criteria and very bad on the remaining ones.

Hence we obtain the following constraints for the interpretation of the w rank weights:

$$\forall s \in \{1, \ldots, n-1\} \qquad \frac{1}{\binom{n}{s}} \sum_{S \subseteq N \,:\, |S|=s} \mu(S) = \sum_{k=1}^{s} w_k. \tag{4}$$

Note that (4) with $s = n$ has been removed as it is trivially satisfied by every capacity.

3.3 Set of Capacities Consistent with the Semantics of p and w

According to the previous subsections, we are interested to the set of capacities that are consistent with the semantics given to p and w (relations (3) and (4)):

$$\begin{aligned}\mathcal{M}_N(p,w) := \Big\{ \mu \in \mathcal{M}_N \ : \\ &\forall i \in N \quad \sum_{S \subseteq N \setminus \{i\}} \frac{|S|!(n-|S|-1)!}{n!} \left(\mu(S \cup \{i\}) - \mu(S)\right) = p_i \ , \\ &\forall s \in \{1, \ldots, n-1\} \ \frac{1}{\binom{n}{s}} \sum_{S \subseteq N \,:\, |S|=s} \mu(S) = \sum_{k=1}^{s} w_k \ \Big\}\end{aligned}$$

4 Satisfaction of the Constraints with the Existing Proposals

The aim of this section is to analyze whether the WOWA and SUOWA operators satisfy relations (3) and (4), and if not, how far they are from fulfilling it.

Experimental Setting. We consider a family $\mu_{p,w}$ of capacities parametrized by the two weight vectors p and w.

Given two weight vectors p and w, we first construct $\mu_{p,w}$, then we compute $\phi_1(\mu_{p,w}), \ldots, \phi_n(\mu_{p,w}), A_1(\mu_{p,w}), \ldots, A_{n-1}(\mu_{p,w})$. In order to measure how far we are to fulfill relations (3) and (4), we introduce the following errors

$$E_P(p,w) := \sqrt{\sum_{i \in N} \left(\phi_i(\mu_{p,w}) - p_i\right)^2}$$

$$E_W(p,w) := \sqrt{\sum_{1 \le k \le n-1} \left(A_k(\mu_{p,w}) - \sum_{j=1}^{k} w_j \right)^2}$$

These two metrics are equal to zero when (3) and (4) are fulfilled.

We generated 100 000 random vectors p and w. The L^2 average and max value of $E_P(p,w)$ over the 100 000 runs are denoted by L^2_P and L^∞_P respectively. The L^2 average and max value of $E_W(p,w)$ over the 100 000 runs are denoted by L^2_W and L^∞_W respectively.

Satisfaction of the Constraints with the WOWA Operator. First of all, relations (3) and (4) are almost never satisfied by the WOWA operator. The numerical results are given in the following table:

n	3	4	5	6	8	10
L^2_P	0.040	0.026	0.022	0.016	0.010	0.007
L^∞_P	0.400	0.284	0.312	0.219	0.156	0.099
L^2_W	0.082	0.081	0.079	0.077	0.074	0.071
L^∞_W	0.583	0.493	0.469	0.384	0.271	0.242

We note that the average values L^2_P and L^2_W are rather small. Hence relations (3) and (4) are not far from being satisfy on average. However, the error is quite large in the worst cases. In these situations, there is a major discrepancy between the interpretation that the decision maker would expect and what the WOWA will provide.

To illustrate the worst cases, consider the following values: $n = 4$, $p = (0.01, 0.07, 0.12, 0.7)$ and $w = (0.02, 0.42, 0.53, 0.03)$. The capacity corresponding to the WOWA is given by:

$$\begin{aligned}
&\mu(\{1\}) = 0.0004, \quad \mu(\{2\}) = 0.0056, \quad \mu(\{3\}) = 0.0096, \quad \mu(\{4\}) = 0.864\\
&\mu(\{1,2\}) = 0.0064, \quad \mu(\{1,3\}) = 0.0104, \quad \mu(\{2,3\}) = 0.0152\\
&\mu(\{1,4\}) = 0.8852, \quad \mu(\{2,4\}) = 0.9724, \quad \mu(\{3,4\}) = 0.9784\\
&\mu(\{1,2,3\}) = 0.016, \quad \mu(\{1,2,4\}) = 0.9736, \quad \mu(\{1,3,4\}) = 0.9796\\
&\mu(\{2,3,4\}) = 0.9868.
\end{aligned}$$

For this capacity, we have

$$\begin{aligned}
\phi(\mu) &= (0.00567, 0.025, 0.03, 0.93933)\\
A_*(\mu) &= (0.22, 0.478, 0.739, 1.0).
\end{aligned}$$

We obtain the following errors

$$E_P(p,w) = 0.2596, \quad E_W(p,w) = 0.3079.$$

These large errors can be explained somehow by the fact that the information contained in p and w is conflicting.

	with $U_{\min}$					
n	3	4	5	6	8	10
L_P^2	0.124	0.108	0.094	0.0846	0.07	0.0611
L_P^∞	0.755	0.651	0.576	0.479	0.347	0.288
L_W^2	0.111	0.113	0.112	0.111	0.108	0.106
L_W^∞	0.685	0.746	0.707	0.5088	0.456	0.398

	with $U_{\max}$					
n	3	4	5	6	8	10
L_P^2	0.126	0.112	0.100	0.092	0.080	0.071
L_P^∞	0.717	0.716	0.550	0.477	0.364	0.314
L_W^2	0.118	0.125	0.129	0.130	0.133	0.134
L_W^∞	0.704	0.857	0.826	0.829	0.859	0.865

	with U_{T_M}					
n	3	4	5	6	8	10
L_P^2	0.081	0.075	0.068	0.063	0.055	0.049
L_P^∞	0.465	0.387	0.315	0.299	0.206	0.164
L_W^2	0.066	0.070	0.071	0.072	0.074	0.074
L_W^∞	0.424	0.369	0.319	0.399	0.352	0.331

	with U_{T_L}					
n	3	4	5	6	8	10
L_P^2	0.057	0.053	0.049	0.045	0.040	0.036
L_P^∞	0.436	0.401	0.313	0.264	0.202	0.175
L_W^2	0.048	0.053	0.056	0.057	0.060	0.061
L_W^∞	0.394	0.348	0.347	0.301	0.305	0.357

Satisfaction of the Constraints with the SUOWA Operator. As for the WOWA, relations (3) and (4) are almost never satisfy with the SUOWA operator. The following numerical results – using different semi-uninorms – are of the same kind as for the WOWA.

We notice that U_{T_L} gives the smallest errors on average.

5 Conclusion

In order to provide a clear interpretation of the two weight vectors p and w, we consider the set $\mathcal{M}_N(p, w)$ of capacities that fulfill both (3) and (4). These two relations provide a clear semantics to p and w respectively.

We note that neither WOWA nor SUOWA satisfy these properties in the general case. Even though we have seen that these classes of aggregation functions almost satisfy (3) and (4), there are situations for which (3) and (4) are very far from being fulfilled. In these situations, there is a major discrepancy between the interpretation that the decision maker would expect and what these aggregations functions would provide.

Among the elements of $\mathcal{M}_N(p,w)$, one may adopt different strategies. One may for instance propose the element of $\mathcal{M}_N(p,w)$ which maximizes the entropy, as in [18]. It is indeed a safe strategy as the available preference information is very sparse. This translates into a convex optimization problem under linear constraints.

Acknowledgments. This work has been supported by the European project FP7-SEC-2013-607697, PREDICT "PREparing the Domino effect In crisis siTuations".

References

1. Yager, R.R.: On ordered weighted averaging aggregation operators in multicriteria decision making. IEEE Trans. Syst. Man Cybern. **18**, 183–190 (1988)
2. Yager, R.R.: Quantifier guided aggregation using OWA operators. Int. J. Intell. Syst. **11**, 49–73 (1996)
3. Labreuche, C., Mayag, B., Duqueroie, B.: Extension of the MACBETH approach to elicit an owa operator. EURO J. Decis. Processes **3**, 65–105 (2015)
4. Torra, V.: The weighted owa operator. Int. J. Intell. Syst. **12**, 153–166 (1997)
5. Xu, Z., Dai, Q.: An overview of operators for aggregating information. Int. J. Intell. Syst. **18**, 953–969 (2003)
6. Llamazares, B.: An analysis of some functions that generalize weighted means and OWA operators. Int. J. Intell. Syst. **28**, 380–393 (2013)
7. Merigo, J.: A unified model between the weighted average and the induced owa operator. Expert Syst. Appl. **38**, 11560–11572 (2011)
8. Choquet, G.: Theory of capacities. Ann. Inst. Fourier **5**, 131–295 (1953)
9. Sugeno, M.: Fuzzy measures and fuzzy integrals. Trans. SICE **8**(2), 95–102 (1972)
10. Shapley, L.S.: A value for n-person games. In: Kuhn, H.W., Tucker, A.W. (eds.) Contributions to the Theory of Games, Vol. II. Annals of Mathematics Studies, vol. 28, pp. 307–317. Princeton University Press, Princeton (1953)
11. Marichal, J.L.: Tolerant or intolerant character of interacting criteria in aggregation by the choquet integral. Eur. J. Oper. Res. **155**(3), 771–791 (2004)
12. Marichal, J.L.: k-intolerant capacities and choquet integrals. Eur. J. Oper. Res. **177**(3), 1453–1468 (2007)
13. Grabisch, M.: The application of fuzzy integrals in multicriteria decision making. Eur. J. Oper. Res. **89**, 445–456 (1996)
14. Grabisch, M., Marichal, J., Mesiar, R., Pap, E.: Aggregation Functions. Cambridge University Press, Cambridge (2009)
15. Yager, R.R.: Families of OWA operators. Fuzzy Sets Syst. **55**, 255–271 (1993)
16. Llamazares, B.: Constructing choquet integral-based operators that generalize weighted means and OWA operators. Information Fusion **23**, 131–138 (2015)
17. Llamazares, B.: SUOWA operators: Constructing semi-uninorms and analyzing specific cases. Fuzzy Sets and Systems
18. Kojadinovic, I., Marichal, J.L., Roubens, M.: An axiomatic approach to the definition of the entropy of a discrete Choquet capacity. Inform. Sci. **172**, 131–153 (2005)

Computing Superdifferentials of Lovász Extension with Application to Coalitional Games

Lukáš Adam[1] and Tomáš Kroupa[2(✉)]

[1] Institute of Information Theory and Automation, Czech Academy of Sciences, Pod Vodárenskou věží 4, 182 08 Prague, Czech Republic
adam@utia.cas.cz
[2] Dipartimento di Matematica "Federigo Enriques", Università degli Studi di Milano, Via Cesare Saldini 50, 20133 Milano, Italy
tomas.kroupa@unimi.it

Abstract. Every coalitional game can be extended from the powerset onto the real unit cube. One of possible approaches is the Lovász extension, which is the same as the discrete Choquet integral with respect to the coalitional game. We will study some solution concepts for coalitional games (core, Weber set) using superdifferentials developed in non-smooth analysis. It has been shown that the core coincides with Fréchet superdifferential and the Weber set with Clarke superdifferential for the Lovász extension, respectively. We introduce the intermediate set as the limiting superdifferential and show that it always lies between the core and the Weber set. From the game-theoretic point of view, the intermediate set is a non-convex solution containing the Pareto optimal payoff vectors, which depend on some ordered partition of the players and the marginal coalitional contributions with respect to the order.

Keywords: Coalitional game · Lovász extension · Choquet integral · Core · Weber set · Superdifferential

1 Introduction

Many important solution concepts for transferable-utility n-person coalitional games can be expressed in terms of formulas involving gradients or generalized gradients of a suitable extension of the game. The purpose of such a "differential representation" is not only computational, but it is also to provide a new interpretation of the corresponding payoff vectors, which usually revolves around the idea of marginal contributions to a given coalition.

In this contribution we will build a bridge between the class of solution concepts involving the core and the Weber set by applying certain generalized derivatives, namely the supergradients, which are studied in variational analysis [8,12]. Among the main superdifferentials count the Fréchet, the limiting and the Clarke superdifferential, respectively. By adopting the idea proposed in [13] we employ the limiting superdifferential to define directly a new solution concept for coalitional games, the so-called intermediate set. Specifically, the

J.P. Carvalho et al. (Eds.): IPMU 2016, Part I, CCIS 610, pp. 35–45, 2016.
DOI: 10.1007/978-3-319-40596-4_4

intermediate set is the limiting superdifferential of the Lovász extension [6] of the game v (or, equivalently, the discrete Choquet integral with respect to v [4]) calculated at the grand coalition. The associated payoff vectors are thus marginal contributions to the grand coalition in the sense conveyed by the limiting superdifferential.

It turns out that the newly constructed solution is meaningful and interesting from many viewpoints. The intermediate set can be seen as a nonempty interpolant between the core and the Weber set, which makes it applicable especially when the former is empty or small and the latter is huge. Theorem 2 provides a combinatorial description of the payoff vectors from the intermediate set in the following sense. For some ordered partition of the player set, each such vector is a Weber-style marginal vector on the level of blocks of coalitions and, at the same time, no coalition inside each block can improve upon this payoff vector in the sense of marginal coalitional contributions. The intermediate set is thus a solution concept that looks globally like the Weber set, but behaves locally like the core concept.

The paper is structured as follows. Section 2 introduces the basic notions and results from cooperative game theory and non-smooth analysis needed throughout the paper. The intermediate set is introduced in Sect. 3, where we formulate its equivalent characterization using ordered partitions of the player set and discuss its properties together with some examples.

The proofs are omitted for the space restrictions in this paper. The interested reader is invited to consult the authors' paper [1], which provides full details and further arguments in favor of the solution concept presented in this proceedings paper.

2 Basic Notions

We recall basic notions and results from cooperative game theory [10] and non-smooth variational analysis [8,12].

2.1 Coalitional Games

Let $N = \{1, \ldots, n\}$ be a finite set of *players*, where n is a positive integer. By 2^N we denote the powerset of N whose elements $A \subseteq N$ are called *coalitions*. A *(transferable utility coalitional) game* is a function $v\colon 2^N \to \mathbb{R}$ with $v(\emptyset) = 0$. Any $\mathbf{x} = (x_1, \ldots, x_n) \in \mathbb{R}^n$ is called a *payoff vector*. We introduce the following notation:

$$\mathbf{x}(A) = \sum_{i \in A} x_i, \quad \text{for every } A \subseteq N.$$

We say that a payoff vector $\mathbf{x}$ is *feasible* in a game v whenever $\mathbf{x}(N) \leq v(N)$. The set of all feasible payoff vectors in v is denoted by $\mathcal{F}(v)$.

Let $\Gamma(N)$ be the set of all games and $\Omega \subseteq \Gamma(N)$. A *solution* on Ω is a set-valued mapping $\sigma\colon \Omega \to 2^{\mathbb{R}^n}$ that maps every game $v \in \Omega$ to a set $\sigma(v) \subseteq \mathcal{F}(v)$.

We recall the core solution and the Weber set. The *core* of a game v is the convex polytope $\mathcal{C}(v) = \{\mathbf{x} \in \mathbb{R}^n \mid \mathbf{x}(N) = v(N),\ \mathbf{x}(A) \geq v(A) \text{ for every } A \subseteq N\}$.

Let Π_n be the set of all the permutations π of the player set N. Let $v \in \Gamma(N)$ and $\pi \in \Pi_n$. A *marginal vector* of a game v with respect to π is the payoff vector $\mathbf{x}^v(\pi) \in \mathbb{R}^n$ with coordinates

$$x_i^v(\pi) = v\left(\bigcup_{j \leq \pi^{-1}(i)} \{\pi(j)\}\right) - v\left(\bigcup_{j < \pi^{-1}(i)} \{\pi(j)\}\right), \quad i \in N. \tag{1}$$

The *Weber set* of v is defined as

$$\mathcal{W}(v) = \operatorname{conv}\{\mathbf{x}^v(\pi) \mid \pi \in \Pi_n\}.$$

Since $\mathbf{x}^v(\pi)(N) = v(N)$, the Weber set is a solution on $\Gamma(N)$ in the sense defined above. Moreover, it always contains the core solution; see [15, Theorem 14].

Proposition 1. *$\mathcal{C}(v) \subseteq \mathcal{W}(v)$ for every $v \in \Gamma(N)$.*

The fundamental tool in this paper is the concept of Lovász extension [6]. For every set $A \subseteq N$ let χ_A denote the incidence vector in $\mathbb{R}^n$ whose coordinates are given by

$$(\chi_A)_i = \begin{cases} 1 & \text{if } i \in A, \\ 0 & \text{otherwise.} \end{cases}$$

We write 0 in place of $\chi_\emptyset$. The embedding of 2^N into $\mathbb{R}^n$ by means of the mapping $A \mapsto \chi_A$ makes it possible to interpret a game on 2^N as a real function on $\{0,1\}^n$. Indeed, it suffices to define $\hat{v}(\chi_A) = v(A)$, for every $A \subseteq N$. We will extend the function $\hat{v}$ onto $\mathbb{R}^n$. For every $\mathbf{x} \in \mathbb{R}^n$, put

$$\Pi(\mathbf{x}) = \{\pi \in \Pi_n \mid x_{\pi(1)} \geq \cdots \geq x_{\pi(n)}\}.$$

Given $i \in N$ and $\pi \in \Pi(\mathbf{x})$, define $V_i^\pi(\mathbf{x}) = \{j \in N \mid x_j \geq x_{\pi(i)}\}$. Note that $V_i^\pi(\mathbf{x}) = V_i^\rho(\mathbf{x})$ for every $\pi, \rho \in \Pi(\mathbf{x})$. This implies that any vector $\mathbf{x} \in \mathbb{R}^n$ can be unambiguously written as a linear combination

$$\mathbf{x} = \sum_{i=1}^{n-1} (x_{\pi(i)} - x_{\pi(i+1)}) \cdot \chi_{V_i^\pi(\mathbf{x})} + x_{\pi(n)} \cdot \chi_N. \tag{2}$$

Using the convention $V_0^\pi(\mathbf{x}) = \emptyset$, we can rewrite (2) as

$$\mathbf{x} = \sum_{i=1}^{n} x_{\pi(i)} \cdot \left(\chi_{V_i^\pi(\mathbf{x})} - \chi_{V_{i-1}^\pi(\mathbf{x})}\right). \tag{3}$$

The *Lovász extension* $\hat{v}$ of $v \in \Gamma(N)$ is the function $\mathbb{R}^n \to \mathbb{R}$ defined linearly with respect to the decomposition (3):

$$\hat{v}(\mathbf{x}) = \sum_{i=1}^{n} x_{\pi(i)} \cdot \left(v(V_i^\pi(\mathbf{x})) - v(V_{i-1}^\pi(\mathbf{x}))\right), \quad \text{for any } \mathbf{x} \in \mathbb{R}^n.$$

Observe that the definition of $\hat{v}(\mathbf{x})$ is independent on the choice of $\pi \in \Pi(\mathbf{x})$. Clearly $\hat{v}(\chi_A) = v(A)$ for every coalition $A \subseteq N$. It is easy to see that the Lovász extension $\hat{v}$ of any game v fulfills these properties:

- $\hat{v}$ is continuous and piecewise affine on $\mathbb{R}^n$;
- $\hat{v}$ is positively homogeneous: $\hat{v}(\lambda \cdot \mathbf{x}) = \lambda \cdot \hat{v}(\mathbf{x})$ for every $\lambda \geq 0$ and $\mathbf{x} \in \mathbb{R}^n$;
- the mapping $v \in \Gamma(N) \mapsto \hat{v}$ is linear.

The following lemma says that the local behavior of $\hat{v}$ is the same around χ_N as in the neighborhood of 0.

Lemma 1. *For any* $\mathbf{x} \in \mathbb{R}^n$ *it holds true that* $\hat{v}(\mathbf{x} + \chi_N) = \hat{v}(\mathbf{x}) + \hat{v}(\chi_N)$.

A game $v \in \Gamma(N)$ is called *supermodular* (or *convex*) if the following inequality is satisfied: $v(A \cup B) + v(A \cap B) \geq v(A) + v(B)$, for every $A, B \subseteq N$. A *submodular* game v is such that $-v$ is supermodular. A game v is called *additive* when $v(A \cup B) = v(A) + v(B)$ for every $A, B \subseteq N$ with $A \cap B = \emptyset$. We will make an ample use of several characterizations of supermodular games appearing in the literature; see [5,6,14,15].

Theorem 1. *Let* $v \in \Gamma(N)$. *Then the following assertions are equivalent:*

1. *v is supermodular;*
2. $\{\mathbf{x}^v(\pi) \mid \pi \in \Pi_n\} \subseteq \mathcal{C}(v)$;
3. $\mathcal{C}(v) = \mathcal{W}(v)$;
4. *The Lovász extension $\hat{v}$ of v is a concave function.*

2.2 Superdifferentials

In this section we will define the selected concepts of variational (nonsmooth) analysis, namely various superdifferentials which generalize the superdifferential of convex functions. Since the superdifferentials will be computed only for the Lovász extension, we will confine to defining superdifferentials only for piecewise affine functions at a point $\bar{\mathbf{x}} \in \mathbb{R}^n$. This assumption enables us to neglect the term $o(\|\mathbf{x} - \bar{\mathbf{x}}\|)$ present in the more general definitions; see [12, Definition 8.3], for example. We refer the reader to [12] for the general framework involving upper semicontinuous functions.

While the standard monographs on variational analysis [8,11,12] deal with subdifferentials instead of superdifferentials, most of the results can be readily transformed to the setting of superdifferentials, usually by reversing inequalities only.

Definition 1. *Let* $f\colon \mathbb{R}^n \to \mathbb{R}$ *be a piecewise affine function and* $\bar{\mathbf{x}} \in \mathbb{R}^n$. *We say that* $\mathbf{x}^* \in \mathbb{R}^n$ *is a*

1. Fréchet supergradient *of f at $\bar{\mathbf{x}}$ if there exists neighborhood $\mathcal{X}$ of $\bar{\mathbf{x}}$ such that for all $\mathbf{x} \in \mathcal{X}$ we have*

$$f(\mathbf{x}) - f(\bar{\mathbf{x}}) \leq \langle \mathbf{x}^*, \mathbf{x} - \bar{\mathbf{x}} \rangle;$$

2. limiting supergradient *of f at $\bar{\mathbf{x}}$ if for every neighborhood $\mathcal{X}$ of $\bar{\mathbf{x}}$ there exists $\mathbf{x} \in \mathcal{X}$ such that $\mathbf{x}^*$ is a Fréchet supergradient of f at $\mathbf{x}$;*
3. Clarke supergradient *of f at $\bar{\mathbf{x}}$ if*

$$\mathbf{x}^* \in \operatorname{conv}\{\mathbf{y} | \ \forall \textit{ neighborhood } \mathcal{X} \textit{ of } \bar{\mathbf{x}} \ \exists \mathbf{x} \in \mathcal{X} \cap D \textit{ with } \mathbf{y} = \nabla f(\mathbf{x})\},$$

where $D := \{\mathbf{x} \in \mathbb{R}^n | \ f$ is differentiable at $\mathbf{x}\}$.

The collection of all (Fréchet, limiting, Clarke) supergradients of f at $\bar{\mathbf{x}}$ is called (Fréchet, limiting, Clarke) superdifferential *and it is denoted by $\hat{\partial} f(\bar{\mathbf{x}})$, $\partial f(\bar{\mathbf{x}})$ and $\overline{\partial} f(\bar{\mathbf{x}})$, respectively.*

It is easy to see that

$$\hat{\partial} f(\bar{\mathbf{x}}) \subseteq \partial f(\bar{\mathbf{x}}) \subseteq \overline{\partial} f(\bar{\mathbf{x}}), \quad \bar{\mathbf{x}} \in \mathbb{R}^n,$$

where all the inequalities may be strict. Moreover, [12, Theorem 8.49] yields that the limiting and the Clarke superdifferential of a piecewise affine function f are related as follows: $\overline{\partial} f(\bar{\mathbf{x}}) = \operatorname{conv} \partial f(\bar{\mathbf{x}})$. The following two examples show that the three superdifferentials can differ significantly.

Example 1 ([2, Example 10.28]). Consider the function $\mathbb{R}^2 \to \mathbb{R}$ defined as $f(x_1, x_2) = \max(\min(2x_1 + x_2, x_1), 2x_2)$. This function is piecewise affine and it can be expressed as follows:

$$f(x_1, x_2) = \begin{cases} 2x_1 + x_2 & \text{if } x_2 \leq 2x_1 \text{ and } x_2 \leq -x_1, \\ x_1 & \text{if } x_2 \leq \frac{x_1}{2} \text{ and } x_2 \geq -x_1, \\ 2x_2 & \text{if } x_2 \geq 2x_1 \text{ or } x_2 \geq \frac{x_1}{2}. \end{cases}$$

Let us compute all the three superdifferentials of f at $\bar{\mathbf{x}} = 0$:

$$\begin{aligned} \hat{\partial} f(\bar{x}) &= \emptyset, \\ \partial f(\bar{x}) &= \operatorname{conv}\{(2,1), (1,0)\} \cup \{(0,2)\}, \\ \overline{\partial} f(\bar{x}) &= \operatorname{conv}\{(2,1), (1,0), (0,2)\}. \end{aligned}$$

Example 2. Let

$$g(x_1, x_2) = \begin{cases} 0 & \text{if } x_1 \leq 0 \text{ or } x_2 \leq 0, \\ -x_1 & \text{if } x_2 \geq x_1 \geq 0, \\ -x_2 & \text{if } x_1 \geq x_2 \geq 0, \end{cases} \qquad \text{for every } (x_1, x_2) \in \mathbb{R}^2.$$

Function g is piecewise affine and the three superdifferentials of g at $\bar{\mathbf{x}} = 0$ are, respectively,

$$\begin{aligned} \hat{\partial} g(\bar{x}) &= \{(0,0)\}, \\ \partial g(\bar{x}) &= \operatorname{conv}\{(0,0), (-1,0)\} \cup \operatorname{conv}\{(0,0), (0,-1)\}, \\ \overline{\partial} g(\bar{x}) &= \operatorname{conv}\{(0,0), (-1,0), (0,-1)\}. \end{aligned}$$

3 Intermediate Set

The Lovász extension $\hat{v}$ of a coalitional game v is instrumental in characterizing the core solution and the Weber set by the tools of nonsmooth calculus. Specifically, it was shown that the core coincides with the Fréchet superdifferential of $\hat{v}$ at 0 [3, Proposition 3] and that the Weber set is the Clarke superdifferential of $\hat{v}$ at 0 [13, Proposition 4.1]. It may be more natural to use the grand coalition N in place of the empty coalition $\emptyset$ in those formulas. Lemma 1 says that this is always possible.

Proposition 2. *For every game* $v \in \Gamma(N)$, $\mathcal{C}(v) = \hat{\partial}\hat{v}(\chi_N) = \hat{\partial}\hat{v}(0)$ *and* $\mathcal{W}(v) = \overline{\partial}\hat{v}(\chi_N) = \overline{\partial}\hat{v}(0)$.

It can easily be shown that the gap between the core and the Weber set can be too large. Indeed, the core can be empty, while the Weber set can be a large convex polytope. Taking into account the hierarchy of superdifferentials introduced in the previous section, we will pursue an idea mentioned in [13] and by analogy with Proposition 2 we define a new solution concept as $\partial\hat{v}(\chi_N)$, where ∂ is the limiting superdifferential. This leads to the following notion.

Definition 2. *The* intermediate set $\mathcal{M}(v)$ *of* $v \in \Gamma(N)$ *is the set*

$$\mathcal{M}(v) := \partial\hat{v}(\chi_N).$$

Similarly as in Proposition 2, we can show that for every game $v \in \Gamma(N)$, $\mathcal{M}(v) = \partial\hat{v}(0)$. Lemma 2 explains why the solution concept $\mathcal{M}(v)$ was termed the "intermediate set".

Lemma 2. *Let* $v \in \Gamma(N)$. *Then:*

1. $\mathcal{M}(v) \neq \emptyset$.
2. *We have*
$$\mathcal{C}(v) \subseteq \mathcal{M}(v) \subseteq \mathcal{W}(v),$$
where both inclusions may be strict.
3. $\mathcal{W}(v) = \operatorname{conv} \mathcal{M}(v)$.
4. v *is supermodular if and only if* $\mathcal{C}(v) = \mathcal{M}(v) = \mathcal{W}(v)$.

Example 3. [3-player glove game] Let $N = \{1, 2, 3\}$. The first player owns a single left glove and the remaining two players possess one right glove each. The profit of a coalition is a total of glove pairs the coalition owns:

$$v(A) = \begin{cases} 1 & \text{if } A \in \{\{1,2\}, \{1,3\}, N\}, \\ 0 & \text{otherwise.} \end{cases}$$

It is not difficult to compute $\mathcal{C}(v), \mathcal{M}(v)$ and $\mathcal{W}(v)$ directly:

$$\begin{aligned} \mathcal{C}(v) &= \{(1,0,0)\}, \\ \mathcal{M}(v) &= \operatorname{conv}\{(1,0,0),(0,1,0)\} \cup \operatorname{conv}\{(1,0,0),(0,0,1)\}, \\ \mathcal{W}(v) &= \operatorname{conv}\{(1,0,0),(0,1,0),(0,0,1)\}. \end{aligned}$$

3.1 Characterization by Ordered Partitions

In this section we are going to show an alternative expression for the intermediate set using the concept of an ordered partition. Thus the purely analytic definition of intermediate set can be equivalently stated in terms of the combinatorial and order-theoretic properties of a coalitional game.

Let $K \geq 1$. An *ordered partition* of the player set N is a K-tuple

$$P := (B_1, \ldots, B_K)$$

of coalitions $\emptyset \neq B_i \subseteq N$ such that $B_i \cap B_j = \emptyset$ $(i \neq j)$ and $B_1 \cup \cdots \cup B_K = N$. Let

$$\mathcal{P} = \{P \mid P \text{ is an ordered partition of } N\}.$$

The family $\mathcal{P}$ is associated with the following scheme of allocating profits $\mathbf{x}$ among the players in a game v:

1. The players may be split into any ordered partition $P = (B_1, \ldots, B_K) \in \mathcal{P}$.
2. Each block of players B_k can distribute the total amount

$$\mathbf{x}(B_k) = v(B_1 \cup \cdots \cup B_{k-1} \cup B_k) - v(B_1 \cup \cdots \cup B_{k-1})$$

   to its members, which can be interpreted as the marginal contribution of coalition B_k to the coalition $B_1 \cup \cdots \cup B_{k-1}$ with respect to P.
3. No coalition B in a block B_k may improve upon $\mathbf{x}$, while respecting the given order of coalition blocks, that is,

$$\mathbf{x}(B) \geq v(B_1 \cup \cdots \cup B_{k-1} \cup B) - v(B_1 \cup \cdots \cup B_{k-1}).$$

Note that the players share total of $v(N)$ among them as a consequence of the second principle. The distribution procedure explained above has two extreme cases. Assume that the ordered partition P is the finest possible, that is, $P = (\{\pi(1)\}, \ldots, \{\pi(n)\})$ for some permutation $\pi \in \Pi_n$. In this case the allocation scheme in a game v leads to the marginal vectors $\mathbf{x}^v(\pi)$ defined by (1). On the contrary, if the partition contains one block only, $P = (N)$, then all the players (and coalitions) are treated equally, which results in distributing payoffs according to the definition of core. Any ordered partition $P = (B_1, \ldots, B_K)$ different from the two extreme cases generates a combination of the principle of marginal distribution on the level of blocks with the core-like stability inside each block of the partition, while respecting the given order of coalitions. Such a distribution process is thus always a mixture of the considerations endogenous to B_i and those which are exogenous to B_i.

Our characterization says that $\mathbf{x} \in \mathcal{M}(v)$ if and only if there is an ordered partition P such that $\mathbf{x}$ is allocated to the players according to the above distribution principles.

Theorem 2. *For every game $v \in \Gamma(N)$,*

$$\mathcal{M}(v) = \bigcup_{P \in \mathcal{P}} \mathcal{M}_P(v),$$

where $\mathcal{M}_P(v)$ *with* $P = (B_1, \ldots, B_K)$ *is the set of all* $\mathbf{x} \in \mathbb{R}^n$ *such that the following two conditions hold for every* $k = 1, \ldots, K$ *and for each* $B \subseteq B_k$*:*

$$\mathbf{x}(B_k) = v(B_1 \cup \cdots \cup B_{k-1} \cup B_k) - v(B_1 \cup \cdots \cup B_{k-1}),$$
$$\mathbf{x}(B) \geq v(B_1 \cup \cdots \cup B_{k-1} \cup B) - v(B_1 \cup \cdots \cup B_{k-1}).$$

Example 4. Let $N = \{1, 2, 3\}$ and

$$v(A) = \begin{cases} 0 & \text{if } |A| = 1, \\ 2 & \text{if } |A| = 2, \\ 3 & \text{if } A = N. \end{cases}$$

It is easy to see that v is not supermodular but only superadditive, that is, $v(A \cup B) \geq v(A) + v(B)$ for every $A, B \subseteq N$ with $A \cap B = \emptyset$.

The core of this game is $\mathcal{C}(v) = \{(1, 1, 1)\}$, while the Weber set $\mathcal{W}(v)$ coincides with the hexagon whose 6 vertices are all the permutations of the payoff vector $(0, 1, 2)$. The intermediate set is the union of three line segments; see Fig. 1. We obtain that $\mathcal{M}_{(i,jk)}(v) = \emptyset$ for every ordered partition (i, jk) of N.[1] On the other hand, $\mathcal{M}_{(ij,k)}(v)$ is the line segment whose endpoints are the two marginal vectors $\mathbf{x}$ with $x_k = 1$. Thus $\mathbf{x} \in \mathcal{M}(v)$ iff it belongs to $\mathcal{M}_{(ij,k)}(v)$ for some ordered partition (ij, k) of N. The example shows that, in general, the intermediate set is not a union of selected faces of the Weber set.

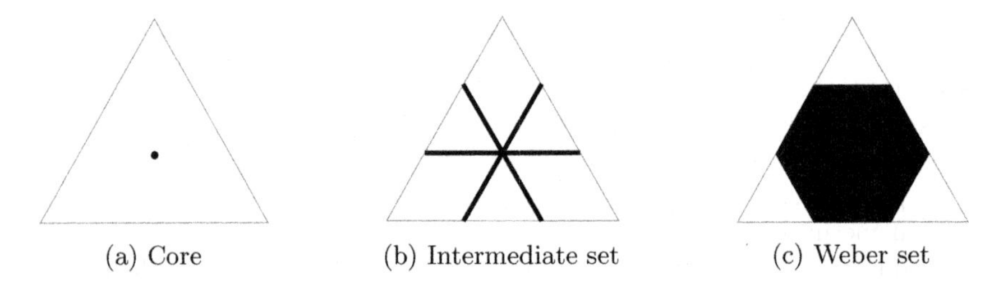

Fig. 1. The solutions from Example 4 in the barycentric coordinates

3.2 Properties

It was proved in [1] that the core, the intermediate set, and the Weber set share many properties of solution concepts for coalitional games. Namely each of the three solutions is Pareto optimal, anonymous, covariant, and has both the null player property and the dummy player property.

In sharp contrast to the core and the Weber set, the previous examples showed that the intermediate set is typically non-convex. Indeed, the Weber set

[1] We may occasionally switch to a simplified notation for coalitions, writing ij in place of $\{i, j\}$ and similarly.

is always the convex hull of the intermediate set. Moreover, the core can be void, while the intermediate set is always non-empty. Individual rationality is not fulfilled by the intermediate set, in general. However, the intermediate set satisfies this property on the class of all weakly superadditive games, that is, the coalitional games v for which the following property holds true:

$$v(A \cup \{i\}) \geq v(A) + v(\{i\}), \quad \text{for every } A \subseteq N \text{ and } i \in N \setminus A.$$

3.3 Example: Simple Games

We will compute the intermediate set for the class of all simple games and compare the achieved results to the shape of the core. A game $v \in \Gamma(N)$ is *monotone* if $v(A) \leq v(B)$ whenever $A \subseteq B \subseteq N$. A *simple game* is a monotone game v with $v(A) \in \{0,1\}$ and $v(N) = 1$. Every simple game v over the player set N can be identified with the family of *winning coalitions*

$$\mathcal{V} = \{A \subseteq N \mid v(A) = 1\}.$$

Conversely, any system of coalitions $\mathcal{V}$ such that

1. $N \in \mathcal{V}$, $\emptyset \notin \mathcal{V}$ and
2. $A \subseteq B \subseteq N,\ A \in \mathcal{V} \Rightarrow B \in \mathcal{V}$,

gives rise to a simple game v by putting $v(A) = 1$ if $A \in \mathcal{V}$ and $v(A) = 0$, otherwise. The family of *minimal winning coalitions* in v is given by

$$\mathcal{V}^m = \{A \in \mathcal{V} \mid B \subsetneq A \Rightarrow B \notin \mathcal{V}, \text{ for every } B \subseteq N\}.$$

The core of a simple game v is fully determined by the minimal winning coalitions in v. Indeed, it is well-known that

$$\mathcal{C}(v) = \bigcap_{E \in \mathcal{V}^m} \{\mathbf{x} \in \mathcal{I}(v) \mid x_i = 0 \text{ for every } i \in N \setminus E\},$$

where $\mathcal{I}(v) := \{\mathbf{x} \in \mathbb{R}^n \mid \mathbf{x}(N) = v(N),\ x_i \geq v(i),\ i \in N\}$ is the set of imputations in v. Using our Theorem 2 we can show that an analogous formula exists for the intermediate set. It states that $\mathcal{M}(v)$ arises as a union of faces of the standard simplex, where each face corresponds to one minimal winning coalition.

Theorem 3. *Let $v \in \Gamma(N)$ be a simple game. Then*

$$\mathcal{M}(v) = \bigcup_{E \in \mathcal{V}^m} \left\{ \mathbf{x} \in \mathbb{R}^n \;\middle|\; \begin{array}{ll} x_i = 0 & \text{if } i \in N \setminus E \\ x_i \geq 0 & \text{if } i \in E \\ \sum_{i \in E} x_i = 1 & \end{array} \right\}.$$

As an example we will compute the intermediate set of the UN Security Council voting scheme.

Example 5. The UN Security Council contains 5 permanent members with veto power and 10 non–permanent members. To pass a resolution, all the permanent members and at least 4 non–permanent members have to vote for the proposal. This is a mildly simplified version of the real voting process, in which abstention of a permanent member is not usually regarded as a veto. However, this assumption is usually accepted in game-theoretic literature; see e.g. [9, Example XI.2.9] or [7, Example 16.1.3].

We assume that the players $N = \{1, \ldots, 15\}$ are ordered in such a way that the first five are the permanent members and the last ten are the non–permanent members. Then it is easy to show that the core and and the Weber set of the corresponding simple game v are, respectively,

$$\mathcal{C}(v) = \left\{ \mathbf{x} \in \mathbb{R}^{15} \;\middle|\; \mathbf{x} \geq 0,\ \sum_{i=1}^{5} x_i = 1,\ x_i = 0 \text{ for } i = 6, \ldots, 15 \right\} \text{ and}$$

$$\mathcal{W}(v) = \left\{ \mathbf{x} \in \mathbb{R}^{15} \;\middle|\; \mathbf{x} \geq 0,\ \sum_{i=1}^{15} x_i = 1 \right\}.$$

Since core allocations are stable, any payoff $\mathbf{x} \in \mathcal{C}(v)$ is distributed only among the permanent members (the vetoers). By contrast, the Weber set is the whole 14-dimensional standard simplex in $\mathbb{R}^{15}$, which contains some payoff vectors whose meaning is problematic. For instance, it is not entirely clear how to interpret a vector $\left(0, \ldots, 0, \frac{1}{10}, \ldots, \frac{1}{10}\right) \in \mathcal{W}(v)$. As we will see, this vector is not contained in $\mathcal{M}(v)$.

Given $i \in N$, denote by $\mathbf{e}_i \in \mathbb{R}^{15}$ the vector whose coordinates are $e_j = 1$ if $j = i$ and $e_j = 0$ otherwise. Put $\mathcal{D} = \{D \subseteq \{6, \ldots, 15\} \mid |D| = 4\}$. Theorem 3 yields

$$\mathcal{M}(v) = \bigcup_{D \in \mathcal{D}} \operatorname{conv}\left(\{\mathbf{e}_1, \mathbf{e}_2, \mathbf{e}_3, \mathbf{e}_4, \mathbf{e}_5\} \cup \{\mathbf{e}_i \mid i \in D\}\right).$$

In other words, $\mathcal{M}(v)$ is a union of $\binom{10}{4}$ 8-dimensional standard simplices, each of which is a convex hull of $\mathbf{e}_i$s corresponding to the five permanent members and four other non–permanent members. Each such simplex is associated with the ordered partition having two blocks, $(\{1, \ldots, 5\} \cup D, N \setminus (\{1, \ldots, 5\} \cup D))$ where $D \in \mathcal{D}$.

4 Conclusions

Not every solution concept is usually suitable for the entire class of coalitional games. In our future research we plan to study if the intermediate set is well-tailored for some subclass of games. The intuition says that such a class of games has the small core and the large Weber set since this makes the interpolation by the intermediate set between the two solutions especially appealing.

An interesting open question is based on the behavior of the core and the components of the intermediate set $\mathcal{M}_{\mathcal{P}}(v)$ observed in Example 4 and Theorem 3: Can we recover the core of a coalitional game v as an intersection of selected components $\mathcal{M}_{\mathcal{P}}(v)$?

Acknowledgments. L. Adam gratefully acknowledges the support from the Grant Agency of the Czech Republic (15-00735S). The work of T. Kroupa was supported by Marie Curie Intra-European Fellowship OASIG (PIEF-GA-2013-622645).

References

1. Adam, L., Kroupa, T.: The intermediate set and limiting superdifferential for coalition games: between the core and the Weber set. Submitted to International Journal of Game Theory (2015). http://arxiv.org/abs/1504.08195
2. Clarke, F.: Functional Analysis, Calculus of Variations and Optimal Control. Graduate Texts in Mathematics, vol. 264. Springer, London (2013)
3. Danilov, V., Koshevoy, G.: Cores of cooperative games, superdifferentials of functions, and the Minkowski difference of sets. J. Math. Anal. Appl. **247**(1), 1–14 (2000)
4. Grabisch, M.: Set functions over finite sets: transformations and integrals. In: Pap, E. (ed.) Handbook of Measure Theory, vol. II, pp. 1381–1401. Elsevier, Amsterdam (2002)
5. Ichiishi, T.: Super-modularity: applications to convex games and to the greedy algorithm for LP. J. Econ. Theory **25**(2), 283–286 (1981)
6. Lovász, L.: Submodular functions and convexity. In: Bachem, A., Korte, B., Grötschel, M. (eds.) Mathematical Programming: The State of the Art, pp. 235–257. Springer, Berlin (1983)
7. Maschler, M., Solan, E., Zamir, S.: Game Theory. Cambridge University Press, Cambridge (2013)
8. Mordukhovich, B.S.: Variational Analysis and Generalized Differentiation I. Springer, Heidelberg (2006)
9. Owen, G.: Game Theory, 3rd edn. Academic Press Inc., San Diego (1995)
10. Peleg, B., Sudhölter, P.: Introduction to the Theory of Cooperative Games. Theory and Decision Library. Series C: Game Theory, Mathematical Programming and Operations Research, vol. 34, 2nd edn. Springer, Berlin (2007)
11. Rockafellar, R.T.: Convex Analysis. Princeton University Press, Princeton (1970)
12. Rockafellar, R.T., Wets, R.J.-B.: Variational Analysis. Springer, Heidelberg (1998)
13. Sagara, N.: Cores and Weber sets for fuzzy extensions of cooperative games. Fuzzy Sets Syst. **272**, 102–114 (2015)
14. Shapley, L.S.: Cores of convex games. Int. J. Game Theory **1**, 11–26 (1972)
15. Weber, R.J.: Probabilistic values for games. In Roth, A. E., editor, The Shapley Value. Essays in Honor of Lloyd S. Shapley, 101–120. Cambridge University Press (1988)

Conjoint Axiomatization of the Choquet Integral for Heterogeneous Product Sets

Mikhail Timonin(✉)

Queen Mary University of London, London E1 4NS, UK
m.timonin@qmul.ac.uk

Abstract. We propose an axiomatization of the Choquet integral model for the general case of a heterogeneous product set $X = X_1 \times \ldots \times X_n$. In MCDA elements of X are interpreted as alternatives, characterized by criteria taking values from the sets X_i. Previous axiomatizations of the Choquet integral have been given for particular cases $X = Y^n$ and $X = \mathbb{R}^n$. However, within multicriteria context such indenticalness, hence commensurateness, of criteria cannot be assumed a priori. This constitutes the major difference of this paper from the earlier axiomatizations. In particular, the notion of "comonotonicity" cannot be used in a heterogeneous structure, as there does not exist a "built-in" order between elements of sets X_i and X_j. However, such an order is implied by the representation model. Our approach does not assume commensurateness of criteria. We construct the representation and study its uniqueness properties.

Keywords: Choquet integral · Decision theory · MCDA · Multi-criteria decision making

1 Introduction

The Choquet integral is widely used in decision analysis and, in particular, MCDA [5], although its use is still somewhat restricted due to both methodological problems and difficulties in practical implementation. Rank-dependent models first appeared in the axiomatic decision theory in reply to the criticism of Savage's postulates of rationality [12]. The renowned Ellsberg paradox [3] has shown that people can violate Savage's axioms and still consider their behaviour rational. First models accounting for the so-called uncertainty aversion observed in this paradox appeared in the 1980s, in the work [11] and others (see [17] for a review). One particular generalization of the expected utility model (EU) characterized by Schmeidler [13] is the Choquet expected utility (CEU), where probability is replaced by a non-additive set function (called capacity) and integration is performed using the Choquet integral.

Since Schmeidler's paper, various versions of the same model have been characterized in the literature (e.g. [4,16]). CEU has gained some momentum in both theoretical and applied economic literature, being used mainly for analysis

J.P. Carvalho et al. (Eds.): IPMU 2016, Part I, CCIS 610, pp. 46–57, 2016.
DOI: 10.1007/978-3-319-40596-4_5

of problems involving Knightian uncertainty. At the same time, rank-dependent models, in particular the Choquet integral, were adopted in multiattribute utility theory (MAUT) [7]. Here the integral gained popularity due to the tractability of non-additive measures in this context (see [5] for a review). The model permitted various preferential phenomena, such as criteria interaction, which were impossible to reflect in the traditional additive models.

The connection between MAUT and decision making under uncertainty has been known for a long time. In the case when the number of states is finite, which is assumed hereafter, states can be associated with criteria. Accordingly, acts correspond to multicriteria alternatives. Finally, the sets of outcomes at each state can be associated with the sets of criteria values. However, this last transition is not quite trivial. It is commonly assumed that the set of outcomes is the same in each state of the world [12,13]. In multicriteria decision making the opposite is true. Indeed, consider preferences of consumers choosing cars. Each car is characterized by a number of features (criteria), such as colour, maximal speed, fuel consumption, comfort, etc. Apparently, sets of values taken by each criterion can be completely different from those of the others. In such context the ranking stage of rank-dependent models, which in decision under uncertainty involves comparing outcomes attained at various states, would amount to comparing colours to the level of fuel consumption, and maximal speed to comfort. Indeed, the traditional additive model [2,9] only implies meaningful comparability of units between goods in the bundle, but not of their absolute levels. However, in rank-dependent models such comparability seems to be a necessary condition.

We propose a representation theorem for the Choquet integral model in the MCDA context. Binary relation $\succcurlyeq$ is defined on a heterogeneous product set $X = X_1 \times \ldots \times X_n$. In multicriteria decision analysis (MCDA), elements of the set X are interpreted as alternatives, characterized by criteria taking values from sets X_i. Previous axiomatizations of the Choquet integral model have been given for the special cases of $X = Y^n$ (see [8] for a review of approaches) and $X = \mathbb{R}^n$ (see [5] for a review). One related result is the recent axiomatization of the Sugeno integral model ([1,6]). Another approach using conditions on the utility functions was proposed in [10]. The "conjoint" axiomatization of the Choquet integral for the case of a general X was an open problem in the literature. The crucial difference with the previous axiomatizations is that the notion of "comonotonicity" cannot be used in the heterogeneous case, due to the fact that there does not exist a meaningful "built-in" order between elements of sets X_i. New axioms and modifications of proof techniques had to be introduced to account for that.

Our first axiom shows, roughly, how the set X can be partitioned into subsets based on properties necessary for existence of an additive representation. The axiom (**A3**) we introduce is similar to the "2-graded" condition previously used for characterizing of MIN/MAX and the Sugeno integral ([1,6]). At every point $z \in X$ for every pair of coordinates $i, j \in N$ it is possible to build two "rectangular cones" - one made up of points from X_i which are "greater" than

z_i and points from X_j which are "less" than z_j, and the second for the opposite case. The axiom states that triple cancellation for $\succcurlyeq$ restricted to i, j must then hold on at least one of these cones. This allows to partition X into subsets by using intersection of such cones for various pairs i, j.

The second property is that the additive representations on different subsets are interrelated, in particular "trade-offs" between criteria values are consistent across partition elements both within the same dimension and across different ones. This is reflected by two axioms (**A4, A5**), similar to the ones used in [9,16] (Sect. 8.2). One, roughly speaking, states that triple cancellation holds across subsets, while the other says that ordering of intervals on any dimension must be preserved when they are projected onto another dimension by means of equivalence relations. These axioms are complemented by a new condition called bi-independence (**A6**) and weak separability (**A2**) [1] - which together reflect the monotonicity property of the integral, and also the standard essentiality, "comonotonic" Archimedean axiom and restricted solvability (**A7,A8,A9**). Finally, $\succcurlyeq$ is supposed to be a weak order (**A1**), and X is order dense.

2 Choquet Integral in MCDA

Definition 1. *Let $N = \{1, \dots, n\}$ be a finite set and 2^N its power set. A* capacity *(non-additive measure, fuzzy measure) is a set function $\nu : 2^N \to \mathbb{R}_+$ such that:*

1. $\nu(\varnothing) = 0$;
2. $A \subseteq B \Rightarrow \nu(A) \leq \nu(B),\ \forall A, B \in 2^N$.

In this paper, it is also assumed that capacities are normalized, i.e. $\nu(N) = 1$.

Definition 2. *The* Choquet integral *of a function $f : N \to \mathbb{R}$ with respect to a capacity ν is defined as*

$$C(\nu, f) = \int_0^{\infty} \nu(\{i \in N \colon f(i) \geq r\})dr + \int_{-\infty}^{0} [\nu(\{i \in N \colon f(i) \geq r\}) - 1]dr$$

Denoting the vector of values of $f : N \to \mathbb{R}$ as $(f_1, \dots, f_n)$, the definition can be written down as:

$$C(\nu, (f_1, \dots, f_n)) = \sum_{i=1}^{n} (f_{(i)} - f_{(i-1)})\nu(\{j \in N \colon f_j \geq f_{(i)}\})$$

where $f_{(1)}, \dots, f_{(n)}$ is a permutation of $f_1, \dots, f_n$ such that $f_{(1)} \leq f_{(2)} \leq \dots \leq f_{(n)}$, and $f_{(0)} = 0$.

One of the most useful tools for analysis of the capacity is the so-called Möbius transform. It's a linear transformation of the capacity which is given by:

$$m(A) = \sum_{B \subset A} (-1)^{|A \setminus B|} \nu(B).$$

2.1 The Model

Let $\succcurlyeq$ be a binary relation on the set $X = X_1 \times \ldots \times X_n$. $\succ, \prec, \preccurlyeq, \sim, \not\succ$ are defined in the usual way. In MCDA, elements of set X are interpreted as alternatives characterized by criteria from the set $N = \{1, \ldots, n\}$. Set X_i contains criteria values for criterion i. We say that $\succcurlyeq$ can be represented by a Choquet integral, if there exists a capacity ν and functions $f_i : X_i \to \mathbb{R}$, called value functions, such that for all $x, y \in X$:

$$x \succcurlyeq y \iff C(\nu, (f_1(x_1), \ldots, f_n(x_n))) \geq C(\nu, (f_1(y_1), \ldots, f_n(y_n))).$$

As seen in the definition of the Choquet integral, its calculation involves comparison of f_i's to each other. It is not immediately obvious how this operation can have any meaning in the MCDA decision framework. It is well-known that direct comparison of value functions for various attributes is meaningless in the additive model [9] (recall that the origin of each value function can be changed independently). In the homogeneous case $X = Y^n$ this problem is readily solved, as we have a single set of "consequences" Y (in the context of decision making under uncertainty). The required order is either assumed as given [17] or is readily derived from the ordering of "constant" acts $(y, \ldots, y)$ [16]. Since there is a single "consequence" set, we also only have one value function $U : Y \to \mathbb{R}$, and thus comparing $U(y_i)$ to $U(y_j)$ is perfectly sensible, since U represents the order on the set Y. None of these methods can be readily applied in the heterogeneous case.

2.2 Properties of the Choquet Integral

Below are given some important properties of the Choquet integral:

1. Functions $f : N \to \mathbb{R}$ and $g : N \to \mathbb{R}$ are comonotonic if for no $i, j \in N$ holds $f(i) > f(j)$ and $g(i) < g(j)$. For all comonotonic f the Choquet integral reduces to a usual Lebesgue integral. In the finite case, the integral is accordingly reduced to a weighted sum.
2. Particular cases of the Choquet integral (e.g. [5]), where m is a Möbius transform of the capacity:
   - If $m(\{1\}) = \ldots = m(\{n\}) = 1$, then $C(\nu, (f_1, \ldots, f_n)) = \max(f_1, \ldots, f_n)$.
   - If $m(N) = 1, m(A) = 0, A \neq N$, then $C(\nu, (f_1, \ldots, f_n)) = \min(f_1, \ldots, f_n)$.
   - If $m(A) = 0$, for all $A \subset N : |A| \geq 2$, then $C(\nu, (f_1, \ldots, f_n)) = \sum_{i \in N} \nu(\{i\}) f_i$

Property 1 states that the set X can be partitioned into subsets corresponding to particular ordering of the value functions. There are $n!$ such sets. Since the integral on each of the sets is reduced to a weighted sum, i.e. an additive representation, we should expect many of the axioms of the additive conjoint model to be valid on this subsets. This is the intuition behind several of the axioms given in the following section.

3 Axioms and Definitions

A1 - Weak order. $\succsim$ is a weak order.
A2 - Weak separability. For all i, if $a_i x_{-i} \succ b_i x_{-i}$ for some $a_i, b_i \in X_i$,$x_{-i} \in X_{-i}$, then $a_i y_{-i} \succsim b_i y_{-i}$ for all $y_{-i} \in X_{-i}$.

Note, that from this follows, that for any $a_i, b_i \in X_i$ either $a_i x_{-i} \succsim b_i x_{-i}$ or $b_i x_{-i} \succsim a_i x_{-i}$ for all $x_{-i} \in X_{-i}$. This allows to introduce the following definition:

Definition 3. *For all $a_i, b_i \in X_i$ define $\succsim_i$ as $a_i \succsim_i b_i \iff a_i x_{-i} \succsim b_i x_{-i}$ for all $x_{-i} \in X_{-i}$.*

Definition 4. *For any $z \in X$ define $\mathbf{SE}^z_{ij} = \{x_i x_j z_{-ij} \in X \colon x_i \succsim_i z_i, z_j \succsim_j x_j\}$, and $\mathbf{NW}^z_{ij} = \{x_i x_j z_{-ij} \in X \colon z_i \succsim_i x_i, x_j \succsim_j z_j\}$.*

Definition 5. *Given $i, j \in N$, a relation $\succsim$ on $X_1 \times \ldots \times X_n$ satisfies ij*-triple cancellation **(ij-3C)**, *if for all $a_i, b_i, c_i, d_i \in X_i$, $p_j, q_j, r_j, s_j \in X_j$, and all $z_{-ij} \in X_{-ij}$ holds:*

$$\left.\begin{array}{l} a_i p_j z_{-ij} \precsim b_i q_j z_{-ij} \\ a_i r_j z_{-ij} \succsim b_i s_j z_{-ij} \\ c_i p_j z_{-ij} \succsim d_i q_j z_{-ij} \end{array}\right\} \Rightarrow c_i r_j z_{-ij} \succsim d_i s_j z_{-ij}.$$

We can introduce the following binary relations:

Definition 6. *We write:*

1. *$i\,\mathbf{R}^z\,j$ if ij-triple cancellation holds on the set $\mathbf{SE}^z_{ij}$.*
2. *$i\mathbf{S}^z j$ if [NOT $j\,\mathbf{R}^z\,i$].*
3. *$i\,\mathbf{E}^z\,j$ if [$i\,\mathbf{R}^z\,j$ AND $j\,\mathbf{R}^z\,i$].*

The following axiom has two parts.

A3 - Coordinate Ordering Completeness. $\mathbf{R}^z$ is complete, i.e. for any $z \in X$, and all $i, j \in N$,

$$[i\,\mathbf{R}^z\,j] \text{ OR } [j\,\mathbf{R}^z\,i].$$

This new property would allow us to divide X into subsets without the need to use the notion of comonotonicity. Note that while $\mathbf{R}^z$ is complete, $\mathbf{S}^z$ is partial.[1] Since N is finite, there is only a finite number of various partial orders $\mathbf{S}^z$, so we can index them $(\mathbf{S}_a, \mathbf{S}_b, \ldots)$ and drop the superscripts when not needed. Also, each of the partial orders $\mathbf{S}_k$ uniquely defines the corresponding $\mathbf{R}_k$ - $i\,\mathbf{R}_k\,j$ if [NOT $j\mathbf{S}_k i$].

In contrast to the case with two variables, this property alone is not sufficient to construct a representation. Comparing value functions for different attributes suggests some sort of transitivity. For example, $f_i(x_i) > f_j(x_j)$ and $f_j(x_j) > f_k(x_k)$ imply $f_i(x_i) > f_k(x_k)$. The property we introduce is weaker - it is acyclicity (for some pairs of coordinates we might have $i\,\mathbf{E}^z\,j$ for all $z \in X$).

[1] If it is empty for all z, other axioms entail the existence of an additive representation on X.

A3 - Coordinate Ordering Acyclicity. For all $z \in X$, $\mathbf{S}^z$ is acyclic. In other words,

$$[i\mathbf{S}^z j\mathbf{S}^z \ldots \mathbf{S}^z k] \Rightarrow i\,\mathbf{R}^z\,k.$$

This axiom effectively defines how the set X is partitioned. It is required for the Choquet integral representation to exist.

We also introduce the following notions:

Definition 7. *Define* $\mathbf{SE}_{ij}$ *as a union of the following three sets:*

- *All* $z \in X$ *such that* $i\,\mathbf{R}^z\,j$, *if* z_i *is not maximal and* z_j *is not minimal;*
- *All* $z \in X$ *such that* z_i *is maximal and for no* $x_j, y_j \in X_j : z_j \succcurlyeq_j x_j \succcurlyeq_j y_j$ *we have* $j\,\mathbf{R}^{x_j z_{-j}}\,i$ *and NOT* $j\,\mathbf{R}^{y_j z_{-j}}\,i$;
- *All* $z \in X$ *such that* z_j *is minimal and for no* $x_i, y_i \in X_i : y_i \succcurlyeq_i x_i \succcurlyeq_i z_i$ *we have* $j\,\mathbf{R}^{x_i z_{-i}}\,i$ *and NOT* $j\,\mathbf{R}^{y_i z_{-i}}\,i$.

Define $\mathbf{NW}_{ij}$ *as a union of the following three sets:*

- *All* $z \in X$ *such that* $j\,\mathbf{R}^z\,i$, *if* z_j *is not maximal and* z_i *is not minimal;*
- *All* $z \in X$ *such that* z_i *is minimal and for no* $x_j, y_j \in X_j : y_j \succcurlyeq_j x_j \succcurlyeq_j z_j$ *we have* $i\,\mathbf{R}^{x_j z_{-j}}\,j$ *and NOT* $i\,\mathbf{R}^{y_j z_{-j}}\,j$;
- *All* $z \in X$ *such that* z_j *is maximal and for no* $x_i, y_i \in X_i : z_i \succcurlyeq_i x_i \succcurlyeq_i y_i$ *we have* $i\,\mathbf{R}^{x_i z_{-i}}\,j$ *and NOT* $i\,\mathbf{R}^{y_i z_{-i}}\,j$.

Presence of maximal and minimal points significantly complicates the definitions of $\mathbf{SE}_{ij}$ and $\mathbf{NW}_{ij}$, since at such points some of the sets $\mathbf{SE}^z_{ij}$ and $\mathbf{NW}^z_{ij}$ become degenerate and condition **ij-3C** trivially holds. If sets X_i and X_j do not contain minimal or maximal points, we can drop the corresponding conditions in each definition and simply state that $\mathbf{SE}_{ij} = \{z : i\,\mathbf{R}^z\,j\}$ and $\mathbf{NW}_{ij} = \{z : j\,\mathbf{R}^z\,i\}$.

Partial orders $\mathbf{S}_i$ define subsets of the set X as follows.

Definition 8. *We write* $X^{\mathbf{S}_i} = \bigcap_{(k,j):k\,\mathbf{R}_i\,j} \mathbf{SE}_{kj}$

It is well known that the sufficient property for an additive representation to exist on a Cartesian product is strong independence [9]. In the $X = Y^n$ case, the Choquet integral was previously axiomatized using comonotonic strong independence (or comonotonic trade-off consistency [16]). In this paper we will be using sets $X^{\mathbf{S}_i}$ to formulate a similar condition.

Definition 9. *We say that* $i \in N$ *is essential on* $A \subset X$ *if there exist* $x_i x_{-i}, y_i x_{-i} \in A$, *such that* $x_i x_{-i} \succ y_i x_{-i}$.

A4 - Intra-coordinate trade-off consistency

$$\left.\begin{array}{r} a_i x_{-i} \preccurlyeq b_i y_{-i} \\ a_i w_{-i} \succcurlyeq b_i z_{-i} \\ c_i x_{-i} \succcurlyeq d_i y_{-i} \end{array}\right\} \Rightarrow c_i w_{-i} \succcurlyeq d_i z_{-i},$$

provided that either:

(a) Exists $X^{\mathbf{S}_j}$ such that $a_ix_{-i}, b_iy_{-i}, a_iw_{-i}, b_iz_{-i}, c_ix_{-i}, d_iy_{-i}, c_iw_{-i}, d_iz_{-i} \in X^{\mathbf{S}_j}$
(b) Exist $X^{\mathbf{S}_j}, X^{\mathbf{S}_k}$ such that $a_ix_{-i}, b_iy_{-i}, a_iw_{-i}, b_iz_{-i} \in X^{\mathbf{S}_j}$, i is essential on $X^{\mathbf{S}_j}$, and $c_ix_{-i}, d_iy_{-i}, c_iw_{-i}, d_iz_{-i} \in X^{\mathbf{S}_k}$, or;
(c) Exist $X^{\mathbf{S}_j}, X^{\mathbf{S}_k}$ such that $a_ix_{-i}, b_iy_{-i}, c_ix_{-i}, d_iy_{-i} \in X^{\mathbf{S}_j}$, i is essential on $X^{\mathbf{S}_j}$, and $a_iw_{-i}, b_iz_{-i}, c_iw_{-i}, d_iz_{-i} \in X^{\mathbf{S}_k}$.

Informally, the meaning of the axiom is that ordering between preference differences ("intervals") is preserved irrespective of the "measuring rods" used to measure them. However, contrary to the additive case this does not hold on all X, but only when either points involved in all four relations lie in the same "3C-set" X^{S_j}, or points involved in two relations lie in one such set and those involved in the other two in another.

A5 - Inter-coordinate trade-off consistency

$$\left.\begin{aligned} a_ix_{-i} &\preccurlyeq b_iy_{-i} \\ c_ix_{-i} &\succcurlyeq d_iy_{-i} \\ a_iy^0_{-i} &\sim p_jx^0_{-j} \\ b_iy^0_{-i} &\sim q_jx^0_{-j} \\ c_iy^1_{-i} &\sim r_jx^1_{-j} \\ d_iy^1_{-i} &\sim s_jx^1_{-j} \\ p_je_{-j} &\succcurlyeq q_jf_{-j} \end{aligned}\right\} \Rightarrow r_je_{-j} \succcurlyeq s_jf_{-j}$$

for all $a_ix_{-i}, b_iy_{-i}, c_ix_{-i}, d_iy_{-i} \in X^{\mathbf{S}_j}$ provided i is essential on $X^{\mathbf{S}_j}$, $a_iy^0_{-i}, b_iy^0_{-i}, c_iy^1_{-i}, d_iy^1_{-i} \in X^{\mathbf{S}_k}$, $p_jx^0_{-j}, q_jx^0_{-j}, r_jx^1_{-j}, s_jx^1_{-j} \in X^{\mathbf{S}_l}$ provided j is essential on $X^{\mathbf{S}_l}$, $p_je_{-j}, q_jf_{-j}, r_je_{-j}, s_jf_{-j} \in X^{\mathbf{S}_m}$.

The formal statement of the **A5** is rather complicated, but it simply means that the ordering of the "intervals" is preserved across dimensions. Together with **A4** the conditions are similar to Wakker's trade-off consistency condition [17]. The axiom bears even stronger similarity to Axiom 5 (compatibility) from Sect. 8.2.6 of [9]. Roughly speaking, it says that if the "interval" between c_i and d_i is "larger" than that between a_i and b_i, then "projecting" these intervals onto another dimension by means of the equivalence relations must leave this order unchanged. We additionally require the comparison of intervals and "projection" to be consistent - meaning that each quadruple of points in each part of the statement belongs to the same $X^{\mathbf{S}_i}$. This axiom can be also conveniently formulated in terms of standard sequences - if we map all members of a sequence on some dimension onto another dimension via equivalence relations, it will be a standard sequence as well, provided the above restrictions hold.

A6 - Bi-independence Let $a_ix_{-i}, b_ix_{-i}, c_ix_{-i}, d_ix_{-i} \in X^{\mathbf{S}_i}$ and $a_ix_{-i} \succ b_ix_{-i}$. If for some $y_{-i} \in X_{-i}$ we have $c_iy_{-i} \succ d_iy_{-i}$, then $c_ix_{-i} \succ d_ix_{-i}$ for all $i \in N$.

This axiom is similar to "strong monotonicity" in [17]. In simple terms it means that if a coordinate is essential "somewhere" within a set X^{S_i}, then it is essential everywhere on this set.

A7 - Essentiality All coordinates are essential on X.

A8 - Restricted solvability If $a_i x_{-i} \succcurlyeq y \succcurlyeq b_i x_{-i}$, then there exists $c : c_i x_{-i} \sim y$ for $i \in N$.

A9 - Archimedean axiom Every bounded standard sequence contained in some $X^{\mathbf{S}_i}$ is finite, and in the case of only one essential coordinate, there exists a countable order-dense subset of $X^{\mathbf{S}_i}$.

Finally, we can introduce a notion of *interacting* coordinates.

Definition 10. *Coordinates i and j are* interacting *if exists $z \in X$, such that $i\mathbf{S}^z j$ or $j\mathbf{S}^z i$. We call a set $A \subset N$ an* interaction clique *if for each $i, j \in A$ we can build a chain of coordinates $i, k, \ldots, j$, such that every two subsequent coordinates in the chain are interacting.*

Interaction cliques play an important role in the uniqueness properties of the representation. In what follows we will be considering only cliques of maximal possible size if not specified otherwise.

3.1 Additional Assumptions

The following additional assumptions are made. The reasoning behind each one is explained below. They are not required for the construction of the representation in general.

"Collapsed" equivalent points along dimensions. For no $i \in N$ and no $a_i, b_i \in X_i$ holds $a_i x_{-i} \sim b_i x_{-i}$ for all $x_{-i} \in X_{-i}$.

If this wasn't true, we could have value functions assigning the same value to several points in the same set X_i. To simplify things we exclude such case, however, it can be easily reconstructed once the representation is built.

Density. We assume that for all $i \in N$, whenever $a_i x_{-i} \succ b_i x_{-i}$, there exists $c_i \in X_i$ such that $a_i x_{-i} \succ c_i x_{-i} \succ b_i x_{-i}$ (X is order dense).

"Closedness". For every i and j, if there exist $x_i x_j z_{-ij}$ such that $i\mathbf{S}^{x_i x_j z_{-ij}} j$ and $y_i x_j z_{-ij}$ such that $j\mathbf{S}^{y_i x_j z_{-ij}} i$, then exists $z_i \in X_i$ such that $i\,\mathbf{E}^{z_i x_j z_{-ij}}\, j$.

This assumption says that sets $\mathbf{SE}_{ij}$ and $\mathbf{NW}_{ij}$ are "closed". In the representation this translates into existence of the inverse for all points where value functions f_i and f_j are equal, provided i and j are interacting and the inverse exists for at least one of the coordinates (i.e. there is a corresponding point in X_i or X_j). This is a technical simplifying assumption and the proof can be done without it.

Geometry of X. For every clique of interacting variables $A \subset N$, there exist at least two points $r_A^0, r_A^1 \in X$ such, that for every pair $i, j \in A$, we have $i\,\mathbf{E}^{r_A^0}\,j$ and $i\,\mathbf{E}^{r_A^1}\,j$.

Again, this is a simplifying assumption, but this time a restrictive one. It takes the proof somewhat closer to the homogeneous case (see Sect. 5). Without it we can have a situation, where the smallest value of $f_i : X_i$ is larger then the greatest value of $f_j : X_j$ for some $i, j \in N$. This in turn does not allow to construct the capacity in a unique way. Another way to stating this assumption, is to say that X must contain points corresponding to all possible acyclic partial orders on N, generated by interacting pairs $i\mathbf{S}j$. Work to remove this assumption is still in progress.

4 Representation Theorem

As follows from the definition of the Choquet integral (Sect. 2), every point $x \in X$ uniquely corresponds to a set of weights $p_i^x : p_i^x \geq 0, \sum_{i \in N} p_i^x = 1$. This notation is used to simplify the statement of the following theorems.

Theorem 1. *Let $\succcurlyeq$ be an order on X and the structural assumption hold. Then, if axioms* ***A1-A9*** *are satisfied, there exists a capacity ν and value functions $f_1 : X_1 \to \mathbb{R}, \ldots, f_n : X_n \to \mathbb{R}$, such that $\succcurlyeq$ can be represented by the Choquet integral:*

$$x \succcurlyeq y \iff C(\nu, (f_1(x_1), \ldots, f_n(x_n))) \geq C(\nu, (f_1(y_1), \ldots, f_n(y_n))), \tag{1}$$

for all $x, y \in X$.

Capacity and value functions have the following uniqueness properties. Let $\mathcal{I} = \{A_1, \ldots, A_k\}$ be a partition of N, such that $m(B) = 0$ for all $B \subset N$ such that $B \cap A_i \neq \emptyset, B \cap A_j \neq \emptyset$. If no such partition exists, let $\mathcal{I} = \{N\}$.

Theorem 2. *Let $g_1 : X_1 \to \mathbb{R}, \ldots, g_n : X_n \to \mathbb{R}$ be such that (1) holds with f_i substituted by g_i. Then, at all $x_i \in X_i$, such that for some z_{-i} we have $p_i^{x_i z_{-i}} > 0$, and also $p_j^{x_i z_{-i}} > 0, j \neq i$, value functions f_i and g_i are related in the following way:*

$$f_i(x_i) = \alpha_{A_j} g_i(x_i) + \beta_{A_j},$$

Capacity changes as follows

$$m'(B) = \frac{\alpha_{A_j} m(B)}{\sum_{C \subset A_i, A_i \in \mathcal{I}} \alpha_{A_i} m(C)}.$$

At the remaining points of X, i.e. for x_i such that for any $z_{-i} \in X_{-i}$ we have $p_i^{x_i z_{-i}} = 1$, and $p_j^{x_i z_{-i}} = 0$ for all $j \neq i$, value functions f_i have the following uniqueness properties[2]*:*

$$f_i(x_i) = \psi_i(g_i(x_i)),$$

[2] Due to our assumption that for no a_i, b_i we have $a_i z_{-i} \sim b_i z_{-i}$ for all z_{-i}, we can't have $p^{x_i z_{-i}} = 0$ for all z_{-i}.

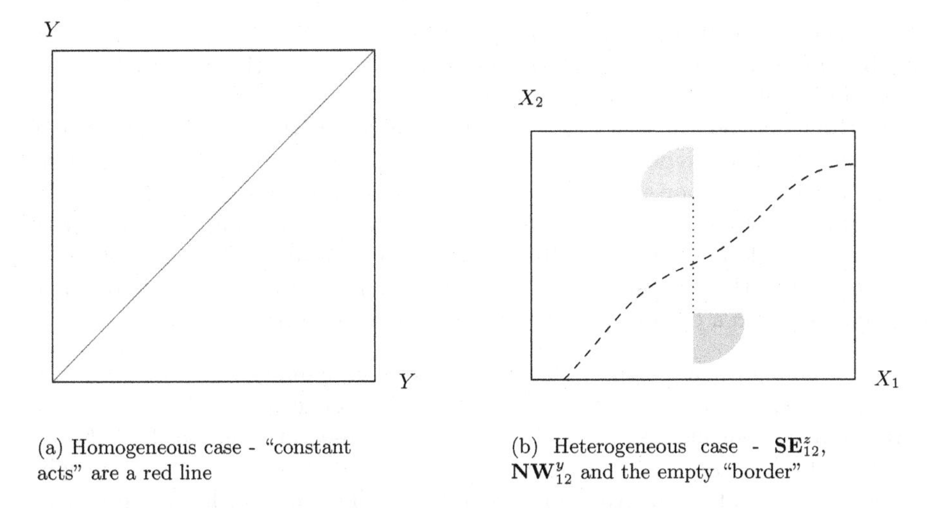

(a) Homogeneous case - "constant acts" are a red line

(b) Heterogeneous case - $\mathbf{SE}^z_{12}$, $\mathbf{NW}^y_{12}$ and the empty "border"

Fig. 1. Differences with the homogeneous case - border

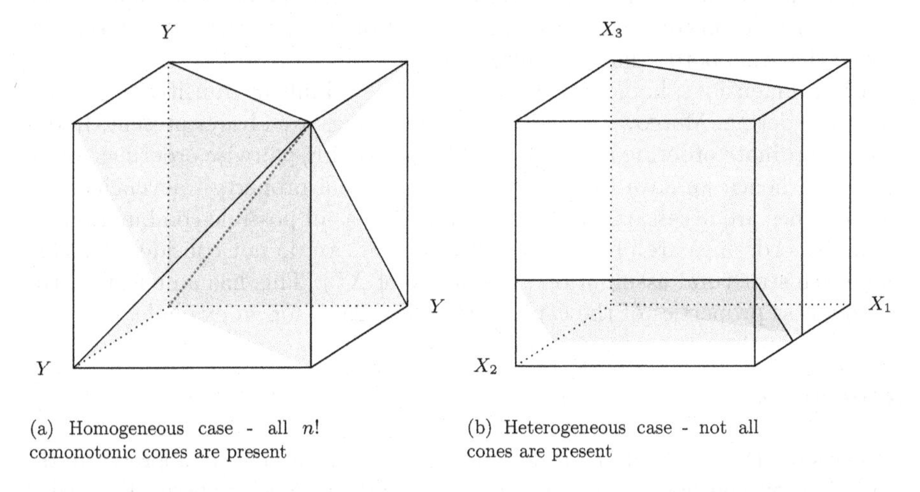

(a) Homogeneous case - all $n!$ comonotonic cones are present

(b) Heterogeneous case - not all cones are present

Fig. 2. Differences with the homogeneous case - geometry of X

where ψ_i is an increasing function, and for all $j \in N, j \neq i$, such that exists $A \in N : i, j \in A, m(A) > 0$, we additionally have

$$f_i(x_i) = f_j(x_j) \iff g_i(x_i) = g_j(x_j).$$

5 Constructing the Representation

The axiomatization rests on three main pillars: a weakened form of comonotonicity (**A3**), trade-off consistency conditions (**A4**,**A5**), and strong monotonicity

(**A6**). The main difference from the previous results is **A3**, other axioms are notably quite similar to the homogeneous case. Details of the construction can be found in [14,15].

In the homogeneous case, the characterization is performed by stating that there exist additive representations on pre-defined subsets of X, in particular on the comonotonic cones, and the representations are related to each other. In this paper we state a weakened form of this property, showing that additivity holds on some loosely-defined subsets of X, and then use additional axioms to guarantee that these subsets have a required shape in the representation. The differences between comonotonicity and **A3** are as follows:

- In the homogeneous case we have a hypercube, symmetrically divided into conical subsets. In our case, the symmetry is no longer present.
- In the homogeneous case there always exists a non-empty thin border between comonotonic subsets (so-called "constant acts"). In some sense, it is a "straight line" (see Fig. 1a, b). **A3** alone guarantees that the border is only non-decreasing, in the sense that an increase along X_i cannot be accompanied only by a decrease along X_j. Moreover, the boundary does not have to be non-empty or, on contrary, be a "line". Additional axioms (**A4**,**A5**,**A6**) guarantee that this border has a required shape.
- With n dimensions, homogeneous case provides a built-in transitivity of coordinate ordering. Moreover, all $n!$ possible orderings are always present. In our case coordinate ordering is constructed by combining pairwise orderings for all $i, j \in N$, hence we have to introduce an additional property - acyclicity. Not all $n!$ cones are necessarily present, moreover not all possible combinations of pairwise orderings are present as well (however, we do not consider this case here, see structural assumption "Geometry of X"). This has an effect on the uniqueness properties of the capacity. See Fig. 2a, b for an example.

References

1. Bouyssou, D., Marchant, T., Pirlot, M.: A conjoint measurement approach to the discrete Sugeno integral. In: Brams, S.J., Gehrlein, W.V., Roberts, F.S. (eds.) The Mathematics Preference, Choice and Order, pp. 85–109. Springer, Heidelberg (2009)
2. Debreu, G.: Topological methods in cardinal utility theory. Cowles Foundation Discussion Papers (1959)
3. Ellsberg, D.: Risk, ambiguity, and the savage axioms. Q. J. Econ. **75**(4), 643–669 (1961)
4. Gilboa, I.: Expected utility with purely subjective non-additive probabilities. J. Math. Econ. **16**(1), 65–88 (1987)
5. Grabisch, M., Labreuche, C.: A decade of application of the Choquet and Sugeno integrals in multi-criteria decision aid. 4OR Q. J. Oper. Res. **6**(1), 1–44 (2008)
6. Greco, S., Matarazzo, B., Słowiński, R.: Axiomatic characterization of a general utility function and its particular cases in terms of conjoint measurement and rough-set decision rules. Eur. J. Oper. Res. **158**(2), 271–292 (2004)

7. Keeney, R.L., Raiffa, H.: Decisions with Multiple Objectives: Preferences and Value Tradeoffs. Cambridge University Press, Cambridge (1976)
8. Köbberling, V., Wakker, P.P.: Preference foundations for nonexpected utility: a generalized and simplified technique. Math. Oper. Res. **28**(3), 395–423 (2003)
9. Krantz, D.H., Luce, R.D., Suppes, P., Tversky, A.: Foundation of Measurement. Additive and Polynomial Representations, vol. 1. Academic Press, New York (1971)
10. Labreuche, C.: An axiomatization of the Choquet integral and its utility functions without any commensurability assumption. In: Greco, S., Bouchon-Meunier, B., Coletti, G., Fedrizzi, M., Matarazzo, B., Yager, R.R. (eds.) IPMU 2012, Part IV. CCIS, vol. 300, pp. 258–267. Springer, Heidelberg (2012)
11. Quiggin, J.: A theory of anticipated utility. J. Econ. Behav. Organ. **3**(4), 323–343 (1982)
12. Savage, L.: The Foundations of Statistics. Wiley Publications in Statistics. Wiley, New York (1954)
13. Schmeidler, D.: Subjective probability and expected utility without additivity. Econometrica J. Econometric Soc. **57**(3), 571–587 (1989)
14. Timonin, M.: Axiomatization of the Choquet integral for 2-dimensional heterogeneous product sets (2015). arXiv:1507.04167
15. Timonin, M.: Axiomatization of the Choquet integral for heterogeneous product sets (2016). arXiv:1603.08142
16. Wakker, P.: Additive representations of preferences, a new foundation of decision analysis; the algebraic approach. In: Doignon, J.-P., Falmagne, J.-C. (eds.) Mathematical Psychology. Recent Research in Psychology, pp. 71–87. Springer, Heidelberg (1991)
17. Wakker, P.: Additive representations on rank-ordered sets. i. the algebraic approach. J. Math. Psychol. **35**(4), 501–531 (1991)

Aggregation of Choquet Integrals

Radko Mesiar, Ladislav Šipeky(✉), and Alexandra Šipošová

Faculty of Civil Engineering, Department of Mathematics and Descriptive Geometry, Slovak University of Technology, Bratislava, Slovakia
ladislav.sipeky@stuba.sk

Abstract. Aggregation functions acting on the lattice of all Choquet integrals on a fixed measurable space $(\mathrm{X}, \mathcal{A})$ are discussed. The only direct aggregation of Choquet integrals resulting into a Choquet integral is linked to the convex sums, i.e., to the weighted arithmetic means. We introduce and discuss several other approaches, for example one based on compatible aggregation systems. For X finite, the related aggregation of *OWA* operators is obtained as a corollary. The only exception, with richer structure of aggregation functions, is the case *card* $\mathrm{X} = 2$, when the lattice of all *OWA* operators forms a chain.

Keywords: Aggregation function · Capacity · Choquet integral · *OWA* operator

1 Introduction

Consider a fixed measurable space $(\mathrm{X}, \mathcal{A})$, i.e., X is a non-empty set and $\mathcal{A}$ is a σ-algebra of subsets of X. If X is finite, then $\mathcal{A} = 2^X$ is considered, by convention. A set function $m : \mathcal{A} \to [0, 1]$ is called a capacity whenever it is monotone and satisfies the boundary conditions $m(\emptyset) = 0$ and $m(\mathrm{X}) = 1$. The set of all capacities on $(\mathrm{X}, \mathcal{A})$ will be denoted as $\mathcal{M}_{(\mathrm{X},\mathcal{A})}$. Similarly, $\mathcal{F}_{(\mathrm{X},\mathcal{A})}$ is the set of all $\mathcal{A}$-measurable functions $f : \mathrm{X} \to [0, \infty[$. Recall that the Choquet integral [1,4] of $f \in \mathcal{F}_{(\mathrm{X},\mathcal{A})}$ with respect to $m \in \mathcal{M}_{(\mathrm{X},\mathcal{A})}$ is given by

$$Ch_m(f) = \int_0^\infty m(f \geq x)\, dx, \tag{1}$$

where the right-hand side of (1) is the classical (possibly improper) Riemann integral. Observe that due to Schmeidler [13,14], a functional $\mathrm{I} : \mathcal{F}_{(\mathrm{X},\mathcal{A})} \to [0, \infty]$ is a Choquet integral, $\mathrm{I} = Ch_m$ for some $m \in \mathcal{M}_{(\mathrm{X},\mathcal{A})}$, if and only if it is comonotone additive and $\mathrm{I}(\mathbf{1}_X) = 1$. Then $m(E) = \mathrm{I}(\mathbf{1}_E)$ for any event $E \in \mathcal{A}$.

Formally, the set $Ch_{(\mathrm{X},\mathcal{A})} = \{Ch_m | m \in \mathcal{M}_{(\mathrm{X},\mathcal{A})}\}$ forms a distributive bounded lattice with respect to the partial ordering $\preceq$ herited from the standard ordering of reals, i.e., $Ch_{m_1} \preceq Ch_{m_2}$ if and only if $Ch_{m_1}(f) \leq Ch_{m_2}(f)$ for each $f \in \mathcal{F}_{(\mathrm{X},\mathcal{A})}$. It is not difficult to check that $Ch_{m_1} \preceq Ch_{m_2}$ if and only if $m_1 \leq m_2$, i.e., if $m_1(E) \leq m_2(E)$ for each event $E \in \mathcal{A}$. Thus the lattices $(Ch_{(\mathrm{X},\mathcal{A})}, \preceq)$ and $(\mathcal{M}_{(\mathrm{X},\mathcal{A})}, \leq)$ are isomorphic, with the respective top and bottom elements given

J.P. Carvalho et al. (Eds.): IPMU 2016, Part I, CCIS 610, pp. 58–64, 2016.
DOI: 10.1007/978-3-319-40596-4_6

by Sup and Inf and m^* and m_*, respectively, where $Sup, Inf\colon \mathcal{F}_{(\mathrm{X},\mathcal{A})} \to [0,\infty]$ are given by $Sup\ (f) = sup\ \{f(x)|x \in \mathrm{X}\}, Inf\ (f) = inf\ \{f(x)|x \in \mathrm{X}\}$, and $m^*, m_* : \mathcal{A} \to [0,1]$ are given by

$$m^*(E) = \begin{cases} 0 \text{ if } E = \emptyset, \\ 1 \text{ otherwise} \end{cases},$$

$$m_*(E) = \begin{cases} 1 \text{ if } E = \mathrm{X}, \\ 0 \text{ otherwise} \end{cases}.$$

The aim of this contribution is a deeper look on the aggregation of Choquet integrals from $Ch_{(\mathrm{X},\mathcal{A})}$. Recall that an aggregation function on a bounded lattice $(L, \leq_L)$ [2,11] is a mapping $A : L^n \to L$ which is monotone with respect to the partial order $\leq_L$ and satisfies the boundary conditions $A(0_L, \ldots, 0_L) = 0_L$, $A(1_L, \ldots, 1_1) = 1_L$. Here n is a fixed integer from the set $\{2, 3, \ldots\}$. Choquet integrals (on finite sets) are frequently applied as utility functions in the area of multicriteria decision support [6]. Here, the considered universe X is the set of criteria, and each evaluating expert can propose his own Choquet integral (i.e., his own capacity m) to evaluate single alternatives. To find a consensus among experts, an aggregated Choquet integral can be one alternative. This fact has motivated us to study aggregations of Choquet integrals. Our approach differs from multi-step procedures considered, e.g., in [12]. Observe that OWA operators [16] are a particular subclass of Choquet integrals, and then several results shown for the aggregation of Choquet integrals can be applied as corollaries for the aggregation of OWA operators, too.

The contribution is organised as follows. In the next section, a direct aggregation of Choquet integrals is considered. Section 3 is devoted to the aggregation of Choquet integrals based on the aggregation of capacities. In Sect. 4, we discuss the aggregation of OWA operators. Finally, some concluding remarks are added.

2 Direct Aggregation of Choquet Integrals

For any aggregation function $A : [0,\infty]^n \to [0,\infty]$, see [7], one can apply A to aggregate Choquet integrals in the form $\mathbb{A}_A : Ch^n_{(\mathrm{X},\mathcal{A})} \to \mathcal{I}_{(\mathrm{X},\mathcal{A})}$,

$$\mathbb{A}_A(Ch_{m_1}, \ldots, Ch_{m_n})(f) = A(Ch_{m_1}(f), \ldots, Ch_{m_n}(f)) \tag{2}$$

where $\mathcal{I}_{(\mathrm{X},\mathcal{A})}$ is the set of all monotone functionals $I : \mathcal{F}_{(\mathrm{X},\mathcal{A})} \to [0,\infty]$, $I(\mathbf{0}) = 0$.

In general, $\mathbb{A}_A$ need not result into a Choquet integral. The positive homogeneity of Choquet integrals forces this property for A in such case, for example. Similarly, the comonotone additivity of Choquet integrals results into the additivity of A. Moreover, $A(\mathbf{1}) = 1$ is also forced. Summarizing, we have the next result.

Theorem 1. *Let $A : [0,\infty]^n \to [0,\infty]$ be an aggregation function. Then the following are equivalent.*

- *for any Choquet integrals* $Ch_{m_1}, \ldots, Ch_{m_n} \in Ch_{(X,\mathcal{A})}$, *also* $\mathbb{A}_A(Ch_{m_1}, \ldots, Ch_{m_n})$ *given by (2) belongs to* $Ch_{(X,\mathcal{A})}$;
- A *is a weighted arithmetic mean,* $A(x_1, \ldots, x_n) = \sum_{i=1}^n w_i x_i$, *where* $(w_1, \ldots, w_n) \in [0,1]^n$ *and* $\sum_{i=1}^n w_i = 1$.

Obviously, under the notation of Theorem 1,

$$\mathbb{A}_A(Ch_{m_1}, \ldots, Ch_{m_n}) = Ch_m, \text{ where } m = \sum_{i=1}^n w_i m_i \in \mathcal{M}_{(X,\mathcal{A})}. \tag{3}$$

3 Aggregation of Choquet Integrals Based on the Aggregation of Capacities

Consider any aggregation function $\mathbb{A} : Ch^n_{(X,\mathcal{A})} \to Ch_{(X,\mathcal{A})}$. Obviously, $\mathbb{A}$ is in a one-to-one correspondence with some aggregation function $\mathbb{B} : \mathcal{M}^n_{(X,\mathcal{A})} \to \mathcal{M}_{(X,\mathcal{A})}$.

For example, the case of direct aggregation of Choquet integrals discussed in Sect. 2 relates convex sums (weighted arithmetic means) of Choquet integrals and of capacities, see formula (3). Maybe the simplest approach to the aggregation of capacities is based on fixed aggregation function $A : [0,1]^n \to [0,1]$, see [7], $\mathbb{B}_A : \mathcal{M}^n_{(X,\mathcal{A})} \to \mathcal{M}_{(X,\mathcal{A})}$ is then given by

$$\mathbb{B}_A(m_1, \ldots, m_n)(E) = A(m_1(E), \ldots, m_n(E)), E \in \mathcal{A} \tag{4}$$

Properties od A are then straightforwardly applicable to $\mathbb{B}_A$, and subsequently to the related aggregation $\mathbb{A}_A$ of Choquet integrals. This fact allows to model the conjunctive, disjunctive or averaging attitude of the global decision maker. One can define also particular aggregation functions on $Ch_{(X,\mathcal{A})}$ in this way, such as triangular norms or conorms [10].

Theorem 2. *Let* $A : [0,1]^n \to [0,1]$ *be a given aggregation function. Then the following are equivalent.*

- $\mathbb{A}_A : Ch^n_{(X,\mathcal{A})} \to Ch_{(X,\mathcal{A})}$ *given by* $\mathbb{A}_A(Ch_{m_1}, \ldots, Ch_{m_n}) = Ch_{\mathbb{B}(m_1, \ldots, m_n)}$, *see (4), is a t-norm on* $Ch_{(X,\mathcal{A})}$;
- A *is a t-norm.*

Example 1. Let $\Pi : [0,1]^n \to [0,1]$ be the standard product, $\Pi(x_1, \ldots, x_n) = \Pi_{i=1}^n x_i$. Observe that Π is a t-norm, i.e., a symmetric associative aggregation function on $[0,1]$ with neutral element 1. For $X = \{1,2\}, \mathcal{A} = 2^X$, it holds $\mathcal{M}_{(X,\mathcal{A})} = \{m_{a,b} \mid a, b \in [0,1]^2\}$, where $m_{a,b}(\{1\}) = a$ and $m_{a,b}(\{2\}) = b$. More, $\mathcal{F}_{(X,\mathcal{A})}$ can be represented as $[0,\infty[^2$, and

$$Ch_{m_{a,b}}(x,y) = \begin{cases} ax + (1-a)y & \text{if } x \geq y, \\ (1-b)x + by & \text{otherwise} \end{cases}.$$

Then

$$\mathbb{A}_{\Pi}(Ch_{m_{a_1,b_1}},\ldots,Ch_{m_{a_n,b_n}})(x,y) = Ch_{m_{\Pi_{i=1}^n a_i,\Pi_{i=1}^n b_i}}(x,y)$$
$$= \begin{cases} (\Pi_{i=1}^n a_i)x + (1-\Pi_{i=1}^n a_i)y & \text{if } x \geq y, \\ (1-\Pi_{i=1}^n b_i)y + (\Pi_{i=1}^n b_i)y & \text{otherwise} \end{cases}.$$

Formula (4) can be generalized in several ways. Observe that for each event E, in (4) the same aggregation function A is applied. In general, one can use, for $E_1 \neq E_2$, different aggregation functions, but not arbitrary. Note that the monotonicity of capacities poses some restrictions here. On the other hand, for any aggregation function A it holds $A(m_1(\emptyset),\ldots,m_n(\emptyset)) = A(0,\ldots,0) = 0$ and $A(m_1(X),\ldots,m_n(X)) = A(1,\ldots,1) = 1$.

Definition 1. *For a fixed measurable space* $(\mathrm{X},\mathcal{A})$ *and* $n \in \{2,3,\ldots\}$, *let* $\mathcal{H}_n = (A_E)_{E\in\mathcal{A}}$ *be a system of n-ary aggregation functions* $A_E : [0,1]^n \to [0,1]$ *such that for any* $\emptyset \neq E_1 \subsetneqq E_2 \subsetneqq X$ *it holds* $A_{E_1} \leq A_{E_2}$. *Then* $\mathcal{H}$ *is called a compatible aggregation system on* $(\mathrm{X},\mathcal{A})$.

Proposition 1. *Let* $\mathcal{H}_n = (A_E)_{E\in\mathcal{A}}$ *be a compatible aggregation system on* $(\mathrm{X},\mathcal{A})$. *For any n-tuple* $(m_1,\ldots,m_n) \in (\mathcal{M}_{(\mathrm{X},\mathcal{A})})^n$ *define a set function* $\mathbb{B}_{\mathcal{H}_n}(m_1,\ldots,m_n) : \mathcal{A} \to [0,1]$ *by*

$$\mathbb{B}_{\mathcal{H}_n}(m_1,\ldots,m_n)(E) = A_E(m_1(E),\ldots,m_n(E)). \tag{5}$$

Then $\mathbb{B}_{\mathcal{H}_n}(m_1,\ldots,m_n)$ *is a capacity on* $(\mathrm{X},\mathcal{A})$, *i.e.,* $B_{\mathcal{H}_n}(m_1\ldots,m_n) \in \mathcal{M}_{(\mathrm{X},\mathcal{A})}$.

Based on (5), one can define a new type of aggregation of Choquet integrals.

Theorem 3. *Let* $\mathcal{H}_n = (A_E)_{E\in\mathcal{A}}$ *be a compatible aggregation system on* $(\mathrm{X},\mathcal{A})$. *Then the mapping* $\mathbb{A}_{\mathcal{H}_n} : Ch^n_{\mathrm{X},\mathcal{A}} \to \mathcal{I}_{(\mathrm{X},\mathcal{A})}$ *given by* $\mathbb{A}_{\mathcal{H}_n}(Ch_{m_1},\ldots,Ch_{m_n}) = Ch_m$, *where* $m = \mathbb{B}_{\mathcal{H}_n}(m_1,\ldots,m_n)$, *is an aggregation of Choquet integrals.*

Example 2. For $X = \{1,2\}$, put $A_\emptyset = A_X = \Pi$, $A_{\{1\}} = Min$ and $A_{\{2\}} = Max$. Then $\mathcal{H}_n = (A_E)_{E\in 2^X}$ is a compatible aggregation system on $(\mathrm{X},\mathcal{A})$. Then $\mathbb{B}_{\mathcal{H}_n}(m_{a_1,b_1},\ldots,m_{a_n,b_n}) = m_{Min\{a_i\},Max\{b_i\}}$, and

$$\mathbb{A}_{\mathcal{H}_n}(Ch_{m_1},\ldots,Ch_{m_n})(x,y) = \begin{cases} Min\{Ch_{m_i}(x,y)\} & \text{if } x \geq y, \\ Max\{Ch_{m_i}(x,y)\} & \text{otherwise} \end{cases}.$$

Note that also Theorem 3 can be further generalized. Indeed, to determine the value $m(E)$ of an aggregated capacity m, we exploit values $m_1(E),\ldots,m_n(E)$, only. This approach is reasonable, and we add the next example as an exotic one, to see how far we can go not considering any reasonable constraints.

Example 3. Let $X = \{1,2\}$ and let $A_1, A_2 : [0,1]^4 \to [0,1]$ be aggregation functions. Define $\mathbb{A} : Ch^2_{(\mathrm{X},\mathcal{A})} \to Ch_{(\mathrm{X},\mathcal{A})}$ by $\mathbb{A}(Ch_{m_{a_1,b_1}}, Ch_{m_{a_2,b_2}}) = Ch_{m_{a,b}}$, where $a = A_1(a_1,b_1,a_2,b_2)$, $b = A_2(a_1,b_1,a_2,b_2)$. Clearly, $\mathbb{A}$ is an aggregation of Choquet integrals.

4 Aggregation of *OWA* Operators

Consider a finite space $X, card\ X = k$. A capacity $m \in \mathcal{M}_{(X,2^X)}$ is symmetric whenever $m(E) = \varphi(\frac{card\ X}{k})$, where $\varphi : [0,1] \to [0,1]$ is some monotone function such that $\varphi(0) = 0,\ \varphi(1) = 1$. Then the corresponding Choquet integral $Ch_m : [0,\infty[^k \to [0,\infty[$ is given by $Ch_m(a_1,\dots,a_k) = \sum_{i=1}^k w_i b_i$, where b_i is the i-th greatest input among $(a_1,\dots,a_k)$, i.e., i-th order statistics, and $w_i = \varphi(\frac{i}{k}) - \varphi(\frac{i-1}{k})$. Hence, $Ch_m = OWA_{\mathbf{w}}$ is the OWA operator based on the weighting vector $\mathbf{w} = (w_1,\dots,w_k)$ as introduced by Yager in [16]. For more details and proofs we recommend [5]. The set OWA_k of all OWA operators on $(\mathrm{X},\mathcal{A})$, $card\ X = k$, is a lattice with top element $OWA_{(1,0,\dots,0)} = Max$ and bottom element $OWA_{(0,\dots,0,1)} = Min$. Note that if $k = 2$, then OWA_2 is a bounded chain. Based on the results highlighted in Sects. 2 and 3, we have the next results for aggregation functions on OWA_k. The next result is a corollary of Theorem 1 if $k > 2$, but not more for $k = 2$.

Corollary 1. *Let $A : [0,\infty]^n \to [0,\infty]$ be an aggregation function. Then the following are equivalent.*

- $\mathbb{A}_A : OWA_k^n \to \mathcal{I}_{(X,2^X)}$ *is an aggregation function on OWA_k, where*

$$\mathbb{A}(OWA_{\mathbf{w}^{(1)}},\dots,OWA_{\mathbf{w}^{(n)}})(a_1,\dots,a_k)$$
$$= A(OWA_{\mathbf{w}^{(n)}}(a_1,\dots,a_k),\dots,OWA_{\mathbf{w}^{(n)}}(a_1,\dots,a_k));$$

- *if $k = 2$, then A is the Choquet integral on $N = \{1,\dots,n\}$;*
  if $k > 2$, A is the weighted arithmetic mean.

Similarly, one can consider Theorems 2 and 3. The next result is related to Theorem 3.

Corollary 2. *Let $A_1,\dots,A_{k-1} : [0,1]^n \to [0,1]$ be aggregation functions such that $A_1 \le \dots \le A_{k-1}$. Then the mapping $\mathbb{A} : OWA_k^n \to OWA_k$ given by $\mathbb{A}(OWA_{\mathbf{w}^{(1)}},\dots,OWA_{\mathbf{w}^{(n)}}) = OWA_{\mathbf{w}}$ is an OWA-aggregation function, where $\mathbf{w} = (w_1,\dots,w_k)$ is defined as follows:*

- *for $\mathbf{w}^{(i)} = (w_1^{(i)},\dots,w_k^{(i)})$ define the corresponding cumulative vector $\mathbf{v}^{(i)} = (v_1^{(i)},\dots,v_k^{(i)})$, $v_j^{(i)} = w_1^{(i)} + \dots + w_j^{(i)}$;*
- *define $\mathbf{v} = (v_1,\dots,v_k)$ by $v_j = A_j(v_j^{(1)},\dots,v_j^{(n)})$, $j = 1,\dots,k-1$, and $v_k = 1$;*
- *then $w_1 = v_1$, $w_2 = v_2 - v_1,\dots,w_k = v_k - v_{k-1}$.*

There are also alternative approaches. We will consider them in our future research, especially bilinear types of aggregation functions.

Note that formally the set OWA_k can be considered as a convex closure of the set of order statistics $OS_1 = Max, OS_2,\dots,OS_k = Min$, and thus, in the case of bilinearity, we need to introduce the aggregation of order statistics only (their aggregation should be an OWA operator but not necessarily an order statistics).

5 Concluding Remarks

We have discussed the aggregation of Choquet integrals, and as a particular case, also the aggregation of *OWA* operators. Observe that some particular aggregation functions on *OWA* operators were introduced in a recent paper [9] where the focus was on the *ORness* parameter characterizing the disjunctive attitude of *OWA* operators. Also in some other distinguished classes of Choquet integrals one can find several approaches to the aggregation (or, equivalently, to the aggregation of capacities). So, for example, in the case of belief measures, i.e., capacities linked to a probability P on $2^X \setminus \{\emptyset\}$, $m(E) = \sum_{F \subseteq E} P(F)$, we have $Ch_m(a_1, \ldots, a_k) = \sum_{E \subseteq X} P(E) \cdot min\{a_i \mid i \in E\}$. For belief measures, already the seminal paper of Dempster [3] brings a combination rule for belief functions, i.e., their aggregation. Some other approaches to aggregation of belief measures can be found, e.g., in [8].

Note that due to the fact that the Šipoš integral [15] based on a capacity m and real valued function vector $\mathbf{x} \in \mathbb{R}^n$ is given by $\check{S}_m(\mathbf{x}) = Ch_m(\mathbf{x}^+) - Ch_m(\mathbf{x}^-)$, where $\mathbf{x}^+ = (max\{x_1, 0\}, \ldots, max\{x_n, 0\})$ and $\mathbf{x}^- = \mathbf{x}^+ - \mathbf{x}$, our results for the Choquet integral can be straightforwardly formulated for the Šipoš integral, too. Note that this is due to the fact that for $\mathbf{x}, \mathbf{y} \in \mathbb{R}^n$, if $\mathbf{x} \leq \mathbf{y}$ then $\mathbf{x}^+ \leq \mathbf{y}^+$ but $\mathbf{x}^- \geq \mathbf{y}^-$.

Acknowledgment. The work on this contribution was supported by the grant APVV-14-0013.

References

1. Choquet, G.: Theory of Capacities. Annales de l'Institute Fourier (Grenoble), vol. 5, pp. 131–295 (1953)
2. De Cooman, G., Kerre, E.: Order norms on bounded partially ordered sets. J. Fuzzy Math. **2**, 281–310 (1994)
3. Dempster, A.P.: Upper and lower probabilities induced by a multi-valued mapping. Ann. Math. Stat. **38**, 325–339 (1967)
4. Denneberg, D.: Non-additive Measure and Integral. Theory and Decision Library (Series B Mathematical and Statistical Methods), vol. 27, p. ix–178. Kluwer Academic Publishers, Dordrecht (1994)
5. Grabisch, M.: Fuzzy integral in multicriteria decision making. Fuzzy Sets Syst. **69**(3), 279–298 (1995)
6. Grabisch, M., Labreuche, C.: A decade of application of the Choquet and Sugeno integrals in multi-criteria decision aid. 4OR **6**(1), 1–44 (2008)
7. Grabisch, M., Marichal, J.L., Mesiar, R., Pap, E.: Aggregation Functions (Encyclopedia of Mathematics and Its Applications). Cambridge University Press, New York (2009)
8. Jiroušek, R., Vejnarová, J., Daniel, M.: Compositional models for belief functions. In: Proceedings of ISIPTA 2007, Prague, 243–252 (2007)
9. Jin, L.: OWA monoid, submitted
10. Klement, E.P., Mesiar, R., Pap, E.: Triangular Norms. Kluwer Academic Publisher, Dordrecht (2000)

11. Komorníková, M., Mesiar, R.: Aggregation functions on bounded partially ordered sets and their classification. Fuzzy Sets Syst. **175**(1), 48–56 (2011)
12. Narukawa, Y., Torra, V.: Twofold integral and multi-step Choquet integral. Kybernetika **40**(1), 39–50 (2004)
13. Schmeidler, D.: Subjective probability and expected utility without additivity. Econometrica **57**, 571–587 (1989)
14. Schmeidler, D.: Integral representation without additivity. Proc. Am. Math. Soc. 2 **97**, 255–261 (1986)
15. Šipoš, J.: Integral with respect to a pre-measure. Math. Slov. **29**, 141–155 (1979)
16. Yager, R.R.: On ordered weighted averaging aggregation operators in multicriteria decision-making. IEEE Trans. Syst. Man Cybern. **18**(1), 183–190 (1988)

Inclusion-Exclusion Integral and t-norm Based Data Analysis Model Construction

Aoi Honda[1(✉)] and Yoshiaki Okazaki[2]

[1] Kyushu Institute of Technology, 680-2 Kawazu, Iizuka, Fukuoka 820-8502, Japan
aoi@ces.kyutech.ac.jp

[2] Fuzzy Logic Systems Institute, 680-41 Kawazu, Iizuka, Fukuoka 820-0067, Japan
okazaki@flsi.or.jp

Abstract. A data analysis model using the inclusion-exclusion integral and a new construction method of a model utilizing t-norms are proposed. This model is based on the integral with respect to the nonadditive measure and is constructed in three steps of specifications of monotone functions, t-norm and of monotone measures. The model has good description ability and can be applied flexibly to real problems. Applying this model to the data set of a multiple criteria decision making problem, the efficiency of the model is verified by comparing it with the classical linear regression model and with the Choquet integral model.

Keywords: Monotone measure · Inclusion-exclusion integral · Möbius transform · Interaction operator · t-norm

1 Introduction

In recent years, data analysis has become increasingly important for us with the development of the storage and data management solutions. Therefore, data analysis models describing relations between several explanatory variables as input data and objective variables are needed in different fields, such as multiple criteria decision making. In dealing with these problems, the linear regression model has been widely used for a long time and plays a central role. There are, however, many cases where the linear model is insufficient. Therefore, more flexible models and methods are needed. As the objective variable is monotone with respect to the explanatory variables, in many cases, using a monotone measure, which is also called a nonadditive measure or fuzzy measure, has been considered. Indeed, the Choquet integral model as the integral with respect to monotone measures has been proposed and has produced good results [5, 7, 14, 15, 19, 20].

We have proposed an integral with respect to a monotone measure and call this the inclusion-exclusion integral. We also propose a data analysis model using this integral [10, 11]. This integral contains the Lebesgue integral and also the Choquet integral. In other words, our model contains the linear regression model and the Choquet integral model. It has a strong model description ability and

J.P. Carvalho et al. (Eds.): IPMU 2016, Part I, CCIS 610, pp. 65–77, 2016.
DOI: 10.1007/978-3-319-40596-4_7

may be flexibly applied to real problems. Later in this paper, a new construction method [11] of our proposed model is given using t-norms to evaluate the mutual interactions among the events. We give experimental results and verify the validity of our proposed method.

2 Preliminaries

Throughout the paper, the whole set is a finite n points set denoted by $\Omega := \{1, 2, \ldots, n\}$ and $\mathcal{P}(\Omega)$ denotes the power set of Ω. For a set A, $|A|$ is the cardinal number of A.

2.1 The Inclusion-Exclusion Integral

Definition 1 (monotone measure [1,16,18]). *A set function $v : \mathcal{P}(\Omega) \to [0, +\infty)$ is a monotone measure if v satisfies the following conditions:*

1. $v(\emptyset) = 0, v(\Omega) < +\infty$.
2. *For any $A, B \in \mathcal{P}(\Omega)$, $A \subset B$ implies $v(A) \leq v(B)$.*

The monotone measure space, denoted by $(\Omega, \mathcal{P}(\Omega), v)$, means that v is a monotone measure on $\mathcal{P}(\Omega)$. We denote a function on Ω by $f := (f(1), \ldots, f(n))$ or also by $f := (f_1, \ldots, f_n)$ for simplicity.

Definition 2 (interaction operator). *Let Ω be a nonempty n points set, $K \in (0, +\infty]$ and $I(\boldsymbol{x} \mid A) : [0, K]^n \times \mathcal{P}(\Omega) \to [0, K]$ be a function. We say I is an* interaction operator *on $[0, K]$ if I satisfies the following conditions:*

1. $I(\boldsymbol{x} \mid \emptyset) = K,\ I(\boldsymbol{x} \mid \Omega) = 0$.
2. $I(\boldsymbol{x} \mid \{i\}) := x_i$ *for any $i \in \Omega$, that is, $I(\boldsymbol{x} \mid \{i\})$ is a coordinate function.*
3. $I(\boldsymbol{x} \mid A) \leq \min_{B \subsetneq A}\{I(\boldsymbol{x} \mid B)\}$ *for any $A \in \mathcal{P}(\Omega)$.*

The interactive monotone measure space, denoted by $(\Omega, \mathcal{P}(\Omega), v, I, K)$, means that $\Omega = \{1, \ldots, n\}$, v is a monotone measure on $\mathcal{P}(\Omega)$ and $I(\boldsymbol{x} \mid A) : [0, K]^n \times \mathcal{P}(\Omega) \to [0, K], K \in (0, +\infty]$ is an interaction operator.

Lemma 1. *Let $\Omega = \{1, 2, \ldots, n\}$ and I be an interaction operator on $[0, K]$. Then K is the unit element. In other words, for any $i \in \Omega$, if $\boldsymbol{x} \in [0, K]^n$ satisfies $x_i = K$, then $I(\boldsymbol{x} \mid A) = I(\boldsymbol{x} \mid A \cup \{i\})$ holds for any $A \in \mathcal{P}(\Omega)$ satisfying $A \not\ni i$.*

Definition 3 (Inclusion-exclusion integral [11], Cf. [9,10]). *Let $(\Omega, \mathcal{P}(\Omega), v, I, K)$ be an interactive monotone measure space. For any function $f : \Omega \to [0, K]$, the inclusion-exclusion integral of f with respect to v and I, denoted by $(I) \int f dv$, is defined by*

$$(I) \int f\, dv := \sum_{A \in \mathcal{P}(\Omega)} M^I(f \mid A)\, v(A),$$

where

$$M^I(f \mid A) := \sum_{B \supset A} (-1)^{|B \setminus A|} I(f \mid B).$$

Example 1. A binary operator on $[0,1]$ *satisfying the following (T1-4)*

(T1) $0 \otimes 0 = 0, x \otimes K = x$ for any $x > 0$.
(T2) $x \leq y$ implies $x \otimes z \leq y \otimes z$.
(T3) $x \otimes y = y \otimes x$.
(T4) $x \otimes (y \otimes z) = (x \otimes y) \otimes z$.

is called *t-norm* ([12,17], Cf. [13]) These t-norms can be extended to multi-variable function by (T4) and they are examples of the interaction operators. Specifically, it shall be:

$$I(f \mid A) := \begin{cases} \bigotimes_{i \in A} f_i, & |A| > 1, \\ f_i, & A = \{i\}. \end{cases}$$

Definition 4 (Discrete Choquet integral [1]). *Let* $(\Omega, \mathcal{P}(\Omega), v)$ *be a monotone measure space and* $f := (f_1, \dots, f_n)$ *a non-negative function on* Ω*. The Choquet integral of* f *with respect to* v*, denoted by* $(C) \int f dv$*, is defined by*

$$(C) \int f \, dv := (C) \int_0^\infty \mu(\{f \geq t\}) dt = \sum_{i=1}^n (f_{\sigma(i)} - f_{\sigma(i+1)}) v(\{\sigma(1), \dots, \sigma(i)\}),$$

where σ *is a permutation on* Ω *such that* $f_{\sigma(1)} \geq \dots \geq f_{\sigma(n)}$*, and* $f_{\sigma(n+1)} := 0$.

Theorem 1. *Let* $(\Omega, \mathcal{P}(\Omega), v, I, K)$ *be an interactive monotone measure space. Then it holds that*

$$\sum_{A \ni i} M^I(f \mid A) = f(i).$$

By Theorem 1, the inclusion-exclusion integral can be interpreted as follows. Each $f_i, i \in \Omega$ of the integrand f is divided into $M^I(f \mid A), A \ni i$. $M^I(f \mid A)$ is the quantity of f assigned for A as interaction of elements only in A. In contrast, $I(f \mid A)$ corresponds to the quantity by interactions of elements involving A and others. This derives the name"interaction operator."

We say I is a *nonnegative interaction operator* if for any $\boldsymbol{x} \in [0, K]^n$ we have $M^I(\boldsymbol{x} \mid A) \geq 0$ for any $A \in \mathcal{P}(\Omega)$. We say I is a *monotone interaction operator* if for any $\boldsymbol{x}, \boldsymbol{y} \in [0, K]^n$ satisfying $\boldsymbol{x} \leq \boldsymbol{x} \leq \boldsymbol{y}$, that is, $x_i \leq y_i, i \in \Omega$, we have $M^I(\boldsymbol{x} \mid A) \leq M^I(\boldsymbol{y} \mid A)$ for any $A \in \mathcal{P}(\Omega)$.

Theorem 2. *Let* $(\Omega, \mathcal{P}(\Omega), v, I, K)$ *be an interactive monotone measure space.*
(i) *If* v *is additive. Then we have*

$$(I) \int f dv = \int f dv.$$

(ii) *If* I *is a nonnegative interaction operator, then it holds that*

$$(I) \int f dv \geq 0.$$

(iii) *Let* $f := (f_1, \ldots, f_n)$ *and* $g := (g_1, \ldots, g_n)$ *be functions from* Ω *to* $[0, K]$. *If* I *is a monotone interaction operator, then the inclusion-exclusion integral is monotone. In other words, for any pair of* f *and* g *that satisfy* $f_i \leq g_i$ *for any* $i \in \Omega$, *we have*

$$(I)\int f dv \leq (I)\int g dv.$$

Definition 5 (Möbius transform). *Let* $(\Omega, \mathcal{P}(\Omega), v)$ *be a monotone measure space. The Möbius transform of* v, *denoted by* m^v, *is defined by*

$$m^v(A) := \sum_{B \subset A} (-1)^{|A \setminus B|} v(B)$$

for any $A \in \mathcal{P}(\Omega)$. *There is one-to-one correspondence between* v *and* m^v *with Möbius inversion formula:*

$$v(A) = \sum_{B \subset A} m^v(B)$$

for any $A \in \mathcal{P}(\Omega)$.

Theorem 3. *Let* $(\Omega, \mathcal{P}(\Omega), v, I, K)$ *be an interactive monotone measure space. The inclusion-exclusion integral can be represented by using the Möbius transform as*

$$(I)\int f\ dv = \sum_{A \in \mathcal{P}(\Omega)} I(f \mid A)\ m^v(A).$$

Example 2 (Cf. Definition 4). The Choquet integral is an inclusion-exclusion integral. In fact, it is known that the Choquet integral is represented by the Möbius transform [6]:

$$(C)\int f\ dv = \sum_{A \in \mathcal{P}(\Omega)} \min\{f_i \mid i \in A\}\ m^v(A).$$

It is known that the Lebesgue integral and the Choquet integral both have the following property. The inclusion-exclusion integral also has this property.

Theorem 4. *Let* $(\Omega, \mathcal{P}(\Omega), v, I, K)$ *be an interactive monotone measure space and* K *be the unit element. Define the characteristic function* χ_A *for* $A \in \mathcal{P}(\Omega)$ *on* Ω *by*

$$\chi_A(i) := \begin{cases} K, & i \in A, \\ 0, & i \notin A. \end{cases}$$

Then we have

$$(I)\int \chi_A dv = Kv(A).$$

In application to multivariate analysis, the whole set Ω corresponds to the set of, for example, criteria or attributes of objects and integrand f corresponds to a set of values assigned to each elements of Ω. In general, the scale or the direction of variables $f = (f_1, \dots, f_n)$ depend on each $i \in \Omega$. Values of f on Ω are measured objectively and each unit and scale depend on each criterion. Some of these criteria are "the higher the better" attributes and the others are "the lower the better". Moreover there are cases where the value range of f is not $\mathbb{R}$ but just totally ordered sets. Therefore, the following generalization of the interaction operator is useful in a practical sense.

Definition 6. *Let Ω be a nonempty n points set, $\mathcal{R}_1, \dots, \mathcal{R}_n$ be totally ordered sets, and $K \in (0, +\infty]$ and $I(\boldsymbol{x} \mid A) : \mathcal{R}^n \times \mathcal{P}(\Omega) \to [0, K]$ be functions. We say I is a* rescaling interaction operator *on $\mathcal{R} := \mathcal{R}_1 \times \cdots \times \mathcal{R}_n$ if I satisfies the following conditions:*

1. *$I(\boldsymbol{x} \mid \emptyset) = K,\ I(\boldsymbol{x} \mid \Omega) = 0$.*
2. *For each $i \in \Omega$, $I(\boldsymbol{x} \mid \{i\})$ is either monotonically increasing or monotonically decreasing. That is, for any $\boldsymbol{x}, \boldsymbol{y} \in \mathcal{R}$, either $x_i < y_i$ implies $I(\boldsymbol{x} \mid \{i\}) \leq I(\boldsymbol{y} \mid \{i\})$ or $x_i < y_i$ implies $I(\boldsymbol{x} \mid \{i\}) \geq I(\boldsymbol{y} \mid \{i\})$.*
3. *$I(\boldsymbol{x} \mid A) \leq \min_{B \subsetneq A}\{I(\boldsymbol{x} \mid B)\}$ for any $A \in \mathcal{P}(\Omega)$.*

3 Modeling Method

We propose concrete methods for constructing the Möbius type inclusion-exclusion models using t-norms. To identify an inclusion-exclusion model, it is needed to specify two functions need to specified; an interaction operator I and a monotone measure v.

Let $f = (f_1, \dots, f_n)$ be an input data set consisting n attributes and let y be its objective variable. Suppose $\varphi_i, i = 1, \dots, n$ are monotone functions from $\mathcal{R}_i$ to $[0, 1]$ and $\bigotimes : \bigcup_{k=2}^{n} [0, 1]^k \to [0, 1]$ be t-norm. We define $I(f \mid \{i\}) := \varphi_i(f_i)^{\otimes}$k $= 2$ and $I(f \mid A := \bigotimes_{i \in A} \varphi_i(f_i)$ for $A \in \mathcal{P}(\Omega), |A| > 1$. Then for a data set $f = (f_1, f_2, \dots, f_n)$, the representation of the inclusion-exclusion integral, which we call the Möbius type inclusion-exclusion model, is written as the following formula:

$$
\begin{aligned}
\hat{y} := (I) \int f dv + a_0 &= \sum_{A \in \mathcal{P}(\Omega)} m^v(A)\ I(f \mid A) + a_0 \\
&= \sum_{A \in \mathcal{P}(\Omega)} \left\{ m^v(A) \bigotimes_{i \in A} \varphi_i(f_i) \right\} + a_0 \qquad (1) \\
&=: a_{\{1\}}\varphi_1 + a_{\{2\}}\varphi_2 + \cdots a_{\{3\}}\varphi_n \\
&\quad + a_{\{1,2\}}(\varphi_1 \otimes \varphi_2) + a_{\{1,3\}}(\varphi_1 \otimes \varphi_3) + \cdots + a_{\{n-1,n\}}(\varphi_{n-1} \otimes \varphi_n) \\
&\quad \vdots \\
&\quad + a_{\{1,2,\dots,n\}}(\varphi_1 \otimes \varphi_2 \otimes \cdots \otimes \varphi_n) + a_0,
\end{aligned}
$$

where $\hat{y}$ denotes an estimated objective variable and φ_i denotes $\varphi_i(f_i)$ in the right side. Therefore the process of constructing an inclusion-exclusion integral model consists of three processes of specification.

1. monotone functions $\varphi_1, \ldots, \varphi_n$ as $I(f \mid \{i\})$,
2. t-norm $\bigotimes$ as $I(f \mid A), |A| > 1$, and
3. coefficients $a_A, A \in \mathcal{P}(\Omega)$ as $m^v(A)$, and a_0 as $(I) \int (0, 0, \ldots, 0) dv$.

3.1 Specification of Monotone Functions

First, it is needed to specify n monotone functions $\varphi_1, \ldots, \varphi_n$ need to be specified. These functions rescale each data f_i, as $I(f \mid \{i\}), i \in \Omega$. This process corresponds to rescaling or pretreatment of data sets. To guarantee the monotonicity of v, for attributes where the larger they take, the larger the objective value takes, φ_i are supposed to be monotone. For attributes where the larger they take, the smaller the objective value takes, φ_i are supposed to be reverse-monotone. Both linear and nonlinear functions are acceptable as φ_i, however the range of φ_i needs to be $[0, 1]$ because we adopt t-norms as $I(f \mid A), |A| > 1$. We suggest a way of determining φ_i a linear rescaling function.

$$\varphi_i(x) := \max\left\{\min\left\{\frac{x - \varphi_i^{\min}}{\varphi_i^{\max} - \varphi_i^{\min}}, 1\right\}, 0\right\},$$

and in the case that f_i needs to be reversed,

$$\varphi_i(x) := 1 - \max\left\{\min\left\{\frac{x - \varphi_i^{\min}}{\varphi_i^{\max} - \varphi_i^{\min}}, 1\right\}, 0\right\}.$$

$\varphi_i^{\min}$ and $\varphi_i^{\max}$ are the minimum and the maximum values of range of φ_i, respectively.

If needed, nonlinear monotone functions can provide a more effective rescaling, for example, where monotone functions which raise the correction coefficient between $\varphi_i(f_i)$ and the objective value. In Sect. 4.1, we use nonlinear monotone functions as φ_i such that

$$\varphi_i := \arg\max_{\varphi_i} \left(\mathrm{cor}(\{y^j\}_{j=1}^M, \{x^j\}_{j=1}^M)\right), \quad i = 1, 2, \ldots, n,$$

where

$$\mathrm{cor}(\{y^j\}_{j=1}^M, \{x^j\}_{j=1}^M) := \frac{\sum_j \left(y^j - \frac{1}{M}\sum_j y^j\right)\left(x^j - \frac{1}{M}\sum_j x^j\right)}{\sqrt{\left(\sum_j \left(y^j - \frac{1}{M}\sum_j y^j\right)^2\right)\left(\sum_j \left(x^j - \frac{1}{M}\sum_j x^j\right)^2\right)}}.$$

3.2 Specification of t-norms

Next, t-norms to be specified. We show the both formulas as binary operator and k-ary operator, $k = 1, \ldots, n$, of non-parametric and parametric t-norms as follows. For $x, y \in [0,1], i = 1, 2, \ldots, k$,

1. the algebraic product: $x \otimes_{\mathrm{a}} y := xy$,
2. the logical product: $x \otimes_{\min} y := \min\{x, y\}$,
3. the bounded product: $x \otimes_{\mathrm{b}} y := \max(x + y - 1, 0)$, xy
4. t-product: $x \otimes_{\mathrm{T}} y := \dfrac{xy}{x + y - xy}$,
5. Dombi's t-norm [2]: $x \otimes_{\mathrm{D}} y := \dfrac{1}{1 + \left(\left(\frac{1}{x} - 1\right)^{\lambda} + \left(\frac{1}{y} - 1\right)^{\lambda}\right)^{\frac{1}{\lambda}}}, \quad \lambda \geq 0,$
6. Dubois and Prade's t-norm [4]: $x \otimes_{\mathrm{DP}} y := \dfrac{xy}{\max\{x, y, \lambda\}}$,
7. Schweizer and Sklar's t-norm [17]:
$$x \otimes_{\mathrm{SS}} y := 1 - \{(1-x)^p + (1-y)^p - (1-x)^p(1-y)^p\}^{\frac{1}{p}}, \quad p > 0,$$
8. Hamacher's t-norm [8]: $x \otimes_{\mathrm{H}} y := \dfrac{xy}{\gamma + (1-\gamma)(x + y - xy)}, \quad \gamma > 0,$
9. Yager's t-norm [21]: $x \otimes_{\mathrm{Y}} y := 1 - (\min\{1, (1-x)^p + (1-y)^p\})^{\frac{1}{p}}, \quad p > 0$

The t-product is obtained by putting $\gamma = 0$ of Hamacher's t-norm and also $\lambda = 1$ of Dombi's t-norm.

The inclusion-exclusion integral using the logical product as I corresponds to the Choquet integral.

t-norms correspond to $I(f \mid A)$ controlling the quantities of the interaction among elements in $A \in \mathcal{P}(\Omega)$.

Remark 1. We may also adopt generalized t-norms under the condition of the interaction operator. For example, for $x_i \in [0,1], i = 1, \ldots, n$, Dubois and Prade's t-norm can be generalized to the interaction operator as follows:

$$\bigotimes_{i \in A} {}_{\mathrm{GDP}}\, x_i := \frac{\prod_{i \in A} x_i \cdot \min\{x_1, x_2, \ldots, x_{|A|}\}}{\left(\prod_{i : x_i > \lambda_{|A|}} x_i\right) \cdot {\lambda_{|A|}}^{|\{i \in A | x_i \leq \lambda_{|A|}\}| - 1} \cdot \min\{x_1, x_2, \ldots, x_{|A|}, \lambda\}},$$
$$0 \leq \lambda_1 \leq \lambda_1 \leq \ldots \lambda_n \leq 1,$$

which is not a t-norm but an interaction operator.

3.3 Specification of Monotone Measures

The last step in the construction process is to select good variables in regression and to determine the values of their coefficients. One of the disadvantages of the inclusion-exclusion model is its large number of terms, so that using the selecting method of the linear regression model is effective for eliminating the defect. This problem can be solved by regarding the Möbius type inclusion-exclusion [11] model as the linear regression model. We propose using methods proposed for the linear regression model to select a subset of variables used in the model [3]. In Sect. 4.3, we use the stepwise regression method which is the most common method of selecting variables in regression models. The only difference point in the case of the inclusion-exclusion integral model is that we have to consider the monotonicity of v.

4 Experiments

In this section, we show the concrete process and the results of an experimental trial to validate the performance of the inclusion-exclusion integral model compared with both the linear multi regression model and the Choquet integral model. We used the Möbius type inclusion-exclusion model (Eq. (1)) as the inclusion-exclusion model and the Choquet integral model. The linear regression model is

$$y = \sum_{i \in \Omega} a_i f_i + a_0.$$

We used the "Car Evaluation" data sets that were provided for open access in the UCI repository[1]. The data set consisted of 1728 data sets containing "overall evaluation" and a further six attributes for evaluating a car: "price", "cost of maintenance", "number of doors", "passenger capacity", "trunk capacity" and "estimated safety". Therefore, Ω is a six-points set consisting of these attributes, $\Omega := \{1, 2, 3, 4, 5, 6\} :=$ {" price", "cost of maintenance", "number of doors", "passenger capacity, "trunk capacity", "estimated safety".

4.1 Monotone Functions

We adopted the monotone functions shown in Table 1. The correlation coefficient is maximized with the objective value, "overall evaluation". We have, for example,

$$\varphi_1 := \varphi_{\text{price}} = \begin{pmatrix} \text{veryhigh} & \text{high} & \text{medium} & \text{low} \\ 0 & 0.1 & 0.7 & 1 \end{pmatrix}$$

[1] http://archive.ics.uci.edu/ml/.

Table 1. Car evaluation data sets and monotone functions

	Attributes	Alternatives	Rescaling
Explanatory variables	price	buying (very high, high, medium, low)	(0, 0.1, 0.7, 1)
	doors number	(2, 3, 4, more)	(0, 0.7, 0.9, 1)
	passenger capacity	(2, 4, more)	(0, 0.8, 1)
	trunk capacity	(big, medium, small)	(1, 0.7, 0)
	estimated safety	(high, medium, low)	(1, 0.6, 0)
Objective variable	overall evaluation	(unacceptable, acceptable, good, very good)	(1, 2, 3, 4)

and for example,

$$I\big((\text{high}, \text{medium}, 3, 2, \text{big}, \text{low}) \mid \{\text{“doors numbers''}\}\big) = 0.7.$$

Selection took place by the exhaust search between 0 and 1 with stepping width 0.1. In other words, in the case that the number of attributes is four, $(0, 0.1, 0.2, 1), (0, 0.1, 0.3, 1), (0, 0.1, 0.4, 1), \ldots, (0, 0.2, 0.3, 1), \ldots, (0, 0.8, 0.9, 1)$.

4.2 t-norms

We adopted the t-norms in Sect. 3.2. The parameters of parametric t-norms are determined by the exhaust search. For example, in the case of Dubois and Prade's t-norm, we try the model with λ between 0 and 1 with stepping width 0.05 and adopt the best parameter that maximizes adjusted R^2. This quantifies how well the model fits the data:

$$\text{adjusted } R^2 := 1 - (1 - R^2)\left(\frac{M-1}{M-K-1}\right),$$

where R^2 is the coefficient of determination,

$$R^2 := 1 - \frac{\sum_{j=1}^{M}\left(y^j - \hat{y}^j\right)^2}{\sum_{j=1}^{M}\left(y^j - \frac{1}{M}\sum_{j=1}^{M} y^j\right)^2},$$

M is the number of the data sets and K is the number of the selected variables in the model.

4.3 Monotone Measures

Regarding the Möbius type inclusion-exclusion model as the linear regression model, we obtained $a_{\{1\}}, a_{\{2\}}, \ldots, a_{\{n\}}, a_{\{1,2\}}, a_{\{1,3\}}, \ldots, a_{\{n-1,n\}}, \ldots, a_{\{1,\ldots,n\}}$ as the least-square method and selected them using the stepwise regression

Table 2. Results

Regression model	t-norm	R	Adjusted R^2	#variables	Parameter
Linear regression	—	0.715	0.510	6	—
Choquet integral	logical product	0.920	0.846	13	—
Möbius type	algebraic product	0.928	0.861	8	—
inclusion-exclusion	bounded product	0.907	0.821	12	—
integral	t-product	0.928	0.861	9	—
	Dombi's t-norm	0.930	0.864	10	$\lambda = 0.9$
	Dubois's t-norm	0.930	0.864	11	$\lambda = 0.95$
	Schweizer's t-norm	0.930	0.864	11	$p = 1.3$
	Hamacher t-norm	0.930	0.864	10	$p = 0.4$
	Yager's t-norm	0.927	0.858	10	$p = 1.7$

method. We also took account of the variance inflation factor (VIF). In general, a value of 10 is recommended as the maximum level for VIF in order to avoid trouble with the stability of the coefficients. We also assumed that v is monotone when using the stepwise selection method. Therefore, the conditions to enter came to be that

1. F-value is greater than 3.84,
2. v is monotone, and
3. VIF is less than or equal to 10,

and the conditions to remove are that

1. F-value is less than 2.71,

where 3.84 is based on the upper critical value of the F-distribution at 5 percent significant level and 2.71 is at 10 percent significant level, and the column "#variables" in Table 3 show the number of selected variables using this selection.

4.4 Results

The results of the experiment are shown in Tables 2 and 3. Table 2 shows "R" which is the correlation coefficient of observed values and the fitted values by the regression models. The closer to 1 they are, the better the regression model is. Table 3 shows the selected variables of each regression models. The results shows validity of the proposed method using the inclusion-exclusion integral model compared with the classical linear regression model. The concrete formula of one of the best model with Dubois and Prade's t-norm was

Table 3. Selected variables of each model

Regression model	t-norm	Selected variables
Linear regression	—	{1},{2},{3},{4},{5},{6}
Choquet integral	logical product	{3},{5},{6},{1,4,6},{2,4,6},{3,4,6},{1,4,5,6},{2,3,4,6},{2,4,5,6},{1,3,4,5,6},{1,2,3,4,6},{1,2,4,5,6},{1,2,3,4,5,6}
Möbius inclusion-exclusion integral	algebraic product	{4,6},{1,4,6},{1,2,4,6},{1,3,4,6},{1,4,5,6},{2,3,4,6},{2,4,5,6},{2,3,4,5,6}
	bounded product	{6},{4,5},{1,2,3},{1,3,6},{1,4,6},{2,4,6},{3,4,6},{1,2,3,4}, {1,4,5,6},{2,4,5,6},{2,3,4,5,6},{1,2,4,5,6},{1,2,3,4,5,6}
	t-product	{1,4,6},{2,3,6},{2,4,6},{3,4,6},{1,3,4,6},{1,4,5,6},{2,4,5,6},{1,2,3,4,6},{2,3,4,5,6}
	Dombi's t-norm	{6},{1,4,6},{2,4,6},{1,3,4,6},{1,4,5,6},{2,3,4,6},{2,4,5,6},{1,2,3,4,6}, {1,2,4,5,6},{1,2,3,4,5,6}
	Dubois's t-norm	{6},{4,5},{1,4,6},{2,4,6},{1,3,4,6},{1,4,5,6},{2,3,4,6},{2,4,5,6},{1,2,3,4,6},{1,2,4,5,6},{1,2,3,4,5,6}
	Schweizer's t-norm	{6},{1,4,6},{2,4,6},{3,4,5},{1,3,4,6},{1,4,5,6},{2,3,4,6},{2,4,5,6},{1,2,3,4,6},{1,2,4,5,6},{1,2,3,4,5,6}
	Hamacher's t-norm	{6},{45},{146},{246},{1346},{1456},{2346},{2456},{12346},{12456},{123456}
	Yager's t-norm	{6},{4,5},{1,4,6},{3,4,6},{1,2,4,6},{1,3,4,6},{1,4,5,6},{2,3,4,6},{2,4,5,6},{2,3,4,5,6}...

$$
\begin{aligned}
F_{\mathrm{DP}}(f) = {} & 0.075\ \varphi_6 + 0.059 \left(\bigotimes_{i\in\{4,5\}}{}_{\mathrm{DP}}\ \varphi_i \right) + 0.462 \left(\bigotimes_{i\in\{1,4,6\}}{}_{\mathrm{DP}}\ \varphi_i \right) \\
& + 0.384 \left(\bigotimes_{i\in\{2,4,6\}}{}_{\mathrm{DP}}\ \varphi_i \right) + 0.509 \left(\bigotimes_{i\in\{1,3,4,6\}}{}_{\mathrm{DP}}\ \varphi_i \right) + 0.937 \left(\bigotimes_{i\in\{1,4,5,6\}}{}_{\mathrm{DP}}\ \varphi_i \right) \\
& + 0.334 \left(\bigotimes_{i\in\{2,3,4,6\}}{}_{\mathrm{DP}}\ \varphi_i \right) + 0.541 \left(\bigotimes_{i\in\{2,4,5,6\}}{}_{\mathrm{DP}}\ \varphi_i \right) + 0.582 \left(\bigotimes_{i\in\{1,2,3,4,6\}}{}_{\mathrm{DP}}\ \varphi_i \right) \\
& + 0.826 \left(\bigotimes_{i\in\{1,2,4,5,6\}}{}_{\mathrm{DP}}\ \varphi_i \right) - 0.911 \left(\bigotimes_{i\in\{1,2,3,4,5,6\}}{}_{\mathrm{DP}}\ \varphi_i \right) + 0.967,
\end{aligned}
$$

where φ_i denotes $\varphi_i(f_i)$.

The model gives additional information. We can read the strengths of the interaction between attributes by m^v and I and obtain the importance of each criterion by the Shapley value of v. In the case of F_{DP}, we obtain $\Phi(v) = (\phi_1, \phi_2, \phi_3, \phi_4, \phi_5, \phi_6) = (0.175, 0.137, 0.052, 0.230, 0.120, 0.286)$ which shows that "estimated safety" is the most significant attribute.

Acknowledgment. This work was supported by JSPS KAKENHI Grant Number 50271119.

References

1. Choquet, G.: Theory of capacities. Ann. Inst. Fourier **5**, 131–295 (1953)
2. Dombi, J.: A general class of fuzzy operators, the DeMorgan class of fuzzy operators and fuzziness measures induced by fuzzy operators. Fuzzy Sets Syst. **8**, 149–163 (1982)
3. Draper, N.R., Smith, H.: Applied Regression Analysis. Wiley Series in Probability and Statistics, 2nd edn. Wiley, New York (1998)
4. Dubois, S., Prade, H.: New results about properties, semantics of fuzzy set-theoretic operators. In: Wang, P.P., et al. (eds.) Fuzzy Sets, pp. 59–75. Plenum Press, New York (1980)
5. Grabisch, M.: Fuzzy integral in multicriteria decision making. Fuzzy Sets Syst. **69**(3), 279–298 (1995)
6. Grabisch, M., Labreuche, C.: A decade of application of the Choquet and Sugeno integrals in multi-criteria decision aid. A Q. J. Oper. Res. **6**, 1–44 (2008). Springer, Verlag
7. Grabisch, M., Murofushi, T., Sugeno, M. (eds.): Fuzzy Measures and Integrals: Theory and Applications. Physica, Heidelberg (2000)
8. Hamacher, H.: Über Logische Aggregation Nicht-binär Explizierter Entscheidnungskriterien. Fischer, Frankfurt (1978)
9. Honda, A., Fukuda, R., Okamoto, J.: Rescaling for evaluations using inclusion-exclusion integral. In: Laurent, A., Strauss, O., Bouchon-Meunier, B., Yager, R.R. (eds.) IPMU 2014, Part I. CCIS, vol. 442, pp. 284–293. Springer, Heidelberg (2014)
10. Honda, A., Okamoto, J.: Inclusion-exclusion integral and its application to subjective video quality estimation. In: Kruse, R., Hoffmann, F., Hüllermeier, E. (eds.) IPMU 2010. CCIS, vol. 80, pp. 480–489. Springer, Heidelberg (2010)
11. Honda, A., Okazaki, Y.: Inclusion exclusion integral. In: Proceedings of the 12th Modeling Decisions for Artificial Intelligence (CD-ROM) (2015)
12. Klement, E.P., Mesiar, R., Pap, E.: Triangular Norms. Kluwer Academic Publishers, Dordrecht (2000)
13. Menger, K.: Statistical metrics. Proc. Nat. Acad. Sci. USA **28**, 535–537 (1942)
14. Mori, D., Honda, A., Uchida, M., Okazaki, Y.: Quality evaluations of network services using a non-additive set function. In: Proceedings of The 5th Modeling Decisions for Artificial Intelligence (CD-ROM) (2008)
15. Mori, T., Murofushi, T.: An analysis of evaluation model using fuzzy measure and the Choquet integral. In: Proceedings of the 5th Fuzzy System Symposium, pp. 207–212. Japan Society for Fuzzy Sets and Systems (1989)
16. Murofushi, T., Sugeno, M.: Fuzzy measures, fuzzy integrals. In: Grabisch, M., Murofushi, T., Sugeno, M. (eds.) Fuzzy Measures and Integrals, pp. 3–41. Physica-Verlag, Heidelberg (2000)

17. Schweizer, B., Sklar, A.: Statistical metric spaces. Pacific. J. Math. **10**, 313–334 (1960)
18. Sugeno, M.: Fuzzy measures and fuzzy integrals-a survey. In: Gupta, M.M., Saridis, G.N., Gaines, B.R. (eds.) Fuzzy Automata and Decision Processes, pp. 89–102. North Holland, Amsterdam (1977)
19. Tehrani, A.F., Cheng, W., Dembczyński, K., Hüllermeier, E.: Learning monotone nonlinear models using the Choquet integral. Mach. Learn **89**, 183–211 (2012)
20. Torra, V.: Learning aggregation operators for preference modeling. In: Fürnkranz, J., Hüllermeier, E. (eds.) Preference Learning, pp. 317–333. Springer, Berlin (2011)
21. Yager, R.R.: On a general class of fuzzy connectives. Fuzzy Sets Syst. **4**, 235–242 (1980)

Fuzzy Integral for Rule Aggregation in Fuzzy Inference Systems

Leary Tomlin[1(✉)], Derek T. Anderson[1], Christian Wagner[2], Timothy C. Havens[3], and James M. Keller[4]

[1] Electrical and Computer Engineering, Mississippi State University, Starkville, USA
lt648@msstate.edu, anderson@ece.mstate.edu

[2] School of Computer Science, University of Nottingham, Nottingham, UK
christian.wagner@nottingham.ac.uk

[3] Electrical and Computer Engineering, Computer Science, Michigan Technological University, Houghton, USA
thavens@mtu.edu

[4] Electrical and Computer Engineering, Computer Science, University of Missouri, Columbia, USA
kellerj@missouri.edu

Abstract. The *fuzzy inference system* (FIS) has been tuned and revamped many times over and applied to numerous domains. New and improved techniques have been presented for fuzzification, implication, rule composition and defuzzification, leaving one key component relatively underrepresented, rule aggregation. Current FIS aggregation operators are relatively simple and have remained more-or-less unchanged over the years. For many problems, these simple aggregation operators produce intuitive, useful and meaningful results. However, there exists a wide class of problems for which quality aggregation requires non-additivity and exploitation of interactions between rules. Herein, we show how the fuzzy integral, a parametric non-linear aggregation operator, can be used to fill this gap. Specifically, recent advancements in extensions of the fuzzy integral to "unrestricted" fuzzy sets, i.e., subnormal and non-convex, makes this now possible. We explore the role of two extensions, the gFI and the NDFI, discuss when and where to apply these aggregations, and present efficient algorithms to approximate their solutions.

Keywords: Fuzzy inference system · Choquet integral · Fuzzy integral · gFI · NDFI · Fuzzy measure

1 Introduction

In Lofti Zadeh's seminal 1965 paper on *fuzzy set* (FS) theory, a new philosophy was put forth to address uncertain data and/or information [1]. In 1973, Zadeh introduced fuzzy logic and he suggested that it might be a useful mechanism to model higher-level thought and reasoning in humans [2]. The first application selected was a steam engine and boiler control system and the rules were provided

J.P. Carvalho et al. (Eds.): IPMU 2016, Part I, CCIS 610, pp. 78–90, 2016.
DOI: 10.1007/978-3-319-40596-4_8

by the system operators [3]. The Mamdani-Assilian *fuzzy inference system* (FIS) is built on top of Zadeh's *compositional rule of inference* (CRI), a generalization of *modus ponens*, *modus tollens*, etc. The CRI is a way to calculate a FS-valued output based on crisp or FS-valued inputs and an implication function. Other well-known FISs that generalize the CRI are the *Takagi-Sugeno-Kang* (TSK) [4], Tsukamoto [5], and *single input rule modules* (SIRM) FISs [6,7]. Nearly all FISs consist of some subset of fuzzification, CRI, rule firing, rule combination and defuzzification. Of particular importance to this paper is the aggregation step in an FIS, which is responsible for combining the output of different rules and deriving a final comprehensive decision. Specifically, we are concerned with the identification of functions that can take multiple FS-valued inputs and produce a FS-valued output. The Mamdani-Assilian FIS typically uses FS aggregation strategies such as maximum and summation per individual element in the discrete output domain. The aggregations used in most FISs share something in common, they are relatively simple and they do not model nor exploit interactions (when/if available) between rules. Herein, we investigate the role of the *fuzzy integral* (FI) [8], namely the *Choquet integral* (CI), for non-linear rule combining in fuzzy logic. In particular, we explore our two recent extensions, the *generalized FI* (gFI) and the *non-direct FI* (NDFI) [9], that are capable of aggregating any type of FS in contrast to prior extensions for interval-valued data and fuzzy numbers. The *fuzzy measure* (FM), introduced by Sugeno in 1974 [8], is used to model interactions (when/if available) between rules. Specifically, for the Mamdani-Assilian FIS, let X be the set of N rules, let $x_i \in X$ be the ith rule and $H(x_i)$, or H_i for short, be the ith FS-valued rule output. Herein, we address how to use the NDFI and gFI for fuzzy logic and when and where to use one or the other. Figure 1 illustrates the role of the FI in fuzzy logic.

In 1992, Yager explored rule aggregation in fuzzy logic [10]. He considered two rule representations, the Mamdani model and the logical model, and obtained a general rule representation. For aggregation, he explored two soft classes of rule combining, *or-like* and *and-like*. Ultimately, Yager's work is subsumed by our

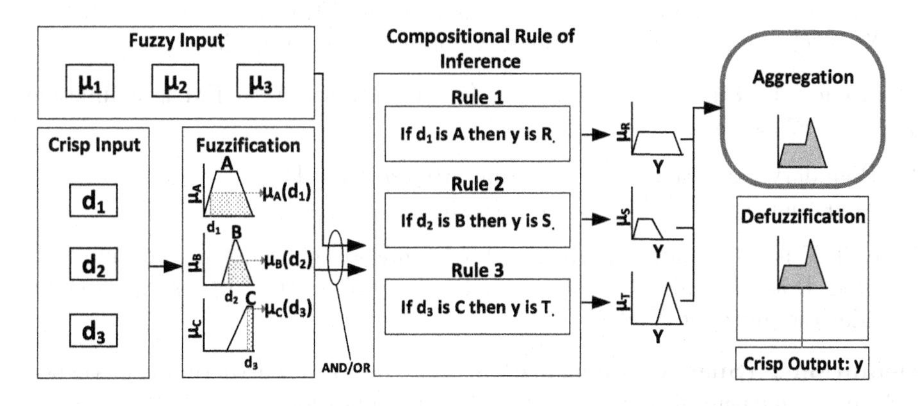

Fig. 1. Block diagram of a fuzzy inference system and where the FI fits in.

current article. First, Yager's operators are performed per-element in a discrete output domain. This is how the NDFI operates, although not the gFI (which is based on the *extension principle* (EP)). Second, Yager's two classes of operators, and even his related *ordered weighted averages* (OWAs) [11], are special cases of the CI (with respect to particular FMs) [12]. Third, and related to the second point, the NDFI and gFI are not restricted to being additive (they only need to satisfy the more general rule of monotonicity) and they can model and exploit rich interactions between rules when/if available. This allows us to achieve a much wider, sophisticated and custom set of non-linear aggregation operators.

The organization of this paper is as follows. In Sect. 2, the FI, NDFI and gFI are reviewed and their role in FIS rule output aggregation is addressed. In Sect. 3, we provide an example of the proposed theory. The goal of this article is to make the necessary introduction to different relevant FI extensions, propose algorithms for computing the integrals, and present discussions about specifying or learning g in the context of fuzzy logic.

2 Fuzzy Measure and Integral

2.1 Fuzzy Measure

The concepts of a measure and integral are two of the most fundamental aspects of mathematics. A famous example is the Lebesgue measure and integral with respect to that measure. The Lebesgue measure is the typical way of assigning a value to subsets of a Euclidean space, i.e., length, area, volume. However, measures can capture a wide range of concepts outside of n-dimensional Euclidean spaces. In the context of the FI, the FM often is a subjective assessment of the *worth* or *importance* of a subset of data or information sources. In general, the FM requires the property of monotonicity with respect to set inclusion, a weaker property than the additive property of a probability measure. In [8], the initial focus was an integrand (h) and FM (g) in the interval $[0,1]$, $h : X \rightarrow [0,1]$ and $g : 2^X \rightarrow [0,1]$, where 2^X is all subsets of X. However, both integrand and FM have been defined more generally and used in the FI, e.g., $h : X \rightarrow [-\infty, \infty]$, $h : X \rightarrow \Re_0^+$, $g : 2^X \rightarrow \Re_0^+$, etc. Next, we briefly review the FM.

Definition 1 (Fuzzy measure) [8]. For a finite set X, a FM is a function $g : 2^X \rightarrow \Re_0^+$, such that

1. (Boundary Condition) $g(\phi) = 0$ (and often $g(X) = 1$);
2. (Monotonicity) If $A, B \subseteq X$, $A \subseteq B$, then $g(A) \leq g(B)$.

Note, if X is an infinite set, then a third condition guaranteeing continuity is required. However, this is a moot point for finite X. In the FIS system, we only consider the finite set X of N rules.

Definition 2 (Sugeno λ-measure) [8]. Let g be a FM. Furthermore, g is called a Sugeno λ-measure if it satisfies the following condition,

$$g_\lambda(A \cup B) = g_\lambda(A) + g_\lambda(B) + \lambda g_\lambda(A) g_\lambda(B), \tag{1}$$

for some $\lambda > -1$. The Sugeno λ-measure is built using the singleton values (aka densities), $g^i = g(\{x_i\})$. Sugeno showed that λ can be found by solving

$$\lambda + 1 = \prod_{i=1}^{N} (1 + \lambda g^i), \lambda > -1, \tag{2}$$

where there exists one real solution that satisfies $\lambda > -1$.

Definition 3 (S-Decomposable Measure). Let S be a *triangular-conorm* (t-conorm). A FM g is called an S-decomposable measure if $g(\phi) = 0$, $g(X) = 1$, and for all A, B such that $A \cap B = \phi$,

$$g(A \cup B) = S(g(A), g(B)). \tag{3}$$

2.2 Choquet Integral

Next, we briefly review the real-valued discrete CI; see [8,9,13,14] for additional information (proofs, properties, etc.).

Definition 4 ($\Re$-Valued Discrete Choquet FI) [8]. Given a finite set X, FM g and function $h : X \to \Re$, the CI of h with respect to g is

$$\int h \circ g = C_g(h) = \sum_{i=1}^{N} h_{\Delta(i)} \left(g(A_{\Delta(i)}) - g(A_{\Delta(i-1)}) \right), \tag{4}$$

where $h_{\Delta(i)} = h(\{x_{\Delta(i)}\})$ and Δ is a permutation on X such that $h_{\Delta(1)} \geq h_{\Delta(2)} \geq ... \geq h_{\Delta(N)}$, $A_{\Delta(i)} = \{x_{\Delta(1)}, ..., x_{\Delta(i)}\}$ and $g(A_{\Delta(0)}) = 0$.

2.3 Generalized Fuzzy Integral

In [9], we introduced the gFI, an extension of the FI based on Zadeh's *extension principle* (EP) for "unrestricted" (potentially non-convex and subnormal) FSs. Initially, the gFI was created for skeletal age-at-death estimation in forensic anthropology [15,16]. Due to space considerations herein, see [9] for full mathematical detail, proofs and in-depth analysis. Algorithm 1 is an algorithmic description of how to calculate the gFI.

Note, that at a particular α-cut we obtain, for non-convex FSs, a *discontinuous interval*, e.g., an ${}^{\alpha}H_i$ that yields more than one interval, like $[0, 0.1]$ and $[0.7, 0.9]$, versus a single continuous interval, such as $[0, 0.9]$. The gFI plays off the fact that a discontinuous interval can be represented as the union of its continuous sub-intervals [9]. At each α-cut, we first decompose the discontinuous intervals into their corresponding continuous interval counterparts. Computationally, that is for a finite number of α-cut approximation, this only makes sense when we consider a finite number of continuous intervals. Next, we compute the continuous interval-valued FI on each combination ($\mathcal{M}_\alpha$ in total) of continuous intervals with respect to our different inputs/sources. For example,

Algorithm 1. Algorithm to calculate the *generalized FI* (gFI) [9]

1: Input FM g ▷ *e.g., Sugeno* λ*-FM, learn* g *from data, specify* g*, etc.*
2: Input FS-valued partial support function H ▷ *i.e.,* H_i *for* $i = \{1, ..., N\}$
3: Calculate $\beta = \bigwedge_{i=1}^{N} Height(H_i)$ ▷ *minimum height of partial support FSs*
4: **for** each $\alpha \in (0, \beta]$ **do** ▷ *note, this step is discretized in practice*
5: $[\int H \circ g]_\alpha = (\int [^\alpha H]_1 \circ g)$ ▷ *first gFI calculation; interval-valued FI*
6: **for** k = 2 to $\mathcal{M}_\alpha$ **do** ▷ *each continuous interval combo* ($\mathcal{M}_\alpha$) *at* α
7: $[\int H \circ g]_\alpha = [\int H \circ g]_\alpha \bigcup (\int [^\alpha H]_k \circ g)$ ▷ *kth gFI calculation*
8: **end for**
9: **end for**

for two inputs, if the first has three continuous intervals and the second has two then $\mathcal{M}_\alpha$ equals six. Let $[^\alpha H]_{k=1}$ be the first such combination, i.e., the first continuous interval from source one, first continuous interval from source two, etc. The gFI is simply the union of these continuous-valued interval FI results. At each α-cut, the beauty is that the gFI breaks down into a set of interval-valued FI calculations which breaks down further into two $\Re$-valued FIs [9,17]—one for the left interval endpoints and one for the right interval endpoints.

2.4 Non-Direct Fuzzy Integral

Unlike the gFI, the NDFI does not extend the FI according to Zadeh's EP. It should be noted that the EP is not the true and only way to address extensions of the FI, it is just one well grounded and heavily studied approach. Due to space considerations herein, see [9] for full mathematical detail, proofs and indepth analysis of the NDFI. Algorithm 2 is a description of the NDFI.

Algorithm 2. Algorithm to calculate the *non-direct FI* (NDFI) [9]

1: Input FM g ▷ *e.g., Sugeno* λ*-FM, learn* g *from data, specify* g*, etc.*
2: Input the FS-valued partial support function H ▷ *i.e.,* H_i *for* $i = \{1, ..., N\}$
3: Discretize the output domain, $D = \{d_1, ..., d_{|D|}\}$ ▷ *e.g.,* $D = \{0, 0.01, ..., 1\}$
4: Initialize the (FS) result to $R[d_k] = 0$
5: **for** each $d_k \in D$ **do** ▷ *for each output domain location*
6: **for** each $i \in \{1, ..., N\}$ **do** ▷ *for each input*
7: Let $z_i = H_i(d_k)$ ▷ *i.e.,* $\mathbf{z}$ *is the vector of memberships at* d_k
8: **end for**
9: $R[d_k] = \int \mathbf{z} \circ g$ ▷ $\Re$*-valued FI of* $\mathbf{z}$ *with* g
10: **end for**

Whereas the gFI decomposes the FI into a sequence of interval-based FI calculations across the membership domain, NDFI decomposes the FI into a sequence of $\Re$-valued FI calculations across the input/element domain. The gFI always produces FSs whose cardinality, i.e., number of elements of the set, is greater than zero, as long as the minimum input FS height is greater than zero.

However, the NDFI is not guaranteed to do this. For example, consider when the FI becomes the minimum operator, i.e., when $g(A) = 0, \forall A \subset X, g(X) = 1$. Furthermore, assume the input FSs are completely disjoint. The NDFI therefore produces a value of 0 everywhere.

2.5 FI for Combining Rules in a FIS

As the reader can clearly see, the NDFI and gFI are two ways to aggregate FS-valued inputs. Furthermore, they fit naturally into existing FISs. The first question we explore is, when to use the NDFI and when to use the gFI. Table 1 summarizes a few important properties of these two extensions.

Table 1. Summary of important properties of the NDFI and the gFI

FI extension	domain	order	"types" of sets
NDFI	discrete or continuous	no assumption	any
gFI	real-number domain	total order	any

In general, if the domain is naturally discrete and no total order exists then the gFI is not applicable but the NDFI is. The NDFI "aggregates in place"—it does not allow for interaction or cross-pollination between elements in the domain. On the other hand, the gFI allows for interaction between elements (thus why it works on $\Re$). Additionally, the gFI is based on the EP and it extends a function to FS-valued inputs. In the next section we provide numeric examples that illustrates these concepts.

Next, we explore common FMs encountered in practice (and FISs in the case of the NDFI). In Yager's approach [10], he explored *and-like* and *or-like* aggregation operators on a per-element basis. The NDFI yields Yager's *and-like* and *or-like* operators, OWA operators (specifically, all linear combinations of order statistics) and numerous other aggregation operators outside this set. It is well-known that the FI, specifically the CI, turns into a particular aggregation operator based on g (Table 2 provides an example for a few well-known FMs).

Table 2. Example FMs and aggregation operators induced

Resultant CI operator	FM properties
OWA	$g(\phi) = 0$, $g(X) = 1$, $g(A) = g(B)$ when $\lvert A\rvert = \lvert B\rvert, \forall A, B \subseteq X$
Maximum	$g(A) = 1, \forall A \subseteq X, A \neq \phi, g(\phi) = 0$
Minimum	$g(X) = 1$, $g(A) = 0, \forall A \subset X$
Mean	$g(A) = \frac{\lvert A\rvert}{N}, \forall A \subseteq X, A \neq \phi, g(\phi) = 0$
Sum	$g(A) = \lvert A\rvert, \forall A \subseteq X, A \neq \phi, g(\phi) = 0$

Of course, the FMs in Table 2 (minus "sum") are all OWAs and it is well-known that the CI can yield a class of custom non-linear aggregation operators outside of OWAs. The FM can be a number of things from a belief measure to a Sugeno λ-FM to much more. This is not counting the vast sea of custom and unique FMs that one might need and attempt to learn from data. The NDFI can also reproduce the *sum*, used by many FISs, if one discards the upper $g(X) = 1$ boundary condition. For example, the FM $g(A) = |A|$ makes the CI yield the summation operator, e.g., for $N = 4$, $h_{\Delta(1)}(1-0) + h_{\Delta(2)}(2-1) + h_{\Delta(3)}(3-2) + h_{\Delta(4)}(4-3) = h_{\Delta(1)} + h_{\Delta(2)} + h_{\Delta(3)} + h_{\Delta(4)}$.

Instead, if we do not want to use a standard aggregation operator then we can specify the full FM. However, this can be incredibly complex as g grows quickly in size—2^N terms for N inputs. If we know the "worth" of the singletons (aka densities) then one can refer to techniques such as the Sugeno λ-measure and S-decomposable measure to impute the remainder of the FM. We can also look to a technique such as the k-additive FI.

However, if one wants to learn the FM, densities or the full FM, from data then a number of methods can be used, e.g., an optimization algorithm such as a genetic algorithm [18], linear and quadratic programming [13,14,19], gradient descent [20], etc. See [21] for FI learning works prior to 2008. In the case of a FIS, learning depends on if one has labeled data that is the defuzzified or FS-valued target outputs. In the latter case, we must select a measure of distance or divergence between FSs, such as one that calculates the sum of differences of interval endpoints across α-cuts [18,22]. While it is possible to learn the FM in fuzzy logic, we note that in practice it is likely not a trivial task as it has to be achieved in combination with FIS parameter specification.

Next, we explore the complexity of the NDFI versus the gFI. If the output domain has been discretized (or is already discrete) and is $D = \{d_1, d_2, ..., d_{|D|}\}$, then the NDFI requires $|D|$ FIs, each of which take $3N-1$ operations (specifically N multiplications, N subtractions and $N-1$ additions), therefore $3N|D| - |D|$ total calculations (not counting the H_i sorting step, which occurs for each element in D). The NDFI has more-or-less the same cost as current FIS operations, e.g., maximum, average, sum, etc. The only added complexity is determining the FM. Depending on the number of desired α-cuts, the gFI may or may not be more expensive than the NDFI (depends on the discretization level of the output domain). Assuming P α-cuts, we must first determine each continuous interval, all combinations of continuous intervals must be computed ($\mathcal{M}_\alpha$) and subjected to the interval-valued FI (which is the invocation of two $\Re$-valued FIs, thus N operations), and we must take the union of the results. The gFI cost formula is not as easily expressible (as the NDFI) because each α-cut on each FS can yield a different number of discontinuous intervals. We do not see complexity as the driving factor in selecting the NDFI or the gFI. Instead, it seems more important that a user prefers the EP (gFI) or "aggregation in place" (NDFI).

3 Numeric Example

In this section, we consider an augmented fuzzy logic tipping problem. We consider four arbitrary rules (Table 3) and an input of $(0.58, 0.75, 0.96)$, for (responsiveness, satisfaction, food), which results in four FSs to aggregate. Note, the particular vocabulary and their parameters is not as important as the trends of FSs. We want to show the case of one "outlier" (FS in extreme disagreement, i.e., H_1) and three overlapping (partially agreeing) FSs. Table 4 contains the terms and parameters used in the four rules. For comparison's sake, Table 5 contains five different FMs (whose full set of values are reported) used to aggregate our rule FS outputs. The particular FMs in Table 5 were selected in order to demonstrate the behavior and differences between the NDFI and gFI. In particular, we compute OWAs (maximum, minimum, average), the sum, the Sugeno λ-FM (for a set of densities) and an S-Decomposable FM (labeled possibility in Table 5), where S=max, for a given set of densities. In addition, we report two "binary FMs"—a FM (labeled $g_{2,3}$ in Table 5 for compactness sake) that is 0 except for $g(\{x_2, x_3\}) = 1$ (and satisfying monotonicity) and a FM (called $g_{1,4}$ in Table 5) that is 0 except for $g(\{x_1, x_4\}) = 1$ (and satisfying monotonicity).

Table 3. Example toy rules for tipping

Rule	If	Then
1	responsiveness is slow	tip is very low
2	satisfaction is high	tip is high
3	responsiveness is moderate AND satisfaction is high	tip is very high
4	responsiveness is moderate AND food is delicious	tip is stellar

Table 4. Fuzzy set parameters for FIS example[1]

Variable	Term	Tri. Membership	Variable	Term	Tri. Membership
responsiveness	slow	(0.3,0.45,0.6)	Tip	very low	(0,0.15,0.3)
	moderate	(0.35,0.5,0.65)		high	(0.55,0.7,0.85)
satisfaction	high	(0.7,0.8,0.9)		very high	(0.65,0.8,0.95)
food	delicious	(0.8,0.9,1)		stellar	(0.8,0.9,1)

[1]note, this could be scaled to a level of tip, e.g. in [0,30]%

Our tipping example exposes a few interesting things. First, the gFI results look rather extreme as they are truncated according to the minimum height input FS (due to the formulation of the EP). Truncation is extremely frequent as each rule fires to some degree so the FSs are almost always subnormal. Next, the max, min and average look more like what we might expect, if our desire is fuzzy arithmetic. On the other hand, the NDFI aggregates in place—restricting

Table 5. List of FMs used for examples

FM	max	min	sum	average	Sugeno λ-FM ($g_{\lambda 1}$)	$g_{2,3}$	$g_{1,4}$	possibility
$g(\{x_1\})$	1	0	1	0.25	0.001	0	0	0.1
$g(\{x_2\})$	1	0	1	0.25	0.25	0	0	0.8
$g(\{x_3\})$	1	0	1	0.25	0.001	0	0	0.2
$g(\{x_4\})$	1	0	1	0.25	0.25	0	0	0.7
$g(\{x_1, x_2\})$	1	0	2	0.5	0.253	0	0	0.8
$g(\{x_1, x_3\})$	1	0	2	0.5	0.002	0	0	0.2
$g(\{x_1, x_4\})$	1	0	2	0.5	0.253	0	1	0.7
$g(\{x_2, x_3\})$	1	0	2	0.5	0.253	1	0	0.8
$g(\{x_2, x_4\})$	1	0	2	0.5	0.983	0	0	0.8
$g(\{x_3, x_4\})$	1	0	2	0.5	0.253	0	0	0.7
$g(\{x_1, x_2, x_3\})$	1	0	3	0.75	0.256	1	0	0.8
$g(\{x_1, x_2, x_4\})$	1	0	3	0.75	0.993	0	1	0.8
$g(\{x_1, x_3, x_4\})$	1	0	3	0.75	0.256	0	1	0.7
$g(\{x_2, x_3, x_4\})$	1	0	3	0.75	0.991	1	0	0.8
$g(X)$	1	1	4	1	1	1	1	1

fusion to just the set of available evidence at a particular output domain element. This produces different results for the same FM. Next, each FI extension and FM yields different FSs, which leads to different defuzzified results (when/if a crisp overall output is desired). The results of firing the rules with the given inputs and utilizing each measure in Table 5 can be found in Fig. 2.

In the case of the sum FM and the gFI, if the output domain is restricted to $[0, 1]$ then the result is officially "out of range". However, the gFI and the NDFI yield results that do not exceed the min and max input FSs, which therefore is in the valid interval as long as the inputs are in the valid interval and the FM has at max $g(X) = 1$ [9]. Additionally, the NDFI also yields (due to boundedness property of the FI) results between $[0, 1]$ as long as the sets have memberships in $[0, 1]$ and the FM has at max $g(X) = 1$. Obviously, the sum FM violates these conditions. One rationalization of the usefulness of an NDFI operator that produces membership values greater than 1 is it still leads to a technically valid defuzzified result (but not a valid FS).

Next, we explore a binary FM. A binary FM is a FM whose values are restricted to $\{0, 1\}$. Consider the set of all maximal chains of the Hasse diagram $(2^X, \subseteq)$. A maximal chain in $(2^X, \subseteq)$ is a sequence ϕ, $\{x_{\pi(1)}\}$, $\{x_{\pi(1)}, x_{\pi(2)}\}$, ..., $\{x_{\pi(1)}, ..., \pi_{\pi(N)}\}$ (for a given permutation π). In the discrete case, the CI is nothing more than a weighted sum—the sorted integrands times the differences in their respective values in the maximal chain. For a binary FM, one of these difference-in-g's is 1 and the other values are 0 (trivial to prove due to monotonicity). Now, consider the FM in Fig. 2 labeled $g_{2,3}$. This FM has value

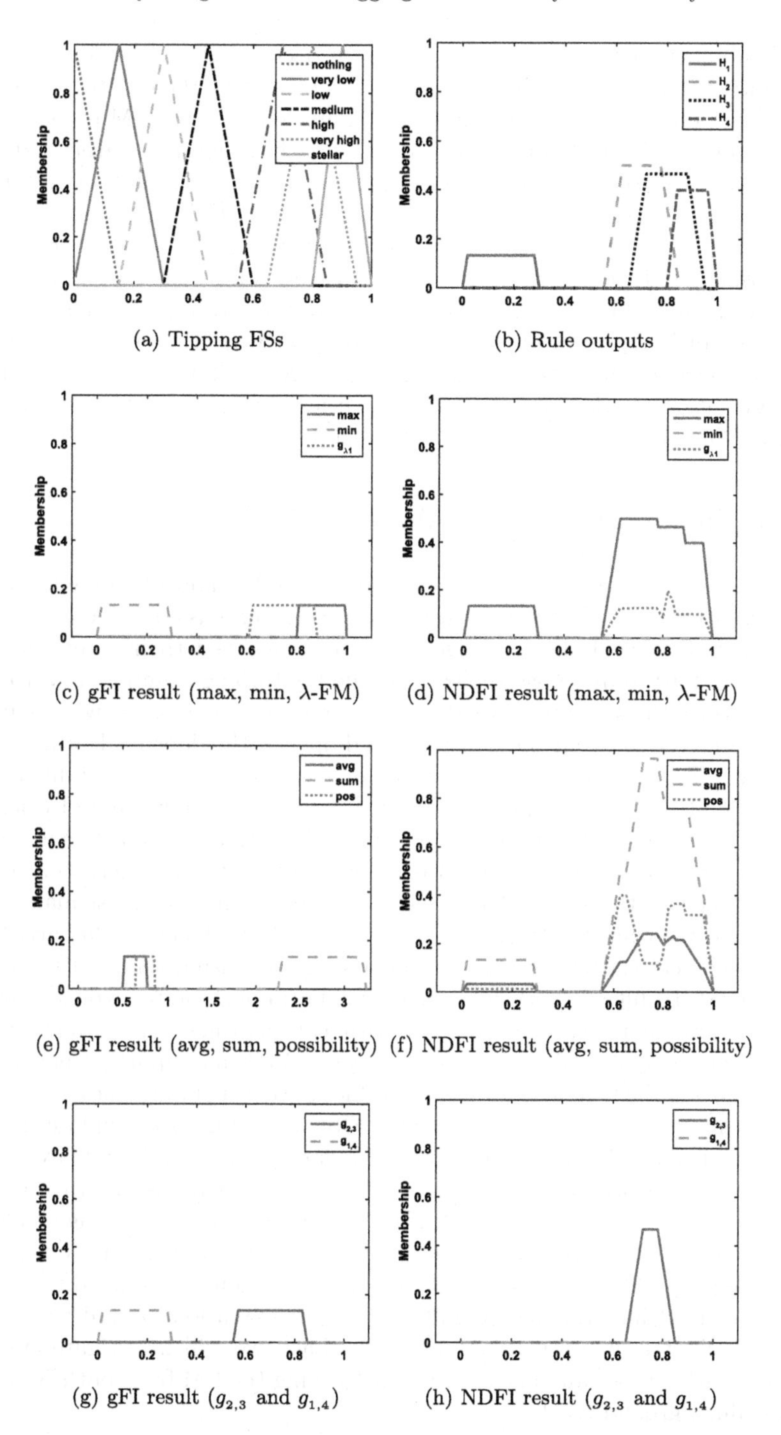

(a) Tipping FSs (b) Rule outputs

(c) gFI result (max, min, λ-FM) (d) NDFI result (max, min, λ-FM)

(e) gFI result (avg, sum, possibility) (f) NDFI result (avg, sum, possibility)

(g) gFI result ($g_{2,3}$ and $g_{1,4}$) (h) NDFI result ($g_{2,3}$ and $g_{1,4}$)

Fig. 2. Graphical illustration of the gFI and NDFI outputs for the FMs reported in Table 5, terms in Table 4 and rules in Table 3.

0 for all subsets except 1 for $\{x_2, x_3\}$ and all respective subsets needed to ensure monotonicity. The result is the minimum of the two sets H_2 and H_3 (shown in Fig. 2). The same argument holds for the FM labeled $g_{1,4}$ (the FM is computing the minimum of the two inputs H_1 and H_4). While this is a rather extreme (binary) example to illustrate the importance and resulting effect of specification and/or learning of higher-order tuple terms in the FM, the point is we can achieve a wealth of custom aggregation strategies for combinations of rules.

In closing, while it would be nice to definitively state that the NDFI or the gFI is better, we feel this is not the case. Instead, the ultimate decision of which extension to pick appears to break down into either what the problem demands (e.g., a discrete domain where elements cannot cross-pollinate) or the desire of what aggregation should do to the outputs, either aggregate in place or compute a function that has been extended to FS-valued inputs.

4 Conclusion and Future Work

Herein, we proposed and started to explore the NDFI and gFI extensions for rule aggregation in a FIS. We discussed how these extensions fit into existing FISs, reviewed efficient discrete algorithms and analyzed their computational complexities. Furthermore, we showed, via the toy tipping example, how to pick FMs that turn the CI into existing FIS aggregation operators, we discussed imputation methods (from the densities) and opened the door for learning. The example and Table 1 gave the reader a feel for the inner workings and differences between the gFI and NDFI and it helps with understanding when and where to use one extension over the other. We also highlighted that the NDFI "aggregates in place" (per-element) while the gFI is based on the EP. Ultimately, this makes a big difference, in terms of output, and it is a choice the user must make.

Due to space considerations, we had to restrict the focus of this article to different technical aspects of rule output aggregation using the NDFI and the gFI. However, in future work we feel it will be important to explore various applications, in particular in conjunction with learning rules from data and going beyond theories like the Wang-Mendel approach. While an important part of the gFI and NDFI is the fact that they can produce a wealth of different aggregation operators and extend the capabilities of a FIS, we will need to explore if there are any particular FMs (and therefore aggregation operators) that are "more appropriate" than others for different scenarios. The point is, a FIS user will need to either select or learn the FM. If no data is available then a big challenge they will have to face is what FM to select for their given application. Next, while we discussed type-1 fuzzy logic herein, in [23] we previously extended the FI for type-2 FS-valued integrands. We will explore the extension of NDFI and gFI, and thus FISs, for type-2 valued integrands. Last, we will investigate efficient methods (mathematics and algorithms) for learning the FM from both $\Re$-valued and FS-valued ground truth.

Acknowledgments. This work has been partially supported by the U.S. Army Research Office, the U.S. Army and RDECOM CERDEC NVESD via W911NF-16-1-0017, W911NF-14-1-0114, W911NF-14-1-0673, W909MY-13-C-0013 and 57940-EV. This work was also partially funded by RCUK's EP/M02315X/1.

References

1. Zadeh, L.: Fuzzy sets. Inf Control **8**(3), 338–353 (1965)
2. Zadeh, L.: Outline of a new approach to the analysis of complex systems and decision processes. IEEE Trans. Syst. Man Cybern. **3**(1), 28–44 (1973)
3. Mamdani, E., Assilian, S.: An experiment in linguistic synthesis with a fuzzy logic controller. Int. J. Man-Mach. Stud. **7**(1), 1–13 (1975)
4. Sugeno, M.: Industrial Applications of Fuzzy Control. Elsevier Science Inc., New York (1985)
5. Tsukamoto, Y.: An approach to fuzzy reasoning method. Adv. Fuzzy Set Theory Appl. **137**, 149 (1979)
6. Yubazaki, N., Yi, J., Hirota, K.: Sirms (single input rule modules) connected fuzzy inference model. JACIII **1**(1), 23–30 (1997)
7. Seki, H., Mizumoto, M.: Sirms connected fuzzy inference method adopting emphasis and suppression. Fuzzy Sets Syst. **215**, 112–126 (2013)
8. Sugeno, M.: Theory of Fuzzy Integrals and Its Applications. Tokyo Institute of Technology, Tokyo (1974)
9. Anderson, D.T., Havens, T.C., Wagner, C., Keller, J.M., Anderson, M.F., Wescott, D.J.: Extension of the fuzzy integral for general fuzzy set-valued information. IEEE Trans. Fuzzy Syst. **22**(6), 1625–1639 (2014)
10. Yager, R.: A general approach to rule aggregation in fuzzy logic control. Appl. Intell. **2**(4), 333–351 (1992)
11. Yager, R.R., Filev, D.P.: Induced ordered weighted averaging operators. IEEE Trans. Syst. Man Cybern. Part B Cybern. **29**(2), 141–150 (1999)
12. Tahani, H., Keller, J.: Information fusion in computer vision using the fuzzy integral. IEEE Trans. Syst. Man Cybern. **20**, 733–741 (1990)
13. Grabisch, M., Nguyen, H., Walker, E.: Fundamentals of Uncertainty Calculi, with Applications to Fuzzy Inference. Kluwer Academic, Dordrecht (1995)
14. Beliakov, G., Pradera, A., Calvo, T.: Aggregation Functions: A Guide for Practitioners, 1st edn. Springer, Heidelberg (2008)
15. Anderson, M., Anderson, D.T., Wescott, D.: Estimation of adult skeletal age-at-death using the sugeno fuzzy integral. Am. J. Phys. Anthropol. **142**(1), 30–41 (2010)
16. Anderson, D.T., Keller, J.M., Anderson, M.F., Wescott, D.J.: Linguistic description of adult skeletal age-at-death estimations from fuzzy integral acquired fuzzy sets. In: 2011 IEEE International Conference on Fuzzy Systems (FUZZ), pp. 2274–2281. IEEE (2011)
17. Grabisch, M., Murofushi, T., Sugeno, M.: Fuzzy Measures and Integrals: Theory and Applications. Physica-Verlag, Heidelberg (2000)
18. Anderson, D.T., Keller, J.M., Havens, T.C.: Learning fuzzy-valued fuzzy measures for the fuzzy-valued Sugeno fuzzy integral. In: Hüllermeier, E., Kruse, R., Hoffmann, F. (eds.) IPMU 2010. LNCS, vol. 6178, pp. 502–511. Springer, Heidelberg (2010)
19. Beliakov, G.: Construction of aggregation functions from data using linear programming. Fuzzy Sets Syst. **160**, 65–75 (2009)

20. Mendez-Vazquez, A., Gader, P., Keller, J., Chamberlin, K.: Minimum classification error training for choquet integrals with applications to landmine detection. IEEE Trans. Fuzzy Syst. **16**(1), 225–238 (2008)
21. Grabisch, M., Kojadinovic, I., Meyer, P.: A review of methods for capacity identification in choquet integral based multi-attribute utility theory: applications of the kappalab r package. Eur. J. Oper. Res. **186**(2), 766–785 (2008)
22. McCulloch, J., Wagner, C., Aickelin, U.:Analysing fuzzy sets through combining measures of similarity and distance. In: 2014 IEEE International Conference on Fuzzy Systems (FUZZ-IEEE), pp. 155–162, July 2014
23. Havens, T.C., Anderson, D.T., Keller, J.M.: A fuzzy choquet integral with an interval type-2 fuzzy number-valued integrand. In: 2010 IEEE International Conference on Fuzzy Systems (FUZZ), pp. 1–8, July 2010

On a Fuzzy Integral as the Product-Sum Calculation Between a Set Function and a Fuzzy Measure

Eiichiro Takahagi(✉)

School of Commerce, Senshu University, 2-1-1, Higashimita, Tamaku, Kawasaki 214-8580, Japan
takahagi@isc.snshu-u.ac.jp
http://www.isc.senshu-u.ac.jp/~thc0456

Abstract. We propose the Choquet integral with respect set to a function defined as the product-sum calculation between a set function and a fuzzy measure. The fuzzy integral is an extension of the Choquet integral. The Choquet integral assumes that the interactions among input values are interact fully but the extension assumes the values partially interaction. In this paper, we define another integral expression and analyze its properties. For an input vector the optimal set function is calculated through linear programming. Lastly, we analyze coalitions among set functions that are a cooperative game using the proposed integral.

Keywords: Set function · Choquet integral · Fuzzy measure · Möbius transformation · co-Möbius transformation · Linear programming · Supermodular · Cooperative game

1 Introduction

In the ordinal Choquet Integral model [3,5], the input functions had to be combined if the input values could be combined. If worker A works 8 h and B works 10 h, the Choquet integral model interprets this as the workers work 8 h together and worker B works 2 h alone. In this paper, we propose a model to deal with the case that A works 1 h alone, B works 3 h alone, and both work 7 h together.

To represent such a case, the input values are represented as a set function η. In this case, $\eta(\{A\}) = 1$, $\eta(\{B\}) = 3$, $\eta(\{A, B\}) = 7$. In the proposed model, for a fuzzy measure μ, the output value is the product sum values between η and μ.

In this paper, we show the properties of the proposed model such as its representation using the Möbius transformation and co-Möbius transformation and analyze the coalition among η.

J.P. Carvalho et al. (Eds.): IPMU 2016, Part I, CCIS 610, pp. 91–100, 2016.
DOI: 10.1007/978-3-319-40596-4_9

2 Choquet Integral with Respect to a Set Function

2.1 Choquet Integral and Choquet Integral with Respect to a Set Function

Let $X = \{1, \ldots, n\}$ be the set of criteria (n: number of criteria), and $\boldsymbol{x} = (x_1, \ldots, x_n)$ be the individual score vector where $x_i \geq 0, \forall i$.

Definition 1. *A fuzzy measure μ is defined as*

$$\mu : 2^X \to \mathbb{R}^+ \text{ where } \mu(\emptyset) = 0 \text{ and } \mu(A) \geq \mu(B) \text{ if } A \supseteq B. \tag{1}$$

Definition 2. *The Choquet integral ([3,5]) is defined as*

$$f_\mu^C(\boldsymbol{x}) \equiv \sum_{i=1}^{n} [x_{\sigma(i)} - x_{\sigma(i+1)}] \mu(\{\sigma(1), \ldots, \sigma(i)\}) \tag{2}$$

where $\sigma(i)$ is the permutation on X, that is, $x_{\sigma(1)} \geq \ldots \geq x_{\sigma(n)}$, $x_{\sigma(n+1)} = 0$ and $X = \{\sigma(1), \ldots, \sigma(n)\}$.

Definition 3. *Input of the Choquet integral with respect to a set function is a set function η defined as*

$$\eta : 2^X \to \mathbb{R}^+, \ \eta(\emptyset) = 0. \tag{3}$$

Definition 4. *The Choquet integral with respect to a set function $f_\mu^{\mathrm{PS}}(\eta)$ is defined as the product-sum (PS) calculation between the two set functions, that is*

$$f_\mu^{\mathrm{PS}}(\eta) \equiv \sum_{A \in 2^X} [\eta(A)\mu(A)]. \tag{4}$$

Definition 5 (Maximal Chain). *Maximal Chain θ is defined as a family of sets that $\theta = \{C_0, \ldots, C_n\}$ where $C_i \in 2^X, C_0 \subsetneq \ldots \subsetneq C_n$, $C_0 = \emptyset$, $C_n = X$.*

Definition 6 ($g : \boldsymbol{x} \to \eta$). *For a $\boldsymbol{x}$, the maximal chain $\theta = \{C_i \mid C_i = \{\sigma(1), \ldots, \sigma(i)\}, i = 1, \ldots, n\} \cup \emptyset$. $\eta = g(\boldsymbol{x})$ is defined as, $\forall A \in 2^X$*

$$\eta(A) = \begin{cases} x_{\sigma(i)} - x_{\sigma(i+1)} & \text{if } A = C_i \in (\theta \setminus \emptyset) \\ 0 & \text{otherwise.} \end{cases} \tag{5}$$

For example, if $n = 3$ and $\boldsymbol{x} = (2, 5, 3)$ then $\eta(\{2\}) = 5 - 3$, $\eta(\{2,3\}) = 3 - 2$, $\eta(\{1,2,3\}) = 2$ and $\eta(A) = 0$, $\forall A \in (2^X \setminus \{\{2\}, \{2,3\}, \{1,2,3\}\})$.

Theorem 1. *The Choquet integral with respect to a set function is an extension of the Choquet integral.*

Proof. For any $\boldsymbol{x}$ and μ, $\eta = g(\boldsymbol{x})$ and θ is the maximal chain of $g(\boldsymbol{x})$

$$\begin{aligned} f_\mu^{\mathrm{C}}(\mathbf{x}) &= \sum_{i=1}^{n} [(x_{\sigma(i)} - x_{\sigma(i+1)})\mu(\{\sigma(1), \ldots, \sigma(i)\})] \\ &= \sum_{A \in \theta} [\eta(A)\mu(A)] + \sum_{A \notin \theta} [\eta(A)\mu(A)] = \sum_{A \in 2^X} [\eta(A)\mu(A))] = f_\mu^{\mathrm{PS}}(\eta) \end{aligned} \tag{6}$$

□

Definition 7 ($h : \eta \to \boldsymbol{x}$)**.** *For a* η, $\boldsymbol{x} = h(\eta)$ *is defined as,* $\forall i \in X$,

$$x_i = \sum_{A \ni i} \eta(A), \quad i = 1, \ldots, n. \tag{7}$$

$A (\in X)$ of η is an active set when $\eta(A) > 0$.

Theorem 2. *If all active* A *of* η *are in a maximal chain,* θ*, that is* $\eta = g(h(\eta))$ *then*

$$f_\mu^{\mathrm{PS}}(\eta) = f_\mu^{\mathrm{C}}(h(\eta)) \tag{8}$$

Proof

$$f_\mu^{\mathrm{PS}}(\eta) = \sum_{A \in \theta} [\eta(A)\mu(A)] + \sum_{A \notin \theta} [\eta(A)\mu(A)] = \sum_{A \in \theta} [\eta(A)\mu(A)] = f_\mu^{\mathrm{C}}(h(\eta)) \tag{9}$$

□

Example 1 (2 workers). Two workers ($X = \{1, 2\}$) work in a workshop. Worker W1 produces 20 kg per hour ($\mu(\{1\}) = 20$) and W2 produces 30 kg per hour ($\mu(\{2\}) = 30$). If the two workers work in cooperation, they produce 60 kg per hour ($\mu(\{1, 2\}) = 60$). One day, W1 starts at 9 and works 7 h and W2 start at 10 and works 8 h. From 9 am to 10 am, W1 works 1 h alone ($\eta(\{1\}) = 1$). From 10 am to 4 pm, W1 and W2 work in cooperation ($\eta(\{1, 2\}) = 6$). From 4 pm to 6 pm, W2 works 2 h alone ($\eta(\{2\}) = 2$). The output of the workshop is

$$f_\mu^{\mathrm{PS}}(\eta) = \eta(\{1\})\mu(\{1\}) + \eta(\{2\})\mu(\{2\}) + \eta(\{1, 2\})\mu(\{1, 2\}) = 440. \tag{10}$$

If the two workers start at the same time, the output is the Choquet integral with $x_1 = 7$ and $x_2 = 8$, $f_\mu^{\mathrm{C}}(\boldsymbol{x}) = (x_2 - x_1)\mu(\{2\}) + x_1\mu(\{1, 2\}) = 450$.

The Choquet integral is based on the assumption that the input values are combined if they can combine.

2.2 Sum of Choquet Integral Outputs

For any two input vectors $\boldsymbol{x}^1$ and $\boldsymbol{x}^2$, $\boldsymbol{x}^3 = \boldsymbol{x}^1 + \boldsymbol{x}^2$. Generally, the sum of two Choquet integral output values are not the same, that is,

$$f_\mu^{\mathrm{C}}(\boldsymbol{x}^3) \lesseqgtr f_\mu^{\mathrm{C}}(\boldsymbol{x}^1) + f_\mu^{\mathrm{C}}(\boldsymbol{x}^2).$$

The addition of two set functions, $\eta^3 = \eta^1 + \eta^2$ is defined as $\eta^3(A) = \eta^1(A) + \eta^2(A), \forall A \in 2^X$.

Proposition 1 (Sum of two η)

$$f_\mu^{\mathrm{PS}}(\eta^1 + \eta^2) = f_\mu^{\mathrm{PS}}(\eta^1) + f_\mu^{\mathrm{PS}}(\eta^2). \tag{11}$$

Proof. Trivial. □

From Proposition 1, for any $\boldsymbol{x}^1$ and $\boldsymbol{x}^2$,

$$f_\mu^{\mathrm{PS}}(g(\boldsymbol{x}^1) + g(\boldsymbol{x}^2)) = f_\mu^{\mathrm{PS}}(g(\boldsymbol{x}^1)) + f_\mu^{\mathrm{PS}}(g(\boldsymbol{x}^2)). \tag{12}$$

Let's $(\theta_1, \ldots, \theta_K)$ be a maximal chain group which covers the union of all of active A, that is, $(\theta_1 \cup \ldots \cup \theta_K) \supseteq \{A \mid \eta(A) > 0\}$ where θ_i is the maximal chain of η. For $(\theta_1, \ldots, \theta_K)$, it is easy to assign $\eta_{\theta_1}, \ldots, \eta_{\theta_K}$ such that,

$$\eta = \sum_{i=1}^{K} \eta_{\theta_i}, \text{ where } \eta_{\theta_i}(A) \begin{cases} \geq 0 & \text{if } A \in \theta_i \\ = 0 & \text{otherwise.} \end{cases} \tag{13}$$

Theorem 3 (Decomposition to Choquet Integrals). *For any η and μ, $f_\mu^{\mathrm{PS}}(\eta)$ is represented as the sum of Choquet integral outputs.*

Proof

$$f_\mu^{\mathrm{PS}}(\eta) = f_\mu^{\mathrm{PS}}(\sum_{i=1}^{K} \eta_{\theta_i}) = \sum_{i=1}^{K} [f_\mu^{\mathrm{PS}}(\eta_{\theta_i})] = \sum_{i=1}^{K} [f_\mu^{\mathrm{C}}(h(\eta_{\theta_i}))] \tag{14}$$

□

Generally, η_{θ_i} of Eq. (13) and the decomposition $f_\mu^{\mathrm{C}}(h(\eta_{\theta_i}))$ of Theorem 3 are not unique.

Example 2 (2 Workers and 2 days). The 2 workers work 2 days and every day they start work at the same time. The first day, W1 works 8 h and W2 works 10 h, $\boldsymbol{x}^1 = (8, 10)$ and $\eta^1 = g(\boldsymbol{x}^1)$. The second day, W1 works 10 h and W2 works 8 h. The fuzzy measure μ would be the same as in example 1.

Table 1. Calculations of Example 2

i		$\boldsymbol{x}^i$	$f_\mu^{\mathrm{C}}(\boldsymbol{x}^i)$	$\eta^i(\{1\})$	$\eta^i(\{2\})$	$\eta^i(\{1,2\})$	$f_\mu^{\mathrm{PS}}(\eta^i)$
1	First day	$(8, 10)$	540	0	2	8	540
2	Second day	$(10, 8)$	520	2	0	8	520
	Sum(f^{C} and f^{PS})		1060				1060
3	Sum($\boldsymbol{x}$ and η)	$(18, 18)$	1080	2	2	16	1060

Table 1 shows the difference between the two Choquet integral models, that is, the Choquet integral did not represent the two day case because $f_\mu^{\mathrm{C}}(\boldsymbol{x}^1+\boldsymbol{x}^2) \neq f_\mu^{\mathrm{C}}(\boldsymbol{x}^1) + f_\mu^{\mathrm{C}}(\boldsymbol{x}^2)$. However, the Choquet integral with respect to a set function can represent this case, that is, $f_\mu^{\mathrm{PS}}(\eta^1+\eta^2) = f_\mu^{\mathrm{PS}}(\eta^1) + f_\mu^{\mathrm{PS}}(\eta^2)$.

2.3 Möbius and Co-Möbius Transformation

The Möbius and the co-Möbius transformations [4,7] are equivalent representations of set functions [8].

Definition 8 (Möbius and Co-Möbius Transformation). *The Möbius transformation μ^m of a set function μ and the co-Möbius transformation μ^c of a set function μ are defined as, $\forall A \in 2^X$,*

$$\mu^m(A) \equiv \sum_{B \subseteq A} [(-1)^{|A|-|B|}\mu(B)], \;\; \mu^c(A) \equiv \sum_{B \subseteq A} [(-1)^{|B|}\mu(X \setminus B)]. \tag{15}$$

The inverse transformations are, $\forall A \in 2^X$,

$$\mu(A) = \sum_{B \subseteq A} \mu^m(A), \;\; \mu^c(A) = \sum_{B \supseteq A} \mu^m(B). \tag{16}$$

Using those transformations, we can represent the Choquet integral with respect to a set the function as another expression. In this transformation, η is the Möbius transformation of δ, that is $\eta = \delta^m$.

Theorem 4 (Another Expression of the Choquet Integral with respect to a set function). *For any η and μ,*

$$f_{\mu^m}^{\mathrm{PS}}(\delta^c) = f_\mu^{\mathrm{PS}}(\eta) \tag{17}$$

where $\delta^c(A) = \sum_{B \supseteq A} \eta(B), \forall A \in 2^X$.

Proof

$$\begin{aligned} f_{\mu^m}^{\mathrm{PS}}(\delta^c) &= \sum_{A \in 2^X} [(\delta^c(A)\mu^m(A)] = \sum_{A \in 2^X} [(\sum_{B \supseteq A} \eta(B))\mu^m(A)] \\ = \sum_{B \supseteq A, A, B \in 2^X} \eta(B)\mu^m(A) &= \sum_{B \in 2^X} [\eta(B) \sum_{A \subseteq B} \mu^m(A)] = \sum_{B \in 2^X} \eta(B)\mu(B) = f_\mu^{\mathrm{PS}}(\eta) \end{aligned} \tag{18}$$

□

Example 3. Table 2 shows calculation process used in example 2. The upper line is $f_\mu^{\mathrm{PS}}(\eta)$. η shows the two day's total hours, that is, the two workers works 16 h in corporation and each worker works 2 h alone. The output value is the product sum of the two set functions. The line below shows $f_{\mu^m}^{\mathrm{PS}}(\delta^c)$. $\delta^c(\{1\})$ shows W1's total work time (18) and $\delta^c(\{1,2\})$ shows total work time in corporation (16). $\mu^m(\{1,2\})$ shows incremental value from the additive value. The output of the two Choquet integrals with respect to a set function is equivalent.

Table 2. Two Choquet Integrals with respect set function

	{1}	{2}	{1, 2}		{1}	{2}	{1, 2}	product-sum
$\eta(A)$:	2	2	16	$\mu(A)$:	20	30	60	$f_{\mu}^{\mathrm{PS}}(\eta) = 1060$
$\delta^c(A)$:	18	18	16	$\mu^m(A)$:	20	30	10	$f_{\mu^m}^{\mathrm{PS}}(\delta^c) = 1060$

3 Optimization and Coalition of η

3.1 Range of η

For some $\boldsymbol{x}$, we discuss the case that η is only assigned suitable values. The range of $\eta(A)$ is restricted by the following conditions.

Sum of input. The sum of the input values related to input i is x_i, that is,

$$\sum_{A \ni i} \eta(A) = x_i, i = 1, \ldots, n \tag{19}$$

Non-negative constraint. $\eta(A), \forall A$ are non-negative.

$$\eta(A) \geq 0, \forall A \in 2^X \tag{20}$$

3.2 Optimization

For a $\boldsymbol{x}$ and a μ, we assign a η^* that maximizes or minimizes $f_{\mu}^{\mathrm{PS}}(\eta^*)$, according to the model. Our model is based on the linear programming (LP) model. For a given $\boldsymbol{x}$ and μ,

$$\begin{aligned} \text{(Maximize)} \quad & f_{\mu}^{\mathrm{PS}}(\eta^*) = \sum_{A \in 2^X} \eta^*(A)\mu(A) \\ \text{where} \quad & \sum_{A \ni i} \eta^*(A) = x_i, i = 1, \ldots, n \\ & \eta^*(A) \geq 0, \forall A \end{aligned} \tag{21}$$

In the minimum model, Eq. (21) is changed to minimize the equation. For an LP model, the optimal η^* is not always unique, for example, when μ is an additive fuzzy measure.

Wang [9] defined this optimization as the new integral with respect to μ and $\boldsymbol{x}$. The output value of the Wang's integral is calculated using the computational algorithm.

Example 4 (3 workers, super-additive). The fuzzy measure of this example in Table 3 is super-additive, $\mu(A \cup B) \geq \mu(A) + \mu(B)$ if $A \cap B = \emptyset$. For the input $\boldsymbol{x} = (8, 6, 7)$, the maximum $f_\mu^{\mathrm{PS}}(\eta^*)$ is shown by the following LP model.

$$\text{(Maximize)} \ f_\mu^{\mathrm{PS}}(\eta^*) = \sum_{A \in 2^X} \eta^*(A)\mu(A)$$

$$\begin{aligned} \text{where } & \eta^*(\{1\}) + \eta^*(\{1,2\}) + \eta^*(\{1,3\}) + \eta^*(\{1,2,3\}) = x_1 \\ & \eta^*(\{2\}) + \eta^*(\{1,2\}) + \eta^*(\{2,3\}) + \eta^*(\{1,2,3\}) = x_2 \\ & \eta^*(\{3\}) + \eta^*(\{1,3\}) + \eta^*(\{2,3\}) + \eta^*(\{1,2,3\}) = x_3 \\ & \eta^*(A) \geq 0, \forall A \in 2^X \end{aligned}$$

Table 3. Example 4

	A	$\{1\}$	$\{2\}$	$\{3\}$	$\{1,2\}$	$\{1,3\}$	$\{2,3\}$	$\{1,2,3\}$
Paramiters: $\boldsymbol{x} = (8,6,7)$	$\mu(A)$:	10	12	15	30	35	40	60
Optimal soulution $f_\mu^{\mathrm{PS}}(\eta^*) = 405$	$\eta^*(A)$:	1	0	0	0	1	0	6

The optimal solution is $f_\mu^{\mathrm{PS}}(\eta^*) = 405$ and η^* is on the bottom line of Table 3. The active elements A, that is $\eta^*(A) > 0$, are $\{1\} \subset \{1,3\} \subset \{1,2,3\}$, which is a maximal chain. Therefore, in this case, $f^{\mathrm{PS}}(\eta^*) = f^{\mathrm{C}}(\boldsymbol{x}^*)$ where $\boldsymbol{x}^* = g(\eta^*)$.

Example 5 (pizzeria). A pizzeria offers 'pizza'(1), 'salad'(2) and 'dessert'(3). The pizzeria sells them either as single items or set items. Their price is a fuzzy measure μ in Table 4. This fuzzy measure is subadditive, $\mu(A \cup B) \leq \mu(A) + \mu(B)$, if $A \cap B = \emptyset$.

Table 4. Example 5

	A	$\{1\}$	$\{2\}$	$\{3\}$	$\{1,2\}$	$\{1,3\}$	$\{2,3\}$	$\{1,2,3\}$
Parameters: $\boldsymbol{x} = (10,6,7)$	$\mu(A)$:	7	2	5	8	10	7	12
Optimal solution $f_\mu^{\mathrm{PS}}(\eta^*) = 100$	$\eta^*(A)$:	0	0	0	3	4	0	3

A group buys 10 pizzas, 6 salads and 7 desserts. The minimum expenditure of the group is the minimum value of the LP model, $f_\mu^{\mathrm{PS}}(\eta^*) = 100$ and η^* is shown in Table 4. The deals available are the 3 'pizza and salad' set ($\eta^*(\{1,2\}) = 3$), 4 'pizza and dessert' set($\eta^*(\{1,3\}) = 4$) and 3 'pizza, salad, and deserts' set($\eta^*(\{1,2,3\}) = 3$). The solution cannot be represented by one Choquet integral because active sets are not constructed by a maximal chain, for example, $\theta_1 = \{\emptyset, \{1\}, \{1,2\}, \{1,2,3\}\}$ and $\theta_2 = \{\emptyset, \{1\}, \{1,3\}, \{1,2,3\}\}$. Therefore the solution is represented by the sum of two Choquet integral outputs, that is, $f_\mu^{\mathrm{PS}}(\eta^*) = f_\mu^{\mathrm{C}}((6,6,3)) + f_\mu^{\mathrm{C}}((4,0,4))$.

3.3 Supermodular, Submodular and Optimization

Definition 9 (Supermodular and Submodular [3])

$$\text{(Supermodular)}\ \mu(A \cup B) + \mu(A \cap B) \geq \mu(A) + \mu(B),\ \forall A, B \in 2^X \tag{22}$$

$$\text{(Submodular)}\ \mu(A \cup B) + \mu(A \cap B) \leq \mu(A) + \mu(B),\ \forall A, B \in 2^X. \tag{23}$$

Theorem 5 (Superadditivity and Subadditivity Theorems [6]). *Given a supermodular set function μ and $\boldsymbol{x}^A$ and $\boldsymbol{x}^B$,*

$$f_\mu^{\mathrm{C}}(\boldsymbol{x}^A + \boldsymbol{x}^B) \geq f_\mu^{\mathrm{C}}(\boldsymbol{x}^A) + f_\mu^{\mathrm{C}}(\boldsymbol{x}^B) \tag{24}$$

and given a submodular set function μ and $\boldsymbol{x}^A$ and $\boldsymbol{x}^B$,

$$f_\mu^{\mathrm{C}}(\boldsymbol{x}^A + \boldsymbol{x}^B) \leq f_\mu^{\mathrm{C}}(\boldsymbol{x}^A) + f_\mu^{\mathrm{C}}(\boldsymbol{x}^B). \tag{25}$$

Theorem 6. *Given a supermodular set function μ and a set function η, $\eta^\sharp = g(h(\eta))$,*

$$f_\mu^{\mathrm{PS}}(\eta^\sharp) \geq f_\mu^{\mathrm{PS}}(\eta) \tag{26}$$

and given a submodular set function μ and η, $\eta^\sharp = g(h(\eta))$,

$$f_\mu^{\mathrm{PS}}(\eta^\sharp) \leq f_\mu^{\mathrm{PS}}(\eta). \tag{27}$$

Proof. First, we proof the supermodular case. We decompose η into $\eta_{\theta_1}, \ldots, \eta_{\theta_K}$ of Eq. (13). As $\sum_{i=1}^K h(\eta_{\theta_i}) = h(\eta)$, Theorems 3 and 5,

$$f_\mu^{\mathrm{PS}}(\eta) = f_\mu^{\mathrm{PS}}(\sum_{i=1}^K \eta_{\theta_i}) = \sum_{i=1}^K [f_\mu^{\mathrm{PS}}(\eta_{\theta_i})] = \sum_{i=1}^K [f_\mu^{\mathrm{C}}(h(\eta_{\theta_i}))]$$

$$\leq f_\mu^{\mathrm{C}}(\sum_{i=1}^K h(\eta_{\theta_i})) = f_\mu^{\mathrm{C}}(g(h(\eta))) = f_\mu^{\mathrm{PS}}(\eta^\sharp).$$

The submodular case can be proven in a similar manner. □

This theorem shows that if the fuzzy measure is supermoduler (submodular), then the maximum (minimum) output of the Choquet integral with respect to a set function is the Choquet integral.

In example 4 (Table 3), the fuzzy measure μ satisfies the supermoudular condition. Therefore the active elements in the maximal output is a subset of the maximal chain $\{\emptyset, \{1\}, \{1,3\}, \{1,2,3\}\}$. In example 5 (Table 4), the fuzzy measure μ satisfies subadditive but not the submodular condition, for example $\mu(\{1,2\} \cup \{1,3\}) + \mu(\{1,2\} \cap \{1,3\}) \not\leq \mu(\{1,2\}) + \mu(\{1,3\})$. The active elements of the maximal output $\{\{1,2\}, \{1,3\}, \{1,2,3\}\}$ are covered by a union of two maximal chains $\{\emptyset, \{1\}, \{1,2\}, \{1,2,3\}\} \cup \{\emptyset, \{1\}, \{1,3\}, \{1,2,3\}\}$.

3.4 Coalition of Cooperative Game

In this section we define a cooperative game [1] using the Choquet integral with respect to a set function. Let $Y = \{1, \ldots, m\}$ be the set of players and $X = \{1, \ldots, n\}$ be the set of inputs (resources). $\boldsymbol{x}^{\{j\}}, j = 1, \ldots, m$ are player j's input vector of resources and coalition $B \subseteq Y$'s vector is $\boldsymbol{x}^B = \sum_{j \in B} \boldsymbol{x}^{\{j\}}$. Each player and coalition produces output by combining their resources $\boldsymbol{x}^B$. The output is calculated by the Choquet integral with respect to a set function. The fuzzy measure μ is the common to all players and coalitions. The characteristic function π is defined as

$$\pi(B) \equiv f_{\mu}^{\mathrm{PS}}(\eta^B), \forall B \in 2^Y \tag{28}$$

where η^B are the maximum outputs within their resources $\boldsymbol{x}^B$ optimized by the LP model (Eq. (21)).

Theorem 7 (Superadditive of π). *The characteristic function π is superadditive, that is*

$$\pi(A \cup B) \geq \pi(A) + \pi(B), \quad A \cap B = \emptyset, A, B \in 2^Y. \tag{29}$$

Proof. $\eta' = \eta^A + \eta^B$ is a feasible solution of the coalition $A \cup B$. As $f_{\mu}^{\mathrm{PS}}(\eta') = f_{\mu}^{\mathrm{PS}}(\eta^A) + f_{\mu}^{\mathrm{PS}}(\eta^B)$, and the maximization LP model, $f_{\mu}^{\mathrm{PS}}(\eta^{A \cup B}) \geq f_{\mu}^{\mathrm{PS}}(\eta^A) + f_{\mu}^{\mathrm{PS}}(\eta^B)$. □

Example 6 (4 players and 3 resource models). Table 5 shows each player's resources. The resources are unevenly distributed for player 1,2,3 but evenly distributed for player 4. Fuzzy measure μ is $\mu(\{1\}) = 0.2$, $\mu(\{2\}) = 0.1$, $\mu(\{3\}) = 0.1$, $\mu(\{1,2\}) = 0.7$, $\mu(\{1,3\}) = 0.8$, $\mu(\{2,3\}) = 0.2$, $\mu(\{1,2,3\}) = 1.0$, that is superadditive, but not supermodular, $\mu(\{1,2\} \cup \{1,3\}) + \mu(\{1,2\} \cap \{1,3\}) \not\geq \mu(\{1,2\}) + \mu(\{1,3\})$.

Table 5. Resources and shapley value of π

	Player 1($\boldsymbol{x}^{\{1\}}$)	Player 2($\boldsymbol{x}^{\{2\}}$)	Player 3($\boldsymbol{x}^{\{3\}}$)	Player 4($\boldsymbol{x}^{\{4\}}$)
Resource 1 (x_1)	10	4	3	4
Resource 2 (x_2)	1	8	0	4
Resource 3 (x_3)	1	1	9	4
Shapley Value of π	5.175	4.125	4.075	4.225

Table 6 shows the results of all coalitions. Player 4 does not benefit greatly from the coalitions. Player 1 takes a large role when it combines with another player, for example $\mu(\{1,2\}) - \mu(\{1\}) = 0.5$. As player 1 has large amount of resource 1, player 1 has the largest Shapley value [2] (Table 5).

Table 6. Coalitions

$A \setminus B$	$\{1\}$	$\{2\}$	$\{3\}$	$\{4\}$	$\{1,2\}$	$\{1,3\}$	$\{1,4\}$	$\{2,3\}$	$\{2,4\}$	$\{3,4\}$	$\{1,2,3\}$	$\{1,2,4\}$	$\{1,3,4\}$	$\{2,3,4\}$	Y
$\{1\}$	8	0	0	0	3	2	4	0	0	0	0	0	0	0	0
$\{2\}$	0	4	0	0	0	0	0	0	4	0	0	0	0	0	0
$\{3\}$	0	0	6	0	0	0	0	2	0	6	0	0	0	2	0
$\{1,2\}$	1	3	0	0	9	1	5	0	3	0	6	12	3	0	6
$\{1,3\}$	1	0	3	0	2	10	5	0	0	3	8	5	12	0	8
$\{2,3\}$	0	0	0	0	0	0	0	1	0	0	0	0	0	1	0
X	0	1	0	4	0	0	0	7	5	4	3	1	2	11	7
$\pi(B)$	3.1	3.5	3.0	4.0	8.5	9.1	8.3	7.4	7.5	7.0	13.6	13.4	13.7	11.4	17.6

4 Conclusion

We have defined the Choquet integral with respect to a set function and have analyzed the properties, including another representation using the Möbius transformation, supermodularity and so on. We also define a cooperative game model using the integral but the properties of the game have not explored.

References

1. Von Neumann, J., Morgenstern, O.: Theory of Games and Economic Behavior. Princeton Univ Press, Princeton (1944)
2. Shapley, L.: A value for n-person games. Contribution Theory of Games, II. Ann. Math. Stud. **28**, 307–317 (1953)
3. Choquet, G.: Theory of capacities. Ann. Inst. Fourier **5**, 131–295 (1954)
4. Rota, G.-L.C.: On the foundations of combinatorial theory: I.Theory of Möbius functions. Z Wahrscheinlichkeitstheorie und Verwandte Gebiete **2**, 340–368 (1964)
5. Murofushi, T., Sugeno, M.: A theory of fuzzy measure: Representation, the Choquet integral and null sets. J. Math. Anal. Appl. **159**, 532–549 (1991)
6. Denneberg, D.: Non-additive Measure and Integral. Kluwer Academic Publishers, Dordrecht (1994)
7. Fujimoto, K., Murofushi, T.: Some characterizations of the systems represented by choquet and multi-linear functionals through the use of möbius inversion. Int. J. Uncertainty Fuzziness Knowl. Based Syst. **5**, 547–561 (1997)
8. Grabisch, M., Marichal, J., Roubens, M.: Equivalent representation of set functions. Math. Oper. Res. **25**(2), 157–178 (2000)
9. Wang, Z., Leung, K.S., Wong, M., Fang, J.: A new type of nonlinear integrals and the computational algorithm. Fuzzy Sets Syst. **112**, 223–231 (2000)

A 2-Additive Choquet Integral Model for French Hospitals Rankings in Weight Loss Surgery

Brice Mayag(✉)

University Paris-Dauphine, PSL Research University, LAMSADE, CNRS, UMR 7243, Place du Maréchal de Lattre de Tassigny, 75775 Paris Cedex 16, France
brice.mayag@dauphine.fr
http://www.lamsade.dauphine.fr/~mayag/

Abstract. In a context of Multiple Criteria Decision Aid, we present a decision model explaining some French hospitals rankings in weight loss surgery. To take into account interactions between medical indicators, we elaborated a model based on the 2-additive Choquet integral. The reference subset, defined during the elicitation process of this model, is composed by some specific alternatives called binary alternatives. To validate our approach, we showed that the proposed 2-additive Choquet integral model is able to approximate the hospitals ranking, in weight loss surgery, published by the French magazine "Le Point" in August 2013.

Keywords: MCDA · Binary alternatives · Hospitals rankings · Choquet integral · Capacity

1 Introduction

MultiCriteria Decision Aid (MCDA) aims at representing the preferences of a Decision-Maker (DM), or a group of Decision-Makers, over a set of alternatives X evaluated on a finite set of n criteria $N = \{1, \ldots, n\}$ often conflicting. Many softwares implementing MCDA methods have been developed and most of them have proved their efficiency in real applications, e.g. MACBETH [2], MYRIAD [12]. One of the problem statement treated by MCDA is the elaboration of rankings from a set of alternatives. In general, this ranking is assumed to be a weak order $\succsim$.

Since many years, there exist some hospitals rankings published by newspapers. In France, three newspapers publish every year their hospital rankings per surgery specialties. In our knowledge, two other countries publish regularly hospitals rankings:

- *United Stated of America*: these rankings are published each year by a news paper called Usnews[1]. The methodology used is based on the weighted sum and developed by the Research Triangle Institute (RTI international), a scientific organism. The report of 129 pages about this methodology is free available[2].

[1] http://health.usnews.com/best-hospitals.

[2] http://www.usnews.com/pubfiles/BH_2014_Methodology_Report_Final_Jul14.pdf.

J.P. Carvalho et al. (Eds.): IPMU 2016, Part I, CCIS 610, pp. 101–116, 2016.
DOI: 10.1007/978-3-319-40596-4_10

- *United Kingdom*: the rankings are elaborated by the National Health Service (NHS)[3].

From the view of MCDA, we were interested in the methodologies used in French hospital rankings. We studied them in details, but we were disappointed because all the French methodologies are just presented in few lines (not more than a half page) compared to the Usnews methodology which is presented in more than 100 pages. Furthermore there is no relevant information concerning MCDA aspects. The main reason is that, behind these rankings, there are only journalists (François Malye and Jérôme Vincent for "Le point") and some very small consulting companies (Le Guide santé for "Le Figaro Magazine" and Santé Value for "Le Nouvel Observateur") without knowledge about good best practices of MCDA. In general, to improve their reputation, the hospitals need and wish to know each year their rank in the published hospital rankings. Most of these hospitals choose to advertise this rank, when they are good, in their website. Health governments agencies also can use these rankings to identify which are the "weak" hospitals.

We propose here a way to improve a ranking in a given medical specialty. We choose the weight loss surgery as an example for which it can be important to take into account interactions between criteria. The elaborated model is based on the 2-additive Choquet integral, and it was built independently of the data used by the magazines (except the definition of criteria). To do this, we assume that the DM is able to give his preferences on the set of fictitious hospitals called binary alternatives. This latter is an alternative which takes either the neutral value **0** for all criteria, or the neutral value **0** for all criteria except for one or two criteria for which it takes the satisfactory value **1**. Our aim is only to show the possibility to elaborate a transparent model, which is understandable by the readers or by the experts.

The paper is organized as follows: we present in Sect. 2 the three main French hospitals rankings, especially in weight loss surgery. The 2-additive Choquet integral is introduced in Sect. 3 and our model related to the weight loss surgery is proposed in Sect. 4.

2 About French Hospital Rankings

In France, hospitals rankings are published each year by three magazines: "Le Nouvel observateur"[4], "Le Point"[5] and "Le Figaro Magazine"[6]. To establish these rankings, they manipulate data coming from some official databases like HOSPIDIAG[7]. This latter, a tool developed by the national performance support agency (Agence Nationale d'Appui à la Performance : ANAP), sheds

[3] http://www.nhs.uk.
[4] http://classement-hopitaux.nouvelobs.com/.
[5] http://hopitaux.lepoint.fr/.
[6] http://sante.lefigaro.fr.
[7] http://hospidiag.atih.sante.fr.

light on a given facility, bringing together data from different databases (PMSI, annual institutional statistics, etc.) in a single tool [3]. The databases contain around eighty indicators which are likely to be filled each year by all the hospitals. In French health system, there are approximately 1600 hospitals classified as public, nonprofit private and commercial private.

All the three newspapers propose a ranking per surgery specialty, for instance a ranking of weight loss surgery. Our analysis in this paper is focused on weight loss surgery. The remarks and comments developed in this section are valid for all the specialties.

2.1 Weight Loss Surgery

Bariatric surgery[8] (weight loss surgery) includes a variety of procedures performed on people who are obese. Weight loss is achieved by reducing the size of the stomach with a gastric band or through removal of a portion of the stomach (sleeve gastrectomy or biliopancreatic diversion with duodenal switch) or by resecting and re-routing the small intestines to a small stomach pouch (gastric bypass surgery).

To identify the "best" hospitals in weight loss surgery, the magazines combine a part of the following indicators:

1. (CR_1) *Volume of activity:* it is the number of stays of all patients with respect to the value of care and some homogeneous price. This criterion has to be maximized.
2. (CR_2) *Activity*: number of procedures performed during one year. "Le Point" supposes that if a hospital has a good score on activity then its teams are more trained and often have good results. Therefore this criterion has to be maximized. This opinion is not totally shared by some other experts who estimate that a good score on the activity of a hospital does not imply necessarily that its teams are best. In this case, one should also investigate if this hospital does not focus on getting grants of the government because in France some grants depend on the activity.
3. (CR_3) *Average Length Of Stay (ALOS)*: a mean calculated by dividing the sum of inpatient days by the number of patients admissions with the same diagnosis-related group classification. A variation in the calculation of ALOS considers only the length of stay during the period under analysis. If a hospital is more organized in terms of resources then its ALOS score should be low.
4. (CR_4) *Notoriety*: Its corresponds to the reputation and attractiveness of the hospital.
   For "the Nouvel Observateur", the attractiveness of the hospital depends on the distance between the hospital and the patient's home. This distance is considered significant if it is more than fifty kms. Its reputation reflects the gradual isolation of patients: the more they come from far away, the more the reputation of the institution is important.

[8] http://en.wikipedia.org/wiki/Bariatric_surgery.

The notoriety indicator of "Le Point" is a percentage of patients treated in the hospital but living in another French administrative department. More the percentage increases, more the hospital is attractive.

5. (CR_5) *Heaviness*: it is a percentage measuring the level of resources consumed (equipment, staff, ...) in the hospital.
6. (CR_6) *Quality score of French National Authority for Health (HAS)*[9]: It is the score (between • and •••••) obtained by the hospital after the accreditation and quality visit made by the experts of HAS.
7. (CR_7) *% of By-Pass*: It is the percentage of surgical procedures using gastric bypass system. This criterion has to be maximized.

Table 1. The best 20 hospitals in Weight loss surgery (2013). Source: "Le Nouvel Observateur" [14]

Hospitals	CR_1	CR_3	CR_7	CR_5	CR_4	F_O
Georges-Pompidou	406	5.2	55	77	95	**19.3**
Bichat	203	7.8	75	83	94	**18.9**
Ambroise-Paré	193	6.6	90	83	94	**18.7**
Strasbourg	330	6.2	84	79	45	**18.2**
Nice	351	6.5	94	79	20	**18.1**
Nancy	230	6.9	87	81	76	**17.9**
Louis-Mourier	154	5.0	81	81	27	**17.9**
Pitié-Salpetrière	127	6.0	75	79	92	**17.8**
Laon	299	1.8	0	54	58	**17.7**
Lille	233	6.2	68	83	30	**17.4**
Colmar	192	3.5	97	77	19	**17.4**
Conception	287	3.1	28	63	22	**17.3**
Caen	152	6.7	89	79	63	**17.1**
Toulouse	173	4.3	63	77	87	**17.0**
Antibes	181	5.6	96	77	23	**16.9**
Edouard-Herriot	89	4.9	52	81	38	**16.9**
Havre	115	2.7	78	74	9	**16.5**
Jean-Verdier	116	6.7	44	79	32	**16.4**
Timone adultes	69	5.0	32	81	36	**16.3**
Orleans	131	6.1	69	81	41	**16.4**

[9] French National Authority for Health (HAS) aims to improve quality and safety of healthcare. The objectives are to accredit health care organizations and health professionals, to produce guidelines for health professionals (practices, public health, patient safety), to develop disease management for chronic conditions, to advise decision makers on health technologies (drugs, devices, procedures), and to inform professionals, patients, and the public.

8. (CR_8) *Technicality*: this particular indicator measures the ratio of procedures performed with an efficient technology compared to the same procedures performed with obsolete technology. The higher the percentage is, the more the team is trained in advanced technologies or complex surgeries.

Remark 1

- *Nothing is said about how and why these eight indicators are selected.*
- *"Le Nouvel Observateur" uses the term activity as a composite indicator of ALOS (CR_3) and Volume of activity (CR_1).*

2.2 The 2013 Results

The rankings given by "Le Nouvel observateur" [14] take into account, in the same tables, both public and private hospitals. They argue that this logic is in spirit of their readers. In terms of MCDA, this justification of the choice of this set of alternatives appears weak and seems to be only a "marketing argument". Table 1 presents the ranking of only twenty hospitals (among the first hundred hospitals evaluated)

Table 2. The best 20 hospitals in Weight loss surgery (2013). Source: "Le Point" [15]

Hospitals	CR_2	CR_4	CR_3	CR_8	F_P
Bichat	372	80	7.8	94	**17.84**
Nice	253	19	8.2	95	**17.59**
Nancy	208	60	8	90	**17.37**
Ambroise-Paré	140	85	6.5	96	**17.23**
Colmar	165	14	3.8	99	**17.20**
Caen	167	47	6.7	96	**17.14**
Strasbourg	289	25	6.3	82	**17.13**
Georges-Pompidou	394	80	5.5	56	**17.06**
Lille	247	18	4.8	63	**17.02**
Antibes	156	13	5.5	96	**16.75**
Orleans	167	35	6.7	86	**16.66**
Rouen	237	29	5.1	48	**16.55**
Jean-Verdier	174	40	9.7	82	**16.45**
Conception	332	19	3.8	24	**16.44**
Louis-Mourier	166	51	5.3	86	**16.36**
Poissy/St Germain	192	34	4.1	60	**16.30**
Montpellier	297	25	5.6	33	**16.24**
Toulouse	181	73	4.6	50	**15.94**
Amiens	170	28	3.8	10	**15.63**
Laon	242	23	1.4	0	**15.54**

in weight loss surgery published by "Le Nouvel observateur" in 2013. These hospitals are evaluated on five indicators: Volume of activity (CR_1), ALOS (CR_3), % of By-Pass (CR_7), Heaviness (CR_5) and Notoriety (CR_4). In their methodology, they mention that they chose indicators which are most significant in terms of medical innovation, but nothing is said about the concrete selection of such indicators. The last column, F_O, concerns the overall score given to each hospital by an aggregation function not defined by the magazines. Indeed, nothing is said about this function and how the overall score of each hospital is computed. We think that it could be a simple weighted sum.

"Le Point" [15] have analyzed 952 hospitals in their rankings. Just 50, 40, 30, 25 or 20 best hospitals per specialty were published. In Table 2, the ranking published in 2013 concerns the 20 best hospitals in weight loss surgery evaluated on Activity (CR_2), (Notoriety) (CR_4); ALOS (CR_3) and Technicality: (CR_8). The last column of the table refers to the overall score assigned to each hospital and obtained by using an aggregation function F_P. Like the previous newspaper, nothing is said about this function and nothing about the elaboration of criteria.

Table 3. The best 20 hospitals in Weight loss surgery (2013). Source: "Le Figaro Magazine" [13]

Hospitals	CR_2	CR_6
Georges-Pompidou	878	•••••
Bichat	384	••••
Saint-Louis	285	••••
Rouen	300	••••
Laon	277	••
Lille	271	••••
Caen	179	••
Nantes	175	••
Limoges	103	•••
Rennes	89	••
Montpellier	353	••
Nice	263	••
Orleans	206	••••
Tours	122	•••
Jean-Mermoz Lyon	312	••
Sens	140	•••
Nancy	305	••
Colmar	169	••
Toulouse	352	••••
Bordeaux	133	••

They only indicate that it is a weighted sum. There is no information concerning the weights associated to the criteria.

Among 1308 hospitals analyzed by the last newspaper, “Le Figaro Magazine” [13], only 830 have been evaluated. The rankings published concern the 10 best hospitals per specialty and per French region. We show in Table 3 some best hospitals in eight regions. The criteria used are: Activity ($CR_{\mathbf{2}}$) and Quality score of French National Authority for Health (HAS) ($CR_{\mathbf{6}}$). The ranking is based on a lexicographic order ($CR_{\mathbf{6}} \ll CR_{\mathbf{2}}$), but nothing about how these rankings were elaborated.

We are not really surprised if the information about the methodologies used by these three newspaper are poor and not available. Indeed, in France, the sales of newspapers devoted to hospitals ranking are often the best of the year. So there exist a real competition between the three organisms. Therefore, each of magazine tries to keep secret its methodology.

3 MCDA Concepts

We think that, the *elaboration of a hospital ranking* is a practical application where a MCDA process could be applied successfully. Let us give below some suggestions indicating how to proceed.

As indicated in [4], we have to start with a a number of crucial questions when trying to build an evaluation (ranking) model in MCDA [6,7]. These questions, known as good practices, are:

1. What is the definition of objects to be evaluated?
2. What is the purpose of the model? Who will use it?
3. How to structure objectives?
4. How to achieve a “consistent family of criteria”?
5. How to take uncertainty, imprecision, and inaccurate definition into account? All the French hospital ranking fail this last point.

After answering these questions, the choice of the suitable MCDA method will be another problem. Some methodologies are based on an additive model, in particular the weighted sum (e.g. methodologies of “Le Point”), because this function is simple and understandable by many persons who are not experts in MCDA.

We can observe that the ranking $\succsim_{LP}$ proposed, from the set of the 20 hospitals treating the weight loss surgery, by the magazine “Le Point” (see Table 2) can be obtained by using an additive model i.e. there exist partial utility functions $U_{CR_i} : CR_i \to \mathbb{R}$, $i \in \{2,3,4,8\}$ such that

$$\text{hospital } a \succsim_{LP} \text{hospital } b \Leftrightarrow \sum_{i \in \{2,3,4,8\}} U_{CR_i}(\text{hospital } a_{CR_i}) \geq \sum_{i \in \{2,3,4,8\}} U_{CR_i}(\text{hospital } a_{CR_i}) \quad (1)$$

where hospital a_{CR_i} is the value of the hospital a on the criterion CR_i.

Table 4. Partial utility functions for an additive model associated to $\succsim_{LP}$

CR_2	CR_4	CR_3	CR_8
$U_{CR_2}(156) = 0.01$	$U_{CR_4}(13) = 0.3575$	$U_{CR_3}(1.4) = 0.15$	$U_{CR_8}(0) = 0$
$U_{CR_2}(165) = 0.02$	$U_{CR_4}(14) = 0.395$	$U_{CR_3}(3.8) = 0.14$	$U_{CR_8}(10) = 0.1425$
$U_{CR_2}(166) = 0.03$	$U_{CR_4}(18) = 0.405$	$U_{CR_3}(4.1) = 0.13$	$U_{CR_8}(24) = 0.1525$
$U_{CR_2}(167) = 0.15$	$U_{CR_4}(19) = 0.415$	$U_{CR_3}(4.6) = 0.12$	$U_{CR_8}(33) = 0.1625$
$U_{CR_2}(170) = 0.16$	$U_{CR_4}(23) = 0.425$	$U_{CR_3}(4.8) = 0.11$	$U_{CR_8}(48) = 0.2975$
$U_{CR_2}(174) = 0.17$	$U_{CR_4}(25) = 0.4825$	$U_{CR_3}(5.1) = 0.1$	$U_{CR_8}(50) = 0.3075$
$U_{CR_2}(181) = 0.18$	$U_{CR_4}(28) = 0.4925$	$U_{CR_3}(5.3) = 0.09$	$U_{CR_8}(56) = 0.3175$
$U_{CR_2}(192) = 0.23$	$U_{CR_4}(29) = 0.5025$	$U_{CR_3}(5.5) = 0.08$	$U_{CR_8}(60) = 0.3275$
$U_{CR_2}(208) = 0.33$	$U_{CR_4}(34) = 0.5125$	$U_{CR_3}(5.6) = 0.07$	$U_{CR_8}(63) = 0.3375$
$U_{CR_2}(237) = 0.34$	$U_{CR_4}(35) = 0.5225$	$U_{CR_3}(6.3) = 0.06$	$U_{CR_8}(82) = 0.5275$
$U_{CR_2}(242) = 0.35$	$U_{CR_4}(40) = 0.5325$	$U_{CR_3}(6.5) = 0.05$	$U_{CR_8}(86) = 0.5375$
$U_{CR_2}(247) = 0.4175$	$U_{CR_4}(47) = 0.5425$	$U_{CR_3}(6.7) = 0.04$	$U_{CR_8}(90) = 0.5475$
$U_{CR_2}(253) = 0.455$	$U_{CR_4}(51) = 0.5525$	$U_{CR_3}(7.8) = 0.03$	$U_{CR_8}(94) = 0.5575$
$U_{CR_2}(289) = 0.465$	$U_{CR_4}(60) = 0.5625$	$U_{CR_3}(8.0) = 0.02$	$U_{CR_8}(95) = 0.8025$
$U_{CR_2}(297) = 0.475$	$U_{CR_4}(73) = 0.5725$	$U_{CR_3}(8.2) = 0.01$	$U_{CR_8}(96) = 0.8125$
$U_{CR_2}(332) = 0.5125$	$U_{CR_4}(80) = 0.5825$	$U_{CR_3}(9.7) = 0$	$U_{CR_8}(99) = 1$
$U_{CR_2}(372) = 0.5225$	$U_{CR_4}(85) = 0.8$		
$U_{CR_2}(374) = 0.535$			
$U_{CR_2}(394) = 0.545$			

Indeed the overall score associated to each hospital in this ranking $\succsim_{LP}$ (see Table 5) can be computed by choosing the partial utility functions U_{CR_i}, $i \in \{2, 3, 4, 8\}$ given in Table 4. We used a simple linear programming to obtain these parameters of an additive model, but there exist some MCDA methodologies like UTA methods [21] which can help to find a suitable additive model corresponding to a given ranking.

It is usually known that, an additive model like the weighted sum is unable to take into account some phenomena like interactions between criteria described in the following Example:

Example 1. *Four hospitals are evaluated on three criteria Notoriety (CR_4), Average Lenght of Stay (ALOS) (CR_3) and Activity (CR_2) defined above in Sect. 2.1. The performances of each hospital on the three criteria are given by the following Table 6. We suppose that the DM gives these preferences:*

- *Hospital 1 is strictly preferred to the hospital 2 (if ALOS is "weak", it is preferable to have a hospital with good evaluation in Activity).*
- *Hospital 4 is strictly preferred to hospital 3 (If ALOS is "good", he prefers in this case a hospital with good evaluation in Notoriety).*

Table 5. The ranking $\succsim_{LP}$ computed with the utility functions given in Table 4

Hospitals	CR_2	CR_4	CR_3	CR_8	F_P
1: Bichat	372	80	7.8	94	**1.6925**
2: Nice	253	19	8.2	95	**1.6825**
3: Nancy	208	60	8	90	**1.6725**
4: Ambroise-Paré	140	85	6.5	96	**1.6625**
5: Colmar	165	14	3.8	99	**1.555**
6: Caen	167	47	6.7	96	**1.545**
7: Strasbourg	289	25	6.3	82	**1.535**
8: Georges-Pompidou	394	80	5.5	56	**1.525**
9: Lille	247	18	4.8	63	**1.27**
10: Antibes	156	13	5.5	96	**1.26**
11: Orleans	167	35	6.7	86	**1.25**
12: Rouen	237	29	5.1	48	**1.24**
13: Jean-Verdier	174	40	9.7	82	**1.23**
14: Conception	332	19	3.8	24	**1.22**
15: Louis-Mourier	166	51	5.3	86	**1.21**
16: Poissy/St Germain	192	34	4.1	60	**1.2**
17: Montpellier	297	25	5.6	33	**1.19**
18: Toulouse	181	73	4.6	50	**1.18**
19: Amiens	170	28	3.8	10	**0.9350**
20: Laon	242	23	1.4	0	**0.9250**

Table 6. Evaluations of four hospitals on Notoriety, ALOS and Technicality.

	$CR_4 \equiv$ Notoriety	$CR_3 \equiv$ ALOS	CR_2 Activity
Hospital 1	35	8	9000
Hospital 2	37	8	8900
Hospital 3	35	3	9000
Hospital 4	37	3	8900

If w_i and u_i are respectively the weight and the partial function associated to the criterion CR_i, $i \in \{2,3,4\}$, then we get the following system:

$$w_1\ u_1(35) + w_2\ u_2(8) + w_3\ u_3(9000) > w_1\ u_1(37) + w_2\ u_2(8) + w_3\ u_3(8900) \quad (2)$$

$$w_1\ u_1(37) + w_2\ u_2(3) + w_3\ u_3(8900) > w_1\ u_1(35) + w_2\ u_2(3) + w_3\ u_3(9000) \quad (3)$$

The combination of these two equations leads to a contradiction.

To model these type of interactions, we propose to elaborate a 2-additive Choquet integral model [5,17,18] in the next section by using the classic Multi Attribute Utility Theory (MAUT) approach [8]. The 2-additive Choquet integral is a particular case of the well known Choquet integral [17,18]. Its main property is to model interactions between two criteria. These interactions are simple and more meaningful than those produced by using the general Choquet integral. This aggregation function is based on the notion of *capacity* μ defined as a set function from the powerset of criteria 2^N to $[0,1]$ such that:

1. $\mu(\emptyset) = 0$
2. $\mu(N) = 1$
3. $\forall A, B \in 2^N$, $[A \subseteq B \Rightarrow \mu(A) \leq \mu(B)]$ (monotonicity).

A capacity μ on N is said to be *2-additive* if its *Möbius transform* $m : 2^N \rightarrow \mathbb{R}$ defined by

$$m(T) := \sum_{K \subseteq T} (-1)^{|T \setminus K|} \mu(K), \forall T \in 2^N. \tag{4}$$

satisfies the following two conditions:

- For all subset T of N such that $|T| > 2$, $m(T) = 0$;
- There exists a subset B of N such that $|B| = 2$ and $m(B) \neq 0$.

Given an alternative
$x := (x_1, ..., x_n) \in X$, the expression of the 2-additive Choquet integral is given by [11]:

$$C_\mu(u(x)) = \sum_{i=1}^{n} v_i u_i(x_i) - \frac{1}{2} \sum_{\{i,j\} \subseteq N} I_{ij} |u_i(x_i) - u_j(x_j)| \tag{5}$$

where

- For all $i \in N$, $u_i : X_i \rightarrow \mathbb{R}_+$ is an utility function associated to the attribute X_i;
- $u(x) = (u_1(x_1), \ldots, u_n(x_n))$ for $x = (x_1, ..., x_n) \in X$;
- $v_i = \sum_{K \subseteq N \setminus i} \frac{(n - |K| - 1)!|K|!}{n!} (\mu(K \cup i) - \mu(K))$ is the importance of criterion i corresponding to the Shapley value of μ [20];
- $I_{ij} = \mu_{ij} - \mu_i - \mu_j$ is the interaction index between the two criteria i and j [10,19].

Therefore the 2-additive Choquet integral appears as a good compromise between the arithmetic mean and the Choquet integral.

We simplify our notation for a capacity μ by using the following shorthand: $\mu_i := \mu(\{i\})$, $\mu_{ij} := \mu(\{i, j\})$ for all $i, j \in N$, $i \neq j$. Whenever we use i and j together, it always means that they are different.

4 A 2-Additive Choquet Integral Model Explaining the Hospitals Rankings

Let us consider as the set of alternatives X, the set of French hospitals evaluated, like in "Le Point", on the following four criteria:

- *Criterion 1:* Activity (CR_2);
- *Criterion 2:* Notoriety (CR_4);
- *Criterion 3:* Average Lenght of Stay (ALOS) (CR_3);
- *Criterion 4:* Technicality (CR_8);

We denote this set of criteria by $N = \{1, 2, 3, 4\}$ and the attribute associated to the criterion i by X_i, $i = 1, \ldots, 4$. We assume that each attribute X_i is defined by a scale $[L_i; U_i]$ with L_i the lower bound of the attribute (the worst element), and U_i the upper bound (the best element). We set $X_1 = [0; 500]$, $X_2 = [0; 100]$; $X_3 = [0; 10]$ and $X_1 = [0; 100]$. In our model, the set of alternatives corresponds to the Cartesian product $X = X_1 \times X_2 \times X_3 \times X_4$, i.e. a hospital h is equivalent to an alternative $h = (h_1, h_2, h_3, h_4)$ where h_i is the evaluation of h on the criterion i.

To compute the partial utility function $u_i : X_i \to \mathbb{R}$ associated to the criterion i, we choose the monotone normalization formula described as follows: Given a hospital $h = (h_1, h_2, h_3, h_4)$,

$$\begin{cases} u_i(h_i) = \dfrac{h_i}{U_i} & \text{if } i \text{ is to be maximized } (\textit{criteria } 1,\ 2 \text{ and } 4) \\ u_i(h_i) = 1 - \dfrac{h_i}{U_i} & \text{if } i \text{ is to be minimized } (\textit{criterion } 3) \end{cases} \tag{6}$$

By using an appropriate linear program, it is not difficult to check that, the above ranking $\succsim_{LP}$ is not representable by a weighted sum based on these normalization transformation. There exist MCDA methodologies like the MACBETH method [1] computing utility functions between 0 and 1, and ensuring to obtain commensurable scales, which is an important property for the use of Choquet integral [11].

Let us now assume that, the DM is able to identify on each attribute X_i, two reference levels $\mathbf{1}_i$ and $\mathbf{0}_i$ corresponding respectively to his satisfactory and neutral levels on the criterion i. We set for the convenience $u_i(\mathbf{1}_i) = 1$ and $u_i(\mathbf{0}_i) = 0$. The Table 7 below gives the reference levels associated to each criterion. These reference elements could be different to the bounds defined in X_i, even if it is not the case in our model.

The next step of our approach consists to ask to the DM some preference information on a reference subset of alternatives, possibly fictitious, called the set of binary alternatives. A *binary action* is an element of the set $\mathcal{B} = \{\mathbf{0}_N,\ (\mathbf{1}_i, \mathbf{0}_{N-i}),\ (\mathbf{1}_{ij}, \mathbf{0}_{N-ij}),\ i, j \in N\} = \{a_0,\ a_i,\ a_{ij},\ i, j \in N\} \subseteq X$ where

- $\mathbf{0}_N = (\mathbf{1}_\emptyset, \mathbf{0}_N) := a_0$ is a hospital considered neutral on all the four criteria.

Table 7. The reference levels associated to each criterion

	1-Activity	2-Notoriety	3-ALOS	4-Technicality
Satisfactory level $\mathbf{1}_i$	500	100	0	100
Neutral level $\mathbf{0}_i$	0	0	10	0

- $(\mathbf{1}_i, \mathbf{0}_{N-i}) := a_i$ is a hospital considered satisfactory on criterion i and neutral on the other criteria.
- $(\mathbf{1}_{ij}, \mathbf{0}_{N-ij}) := a_{ij}$ is a hospital considered satisfactory on criteria i and j and neutral on the other criteria.

From $\mathcal{B} = \{a_0; a_1; a_2; a_3; a_4; a_{12}; a_{13}; a_{14}; a_{23}; a_{24}; a_{34}\}$, we get the following preferences of the DM called in [18] an ordinal information:

- A satisfactory hospital on Activity and ALOS (neutral on the other criteria) is better than a satisfactory hospital on Notoriety and Technicality (neutral on the other criteria), i.e. a_{13} is preferred to a_{24}.
- A satisfactory hospital on Activity (neutral on the other criteria) is better that a satisfactory hospital only on Notoriety and ALOS (neutral on the other criteria), i.e. a_1 is preferred to a_{23}.
- A hospital only better in Activity (neutral on the other criteria) is judged indifferent to a hospital better on Activity and ALOS (neutral on the other criteria), i.e. a_1 is indifferent to a_{13}.
- If a hospital is fully satisfying on the criterion Technicality (neutral on the other criteria), then it will be preferred to a hospital satisfactory on Notoriety (neutral on the other criteria), i.e. a_4 is preferred to a_2.
- A satisfactory hospital on Activity and Technicality (neutral on the other criteria) is better than a satisfactory hospital on ALOS and Technicality (neutral on the other criteria), i.e. a_{14} is preferred to a_{34}.

All these preferences are then translated in a linear programming in order to test their compatibility with a 2-additive Choquet integral. The constraints of this program are the following:

$$\begin{cases} C_\mu(u(a_{13})) > C_\mu(u(a_{24})) \\ C_\mu(u(a_1)) > C_\mu(u(a_{23})) \\ C_\mu(u(a_1)) = C_\mu(u(a_{13})) \\ C_\mu(u(a_4)) > C_\mu(u(a_2)) \\ C_\mu(u(a_{14})) > C_\mu(u(a_{34})) \\ \text{The 2-additive constrains of } \mu \end{cases} \tag{7}$$

If this preference information is not consistent, the axiomatization of a 2-additive Choquet integral proposed in [18] could be used as an alternative to restore the consistency of the preferences (see [16] for more details).

Table 8. The parameters computed for our 2-additive Choquet integral model in weight loss surgery

A 2 additive capacity μ	$\mu_1 = 0.2$; $\mu_2 = 0$; $\mu_3 = 0.1$; $\mu_4 = 0.1$;
	$\mu_{12} = 0.9$; $\mu_{13} = 0.2$; $\mu_{14} = 0.3$; $\mu_{23} = 0.1$;
	$\mu_{24} = 0.1$; $\mu_{34} = 0.2$
Importance index v_i	$v_1 = 0.5$; $v_2 = 0.35$; $v_3 = 0.05$; $v_4 = 0.1$
interaction index I_{ij}	$I_{12} = 0.7$; $I_{13} = -0.1$; $I_{14} = I_{23} = I_{24} = I_{34} = 0$

The preference information on $\mathcal{B}$ introduced previously is consistent with a 2-additive Choquet integral model with the following parameters (the values of the capacity on pairs and singletons, importance index of criteria v_i and the interaction index I_{ij}, $i, j = 1, 2, 3, 4$) given in Table 8.

In this 2-additive Choquet integral model, Activity and Notoriety are the most important criteria, while the criteria ALOS and Technicality are not very

Table 9. The ranking of the best 20 hospitals in Weight loss surgery obtained by a 2-additive model

Hospitals	C_μ
1: Bichat	**4.212**
2: Ambroise-Paré	**4.12**
3: Colmar	**3.9**
4: Nancy	**3.658**
5: Georges-Pompidou	**3.637**
6: Antibes	**3.566**
7: Nice	**3.539**
8: Caen	**3.507**
9: Laon	**3.391**
10: Louis-Mourier	**3.384**
11: Strasbourg	**3.377**
12: Jean-Verdier	**3.209**
13: Toulouse	**3.104**
14: Orleans	**3.077**
15: Conception	**3.015**
16: Lille	**2.912**
17: Poissy/St Germain	**2.807**
18: Montpellier	**2.585**
19: Amiens	**2.550**
20: Rouen	**2.437**

important. Furthermore Activity and Notoriety interact positively i.e. the DM is satisfied when a hospital has a good evaluations on these two criteria. The criteria Activity and ALOS are judged redundant (negative interaction) i.e. the DM is satisfied if one of these two criteria is fully satisfactory. By applying this model to the the data of the magazine "Le Point", we get the ranking presented in Table 9.

We can noticed that Bichat remains the best hospital in weight loss surgery because it has good scores in Activity and Notoriety. The hospital Amiens remains one of the last hospitals in the ranking, due to its low scores in all the criteria. The hospital Laon is now at the middle of the ranking because it has better evaluation on the criterion ALOS and an average score on Activity. Since these two criteria are redundant, it not surprising to have this rank.

The Kendall tau distance[10] between the ranking of the magazine "Le Point" (see Table 2) and the ranking produced by our model (see Table 9) is 0.2263. This value indicates that 23 % of pairs differ in ordering between the two rankings. Therefore we think that a 2-additive Choquet integral is able to better explain this ranking in weight loss surgery provided by "Le Point".

5 Conclusion

The model we proposed is independent to any published data of hospitals rankings in weight loss surgery. It suppose that, in order to express his preferences, the DM is able to understand the meaning of binary alternatives related to the four chosen criteria. Of course, this model can be improved by the DM and the assumptions we made (transformation scale, reference subset, ...) can be modified by the experts of this specialty (doctors, nurses, ...).

Since the 2-additive Choquet integral is a good compromise between the weighted sum and the general Choquet integral, we suggest to use this type of model in the elaboration of hospitals rankings when the DM suspects the existence of some interactions between criteria. The study of the robustness and validation of our model will be the next steps in the future works. We think that, these two aspects must include the real expert in weight loss surgery.

We recommend to the French magazines to publish hospitals rankings in a transparent way by using a suitable model such as a 2-additive Choquet integral where the DM is the reader. Hence, this model can be viewed as an interactive decision process where each reader can add in the proposed model its preferences (in terms of Notoriety for instance), in order to get its own hospital ranking.

[10] The Kendall tau ranking distance between two rankings R_1 and R_2 is the quantity $K(R_1, R_2) = \frac{|\{(i,j) : i < j, (\tau_1(i) < \tau_1(j) \wedge \tau_2(i) > \tau_2(j)) \vee (\tau_1(i) > \tau_1(j) \wedge \tau_2(i) < \tau_2(j))\}|}{n/2(n-1)}$ where $\tau_1(i)$ et $\tau_2(i)$ are the ranks of the element i in the rankings R_1 and R_2 respectively.

References

1. Bana e Costa, C.A., De Corte, J.M., Vansnick, J.C.: On the mathematical foundations of MACBETH. In: Figueira, J., Greco, S., Ehrgott, M. (eds.) Multiple Criteria Decision Analysis: State of the Art Surveys, pp. 409–437. Springer, New York (2005)
2. Bana e Costa, C.A., Vansnick, J.C.: The MACBETH approach: basic ideas, software and an application. In: Meskens, N., Roubens, M. (eds.) Advances in DecisionAnalysis, pp. 131–157. Kluwer Academic Publishers, Dordrecht (1999)
3. Baron, S., Duclos, C., Thoreux, P.: Orthopedics coding and funding. Orthop. Traumatol. Surg. Res. **100**(1 Supplement), 99–106 (2014). 2013 Instructional Course Lectures (SoFCOT)
4. Billaut, J.-C., Bouyssou, D., Vincke, P.: Should you believe in the shanghai ranking? - an mcdm view. Scientometrics **84**(1), 237–263 (2010)
5. Bouyssou, D., Couceiro, M., Labreuche, C., Marichal, J.-L., Mayag, B.: Using choquet integral in machine learning: What can MCDA bring? In: DA2PL 2012 Workshop: From Multiple Criteria Decision Aid to Preference Learning, Mons, Belgique (2012)
6. Bouyssou, D., Marchant, T., Pirlot, M., Tsoukiàs, A.: Evaluation and Decision Models: A Critical Perspective. Kluwer Academic, Dordrecht (2000)
7. Bouyssou, D., Marchant, T., Pirlot, M., Tsoukiàs, A., Vincke, P.: Evaluation and Decision Models: Stepping Stones for the Analyst. Springer Verlag, New York (2006)
8. Dyer, J.S.: MAUT - multiattribute utility theory. In: Figueira, J., Greco, S., Ehrgott, M. (eds.) Multiple Criteria Decision Analysis: State of the Art Surveys, pp. 265–285. Springer Verlag, Boston, Dordrecht, London (2005)
9. Fürnkranz, J., Hüllermeier, E.: Preference Learning. Springer, New York (2011)
10. Grabisch, M.: k-order additive discrete fuzzy measures and their representation. Fuzzy Sets Syst. **92**, 167–189 (1997)
11. Grabisch, M., Labreuche, C.: A decade of application of the Choquet, Sugeno integrals in multi-criteria decision aid. 4OR **6**, 1–44 (2008)
12. Labreuche, C. Le Huédé, F.: Myriad: a tool suite for MCDA. In: International Conference of the Euro Society for Fuzzy Logic and Technology(EUSFLAT), Barcelona, Spain, September, 7–9 2005, pp. 204–209 (2005)
13. De Linares, J.: Hôpitaux et cliniques: Le palmarès 2013. Le Figaro Magazine, pp. 39–49, 21 June 2013. http://sante.lefigaro.fr/actualite/2013/06/23/20819-palmares-2013-hopitaux-cliniques
14. De Linares, J.: Le palmarès national 2013: Hôpitaux et cliniques. Le Nouvel Observateur, pp. 77–117, 28 November 2013. http://classement-hopitaux.nouvelobs.com/
15. Malye, F., Vincent, J.: Hôpitaux et cliniques: Le palmarès 2013, pp. 86–142, 22 August 2013. http://hopitaux.lepoint.fr/
16. Mayag, B., Grabisch, M., Labreuche, C.: An interactive algorithm to deal with inconsistencies in the representation of cardinal information. In: Hüllermeier, E., Kruse, R., Hoffmann, F. (eds.) IPMU 2010. CCIS, vol. 80, pp. 148–157. Springer, Heidelberg (2010)
17. Mayag, B., Grabisch, M., Labreuche, C.: A characterization of the 2-additive Choquet integral through cardinal information. Fuzzy Sets Syst. **184**(1), 84–105 (2011)
18. Mayag, B., Grabisch, M., Labreuche, C.: A representation of preferences by the Choquet integral with respect to a 2-additive capacity. Theory Decis. **71**(3), 297–324 (2011)

19. Murofushi, T., Soneda, S., Techniques for reading fuzzy measures (III): interaction index.In: 9th Fuzzy System Symposium, Sapporo, Japan, pp. 693–696, May 1993. (in Japanese)
20. Shapley, L.S.: A value for n-person games. In: Kuhn, H.W., Tucker, A.W. (eds.) Contributions to the Theory of Games, Vol. II. Annals of Mathematics Studies, vol. 28, pp. 307–317. Princeton University Press, Princeton (1953)
21. Siskos, Y., Grigoroudis, E., Matsatsinis, N.F.: UTA methods. In: Figueira, J., Greco, S., Ehrgott, M. (eds.) Multiple Criteria Decision Analysis: State of the Art Surveys, pp. 297–343. Springer, New York (2005)

Benchmarking over Distributive Lattices

Marta Cardin(✉)

Department of Economics, Università Ca' Foscari Venezia, Venice, Italy
mcardin@unive.it

Abstract. We provides an axiomatic characterization of preorders in lattices that are representable as benchmarking procedure. We show that the key axioms are related to compatibility with lattice operations.

This paper propose also a characterization and a generalization of Sugeno integral in a ordinal framework.

Keywords: Lattice · Benchmark · Congruence · Compatible preorder · Aggregation function · Sugeno integral

1 Introduction

Benchmarking is the process of comparing objects according to a set of benchmarks where a benchmark is an important accomplishment. An object is considered better than another object in a benchmarking method if and only if it achieves more benchmarks.

A benchmark-based approach is a very natural approach in many situations and it is used in many settings as noted by Chambers and Miller in [5,6].

It may be used when comparing candidates in a selection process, in this case benchmarks can refer to qualifications or to experience. When comparing scholars by their citation profiles as in [5] an accomplishment is a pair of numbers (x, y) where x is the number of publications with at least y citations.

Benchmarking provides information about quality of services perceived by customers. As an example hotel classification systems are based on many characteristics recorded through a benchmarking process. This approach can also be used when we consider the question of how one can construct an ordering over subsets of a set A given an ordering over the elements of A (see [9] for example).

In fact a lot of problems in individual and collective decision making involve the comparison of sets of alternatives (e.g.,when a firm have to hire not just one person but a team).[1]

It can be proved that a benchmark rule defines a preorder in a set of objects.

In the present note we study and characterize preferences in lattices that are defined with respect to a set of benchmarks and we extend in a natural way the results in [6]. The structure of the paper is as follows. To make this work self-contained in Sect. 2 we briefly mention some basic concepts on lattices theory

[1] I would like to thank Esteban Indurain for pointing out this problem.

J.P. Carvalho et al. (Eds.): IPMU 2016, Part I, CCIS 610, pp. 117–125, 2016.
DOI: 10.1007/978-3-319-40596-4_11

and we provide the necessary definitions. Section 3 is devoted to introduce compatible preorders on a lattice. In Sect. 4 we focus on benchmarking preferences in finite lattices while in Sect. 5 we consider the case of distributive and bounded lattices and we prove characterization theorems for benchmarking preferences. We then focus in Sect. 6 on compatible lattice functions and we study the class of aggregation functions that are componentwise compatible with every congruence or every preference and we obtain a characterization of Sugeno integral that is a noteworthy aggregation function that plays a relevant role as a preference functional in multicriteria decision making.

2 Basic Notions and Terminology

First we recall some basic notions in lattice and ordered set theory. For further background in ordered set theory we refer the reader to, e.g., Birkhoff [1], Davey and Priestley [8] Caspard, Leclerc and Monjardet [7] or Grätzer [11].

A *partially ordered set* (*poset* for short)$(P, \geq)$ is a set P with a reflexive, antisymmetric and transitive binary relation $\geq$. We will write $(x, y) \in R$ as $x \geq y$ (or equivalently, $y \leq x$) and we will use $x > y$ to mean that $x \geq y$ and $x \neq y$.

The word "partial" indicates that there's no guarantee that all elements can be compared to each other i.e. we don't know that for all $x, y \in P$, at least one of $x \leq y$ and $x \geq y$ holds. A poset in which this is guaranteed is a *complete* poset and it is called a *totally ordered set.*

A relation that is reflexive and transitive is said to be a *preorder.* This is a rather general concept, as every partial order and every equivalence order is a preorder.

We say that y *covers* x if $x \neq y$ and no elements of the poset lie strictly between x and y i.e. if $x \leq z \leq y$ implies that $z = x$ or $z = y$.

Our general approach to representation of preferences is based on lattice theory and in this framework we extend the results in [6].

A poset L is a *lattice* if every pair of elements x, y has

(i) a *least upper bound* $x \vee y$ (called *join*), and
(ii) a *greatest lower bound* $x \wedge y$ (called *meet*);

that is

$$z \geq x \vee y \iff z \geq x \text{ and } z \geq y$$
$$z \leq x \wedge y \iff z \leq x \text{ and } z \leq y.$$

A lattice L is said to be *distributive* if, for every $a, b, c \in L$,

$$a \vee (b \wedge c) = (a \vee b) \wedge (a \vee c) \quad \text{or, equivalently,}$$

$$a \wedge (b \vee c) = (a \wedge b) \vee (a \wedge c).$$

A *chain* is a lattice such that for every $a, b \in L$ we have $x \geqslant y$ or $y \geqslant x$. Clearly, every chain is distributive. A lattice L is said to be *bounded* if it has a least and a greatest element, usually denoted by 0 and 1, respectively.

If L is a lattice the cartesian product L^n also constitutes a lattice by defining the lattice operations componentwise. Observe that if L is bounded (distributive), then L^n is also bounded (resp. distributive). We denote by $\mathbf{0}$ and $\mathbf{1}$ the least and the greatest elements, respectively, of L^n.

If L, M are two lattices a mapping $f: L \to L$ is said to be a lattice *homomorphism* if it preserves meets and joins, i.e., $f(x \wedge y) = f(x) \wedge f(y)$ and $f(x \vee y) = f(x) \vee f(y)$ for all $x, y \in L$. Every lattice homomorphism is clearly order-preserving.

We will need the following definition. An element z of a lattice L is called *join irreducible* if $z = x \vee y$ for $x, y \in L$ implies that $z = x$ or $z = y$. The notion of meet-irreducible element is defined dually. The set of join-irreducible elements of a lattice L is denoted by $J(L)$.

An element z of a lattice L is called *join prime* if $z \leq x \vee y$ for $x, y \in L$ implies that $z \leq x$ or $z \leq y$. The notion of meet-prime element is defined dually. The set of join-prime elements of a lattice L is denoted by $JI(L)$. It can be proved that a join-prime element is join-irreducible and if L is a finite distributive lattice we have that $J(L) = JI(L)$.

A *filter* of a lattice L is a nonempty subset F such that

(i) if $x \in F$ and $x \leq y$ then $y \in F$,
(ii) $x, y \in F$ then $x \wedge y \in F$.

Sets satisfying Condition (i) of a filter are called upsets. The dual notation is that of an *ideal*. If $a \in L$ we define the *principal filter* generated by x as $\uparrow x = \{y \in L : y \geq x\}$. It is easy to prove that $\uparrow x$ is a filter for every $x \in L$. It can be proved that in a finite lattice each filter and each ideal are principal.

A *proper filter* is a filter that is neither empty nor the whole lattice while a *prime filter* is a proper filter P such that if $x \vee y \in P$ then $x \in P$ or $y \in P$. An element x of a lattice L is join-prime if and only if $\uparrow x$ is prime. A filter F is prime if and only if $L \setminus F$ is an ideal, which is then a prime ideal.

Throughout this paper lattice means bounded and distributive lattice.

3 Benchmarking Rules on a Lattice

We introduce a formal definition of a preorder defined in a lattice by a set of benchmarks. If L is a lattice and $B \subseteq L$ for every $x \in L$ we consider the set $B(x) = \{b \in B : b \leq x\}$ and the relation $\trianglerighteq$ in L defined by

$$x \trianglerighteq y \quad \text{if and only if} \quad B(x) \supseteq B(y) \tag{1}$$

is said to be a *benchmarking rule*(as in [6]). According to this definition a benchmark rule is a comparison according to set-inclusion.

Employers commonly consider benchmarks when making hiring decisions and often benchmarks are included on résumés and it is meaningful to note that in a résumé there may be accomplishments that are not considered as benchmarks.

A binary relation $R \subseteq L \times L$ in a lattice L is said to be *monotone* if when $x \geq y$ then $(x, y) \in R$. Benchmark rules are not necessarily complete but are transitive and monotone. It is easy to prove that if we have three objects x, y, z in a lattice L if x is considered better than y in a benchmarking rule and z has exactly the accomplishments common to x and y then y and z are equivalent. This property requires that the preorder is compatible with the lattice structure and we can note that a benchmarking rule respects the lattice operations as in the following definition.

Let us say that a binary relation $R \subseteq L \times L$ in a lattice L is *compatible* whenever it preserves the join and the meet i.e.

(i) if $(x, y) \in R$ then $(x \wedge z, y \wedge z) \in R$ for each $z \in L$;
(ii) if $(x, y) \in R$ then $(x \vee z, y \vee z) \in R$ for each $z \in L$.

It is straightforward to prove that a compatible binary relation on a lattice L is monotone if and only if for every $x \in L$, $(x, 0) \in R$. An equivalence relation defined on a lattice and compatible with the two lattice operations is said to be a *congruence*. A *preference* on a lattice L is a monotone and compatible preorder on L.

There is a strong link between preferences and congruences in a lattice.

If $\trianglerighteq$ is a preference on a lattice L then the relation $\sim$ defined by

$$x \sim y \quad \text{if} \quad x \trianglerighteq y \quad \text{and} \quad y \trianglerighteq x$$

is a congruence on L. It can be proved also that

$$x \trianglerighteq y \quad \text{if and only if} \quad x \sim x \vee y \quad \text{and} \quad x \wedge y \sim y.$$

Moreover if $\sim$ is a congruence on a lattice L we can define a preference on L by

$$x \trianglerighteq y \quad \text{if and only if} \quad x \sim x \vee y \quad \text{and} \quad x \wedge y \sim y.$$

The following proposition considers some properties of benchmarking rules.

Proposition 1. *If $B \subseteq L$ and $\trianglerighteq$ is defined by*

$$x \trianglerighteq y \quad \textit{if and only if} \quad B(x) \supseteq B(y) \tag{2}$$

then $\trianglerighteq$ is a monotone preorder. If $B \subseteq JI(L)$ then $\trianglerighteq$ is a preference on L.

Proof. The transitivity of $\trianglerighteq$ follows from the transitivity of $\supseteq$ and the relation is obviously reflexive. If $x, y \in L$ are such that $x \geq y$ then if $y \geq b$ for $b \in B$ it follows that $x \geq y \geq b$ so $B(x) \supseteq B(y)$.

Now we suppose that $B \subseteq JI(L)$ and that $x \trianglerighteq y$. If $z \in L$, $b \in B$ and $b \leq y \vee z$ since b is join-prime we have that $b \leq y$ or $b \leq z$. If $b \leq y$ we have also that $b \leq x$ by $x \trianglerighteq y$ and then we have proved that $b \leq x \vee z$. It follows that $B(x \vee z) \supseteq B(y \vee z)$ and then $x \vee z \trianglerighteq y \vee z$.

Let $b \leq y \wedge z$ with $z \in L$, $b \in B$. Then $b \leq y$ and $b \leq z$ and we can prove that $b \leq x \wedge z$. We have that $B(x \wedge z) \supseteq B(y \wedge z)$ and so $x \wedge z \trianglerighteq y \wedge z$.

If $B \subseteq JI(L)$ a *benchmarking preference* $\trianglerighteq$ is a preference on L defined by (1) with respect to the subset B of $JI(L)$.

The following result characterizes complete benchmarking preferences.

Proposition 2. *If $B \subseteq JI(L)$ and $\trianglerighteq$ is a benchmarking preference on L defined with respect to the set B, $\trianglerighteq$ is a total order if and only if the set B is a chain.*

Proof. If $\trianglerighteq$ is a complete preorder and b_1, b_2 are elements of B then or $B(b_1) \supseteq B(b_2)$ or $B(b_1) \subseteq B(b_2)$. Then we can conclude that $b_1 \leq b_2$ or $b_1 \geq b_2$.

Now we assume that B is a totally ordered set. If $x, y \in L$ and $x \ntrianglerighteq y$ there exists $b_1 \in B$ such that $y \geq b_1$ and $x \ngeq b_1$. Let b_2 an element in $B(x)$. Since $b_2 \leq x$ we can prove that $b_1 \nleq b_2$ therefore $b_1 > b_2$. Then $y \geq b_1 \geq b_2$ and so we have that $B(x) \subseteq B(y)$ and we can conclude that $x \trianglelefteq y$.

4 Compatible Relations on a Finite Lattice

Our aim is to characterize benchmarking preference on a finite lattice as compatible relations. The following proposition proves that every compatible and monotone preorder can be defined with respect to a set of benchmarks.

Theorem 1. *If L is a finite lattice every preference relation on L is a benchmarking preference with respect to a set $B \subseteq J(L)$.*

Proof. We note that in a finite lattice $JI(L) = J(L)$. If $\trianglerighteq$ is a preference in L we define the set $I(L) = \{z \in J(L) : \text{if } \ w \in L, \ w \trianglelefteq z \ \text{ then } \ w \sim z\}$ and the set $B = J(L) \setminus I(J)$. We prove that if $x, y \in L$ the $x \trianglerighteq y$ if and only if $B(x) \supseteq B(y)$. If there is an element $z \in L$ such that $z \leq y$ and $z \nleq x$ we consider $w \in L$ such that z covers w. We have that $z \trianglerighteq w$ and since $\trianglerighteq$ is a preference and $x \trianglerighteq y$ we have that $x \wedge w \trianglerighteq y \wedge w$. Since $w = (x \wedge z) \vee w \trianglerighteq (y \wedge z) \vee w = z$ and thus $w \sim z$. Therefore $z \in I(L)$ and we have proved that if $x \trianglerighteq y$ then $B(x) \supseteq B(y)$.

Let $x, y \in L$ such that $B(x) \supseteq B(y)$. Since L is a finite lattice $y = y_1 \vee \ldots \vee y_n$ where y_i is a join-irreducible element of L for every $i, 1 \leq i \leq n$. If $y_i \nleq x$ there exists an element z such that $z \leq x, z < y_i$. Since $y_i \notin B(x)$ we have that $y_i \notin B(y)$ it follows that $z \sim y_i$ then $y_i \sim z \trianglelefteq x$. We have proved that for every $i, 1 \leq i \leq n$, $y_i \trianglelefteq x$ and then $y \trianglelefteq x$.

As a consequence of Theorem 1 we can give the following characterization of congruences on finite lattices.

Proposition 3. *If L is a finite lattice a relation on L is a congruence if and only if there exists a set B, $B \subseteq J(L)$ such that*

$$x \sim y \quad \textit{if and only if} \quad B(x) = B(y) \tag{3}$$

Proof. If we define a relation $\sim$ by (3) it is easy to see that $\sim$ is a congruence. The converse follows directly from Theorem 1.

5 Generalized Benchmarking Preferences

The results in Sect. 4 depend on the existence of many join-irreducible elements in a finite distributive lattice. Note that in an infinite distributive lattice there may be no join-irreducible element. In the infinite case, the role of join-irreducible elements is taken by prime filters.

Let $\mathcal{P}$ be the set of prime filters of a lattice L. If $\mathcal{B} \subseteq \mathcal{P}$ and $x \in L$ we define $\mathcal{B}(x) = \{F \in \mathcal{B} : x \in F\}$.

Proposition 4. *If $\mathcal{B} \subseteq \mathcal{P}$ and $\trianglerighteq$ is defined by*

$$x \trianglerighteq y \quad \text{if and only if} \quad \mathcal{B}(x) \supseteq \mathcal{B}(y) \tag{4}$$

then $\trianglerighteq$ is is a preference on L.

Proof. As in Proposition 1 we can prove that $\trianglerighteq$ is a reflexive, transitive and monotone relation. Let $x, y, z \in L$ and $x \trianglerighteq y$. If $F \in \mathcal{P}$ and $y \vee z \in F$ then either $y \in F$ or $z \in F$. If $y \in F$ then $x \in F$ so we can conclude that $x \vee z \in F$. It follows that $\mathcal{B}(x \vee z) \supseteq \mathcal{B}(y \vee z)$ and then $x \vee z \trianglerighteq y \vee z$.

Moreover if $F \in \mathcal{B}$ and $y \wedge z \in F$ then $y, z \in F$. Since $\mathcal{B}(x) \supseteq \mathcal{B}(y)$ we have also that $x \in F$. Then $x \wedge z \in F$ hence $\mathcal{B}(x \wedge z) \supseteq \mathcal{B}(y \wedge z)$ and $x \wedge z \trianglerighteq y \wedge z$.

If $\mathcal{B} \subseteq \mathcal{P}$ a *generalized benchmarking preference* $\trianglerighteq$ is a preference on L defined by (4) with respect to $\mathcal{B}$.

We present the following important result (see [11]).

Proposition 5. *If x, y are two elements of the distributive lattice L and $x \not\geq y$ there exists a prime filter F with $y \in F$ and $x \notin F$.*

The following theorem which characterizes preference relations as generalized benchmarking preferences can now be proved.

Theorem 2. *Every preference relation on a distributive lattice L is a generalized benchmarking preference with respect to a set $\mathcal{B} \subseteq \mathcal{P}$.*

Proof. First we consider the preorder $\geq$ defined on the lattice L and we prove that

$$x \geq y \quad \text{if and only if} \quad \mathcal{P}(x) \supseteq \mathcal{P}(y).$$

We can easily prove that if $x \geq y$ then $\mathcal{P}(x) \supseteq \mathcal{P}(y)$.

By Proposition 5 if x, y are two elements of the lattice L and $x \not\geq y$ there exists a prime filter F with $y \in F$ and $x \notin F$ and then $\mathcal{P}(x) \not\supseteq \mathcal{P}(y)$.

Let $\trianglerighteq$ a preference on L an $\sim$ the associated congruence.

We denote by $\overline{L}$ the set of the equivalence classes with respect to $\sim$ and by $\overline{x}$ the equivalence class of the element x. It is well known that $\overline{L}$ is a lattice with the operations defined naturally on the equivalence classes and that there is a surjective homomorphism f of L in $\overline{L}$ such that $f(x) = \overline{x}$. Moreover we can define a preference $\trianglerighteq$ in $\overline{L}$ by $\overline{x} \trianglerighteq \overline{y}$ if and only if $x \trianglerighteq y$.

It is straightforward to prove that $\trianglerighteq$ is a well defined preference in $\overline{L}$.

If $\mathcal{P}$ is the set of prime filter in $\overline{L}$ the relation $\trianglerighteq$ in $\overline{L}$ is a generalized preference relation on $\overline{L}$ with respect to $\mathcal{P}$. It can be proved that the inverse image of a prime filter with respect to a lattice homomorphism is a prime filter and then the set $\mathcal{B} = \{f^{-1}(F) : F \in \mathcal{P}\}$ is a set of prime filters in L. It follows that the relation $\trianglerighteq$ in L is a generalized benchmarking preference with respect to $\mathcal{B}$.

Using Theorem 2 one can easily prove the following result.

Proposition 6. *A relation on a distributive lattice L is a congruence if and only if there exists a set $\mathcal{B}$, $\mathcal{B} \subseteq \mathcal{P}$ such that*

$$x \sim y \quad \textit{if and only if} \quad \mathcal{B}(x) = \mathcal{B}(y) \tag{5}$$

By Proposition 6 we can prove that if F is a prime filter of the lattice L we can define a congruence such that the equivalence classes are F and $L \setminus F$.

Proposition 7. *If F is a prime filter of the distributive lattice L the relation $\sim_F$ defined by*

$$x \sim_F y \quad \textit{if and only if} \quad x, y \in F \quad \textit{or} \quad x, y \notin F$$

is a congruence in L.

6 Compatible Aggregation Function on a Lattice

Aggregation operators are mathematical functions that are used to combine several inputs into a single representative outcome;see [10] for a comprehensive overview on aggregation theory. Aggregation operators play an important role in several fields such as decision sciences, computer and information sciences, economics and social sciences. There are a large number of different aggregation operators that differ on the assumptions on the inputs and about the information that we want to consider in the model. There are many situations where inputs to be aggregated are qualitative and numerical values are used by convenience. Sometimes we need to evaluate objects with a scale that is not totally ordered. In this paper we focus on polynomial functions and on Sugeno integral in a bounded and distributive lattice(see [2–4]). In fact the definition of Sugeno integral primarily introduced on real intervals can be extended to ordered domains.

We briefly introduce lattice functions that are compatible with lattice operations.

We follow the approach in [12] and we consider componentwise compatibility with respect to a relation defined on a lattice L.

A lattice function $f\colon L^n \to L$ is said to be *compatible with the preference* $\trianglerighteq$ if when $\mathbf{x}, \mathbf{y} \in L^n$ and for every $i, 1 \leq i \leq n$, $x_i \trianglerighteq y_i$ then $f(\mathbf{x}) \trianglerighteq f(\mathbf{y})$.

A lattice function $f\colon L^n \to L$ is said to be *compatible with the congruence* $\sim$ if when $\mathbf{x}, \mathbf{y} \in L^n$ and for every i, $1 \leq i \leq n$, $x_i \sim y_i$ then $f(\mathbf{x}) \sim f(\mathbf{y})$.

A function $f\colon L^n \to L$ is said to be *monotone* if, for every $\mathbf{x}, \mathbf{y} \in L^n$ such that for every i, $1 \leq i \leq n$, $x_i \geq y_i$ then $f(\mathbf{x}) \geq f(\mathbf{y})$.

The following proposition introduces a property of compatible aggregation functions.

Proposition 8. *If L is a distributive lattice and $f\colon L^n \to L$ is a monotone lattice function, f is compatible with a preference $\trianglerighteq$ if and only if it is compatible with the associated congruence $\sim$.*

Proof. By the definition of $\sim$ it is straightforward to prove that if f is compatible with a preference $\trianglerighteq$ it is also compatible with the associated congruence $\sim$.

Conversely we consider a lattice function compatible with a congruence $\sim$. Then if for every $i, 1 \leq i \leq n$, $x_i \trianglerighteq y_i$ hence $x_i \sim x_i \vee y_i$ and $y_i \sim x_i \wedge y_i$.

Since F is compatible with $\sim$ we can prove that $f(x) \sim f(x \vee y)$ and $f(y) \sim f(x \wedge y)$. Being f monotone it follows that $f(x) \vee f(y) \sim f(x)$ and $f(x) \wedge f(y) \sim f(y)$ and then $f(x) \trianglerighteq f(y)$.

A lattice function $f\colon L^n \to L$ is called a *polynomial function* if can be obtained by composition of the binary operations $\vee$ and $\wedge$, the projections and the constants functions (see [2–4]). If $f(\mathbf{0}) = 0$ and $f(\mathbf{1}) = 1$ a polynomial function is a *Sugeno integral* (see [4]).

We want to show that the class of polynomial functions coincides with the class of multivariate functionals compatible with every congruence defined in the lattice L. So if we suppose that $f(\mathbf{0}) = 0$ and $f(\mathbf{1}) = 1$ we obtain a characterization of Sugeno integral as in [12].

Theorem 3. *If L is a distributive lattice, a monotone lattice function $f\colon L^n \to L$ is compatible with every preference (congruence) if and only if it is a polynomial function.*

Proof. Since binary operations $\vee$ and $\wedge$, the projections and the constants functions are compatible with every congruence it is clear that a polynomial function is compatible with every congruence.

Suppose that f is a monotone lattice function $f\colon L \to L$ compatible with every congruence. Our goal is to prove that for every $x, y \in L$, $f(x \vee y) = f(x) \vee f(y)$ and $f(x \wedge y) = f(x) \wedge f(y)$ and then we can conclude that f is an unary polynomial (see [2]). If $x, y \in L$, obviously $f(x \vee y) \geq f(x) \vee f(y)$ and we suppose by contradiction that $f(x) \vee f(y) \ngeq f(x \vee y)$. By Proposition 5 there exists a prime filter F such that $f(x \vee y) \in F$ and $f(x) \vee f(y) \notin F$ and then $f(x) \notin F$ and $f(y) \notin F$. Now we consider the congruence $\sim_F$ defined by the prime filter F that is characterized by two class of equivalence namely the class of elements that belongs to F and the class of elements that do not belong to the filter F. It is clear that $f(x) \sim_F f(y)$ and so we have that $x \sim_F y$. Therefore $x \sim_F x \vee y$ and then $f(x) \sim_F f(x \vee y)$ that is a contradiction since $f(x) \notin F$ and $f(x \vee y) \in F$. We have proved that $f(x \vee y) = f(x) \vee f(y)$ and in a similar way we obtain that $f(x \wedge y) = f(x) \wedge f(y)$ for every $x, y \in L$.

We note that any unary function obtained from an n-ary function compatible with every congruence substituting constants for $n-1$-variables is compatible with every congruence and then by Lemma 9 of [2] we can conclude that every function f, $f\colon L^n \to L$ compatible with every congruence in L is a polynomial function.

7 Concluding Remarks

This paper considers a qualitative approach to preferences representation. We introduce a very general framework where a preference relation is not assumed to be complete. We characterize preferences that respect lattice operations in a bounded and distributive lattice. It can be proved that these preference relations are comparisons with respect to set inclusion and that these preference are benchmarks rules. A benchmark rule considers an object better than another object if it satisfies more benchmarks. So benchmarks represent relevant aspects of the considered comparison.

Moreover in [12] it is proved that if L is a bounded chain a function that an n-ary monotone function that preserves every congruence it is a polynomial function. We extend the result to the case of a bounded and distributive lattice.

References

1. Birkhoff, G.: Lattice Theory, Colloquium Pub, vol. 25. American Mathematical Society, Providence, R.I. (1967)
2. Couceiro, M., Marichal, J.-L.: Polynomial functions over bounded distributive lattices. J. Mul-Valued Log S **18**, 247–256 (2012)
3. Couceiro, M., Marichal, J.L.: Characterizations of discrete Sugeno integrals as lattice polynomial functions. In: Proceedings of the 30th Linz Seminar on Fuzzy Set Theory (LINZ2009), pp. 17–20 (2009)
4. Couceiro, M., Marichal, J.L.: Characterizations of discrete Sugeno integrals as polynomial functions over distributive lattices. Fuzzy Set Syst. **161**, 694–707 (2010)
5. Chambers, C.P., Miller, A.D.: Scholarly influence. J. Econ. Theory **151**(1), 571–583 (2014)
6. Chambers, C.P., Miller, A.D.: Benchmarking, working paper (2015)
7. Caspard, N., Leclerc, B., Monjardet, B.: Finite Ordered Sets. Encyclopedia of Mathematics and its Applications. Cambridge University Press, Cambridge (2012)
8. Davey, B.A., Priestley, H.A.: Introduction to Lattices and Order. Cambridge University Press, New York (2002)
9. Fishburn, P.C.: Signed order and power set extensions. J. Econ. Theory **56**, 1–19 (1992)
10. Grabisch, M., Marichal, J.L., Mesiar, R., Pap, E.: Aggregation Functions. Encyclopedia of Mathematics and its Applications. Cambridge University Press, Cambridge (2009)
11. Grätzer, G.: General Lattice Theory. Birkhäuser Verlag, Berlin (2003)
12. Halaš R., Mesiar, R., Pócs, J.: A new characterization of the discrete Sugeno integral. Inform. Fusion **29**, 84–86 (2016)

Uncertainty Quantification with Imprecise Probability

Efficient Simulation Approaches for Reliability Analysis of Large Systems

Edoardo Patelli(✉) and Geng Feng

Institute for Risk and Uncertainty, University of Liverpool, Chadwick Building, Peach Street, Liverpool L69 7ZF, UK
{edoardo.patelli,fenggeng}@liverpool.ac.uk
http://www.liverpool.ac.uk/risk-and-uncertainty

Abstract. Survival signature has been presented recently to quantify the system reliability. However, survival signature-based analytical methods are generally intractable for the analysis of realistic systems with multi-state components and imprecisions on the transition time. The availability of numerical simulation methods for the analysis of such systems is required. In this paper, novel simulation methods for computing system reliability are presented. These allow to estimate the reliability of realistic and large-scale systems based on survival signature including parameter uncertainties and imprecisions. The simulation approaches are generally applicable and efficient since only one estimation of the survival signature is needed while Monte Carlo simulation is used to generate component transition times. Numerical examples are presented to show the applicability of the proposed methods.

Keywords: Reliability analysis · Survival signature · Monte Carlo simulation

1 Introduction

The structure of a complex system cannot be sequentially reduced considering alternative series and parallel sections. Consequentially, the study of the reliability of such systems is still a topical subject in the literature and it has obvious importance in many application areas [11]. Traditionally, the reliability analysis of systems is performed adopting different well-known tools such as reliability block diagram, fault tree and success tree methods, failure mode and effect analysis, and master logic diagram [12]. The main limitation of these traditional approaches is their lack of applicability for very large systems.

In recent years, the system signature [16] has been recognized as an important tool to quantify the reliability of systems consisting of independent and identically distributed (*iid*) or exchangeable components with respect to the random failure times. System signature separates the system structure from the component probabilistic failure distribution. However, when it is adopted to solve a

J.P. Carvalho et al. (Eds.): IPMU 2016, Part I, CCIS 610, pp. 129–140, 2016.
DOI: 10.1007/978-3-319-40596-4_12

complex systems with more than one component type, it requires the computation of the probabilities of all possible different ordering statistics of each component failure lifetime distributions, which is an intractable and tedious procedure. In order to overcome the limitations of the system signature, Coolen and Coolen-Maturi [7] proposed the use of survival signature for analysing complex systems consisting of more than one single component type increasing the applicability of such approach to characterise complex systems and networks. System survival signature can also be derived from the subsystems survival signature [8], which provides a theory for reliability analysis on real world systems of non-trivial size. Based on the above concepts, Feng et al. [10] developed an analytical method to calculate survival functions of systems with uncertain components parameters which belong to the exponential family. The analytical solutions are exact within the assumptions made, but they are sometimes hard or impossible to derive for large complex systems. Hence, general applicable numerical solutions are required to perform numerical experiments, analysing the effect of different distribution types and to overcome the limitation of analytical methods that are usually based on ad-hoc solutions and limited to some specific families of probability distribution functions. Simulation methods can be used for the sensitivity analysis of multi-criteria decision models [6], optimise models with rare events [15] and perform multi-attribute decision making [18].

In this paper, efficient simulation approaches are proposed to estimate the reliability of large systems based on survival signature without the calculation of all the cut-sets, which is a challenging and error prone task. The proposed simulation approaches are applicable to any system configuration and able to consider different representations of the uncertainties. The numerical implementation of the proposed approaches is based on two open source packages: the R package "ReliabilityTheory" [3,4] adopted to calculate the survival signature and OpenCossan [14] a Matlab toolbox for uncertainty quantification and reliability analysis. The applicability and efficiency of the proposed approaches are demonstrated by solving numerical examples.

2 Background

2.1 Survival Signature

Suppose there is one system formed by M components. Let the state vector of components be $\underline{x} = (x_1, x_2, ..., x_M) \in \{0,1\}^M$ with $x_i = 1$ if the i-th component is in working state and $x_i = 0$ if not. $\Phi = \Phi(\underline{x}) : \{0,1\}^m \rightarrow \{0,1\}$ defines the system structure function, i.e., the system status based on all possible $\underline{x}$. Φ is 1 if the system functions for state vector $\underline{x}$ and 0 if not.

Now consider a system with $K \geq 2$ types of M components, with m_k indicating the number of components of each type and $\sum_{k=1}^{K} m_k = M$. It is assumed that the failure times of the same component type are independently and identically distributed (*iid*) or exchangeable. Coolen et al. [8] introduced the survival signature for such a system, denoted by $\phi(l_1, l_2, ..., l_K)$, with $l_k = 0, 1, ..., m_k$ for $k = 1, 2, ..., K$, which is defined to be the probability that the system functions given that l_k of its m_k components of type k work, for each $k \in \{1, 2, ..., K\}$. There are $\binom{m_k}{l_k}$ state vectors $\underline{x}^k$ with precisely l_k components x_i^k equal to 1, so with $\sum_{i=1}^{m_k} x_i^k = l_k$. x_i^k denotes the state of the i-th component of type k.

Let $S_{l_1,l_2,...,l_K}$ denote the set of all state vectors for the whole system for which $\sum_{i=1}^{m_k} x_i^k = l_k$, $k = 1, 2, ..., K$. Assume that the random failure times of components of the different types are fully independent, and in addition the components are exchangeable within the same component types, the survival signature can be rewritten as:

$$\phi(l_1, ..., l_K) = \left[\prod_{k=1}^{K} \binom{m_k}{l_k}^{-1} \right] \times \sum_{\underline{x} \in S_{l_1,l_2,...,l_K}} \phi(\underline{x}), \tag{1}$$

$C_k(t) \in \{0, 1, ..., m_k\}$ denotes the number of type k components working at time t. Assume that the components of the same type have a known CDF, $F_k(t)$ for type k. Moreover, the failure times of different component types are assumed independent, then:

$$P(\bigcap_{k=1}^{K} \{C_k(t) = l_k\}) = \prod_{k=1}^{K} P(C_k(t) = l_k) = \prod_{k=1}^{K} \binom{m_k}{l_k} [F_k(t)]^{m_k - l_k} [1 - F_k(t)]^{l_k} \tag{2}$$

Hence, the survival function of the system with K types of components becomes:

$$P(T_s > t) = \sum_{l_1=0}^{m_1} ... \sum_{l_K=0}^{m_K} \phi(l_1, ..., l_K) P(\bigcap_{k=1}^{K} \{C_k(t) = l_k\}) \tag{3}$$

Equation (3) separates the structure of the system from the failure time distribution of its components, which is the main advantage of the system signature. The survival signature only needs to be calculated once for any system, which is similar to the system signature for systems with only single type of components. For a special case of a system with only one type ($K = 1$) of components, the survival signature and the system signature [16] are directly linked to each other through a simple equation, however, the latter cannot be easily generalized for systems with multiple types ($K \geq 2$) of components [7]. This implies that all attractive properties of the system signature also hold for the method using the survival signature. The survival signature is easy to apply for systems with multiple types of components, and one could argue it is much easier to interpret than the system signature.

2.2 Modelling the Uncertainties

Multiple mathematical concepts can be used to characterize variability and uncertainty. Often in practical situations very limited data are available, and to avoid the inclusion of subjective and often unjustified hypothesis, the imprecision and vagueness of the data can be treated combining probabilistic and set theoretical components in a unified construct allowing the identification of bounds on probabilities for the events of interest in order to give a different prospective to the results [5]. Random Set theory is a general framework suited to model uncertainty represented as cumulative distribution functions (CDFs), intervals, probability boxes, normalized fuzzy sets (also known as possibility distributions) and Dempster-Shafer [9,17] structures without making any implicit or explicit assumption at all. Explanatory examples of such flexible frameworks are provided in [1,2,13].

Without entering in the mathematical formalism, the Random Sets can be understood as random variables that sample sets and not points as realizations. These realization are called focal elements. When all focal elements are singletons, then the Random Set becomes a random variable.

Focal elements propagated through a model produce a collection of sets and not points. These sets are generally identified by means of an opportune optimization strategy. The collection of these set produces the so called Dempster-Shafer structure [9]. The upper and lower bounds of these structures form the distribution bounds of model output.

3 Simulation Approaches

The survival signature presented in the Sect. 2.1 can be adopted in a Monte Carlo based simulation method to estimate the system reliability in a simple and efficient way. A possible system evolution is simulated by generating random events (i.e. the random transition time such as failures of the system components) and then estimating the status of the system based on the survival signature (Eq. 1). Then, counting the occurrence number of a specific condition (i.e. number of system failures), it is possible to estimate the reliability of the system.

3.1 Reliability Analysis of System Without Imprecision

The simulation approach is based on the realizations of failure events of the system's components. Then, for each failure event the status of the system is generated based on the probability that the system is working knowing that a specific number of components are working. Such probability is given by the survival signature as defined in Eq. 1. Suppose there is a system with M components, K component types and m_k components of type k. Hence, $M = \sum_{k=1}^{K} m_k$. The survival signature is computed only once before starting the Monte Carlo simulation. Without loss of generality, the lifetime distributions of components are irrespective of the time they enter into service and once failed they can not be repaired. The reliability of the system can be estimated adopting the following procedure:

1 Sample the failure times for each component. The failure time of component type k is obtained sampling from the CDF F_k corresponding to it.
2 Order the failure times $t_i \leq t_{i+1}$ for $i = 1, 2, \ldots, M$. Hence, t_1 represents the first failure of a system component, t_2 the second failure and so on.

3 At each failure time, it is easy to calculate the number of components working for each component type: $C_k(t_i)$.
4 Evaluate the survival signature which applies immediately after the corresponding failure indicated as $\phi_{t_i} \equiv \phi(C_1(t_i), C_2(t_i), \ldots, C_K(t_i))$.
5 Set $i = 1$ and draw from a Bernoulli distribution with probability $1 - \phi_{t_i}$ the system status X_1 at time t_i, if $X_i = 1$ the system fails.
6 If the system does not fail at t_i, then consider t_{i+1}. The probability that the system functions at time t_{i+1} is $\phi_{t_{i+1}}/\phi_{t_i} = q_{i+1}$, given that it has survived at time t_i. So the system failure at time t_{i+1} X_{i+1} is drawn from a Bernoulli distribution with the probability $1 - q_{i+1}$.
7 Repeat Step 6 to process other failure times ($i \leftarrow i + 1$).

The above procedure is repeated for N samples and the status of the system over the time is collected in appropriated counters. It should be noted that with the assumption that the system fails if no component functions, this implies that $q_{i^*} = 0$ for $i^* \leq M$. Hence the system fails certainly at this t_{i^*} if it has not failed before. The proposed algorithm is applicable to any system and requires the calculation of the survival signature only once.

It is also possible to estimate the system reliability without the necessity to sample the system status at each component failure time. The idea is to interpret the survival signature as a normalised "production capability" of the system defined by the Eq. 1. For instance, if all the components are working, the system output is 1. If all components are in failure status, the system output is 0. Hence, instead of sampling the system state at each failure time, the survival signature is evaluated immediately after each sampled component failure time and collected in proper counters. In other words, for each Monte Carlo simulation this method generates a random grid of time points at which to evaluate the probability of survival to those times that represents the "production level of the system". Finally, the survival function is obtained by directly averaging of the survival signature over the time, i.e. computing the expected production level of the system adopting an algorithm derived from the approach proposed in [19]. Hence, the reliability of the system can be estimated modifying the steps 5–7 of the proposed approach as follows:

5' Compute the probability of survival (production level) of the system by evaluating the survival signature ϕ at each sampled component failure time t_i. For instance, the probability that the system survivals at time t_1 is ϕ_{t_1}.
6' Collect the value of the survival signature in appropriate counters (i.e. $Y(j) = Y(j) + \phi_{t_i}$ for $j : t_{i-1} \leq j * dt < t_i$) where Y represents the counter and dt the discretisation time used to store the results.

The above procedure is repeated for N samples and the reliability of the system is computed averaging the values of the survival signatures ($P(T_s > t) \approx \frac{1}{N}Y(t)$). The uses of the survival signature makes this approach extremely efficient since it does not require to sample the system output at each component transition time (i.e. component failures). The flow chart of the simulation methods proposed for estimating the reliability of non-repairable systems is shown in Fig. 1.

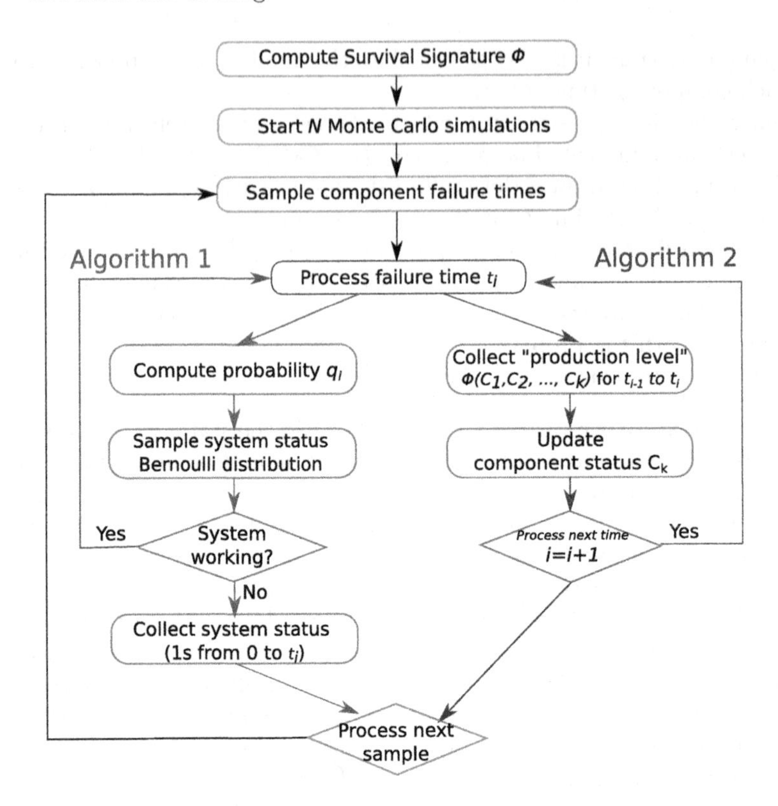

Fig. 1. Flow chart of the proposed numerical approaches.

3.2 Reliability Analysis of Systems with Imprecision

Reliability analysis of complex systems requires the probabilistic characterization of all the possible component transitions. This usually requires a large data-set that is not always available. In fact, it might not be possible to unequivocally characterize some component transitions due to lack of data or ambiguity. As mentioned in Sect. 2.2, to avoid the inclusion of subjective and often unjustified hypotheses, the imprecision and vagueness of the data can be treated by using concepts of imprecise probabilities. Randomness and imprecision are considered simultaneously but viewed separately at any time during the analysis and in the results. The probabilistic analysis is carried out conditional on the elements from the sets, which leads eventually to sets of probabilistic results.

Considering the imprecisions in the component parameters will lead to bounds of the survival function of the systems and it can therefore be seen as a conservative analysis, in the sense that it does not make any additional hypothesis with regard to the available information. In some instances, analytical methods will not be appropriate means to analyse a system. Again, simulation methods based on survival signature can be adopted to study systems considering parameter imprecision. A naive approach consists in adopting a double loop sampling where the outer loop is used to sample realizations in the epistemic space. In other words, each realization in the epistemic space defines a new probabilistic model that needs to be solved adopting the simulation methods proposed above. Then the envelop of the system reliability is identified. However, since almost all the systems are coherent (a system is coherent if each component is relevant, and the structure function is non decreasing) it is only necessary to compute the analysis reliability twice using the lower and upper bounds for all the parameters, respectively.

4 Numerical Examples

4.1 Bridge System

The purpose of this numerical example is to verify the proposed algorithms since analytical solutions are available. The bridge system comprises six components, which belong to two types. It has no series section or parallel section which can enable simplification (see Fig. 2). The survival signature can easily be computed either manually or using the R-package *ReliabilityTheory* [4]. The values of the survival signature are reported in Table 1. In this example the failure times of component type 1 and 2 both obey exponential distributions with parameters shown in Table 2. It is also assumed that the component once failed can not be repaired. The survival function of the bridge system is then calculated by means of the two proposed methods and compared with the analytical solution as shown in Fig. 3. The simulations have been performed using $N = 5000$ samples and collecting the results in 2000 counters. The discretisation time is only required to collect the numerical results (i.e. survival function) although the simulation of the system is continuous with respect to the time. The variance of the estimators evaluated at time $t = 0.8$ and calculated by repeating the Monte Carlo simulation 20 times, is $3.0 \cdot 10^{-5}$ using the Algorithm 1 and $1.4 \cdot 10^{-5}$ with the Algorithm 2, respectively.

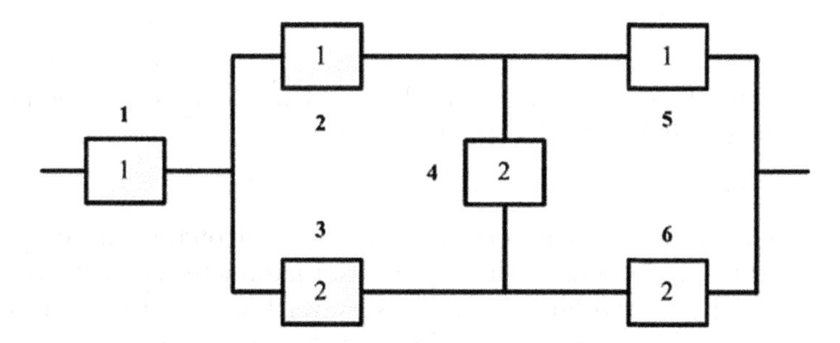

Fig. 2. Block diagramme of the bridge system with two types of components. The number inside the boxes represent the component type.

Table 1. Survival signature of the bridge system of Fig. 2

l_1	l_2	$\phi(l_1, l_2)$
0	[0, 1, 2, 3]	0
[1, 2]	[0, 1]	0
1	2	1/9
1	3	1/3
2	2	4/9
2	3	2/3
3	[0, 1, 2, 3]	1

Table 2. Failure rates of the components in the bridge system.

Component type	Distribution type	λ	λ (with imprecision)
1	Exponential	0.8	$[\underline{0.4}, \overline{1.2}]$
2	Exponential	1.5	$[\underline{1.3}, \overline{2.1}]$

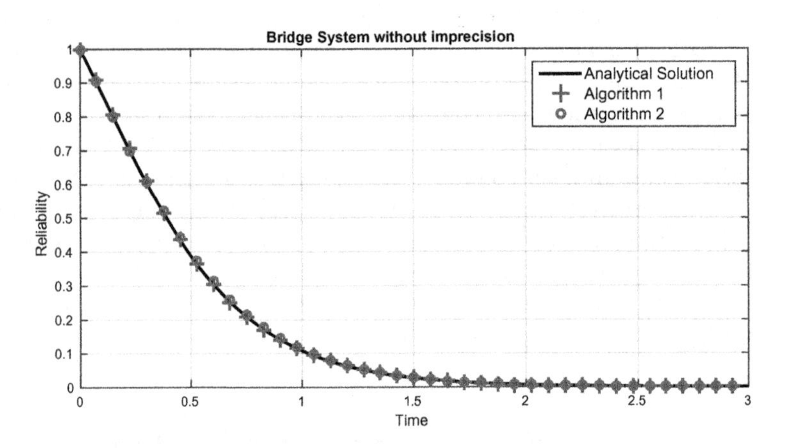

Fig. 3. Survival function of the bridge system calculated by means of the proposed simulation methods and analytical method, respectively.

The system is also analysed in presence of imprecision on the parameter value. In particular, the only bounds of the exponential parameter is known as shown in Table 2. To estimate the bounds of the survival function, the analysis have been performed twice, using the lower and upper bound for all the parameters as explain in Sect. 3.2. The results are shown in Fig. 4 and compared with analytical solutions estimated adopting the method presented in [10] showing a perfect agreement.

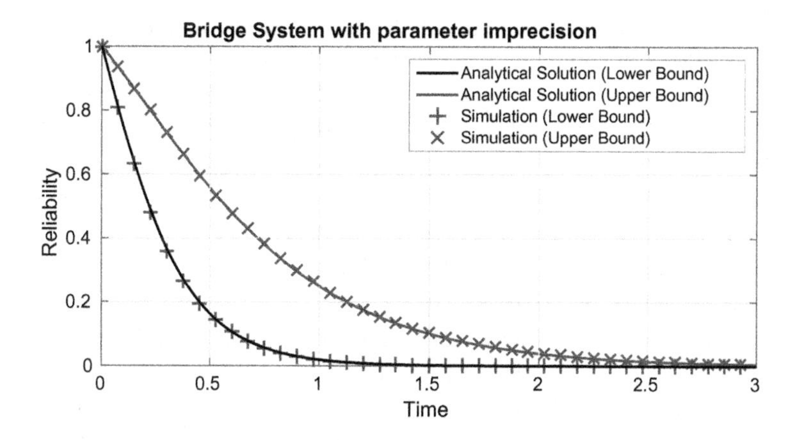

Fig. 4. Bounds of the survival function of the bridge system calculated by means of the simulation methods and compared with analytical solution.

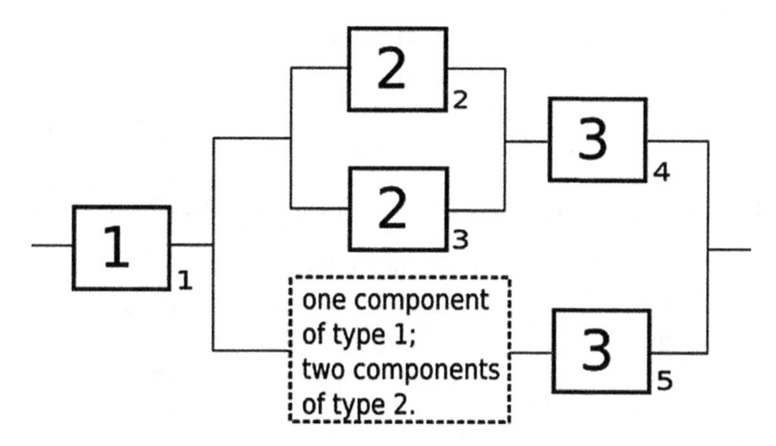

Fig. 5. An eight components system without knowing the exact configuration.

4.2 Imprecise System

In order to illustrate the efficiency and the applicability of the proposed simulation approaches a complex system composed by 8 components of 5 types is analysed. The component failure types and distribution parameters are shown in Table 4. In addition, it is assumed that the exact configuration of part of the system is unknown as shown in Fig. 5. However, the system can still be described using the survival signature although affected by imprecision as shown in Table 3. Algorithm 2 is used to estimated the bounds of the survival function of the system by collecting the values (i.e. intervals) of the survival signature during the Monte Carlo simulation. The collected bounds are then used to estimate the bounds of the survival function as details in Ref [13]. For coherent systems as explain in Sect. 3.2. In principle, Algorithm 1 can also be used for the estimation

Table 3. Imprecise survival signature of the system of Fig. 5, $\phi(l_1, l_2, l_3) = 0$ and $\phi(l_1, l_2, l_3) = 1$ for both lower and upper bounds are omitted.

l_1	l_2	l_3	$[\phi(l_1, l_2, l_3)]$
1	1	1	[1/8,1/8]
1	1	2	[1/4,1/4]
1	2	1	[1/5,1/4]
1	2	2	[3/7,1/2]
1	3	1	[1/4,3/8]
1	3	2	[1/2,1/2]
1	4	1	[1/4,1/2]
1	4	2	[1/2,1/2]
2	0	1	[0,1/2]
2	0	2	[0,1]
2	1	1	[1/4,3/4]
2	1	2	[1/2,1]
2	2	1	[1/2,1]
2	3	1	[3/4,1]

Table 4. Components failure types and distribution parameters for system of Fig. 5

Component type	Distribution	Parameters
1	Weibull	$([1.6, 1.8], [3.3, 3.9])$
2	Exponential	$([2.1, 2.5])$
3	Weibull	$([3.1, 3.3], [2.3, 2.7])$

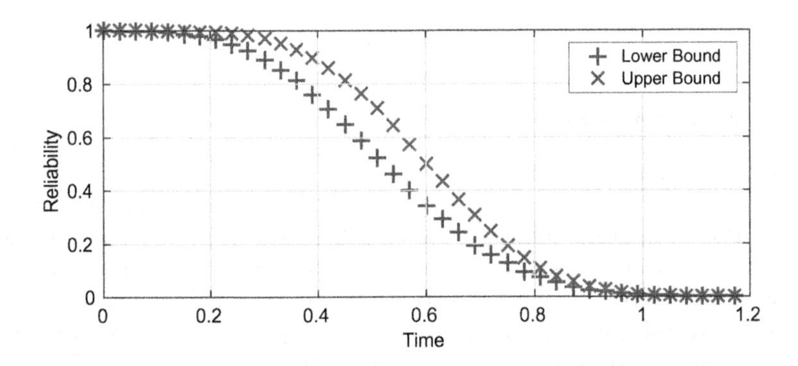

Fig. 6. Upper and lower bounds of survival function for the system in Fig. 5.

of the reliability bounds although it requires some modifications in the sampling of the system status. The upper and lower bounds of survival function for the system with imprecision both in the survival signature and on the component distribution parameters are shown in Fig. 6. The simulation have been performed using 5000 samples. This example shows the flexibility and the applicability of the simulation approaches proposed for analysing complex systems affected by imprecision where no analytical solutions are available.

5 Conclusions

Complex systems occur in various engineering applications, the survival signature has been shown to be a practical method for performing reliability analysis of complex systems with multiple component types. However, there always exist maintainability and imprecision within systems and the analytical methods are only applicable in a few cases (e.g., for components only with exponential distribution types).

In this paper, efficient simulation methods have been proposed for system reliability estimation. In principle, the simulation methods proposed in this paper have the ability to analyse system reliability by using only component failure time simulation and the survival signature, which is of great value for many systems in real world. The proposed simulation methods are generally applicable and able to deal with imprecision in the component distribution parameters as well in the system configuration (i.e. in the survival signature). Furthermore, they can easily be used to analyse non-repairable and repairable systems. The simulation methods are extremely fast and they can be adopted to analyse realistic and complex systems. The feasibility and effectiveness of the presented approaches have been illustrated with two numerical examples, the results indicate that simulation methods based on survival signature are efficient for analysing reliability on complex systems.

References

1. Alvarez, D.A.: Infinite random sets and applications in uncertainty analysis. Ph.D. thesis, Arbeitsbereich für Technische Mathematik am Institut für Grundlagen der Bauingenieurwissenschaften. Leopold-Franzens-Universität Innsbruck, Innsbruck, Austria (2007) https://sites.google.com/site/diegoandresalvarezmarin/RSthesis.pdf
2. Alvarez, D.A., Hurtado, J.E.: An efficient method for the estimation of structural reliability intervals with random sets, dependence modelling and uncertain inputs. Comput. Struct. **142**, 54–63 (2014)
3. Aslett, L.J.M. : MCMC for Inference on Phase-type and Masked System Lifetime Models. Ph.D. thesis, Trinity College Dublin (2012)
4. Aslett, L.J.M.: Reliabilitytheory: Tools for structural reliability analysis. r package (2012)
5. Beer, M., Patelli, E.: Editorial: Engineering analysis with vague and imprecise information. Structural Safety Special Issue: Engineering Analyses with Vague and Imprecise Information 52, Part B(0), 143 (2015)

6. Butler, J., Jia, J., Dyer, J.: Simulation techniques for the sensitivity analysis of multi-criteria decision models. Eur. J. Oper. Res. **103**(3), 531–546 (1997)
7. Coolen, F.P.A., Coolen-Maturi, T.: Generalizing the signature to systems with multiple types of components. In: Zamojski, W., Mazurkiewicz, J., Sugier, J., Walkowiak, T., Kacprzyk, J. (eds.) Complex Systems and Dependability. AISC, vol. 170, pp. 115–130. Springer, Heidelberg (2012)
8. Coolen, F.P.A., Coolen-Maturi, T.: Modelling uncertain aspects of system dependability with survival signatures. In: Zamojski, W., Sugier, J. (eds.) Dependability Problems of Complex Information Systems. AISC, vol. 307, pp. 19–34. Springer, Heidelberg (2014)
9. Arthur, P.: Dempster.: Upper and lower probabilities induced by a multivalued mapping. Annal. Math. Stat. **38**, 325–339 (1967)
10. Feng, G., Patelli, E., Beer, M., Coolen, F.: Imprecise system reliability and component importance based on survival signature. Reliab. Eng. Syst. Saf. **150**, 116–125 (2016)
11. Kołowrocki, K., Kwiatuszewska-Sarnecka, B.: Reliability and risk analysis of large systems with ageing components. Reliab. Eng. Syst. Saf. **93**(12), 1821–1829 (2008)
12. Modarres, M.: What Every Engineer Should Know About Reliability and Risk Analysis, vol. 30. CRC Press, Boca Raton (1992)
13. Patelli, E., Alvarez, D.A., Broggi, M., de Angelis, M.: Uncertainty management in multidisciplinary design of critical safety systems. J. Aerosp. Inf. Syst. **12**, 140–169 (2015)
14. Patelli, E., Broggi, M., de Angelis Marco, M.B.: OpenCossan: An efficient open tool for dealing with epistemic and aleatory uncertainties. In: Vulnerability, Uncertainty, and Risk, pp. 2564–2573. American Society of Civil Engineers, June 2014
15. Rubinstein, R.Y.: Optimization of computer simulation models with rare events. Eur. J. Oper. Res. **99**(1), 89–112 (1997)
16. Samaniego, F.J.: System Signatures and Their Applications in Engineering Reliability, vol. 110. Springer Science & Business Media, New York (2007)
17. Shafer, G.: A Mathematical Theory of Evidence. Princeton University Press, Princeton (1976)
18. Zanakis, S.H., Solomon, A., Wishart, N., Dublish, S.: Multi-attribute decision making: A simulation comparison of select methods. Eur. J. Oper. Res. **107**(3), 507–529 (1998)
19. Zio, E., Baraldi, P., Patelli, E.: Assessment of the availability of an offshore installation by Monte Carlo simulation. Int. J. Press. Vessels Pip. **83**(4), 312–320 (2006)

Bivariate p-boxes and Maxitive Functions

Ignacio Montes[1] and Enrique Miranda[2(✉)]

[1] Department of Statistics, Carlos III University of Madrid, Getafe, Spain
igmontes@est-econ.uc3m.es
[2] Department of Statistics and Operations Research,
University of Oviedo, Oviedo, Spain
mirandaenrique@uniovi.es
http://unimode.uniovi.es

Abstract. We investigate the properties of the upper probability associated with a bivariate p-box, that may be used as a model for the imprecise knowledge of a bivariate distribution function. We give necessary and sufficient conditions for this upper probability to be maxitive, characterize its focal elements, and study which maxitive functions can be obtained as upper probabilities of bivariate p-boxes.

Keywords: Coherent lower and upper probabilities · Uni- and bivariate p-boxes · Maxitive functions · Focal sets · Random sets

1 Introduction

Probability boxes can be given slightly different interpretations, such as confidence bands or the result of interval measurements; our main interpretation here will be that of envelopes of a set of probability distributions [4]. We focus on *bivariate* p-boxes, that were recently introduced in [6,8] as a generalization of p-boxes to the multivariate case. They arise in cases where we have imprecise knowledge about a bivariate distribution function, either because we cannot determine precisely its marginals, because we have uncertain information about the dependence between the underlying variables, or both.

In order to be able to take advantage of all the machinery that has already been developed within imprecise probability theory, it is important to clarify the connection between p-boxes and other imprecise probability models. This was done in [10] by Troffaes and Destercke, who showed that any univariate p-box is a particular case of a plausibility function. This was complemented in [11], where it was determined under which conditions a p-box can be regarded as a particular case of a maxitive function or a possibility measure. The interest of maxitive models is that they are determined by the restrictions to singletons, being thus more attractive from the computational point of view. They are linked in particular to fuzzy models.

Here we study to which extent the aforementioned results can be established for the bivariate case. This continues our work in [8], where we investigated

J.P. Carvalho et al. (Eds.): IPMU 2016, Part I, CCIS 610, pp. 141–152, 2016.
DOI: 10.1007/978-3-319-40596-4_13

when the lower/upper probability models associated with a bivariate p-box satisfied the properties of avoiding sure loss and coherence. We shall focus on finite spaces, and shall determine under which conditions the upper probability that summarizes the set of probability measures of a bivariate p-box is maxitive. As we shall show in Sect. 3, this only holds when one of the bounds of the bivariate p-box is non-informative: either the lower distribution function is constant on 0 or the upper distribution function is constant on 1. In addition to the necessary and sufficient conditions we shall establish, we shall determine the focal elements of the upper probability of a p-box when this upper probability is a maxitive measure. These focal elements can be used to ease the calculations in practical problems.

On the other hand, although in Sect. 3 we characterize those p-boxes that induce a maxitive measure, not all maxitive measures can be obtained as the upper probability of a p-box; in other words, maxitive measures are not a particular case of probability boxes. In Sect. 4, we characterize those maxitive measures that can be represented by means of bivariate p-boxes. Our paper concludes with some additional discussion in Sect. 5. Due to space limitations, proofs have been omitted.

2 Preliminary Concepts

Consider two finite ordered spaces $\mathcal{X} = \{x_1, \ldots, x_n\}, \mathcal{Y} = \{y_1, \ldots, y_m\}$, where $x_i < x_{i+1}$ and $y_j < y_{j+1}$ for every $i = 1, \ldots, n-1$ and every $j = 1, \ldots, m-1$, and let $\mathcal{X} \times \mathcal{Y}$ denote their product. Following the notation introduced in [8], we shall consider the cumulative sets $A_{x_i} = \{x \in \mathcal{X} \mid x \leq x_i\}$ and $A_{y_j} = \{y \in \mathcal{Y} \mid y \leq y_j\}$ for $x_i \in \mathcal{X}$ and $y_j \in \mathcal{Y}$, as well as the *cumulative rectangles* $A_{x_i,y_j} = \{(x,y) \in \mathcal{X} \times \mathcal{Y} \mid x \leq x_i, y \leq y_j\}$ for $(x_i, y_j) \in \mathcal{X} \times \mathcal{Y}$.

We shall consider probability measures P defined on either $\mathcal{X}, \mathcal{Y}$ or the product space $\mathcal{X} \times \mathcal{Y}$. Since these are finite spaces, any such P is determined by its restriction to singletons (its probability mass function). It is also determined by its cumulative distribution function (cdf for short).

Definition 1. *A* (univariate) cdf *on $\mathcal{X}$ is a non-decreasing function $F : \mathcal{X} \to [0,1]$ satisfying $F(x_n) = 1$.*

In the bivariate case, cdfs are defined as follows.

Definition 2. *A function $F : \mathcal{X} \times \mathcal{Y} \to [0,1]$ is a* bivariate cdf *when it is componentwise non-decreasing, $F(x_n, y_m) = 1$ and it satisfies the* rectangle inequality*:*

$$\Delta_F^{i,j} = F(x_i, y_j) + F(x_{i-1}, y_{j-1}) - F(x_{i-1}, y_j) - F(x_i, y_{j-1}) \geq 0$$

for any $i = 2, \ldots, n$, $j = 2, \ldots, m$.

In both cases there is a one-to-one correspondence between probability measures and distribution functions, by means of the cumulative sets.

Imprecise probabilities [1] is a generic term that may be used to refer to most of the mathematical models that serve as an alternative to probability theory in cases of imprecise or ambiguous information. In this paper we shall work with *lower* and *upper* probabilities and some particular cases of interest, such as plausibility or maxitive functions.

Definition 3 *[12, Sect. 2.7]. Consider a universe Ω. An* upper probability $\overline{P}$ *is a map $\overline{P} : \mathcal{K} \subseteq \mathcal{P}(\Omega) \to [0,1]$. Its conjugate* lower probability $\underline{P} : \mathcal{K}^c \to [0,1]$ *is defined on $\mathcal{K}^c = \{A^c \mid A \in \mathcal{K}\}$ by*

$$\underline{P}(A) = 1 - \overline{P}(A^c) \quad \forall A \in \mathcal{K}^c. \tag{1}$$

Any upper or lower probability determines a set of probabilities, or *credal set*, by $\mathcal{M}(\overline{P}) = \{P \text{ prob.} \mid P(A) \leq \overline{P}(A)\ \forall A \in \mathcal{K}\}$ and $\mathcal{M}(\underline{P}) = \{P \text{ prob.} \mid P(A) \geq \underline{P}(A)\ \forall A \in \mathcal{K}^c\}$. Taking into account Eq. (1), $\mathcal{M}(\underline{P}) = \mathcal{M}(\overline{P})$. Because of this, $\underline{P}, \overline{P}$ carry the same probabilistic information, and as a consequence it suffices to work with one of them. In this paper we will focus on upper probabilities. An upper probability $\overline{P}$ is called *coherent* when it is the upper envelope of the set $\mathcal{M}(\overline{P})$. In that case, it can be extended to the power set by means of the procedure of *natural extension*, which gives $\overline{E}(A) = \sup\{P(A) \mid P \in \mathcal{M}(\overline{P})\}$, $\forall A \in \mathcal{P}(\Omega)$.

In this paper, we shall focus on a particular family of coherent upper and lower probabilities: maxitive and minitive functions.

Definition 4. *A* maxitive function $\Pi : \mathcal{P}(\Omega) \to [0,1]$ *is a function satisfying $\Pi(A \cup B) = \max\{\Pi(A), \Pi(B)\}$, $\forall A, B \subseteq \Omega$. Its conjugate function N obtained by* (1) *is called a* minitive function*: $N(A \cap B) = \min\{N(A), N(B)\}$, $\forall A, B \subseteq \Omega$.*

Conjugate maxitive and minitive measures satisfy for instance $N(A) = 0$ if $\Pi(A) < 1$ and therefore $\Pi(A) = 1$ when $N(A) > 0$. Since we are dealing with finite spaces, maxitive and minitive functions are also *possibility* and *necessity* measures [3]. In particular, a maxitive measure is determined by its restriction to singletons, called its *possibility distribution*.

Maxitive and minitive functions are particular cases of *plausibility* and *belief* functions from evidence theory [9], and as a consequence they are uniquely determined by a *basic probability assignment* $m : \mathcal{P}(\Omega) \to [0,1]$, by means of the formula $\Pi(A) = \sum_{B \cap A \neq \emptyset} m(B)$. It holds that $\sum_{A \subseteq \Omega} m(A) = 1$. Any set A such that $m(A) > 0$ is called a *focal element* of m, and if we denote by $\mathcal{F}$ the set of focal elements, it holds that $\Pi(A) = \sum_{B \cap A \neq \emptyset, B \in \mathcal{F}} m(B)$.

Moreover, for maxitive functions the focal elements are nested, and can be determined by the possibility distribution: if we consider the family of sets $E_\alpha := \{\omega : \Pi(\{\omega\}) \geq \alpha\}$ for $\alpha \in [0,1]$, the finiteness of Ω allows us to conclude that $\{E_\alpha : \alpha \in [0,1]\} = \{E_{\alpha_1}, \ldots, E_{\alpha_n}\}$ for some $0 = \alpha_1 < \alpha_2 < \cdots < \alpha_n = 1$. Then the focal elements of Π are

$$\Omega = E_{\alpha_1} \supset E_{\alpha_2} \supset \cdots \supset E_{\alpha_n}, \tag{2}$$

and their mass functions are $m(E_{\alpha_1}) = \alpha_1, m(E_{\alpha_i}) = \alpha_i - \alpha_{i-1}$ for $i = 2, ..., n$.

In the same way as lower and upper probabilities can be used to model the imprecise information about a probability measure, there exist models for the imprecise information about a cumulative distribution function: probability boxes.

Definition 5 [4,8]. *A* (univariate) probability box *(or p-box)* $(\underline{F}_X, \overline{F}_X)$ *defined on* $\mathcal{X}$ *is a pair of ordered cdfs* $\underline{F}_X \leq \overline{F}_X$*, where* $\underline{F}_X, \overline{F}_X : \mathcal{X} \to [0,1]$*. A* bivariate p-box $(\underline{F}, \overline{F})$ *defined on* $\mathcal{X} \times \mathcal{Y}$ *is a pair of ordered componentwise non-decreasing functions* $\underline{F}, \overline{F} : \mathcal{X} \times \mathcal{Y} \to [0,1]$ *such that* $\underline{F} \leq \overline{F}$ *and* $\underline{F}(x_n, y_m) = \overline{F}(x_n, y_m) = 1$.

In particular, we shall say that $\underline{F}_X$ (resp., $\underline{F}_Y, \underline{F}$) is *vacuous* when $\underline{F}_X = I_{x_n}$ (resp., $\underline{F}_Y = I_{y_m}, \underline{F} = I_{(x_n, y_m)}$). Note that the lower and upper bounds of a bivariate p-box are not required to satisfy the rectangle inequality. The reason is that, as shown in [8, Example 1], the lower and upper envelopes of a set of bivariate cdfs may not satisfy such inequality in general.

As we said before, there is a one-to-one correspondence between probability measures and cdfs, both in the univariate and in the bivariate case. It is also possible to make a connection in the imprecise case. To see this, consider an upper probability $\overline{P}$ on $\mathcal{K} \supseteq \mathcal{K}_1$, where $\mathcal{K}_1 = \{A_x \mid x \in \mathcal{X}\} \cup \{A_x^c \mid x \in \mathcal{X}\}$. It defines a univariate p-box $(\underline{F}_X, \overline{F}_X)$ by $\overline{F}_X(x) = \overline{P}(A_x)$ and $\underline{F}_X(x) = 1 - \overline{P}(A_x^c)$ for any $x \in \mathcal{X}$. Conversely, we also have the following correspondence:

Theorem 1 [10,12]. *Let* $(\underline{F}_X, \overline{F}_X)$ *be a univariate p-box defined on* $\mathcal{X}$*. It defines a coherent upper probability on* $\mathcal{K}_1$ *by:*

$$\overline{P}(A_x) = \overline{F}_X(x) \text{ and } \overline{P}(A_x^c) = 1 - \underline{F}_X(x) \quad x \in \mathcal{X}. \tag{3}$$

Its associated coherent lower probability is $\underline{P}(A_x) = \underline{F}_X(x)$ *and* $\underline{P}(A_x^c) = 1 - \overline{F}_X(x)$.

Let us now turn to the bivariate case. Consider $\mathcal{K}_2 = \{A_{x,y} \mid (x,y) \in \mathcal{X} \times \mathcal{Y}\} \cup \{A_{x,y}^c \mid (x,y) \in \mathcal{X} \times \mathcal{Y}\}$. Any coherent upper probability on $\mathcal{K} \supseteq \mathcal{K}_2$ defines a bivariate p-box $(\underline{F}, \overline{F})$ by $\overline{F}(x,y) = \overline{P}(A_{x,y})$ and $\underline{F}(x,y) = 1 - \overline{P}(A_{x,y}^c)$ for any $(x,y) \in \mathcal{X} \times \mathcal{Y}$. Conversely [8, Definition 8], given a bivariate p-box $(\underline{F}, \overline{F})$ on $\mathcal{X} \times \mathcal{Y}$, its associated upper probability $\overline{P}_{(\underline{F}, \overline{F})}$ on $\mathcal{K}_2$ is given by:

$$\overline{P}_{(\underline{F}, \overline{F})}(A_{x,y}) = \overline{F}(x,y) \text{ and } \overline{P}_{(\underline{F}, \overline{F})}(A_{x,y}^c) = 1 - \underline{F}(x,y) \tag{4}$$

for any $(x,y) \in \mathcal{X} \times \mathcal{Y}$, and its lower probability is $\underline{P}_{(\underline{F}, \overline{F})}(A_{x,y}) = \underline{F}(x,y)$ and $\underline{P}_{(\underline{F}, \overline{F})}(A_{x,y}^c) = 1 - \overline{F}(x,y)$.

In general, $\overline{P}$ may not be coherent ([8, Example 2]). We will call a bivariate p-box *coherent* when its associated upper probability given by Eq. (4) is. In what follows, we shall assume that the upper probability derived from a (uni or bivariate) p-box by means of Eqs. (3) or (4) is coherent and that it is defined on the power set, using the notion of natural extension. There may be other upper probabilities compatible with the p-box by means of Eq. (4); here we consider the one determined by the values in $\mathcal{K}_2$ only.

Given a bivariate p-box $(\underline{F}, \overline{F})$ defined on $\mathcal{X} \times \mathcal{Y}$, its *marginal* univariate p-boxes $(\underline{F}_X, \overline{F}_X)$ and $(\underline{F}_Y, \overline{F}_Y)$ are given by:

$$\underline{F}_X(x_i) = \underline{F}(x_i, y_m) \text{ and } \overline{F}_X(x_i) = \overline{F}(x_i, y_m) \; \forall i = 1, \ldots, n.$$
$$\underline{F}_Y(y_j) = \underline{F}(x_n, y_j) \text{ and } \overline{F}_Y(y_j) = \overline{F}(x_n, y_j) \; \forall j = 1, \ldots, m.$$

The connection between maxitive functions and p-boxes was established in the univariate case in [11]:

Theorem 2 [11, Corollary 13]. *Let $(\underline{F}_X, \overline{F}_X)$ be a univariate p-box defined on $\mathcal{X}$, and denote by $\overline{P}$ its associated upper probability given by Eq.* (3) *and extended to $\mathcal{P}(\mathcal{X})$ using natural extension. Then, $\overline{P}$ is maxitive if and only if one of the following conditions holds:*

(a) $\underline{F}_X$ is 0-1 valued.
(b) $\overline{F}_X$ is 0-1 valued.

3 Bivariate p-boxes Inducing a Maxitive Function

The aim of this section is to investigate which conditions a bivariate p-box must satisfy in order to generate a maxitive function, trying to establish a characterization similar to the one Theorem 2 gives for the univariate case. Recall that we will consider a coherent bivariate p-box $(\underline{F}, \overline{F})$ defined on the product space $\mathcal{X} \times \mathcal{Y}$ of two finite and ordered spaces. $\overline{P}_{(\underline{F},\overline{F})}$ will denote the upper probability defined from $(\underline{F}, \overline{F})$ using Eq. (4) and extended to $\mathcal{P}(\mathcal{X} \times \mathcal{Y})$ by means of natural extension, and $\underline{P}_{(\underline{F},\overline{F})}$ will denote its conjugate lower probability.

Definition 6. *We shall say that $(\underline{F}, \overline{F})$ is a* maxitive bivariate p-box *when the natural extension of the upper probability $\overline{P}_{(\underline{F},\overline{F})}$ it induces by means of Eq.* (4) *is a maxitive function.*

We shall interpret $(\underline{F}, \overline{F})$ as a model for the probabilistic information on the joint behavior of two random variables X, Y. We shall assume that these two variables are *logically independent*, in the sense that any pair (x_i, y_j) in the product space $\mathcal{X} \times \mathcal{Y}$ is assumed to be possible. In addition, taking into account that both $\mathcal{X}, \mathcal{Y}$ are finite spaces, we shall assume that each pair has strictly positive upper probability:

$$\overline{P}_{(\underline{F},\overline{F})}(\{(x_i, y_j)\}) > 0 \; \forall i = 1, \ldots, n; j = 1, \ldots, m. \tag{5}$$

This means in particular that there is at least one precise model P that is compatible with our p-box $(\underline{F}, \overline{F})$, in the sense that $\underline{F} \leq F_P \leq \overline{F}$, and that gives positive probability to any pair (x_i, y_j). In other words, any singleton is a focal element of at least one of the compatible precise models. The condition will be instrumental for the technical results that follow.

3.1 Characterization of Maxitive Bivariate p-Boxes

We begin with a simple necessary condition for a p-box to be maxitive:

Proposition 1. *Let $(\underline{F}, \overline{F})$ be a maxitive bivariate p-box. Then, at least one of the following conditions holds:*

(a) $\underline{F}_X, \underline{F}_Y$ are vacuous.
(b) $\overline{F}_X, \overline{F}_Y$ are constantly 1.

Moreover, $\underline{F}(x_i, y_j) = \min\{\underline{F}(x_n, y_j), \underline{F}(x_i, y_m)\} = \min\{\underline{F}_Y(y_j), \underline{F}_X(x_i)\}\ \forall i, j$.

In the remainder of this section, we shall characterize p-boxes inducing a maxitive function. We begin by considering the following notion:

Definition 7. *We shall say that a bivariate p-box $(\underline{F}, \overline{F})$ is of* type-1 *when its lower distribution function $\underline{F}$ satisfies $\underline{F} = I_{(x_n, y_m)}$, i.e., when it is vacuous.*

The following proposition gives a useful property of this type of p-boxes:

Proposition 2. *Let $(\underline{F}, \overline{F})$ be a bivariate p-box. It is of type-1 if and only if both $\underline{F}_X, \underline{F}_Y$ are vacuous. Moreover, in such a case it holds that $\overline{P}_{(\underline{F}, \overline{F})}(\{(x_i, y_j)\}) = \overline{F}(x_i, y_j)$ for every $i = 1, \ldots, n; j = 1, \ldots, m$.*

Thus, type-1 p-boxes correspond to the first possibility determined in Proposition 1. Next we characterize under which conditions they are maxitive:

Proposition 3. *Let $(\underline{F}, \overline{F})$ be a bivariate p-box of type-1. It is maxitive if and only if one of the following conditions is satisfied:*

(a) $\overline{F}_X = 1$ and $\overline{F}(x_i, y_j) = \max\{\overline{F}(x_i, y_1), \overline{F}(x_1, y_j)\}\ \forall i, j$.
(b) $\overline{F}_Y = 1$ and $\overline{F}(x_i, y_j) = \max\{\overline{F}(x_i, y_1), \overline{F}(x_1, y_j)\}\ \forall i, j$.

The conditions established in this result are interesting because they mean that for maxitive type-1 p-boxes at least one of the marginal p-boxes must be completely uninformative: its lower bound is vacuous and its upper bound is constant on 1.

Next we consider a second type of bivariate p-boxes:

Definition 8. *We shall say that a bivariate p-box $(\underline{F}, \overline{F})$ is of* type-2 *if it is not of type-1 and its upper distribution function is constant on 1.*

These cover the second possibility depicted in Proposition 1: the case where both $\overline{F}_X, \overline{F}_Y$ are constant on 1.

Proposition 4. *Let $(\underline{F}, \overline{F})$ be a maxitive bivariate p-box with $\overline{F}_X, \overline{F}_Y$ constant on* 1 *and where $\underline{F}$ is non-vacuous. Then $\overline{F}$ is constantly 1.*

Taking this result into account, any bivariate distribution inducing a maxitive measure is of either type-1 or type-2. For type-1 bivariate p-boxes, we have characterized in Proposition 3 under which conditions they induce a maxitive measure. Our next result provides a characterization in the case of type-2 bivariate p-boxes.

Proposition 5. *Let $(\underline{F}, \overline{F})$ be a coherent bivariate p-box of type-2. Then, $\overline{P}_{(\underline{F}, \overline{F})}$ is maxitive if and only if $\underline{F}(x_i, y_j) = \min\{\underline{F}_X(x_i), \underline{F}_Y(y_j)\}\ \forall i = 1, \ldots, n; j = 1, \ldots, m$.*

3.2 Focal Sets of Maxitive Bivariate p-boxes

Next we determine the focal elements of maxitive bivariate p-boxes. We split our study in type-1 and type-2 bivariate p-boxes.

Given a maxitive bivariate p-box of type-1, Proposition 2 implies that the possibility distribution of $(\underline{F}, \overline{F})$ is given by $\overline{P}_{(\underline{F},\overline{F})}(\{(x_i, y_j)\}) = \overline{F}(x_i, y_j)\ \forall i, j$. Thus, the focal sets associated with $\overline{P}_{(\underline{F},\overline{F})}$ are defined by the different values of $\overline{F}$. From Eq. (2), if $\overline{F}$ takes the values $\gamma_1 < \ldots < \gamma_k = 1$, there are k focal sets given by $E_i = \{(x, y) : \overline{F}(x, y) \geq \gamma_i\}$, and their masses are $m(E_i) = \gamma_i - \gamma_{i-1}$ for any $i = 1, \ldots, k$ (we assume $\gamma_0 = 0$). Obviously, $E_i \supseteq E_{i+1}$ for any $i = 1, \ldots, k-1$ and $E_0 = \mathcal{X} \times \mathcal{Y}$. Our next result gives the form of the focal sets.

Proposition 6. *Let $(\underline{F}, \overline{F})$ be a maxitive bivariate p-box of type-1. Consider the set $S = \{(x_i, y_j) \mid \overline{F}(x_i, y_j) < \min\{\overline{F}(x_{i+1}, y_j), \overline{F}(x_i, y_{j+1})\}\}$. The following statements hold:*

1. *The set S can be expressed as $S = \{(x_{i_1}, y_{j_1}), \ldots, (x_{i_k}, y_{j_k})\}$, where $x_{i_1} \leq \ldots \leq x_{i_k}$ and $y_{j_1} \leq \ldots \leq y_{j_k}$.*
2. *The focal sets of $\overline{P}_{(\underline{F},\overline{F})}$ are $E_0 = \mathcal{X} \times \mathcal{Y}$, $E_l = A^c_{x_{i_l}, y_{j_l}}$, $\forall l = 1, \ldots, k$, with $m(E_l) = \overline{F}(x_{i_{l+1}}, y_{j_{l+1}}) - \overline{F}(x_{i_l}, y_{j_l})$, where $\overline{F}(x_{i_{k+1}}, y_{j_{k+1}}) = 1$, and $m(\mathcal{X} \times \mathcal{Y}) = \overline{F}(x_1, y_1)$.*

Example 1. Consider $\mathcal{X} = \{x_1, x_2, x_3\}$, $\mathcal{Y} = \{y_1, y_2, y_3\}$ and the bivariate p-box:

	(x_1, y_1)	(x_1, y_2)	(x_1, y_3)	(x_2, y_1)	(x_2, y_2)	(x_2, y_3)	(x_3, y_1)	(x_3, y_2)	(x_3, y_3)
$\underline{F}$	0	0	0	0	0	0	0	0	1
$\overline{F}$	0.1	0.5	0.7	0.8	0.8	0.8	1	1	1

By Proposition 3, it induces a maxitive measure. Using the notation of Proposition 6, the set S is given by $S = \{(x_1, y_1), (x_1, y_2), (x_1, y_3), (x_2, y_3)\}$. Therefore, according to Proposition 6, the focal sets of $\overline{P}_{(\underline{F},\overline{F})}$ and their masses are:

i	0	1	2	3	4
E_i	$\mathcal{X} \times \mathcal{Y}$	$A^c_{x_1,y_1}$	$A^c_{x_1,y_2}$	$A^c_{x_1,y_3}$	$A^c_{x_2,y_3}$
$m(E_i)$	0.1	0.4	0.2	0.1	0.2

They are depicted in Fig. 1. ♦

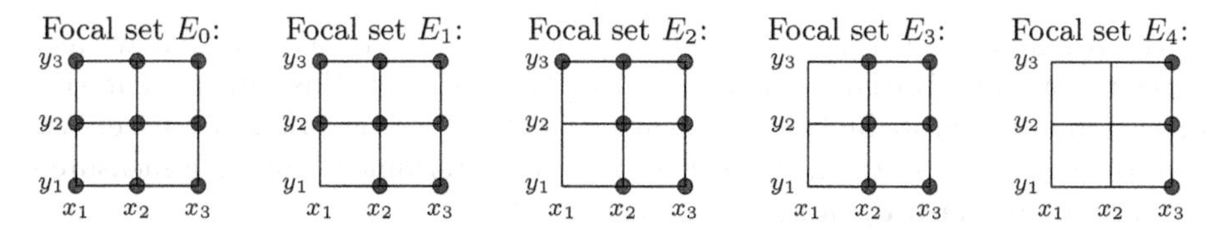

Fig. 1. Description of the focal sets of the upper probability $\overline{P}_{(\underline{F},\overline{F})}$ of Example 1.

Next, we consider type-2 bivariate p-boxes. From Proposition 1, when $\overline{P}_{(\underline{F},\overline{F})}$ is maxitive $\underline{F}(x_i, y_j) = \min\{\underline{F}_Y(y_j), \underline{F}_X(x_i)\}$. This means that $\underline{F}$ is indeed a bivariate cdf (see for Example [7]), and in particular the support of $P_{\underline{F}}$, the probability distribution associated with $\underline{F}$, can be expressed as

$$S = \{(u_1, v_1), \ldots, (u_k, v_k)\} \subset \mathcal{X} \times \mathcal{Y}$$

such that $(u_1, v_1) = (x_1, y_1)$, $(u_k, v_k) = (x_n, y_m)$, $u_i \leq u_{i+1}, v_i \leq v_{i+1}$ for any $i = 1, \ldots, k-1$, which implies that $P_{\underline{F}}(\{(u_i, v_i)\}) > 0$ and $\sum_{i=1}^{k} P_{\underline{F}}(\{(u_i, v_i\}) = 1$. Using S we can determine the focal sets of $\overline{P}_{(\underline{F},\overline{F})}$.

Proposition 7. *Let $(\underline{F}, \overline{F})$ be a maxitive bivariate p-box such that $\overline{F} = 1$, and consider the set S above. Then, the focal sets of $\overline{P}_{(\underline{F},\overline{F})}$ are $E_i = A_{u_i, v_i}$, $m(E_i) = \underline{F}(u_i, v_i) - \underline{F}(u_{i-1}, v_{i-1})$ for any $l = 1, \ldots, k$, where $\underline{F}(u_0, v_0) := 0$.*

Example 2. Consider the bivariate p-box $(\underline{F}, \overline{F})$ given by:

	(x_1, y_1)	(x_1, y_2)	(x_1, y_3)	(x_2, y_1)	(x_2, y_2)	(x_2, y_3)	(x_3, y_1)	(x_3, y_2)	(x_3, y_3)
$\underline{F}$	0.2	0.3	0.3	0.2	0.6	0.9	0.2	0.6	1
$\overline{F}$	1	1	1	1	1	1	1	1	1

The set S is given by $S = \{(x_1, y_1), (x_1, y_2), (x_2, y_2), (x_2, y_3), (x_3, y_3)\}$. Applying the previous proposition, the focal sets of $\overline{P}_{(\underline{F},\overline{F})}$ and their masses are:

E_i	A_{x_1,y_1}	A_{x_1,y_2}	A_{x_2,y_2}	A_{x_2,y_3}	A_{x_3,y_3}
$m(E_i)$	0.2	0.1	0.3	0.3	0.1

We have graphically depicted these focal sets in Fig. 2. ♦

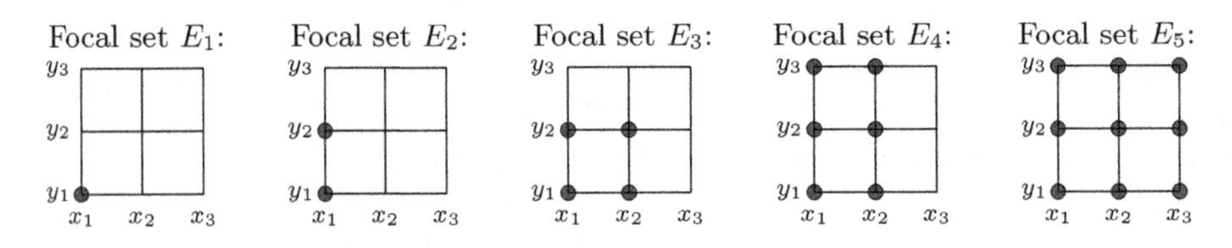

Fig. 2. Description of the focal sets of the upper probability $\overline{P}_{(\underline{F},\overline{F})}$ of Example 2.

3.3 Relevance of Restriction (5)

At the beginning of this section we have mentioned one restriction that we are imposing on the bivariate p-boxes we are considering in this paper: that the associated upper probability of any pair (x_i, y_j) in $\mathcal{X} \times \mathcal{Y}$ is strictly positive. We have then characterized which of these p-boxes determine a maxitive measure, and later determined the associated focal elements.

It is important to stress that Eq. (5) is essential to our results, in the sense that there are bivariate p-boxes that do not satisfy it and still induce a maxitive measure. Those bivariate p-boxes need not satisfy the conditions we have established in Propositions 6, 7, as our next example shows.

Example 3. Consider $\mathcal{X} = \{x_1, x_2\}$ and $\mathcal{Y} = \{y_1, y_2\}$ and the bivariate p-box:

	(x_1, y_1)	(x_1, y_2)	(x_2, y_1)	(x_2, y_2)
$\underline{F}$	0	0.5	0	1
$\overline{F}$	0	1	0.5	1

Note that this bivariate p-box is coherent because it is the lower/upper envelope of the set of distribution functions $\{F_1, F_2\}$, where:

	(x_1, y_1)	(x_1, y_2)	(x_2, y_1)	(x_2, y_2)
F_1	0	0.5	0.5	1
F_2	0	1	0	1

and therefore, according to [8, Proposition 9], $(\underline{F}, \overline{F})$ is coherent.

Secondly, it is easy to check that $\overline{P}_{(\underline{F},\overline{F})}$ is a maxitive function whose associated focal sets are $\{(x_1, y_2)\}$ and $\{(x_1, y_2), (x_2, y_1), (x_2, y_2)\}$, both with mass 0.5. These focal sets are not like those depicted in Propositions 6, 7.

The reason is that the bivariate p-box does not satisfy the necessary condition of Proposition 1, since $\overline{F}_X$ and $\underline{F}_Y$ are 0-1 valued, but $\underline{F}_X, \overline{F}_Y$ are not. The key here is that $\overline{P}_{(\underline{F},\overline{F})}(\{(x_1, y_1)\}) = 0$, and therefore our results from this section cannot be applied. ♦

4 From Maxitive Functions to Bivariate p-boxes

So far, we have been studying the conditions a bivariate p-box must satisfy in order to induce a maxitive function. However, it may be that not every maxitive function can be obtained as the upper probability of a bivariate p-box, meaning that not all maxitive models can be embedded in the framework of p-boxes:

Example 4. Consider $\mathcal{X} = \{x_1, x_2\}, \mathcal{Y} = \{y_1, y_2\}$ and let Π be the maxitive measure determined by the possibility distribution $\pi(x_1, y_1) = 1 = \pi(x_2, y_2)$, $\pi(x_1, y_2) = 0.5 = \pi(x_2, y_1)$. Assume that $(\underline{F}, \overline{F})$ is a bivariate p-box such that $\overline{P}_{(\underline{F},\overline{F})} = \Pi$. Then we should have $\overline{P}_{(\underline{F},\overline{F})}(\{x_1, y_1\}) = 1 = \overline{P}_{(\underline{F},\overline{F})}(\{x_2, y_2\})$, meaning that there must be P_1, P_2 in $\mathcal{M}(\overline{P}_{(\underline{F},\overline{F})})$ such that $P_1(\{x_1, y_1\}) = 1 = P_2(\{x_2, y_2\})$. But this implies that the p-box $(\underline{F}, \overline{F})$ must be vacuous, because $\underline{F}(x_2, y_1) = \underline{F}(x_1, y_2) \leq F_{P_2}(x_1, y_2) = F_{P_2}(x_2, y_1) = 0$, and $\overline{F}(x_1, y_1) \geq F_{P_1}(x_1, y_1) = 1$. As a consequence, $\overline{P}_{(\underline{F},\overline{F})}(\{x_1, y_2\}) = 1 \neq \pi(x_1, y_2)$, and therefore $\overline{P}_{(\underline{F},\overline{F})}$ does not coincide with Π. ♦

Given a maxitive function Π on $\mathcal{P}(\mathcal{X} \times \mathcal{Y})$, its associated bivariate p-box is

$$\overline{F}(x, y) = \Pi(A_{x,y}), \quad \underline{F}(x, y) = 1 - \Pi(A^c_{x,y}) \ \forall (x, y) \in \mathcal{X} \times \mathcal{Y}. \tag{6}$$

One difference with the results in [11, Section 5] is that we shall assume that the order in the universes $\mathcal{X}, \mathcal{Y}$ is fixed, and cannot be adapted to the values of the maxitive measure.

Interestingly, the p-box $(\underline{F}, \overline{F})$ defined above may not satisfy $\overline{P}_{(\underline{F},\overline{F})} = \Pi$, and in fact $\overline{P}_{(\underline{F},\overline{F})}$ may not even be maxitive:

Example 5. Consider $\mathcal{X} = \{x_1, x_2\}$ and $\mathcal{Y} = \{y_1, y_2\}$ and the maxitive function Π whose focal sets are $E_1 = \{(x_2, y_1)\}$, $E_2 = \mathcal{X} \times \mathcal{Y}$, each one with a mass of 0.5. If we denote by $(\underline{F}, \overline{F})$ its associated bivariate p-box by Eq. (6) and by $\underline{P}_{(\underline{F},\overline{F})}, \overline{P}_{(\underline{F},\overline{F})}$ the lower and upper probabilities it induces, we can easily prove that $\underline{P}_{(\underline{F},\overline{F})}, \overline{P}_{(\underline{F},\overline{F})}$ are not minitive and maxitive functions: it holds that $\underline{P}_{(\underline{F},\overline{F})}(A_{x_2,y_1}) = 0.5$ and $\underline{P}_{(\underline{F},\overline{F})}(A^c_{x_1,y_1}) = 0.5$, while $\underline{P}_{(\underline{F},\overline{F})}(A_{x_2,y_1} \cap A^c_{x_1,y_1}) = \underline{P}_{(\underline{F},\overline{F})}(\{(x_2, y_1)\}) = 0$, for instance considering the cdf $F \in (\underline{F}, \overline{F})$ given by $F(x_1, y_1) = F(x_1, y_2) = F(x_2, y_1) = 0.5, F(x_2, y_2) = 1$. Then, $\underline{P}_{(\underline{F},\overline{F})}$ is not minimum-preserving. This can also be seen using the results of the previous section: from Proposition 1, if $\overline{P}_{(\underline{F},\overline{F})}$ is maxitive, either $\overline{F}_X, \overline{F}_Y$ are constantly 1 or $\underline{F}_X, \underline{F}_Y$ are vacuous. However, in this example none of these conditions are satisfied, and therefore $\overline{P}_{(\underline{F},\overline{F})}$ cannot be maxitive. ♦

In what follows, we shall assume that the upper probability $\overline{P}_{(\underline{F},\overline{F})}$ induced by the $(\underline{F}, \overline{F})$ determined by (6) satisfies Eq. (5). We look for conditions in terms of the distribution π of the maxitive function Π to assure that it coincides with the upper probability induced by its associated bivariate p-box. One important remark is that, even if the bivariate p-box induced by Π is maxitive, it may not hold that $\overline{P}_{(\underline{F},\overline{F})} = \Pi$; an example will be given the context of random sets in Example 7 later on.

We begin by establishing a necessary condition for a maxitive measure to be attainable as the upper distribution of a bivariate p-box.

Proposition 8. *Let Π be a maxitive function on $\mathcal{P}(\mathcal{X} \times \mathcal{Y})$, and denote by π its associated possibility distribution. If $\Pi = \overline{P}_{(\underline{F},\overline{F})}$, where $(\underline{F}, \overline{F})$ is the p-box associated with Π by* (6)*, then either $\pi(x_1, y_1) = 1$ or $\max\{\pi(x_1, y_m), \pi(x_n, y_1)\} = 1$.*

Next we give sufficient conditions for the equality between $\overline{P}_{(\underline{F},\overline{F})}$ and Π. In that case, $\overline{P}_{(\underline{F},\overline{F})}$ is maxitive and, as we have explained in the previous result, either $\overline{F}$ or $\underline{F}$ is vacuous. We split now our study in these two cases.

Proposition 9. *Let Π be a maxitive function defined on $\mathcal{P}(\mathcal{X} \times \mathcal{Y})$, and denote by π its associated possibility distribution. The following are equivalent:*

(a) $\Pi = \overline{P}_{(\underline{F},\overline{F})}$ and $\underline{F}$ is vacuous.
(b) π is componentwise increasing, $\max\{\pi(x_n, y_1), \pi(x_1, y_m)\} = 1$ and $\pi(x, y) = \max\{\pi(x_1, y), \pi(x, y_1)\}$ for any (x, y).

Obviously this situation corresponds with the case of type-1 p-boxes. The next result corresponds with the case of $\overline{F} = 1$.

Proposition 10. *Let Π be a maxitive function defined on $\mathcal{P}(\mathcal{X} \times \mathcal{Y})$, and denote by π its associated distribution. The following are equivalent:*

(a) $\Pi = \overline{P}_{(\underline{F},\overline{F})}$ and $\overline{F} = 1$.
(b) π is componentwise decreasing, $\pi(x, y) = \min\{\pi(x_1, y), \pi(x, y_1)\}$ for any (x, y) and $\pi(x_1, y_1) = 1$.

4.1 Maxitive Bivariate p-boxes and Random Sets

One framework where maxitive functions arise naturally is when dealing with nested random sets. A *random set* $\Gamma : \Omega \to \mathcal{P}(\mathcal{X} \times \mathcal{Y})$ defined on the probability space $(\Omega, \mathcal{A}, P)$ is a multivalued map satisfying $\Gamma_*(A) = \{\omega \in \Omega \mid \Gamma(\omega) \subseteq A\} \in \mathcal{A}$ for every $A \subseteq \mathcal{X} \times \mathcal{Y}$. Any random set generates belief and plausibility functions on $\mathcal{P}(\mathcal{X} \times \mathcal{Y})$ by $\underline{P}(A) = P(\{\omega \in \Omega \mid \Gamma(\omega) \subseteq A\})$ and $\overline{P}(A) = P(\{\omega \in \Omega \mid \Gamma(\omega) \cap A \neq \emptyset\})$. When the images of Γ are almost surely nested, that is, when there is some $N \in \mathcal{A}$ with $P(N) = 0$ such that either $\Gamma(\omega_1) \subseteq \Gamma(\omega_2)$ or $\Gamma(\omega_2) \subseteq \Gamma(\omega_1)$ for any $\omega_1, \omega_2 \notin N$, then $\underline{P}$ and $\overline{P}$ are minitive and maxitive functions [5, Theorem 8].

Any random set defines a bivariate p-box $(\underline{F}, \overline{F})$ by means of $\underline{F}(x, y) = \underline{P}(A_{x,y})$ and $\overline{F}(x, y) = \overline{P}(A_{x,y})$. In general, this p-box is not as informative as the lower and upper probabilities of the random set [2]. In this section, we are going to study the case where the upper probability of the random set is maxitive. For this, we are going to study whether $\overline{P}$ and $\overline{P}_{(\underline{F},\overline{F})}$ coincide. Again, we shall assume that the p-box determined by the random set satisfies Eq. (5).

We begin by showing that in general $(\underline{F}, \overline{F})$ is not maxitive.

Example 6. Consider $\Omega = \{\omega_1, \omega_2\}$ with $P(\{\omega_1\}) = 0.6$ and $P(\{\omega_2\}) = 0.4$, and the random set Γ defined by $\Gamma(\omega_1) = \{x_1, x_2\} \times \{y_1, y_2\}, \Gamma(\omega_2) = \{(x_2, y_1)\}$. Since the images of Γ are nested $(\Gamma(\omega_2) \subset \Gamma(\omega_1))$, $\overline{P}$ is maxitive. Let us see that $\overline{P}_{(\underline{F},\overline{F})}$ is not. For this, we compute the bivariate p-box $(\underline{F}, \overline{F})$:

	(x_1, y_1)	(x_1, y_2)	(x_2, y_1)	(x_2, y_2)
$\underline{F}$	0	0	0.4	1
$\overline{F}$	0.6	0.6	1	1

$\overline{P}_{(\underline{F},\overline{F})}(\{(x_1, y_1)\}) = \overline{F}(x_1, y_1) = 0.6$ and $\overline{P}_{(\underline{F},\overline{F})}(\{(x_2, y_2)\}) = 1 - \underline{F}(x_2, y_1) = 1 - 0.4 = 0.6$. However, $\overline{P}_{(\underline{F},\overline{F})}(\{(x_1, y_1), (x_2, y_2)\}) = 1$, for instance considering the cdf F given by $F(x_1, y_1) = F(x_1, y_2) = F(x_2, y_1) = 0.6$ and $F(x_2, y_2) = 1$. Thus, $\overline{P}_{(\underline{F},\overline{F})}$ is not maxitive. ♦

In fact, $\overline{P}$ and $\overline{P}_{(\underline{F},\overline{F})}$ may not coincide even when both are maxitive:

Example 7. Consider $\Omega = \{\omega_1, \omega_2\}$ with $P(\{\omega_1\}) = P(\{\omega_2\}) = 0.5$ and the random set Γ given by $\Gamma(\omega_1) = \{(x_1, y_1), (x_3, y_3)\}$ and $\Gamma(\omega_2) = \{x_1, x_2, x_3\} \times \{y_1, y_2, y_3\}$. It can be easily proven that $\underline{F}$ is vacuous and $\overline{F}$ is constantly 1, and therefore $\overline{P}_{(\underline{F},\overline{F})}$ is maxitive. However, taking the probability distribution with $P(\{(x_2, y_2)\}) = 1$, its associated cdf belongs to $(\underline{F}, \overline{F})$, and therefore $\overline{P}_{(\underline{F},\overline{F})}(\{(x_2, y_2)\}) = 1 > 0.5 = \overline{P}(\{(x_2, y_2)\})$, which shows that $\overline{P}_{(\underline{F},\overline{F})}$ and $\overline{P}$ are different maxitive functions. ♦

5 Conclusions

The connection between p-boxes and maxitive measures established in [11] does not extend straightforwardly to the bivariate case: it does not suffice that either the lower or the upper distribution function is 0-1-valued, but they need moreover satisfy some additional properties: not only the focal elements of a maxitive bivariate p-box must be nested, but they must be nested families of cumulative rectangles (or their complementary sets, depending on whether we are working with type-1 or type-2 bivariate p-boxes). In addition, not every maxitive measure can be obtained as the upper probability of a bivariate p-box, as shown in Sect. 4.

There are several open lines of research that arise from our work. On the one hand, we should extend our results to arbitrary universes, not necessarily finite. In that case, we should distinguish between maxitive and possibility measures, and we envisage that some additional continuity properties should be imposed if a bivariate p-box is to induce a possibility measure, similarly to what has been established in [11]. Related to this, our assumption of positive upper probability on the singletons, that has been instrumental for many of our results, cannot be satisfied when one of the marginal universes is uncountable.

Even if we focus on the finite case, we should also study the connection between bivariate p-boxes and maxitive functions when this positivity condition does not hold. As we have already shown in Example 3, if we lift this restriction we may have other bivariate p-boxes inducing a maxitive measure. Thus, the connection between these two models in the general case should be determined.

Acknowledgements. The research reported in this paper has been supported by project TIN2014-59543-P.

References

1. Augustin, T., Coolen, F., de Cooman, G., Troffaes, M. (eds.): Introduction to Imprecise Probabilities. Wiley, Hoboken (2014)
2. Couso, I., Sánchez, L., Gil, P.: Imprecise distribution function associated to a random set. Inf. Sci. **159**, 109–123 (2004)
3. Dubois, D., Prade, H.: Possibility Theory. Plenum Press, New York (1988)
4. Ferson, S., Kreinovich, V., Ginzburg, L., Myers, D., Sentz, K.: Constructing probability boxes and Dempster-Shafer structures. Technical report, Sandia (2003)
5. Miranda, E., Couso, I., Gil, P.: Relationships between possibility measures and nested random sets. Int. J. Uncertainty Fuzziness Knowl.-Based Syst. **10**, 1–15 (2002)
6. Montes, I., Miranda, E., Pelessoni, R., Vicig, P.: Sklar's theorem in an imprecise setting. Fuzzy Sets Syst. **278**, 48–66 (2015)
7. Nelsen, R.: An Introduction to Copulas. Springer, New York (2006)
8. Pelessoni, R., Vicig, P., Montes, I., Miranda, E.: Bivariate p-boxes. Int. J. Uncertainty Fuzziness Knowl.-Based Syst. **24**, 229–263 (2016)
9. Shafer, G.: A Mathematical Theory of Evidence. Princeton University Press, Princeton (1976)
10. Troffaes, M., Destercke, S.: Probability boxes on totally preordered spaces for multivariate modelling. Int. J. App. Reason. **52**, 767–791 (2011)
11. Troffaes, M., Miranda, E., Destercke, S.: On the connection between probability boxes and possibility measures. Inf. Sci. **224**, 88–108 (2013)
12. Walley, P.: Statistical Reasoning with Imprecise Probabilities. Chapman and Hall, London (1991)

Sets of Priors Reflecting Prior-Data Conflict and Agreement

Gero Walter[1(✉)] and Frank P.A. Coolen[2]

[1] School of Industrial Engineering, Eindhoven University of Technology, Eindhoven, Netherlands
g.m.walter@tue.nl

[2] Department of Mathematical Sciences, Durham University, Durham, UK
frank.coolen@durham.ac.uk

Abstract. In Bayesian statistics, the choice of prior distribution is often debatable, especially if prior knowledge is limited or data are scarce. In imprecise probability, sets of priors are used to accurately model and reflect prior knowledge. This has the advantage that prior-data conflict sensitivity can be modelled: Ranges of posterior inferences should be larger when prior and data are in conflict. We propose a new method for generating prior sets which, in addition to prior-data conflict sensitivity, allows to reflect *strong prior-data agreement* by decreased posterior imprecision.

Keywords: Bayesian inference · Strong prior-data agreement · Prior-data conflict · Imprecise probability · Conjugate priors

1 Introduction

The Bayesian approach to inference [6] offers the advantage to combine data and prior expert knowledge in a unified reasoning process. It combines a parametric *sample model*, denoted by a conditional distribution $f(\boldsymbol{x} \mid \vartheta)$ of data $\boldsymbol{x} = (x_1, \ldots, x_n)$ given parameter ϑ with a *prior distribution* $f(\vartheta)$, expressing expert opinion on ϑ. Given $\boldsymbol{x}$, the prior distribution is updated by Bayes' Rule to obtain the *posterior distribution* $f(\vartheta \mid \boldsymbol{x}) \propto f(\boldsymbol{x} \mid \vartheta) \cdot f(\vartheta)$. The choice of prior distribution is often debatable. One can employ sensitivity analysis to study the effect of different prior distributions on the inferences, as done in robust Bayesian methods [2]. The method presented in this paper also uses sets of priors, with interpretation in line with theory of imprecise probability [1,8], considering sets of posterior distributions as the proper method to express the precision of probability statements themselves: the smaller the set of posteriors, the more precise the probability statements. This relation should hold in particular in case of *prior-data conflict*: From the viewpoint of the prior $f(\vartheta)$, the observed data $\boldsymbol{x}$ seem very surprising, i.e., information from data is in conflict with prior assumptions [4]. This is most relevant when there is not enough data to largely reduce the influence of the prior on the posterior; it is then unclear whether to put

J.P. Carvalho et al. (Eds.): IPMU 2016, Part I, CCIS 610, pp. 153–164, 2016.
DOI: 10.1007/978-3-319-40596-4_14

more trust to prior assumptions or to the observations, and posterior inferences should clearly reflect this state of uncertainty. [11] pointed out that both precise and imprecise models based on conjugate priors can be insensitive to prior-data conflict.

For Bayesian inference based on a precise conjugate prior, learning from data amounts to averaging between prior and data [10, Sect. 1.2.3.1]. This is the root of prior-data conflict insensitivity: When observed data are very different to what is assumed in the prior, this conflict is simply averaged out and not reflected in the variance of the posterior, giving a false sense of certainty: A posterior with small variance indicates that we know what is going on quite precisely, but in case of prior-data conflict we do not. Prior-data conflict is reflected by increased imprecision in inferences, so more cautious probability statements, when using carefully tailored sets of conjugate priors [11]. One approach is to define sets of conjugate priors via sets of canonical parameters which ensure prior-data conflict sensitivity. [11] suggested a parameter set shape that balances tractability and ease of elicitation with desired inference properties. This approach has been applied in common-cause failure modelling [7] and system reliability [12]. We further refine this approach by complementing the increased imprecision reaction to prior-data conflict with further reduced imprecision if prior and data coincide especially well, which we call *strong prior-data agreement*. These desired inference properties are achieved through a novel, more complex parameter set shape. For ease of presentation, we restrict presentation to the Beta-Binomial model, the approach is generalizable to arbitrary canonical conjugate priors. Section 2 gives a quick overview on Bayesian inference with sets of Beta priors. The new shape is defined in terms of a parametrization recently suggested by Bickis [3] and explained in Sect. 3. We suggest a shape in this parametrization that reacts to both prior-data conflict and strong prior-data agreement (Sect. 4). Section 5 discusses generalizations and potential applications.

2 Generalized Bayesian Inference for Binary Data

The Binomial distribution models the probability to observe s successes in n independent trials given p, the success probability in each trial. In a Bayesian setting, information about p is expressed by a prior distribution $f(p)$ and updating is straightforward if one uses a *conjugate* prior distribution, for which the posterior distribution belongs to the same family as the prior, just with updated parameters. The conjugate prior for the Binomial distribution is the Beta distribution,[1]

$$f(p) \propto p^{n^{(0)}y^{(0)}-1}(1-p)^{n^{(0)}(1-y^{(0)})-1}, \tag{1}$$

written here in terms of the *canonical* parameters $n^{(0)} > 0$ and $y^{(0)} \in (0,1)$, where $y^{(0)}$ is the prior expectation for p, and $n^{(0)}$ is a pseudocount or prior

[1] We denote prior parameter values by upper index $^{(0)}$ and posterior parameter values, after n observations, by upper index$^{(n)}$.

strength parameter. The posterior given s successes in n trials is a Beta distribution with updated parameters

$$n^{(n)} = n^{(0)} + n\,, \qquad y^{(n)} = \frac{n^{(0)}}{n^{(0)} + n} \cdot y^{(0)} + \frac{n}{n^{(0)} + n} \cdot \frac{s}{n}\,. \tag{2}$$

The posterior mean $y^{(n)}$ for p is a weighted average of the prior mean $y^{(0)}$ and the observed fraction of successes s/n, with weights proportional to $n^{(0)}$ and n, respectively. This averaging between prior and data is a concern if observed data differ greatly from what is expressed in the prior, as such conflict is averaged out and not reflected in the posterior.

[11] showed that it is possible to obtain a meaningful reaction to prior-data conflict by using sets of priors $\mathcal{M}^{(0)}$ produced through parameter sets $\Pi^{(0)} = [\underline{n}^{(0)}, \overline{n}^{(0)}] \times [\underline{y}^{(0)}, \overline{y}^{(0)}]$. More generally, [10, Sect. 3.1] describes a framework for Bayesian inference using sets of conjugate priors based on arbitrary parameter sets $\Pi^{(0)}$. Here, each prior parameter pair $(n^{(0)}, y^{(0)}) \in \Pi^{(0)}$ corresponds to a Beta prior, so $\mathcal{M}^{(0)}$ can be taken directly as a set of Beta priors. Alternatively, one may take the convex hull of all Beta priors with $(n^{(0)}, y^{(0)}) \in \Pi^{(0)}$ as $\mathcal{M}^{(0)}$; $\mathcal{M}^{(0)}$ then consists of all finite mixtures of Beta distributions with $(n^{(0)}, y^{(0)}) \in \Pi^{(0)}$. It is a modeling decision whether to take $\mathcal{M}^{(0)}$ as containing only Beta priors or also the mixtures. In the first case, bounds for all inferences can be obtained by optimizing over $\Pi^{(0)}$. In the second case, optimizing over $\Pi^{(0)}$ will only yield bounds for all inferences that are *linear functions* of $n^{(0)}$ and $y^{(0)}$, as the linearity ensures that bounds must correspond to the extreme points of the convex set of priors, which are the Beta priors with $(n^{(0)}, y^{(0)}) \in \Pi^{(0)}$. In both cases, the set of posteriors $\mathcal{M}^{(n)}$ is obtained by updating each prior in $\mathcal{M}^{(0)}$ according to Bayes' Rule. This element-by-element updating can be rigorously justified as ensuring coherence [8, Sect. 2.5], and was termed "Generalized Bayes' Rule" by Walley [8, Sect. 6.4]. In the first case, $\mathcal{M}^{(n)}$ is a set of Beta distributions with parameters $(n^{(n)}, y^{(n)})$, obtained by updating $(n^{(0)}, y^{(0)}) \in \Pi^{(0)}$ according to (2), leading to the set of updated parameters

$$\Pi^{(n)} = \Big\{(n^{(n)}, y^{(n)}) \mid (n^{(0)}, y^{(0)}) \in \Pi^{(0)} = [\underline{n}^{(0)}, \overline{n}^{(0)}] \times [\underline{y}^{(0)}, \overline{y}^{(0)}]\Big\}\,. \tag{3}$$

In the second case, the set of Beta distributions corresponding to $(n^{(n)}, y^{(n)}) \in \Pi^{(n)}$ forms the extreme points of the convex set of posteriors $\mathcal{M}^{(n)}$, such that, just like $\mathcal{M}^{(0)}$, $\mathcal{M}^{(n)}$ can be described as a set of all finite mixtures of Beta distributions with $(n^{(n)}, y^{(n)}) \in \Pi^{(n)}$, see [10, pp. 56f].

$\mathcal{M}^{(n)}$ forms the basis for all inferences, leading to probability *ranges* obtained by minimizing and maximizing over $\mathcal{M}^{(n)}$. For example, the posterior predictive probability for the event that a future single draw is a success is equal to $y^{(n)}$; for an imprecise model $\mathcal{M}^{(0)}$ based on $\Pi^{(0)}$, the lower and upper probability are

$$\inf_{\Pi^{(n)}} y^{(n)} = \inf_{\Pi^{(0)}} \frac{n^{(0)} y^{(0)} + s}{n^{(0)} + n} \qquad \text{and} \qquad \sup_{\Pi^{(n)}} y^{(n)} = \sup_{\Pi^{(0)}} \frac{n^{(0)} y^{(0)} + s}{n^{(0)} + n}\,.$$

The relation between $\mathit{\Pi}^{(0)}$ and $\mathcal{M}^{(0)}$, as well as between $\mathit{\Pi}^{(n)}$ and $\mathcal{M}^{(n)}$, allows to characterize model properties through properties of $\mathit{\Pi}^{(0)}$ and $\mathit{\Pi}^{(n)}$, as is done in [10, Sect. 3.1.2–3.1.4]. The well-known Imprecise Dirichlet Model [9] corresponds to a choice of $\mathit{\Pi}^{(0)} = n^{(0)} \times (\underline{y}^{(0)}, \overline{y}^{(0)})$ where $(\underline{y}^{(0)}, \overline{y}^{(0)}) = (0, 1)$. The model proposed by [5] generally assumes $\mathit{\Pi}^{(0)} = n^{(0)} \times [\underline{y}^{(0)}, \overline{y}^{(0)}]$, and was shown to be insensitive to prior-data conflict by [11], who proposed parameter sets $\mathit{\Pi}^{(0)} = [\underline{n}^{(0)}, \overline{n}^{(0)}] \times [\underline{y}^{(0)}, \overline{y}^{(0)}]$ instead. Indeed, for $\mathit{\Pi}^{(0)} = n^{(0)} \times [\underline{y}^{(0)}, \overline{y}^{(0)}]$, we get $\mathit{\Pi}^{(n)} = n^{(n)} \times [\underline{y}^{(n)}, \overline{y}^{(n)}]$, where $\underline{y}^{(n)} = (n^{(0)}\underline{y}^{(0)} + s)/(n^{(0)} + n)$ and $\overline{y}^{(n)} = (n^{(0)}\overline{y}^{(0)} + s)/(n^{(0)} + n)$. The posterior imprecision in the y dimension, denoted by $\Delta_y(\mathit{\Pi}^{(n)})$, is then

$$\Delta_y(\mathit{\Pi}^{(n)}) = \overline{y}^{(n)} - \underline{y}^{(n)} = \frac{n^{(0)}(\overline{y}^{(0)} - \underline{y}^{(0)})}{n^{(0)} + n},$$

and so the same for any fixed n, independent of s. In contrast, parameter sets $\mathit{\Pi}^{(0)} = [\underline{n}^{(0)}, \overline{n}^{(0)}] \times [\underline{y}^{(0)}, \overline{y}^{(0)}]$ provide prior-data conflict sensitivity, since

$$\Delta_y(\mathit{\Pi}^{(n)}) = \frac{\overline{n}^{(0)}(\overline{y}^{(0)} - \underline{y}^{(0)})}{\overline{n}^{(0)} + n} + \inf_{y^{(0)} \in [\underline{y}^{(0)}, \overline{y}^{(0)}]} |s/n - y^{(0)}| \frac{n(\overline{n}^{(0)} - \underline{n}^{(0)})}{(\underline{n}^{(0)} + n)(\overline{n}^{(0)} + n)}.$$

The shape of $\mathit{\Pi}^{(0)}$ poses a trade-off [10, Sect. 3.1.4]: Less complex shapes are easy to handle and lead to tractable models, but will offer less flexibility in expressing prior information and may have undesired inference properties. In contrast, more complex shapes may allow for more sophisticated model behaviour at the cost of more involved handling.

3 A Novel Parametrization for Beta Priors

A conjugate Beta prior is updated by a shift in the parameter space, given by rewriting (2):

$$n^{(0)} \mapsto n^{(0)} + n, \qquad y^{(0)} \mapsto y^{(0)} + \frac{s - ny^{(0)}}{n^{(0)} + n}.$$

The shift for the n coordinate is the same for all elements $(n^{(0)}, y^{(0)})$ of $\mathit{\Pi}^{(0)}$. The shift in the y coordinate depends on $n^{(0)}$, n, s, and the location of $y^{(0)}$ itself (in fact, how far $y^{(0)}$ is from s/n). The shape of $\mathit{\Pi}^{(0)}$ changes during the update step to $\mathit{\Pi}^{(n)}$, the effects on posterior inferences may be difficult to grasp. To isolate the influence of a set shape, we consider a recently proposed parametrization [3], where each coordinate has the same shift in updating, such that updating a prior set corresponds to a shift of the entire set. In this novel parametrization, a conjugate prior is represented by a coordinate $(\eta_0^{(0)}, \eta_1^{(0)})$, related to $(n^{(0)}, y^{(0)})$ by

$$n^{(0)} = \eta_0^{(0)} + 2, \qquad y^{(0)} = \frac{\eta_1^{(0)}}{\eta_0^{(0)} + 2} + \frac{1}{2}. \tag{4}$$

The domain of η_0 and η_1 in case of the Beta-Binomial model is

$$\mathrm{H} = \left\{ (\eta_0, \eta_1) \middle| \eta_0 > -2,\ |\eta_1| < \frac{1}{2}(\eta_0 + 2) \right\}, \tag{5}$$

the Bayes update step in terms of η_0 and η_1 is given by

$$\eta_0^{(n)} = \eta_0^{(0)} + n\,, \quad \eta_1^{(n)} = \eta_1^{(0)} + \frac{1}{2}(s - (n - s)) = \eta_1^{(0)} + s - \frac{n}{2}\,. \tag{6}$$

Each success thus leads to a step of 1 in the η_0 direction and of $+\frac{1}{2}$ in the η_1 direction, while each failure leads to a step of 1 in the η_0 direction and of $-\frac{1}{2}$ in the η_1 direction. While $y^{(0)}$ had the convenient property of being equal to the prior expectation for p, η_1 is only slightly more difficult to interpret. From (4) we can derive that points $(\eta_0, \eta_1) \in \mathrm{H}$ on rays emanating from the coordinate $(-2, 0)$, i.e., coordinates satifying $\eta_1 = (\eta_0 + 2)(y_c - 1/2)$, will have a constant expectation of y_c. The domain H, and these *rays of constant expectation* emanating from the coordinate $(-2, 0)$, can be seen in Fig. 1.

In the parametrization in terms of $(n^{(0)}, y^{(0)})$, posterior inferences based on $y^{(n)}$ become less imprecise with increasing n because $\Delta_y(\mathit{\Pi}^{(n)}) \to 0$ for $n \to \infty$. In the domain H, parameter sets do not change size during update, but the rays of constant expectation fan out for increasing n. The more $\mathrm{H}^{(n)}$ is located to the right, the fewer rays of constant expectation it intercepts, and so imprecision decreases. Imprecision in terms of $y^{(n)}$ can thus be imagined as the size of the 'shadow' that a set $\mathrm{H}^{(n)}$ casts given a light source in $(-2, 0)$. The smaller this shadow, the less imprecise the inferences. Denoting the bounds of this shadow by

$$\underline{y}_{\mathrm{H}}^{(n)} := \min_{(\eta_0^{(n)}, \eta_1^{(n)}) \in \mathrm{H}^{(n)}} \frac{\eta_1^{(n)}}{\eta_0^{(n)} + 2} + \frac{1}{2}\,, \quad \overline{y}_{\mathrm{H}}^{(n)} := \max_{(\eta_0^{(n)}, \eta_1^{(n)}) \in \mathrm{H}^{(n)}} \frac{\eta_1^{(n)}}{\eta_0^{(n)} + 2} + \frac{1}{2}\,,$$

we call the η_0 coordinate of $\arg\min_{(\eta_0, \eta_1) \in \mathrm{H}^{(n)}} y^{(n)}$ and $\arg\max_{(\eta_0, \eta_1) \in \mathrm{H}^{(n)}} y^{(n)}$ the *lower* and *upper touchpoint* of $\mathrm{H}^{(n)}$ responsible for the shadow $[\underline{y}_{\mathrm{H}}^{(n)}, \overline{y}_{\mathrm{H}}^{(n)}]$. Mutatis mutandis, the same definitions can be made for the prior set $\mathrm{H}^{(0)}$. Due to the fanning out of rays, most shapes for $\mathrm{H}^{(0)}$ will lead to decreasing imprecision for increasing n. For example, models with $\mathit{\Pi}^{(0)} = n^{(0)} \times [\underline{y}^{(0)}, \overline{y}^{(0)}]$ are represented by a line segment $\mathrm{H}^{(0)} = \eta_0^{(0)} \times [\underline{\eta}_1^{(0)}, \overline{\eta}_1^{(0)}]$, and imprecision decreases because a line segment of fixed size will cast a smaller shadow when further to the right, as illustrated in Fig. 1.

For prior-data conflict sensitivity, we need sets $\mathrm{H}^{(0)}$ that cover a range of η_0 values, just like sets $\mathit{\Pi}^{(0)}$ with a range of $n^{(0)}$ values are necessary to ensure this property. A set $\mathrm{H}^{(0)}$ that is elongated along a certain ray of constant expectation will behave similar to a rectangular $\mathit{\Pi}^{(0)}$. When shifted along its ray of constant expectation, imprecision will be reduced as the shadow of $\mathrm{H}^{(0)}$ will become smaller just as described above for line segments. When $\mathrm{H}^{(0)}$ is instead shifted away from its ray of constant expectation, imprecision will increase, as a prolonged shape that is now turned away from its ray will cast a larger shadow.

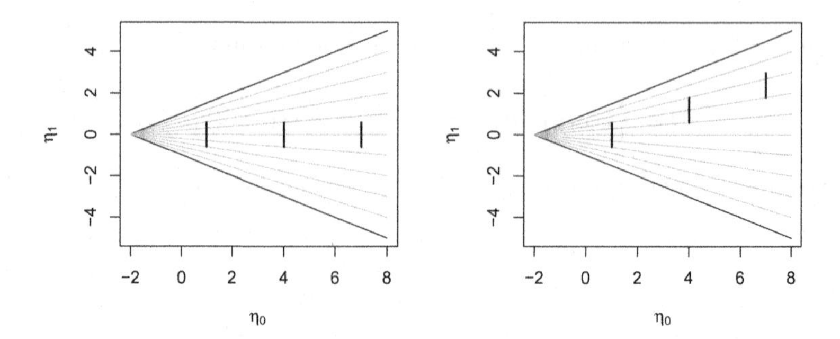

Fig. 1. Prior parameter set $\mathrm{H}^{(0)} = \eta_0^{(0)} \times [\underline{\eta}_1^{(0)}, \overline{\eta}_1^{(0)}]$ and respective posterior sets $\mathrm{H}^{(n)}$ for $s/n = 0.5$ (left) and $s/n = 0.9$ (right). Bounds for the domain H are in black, with rays of constant expectation for $y_c = \{0.1, 0.2, \ldots, 0.9\}$ in grey. Note that all sets have the same size, imprecision decreasing only through their position on the η_0 axis.

4 The Boatshape

The shape for $\mathrm{H}^{(0)}$ that we suggest to obtain both prior-data conflict sensitivity and reduced imprecision in case of strong prior-data agreement looks like a boat with a transom stern (see Fig. 2 below). The curvature along its length in the direction of its constant rays of expectation leads to smaller $\Delta_y(\mathit{\Pi}^{(n)})$ as compared to a rectangular $\mathit{\Pi}^{(0)}$ with the same prior range $\Delta_y(\mathit{\Pi}^{(0)})$, see Fig. 3. The strong prior-data agreement effect is realized through the touchpoints determining $\underline{y}_{\mathrm{H}}^{(n)}$ and $\overline{y}_{\mathrm{H}}^{(n)}$ moving along the shape during updating, see Sect. 4.2. This is advantageous since the spread of the Beta posteriors is determined by $\eta_0 = n^{(0)} - 2$. In case of strong prior-data agreement, variances in the 'critical' distributions at the boundary of the posterior expectation interval $[\underline{y}_{\mathrm{H}}^{(n)}, \overline{y}_{\mathrm{H}}^{(n)}]$ will thus be lower leading to reduced imprecision.

4.1 Basic Definition

We suggest an exponential function for the contours of a boat-shaped parameter set $\mathrm{H}^{(0)}$. We first restrict discussion on prior sets that are symmetric to the η_0 axis, i.e., centered around $y_c = 0.5$. Sets $\mathrm{H}^{(0)}$ with central ray $y_c \neq 0.5$ can be obtained by rotating the set around $(\eta_0, \eta_1) = (-2, 0)$ such that y_c forms the axis of symmetry. Results for sets with $y_c = 0.5$ generalize straightforwardly to the case $y_c \neq 0.5$; an example is given in Fig. 6. The lower and the upper contour functions are defined as

$$\underline{c}^{(0)}(\eta_0) = -a\left(1 - e^{-b(\eta_0 - \underline{\eta}_0)}\right), \qquad \overline{c}^{(0)}(\eta_0) = a\left(1 - e^{-b(\eta_0 - \underline{\eta}_0)}\right), \tag{7}$$

where $a > 0$ and $b > 0$ are parameters controlling the shape of $\mathrm{H}^{(0)}$, which is defined as

$$\mathrm{H}^{(0)} = \{(\eta_0, \eta_1) \colon \underline{\eta}_0 \leq \eta_0 \leq \overline{\eta}_0, \underline{c}^{(0)}(\eta_0) \leq \eta_1 \leq \overline{c}^{(0)}(\eta_0)\}\,. \tag{8}$$

A prior boatshape set, together with corresponding posterior sets for different observations, is shown in Fig. 2. The same prior and posterior sets in terms of $(n^{(0)}, y^{(0)})$ are depicted in Fig. 3.

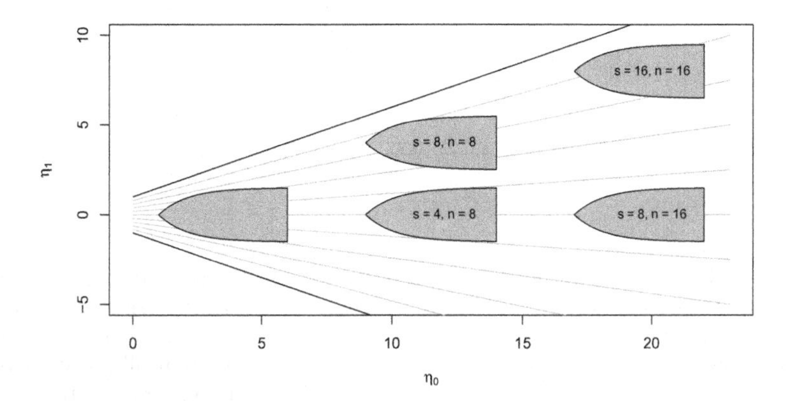

Fig. 2. Boatshape prior and posterior sets for data in accordance and in conflict with the prior set. The parameters for the prior set are $\underline{\eta}_0 = 1$, $\overline{\eta}_0 = 6$, $a = 1.5$, and $b = 0.9$. While the posterior sets for $\frac{s}{n} = 0.5$ move along the ray for $y_c = 0.5$, the posterior sets for $\frac{s}{n} = 1$ are shifted away from the ray for $y_c = 0.5$, resulting in increased posterior imprecision. Note that lower and upper touchpoints are in the middle of the contour for the prior set and the posterior sets resulting for data $\frac{s}{n} = 0.5$, while the lower touchpoint is at the end for the posterior sets for data $\frac{s}{n} = 1$.

The parameter a determines the half-width of the set; the size in the η_1 dimension would be $2a$ if $\overline{\eta}_0 \to \infty$. Parameter b determines the 'bulkyness' of the shape. Together with $\underline{\eta}_0$, a and b determine $[\underline{y}_{\mathrm{H}}^{(0)}, \overline{y}_{\mathrm{H}}^{(0)}]$. Decreasing $\underline{\eta}_0$, or increasing a or b, leads to a wider $[\underline{y}_{\mathrm{H}}^{(0)}, \overline{y}_{\mathrm{H}}^{(0)}]$. $\overline{\eta}_0$ plays only a role in determining when the 'unhappy learning' phase starts (see end of Sect. 4.3).

We see from the prior set in Fig. 3 that the lower and the upper bound for $y^{(0)}$ is attained in the middle of the set contour. To determine $\underline{y}_{\mathrm{H}}^{(0)}$ and $\overline{y}_{\mathrm{H}}^{(0)}$, we need to find the corresponding touchpoints $\eta_0^{l\,(0)}$ and $\eta_0^{u\,(0)}$ by identifying the rays of constant expectation that are tangents to $\mathrm{H}^{(0)}$ and then solving for η_0. Since $\mathrm{H}^{(0)}$ is symmetric to the η_0 axis, we have $\eta_0^{l\,(0)} = \eta_0^{u\,(0)}$ and we will determine $\eta_0^{u\,(0)}$ by considering the upper contour tangent. We get

$$1 + b(\eta_0^{u\,(0)} + 2) \stackrel{!}{=} e^{b(\eta_0^{u\,(0)} - \underline{\eta}_0)} . \tag{9}$$

This equation only has one solution for $\eta_0^{u\,(0)} > \underline{\eta}_0$ that is, however, not available in closed form. Generally, the nearer $\eta_0^{u\,(0)}$ is to $\underline{\eta}_0$, the larger $\frac{d}{d\eta_0}\overline{c}^{(0)}(\eta_0^{u\,(0)})$, such that $\overline{y}_{\mathrm{H}}^{(0)}$ is further away from $\frac{1}{2}$.

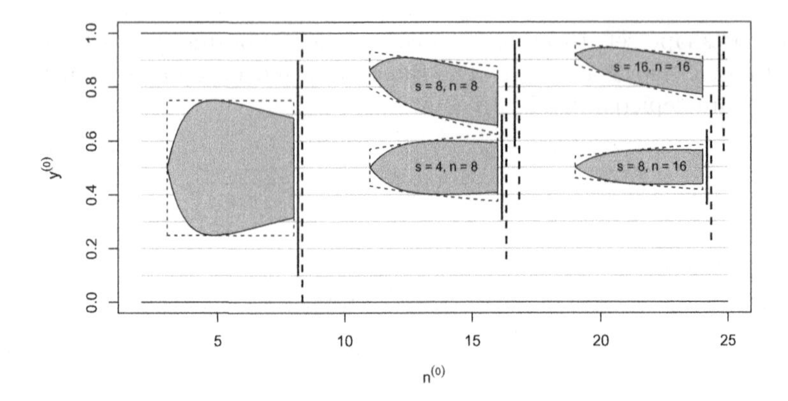

Fig. 3. Boatshape prior and posterior sets from Fig. 2 in the $(n^{(0)}, y^{(0)})$ parametrization. The rectangular prior set with the same range for $y^{(0)}$ as the prior boatshape set and the corresponding posterior sets are drawn with dashed lines. Unions of symmetric credibility intervals ($\gamma = 0.5$) are drawn as vertical bars. Note that all posterior boatshape sets have shorter $y^{(n)}$ ranges than their corresponding posterior rectangle sets, and boatshape credibility regions are especially short for posterior sets where $\frac{s}{n} = 0.5$.

4.2 Strong Prior-Data Agreement Property

Sets (8) lead to reduced imprecision in inferences when data are strongly supporting prior information as the touchpoint moves further to the right in that case. The basic shape is symmetric around the η_0 axis ($\mathrm{H}^{(0)}$ has central ray $y_c = 0.5$), and updating with strong-agreement data $s/n = 0.5$ means that $\mathrm{H}^{(0)}$ is shifted along the η_0 axis by n, such that also $\mathrm{H}^{(n)}$ is symmetric around the η_0 axis. We thus need to consider only one touchpoint. Movement to the right means that the upper posterior touchpoint $\eta_0^{u\,(n)}$ is larger than the updated prior touchpoint $\eta_0^{u\,(0)}$, so we need to show that $\eta_0^{u\,(n)} > \eta_0^{u\,(0)} + n$. The upper contour for the posterior boatshape, updated with $s = \frac{n}{2}$, is $\overline{c}^{(0)}$ from (7) shifted to the right by n, i.e., $\overline{c}^{(n)}(\eta_0) = a - ae^{-b(\eta_0 - n - \underline{\eta}_0)}$. The equation to identify the posterior upper touchpoint is

$$1 + b(\eta_0^{u\,(n)} + 2) \stackrel{!}{=} e^{b(\eta_0^{u\,(n)} - n - \underline{\eta}_0)} . \tag{10}$$

Comparing (10) to (9), both have a linear function with slope b and intercept $1 + 2b$ on the left hand side. The exponential function on the right hand side of (10) is the function on the right hand side of (9) shifted to the right by n. We can picture this situation as in Fig. 4: $\eta_0^{u\,(0)}$ is identified by the intersection of the linear function with the left, non-shifted exponential, whereas $\eta_0^{u\,(n)}$ is at the intersection of the linear function with the right, shifted exponential. Since $b > 0$, we have indeed $\eta_0^{u\,(n)} > \eta_0^{u\,(0)} + n$.

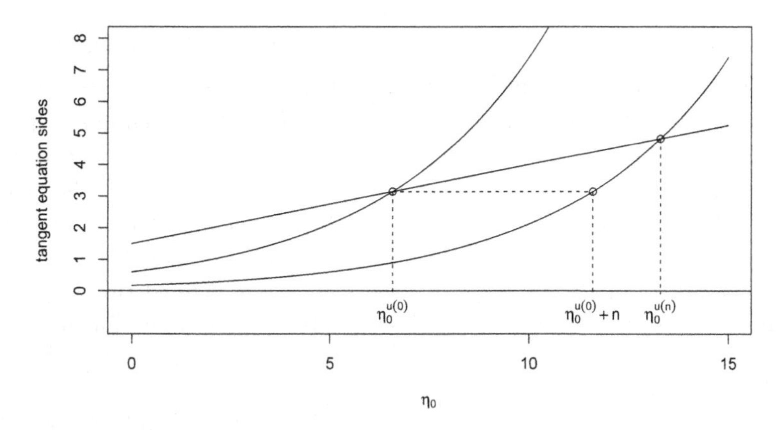

Fig. 4. Illustration for the argument that $\eta_0^{u\,(n)} > \eta_0^{u\,(0)} + n$.

4.3 Touchpoints for Arbitrary Updates

Let us now consider the update of the basic boatshape (8) in the general case $s \neq \frac{n}{2}$, investigating the effect that different values of s for fixed n have on $\eta_0^{l\,(n)}$ and $\eta_0^{u\,(n)}$[2]. For $s \neq \frac{n}{2}$, $\mathrm{H}^{(n)}$ is not symmetric to the η_0 axis, and we have to derive the touchpoints $\eta_0^{l\,(n)}$ and $\eta_0^{u\,(n)}$ separately. The upper and lower contours for $\mathrm{H}^{(n)}$ are

$$\overline{c}^{(n)}(\eta_0) = s - \frac{n}{2} + a - ae^{-b(\eta_0 - n - \underline{\eta}_0)}\,, \quad \underline{c}^{(n)}(\eta_0) = s - \frac{n}{2} - a + ae^{-b(\eta_0 - n - \underline{\eta}_0)}\,,$$

leading to

$$\frac{a}{s - \frac{n}{2} + a}\big(1 + b(\eta_0^{u\,(n)} + 2)\big) \stackrel{!}{=} e^{b(\eta_0^{u\,(n)} - n - \underline{\eta}_0)}\,, \tag{11}$$

$$\frac{a}{\frac{n}{2} - s + a}\big(1 + b(\eta_0^{l\,(n)} + 2)\big) \stackrel{!}{=} e^{b(\eta_0^{l\,(n)} - n - \underline{\eta}_0)}\,. \tag{12}$$

We see that the graph from Fig. 4 holds here as well, except that the linear function on the left hand side of (11) and (12) is changed in slope and intercept by a factor. (Equivalently, we can consider it to be rotated around the root $-2 - \frac{1}{b}$.) For $s = \frac{n}{2}$, this factor is 1 for both (11) and (12), reducing to (10). Due to symmetry of $\mathrm{H}^{(0)}$ we consider, without loss of generality, only the case $s > \frac{n}{2}$.

The factor $\frac{a}{s - \frac{n}{2} + a}$ in (11) is smaller than 1 and decreasing in s to $\frac{a}{\frac{n}{2} + a}$ for $s = n$. As the linear function's slope will be less steep (the intercept is lowered as well), the intersection with the exponential function moves to the left, i.e. $\eta_0^{u\,(n)}(s) < \eta_0^{u\,(n)}(\frac{n}{2})$ for $\frac{n}{2} < s < n$. This means that $\overline{y}_{\mathrm{H}}^{(n)}(s) > \overline{y}_{\mathrm{H}}^{(n)}(\frac{n}{2})$. However, $\eta_0^{u\,(n)}(s)$ can decrease only to $\underline{\eta}_0 + n$: When $\eta_0^{u\,(n)}(s)$ reaches the left end of $\mathrm{H}^{(n)}$

[2] We treat s as a a real-value in $[0, n]$ for convenience of our discussions; this does not affect the conclusions.

at $\underline{\eta}_0 + n$, the gradual increase of $\overline{y}_{\mathrm{H}}^{(n)}$ through the changing tangent slope is replaced by a different change mechanism, where increase of $\overline{y}_{\mathrm{H}}^{(n)}$ is solely due to the shift of $\mathrm{H}^{(n)}$ in the η_1 coordinate. Due to (4), $\overline{y}_{\mathrm{H}}^{(n)}$ is then linear in s.

In (12), the factor to the linear function is $\frac{a}{\frac{n}{2}-s+a}$. Here, we have to distinguish the two cases $\frac{n}{2} \leq s < \frac{n}{2} + a$ and $s \geq \frac{n}{2} + a$. In the first case, the factor is larger than 1 and increasing in s so the intersection of the linear function with the exponential function will move to the right, such that $\eta_0^{l\,(n)}(s)$ becomes larger, and $\underline{y}_{\mathrm{H}}^{(n)}$ increases. In the second case, the factor is undefined (for $s = \frac{n}{2} + a$) or negative (for $s > \frac{n}{2} + a$) and there is no intersection of the linear function with the exponential function for any $\eta_0 > \underline{\eta}_0 + n$. So for $s \geq \frac{n}{2} + a$, the whole set is above the η_0 axis, and the touchpoint must thus be at $\overline{\eta}_0 + n$. Actually, $\eta_0^{l\,(n)}(s) = \overline{\eta}_0 + n$ already for some $\frac{n}{2} \leq s < \frac{n}{2} + a$, when the intersection point reaches $\overline{\eta}_0 + n$. At this point, gradual increase of $\underline{y}_{\mathrm{H}}^{(n)}$ resulting from the movement of $\eta_0^{l\,(n)}(s)$ along the set towards the right is replaced by a linear increase in s. Again, this is because the η_1 coordinate is incremented according to (6), and from (4) we see that $\underline{y}^{(n)}$ is linear in η_1.

4.4 Posterior Imprecision

We now summarize the results from Sect. 4.3 and give two numerical examples. For $s > \frac{n}{2}$, both $\overline{y}^{(n)}$ and $\underline{y}^{(n)}$ will at first increase gradually with s, as $\eta_0^{u\,(n)}$ moves to the left, and $\eta_0^{l\,(n)}$ moves to the right. We will call such updating of the prior parameter set, where both lower and upper posterior touchpoints are in the middle of the set, *happy learning*. At some s^u, $\eta_0^{u\,(n)}$ will reach $\underline{\eta}_0 + n$, and at some s^l, $\eta_0^{l\,(n)}$ will reach $\overline{\eta}_0 + n$. Whether $s^l < s^u$ or vice versa depends on the choice of parameters $\underline{\eta}_0, \overline{\eta}_0, a$ and b. When s is larger than either s^l or s^u, we have *unhappy learning*, where data s is very much out of line with our prior expectations as expressed by $\mathrm{H}^{(0)}$. Ultimately, when $s > s^u$ and $s > s^l$, both $\overline{y}_{\mathrm{H}}^{(n)}$ and $\underline{y}_{\mathrm{H}}^{(n)}$ will increase linearly in s, but with different slopes. $\overline{y}_{\mathrm{H}}^{(n)}$ will increase with slope $\frac{1}{\underline{\eta}_0+n+2}$, whereas $\underline{y}_{\mathrm{H}}^{(n)}$ will increase with the lower slope $\frac{1}{\overline{\eta}_0+n+2}$.

These findings are illustrated in Fig. 5 for a boatshape set with $y_c = 0.5$, $\underline{\eta}_0 = -1$, $\overline{\eta}_0 = 20$, $a = 1$ and $b = 0.4$. These are compared to a rectangular set and two line segment sets with the same $y^{(0)}$ range. Here we see a linear increase of $\overline{y}_{\mathrm{H}}^{(n)}$ for $s < 4$ and a superlinear increase for $s \geq 4$. We have happy learning for $s \in [4, 6]$, and unhappy learning for $s \neq [4, 6]$. For $s \approx 5$, Δ_y for the boatshape set is about half of Δ_y for the rectangle set. The line segment sets lead to very short $y^{(n)}$ ranges, but do not reflect prior-data conflict.

Figure 6 depicts a numerical example for the case $y_c = 0.75$. Notice that the rotated boatshape parameter set is not symmetric in the $(n^{(0)}, y^{(0)})$ space. We see that $[\underline{y}_{\mathrm{H}}^{(n)}, \overline{y}_{\mathrm{H}}^{(n)}]$ is nearly as short as $[\underline{y}^{(n)}, \overline{y}^{(n)}]$ for the line segments sets when $s \approx 0.75$, but that unlike those, the boatshape offers prior-data conflict sensitivity. Interestingly, all four sets lead to a similar $\underline{y}^{(n)}$ for $s < 5$.

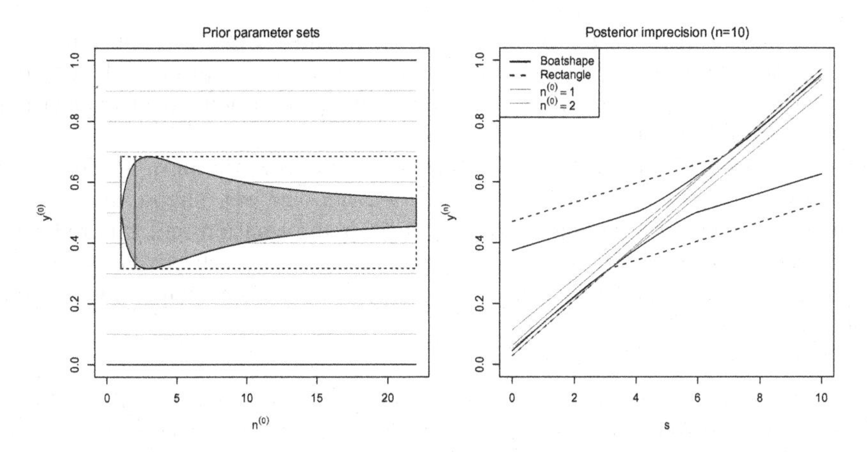

Fig. 5. Boatshape set with $y_c = 0.5$ together with rectangle set, $1 \times [\underline{y}^{(0)}, \overline{y}^{(0)}]$ and $2 \times [\underline{y}^{(0)}, \overline{y}^{(0)}]$ with same prior imprecision (left), and the corresponding lower and upper bounds for $y^{(n)}$ as functions of s (right). (Color figure online)

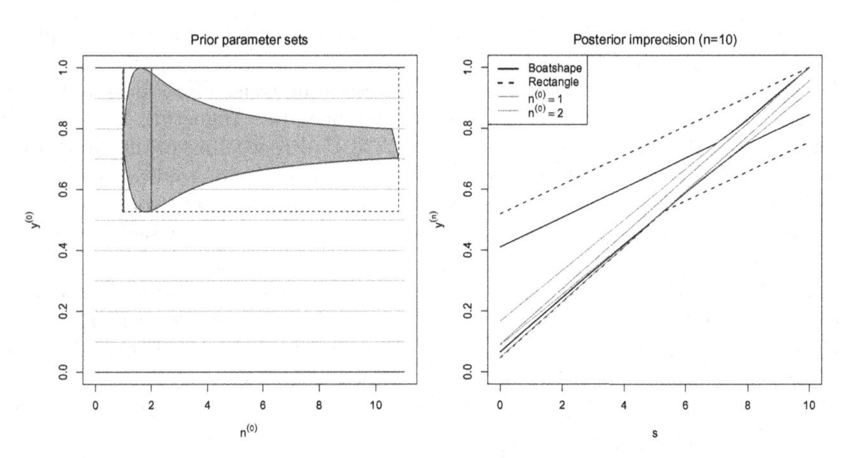

Fig. 6. Boatshape set with $y_c = 0.75$ together with rectangle set, $1 \times [\underline{y}^{(0)}, \overline{y}^{(0)}]$ and $2 \times [\underline{y}^{(0)}, \overline{y}^{(0)}]$ with same prior imprecision (left), and the corresponding lower and upper bounds for $y^{(n)}$ as functions of s (right). (Color figure online)

5 Concluding Remarks

For application of the novel method presented in this paper, elicitation of the boatshape set parameters must be considered, pre-posterior analysis seems useful for this. It will be interesting to investigate whether another way of defining a set aligned to a certain ray could be useful, namely by shifting each part of $\mathrm{H}^{(0)}$ from (8) in the η_1 dimension onto the desired ray (similar to turning a right prism into an oblique prism). Alternatives to the functional form of the contour functions (7) could also be worth of study. The method was presented here for the case of binary data, it can be easily generalized to cover all sample distributions that belong to the exponential family, since for those a conjugate prior in the (η_0, η_1) parametrization can be constructed having a purely data-dependent translation as update step [3, p. 56].

In the parameter space described in Sect. 3, updating the prior set amounts to a purely data-dependent translation, leaving the set shape unchanged. As shown, this enables flexible modeling of prior information and tailored posterior inference properties, while remaining within the generalized Bayesian paradigm, hence opening a wide field of research on prior set shapes for specific inference objectives.

Acknowledgements. Gero Walter was supported by the Dinalog project "Coordinated Advanced Maintenance and Logistics Planning for the Process Industries" (CAMPI).

References

1. Augustin, T., Coolen, F., de Cooman, G., Troffaes, M.: Introduction to Imprecise Probabilities. Wiley, Chichester (2014)
2. Berger, J., et al.: An overview of robust Bayesian analysis. TEST **3**, 5–124 (1994)
3. Bickis, M.: The geometry of imprecise inference. In: Augustin, T., Doria, S., Miranda, E., Quaeghebeur, E. (eds.) ISIPTA 2015: Proceedings of the Ninth International Symposium on Imprecise Probability: Theories and Applications, pp. 47–56. SIPTA (2015). http://www.sipta.org/isipta15/data/paper/31.pdf
4. Evans, M., Moshonov, H.: Checking for prior-data conflict. Bayesian Analysis **1**, 893–914 (2006). http://projecteuclid.org/euclid.ba/1340370946
5. Quaeghebeur, E., de Cooman, G.: Imprecise probability models for inference in exponential families. In: Cozman, F., Nau, R., Seidenfeld, T. (eds.) ISIPTA 2005, Proceedings of the Fourth International Symposium on Imprecise Probabilities and Their Applications, pp. 287–296. SIPTA, Manno (2005)
6. Robert, C.P.: The Bayesian Choice: From Decision-Theoretic Foundations to Computational Implementation. Springer, New York (2007)
7. Troffaes, M., Walter, G., Kelly, D.: A robust Bayesian approach to modelling epistemic uncertainty in common-cause failure models. Reliab. Eng. Syst. Saf. **125**, 13–21 (2014)
8. Walley, P.: Statistical Reasoning with Imprecise Probabilities. Chapman and Hall, London (1991)
9. Walley, P.: Inferences from multinomial data: Learning about a bag of marbles. J. R. Stat. Soc. Ser. B **58**(1), 3–34 (1996)
10. Walter, G.: Generalized Bayesian inference under prior-data conflict. Ph.D. thesis, Ludwig-Maximilians-Universität München (2013). http://nbn-resolving.de/urn:nbn:de:bvb:19-170598
11. Walter, G., Augustin, T.: Imprecision and prior-data conflict in generalized Bayesian inference. J. Stat. Theory Pract. **3**, 255–271 (2009)
12. Walter, G., Graham, A., Coolen, F.P.A.: Robust Bayesian estimation of system reliability for scarce and surprising data. In: Podofillini, L., Sudret, B., Stojadinović, B., Zio, E., Kröger, W. (eds.) Safety and Reliability of Complex Engineered Systems: ESREL 2015, pp. 1991–1998. CRC Press, Boca Raton (2015)

On Imprecise Statistical Inference for Accelerated Life Testing

Frank P.A. Coolen[1(✉)], Yi-Chao Yin[2], and Tahani Coolen-Maturi[3]

[1] Department of Mathematical Sciences, Durham University, Durham, UK
frank.coolen@durham.ac.uk

[2] Institute of Reliability Engineering, University of Electronic Science and Technology of China, Chengdu, Sichuan, China

[3] Durham University Business School, Durham University, Durham, UK

Abstract. Accelerated life testing provides an interesting challenge for quantification of the uncertainties involved, in particular due to the required linking of items' failure times, or failure time distributions, at different stress levels. This paper provides an initial exploration of the use of statistical methods based on imprecise probabilities for accelerated life testing, with explicit emphasis on prediction of a future observation at the actual stress level of interest. We apply nonparametric predictive inference at that stress level, in combination with an estimated parametric form for the function linking different levels. For the latter aspect imprecision is introduced, leading to observations at stress levels other than the actual level of interest, to be transformed to intervals at the latter level. We believe that this is the first attempt to apply imprecise probability methods to accelerated life testing scenarios, and argue in favour of doing so. The paper concludes with a discussion of related research topics.

Keywords: Accelerated life testing · Imprecise probability · Nonparametric predictive inference · Power-weibull model · Right-censored data · Survival functions

1 Introduction

Testing of highly reliable components is often complicated if, under normal conditions, failures tend to occur only after a very long time, e.g. many years. This makes it impossible to infer aspects of the components' failure time distribution at a relatively early stage, for example for comparison of components from different manufacturers. An effective way to still enable data collection for such inferences is provided by so-called *Accelerated Life Testing* (ALT), which is general terminology for a range of test scenarios, which have in common that the components are tested under conditions that differ from the normal conditions.

Yi-Chao Yin gratefully acknowledges financial support from the China Scholarship Council to visit Durham University

J.P. Carvalho et al. (Eds.): IPMU 2016, Part I, CCIS 610, pp. 165–176, 2016.
DOI: 10.1007/978-3-319-40596-4_15

Under the changed conditions, the failure time distribution will change corresponding to reduction of failure times, for example the stress level or temperature at which the components function may be increased for the tests. There is a wide variety of test designs, including constant stress testing, step stress testing and progressive stress testing. These test methods and a variety of statistical methods that can be applied for such methods are in detail described by Nelson [15].

Constant stress testing is the most widely used ALT design: the components are divided into several groups and all components in a group are tested at a constant stress level. In this paper we only consider this relatively straightforward form of ALT. The main challenge for statistical methods for ALT lies in the obvious fact that information from a test with increased stress levels must be transformed to information that can be considered as representative for information about component failure times under the normal conditions. Due to the enormous practical relevance of ALT and the obvious challenges for statistical inference based on ALT data, many statistical models and methods for ALT data have been presented [15]. A standard model for failure time data resulting from ALT is the power-Weibull model, which we use as the first stage in the new approach which we explore in this paper.

The power-Weibull model consists of a Weibull model for failure times at stress level $i = 0, 1, 2, \ldots, k$, where level $i = 0$ is the normal level and levels $i = 1, 2, \ldots, k$ represent k increased stress levels. These Weibull distributions for different stress levels are assumed to have the same shape parameter β, but different scale parameters α_i. Assuming that the stress level is quantified by a single positive measurement V_i for stress level i, which is an increasing function of the stress level i (one can e.g. think of voltage), the different α_i values are assumed to satisfy the equation

$$\alpha_i = \alpha \left(\frac{V_0}{V_i} \right)^p$$

such that $\alpha_0 = \alpha$ is the Weibull scale parameter at the normal stress level and p is the parameter of the power-law which models the links of the different Weibull distributions at different stress levels. For clarity, in this paper we use the parametrization for the Weibull distribution with shape parameter β and scale parameter α corresponding to the survival function

$$P(T > t) = \exp \left\{ - \left(\frac{t}{\alpha} \right)^\beta \right\}$$

A useful alternative way to understand this power-law link between the different stress levels is provided by the fact that, under this model assumption, an observation t^i at stress level i, so subject to stress V_i, can be interpreted as an observation

$$t^i \left(\frac{V_i}{V_0} \right)^p$$

at stress level 0. As the objective of ALT is, obviously, to have a reduction of failure times at higher stress levels, it is necessary to assume that $p > 0$ with $p > 1$ most likely in practical applications. Given failure time data, which can contain right-censored data under the usual assumption that the cause of censoring holds no information about the remaining future time to failure of a right-censored observation, the parameters α, β and p of this model can be estimated by maximising the likelihood function, which requires a quite straightforward numerical optimisation method; computation in this paper was performed with the statistical software R.

Section 2 of this paper provides a short introduction to nonparametric predictive inference (NPI), in particular it provides the NPI lower and upper survival functions for a future observation based on failure time data including right-censored observations, these are used in the new statistical method for ALT data which is presented in Sect. 3. This method consists of two stages. In the first stage, the power-Weibull model is assumed for the observations at all stress levels simultaneously, including the parameter p representing the link between different stress levels. Based on all the data, the parameters in this model are estimated using maximum likelihood estimation. In the second stage, only the point estimate for the link parameter p is used to transform data from the different stress levels to the normal level. Then NPI is used with these combined data to provide lower and upper survival functions for the next observation at the normal stress level. In Sect. 4 this approach is extended by including imprecision in the link parameter, which leads to observations at levels other than the normal stress to be transformed to interval-valued observations at the normal stress level, with the width of these intervals increasing as function of the difference between the corresponding stress level and the normal stress level. These interval-valued observations are then used in the NPI approach to lead to new lower and upper survival functions with increased imprecision. This can be interpreted as a straightforward method to provide robust predictive inferences based on ALT data. This is the first investigation towards developing NPI methods for ALT data and the proposed method is rather ad hoc, but the idea to use imprecision as a safeguard against lack of detailed knowledge in ALT settings seems attractive. Section 5 provides a brief discussion of related research challenges, including performance evaluation of the method proposed here and some suggestions for different imprecise probabilistic methods for ALT data.

2 Nonparametric Predictive Inference

Nonparametric predictive inference (NPI) is a statistical method based on Hill's assumption $A_{(n)}$ (Hill [9]), which gives a direct conditional probability for a future observable random quantity, given observed values of related random quantities [1,3,4,16]. Let $Y_1, \ldots, Y_n, Y_{n+1}$ be positive, continuous and exchangeable random quantities representing event times [8]. Suppose that the values of $Y_1, \ldots, Y_n$ are observed and the corresponding ordered observed values are denoted by $0 < y_1 < \ldots < y_n < \infty$, for ease of notation let $y_0 = 0$ and

$y_{n+1} = \infty$. For ease of presentation, it is assumed that no ties occur among the observed values. It is quite straightforward to deal with tied observations in this setting, by assuming that tied observations differ by small amounts which tend to zero. For the random quantity Y_{n+1} representing a future observation, based on n observations, the assumption $A_{(n)}$ [9] is $P(Y_{n+1} \in (y_{i-1}, y_i)) = 1/(n+1)$ for $i = 1, \ldots, n+1$. $A_{(n)}$ does not assume anything else, and can be interpreted as a post-data assumption related to exchangeability [8]. Inferences based on $A_{(n)}$ are predictive and nonparametric, and can be considered suitable if there is hardly any knowledge about the random quantity of interest, other than the n observations, or if one does not want to use such information, e.g. to study effects of additional assumptions underlying other statistical methods. $A_{(n)}$ is not sufficient to derive precise probabilities for many events of interest, but it provides bounds for probabilities via the 'fundamental theorem of probability' [8], which are lower and upper probabilities in the theory of imprecise probability with strong consistency properties [1,2].

In reliability analyses, events of interest are often failures of units, but such data are often affected by right-censoring, where for a unit it is only known that it has not yet failed by a specific time. Coolen and Yan [7] presented a generalization of $A_{(n)}$, called 'right-censoring $A_{(n)}$', or rc-$A_{(n)}$, which is suitable for NPI with right-censored data and uses the additional assumption that, at the moment of censoring, the residual time to failure of a right-censored unit is exchangeable with the residual times to failure of all other units that have not yet failed or been censored.

Suppose that there are n observations consisting of u failure times, $x_1 < x_2 < \ldots < x_u$, and $n-u$ right-censored observations, $c_1 < c_2 < \ldots < c_{n-u}$. Let $x_0 = 0$ and $x_{u+1} = \infty$. Suppose further that there are s_i right-censored observations in the interval (x_i, x_{i+1}), denoted by $c_1^i < c_2^i < \ldots < c_{s_i}^i$, so $\sum_{i=0}^{u} s_i = n - u$. We introduce notation d_j^i for any observation, either a failure or right-censoring time, with $d_0^i = x_i$ and $d_j^i = c_j^i$ for $j = 1, \ldots, s_i$ and $i = 0, 1, \ldots, u$. Let $\tilde{n}_{c_r}$ and $\tilde{n}_{d_j^i}$ be the number of units in the risk set just prior to time c_r and d_j^i, respectively, with the definition $\tilde{n}_0 = n+1$ for ease of notation. Let $d_{s_i+1}^i = d_0^{i+1} = x_{i+1}$ for $i = 0, 1, \ldots, u-1$, and note that the product taken over an empty set is defined as equal to one. Based on the assumption rc-$A_{(n)}$ [7], the NPI lower and upper survival functions for the failure time of the next unit, $\underline{S}_{X_{n+1}}(t)$ and $\overline{S}_{X_{n+1}}(t)$, respectively, are as follows [13,14]. For $t \in [d_j^i, d_{j+1}^i)$ with $i = 0, 1, \ldots, u$ and $j = 0, 1, \ldots, s_i$,

$$\underline{S}_{X_{n+1}}(t) = \frac{1}{n+1} \tilde{n}_{d_j^i} \prod_{\{r: c_r < d_j^i\}} \frac{\tilde{n}_{c_r} + 1}{\tilde{n}_{c_r}} \tag{1}$$

and for $t \in [x_i, x_{i+1})$ with $i = 0, 1, \ldots, u$,

$$\overline{S}_{X_{n+1}}(t) = \frac{1}{n+1} \tilde{n}_{x_i} \prod_{\{r: c_r < x_i\}} \frac{\tilde{n}_{c_r} + 1}{\tilde{n}_{c_r}} \tag{2}$$

These NPI lower and upper survival functions are step-functions, presented in product forms which lead to relatively straightforward computation. It should be remarked that the Kaplan-Meier (KM) estimate [10] based on such data, which is the classical nonparametric maximum likelihood estimate, always lies between the NPI lower and upper survival functions [7]. Whilst the KM estimate has also been used for accelerated testing data [15], it should be emphasized that its explicit aim is estimation of an underlying population distribution, whilst our NPI approach is explicitly predictive and considers events involving one future observation at the standard stress level.

The assumption rc-$A_{(n)}$ results in lower and upper survival functions, which fit well into the theory of imprecise probability [2]. The imprecision in these inferences therefore results from the limited inferential assumption made, and it reflects the amount of information in the form of observations well. Note, for example, that the upper survival function only decreases at an observed failure time, while the lower survival function decreases both at an observed failure time and, by a smaller amount, at a right-censored observation. This is in line with a useful, albeit somewhat informal interpretation of imprecise probabilities, namely that a lower probability reflects the information in favour of the event of interest, and the difference between 1 and the corresponding upper probability reflects the information against the event of interest. So the decrease of both the NPI lower and upper survival function at an observed failure time reflects a decrease of information supporting survival past this time, while a right-censored observation reduces the information in favour of survival (hence the slight decrease of the lower survival function) but does not provide evidence against survival (so the upper survival function is not affected).

3 NPI with Estimated Link

The new statistical method for ALT data, which we propose in this paper, consists of two stages. In the first stage, the basic power-Weibull model is assumed and its parameters are estimated using maximum likelihood estimation. Other models can be used, as well as other estimation methods. While this stage results in point estimates for all parameters α, β and p, the next stage only uses the estimate for p, which we denote by $\hat{p}$. In the second stage, we transform all observations at stress levels other than the normal level, so $V_1, \ldots, V_k$, to 'equivalent' observations at the normal level, using the estimate $\hat{p}$ within the transformation explained in Sect. 1, leading to

$$t^i \left(\frac{V_i}{V_0}\right)^{\hat{p}}$$

Right-censored observations are transformed similarly, where their status as right-censored observation is maintained. Now, we apply NPI with all these transformed data as well as the original data at the normal level, as explained in Sect. 2. We illustrate the results of this approach in an example using data from the literature; this example will also be used for an extension of this method in Sect. 4.

Example. Lawless [11, p. 341] presents the ALT data set below in an exercise with further reference to an unpublished Master's thesis. The data result from an experiment in which specimens of solid epoxy electrical insulation were studied in an accelerated voltage life test. Twenty specimens are tested at each of three voltage levels, the normal level $V_0 = 52.5$ and increased levels $V_1 = 55.0$ and $V_2 = 57.5$ kilovolts. Most of the sixty specimens actually failed during the experiments, but a few did not, these provide right-censored observations. The failure times, in minutes, are given in Table 1, where a right-censored observation is indicated with a superscript asterisk.

Table 1. Failure times at three voltage levels.

Voltage	Data
$V_0 = 52.5$	245, 246, 350, 550, 600, 740, 745, 1010, 1190, 1225,
	1390, 1458, 1480, 1690, 1805, 2450, 3000, 4690, 6095, 6200*
$V_1 = 55.0$	114, 132, 144, 162, 222, 258, 300, 312, 396, 444,
	498, 520, 745, 772, 1240, 1266, 1464, 1740*, 2440*, 2600*
$V_2 = 57.5$	168, 174, 234, 252, 288, 288, 294, 348, 390, 408,
	444, 510, 528, 546, 558, 690, 696, 714, 900*, 1000*

Maximum likelihood estimation for the power-Weibull model, based on these data, leads to parameter estimates $\hat{\beta} = 1.184$, $\hat{\alpha} = 2038.03$ and $\hat{p} = 15.104$, where it should be remarked that this numerical optimisation appears to be sensitive to the starting point of the algorithm, we noticed some slight variation in the resulting estimates for different starting points.

In the second stage of our procedure, we only use the estimate $\hat{p}$ to transform the data, as explained above. This leads to the values in Table 2, still listed with their corresponding stress level. Of course, the data at the normal stress level $V_0 = 52.5$ have not been transformed, but are also included in the table for ease of comparison with the transformed data from the other stress levels.

Table 2. Failure times transformed to normal voltage level.

Voltage	Data
$V_0 = 52.5$	245, 246, 350, 550, 600, 740, 745, 1010, 1190, 1225,
	1390, 1458, 1480, 1690, 1805, 2450, 3000, 4690, 6095, 6200*
$V_1 = 55.0$	230.2, 266.5, 290.8, 327.1, 448.2, 520.9, 605.7, 630.0, 799.6, 896.5,
	1005.5, 1049.9, 1504.2, 1558.7, 2503.7, 2556.2, 2956.0, 3513.2*, 4926.6*, 5249.6*
$V_2 = 57.5$	663.8, 687.5, 924.6, 995.7, 1138.0, 1138.0, 1161.7, 1375.1, 1541.0, 1612.2,
	1754.4, 2015.2, 2086.3, 2157.4, 2204.9, 2726.4, 2750.1, 2821.3, 3556.2*, 3951.4*

Using these 60 failure observations, either originally observed at the normal voltage level or transformed to it, the NPI approach as described in Sect. 2 provides predictive lower and upper survival functions, which must be interpreted as applying for a further specimen, exchangeable with those in the test, subjected to the normal stress level $V_0 = 52.5$. These lower and upper survival functions are presented in Fig. 1.

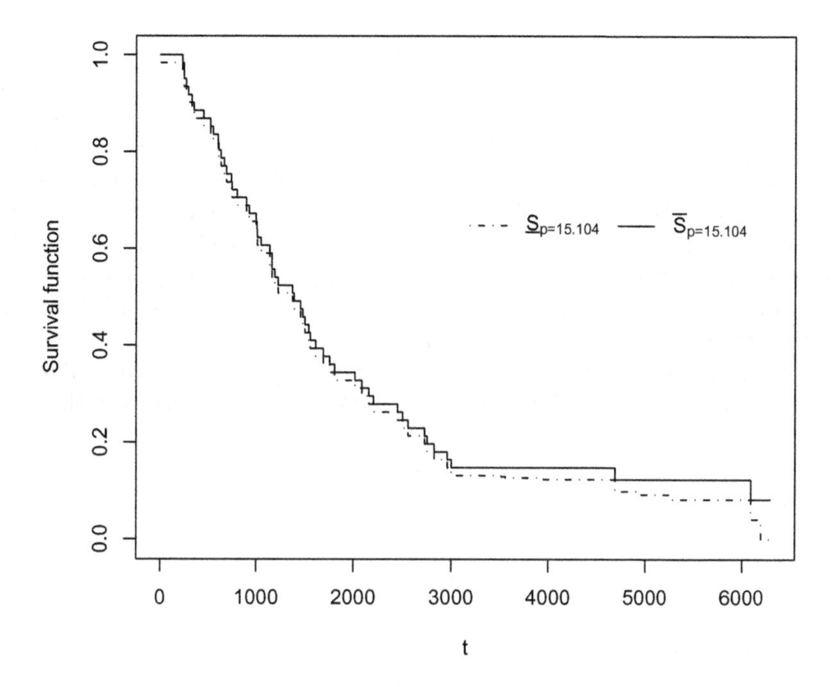

Fig. 1. Lower and upper survival functions with $\hat{p} = 15.104$

4 Imprecision in Link Estimate

The method presented in Sect. 3 is, as will be clear, rather ad-hoc. There are many aspects which will make us doubt the validity of the predictive inference. These include doubt about the model used at stage 1 and, related to this, the fact that the estimation of the Weibull parameters influences the estimate of the parameter $\hat{p}$ but is further neglected at stage 2. As mentioned in the example, there may also be some numerical instability in the estimation computations. One could rebute all such issues by suggesting more detailed modelling, but particularly for ALT data there often remains an element of model-based extrapolation that is difficult, sometimes even impossible, to justify on the basis of available data. Note that the example considered in this paper does have observations at the normal stress level, in other applications this may not be the case or they may only consist of right-censored observations.

We propose a different route to more detailed modelling, although if such modelling can be done on the basis of detailed knowledge of the scenario under study then, of course, this is strongly recommended; it can still be combined by the overall ideas we are presenting here. In an attempt to develop sound predictive inference for ALT data, we suggest to adapt the two stage approach presented in Sect. 3 by replacing the point estimate for p by an interval $[\underline{p}, \overline{p}]$. The use of this interval in the transformation of observations at stress level V_i to the normal level V_0 leads to such observations becoming interval-valued observations at the normal stress level, which we believe is an attractive way for showing the effect of imprecision in line with the absence of perfect information about the link between the different stress levels. Furthermore, these intervals representing observations at other levels will be larger for larger values of i, so an original observation from a stress level that is further away from the normal level is transformed to a wider interval at the normal level than an original observation at a level nearer to the normal level. We believe that, in general, this is also an attractive property of such imprecise inferences for ALT data.

Due to the monotonicity of the transformed data in the power-Weibull model, and the monotonicity of the NPI lower and upper survival functions with regard

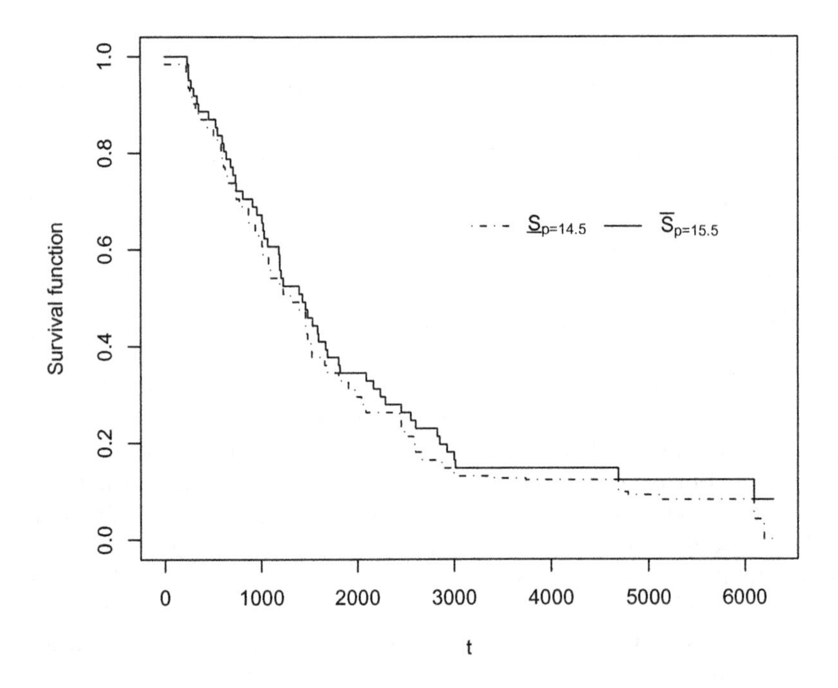

Fig. 2. Lower and upper survival functions with $[\underline{p}, \overline{p}] = [14.5, 15.5]$

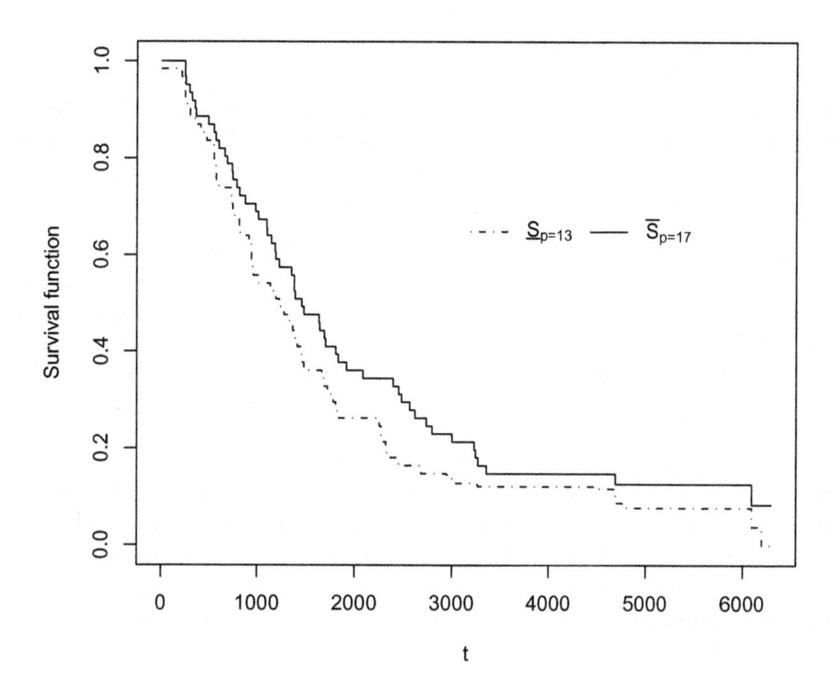

Fig. 3. Lower and upper survival functions with $[\underline{p}, \overline{p}] = [13.0, 17.0]$

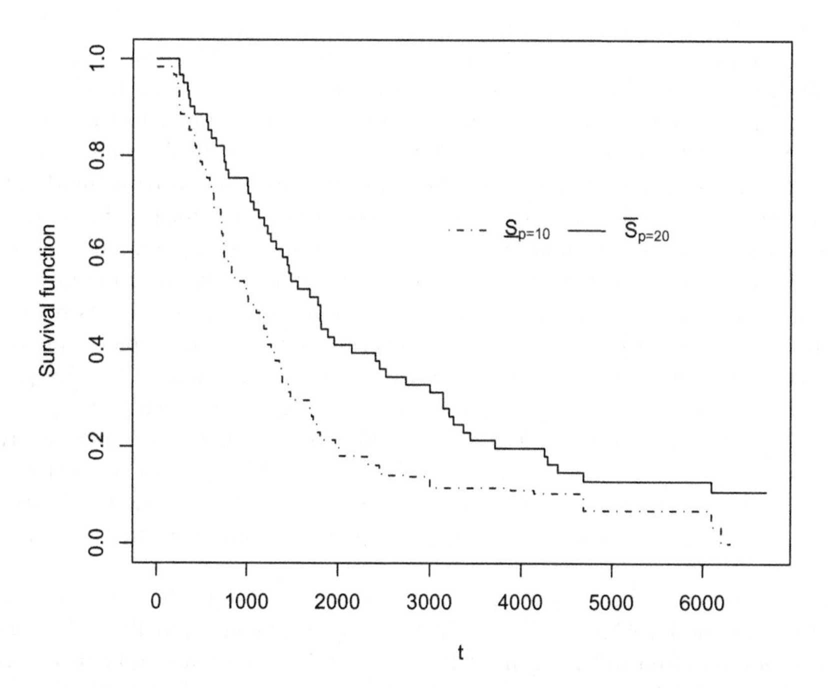

Fig. 4. Lower and upper survival functions with $[\underline{p}, \overline{p}] = [10.0, 20.0]$

to the data on which they are based, this interval $[\underline{p}, \overline{p}]$ straightforwardly leads, in the second stage of our method, to the NPI lower survival function based on the transformed data using $\underline{p}$, and the NPI upper survival function based on the transformed data using $\overline{p}$. Hence, the imprecision in this inferential method, that is the difference between the corresponding NPI upper and lower survival functions, will increase when the width of the interval $[\underline{p}, \overline{p}]$ increases.

Example (ctd). For the example in Sect. 3, the point estimate for p was equal to $\hat{p} = 15.104$. Replacing this by an interval $[\underline{p}, \overline{p}] = [14.5, 15.5]$ leads to the lower and upper survival functions presented in Fig. 2. Increasing the imprecision to $[\underline{p}, \overline{p}] = [13.0, 17.0]$ leads to the lower and upper survival functions presented in Fig. 3, while the substantial further increase in imprecision to $[\underline{p}, \overline{p}] = [10.0, 20.0]$ leads to the lower and upper survival functions in Fig. 4.

These figures show the effect of increased imprecision for the parameter p, where even for very substantial imprecision the effect on the lower and upper survival functions is still quite modest. Of course, this depends on the ratios V_i/V_0 used, which are close to one in this example, but nevertheless it suggests that some concerns, e.g. about some sensitivity to the starting point of the numerical optimisation methods used to derive $\hat{p}$, are not necessary.

5 Discussion

This paper has provided an initial presentation of the possibilities provided by theory of imprecise probabilities [2] for ALT applications. The proposed method combines transformation of failure times at increased stress levels to equivalent observations at the normal stress level, based on an assumed full parametric model, with nonparametric predictive inference based on all the combined data at the normal stress level. As shown, it is quite straightforward to build extra robustness into the inferences by taking an interval of values for the link parameter p instead of a point estimater. One could of course use a confidence interval for this, but we avoided that as the 'confidence' would only be meaningful within the full model, and it would likely lead to far more imprecision than is required for good quality prediction. Instead our overall aim, and the topic of ongoing research, is to get insight into appropriate choice of the interval for the parameter p in order to get predictive inference with good frequentist properties. This is not trivial, as very wide intervals will lead to lower and upper survival functions that are very imprecise, hence will have good frequentist properties for prediction but will be quite meaningless. There has not yet been much attention to performance evaluation for imprecise predictive inference methods in the literature.

There are many unknowns in ALT, with aspects related to the transformation of data from one level to another, or just the simultaneous modelling of data at different levels, often involving modelling assumptions that are difficult or even impossible to justify by observable data. This makes statistical methods using imprecise probabilities attractive for applications in ALT, in particular where these can be kept relatively simple compared to the alternative of using very detailed models. We were slightly surprised that, upon investigation, we did not yet find other contributions of imprecise methods for ALT data, while there is a substantial literature on imprecise methods for other reliability problems [5,6,17] and clearly there are many related research challenges and opportunities for applications ahead. It will also be interesting to consider imprecise statistical methods for ALT data using a different way to deal with the transformation of data from different stress levels. One such an alternative approach that may be of interest for combination with NPI is the linking of data from different stress levels through matching of quantiles [12], which can however only be applied if there are quite substantial numbers of observations at all the stress levels. Another possible approach is by assuming a parametric link function but no further model assumption, and optimising the parameters of the link function through optimal mixing of all the (transformed) data at the stress level of interest.

This initial exploration into development of an NPI-based approach for ALT data will be followed by a detailed simulation study on the predictive performance of our method, which will also provide guidance on suitable levels of imprecision in the parameter linking the data from different stress levels, in particular related to misspecification of the model. Hence, the main justification of the approach should come from the predictive performance, where guidance on suitable levels of imprecision for providing robustness in model specification will be important.

References

1. Augustin, T., Coolen, F.P.A.: Nonparametric predictive inference and interval probability. J. Stat. Plan. Infer. **124**, 251–272 (2004)
2. Augustin, T., Coolen, F.P.A., de Cooman, G., Troffaes, M.C.M.: Introduction to Imprecise Probabilities. Wiley, Chichester (2014)
3. Coolen, F.P.A.: On nonparametric predictive inference and objective bayesianism. J. Logic Lang. Inform. **15**, 21–47 (2006)
4. Coolen, F.P.A.: Nonparametric predictive inference. In: Lovric, M. (ed.) International Encyclopedia of Statistical Science, pp. 968–970. Springer, Berlin (2011)
5. Coolen, F.P.A., Coolen-Schrijner, P., Yan, K.J.: Nonparametric predictive inference in reliability. Reliab. Rng. Syst. Saf. **78**, 185–193 (2002)
6. Coolen, F.P.A., Utkin, L.V.: Imprecise reliability. In: Lovric, M. (ed.) International Encyclopedia of Statistical Science, pp. 649–650. Springer, Berlin (2011)
7. Coolen, F.P.A., Yan, K.J.: Nonparametric predictive inference with right-censored data. J. Stat. Plan. Infer. **126**, 25–54 (2004)
8. De Finetti, B.: Theory of Probability. Wiley, Chichester (1974)
9. Hill, B.M.: Posterior distribution of percentiles: Bayes' theorem for sampling from a population. J. Am. Stat. Assoc. **63**, 677–691 (1968)
10. Kaplan, E.L., Meier, P.: Nonparametric estimation from incomplete observations. J. Am. Stat. Assoc. **53**, 457–481 (1958)
11. Lawless, J.F.: Statistical Models and Methods for Lifetime Data. Wiley, New York (1982)

12. Maciejewski, H.: Accelerated life test data analysis with generalised life distribution function and with no aging model assumption. Microelectron. Reliab. **35**, 1047–1051 (1995)
13. Maturi, T.A.: Nonparametric predictive inference for multiple comparisons. Ph.D. Thesis, Durham University (2010)
14. Maturi, T.A., Coolen-Schrijner, P., Coolen, F.P.A.: Nonparametric predictive inference for competing risks. J. Risk Reliab. **224**, 11–26 (2010)
15. Nelson, W.B.: Accelerated Testing: Statistical Models, Test Plans, and Data Analysis. Wiley, Hoboken (1990)
16. Nonparametric Predictive Inference. http://www.npi-statistics.com
17. Utkin, L.V., Coolen, F.P.A.: Imprecise reliability: an introductory overview. In: Levitin, G. (ed.) Computational Intelligence in Reliability Engineering, Volume 2: New Metaheuristics, Neural and Fuzzy Techniques in Reliability, pp. 261–306. Springer, Berlin (2007)

The Mathematical Gnostics
(Advanced Data Analysis)

Pavel Kovanic(✉)

Institute of Information Theory and Automation of
Czech Academy of Sciences, Prague, Czech Republic
kovanic@tiscali.cz

Abstract. A brief survey of mathematical gnostics is presented. Mathematical gnostics is a tool of advanced data analysis, consisting of

1. theory of individual uncertain data and small samples,
2. algorithms to implement the theory,
3. applications of the algorithms.

The axioms and definitions of the theory are inspired by the Laws of Nature dealt with by physics and the investigation of data uncertainty follows the methods of analysis of physical processes. The first axiom is a reformulation of the measurement theory which mathematically formalizes the empirical cognitive activity of physics. This axiom enables the curvature of the data space to be revealed and quantified. The natural affinity between uncertain data and relativistic mechanics is also shown. Probability, informational entropy and information of individual uncertain data item are inferred from non-statistical Clausius' thermodynamical entropy. The quantitative cognitive activity is modeled as a closed cycle of quantification and estimation, which is proved to be irreversible and maximizes the result's information. A proper estimation of the space's curvature ensure a reliable robustness of the algorithms successfully proven in many applications. Gnostic formulae of data weights and errors, probability and information, which has been proved as valid for small samples of strongly uncertain data converge to statistical ones when uncertainty becomes weak. From this point of view, the mathematical gnostics can be considered as an extension of statistics useful under heavy-duty conditions.

1 A Gnostic System

The notion of the *gnostic system* has been applied in [1] to a general model of recognition characterized as the pairing of a real object and of a subject, its observer. The observation activity *object* $\rightarrow$ *subject* is followed by the feedback *subject* $\rightarrow$ *object* the purpose of which is using the evaluated information in

Motto: This approach is a deterministic theory of indeterminism.
Prof. J.A. Víšek, Charles University, Prague.
P. Kovanic—Retired from the Institute of Information Theory and Automation of Czech Academy of Sciences.

J.P. Carvalho et al. (Eds.): IPMU 2016, Part I, CCIS 610, pp. 177–188, 2016.
DOI: 10.1007/978-3-319-40596-4_16

manipulating, exploiting or control the object. In the special case of *quantitative recognition* , the observation represents the mapping of a real quantity onto numbers called *quantification*, the feed-back being the *estimation* of the true quantity's value. The necessity of quantification originated with the development of the market and the measuring became the task for physics. Mathematical modeling of counting and measuring – the measurement theory [2] – considers the quantification as a consistent mapping of structures of empiric quantities (sets endowed with some relations and operations) onto numeric structures. This theory deals with precise quantification only, leaving the treatment of imprecise quantification to mathematical statistics. Such a quantification process can be named *ideal quantification.*

2 Axiom of Real Quantification

As known from measurement theory, to ensure consistency of the ideal quantification, the relations between quantities and the operations on them must be subject to several logical conditions. This requirement was substituted in [3] by the idea of ideal quantification as the commutative (Abelian) group[1] If the real quantitative observation process would actually be the Abelian group, the estimation would be simply the inverse of this group. Unfortunately, real observations are disturbed by uncertain impacts. But these impacts are as real as the observed quantity. Moreover, their nature is the same: electrical measurements are subject to electrical disturbance. The uncertain impacts can thus be considered as countable or measurable sets and endowed with the same operation as the true observed quantities. Real quantification can be therefore modeled as a pair of two Abelian groups, one of the true and one of disturbing quantities. Considering one single quantitative observation, one actually obtains one single real number of the form of

$$A = A_0 + S\Phi \tag{1}$$

with the *true* real value A_0, real *uncertain* value Φ and a positive dimensionless *scale parameter* S. Both A_0 and Φ are numerical images of elements of empirical structures forming Abelian groups. The multiplicative form of the additive relation 1 is obtained by exponentiation as

$$Z = Z_0 \exp(S\Phi) \tag{2}$$

Quantities A are real numbers and they can have both finite and infinite values when considered as theoretical objects. However, as numeric images of actual quantities, they have values within some finite bounds. This is why the theory involves regular transformations of the actual finite data domains onto the infinite domain to introduce and analyze the corresponding functions of data.

[1] Abel group is a set endowed with a binary operation satisfying the conditions of closedness, associativity, commutativity and existence of an identity element and of the inverse element to each element.

3 Geometry of Real Quantification, Quantifying Error and Weight

The observed data value Z (2) is represented by a point in the bi-dimensional plane $(Z_0, S\Phi)$. Observation is a discrete event, however let us consider the virtual path of a continuous variable z from the true value Z_0 to the observed value Z under the impact of the uncertainty φ changing from the zero starting value to an unknown value Φ. The length of this path is the observation error. A non-trivial question arises, which of many existing geometries is to be applied to quantify the error? Using the identity $\exp(\alpha) = \cosh(\alpha) + \sinh(\alpha)$ and introducing hyperbolic Cartesian coordinates

$$x_Q = Z_0 \cosh(2S\Phi) \qquad y_Q = Z_0 \sinh(2S\Phi) \tag{3}$$

one comes to the relation

$$Z_0 = \sqrt{(x_Q^2 - y_Q^2)} \tag{4}$$

The plane of observed data is thus endowed with the Minkowskian metric and the path of virtual movement is the Minkowskian circle. The number Z_0 is a circle's radius and invariant of the movement. Multiplier 2 of Minkowskian angle $S\Phi$ results from accepting the angular distance between Φ and $-\Phi$ (the mirrored point's angle) as the angular error. Relative coordinates

$$w_Q = \cosh(2S\Phi) \qquad h_Q = \sinh(2S\Phi) \tag{5}$$

called *quantifying weight* and *quantifying irrelevance* have important interpretation in quantification of the uncertainty as data error weight and data error value. These names are motivated by the relation

$$\sinh(2S\Phi) = \int_0^{2S\Phi} \cosh(x)d(x) \tag{6}$$

where w_Q determines the weight of the differential data error $d(2S\Phi)$ thus playing the role of metrical tensor in the sense of Riemannian geometry.

4 Geometry of Estimation, Estimating Error and Weight

An observer aims to use the best available way of measuring for quantification, but he must accept the observed value "as it is" without the chance of choosing the virtual quantification path determined by Nature. However, he knows from geometry, that the length of the quantifying path measured by Minkowskian geometry is an *extremal*: its length between two points exceeds the lengths of each of the other path between the same points. This means that the uncertainty $S\Phi$ makes the observed value as bad as possible by maximizing its distance from the true value. The observer has a chance for his best "countermove" in his game with Nature by choosing the best virtual path of estimation from the known observed value Z back to the unknown true value Z_0 thus minimizing the resulting error. As shown in gnostic theory [4], such a path exists and its points have coordinates

$$x_E = Z_0 \cos(2S\varphi) \qquad y_E = Z_0 \sin(2S\varphi) \tag{7}$$

where the relative coordinates

$$w_E = \cos(2S\varphi) \qquad h_E = \sin(2S\varphi) \tag{8}$$

are *estimating weight* and *estimating irrelevance*, for which an analogue of the 6 exists. Thus this path has the form of Euclidean circle. It means that the observation plane is endowed by two metrics, quantifying (Minkowskian) and estimating (Euclidean) ones. The Euclidean angle φ is related with the Minkowskian ones Φ by

$$\tan(S\varphi) = \tanh(S\Phi) \tag{9}$$

Thus each point of the observation plane has double interpretation, a quantifying and an estimating one.

5 Uncertainty and Curvature

The additive formulae 1 represents the quantity $S\Phi$ as a cause of observed values' uncertainty. It is frequently used as an evaluation of the uncertainty's "size" and its square as an element of the data variance or data "weight". The latter notion has a classic statistical background. As proved in [10], the best asymptotically unbiased and asymptotically normally distributed estimate of the mean of differently dispersed data is a weighted mean where the weights are proportional to the reciprocal value of the data variance. It means, that measurement in different points of the observation space is to be done differently. In terms of Riemannian geometry: the metric tensor is a function of the coordinates of the space and that space is curved. "Locally dependent" metrics have been introduced into statistics as well by using the influence functions to improve robustness of the regression analysis [11]. There are many approaches to this task supported by statistical assumptions and tailored to different data classes. The influence functions derived from gnostic axioms were presented in [6].

Locally dependent metrics are also introduced by quantifying and estimating weights (5) and (8). Their non-linearity with respect to data is obviously exhibiting two types of forms, convex and concave. The scale parameter S is a function of the curvature's radius. It is closely connected with the robustness of the estimation of uncertainty.

6 Entropy of a Datum and Entropy Fields

C.E. Shannon's information entropy is the negative value of L. Boltzmann's statistical entropy. A complete system of probabilities of events is necessary for the evaluation of this entropy. The pre-statistical concept of Clausius' thermodynamic entropy makes use only of the heat amount and the absolute temperature. A Gedanken-experiment helped in [5] to represent the entropy of a single uncertain datum in the Clausius' manner by introduction of the proportional mapping of the squared data value onto the absolute temperature and onto the heat flow. Substitution into Clausius' formula shows that the changes of the thermodynamic entropy of an uncertain datum within quantification and estimation is proportional to the changes of the corresponding data weights,

$$\delta\mathcal{E}_Q = w_Q - 1 \qquad \delta\mathcal{E}_E = w_E - 1 \tag{10}$$

if the coefficients of proportionality of the mapping are suitably chosen. The plane of observation is formed by possible data values, each of which has its quantifying and estimating weight attached. Formulae 10 therefore define two scalar fields of entropy. Gradients of these fields can be shown to be proportional to the corresponding irrelevances h_Q and h_E.

7 Information and Probability of an Individual Datum

The source of a scalar field $\mathcal{E}$ is known to result from the operation $\operatorname{div}\operatorname{grad}\mathcal{E}$, i.e. by application of the Laplace's operator Δ. Looking for the source of the entropy field of $\mathcal{E}_Q$ in the point (x, y) one comes to relation

$$(x^2 + y^2)\Delta\mathcal{E}_Q = \frac{1}{p*(1-p)} \tag{11}$$

where

$$p = (1 - h_E)/2 \tag{12}$$

Introducing the quantity

$$\mathcal{I}(p) = -p * \ln(p) - (1-p) * \ln(1-p) \tag{13}$$

one has the relation

$$\frac{1}{p*(1-p)} = \frac{d^2(\mathcal{I}(1/2) - \mathcal{I}(p))}{dp^2} \tag{14}$$

saying that the right hand side of Eq. 11 is a source of the field of $\mathcal{I}$. The quantity $\mathcal{I}(p)$ would be formally identical with the Shannon's information of an event, the probability of which would be p. Moreover, there is a large set of conditions in [7] under which a quantity is to be accepted as information. As shown in [4], all such conditions are satisfied by $\mathcal{I}$ which thus deserves to be accepted as *information of an individual uncertain datum* and its argument p as the datum's *probability*. Equation 11 can be thus formulated as a general statement: The source of entropy of an individual uncertain datum is proportional to the source of its information. This equation describes the conversion of entropy to information and vice versa. It thus can be considered to be a mathematical model of the Maxwell's demon[2]

8 Ideal Gnostic Cycle and its Features

The observed point, interpreted by quantifying coordinates (x_Q, y_Q) or by estimating ones (x_E, y_E) has its mirrored image $(x_Q, -y_Q)$ and $(x_E, -y_E)$. They are two arcs of virtual paths connecting the observed points with their mirrored images, the "hyperbolic" arc of a Minkowskian circle and an "ordinary"(Euclidean) ones. This closed path is called the *Ideal Gnostic Cycle* (*IGC*). Changes of entropy and information of a datum 10 and 14 enable the important features of the *IGC* to be proved:

[A] Data transformations following the closed path of *IGC* provide the best estimate of the true value in the sense of maximization of results' information and minimization of its entropy.
[B] The closed *IGC* is irreversible: none estimation can completely eliminate the error of an uncertain observation.

Thus the *IGC* according to [A] provides a theoretical model for programs of estimation, but establishes by [B] unsurpassable limits for data analysis like the second law of thermodynamics does for heat transformation.

9 What Should Data Say for Themselves

The ideal of data treatment frequently formulated as "Let data speak for themselves!" resulted from the requirement of maximum objectivity. The more a priori assumptions on data, the more subjectivity is increasing the danger of discrepancy between assumed models and actual features of data. The goal of data treatment is information being brought by data, but reaching it is critically limited by the knowledge of data features. This knowledge requires answering a series of questions:

- What kind of geometry should be applied (Euclidean, Minkowskian, Riemannian)?
- What curvature of the space of uncertain data characterizes the given data?
- Is the data structure additive or multiplicative?
- Are the data homo- or heteroscedastic?
- Is there a data trend?
- Are the data cross- or autocorrelated?
- Are the data homogeneous?
- What is form of the probability and density distribution?

Some of these questions are not asked in statistics, others are answered by assumptions. Mathematical gnostics derives all the answers from data. The crucial point is the robust kernel estimation of probability distributions.

[2] As described in [8], Maxwell introduced in his Gedanken-experiment a virtual creature capable to use the information on movement of molecules to convert it into decrease of entropy.

10 The Unique Kernel for Robust Kernel Estimation

The kernel estimation of a probability density function was introduced in [9] along with five conditions necessary for asymptotic convergence to true density. A lot of kernels can be found in literature satisfying these conditions and giving estimates of different quality dependent on the kernel's form. Kernels are ordinarily defined over the domain of the independent variable using their natural additive or multiplicative scale. Unlike this, the individual data item's probability 12 is defined over the infinite (positive) domain obtained by transformation of the actual data domain. Its density was shown to satisfy all Parzen's conditions. Its application to kernel estimation is not only justified, but advantageous: its form is universally applicable and as a result of the theory, it is unique and optimal. The location of the kernel is determined by the (known) observed value and its "width" by the scale parameter S which is to be estimated by data.

11 Aggregation of Kernels

The Parzen's kernel estimating method creates the density estimate by additive aggregation of kernels without the consideration of any alternatives. It may seem natural, because the historical mathematical forerunners of kernel estimation like Green and Duhamel[3] did essentially the same because of linearity. However, the aggregation of gnostic kernels deserves a special consideration. The space of observed data within the quantification process has been shown as a Minkowskian plane with coordinates proportional to the hyperbolic cosine (w_Q) and the hyperbolic sine (h_Q). But a two-dimensional plane depicting the moment and energy of a relativistic charge-free particle would be endowed by the same geometry. This means that there exists (at least mathematically) a consistent linear mapping of the pair (w_Q, h_Q) onto the pair of *(energy, momentum)* of a relativistic particle moving with velocity corresponding to the argument of said hyperbolic functions. Moreover, this mapping is Lorentz-invariant, i.e. it is valid for all data uncertainties and corresponding particle's velocities. This mapping *uncertain data* $\Leftrightarrow$ *relativistic particle* can be applied to several data. The aggregation law of relativistic particles is known, it is the Momentum-Energy Conservation Law, which is additive with respect to pairs *(energy, momentum)*. To preserve the mapping for a data set, one must aggregate the pairs (w_Q, h_Q) additively as well, although they are nonlinear functions of data. The second axiom of the gnostic theory extends this way of aggregating from quantifying weights and irrelevances to estimating ones to preserve the mapping of quantifying variables to estimating ones and vice versa.

However, a sum of cosine is not a cosine and a sum of sines is not a sine. Therefore, sum of weights (and irrelevances) of a data set will represent the weight (irrelevance) of the whole set but not a pair *(weight, irrelevance)* of a possible single data item. This is why a proper normalization of additively

[3] See Green's function and Duhamel's integral.

aggregated weights and irrelevances should be applied instead of their simple addition.

The form of both quantification and estimation kernels can be shown to be similar, differing only by scale parameters. However, the results of aggregation of kernels depend on metric.

12 Applications of the Gnostic Kernels

The kernel presented above was the derivative of probability. The linearity of this operation allows us to obtain and use kernels of both density and probability. Library of gnostic algorithms includes the following applications of kernel estimation:

12.1 Local Probability and Density Distribution

Local distributions are obtained as means of kernels. They possess a full flexibility controlled by the choice of scale parameter. This feature make them an ideal instrument for revealing the detailed structure of a data set and to perform the marginal analysis showing the data clusters and outliers in a non-homogeneous data set. A special kind of inner robustness of these distributions allows a deep insight into a homogeneous data set to be obtained allowing robust bounds of its important subintervals to be estimated.

12.2 Global Probability and Density Distribution

Each pair *(weight, irrelevance)* has its module determined as the Minkowskian or Euclidean length of the observed point's radius vector. The global probability distribution function is obtained as the mean of integral kernels divided by the module of sums of cosines and sines by using the proper metric. The global density distribution is the first derivative of the global probability distribution. There are two types of the global distributions differing by robustness, the estimating one is robust with respect to outlying data and peripheral clusters while the quantifying distribution is robust with respect to inner disturbances and noises of the treated data sample. Unlike the high flexibility of the local distribution functions, the global ones are more rigid. This feature makes them applicable to robust probability and density estimation, to reliable tests of data homogeneity and to estimation of the observed data's true values and of bounds of data support. Global distribution enable three types of censored data to be estimated: both left- and right censored ones and interval data.

The advantage of gnostic distribution functions lies in their independence on the a priori assumptions, objectivity due to the reliance only on data alone, suitability for small data samples and a much broader application field than standard statistical distributions including small data samples.

12.3 Robust Curve Fitting

Frequently used curve fitting by means of polynomials or by sets of other functions including the orthogonal ones can suffer from un-robustness in the case of application to uncertain data. A careful preliminary gnostic analysis providing reliable estimates of individual data weights, proper geometry and scale parameters of gnostic kernels used for the fit enable the maximum of resulting information to be reached.

12.4 Analysis of Dependencies

The correlation coefficient can say that there exists an interdependence between two vectors, but the interpretation of the interaction is easy only in the case close to the linear relationship. The application of kernel estimates is suitable especially for presentation of non-monotonous dependencies.

13 Robust Regression

The approach to the task of robust multi-dimensional regression modeling based on mathematical gnostics has been demonstrated in [6]. The gist was the choice of a criterion function for the evaluation of model's residuals. Instead of some formal "purely mathematical" functions, natural features of uncertainty were used such as the common source of fields of entropy and information. Results were shown to be applicable as special kinds of influence functions used in robust statistics for the Iterated Weighted Least Squares Method completed by a feed-back filter. Extensive comparisons with statistical models of this type demonstrated the priority of the gnostic approach resulting in better estimates of curvature of the space of uncertain data and in information maximization of the estimation process.

The standard case of a regression model representing the dependent vector as a linear combination of explanatory vectors can be called *explicit*. The *implicit* regression model is obtained from the explicit one by division of all equations of the system by the values of the dependent variable (which must be non-zero). There are some advantages of the implicit regression, e.g. uniqueness of the model independent on the exchange of roles of explanatory/dependent variables, comparability and evaluation of relative impacts of variable.

14 Robust Correlation

The availability of reliable robust regression techniques enabled a new approach to robust correlation coefficients to be introduced. The proportionality between two centered vectors x and y is considered twice, as $x = c \cdot y$ and $y = k \cdot x$ with scalars c and k, which are estimated by the robust regression. The square root of products of estimates can be used as robust estimate of the correlation coefficient.

15 Testing of Hypotheses

The crucial problem of statistical testing of hypotheses is the decision making on the probability distribution of the underlying data. Some statistical tests are based on the Gaussian assumption, but experienced analyst know that the "normal" distribution is not always normal. Relying on a priori assumptions as well as data violation by "normalizing" transformations can lead to incorrect decision making. The availability of robust probability distributions described above allows not only tests on a required significance level to be performed, but actual significance of the required decision to be evaluated.

16 Homogeneity Problem

Problems with data homogeneity can be demonstrated on the task of statistical investigation of political preferences. The careful selection of people used as a data source cannot warrant the homogeneity (similarity, affinity, comparability, closeness) of meaning of all individuals or groups. There is an extensive amount of factors influencing the measurable parameters. All these factors cannot be under the control of the survey's organizers. Increasing the survey's size can be even counter-productive: the more cases, the broader the spectrum of factors. Moreover, it is not always safe to assume that the demanding conditions of Central Limit Theorem are satisfied.

A non-homogeneity of a one-dimensional data set is sensitively and reliably detected by the appearance of a second maximum in gnostic global density distribution. This enables a reliable homogenization to be performed.

17 Robust Cluster Analysis

The local distribution functions enable the homogenization of a one-dimensional non-homogeneous data sample to be implemented by the identification of outliers, inliers and sub-clusters causing the non-homogeneity. Thus a non-homogeneous data set consisting of several homogeneous clusters can be subjected to robust marginal analysis. This approach is efficiently generalized to robust multi-dimensional cluster analysis by a marginal analysis of residuals of an implicit multi-dimensional model. A multi-dimensional non-homogeneous data set is then replaced by several homogeneous clusters.

The robustness of this approach enables the multi-dimensional objects (represented by rows of the model) to be ordered in a rational and reliable way.

18 Implementation

Methods of mathematical gnostics have been implemented as computer programs during the last several decades and the implementation efforts are continuing today, as well. Their application in many fields, including technology, economy, medical, environmental investigation and others, were used not only for tests of their efficacy, but also as motivation and initiation of further development. The long-term experience confirms the usefulness of this approach to uncertainty. Many applications (especially to economic problems like financial statement analysis and financial control, marketing and financial markets) are described in [12]. The gnostic methodology of analysis of environmental parameters was investigated within the framework of two research projects of the European Union [13,14]. Programs based on mathematical gnostics became the main data analytical tool in the Institute of Chemical Process Fundamentals of the Czech Academy of Sciences as documented by series of publications (e.g. [15–18]). Recent results enable a complete automation of the exploratory phase of data analysis providing robust information on actual data model, which offers the rising of the quality assessment control to the level unreachable by other methods ([19].

19 Conclusions

Mathematical gnostics, which is based on the axiomatic theory of individual uncertain data and small samples and supported by laws of physics develops advanced methods for the treatment of strongly uncertain data. These methods maximize the resulting information and are naturally robust. Their applications also extend the range of tasks solvable by statistical methods.

References

1. Rastrigin, P.A., Markov, V.A.: Cybernetic Models of Recognition. Zinatne, Riga (1976). (in Russian)
2. von Helmholtz, H.: Zaehlen und Messen erkentniss-theoretisch betrachtet. In: Philosophische Aufsaetze Eduard Zeller Gewidmet, Leipzig, pp. 17–52(1887). (in German)
3. Kovanic, P.: Gnostical theory of individual data. Prob. Control Inform. Theory **13**, 259–271 (1984)
4. Kovanic P.: Gnostická teorie neurčitých dat, (Gnostic Theory of Uncertain Data), doctor (DrSc.) dissertation, The Institute of Information Theory and Automation, Czechoslovak Academy of Sciences, Prague, 161 pp. (1990). (in Czech)
5. Kovanic, P.: On relations between information and physics. Prob. Control Inf. Theory **13**, 383–399 (1984)
6. Kovanic, P.: A new theoretical and algorithmical basis for estimation. Ident. Control Autom. **V22**(6), 657–674 (1986)
7. Perez, A.: Mathematical theory of information. Appl. Math. **3**(1), 81–99 (1958). (in Czech)
8. von Bayerer, H.C.: Maxwell's Demon. Random House Inc., New York (1998)
9. Parzen, E.: On estimation of a probability density function and mode. Ann. Math. Statist. **33**, 1065–1076 (1962)
10. Linnik, Y.V.: The Least Square Method and Basics of Observation Treatment. GIM-FL, Moscow (1962). (in Russion)

11. Coleman, D., Holland, P., Kaden, N., Klema, V., Peters, S.C.: A system of subroutines for iteratively re-weighted least-squares computation. ACM Trans. Math. Softw. **6**, 327–336 (1980)
12. Kovanic, P., Humber, M.B.: The Economics of Information, 717 p. (2015). www.math-gnostics.com
13. Focus on Key Sources of Environmental Risk. www.projectfoks.eu
14. European Project 2-FUN: Improving Risk Assessment. www.2-fun.org
15. Jacquemin, J., Bendová, M., Sedláková, Z., Holbrey, J.D., Mullan, C.L., Youngs, T.G.A., Pison, L., Wagner, Z., Aim, K., Costa Gomes, M.F., Hardacre, C.: Phase behaviour, interactions, and structural studies of (Amines+Ionic Liquids) binary mixtures. (Eng) Chem. Phys. Chem. **13**(7), 1825–1835 (2012)
16. Borsós, T., Řimnáčová, D., Ždímal, V., Smolík, J., Wagner, Z., Weidinger, T., Burkart, J., Steiner, G., Reischl, G., Hitzenberger, R., Schwarz, J., Salma, I.: Comparison of particulate number concentrations in three central european capital cities. (Eng) Sci. Total Environ. **433**, 418–426 (2012)
17. Setničková, K., Wagner, Z., Noble, R., Uchytil, P.: Semi-empirical model of toluene transport in polyethylene membranes based on the data using a new type of apparatus for determining gas permeability, diffusivity and solubility. (Eng) J. Membr. Sci. **66**(22), 5566–5574 (2011)
18. Andresová, A., Storch, J., Traikia, M., Wagner, Z., Bendová, M., Husson, P.: Branched and cyclic alkyl groups in imidazolium-based ionic liquids: molecular organization and physicochemical properties. (Eng) Fluid Phase Equilib. **371**, 41–49 (2014)
19. Wagner, Z., Kovanic, P.: Advanced Data Analysis for Industrial Applications, Modelling Smart Grids 2015, Prague, September 10–11 (2015). http://www.smartgrids2015.eu/

Textual Data Processing

The Role of Graduality for Referring Expression Generation in Visual Scenes

Albert Gatt[1], Nicolás Marín[2(✉)], François Portet[3], and Daniel Sánchez[2]

[1] Institute of Linguistics, University of Malta, Msida, Malta
albert.gatt@um.edu.mt
[2] Department of Computer Science and A.I., University of Granada, 18071 Granada, Spain
{nicm,daniel}@decsai.ugr.es
[3] Laboratoire d'Informatique de Grenoble, Grenoble Institute of Technology, Grenoble, France
francois.portet@imag.fr

Abstract. Referring Expression Generation (REG) algorithms, a core component of systems that generate text from non-linguistic data, seek to identify domain objects using natural language descriptions. While REG has often been applied to visual domains, very few approaches deal with the problem of fuzziness and gradation. This paper discusses these problems and how they can be accommodated to achieve a more realistic view of the task of referring to objects in visual scenes.

Keywords: Referring expression · Fuzziness · Linguistic description · Visual scenes

1 Introduction

The aim of systems for automatically generating linguistic descriptions of visual scenes is to generate text describing the scene contained in an image, simulating the results provided by humans in the same task. In recent years, scene description has attracted a lot of attention as part of a growing trend to explore the interface between vision and language in Natural Language Generation (NLG) [1], as is the case of other kinds of data [2–4].

Research in this area tends to focus on learning correspondences between parts of an image and textual descriptions to generate descriptive captions [5–7]. A somewhat different focus is offered by work on Referring Expression Generation (REG), one of the most important tasks in NLG. A referring expression is a noun phrase whose communicative purpose is to identify an object to the hearer, for which it must be a distinguishing description, i.e., it must be "an accurate description of the entity being referred to, but not of any other object in the current context set", the context set being "the set of entities that the hearer is currently assumed to be attending to" [8]. The most standard way of approaching REG computationally is by seeing the problem as a content selection problem [9]. If a particular object is to be identified linguistically among a set of other objects, then the challenge is to find a set of properties of the object that

J.P. Carvalho et al. (Eds.): IPMU 2016, Part I, CCIS 610, pp. 191–203, 2016.
DOI: 10.1007/978-3-319-40596-4_17

identifies it uniquely and to do so in a way that matches what human speakers would do in a similar situation.

Formally, this means that there are two important criteria for a REG algorithm to be successful. Let $re = \{p_1, \ldots, p_n\}$ stand for a referring expression (a set of properties) returned by such an algorithm for some target referent o. The description re is considered *accurate* if, for all $p \in re$, $o \in [\![p]\!]$, that is, every property in the description is true of o. Furthermore, re is a *successful* description if it uniquely identifies o, that is:

$$\bigcap_{p \in re} [\![p]\!] = \{o\} \tag{1}$$

The choice problem represented by REG has been explored in a number of algorithms, reviewed in detail in [9], though it has only recently been tackled in the context of complex visual scenes [10,11]. Referring expressions in images use visual properties and concepts for characterizing univocally certain regions in the image, corresponding to objects or areas of interest.

In many cases these concepts and properties are fuzzy or gradable in nature, a factor that has been ignored in many REG algorithms, which typically treat a target's properties as symbols with a crisp extension. While exceptions to this trend exist [12], they have focussed on prototypically gradable properties, such as an object's size. Yet, even properties such as an object's colour admit of boundary cases and gradations. Another simplification in existing REG models has to do with the notion of an object itself: frequently, REG algorithms assume that the relevant components of a visual scene are entities and their properties. However, even the concept of an 'entity' can be argued to be fuzzy. An example where this complication becomes an important issue is a weather forecasting application, which needs to refer to a geographical region in order to describe the weather conditions holding within it. As real-world applications show [13], regions corresponding to a particular set of predicates describing the weather conditions will often not be easily identifiable using place names or simple expressions based on cardinal directions.

Thus, in addition to the problem of modelling the semantics of fuzzy concepts with respect to their correspondence with properties of objects [14,15], referring expressions using fuzzy concepts may also match objects in the scene to a certain extent, and this extent has to be calculated, introducing an additional complexity in the REG problem. In particular, distinguishability of an object with respect to others in terms of a referring expression involving fuzzy concepts turns out to be a matter of degree. In this paper we address this problem.

The paper is organized as follows: in Sect. 2 we briefly discuss fuzzy visual concepts. In Sect. 3 we deal with graduality in referring expressions. Implications of graduality in REG are discussed in Sect. 4. Section 5 concludes the paper.

2 Fuzzy Concepts in Visual Scenes

Concepts related to low-level features like colour, texture, and shape are among the most employed in linguistic description [16], and are usually affected by fuzziness. Colour terms used as adjectives like *red*, *vivid yellow*, etc. are paradigmatic examples of fuzzy concepts, and can be represented by means of fuzzy subsets of crisp colours in some colour space. That is, each crisp colour (usually represented by a triplet of numbers, whose domain and meaning vary from one colour space to another) has a membership degree to each colour term. Several proposals are available in the literature for determining the membership function representing each colour term [17]. Once the terms are defined, they have to be matched to the objects under study. This is not a trivial issue because, unless all the pixels comprising the object have the same colour, approaches for associating colour terms to the whole object are necessary, like for instance the notion of *colour dominance* [18]. Fuzzy approaches for defining shape (round, regular, convex, etc.) and texture properties (coarseness, contrast, orientation, etc.) are also available, see [19,20] and references therein.

Not only properties, but regions themselves can be fuzzy as well, as they are usually defined in terms of their correspondence to properties [16]. In this sense, a fuzzy region is defined as a fuzzy subset of pixels in the image. As an example of this, if we take the homogeneity in colour as the criterion for determining regions, we obtain fuzzy regions like the ones in the image in Fig. 1(a), which shows a collection of cells as seen under the microscope. The regions corresponding to cells are characterized by being *red* but, whilst some of them have clear boundaries, some others do not, since the colour of some pixels has partial membership to *red* for some cells. Dozens of methods for fuzzy segmentation are available [21].

The lack of clear boundaries of fuzzy regions is another source of fuzziness for other properties, which further complicates the task of calculating a region's other properties, including its size, shape, and its spatial relationships with other regions and with the image framework (location), among others. These concepts, which are themselves fuzzy even for crisp regions, have been an object of study

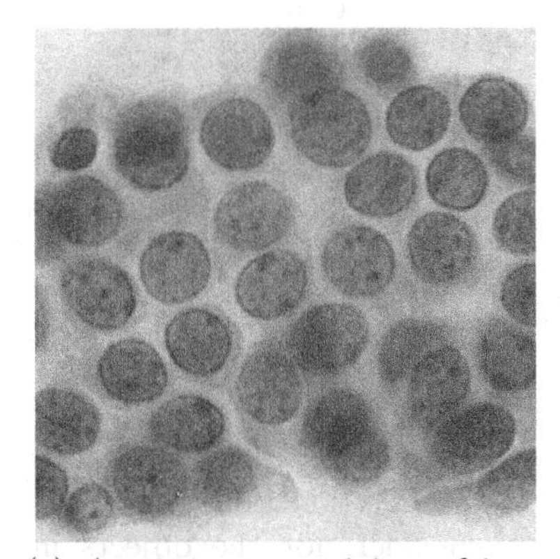

(a) An image containing objects with fuzzy properties

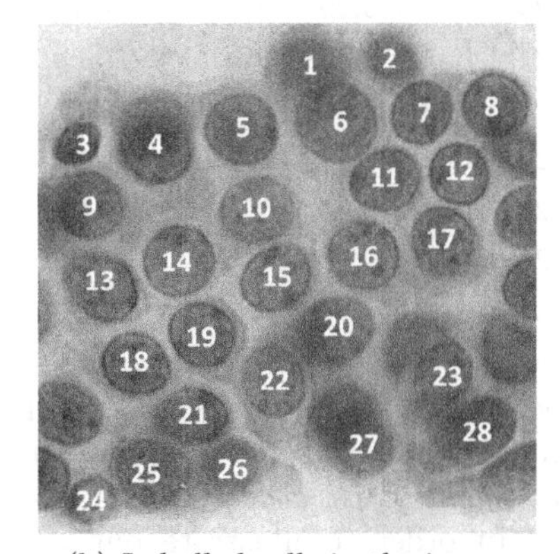

(b) Labelled cells in the image

Fig. 1. An image with fuzzy properties and its labelled counterpart

Table 1. Fulfilment degree of several fuzzy properties and accuracy of three referring expressions for the image in Fig. 1(a).

Cell	Dark	Small	Upperleft	dark & small	dark & upperleft	small & upperleft
1	0,00	0,29	0	0,00	0,00	0,00
2	0,00	0,71	0	0,00	0,00	0,00
3	1,00	1,00	0,96	1,00	0,96	0,96
4	0,83	0,00	0,85	0,00	0,83	0,00
5	0,00	0,00	0,05	0,00	0,00	0,00
6	0,00	0,00	0	0,00	0,00	0,00
7	0,00	0,14	0	0,00	0,00	0,00
8	0,33	0,14	0	0,14	0,00	0,00
9	0,33	0,00	0,33	0,00	0,33	0,00
10	0,00	0,00	0,01	0,00	0,00	0,00
11	0,00	0,14	0	0,00	0,00	0,00
12	0,00	0,43	0	0,00	0,00	0,00
13	0,00	0,00	0	0,00	0,00	0,00
14	0,00	0,00	0	0,00	0,00	0,00
15	0,00	0,00	0	0,00	0,00	0,00
16	0,00	0,00	0	0,00	0,00	0,00
17	0,00	0,14	0	0,00	0,00	0,00
18	0,17	0,14	0	0,14	0,00	0,00
19	0,17	0,14	0	0,14	0,00	0,00
20	0,00	0,00	0	0,00	0,00	0,00
21	0,17	0,43	0	0,17	0,00	0,00
22	0,00	0,00	0	0,00	0,00	0,00
23	0,17	0,29	0	0,17	0,00	0,00
24	0,00	0,86	0	0,00	0,00	0,00
25	0,00	0,00	0	0,00	0,00	0,00
26	0,00	0,57	0	0,00	0,00	0,00
27	0,50	0,43	0	0,43	0,00	0,00
28	0,67	0,00	0	0,00	0,00	0,00

with fuzzy techniques [16,22,23]. It is easy to see examples of concepts that are fulfilled only to some degree by objects in the image depicted in Fig. 1(a), such as *round shape*, *above* (spatial relationship), *in the middle* (location), etc.

In order to illustrate our discussion, let us consider, for the objects in Fig. 1(a), the fuzzy properties *small*, *dark*, and *upperleft*. Table 1 shows example fulfilment degrees of these properties for all the named objects in Fig. 1(a). For the property *small*, a fuzzy subset on the domain of the proportion of pixels in the image has been employed. The property *dark* is calculated as the dominance of the property *low luminance* of the pixels in an object. The property *upperleft* is defined as a fuzzy region with square core and support, core with side 1/4 of the image, and support with side 3/8, with membership decreasing linearly. We omit more specific details about how these degrees have been calculated for reasons of space.

3 Gradation in Referring Expressions

When referring expressions are comprised of fuzzy properties, both their accuracy and their ability to discriminate an object become a matter of degree, contrary to the view usually assumed by REG algorithms (see Sect. 1). In this section we propose different ways to assess these, as well as some other properties which are also important when generating referring expressions and assessing their quality.

3.1 Accuracy

As we have noted, the accuracy of a referring expression with respect to a given object measures the extent to which what the expression says is true for the object, that is, the degree to which *all* the properties that appear in the expression hold for the object. When the referring expression is composed of fuzzy properties, to calculate the accuracy of the expression for a given object we first have to compute the fulfilment degree of each property; subsequently we have to appropriately aggregate these degrees into a single value.

Let us once again consider a referring expression re involving properties $\{p_1, \ldots, p_n\}$ and an object o. Let also $p_i(o)$ be the fulfilment degree of property p_i for object o. The accuracy of re for object o is calculated as follows:

$$a_{re}(o) = \bigotimes_{i=1}^{n} p_i(o) \tag{2}$$

where $\otimes$ is a t-norm.

As an example, consider again Fig. 1(a) (Fig. 1(b) assigns numbers to the cells) and the three referring expressions: *The small object in the upper-left* (re_1), *The dark object in the upper-left* (re_2), and *The small and dark object* (re_3).

According to data in Table 1, and using the minimum as t-norm, we have that:

- $a_{c_3}(re_1) = \min(small(c_3), upperleft(c_3)) = 0.96$
- $a_{c_3}(re_2) = \min(dark(c_3), upperleft(c_3)) = 0.96$
- $a_{c_3}(re_3) = \min(small(c_3), dark(c_3)) = 1$

3.2 Referential Success

As noted in Sect. 1, referential success is the extent to which the referring expression unambiguously identifies the intended referent to the reader or hearer. Consider again a referring expression re and a set of objects $O = \{o_1, \ldots, o_m\}$. The referential success of re with respect to a referent object o_i is the degree to which re is true for o_i and false for the rest of objects, that is:

$$rs_{re}(o_i) = a_{re}(o_i) \otimes \left(\bigotimes_{o_j \in O \wedge j \neq i} \neg(a_{re}(o_j)) \right) \in [0, 1] \tag{3}$$

where $\otimes$ is a t-norm and $\neg$ is a fuzzy negation. Note that this notion of referential success is quite different from the more traditional definition given in REG (see Eq. 1 above). As an example, consider again Fig. 1(a) and the three referring expressions in the previous section. In this case, if we once more use the minimum as t-norm and the standard fuzzy negation operator, we have that:

- $rs_{re_1}(c_3) = \min(a_{re_1}(c_3), \min_{c_j \neq c_3}(1 - a_{re_1}(c_j))) = \min(0.96, 1) = 0.96$
- $rs_{re_2}(c_3) = \min(a_{re_2}(c_3), \min_{c_j \neq c_3}(1 - a_{re_2}(c_j))) = \min(0.96, 0.17) = 0.17$
- $rs_{re_3}(c_3) = \min(a_{re_3}(c_3), \min_{c_j \neq c_3}(1 - a_{re_3}(c_j))) = \min(1, 0.57) = 0.57$

As can be observed, though the three considered referring expressions are almost equally true for cell 3, they have rather different degrees of referential success. The two last expressions fail in the objective of unambiguously identifying the referent object to the reader. This is due to the fact that the accuracy of re_2 regarding cell 4 is also high ($a_{re_2}(c_4) = 0.83$) and the accuracy of re_3 regarding cell 27 is not low ($a_{re_2}(c_27) = 0.43$).

The presented measure is a natural extension of the conventional concept of referential success when the accuracy of the referring expression for a given object is fuzzy. However, this measure produces similar values when evaluating referring expressions independently of the number of objects they *fuzzily* refer to.

For example, consider the following sentences:

- re_4 = The dark object.
- re_5 = The object in the upper-left.

In both cases, the referential success with respect to cell 3 is rather low ($rs_{re_4}(c_3) = 0.17$, $rs_{re_5}(c_3) = 0.15$). However, if we do not have any other alternative, we will probably choose re_5 because, as can be seen in Table 1, there are fewer objects for which re_5 is true to some extent than there are for re_4. That is, the fuzzy sets of objects of which re_5 is to some extent true is *more specific*, i.e. there is less uncertainty about which referent is intended.

To asses this uncertainty, Fuzzy Set Theory offers a wide variety of measures of the specificity of fuzzy sets. If O_{re} is the fuzzy set of objects for which re is true to some extent, with $O_{re}(o) = a_{re}(o)$, then the specificity of re, $Sp(re)$, should satisfy, among others, the following:

- $Sp(re) = 1$ iff O_{re} is a crisp singleton.
- If $O_{re} = \emptyset$, then $Sp(re) = 0$.
- If $card(O_{re}) \leq card(O_{re'})$ and O_{re} and $O_{re'}$ are normal fuzzy sets, then $Sp(re) \geq Sp(re')$, where $card(\cdot)$ is a suitable cardinality measure.

3.3 A Multiobjective Problem

In order to asses the goodness or adequacy of a referring expression, other quality dimensions can be considered, apart from the ones previously discussed, similarly to what happens in complete linguistic descriptions [3]. The REG literature contains a number of such proposals. Thus, *brevity* has been proposed as a criterion of adequacy [24] based on the conversational maxim, due to Grice [25], that a cooperative communicator should say no more than is required for the purposes of the exchange. Similarly, *humanlikeness* has been explored based on a variety of psycholinguistic findings, such as for example the extent to which the properties selected in a referring expression are salient or 'preferred' by speakers [26,27]. Finally, it has been argued that the use of certain properties in a referring expression depends on the relevance to the user given their current communicative task [28] as well as their knowledge state [29].

Though in the example used in this section it has been possible to determine the *best* referring expression for cell 3, in general, the problem of generating a referring expression for a given object does not always lead to a single *best* solution. It is quite normal to find that different measures of quality are correlated to different degrees; for instance, they might exert opposing influences, be contradictory, and/or have a negative correlation. For example, as we add properties to a referring expression, we increase the specificity but we also decrease brevity and probably also accuracy and ease of comprehension. Interestingly, empirical work comparing a large number of REG algorithms against shared datasets suggests that in fact, different evaluation metrics corresponding to different quality dimensions often do not correlate with each other [27].

This means that there is in general no such thing as the optimum or best referring expression for a given object, and the referring expression generation is a multiobjective optimization problem, as explained in [30,31] for complete linguistic descriptions.

4 REG in the Presence of Graduality: A Discussion

As we have noted, gradability and fuzziness have not received much attention in the REG literature. An exception is the work of van Deemter [12], who proposes a semantics for gradable properties, such as those related to the size of an object, and an algorithm to incorporate these properties in a REG procedure. van Deemter proposes to represent gradable properties numerically, preceding the standard REG content determination procedure with a step in which such properties are converted into inequalities. For example, if o is defined as having a value n for its height, this would be 'unpacked' into a set of inequalities of the form $n > m$, for all m which are values of height that are less than n. A standard REG algorithm can now use these inequalities as it would any other property, assuming that content selection is followed by an appropriate realisation step to convert them into an appropriate natural language string.

This approach is arguably less general than the one proposed in this paper. In particular, if we abandon the view that the relationship between objects and their properties is necessarily boolean (in the sense that an object either has a property or it doesn't) and adopt the more graded notion of *degree of fulfilment*, it is possible to treat all properties as potentially graded, or rather, all objects in a domain as having a particular property to different degrees. The contrast between the definition of accuracy and referential success given in Eq. (1) and Eqs. (2-3) makes this explicit.

4.1 Discriminatory Power

From the perspective of knowledge representation, this also eschews a core simplifying assumption made in many approaches to REG, namely, that properties are mutually exclusive. For example, consider an object's colour. In a domain in which some objects are *red* and some are *pink*, it is typically assumed that $[\![\, red \,]\!] \cap [\![\, pink \,]\!] = \emptyset$, so that a REG algorithm that determines, for a target referent o, that $o \in [\![\, red \,]\!]$, can safely ignore all objects in the extension of *pink*, since the latter is a completely different property.

This assumption has had an important implication for the development of heuristics to control the search performed by a REG algorithm during content selection, especially where the notion of *discriminatory power* is concerned. Discriminatory power refers to the informativeness of a property, that is, the extent to which it is true of the target referent, but false of other objects in the domain [24]. Thus, algorithms which define the quality of a referring expression in terms of brevity or informativity make use of this notion in trying to generate brief descriptions, by prioritising properties with higher discriminatory power, under the assumption that this will help the reader or hearer to identify the object more efficiently [24,32,33]. Returning to our example of colour, if the target referent happens to be the only *red* object in the domain, these algorithms will tend to select this property. The problem is that the object in question might also have a non-zero degree of fulfilment with respect to the property *pink*, based on the closeness of this colour to *red*, so that the discriminatory value of the property selected is not as clear-cut as was initially assumed.

4.2 Salience and Contextual Knowledge

Explicitly accounting for fuzziness also has implications for the way in which another important quality heuristic is modelled in a REG algorithm, namely, salience or preference. Many algorithms [26,34] give priority to properties, not based on discriminatory power, but on how salient, relevant or preferred they are. This notion of preference is usually defined by appealing to psycholinguistic findings, for example, the finding that speakers tend to avoid prototypically gradable properties such as size, unless they are absolutely required, in contrast for example to colour which, as we have seen, is used very frequently [35,36]. However, many such findings are based on experimental domains in which colours are maximally salient and distinct. A fuzzy approach can of course handle such cases, given that the fulfilment degree of an object with respect to a colour which is maximally salient in a domain would be higher than it would be in case the property is less salient, that is, closer to other colours. However, this approach would also enable more nuanced models, in which preference or salience can be computed in less straightforward cases, as in the example in Fig. 1(a) above. Indeed, the human tendency to use colour in referring expressions decreases significantly in the presence of colours that are similar, or close to each other [37]. Similarly, the tendency to use size increases, when the size difference between a target referent and its distractors is large [38].

Note that a graded notion of salience is also more faithful to the findings in the vision literature, where many computational models of visual salience identify salient regions based on a combination of the properties we have discussed in this paper, treating these properties as gradable [39,40]. The framework proposed here could potentially be adapted to the findings from the vision literature, by incorporating a psychologically plausible definition of salience into the computation of degree of fulfilment that determines the likelihood for a property to be selected by a REG procedure.

A further trend in computational research on vision is the incorporation of top-down, or contextual, knowledge [15,41]. Certain models [42,43] have successfully modelled human shifts of attention in visual scenes by incorporating both feature-based salience and knowledge of the type of scene being viewed. Thus, the salience of an object depends not only on the distinctiveness of its properties, but on human expectations concerning its location and function in a particular scene. Returning to our running example, when viewing an array of cells, such as those in Fig. 1(a), a biologist's attention might be drawn less by their colour (perhaps because this is relatively unimportant, or expected, where cells and tissues are concerned), but by their other properties. Future research on REG will need to take into account these top-down, expectation-based mechanisms, which are also known to influence the way humans refer to objects [44,45]. It is an open question whether the computation of the degree of fulfilment of a fuzzy property can be modulated by such top-down considerations.

4.3 Spatial Relations

Finally, another area in which fuzzy sets can play an important role in REG is in reference to objects using (spatial) relations to other elements of a scene. Since the early work of Dale and Haddock [46], the problem has been mostly defined in logical terms: the elements of a scene that can serve as potential anchors for a target referent in a spatial relation such as *in front of* are themselves encoded explicitly in the REG input [47], though there has been some work that takes a more realistic perspective in selecting salient landmarks [10,34]. Once again, the simplifying assumptions underlying REG algorithms can be relaxed using a fuzzy approach to spatial relations, especially where these are 'uncertain' and not necessarily mutually exclusive. For instance, in Fig. 1(b) if the object to be identified is cell 1, it can be seen that it is above cell 6 but also overlaps with it to some degree. It is not clear whether a good referring expression in this case should combine the two relations (*the small cell at the top which is above and slightly overlaps another*), select only one, perhaps the most salient one, or simply rely on a non-relational frame of reference (*the one at the top towards the middle*). Many techniques have been described to handle fuzzy spatial relationships, and we believe that REG would benefit from an exploration of this literature. Some related work has shown that fuzzy sets can be useful in generating appropriate natural language descriptions of uncertain temporal relations between events [48]; applying similar strategies to spatial relations is a promising way forward.

5 Conclusions

In this paper, we have tackled a classic problem in the generation of natural language from non-linguistic data, namely, referring expression generation. Focussing on visual scenes, we have sought to extend the remit of REG algorithms, by (a) abandoning some simplifying assumptions related to knowledge representation, especially the assumption that properties are crisp sets and that they are either true or false of an object; and (b) extending the notions of accuracy and referential success to deal with the more complex picture that gradability affords. Our discussion has also pointed to several directions for future work, notably the possibility of integrating the types of models proposed in this paper with findings from the vision literature concerning salience. We believe this is a promising direction in which to take research on REG; indeed, it potentially offers a way to bridge the gap between research on fuzzy sets, research on Natural Language Generation, and work in computer vision.

Acknowledgments. This work has been partially supported by the Spanish Ministry of Economy and Competitiveness and the European Regional Development Fund (FEDER) under project TIN2014-58227-P.

References

1. Reiter, E., Dale, R.: Building Natural Language Generation Systems. Cambridge University Press, Cambridge (2000)
2. Kacprzyk, J., Zadrozny, S.: Computing with words is an implementable paradigm: Fuzzy queries, linguistic data summaries, and natural-language generation. IEEE Trans. Fuzzy Syst. **18**(3), 461–472 (2010)
3. Marín, N., Sánchez, D.: On generating linguistic descriptions of time series. Fuzzy Sets Syst. **285**, 6–30 (2016)
4. Ramos-Soto, A., Bugarín, A., Barro, S.: On the role of linguistic descriptions of data in the building of natural language generation systems. Fuzzy Sets Syst. **285**, 31–51 (2016)

5. Mitchell, M., Dodge, J., Goyal, A., Yamaguchi, K., Stratos, K., Han, X., Mensch, A., Berg, A., Han, X., Berg, T., Daume III., H.: Midge: Generating Image Descriptions From Computer Vision Detections. In: EACL 2012, Avignon, France, pp. 747–756. Association for Computational Linguistics (2012)
6. Kulkarni, G., Premraj, V., Ordonez, V., Dhar, S., Li, S., Choi, Y., Berg, A.C., Berg, T.L.: Baby talk: Understanding and generating simple image descriptions. IEEE Trans. Pattern Anal. Mach. Intell. **35**(12), 2891–2903 (2013)
7. Yatskar, M., Galley, M., Vanderwende, L., Zettlemoyer, L.: See No Evil, Say No Evil : Description Generation from Densely Labeled Images. In: Proceedings of the Third Joint Conference on Lexical and Computation Semantics (*SEM) (2014)
8. Reiter, E., Dale, R.: A fast algorithm for the generation of referring expressions. In: COLING 1992, pp. 232–238 (1992)
9. Krahmer, E., van Deemter, K.: Computational generation of referring expressions: A survey. Comput. Linguist. **38**(1), 173–218 (2012)
10. Elsner, M., Rohde, H., Clarke, A.D.F.: Information structure prediction for visual-world referring expressions. In: EACL 2014, Gothenburg, Sweden, pp. 520–529. Association for Computational Linguistics (2014)
11. Kazemzadeh, S., Ordonez, V., Matten, M., Berg, T.L.: ReferItGame: referring to objects in photographs of natural scenes. In: EMNLP 2014, Doha, Qatar, pp. 787–798. Association for Computational Linguistics (2014)
12. van Deemter, K.: Generating referring expressions that involve gradable properties. Comput. Linguist. **32**(2), 195–222 (2006)
13. Turner, R., Sripada, S., Reiter, E., Davy, I.P.: Selecting the content of textual descriptions of geographically located events in spatio-temporal weather data. In: Ellis, R., Allen, T., Petridis, M. (eds.) Applications and Innovations in Intelligent Systems XV, pp. 75–88. Springer, London (2008)
14. Reiter, E., Sripada, S., Hunter, J., Yu, J., Davy, I.: Choosing words in computer-generated weather forecasts. Artif. Intell. **167**(1–2), 137–169 (2005)
15. Cadenas, J.T., Marín, N., Vila, M.A.: Context-aware fuzzy databases. Appl. Soft Comput. **25**, 215–233 (2014)
16. Castillo-Ortega, R., Chamorro-Martínez, J., Marín, N., Sánchez, D., Soto-Hidalgo, J.M.: Describing images via linguistic features and hierarchical segmentation. In: FUZZ-IEEE 2010, pp. 1–8 (2010)
17. Soto-Hidalgo, J.M., Chamorro-Martínez, J., Sánchez, D.: A new approach for defining a fuzzy color space. In: FUZZ-IEEE 2010, pp. 1–6 (2010)
18. Chamorro-Martínez, J., Medina, J.M., Barranco, C.D., Galán-Perales, E., Soto-Hidalgo, J.M.: Retrieving images in fuzzy object-relational databases using dominant color descriptors. Fuzzy Sets Syst. **158**(3), 312–324 (2007)
19. Chamorro-Martínez, J., Martínez-Jiménez, P.M., Soto-Hidalgo, J.M., León-Salas, A.: A fuzzy approach for modelling visual texture properties. Inf. Sci. **313**, 1–21 (2015)
20. Chamorro-Martínez, J., Martínez-Jiménez, P.M., Soto-Hidalgo, J.M., Prados-Suárez, B.: Fuzzy sets on 2D spaces for fineness representation. Int. J. Approx. Reasoning **62**, 46–60 (2015)
21. Prados-Suárez, B., Chamorro-Martínez, J., Sánchez, D., Abad, J.: Region-based fit of color homogeneity measures for fuzzy image segmentation. Fuzzy Sets Syst. **158**(3), 215–229 (2007)
22. Hudelot, C., Atif, J., Bloch, I.: Fuzzy spatial relation ontology for image interpretation. Fuzzy Sets Syst. **159**(15), 1929–1951 (2008)

23. Buck, A.R., Keller, J.M., Skubic, M.: A memetic algorithm for matching spatial configurations with the histograms of forces. IEEE Trans. Evol. Comput. **17**(4), 588–604 (2013)
24. Dale, R.: Cooking up referring expressions. In: ACL 1989, Vancouver, BC, pp. 68–75. Association for Computational Linguistics (1989)
25. Grice, H.P.: Logic and conversation. In: Cole, P., Morgan, J.L. (eds.) Syntax and Semantics 3: Speech Acts, pp. 41–58. Elsevier, Amsterdam (1975)
26. Dale, R., Reiter, E.: Computational interpretations of the gricean maxims in the generation of referring expressions. Cognitive Sci. **19**(2), 233–263 (1995)
27. Gatt, A., Belz, A.: Introducing shared tasks to NLG: the TUNA shared task evaluation challenges. In: Krahmer, E., Theune, M. (eds.) Empirical Methods in NLG. LNCS, vol. 5790, pp. 264–293. Springer, Heidelberg (2010)
28. Jordan, P.W., Walker, M.A.: Learning content selection rules for generating object descriptions in dialogue. J. Artif. Intell. Res. **24**, 157–194 (2005)
29. Janarthanam, S., Lemon, O.: Adaptive generation in dialogue systems using dynamic user modeling. Computat. Linguist. **40**(4), 883–920 (2014)
30. Castillo-Ortega, R., Marín, N., Sánchez, D., Tettamanzi, A.G.B.: Quality assessment in linguistic summaries of data. In: Greco, S., Bouchon-Meunier, B., Coletti, G., Fedrizzi, M., Matarazzo, B., Yager, R.R. (eds.) IPMU 2012, Part II. CCIS, vol. 298, pp. 285–294. Springer, Heidelberg (2012)
31. Bugarín, A., Marín, N., Sánchez, D., Triviño, G.: Aspects of quality evaluation in linguistic descriptions of data. In: FUZZ-IEEE 2015, pp. 1–8 (2015)
32. Gardent, C.: Generating minimal definite descriptions. In: ACL 2002, pp. 96–103, July 2002
33. Frank, M.C., Goodman, N.D.: Predicting pragmatic reasoning in language games. Science (New York, N.Y.) **336**(6084), 998 (2012)
34. Kelleher, J.D., Kruijff, G.: Incremental generation of spatial referring expressions in situated dialog. In: COLING-ACL 2006, Sydney, Australia, pp. 1041–1048. Association for Computational Linguistics (2006)
35. Pechmann, T.: Incremental speech production and referential overspecification. Linguistics (1989)
36. Belke, E., Meyer, A.S.: Tracking the time course of multidimensional stimulus discrimination: Analyses of viewing patterns and processing times during "same"-"different" decisions. Eur. J. Cognit. Psychol. **14**(2), 237–266 (2002)
37. Viethen, J., Goudbeek, M., Krahmer, E.: The impact of colour difference and colour codability on reference production. In: CogSci 2012, Austin, TX, pp. 1084–1089. Cognitive Science Society (2012)
38. van Gompel, R.P., Gatt, A., Krahmer, E., van Deemter, K.: Overspecification in reference: Modelling size contrast effects. In: AMLAP 2014 (2014)
39. Itti, L., Koch, C.: Computational modelling of visual attention. Nat. Rev. Neurosci. **2**(3), 194–203 (2001)
40. Erdem, E., Erdem, A.: Visual saliency estimation by nonlinearly integrating features using region covariances. J. Vis. **13**(4:11), 1–20 (2013)
41. Oliva, A., Torralba, A.: The role of context in object recognition. Trends Cogn. Sci. **11**(12), 520–527 (2007)
42. Torralba, A., Oliva, A., Castelhano, M.S., Henderson, J.M.: Contextual guidance of eye movements and attention in real-world scenes: The role of global features in object search. Psychol. Rev. **113**(4), 766–786 (2006)
43. Kanan, C., Tong, M.H., Zhang, L., Cottrell, F.W.: SUN: Top-down saliency using natural statistics. Vis. Cogn. **17**(6–7), 979–1003 (2009)

44. Sedivy, J.C.: Pragmatic versus form-based accounts of referential contrast: evidence for effects of informativity expectations. J. Psycholinguist. Res. **32**(1), 3–23 (2003)
45. Westerbeek, H., Koolen, R., Maes, A.: Stored object knowledge and the production of referring expressions: the case of color typicality. Frontiers Psychol. **6**(July), 1–12 (2015)
46. Dale, R., Haddock, N.: Generating referring expressions involving relations. In: EACL 1991, Berlin, Germany, pp. 161–166 (1991)
47. Areces, C., Koller, A., Striegnitz, K.: Referring expressions as formulas of description logic. In: INLG 2008, pp. 42–49 (2008)
48. Gatt, A., Portet, F.: Multilingual generation of uncertain temporal expressions from data: A study of a possibilistic formalism and its consistency with human subjective evaluations. Fuzzy Sets Syst. **285**, 73–93 (2016)

Impact of the Shape of Membership Functions on the Truth Values of Linguistic Protoform Summaries

Akshay Jain, Tianqi Jiang, and James M. Keller(✉)

Department of Electrical and Computer Engineering,
University of Missouri, Columbia, MO, USA
aj4g2@mail.missouri.edu, kellerj@missouri.edu

Abstract. In the recent past, a lot of work has been done on Linguistic Protoform Summaries (LPS). Much of this work focuses on improvement of the ways to compute truth values of LPS as well as on development of different protoforms. However, almost all of the systems using LPS use trapezoidal membership functions. This work investigates the effects of using triangular and pi shaped membership functions and compare their performance when using trapezoids. We start with an experiment using synthetic data and then compare the behavior of the three types of membership functions using real data which is obtained from an eldercare setting.

Keywords: Linguistic protoform summaries · Membership functions · Truth value

1 Introduction

An important aspect of decision support systems based on Fuzzy Logic and Fuzzy Sets is the design of the membership functions employed to represent the linguistic variables. Both the span of the membership functions as well as the shape can be an important factor depending on the application of concern. In many instances of fuzzy systems, there is little sensitivity to the configuration of the underlying membership functions. This work investigates the effect of the shape of membership functions representing the linguistic variables used to generate Linguistic Protoform Summaries (LPS).

Linguistic protoform Summaries are template based Natural Language sentences. In the past they have been used extensively to report the content of data in various fields [1–4]. As a departure from using LPS just to represent data linguistically, in [5] the authors defined a method to compute dissimilarity between two LPS and proved that it is a metric. This enabled them to use LPS as features of data and to find Linguistic Prototypes [6] representing the nightly sleeping patterns of elderly over a month's period of time. Basically, LPS were generated for each of the 31 nights which were then clustered to find Linguistic Prototypes. These prototypes were then deployed to find anomalies in a dataset in [7]. While detecting anomaly nights worked well, describing why a given night deviated from the normal pattern of prototype summaries proved a challenge. One issue that was identified was that the truth values of the

J.P. Carvalho et al. (Eds.): IPMU 2016, Part I, CCIS 610, pp. 204–213, 2016.
DOI: 10.1007/978-3-319-40596-4_18

prototypes and the nightly LPS were very close to 1 in almost all cases. Hence, the truth values played no role in distinguishing between 2 LPS under consideration. In all of these works and most of the work on LPS in the literature, the shape of the membership functions used to represent the linguistic variables of LPS is trapezoidal. Could that be the reason for such high truth values? In the following we compare the performance of trapezoid membership functions with triangular and pi shaped functions [8] related to LPS. We first preset an example with synthetic data and then explore the performance of triangular and pi functions with a set of real data which is generated from the sensors placed in the homes of elderly.

2 Background

In the sense of Yager [9], the most basic form of Linguistic Protoform Summaries (LPS) can be exemplified by '*Q y's are P*' @ *T* where *Q* is called the quantifier (like *few*, *some*, *most*), *P* is the summarizer (like *small*, *big*, *tall*), *y's* are the objects being summarized (like *balls*, *nights*) and *T* is the truth value ranging from 0 to 1. For example, in *Most of the balls are big @ 0.7*, *Most* and *big* are the quantifiers and summarizers respectively, the objects being summarized is size of balls in a bag and the truth value is 0.7. The Quantifier and Summarizers are the linguistic variables and are modelled over Fuzzy Sets. The truth value conveys information about the validity of the summary with respect to the data. Out of the several techniques available in the literature to compute it, we experimented with the methods of [9, 10]. Similar results were obtained for both of these techniques, and therefore, we only present the analysis using [10], in which the truth value is computed using Eq. (1)

$$T(A\,y's\ are\ P) = \max_{\alpha \in [0,1]} (\alpha \wedge A(P_\alpha)) \tag{1}$$

where ^ is the minimum operator, $P(x)$ is the membership function of the summarizer P, $P_\alpha = |\{y_i \in Y | P(y_i) \geq \alpha\}|$ is the proportion of objects whose membership in $P(x)$ is greater than or equal to α (varies from 0 to 1 in small intervals), $|.|$ denotes the cardinality of a set and $A(x)$ is a normal, convex and monotonically non-decreasing membership function of the quantifier A. For a quantifier whose membership function is not monotonically non-decreasing, it is split into two monotonically non-decreasing functions, $A_1(x)$ and $A_2(x)$ (which is used to compute $\overline{A_2}(x)$) and the truth value is computed as shown in (1). Please refer to [10] for more details.

3 Tests with Synthetic Data

As mentioned before, traditionally Quantifiers and Summarizers have been represented by trapezoidal membership functions. In this section we set out to see how changing the shape of the membership functions to triangles and pi shaped functions varies the truth values of Linguistic Protoform Summaries. To this end, we design an experiment with synthetic data.

The synthetic data is comprised of 100 balls inside a bag. We find the LPS summarizing the size of the balls inside this bag. To see the variation of truth values we change the size of the balls in steps and generate LPS at each point. The truth values are computed for trapezoidal, triangular and pi shaped membership functions. Figures 1 and 2 show the summarizer and quantifier membership functions of all three types drawn in the same figures to have a better relative comparison among them.

While converting the trapezoids to triangles we change the span of the functions such that they have the similar core and intersection points in all three shapes. For example, for the case of *mid* in Fig. 1, the core of triangular function is at the center of the core of trapezoid. However, the value of the trapezoid membership function is zero at all points other than from 0.25 to 0.75, while the triangle and pi functions are non-zero from 0.1 to 0.9. This is done in order to have the intersection of the membership function of *mid* and *small*, and *mid* and *big* at same levels for all three types of membership functions. If this modification is not done, then some values on the x axis (that is, the size in this particular case) will have very small memberships in all three functions which is non-intuitive since it would lead to giving less importance to these values in the final truth value calculation. Also, note that the pi function has smaller memberships than the corresponding triangle until 0.5, while it is higher for values above 0.5.

We start with 100 balls, with all sizes having highest membership in the summarizer *Small*. Then step by step, we replace five balls in the bag with balls having highest membership in summarizer *Big*. That is, at the end we have all balls with highest membership in the summarizer *Big*. At each step we calculate the truth values of LPS of the form *Q of the balls are P* where *P* and *Q* are the quantifiers and summarizers, respectively, given in Figs. 1 and 2. For each step we would have twelve summaries (four for each summarizer). Reporting the truth value of all these summaries at every step will result in a lot of clutter. Therefore, in Fig. 3 we only report truth values of 3 sentences, *Almost None of the balls are small*, *Many balls are big* and *Many balls are small* when using trapezoidal, triangular and pi shaped membership functions. Truth

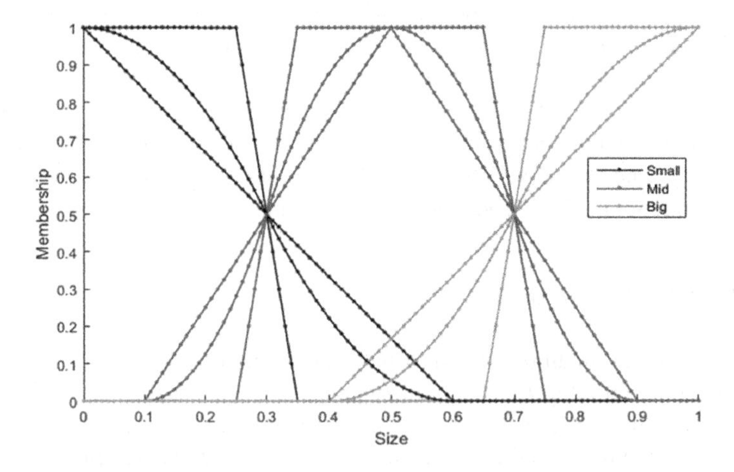

Fig. 1. Trapezoid, triangular and pi shaped membership functions for summarizers used in synthetic data example. The x-axis represents the size of balls. (Color figure online)

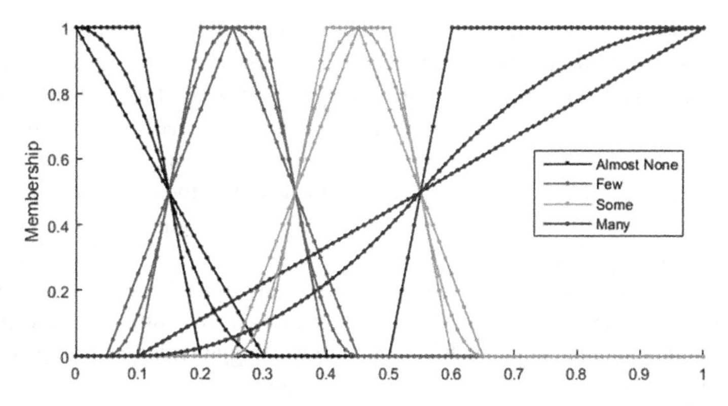

Fig. 2. Trapezoid, triangular and pi shaped membership functions for quantifiers used in synthetic data example. (Color figure online)

values generated from trapezoids are shown with "tick marks", those from triangles have "dots", and truth calculation using pi functions are shown with "open circles". It is easy to see that for the case when triangular and pi shaped functions are used, the truth value varies gradually while it changes in big steps when using trapezoids. This is expected since the trapezoids have longer constant periods of membership 1 while in triangles and pi, the membership gradually reaches 1 and also drops slowly to 0. It is easy to observe this by looking at each set of 3 curves of the same color in Fig. 3. This variation of truth value along with the data the LPS are summarizing looks more intuitive since its better reflecting the changes in the data.

Another important observation that should be noted is that when the truth values computed using trapezoids are below 0.5, the use of triangles and pi functions

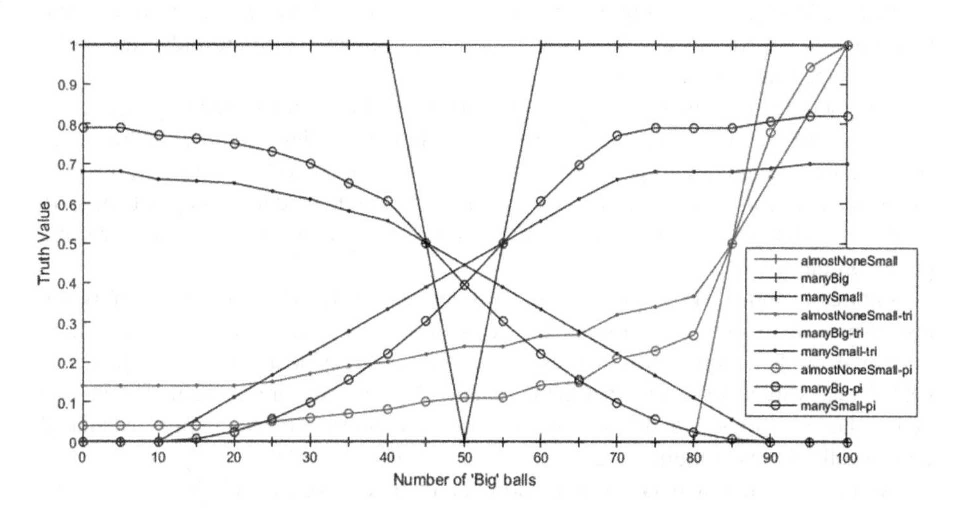

Fig. 3. Truth value vs the number of big balls while using trapezoid, triangle and pi shaped membership functions. The x-axis shows how many of the 100 balls have highest membership in the summarizer big. (Color figure online)

intensifies it, while it suppresses the value when the truth value is greater than 0.5. This is expected, since the triangular and pi membership functions are greater than trapezoids for values below 0.5 while they are less than trapezoids for the values above it. Also, when using pi functions, the truth values are smaller as compared to when using triangles for values less than 0.5, while it is greater with triangles when the value is higher than 0.5. This is also in accordance with how the membership functions are constructed, since the pi functions become higher than triangular functions only after they reach a value of 0.5.

Given these results where intuition can play a significant role, the use of either triangular or pi-shaped membership functions enables truth values to span the entire unit interval. This should allow truth values to play a bigger role in distinguishing one LPS from another in an automated explanation system.

4 Tests with Real Data

In this section we explore how changing the shape of the membership functions effect the results with real data. We use the data which was previously used in [6] to generate Linguistic Prototypes. This data comes from an aging in place facility called TigerPlace in Columbia, MO, USA. The apartments of elderly are equipped with network of sensors such as motion sensors, Kinect or bed sensors to monitor the living patterns of the residents. In that work, nightly bed restlessness data was summarized by summaries of the form:

$$Q \text{ of the } \Delta t \text{ slots had } P \text{ restlessness}$$

where the data during a night was divided into slots of duration Δt (15 min) and Q and P are the quantifiers and summarizers respectively. For example, a summary describing the bed restlessness over a night may be: *Few of the Δt had high bed restlessness*. We use two nights from this dataset to compare the performance of trapezoidal, triangular and pi shaped membership functions.

The membership functions of the Summarizers and Quantifier used to generate the summaries of the form mentioned above are shown in Figs. 4 and 5. While constructing the triangular and pi membership functions, we try to be as close as possible to the trapezoids used in [6]. However, similar to the synthetic data example, their span needs to be changed sometimes to avoid cases having very low memberships for some points on the x axis.

Figures 6 and 7 show the data of two nights. Each point in the graph is the restlessness of Δt slot (15 min time slot). To compute the value of each slot, a night is first divided into 15 min slots, then the restlessness data inside a slot is accumulated to get the final value. Also, note that the number of 15 min slots are not same for the two nights shown in Figs. 6 and 7 because the start and end time of the night is calculated dynamically for each night.

Along with the raw data, we also show the memberships of each data point in the three summarizers shown in Fig. 4 for all three types of membership functions. This is done in order to have a better interpretation of data with respect to the membership functions used.

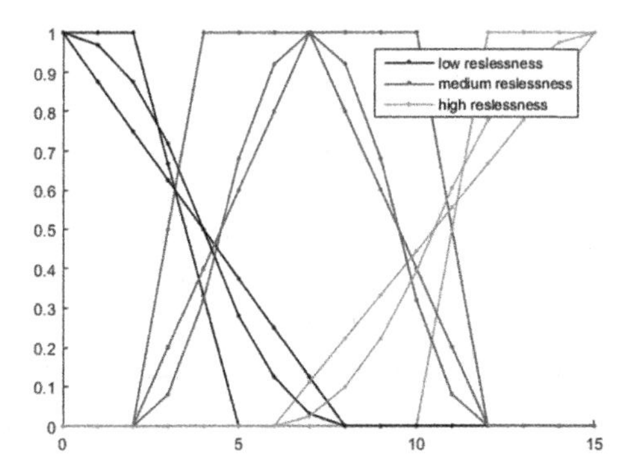

Fig. 4. Trapezoid, triangle and pi shaped membership functions for summarizers used in real data examples. (Color figure online)

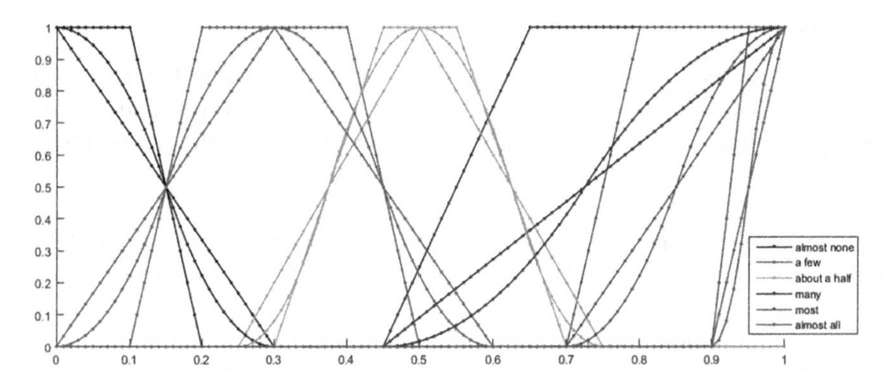

Fig. 5. Trapezoid, triangle and pi shaped membership functions for quantifiers used in real data examples. (Color figure online)

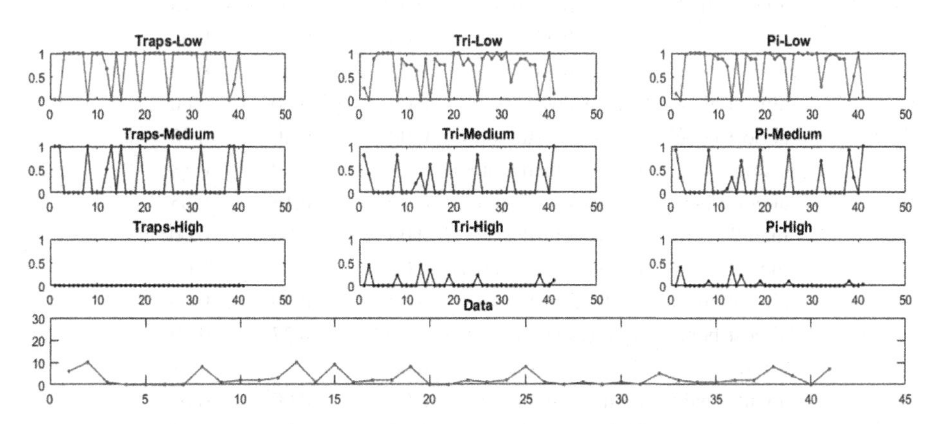

Fig. 6. Case A – night time restlessness. The last row shows the raw data with x-axis showing the number of 15 min slot and y-axis is the restlessness in each slot. The first row shows the membership of the raw data in summarizer low, the second shows membership in medium while the third row for summarizer high, with the first column for trapezoids, the second for triangle and third for pi shaped membership function.

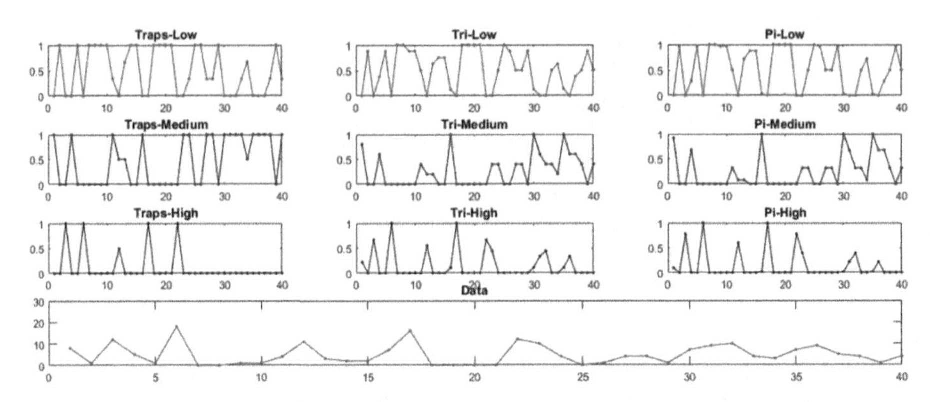

Fig. 7. Case B – night time restlessness.

4.1 Case A

Table 1 lists all the possible LPS generated for the data shown in Fig. 6. For each summary, we compute the truth values with all three types of membership functions. Also, instead of the full sentence of LPS, we only show the quantifier and summarizer

Table 1. Case A-truth values of summaries when using trapezoid, triangular and pi membership functions. The first and second column show the Quantifier and Summarizer respectively, for each summary.

Quantifier	Summarizer	Truth values		
		Trapezoids	Triangles	Pi
'almost none'	'low restlessness'	0.00	0.13	0.04
'a few'	'low restlessness'	0.00	0.25	0.13
'about a half'	'low restlessness'	0.00	0.25	0.13
'many'	'low restlessness'	1.00	0.51	0.52
'most'	'low restlessness'	0.33	0.27	0.14
'almost all'	'low restlessness'	0.00	0.00	0.00
'almost none'	'medium restlessness'	0.00	0.40	0.32
'a few'	'medium restlessness'	1.00	0.60	0.68
'about a half'	'medium restlessness'	0.00	0.16	0.05
'many'	'medium restlessness'	0.00	0.00	0.00
'most'	'medium restlessness'	0.00	0.00	0.00
'almost all'	'medium restlessness'	0.00	0.00	0.00
'almost none'	'high restlessness'	1.00	0.77	0.89
'a few'	'high restlessness'	0.00	0.23	0.11
'about a half'	'high restlessness'	0.00	0.00	0.00
'many'	'high restlessness'	0.00	0.00	0.00
'most'	'high restlessness'	0.00	0.00	0.00
'almost all'	'high restlessness'	0.00	0.00	0.00

Table 2. Case B-truth values of summaries when using trapezoid, triangular and pi memberhsip functions. The first and second column show the Quantifier and Summarizer respectively, for each summary.

Quantifier	Summarizer	Truth values		
		Trapezoids	Triangles	Pi
'almost none'	'low restlessness'	0.00	0.13	0.04
'a few'	'low restlessness'	0.50	0.50	0.50
'about a half'	'low restlessness'	0.67	0.52	0.54
'many'	'low restlessness'	0.33	0.37	0.28
'most'	'low restlessness'	0.00	0.12	0.03
'almost all'	'low restlessness'	0.00	0.00	0.00
'almost none'	'medium restlessness'	0.00	0.40	0.32
'a few'	'medium restlessness'	0.50	0.60	0.68
'about a half'	'medium restlessness'	0.80	0.40	0.32
'many'	'medium restlessness'	0.25	0.09	0.02
'most'	'medium restlessness'	0.00	0.00	0.00
'almost all'	'medium restlessness'	0.00	0.00	0.00
'almost none'	'high restlessness'	0.80	0.56	0.61
'a few'	'high restlessness'	0.20	0.44	0.39
'about a half'	'high restlessness'	0.00	0.11	0.02
'many'	'high restlessness'	0.00	0.00	0.00
'most'	'high restlessness'	0.00	0.00	0.00
'almost all'	'high restlessness'	0.00	0.00	0.00

for each summary of the form, *Q of the Δt slots had P restlessness*. It can be observed that the spread of the truth values is more uniform for the case when triangles and pi membership functions are employed as compared to the trapezoids. For example, for the case with trapezoid membership functions, the truth value of, *Many of the Δt slots had low restlessness* is 1 while it is 0 for all the quantifiers to the left of it in Fig. 5, i.e., *almost none, few, about a half*. However, with triangular shaped membership functions for the same summarizer, the truth values start with 0.13 for the summary with quantifier *almost none*, it progressively rises to 0.51 for *many* and drops gradually to 0 for *almost all*. Similar trends can be observed when using the pi shaped membership functions.

4.2 Case B

LPS for the data displayed in Fig. 7 are shown in Table 2. Similar to Case A, for each possible summary, we present the truth values for all three types of membership functions. The spread of truth values for trapezoidal membership functions is not as non-uniform as in the Case A. For instance, the truth values of summaries with summarizer low restlessness varies gradually from 0 for quantifier almost none to 0.67

for about a half. Nevertheless, the truth values for triangular and pi membership functions are better spaced out among all the LPS.

This type of gradual trend in the truth values of LPS can be considered more informative as compared to the cases where the truth values suddenly rises and drops to some value. Even though it might not be of much value when LPS are used to analyze some data linguistically, they can be considered better when LPS are used as features, like in [6]. Also, it should be noted that in some applications, in order to decide which LPS should be used in the final representation of the data, a threshold based on the truth value is used. In such cases, the threshold would need to be changed for triangular and pi membership functions.

5 Conclusions and Future Work

With the help of synthetic and real data we compared the performance of three types of membership function with respect to computing truth values for Linguistic Protoform Summaries. The contrast between the three was much more evident for synthetic data than for the real data we chose. This is because we had the liberty of designing an experiment tailored to this task in the synthetic data case. As for the real data, it is quite possible that we will find more differences in the performances utilizing the three membership function types as we include more nights into our study. Nevertheless, the two real data examples we presented still pose some interesting comparisons. This naturally leads us to a more rigorous and broad study of comparison of membership functions with different sets of real data. Moreover, evaluations with the help of case studies involving human subjects might present some important insights.

An important exercise that we are currently conducting is to first study the use of triangular and pi shaped membership functions when using LPS as features to compute Linguistic Prototypes. Then from the prototypes, we are building an automated system to explain why particular nights deviate from the prototypes. We believe that using triangular or pi shaped memberships will aid in that endeavor.

Acknowledgments. Akshay Jain is supported by AHRQ Grant 1R01HS023328 and Center for Eldercare and Rehabilitation Technology at the University of Missouri.

References

1. Jain, A., Keller, J.M.: Textual summarization of events leading to health alerts. In: 2015 37th Annual International Conference of the IEEE Engineering in Medicine and Biology Society (EMBC), pp. 7634–7637 (2015)
2. Wilbik, A., Keller, J.M., Alexander, G.L.: Linguistic summarization of sensor data for eldercare. In: 2011 IEEE International Conference on Systems, Man, and Cybernetics (SMC), pp. 2595–2599. IEEE (2011)
3. Castillo-Ortega, R., Marín, N., Sánchez, D.: Time series comparison using linguistic fuzzy techniques. In: Hüllermeier, E., Kruse, R., Hoffmann, F. (eds.) IPMU 2010. LNCS, vol. 6178, pp. 330–339. Springer, Heidelberg (2010)

4. Kacprzyk, J., Wilbik, A., Zadrożny, S.: Linguistic summarization of time series using a fuzzy quantifier driven aggregation. Fuzzy Sets Syst. **159**, 1485–1499 (2008)
5. Wilbik, A., Keller, J.M.: A distance metric for a space of linguistic summaries. Fuzzy Sets Syst. **208**, 79–94 (2012)
6. Wilbik, A., Keller, J.M., Bezdek, J.C.: Linguistic prototypes for data from eldercare residents. IEEE Trans. Fuzzy Syst. **22**, 110–123 (2014)
7. Wilbik, A., Keller, J.M.: Anomaly detection from linguistic summaries. In: 2013 IEEE International Conference on Fuzzy Systems (FUZZ), pp. 1–7. IEEE (2013)
8. Keller, J.M., Liu, D., Fogel, D.: Fundamentals of Computational Intelligence: Neural Networks, Fuzzy Systems, and Evolutionary Computation. Wiley/IEEE Press, Hoboken/New York (2016)
9. Yager, R.R.: A new approach to the summarization of data. Inf. Sci. **28**, 69–86 (1982)
10. Jain, A., Keller, J.M.: On the computation of semantically ordered truth values of linguistic protoform summaries. In: 2015 IEEE International Conference on Fuzzy Systems (FUZZ-IEEE), pp. 1–8 (2015)

A Solution of the Multiaspect Text Categorization Problem by a Hybrid HMM and LDA Based Technique

Sławomir Zadrożny(✉), Janusz Kacprzyk, and Marek Gajewski

Systems Research Institute, Polish Academy of Sciences,
ul. Newelska 6, 01-447 Warszawa, Poland
{zadrozny,kacprzyk,gajewskm}@ibspan.waw.pl

Abstract. In our previous work we introduced a novel concept of the *multiaspect text categorization* (MTC) task meant as a special, extended form of the text categorization (TC) problem which is widely studied in *information retrieval.* The essence of the MTC problem is the classification of documents on two levels: first, on a more or less standard level of thematic categories and then on the level of document sequences which is much less studied in the literature. The latter stage of classification, which is by far more challenging, is the main focus of this paper. A promising way of attacking it requires some kind of modeling of connections between documents forming sequences. To solve this problem we propose a novel approach that combines a well-known techniques to model sequences, i.e., the Hidden Markov Models (HMM) and the Latent Dirichlet Allocation (LDA) technique for the advanced document representation, hence obtaining a hybrid approach. We present details of our proposed approach as well as results of some computational experiments.

Keywords: Multiaspect text categorization · Sequences of documents · HMM · LDA

1 Introduction

We deal with a variant of the *text categorization* (TC) problem. In its basic form, the general TC problem boils down to deciding which of a predefined set of categories a given document belongs to. Thus, usually after adopting an appropriate document representation, e.g., based on the vector space model [2], documents are treated as vectors and one of a multitude of the classification techniques may be employed. In a series of papers [8,16,18,19] (see www.ibspan.waw.pl/~zadrozny/MTC for a complete list of our related papers). we have introduced and studied the concept of the *multiaspect text categorization* (MTC), a novel problem that goes far beyond the usual TC. We have proposed several approaches to solve it. The MTC task may be seen as a special case of the general text categorization problem where two levels of classification are involved. It is inspired by a practical application which may be briefly described as follows.

J.P. Carvalho et al. (Eds.): IPMU 2016, Part I, CCIS 610, pp. 214–225, 2016.
DOI: 10.1007/978-3-319-40596-4_19

Institutions in Poland, as well as in virtually all countries, are obliged to handle documents related to their business processes in a strictly regulated way. First, the documents have to be assigned to some thematic/topical categories arranged in a hierarchy. For example, a document submitted by a citizen while applying for a driving license should be classified as belonging to the top category "Social and civic cases" or, perhaps, within its specialized descendant subcategory at the bottom of a hierarchy, such as "Documentation of a vehicle registration". Second, within such a category this document has to be classified to a specific *case*, i.e., a sequence of documents related to a particular instance of the business process of the driving license issuing of that person. Such a sequence may already exist – for instance, the document under consideration may concern some additional information the applicant has been required to provide – or it may be the first document which initiates a case, e.g., it is the application for the issuing of the driving license of that person. Thus, within the case the documents are sequentially ordered and their order is implied by the logical succession of the documents within a given business process. Each instance of a given process may clearly be associated with a different number of documents of a different type, e.g., some documents may be initially missing and the institution will send a notice to the applicant to complete it which he or she will respond attaching those missing documents or explaining the reasons for their lack, or asking for further information from the institution etc.

The above task is usually dealt with manually, which is costly and time consuming, and our aim is to support the human operator by developing a system automatically generating an advice concerning the proper classification of documents. Thus, on the first level one may apply one of the classification techniques well studied in the classic *text categorization* [13]. The second level classification is more challenging due to several reasons. First of all, there is a limited number of training documents representing particular cases and a straightforward approach of treating each case as a category on its own does not work well. Moreover, the list of cases is growing over time and a classifier has to detect if a document to be classified should start a new case. Hence, grasping the logic of succession of the documents within a case seems to be critical for a classifier to successfully handle the MTC problem. In our previous work we proposed several solutions to the MTC problem. In particular, in [16] we proposed two approaches to model the sequences of the documents using Zaki's sequence mining algorithm [20] and the Hidden Markov Models [12]. In this paper we further develop the latter approach and go beyond the conceptual presentation by making the approach operational. In particular, we follow a widely advocated approach of developing a hybrid system that combines a variant of the well-known HMM technique with one of the modern techniques used to represent textual documents and known as the Latent Dirichlet Allocation (LDA) [4].

We first remind the formal definition of the MTC problem and point out some related works. Next, we present a general scheme of our proposed approach, briefly reminding the basics of the HMM and LDA techniques, focusing on their hybridization. Then, we discuss details of our approach, present the results of some computational experiments and conclude with some final remarks.

2 The Multiaspect Text Categorization Problem

We assume a collection of documents, $D = \{d_1, \ldots, d_n\}$ which is structured as follows. The documents are arranged in a set of predefined *categories* from the set $C = \{c_1, \ldots, c_m\}$ in such a way the each document $d \in D$ belongs to exactly one category $c \in C$. The documents are further arranged within each category in sequences $\sigma \in \Sigma$ which are referred to as *cases*:

$$\sigma_k =< d_{k_1}, \ldots, d_{k_T} > \quad (1)$$

$$\Sigma = \{\sigma_1, \ldots, \sigma_p\} \quad (2)$$

Each document $d \in D$ belongs to exactly one case $\sigma \in \Sigma$. A different rationale and logic assumed for the grouping of the documents into categories and cases is here important. That is, respectively, a topical similarity or the belongingness to the same business process, in relation to its different stages.

Our purpose is to develop a system, following the paradigm of supervised learning, working in such a way that given a collection D structured as above and a new document $d^\star$ the system supports a human user in deciding how to assign $d^\star$ to a category $c \in C$ and to a case $\sigma \in \Sigma$ within this category. For practical reasons, we distinguish between *on-going cases* comprising documents of the business processes still under way, and closed cases related to the business processes which are already completed. The newly incoming documents may be classified only to the cases of the former type while cases of both types may serve as training examples for the construction of a classifier.

The MTC problem formulation and its practical inspirations are original and the literature of this topic basically comprises our recent works only. The most similar problem already known in the literature is *Topic Detection and Tracking* (TDT) [1]. It is inspired by a practical problem of handling a stream of news stories to be organized in a dynamically structured collection. News/documents concerning the same topic/event have to be grouped together and, similarly to our MTC problem, incoming documents may belong to already existing groups or may start new ones. Topics in the TDT are similar to the cases in our MTC problem and, in general, both problems share many points. However, they are inherently different which well justifies the study of the MTC as a separate problem. For example, in the TDT there are no such distinct two levels of classification as in the MTC. Even if the concept of a hierarchical TDT was also considered as an extension to the basic TDT, still the different nature of classes at particular levels of the hierarchy is not considered there. Another important aspect distinguishing both problems is that in the MTC cases are sequences of documents while topics in TDT are just sets of stories. For more discussion of the relation between the TDT and MTC, cf. our paper [8].

The MTC problem may be dealt with in many different ways. Due to a space limit we refer the reader to a number of approaches we proposed in our earlier papers.

3 The Proposed Approach

3.1 The Techniques Employed: HMM, LDA and the Logistic Regression

Hidden Markov Models (HMM) [12]. As we need a model of the sequence (case) of documents, we assume here that the case is a realization of a stochastic process with the Markov property and hidden states, i.e., is a Hidden Markov Model (Chain) (HMM), denoted by $\lambda = (A, B, \pi)$ and characterized by the following parameters:

1. the number of hidden states N_S; the states $S_i \in S = \{S_1, \ldots, S_{N_S}\}$, may be here interpreted as corresponding to the stages of the business process represented by a given case,
2. the number of distinct observation symbols $u_i \in U = \{u_1, \ldots, u_{N_U}\}$; here the observations are the whole documents and we discuss their representation in what follows – in the explanation of the line 5 of the algorithm shown in Fig. 1,
3. the state transition probability distribution, denoted as $A = [a_{ij}]_{1 \le i,j \le N_S}$, i.e., $a_{ij} = P(q_{t+1} = S_j \mid q_t = S_i)$, where q_t denotes the (hidden) state of the stochastic process at time t,
4. the observation symbols probability distribution $b_j \in B = \{b_1, \ldots, b_{N_S}\}$ defined for each state S_j, i.e.:

$$b_j : U \rightarrow [0, 1], \qquad b_j(u_i) = P(O_t = u_i \mid q_t = S_j) \tag{3}$$

   where O_t denotes an observation generated at time t; $O = (O_1, \ldots, O_T)$ will denote the whole sequence generated by the HMM which corresponds here to the sequence of documents (a case),
5. the initial probability distribution π over the state space S, i.e., $\pi(S_i) = P(q_1 = S_i)$.

For our purposes the original basic version of the HMM, as described above, seems to be not adequate. A possible extension [14] consists in adding covariates to condition the probabilities of the transitions and observations. We use a vector of covariates $cov_t = [cov_t^1, \ldots, cov_t^k]$ for the observation distribution conditioning which leads to the following modified form of (3):

$$b_j(u_i) = P(O_t = u_i \mid q_t = S_j, cov_t) \tag{4}$$

There are three basic problems related to the HMMs [12]:

- the evaluation problem, i.e., how to efficiently compute the probability of an observation sequence $O = (O_1, \ldots, O_T)$ given an HMM λ,
- the decoding problem, i.e., given an HMM λ and an observation sequence $O = (O_1, \ldots, O_T)$ what is a most probable (in some sense) sequence of states $(S_1, \ldots, S_T)$ which led to the generation of sequence O,

- the learning problem, i.e., given an HMM $\lambda = (A, B, \pi)$ and a sequence of observations O how to adjust λ's parameters A, B and π so as to maximize the probability of O, i.e., $P(O \mid \lambda)$.

In our algorithm we are dealing mostly with the first and third problem but the second problem is also of interest from the point of view of possible future modifications of our approach.

Thus, we may adopt an HMM λ_c as a rich generative model of sequences $\sigma =< d_1, \ldots, d_T >$ belonging to a given category c. We will discuss this in more detail later, including the form of the covariates involved, in the explanation of the line 5 of the algorithm shown in Fig. 1.

Latent Dirichlet Allocation (LDA) [4]. The Latent Dirichlet allocation (LDA) is a generative probabilistic model of a collection of documents (a corpus). Basically, it assumes that there is a set of k topics[1] $Z = \{z_j\}$ and each document $d \in D$ of the corpus deals with a mixture θ_d of them, i.e., $\theta_d : Z \rightarrow [0, 1]$ such that $\sum_j \theta_d(z_j) = 1$. Each topic z_j is, in turn, a distribution over a set of words (vocabulary) $V = \{w_i\}$, i.e., $z_j : V \rightarrow [0, 1]$ and $\sum_i z_j(w_i) = 1$.

It is assumed that for the whole corpus a parameter denoted by β is fixed and each topic distribution $z_j \in Z$ is sampled from the Dirichlet distribution with parameter β over the space of all multinomial (categorical) distributions over the vocabulary V. Another parameter set for the whole corpus is α which is the parameter of the Dirichlet distribution used to sample the mixtures of topics, to be explained below. Then, a document d, belonging to a corpus characterized by the values of parameters α and β, is assumed to be generated in the following process:

1. First, the length of the document in words, N, is sampled according to the Poisson distribution with the parameter ξ.
2. Second, the mixture of topics θ_d is sampled for the document according to the Dirichlet distribution with the parameter α.
3. Finally, for each of the N positions of words assumed to comprise the document d, first a topic z_j is sampled using the multinomial distribution θ_d and then a word $w \in V$ is chosen using the multinomial distribution related to the topic z_j.

Now, if we are given a corpus of documents we can observe only the values of the variables corresponding to the particular positions of the words within documents. All other random variables mentioned in the description of the generative process above are hidden. There exists a number of approaches to infer the posterior distributions of the hidden variables and to estimate parameters α and β [4]. Using one of them we obtain an LDA model of the corpus. Let us denote its part which will be useful for our further considerations as:

$$L = (\{z_j\}_{j=1,\ldots,k}, \{\theta_d\}_{d \in D}) = (Z, \Theta) \tag{5}$$

[1] To shorten the notation we will denote the topic in the same way as the distribution on the words defining it.

i.e., we have a set of k multinomial distributions z_j over the set of words V for all k topics and for each document $d \in D$ we have a mixture of topics θ_d characterizing it. We are also in a position to determine the representation of a new document $d^\star \notin D$ using the LDA model obtained.

3.2 The Algorithm

Here we assume that the incoming document $d^\star$ has been first properly classified to a category and the algorithm presented assigns a case to $d^\star$. We briefly discuss the question of category assignment in Sect. 3.3.

The general scheme of the proposed algorithm is presented in Fig. 1. Now we will discuss its particular lines, referring to the numbers shown in Fig. 1. In the next section we present the results of the computational experiments carried out using the R environment and its various packages, thus while describing here particular steps of the algorithm we will refer to its more general aspects as well as to the aspects specific for the assumed implementation.

Line 3. The document-term matrix forms a standard representation of the collection of documents in the vector space model [2]. The set of terms (the vocabulary) used to represent the documents is denoted as V. Here we employ the weights of the terms (keywords) in documents equal to the frequencies of their occurrence within those documents, i.e. the `tf` weighting scheme. This is the format preferred for the LDA analysis of the collection.

Line 4. An LDA model $L = (Z, \Theta)$ is constructed for the whole collection of training documents belonging to category c. The number of topics should be

1: **Initialization stage**
2: **for all** categories $c \in C$ **do**
3: *create* a document-term matrix
4: *create* an LDA model, LDA_c, for the collection $D_c \subseteq D$ of the documents belonging to category c
5: *train* an HMM model, λ_c, using all cases belonging to D_c
6: **end for**
7: **Classification stage**
8: $d^\star \leftarrow$ newly arrived document
9: $c^* \leftarrow$ category assigned to $d^\star$
10: *represent* $d^\star$ using the model LDA_c
11: **for all** ongoing cases σ_i **do**
12: *compute*, with respect to the HMM λ_c, the conditional probability of the case σ extended with the document $d^\star$, $< \sigma_i, d^\star >$, under the condition that the sequence σ has been generated, i.e. $P_{\lambda_c}(< \sigma_i, d^\star >| \sigma)$
13: **end for**
14: choose the case σ_i with the highest $P_{\lambda_c}(< \sigma_i, d^\star >| \sigma)$ and assign $d^\star$ to this case.

Fig. 1. A general scheme of the proposed algorithm

chosen experimentally but should not be too large as that number implies the number of parameters that have to be learned during the training of the HMM, in line 5.

Line 5. In this step, first, the representation of each document $d \in D$ provided by the obtained LDA model L in the form of a distribution θ_d is transformed into a binary vector[2], $d = [d_1, \ldots, d_k] \in \{0,1\}^k$, of dimension k in such a way that if the probability of a given topic z_j according to θ_d is greater than a threshold value τ (in the experiments $\tau = 1/k$), then $d_j = 1$ and otherwise $d_j = 0$, i.e.:

$$\theta_d \longrightarrow d : d_j = \begin{cases} 1 & \text{if } \theta_d(z_j) \geq \tau \\ 0 & \text{otherwise} \end{cases} \qquad j = 1, \ldots, k \tag{6}$$

Then, all cases present in collection D are used to train the HMM with a number of states N_S chosen experimentally and observations identified with the binary vectors d_j defined in (6). The observation probability distributions (3)–(4) are assumed to be multivariate Bernoulli distributions, i.e.,

$$b_j(u_i) = P(d \mid q_t = S_j) = \prod_{j=1}^{k} P(d_j = 1 \mid q_t = S_j)^{d_j} * P(d_j = 0 \mid q_t = S_j)^{1-d_j} \tag{7}$$

Actually, we are using a modified form of the formula (4) as we use the covariates for our observation distributions and the logistic regression to take them into account. Thus, in our case the following formula is employed:

$$\text{logit}(P(d_j \mid q_t = S_j)) = \omega_1 cov_t^j + \omega_0 \tag{8}$$

where the vector of covariates $cov = (cov_t^1, \ldots, cov_t^k)$ at time t is defined as follows:

$$cov_t^j = \theta_{d_{t-1}^{tfn}} \cdot z_j = \sum_{i=1}^{|V|} d_{t-1,i}^{tfn} * z_j^i \qquad j = 1, \ldots, k \tag{9}$$

where:

- $|V|$ denotes the size of the vocabulary,
- $d_{t-1}^{tfn} = (d_{t-1,1}^{tfn}, \ldots, d_{t-1,|V|}^{tfn})$ denotes the document occurring in the case at the preceding position (at time $t-1$ in the parlance of the HMM modeling) which is represented by its normalized version present in the document-term matrix created in line 3 of the algorithm shown in Fig. 1; the normalization takes the following form:

$$d_{t-1,i}^{tfn} = \frac{d_{t-1,i}^{tf}}{\max_j d_{t-1,j}^{tf}} \qquad i = 1, \ldots, |V| \tag{10}$$

where $d_{t-1,i}^{tf}$ denotes the i-th coordinate of the vector representing the document in the document-term matrix before normalization,

[2] To simplify notation we denote this vector as d, i.e., in the same way as the document $d \in D$.

- z_j is the probability distribution representing the j-th topic, obtained as a part of the LDA model of the collection, which is here treated as a vector, i.e., $z_j = (z_j^1, \ldots, z_j^{|V|})$, $\sum_{i=1}^{|V|} z_j^i = 1$.

The usage of the covariates defined as above makes it possible to better model the patterns of the similarity/dissimilarity of the documents neighboring in a sequence belonging to a given category. More on that in the discussion provided in Sect. 3.3.

Line 8. A new document $d^\star$ to be classified is first represented both in terms of the document-term matrix mentioned in line 3 as well as in terms of the LDA model mentioned in line 4.

Line 9. As it is mentioned earlier, we assume that the document $d^\star$ is already classified to a category. In our previous work we usually use the k-nearest neighbors algorithm to do that. The current use of the LDA models opens new possibilities and in our further work we will check the efficiency of the method based on the LDA model.

Line 12. In order to select a case to which document $d^\star$ should be classified we compute for each on-going case $\sigma_i =< d_{i_1}, \ldots, d_{i_T} >$ and $d^\star$ the following index:

$$P(d^\star \mid \sigma_i, \lambda) = \frac{P(d_{i_1}, \ldots, d_{i_T}, d^\star|\lambda)}{P(d_{i_1}, \ldots, d_{i_T}|\lambda)} \quad (11)$$

which may be interpreted as the probability of the event that document $d^\star$ makes up the continuation of the case σ_i. In line 14 simply the case for which the probability (11) is highest is selected and the document $d^\star$ is assigned to it.

3.3 Discussion

The essence of the proposed algorithm, shown in Fig. 1, is relatively simple: the succession of the documents within cases is modeled using an HMM whose parameters are learned on the training data and a new document $d^\star$ is suggested to be added to a case for which it is the most probable successor (we do not consider here for simplicity the situation when a new case has to be established; for some solutions of this subproblem the reader is referred to our papers [8,19] as well as, e.g., to [15]). However, a few points do require some extra comments.

It should be noted that several representations of the documents are employed. The first is the standard vector space model based representation using the `tf` weighting scheme which is then employed to create an LDA model of the collection of documents[3]. The LDA based representation is then simplified,

[3] All text processing considered in this paper is carried out separately for each category $c \in C$, which will not be explicitly mentioned again, and, moreover, we will refer to the collection of documents having in mind its subset comprising documents belonging to one category.

namely it is turned into a binary representation, for the purposes of the HMM (see further discussion below). Finally, the original `tf` based representation is normalized/scaled for the purposes of the covariates computation.

The decision on the assumed documents representation is, of course, strongly connected with the form of the observation distributions used for the HMM based cases modeling. The first important assumption we adopted is that about independence of the features representing documents, i.e., terms/keywords in the standard vector space model representation or topics in case of the LDA. While this assumption is obviously incorrect in general, still it is usually assumed as otherwise the number of parameters of multivariate distributions makes effective and efficient learning practically impossible. Then, we have tried several options using both the Boolean representations of documents and their weighted forms, the former combined with the multinomial distribution and the latter combined with the Gaussian distribution. A multinomial distribution becomes cumbersome already for relatively small vocabularies V, requiring $N_S|V|$ parameters to be learned. In our experiments the vocabulary, already aggressively reduced, was composed of ca. 250 terms. The use of the LDA models makes it possible to reduce the number of features and at the same time provides for a more semantic rich representation. The number of parameters to be learned for the observation distributions is now equal $2N_S k$, where N_S is the number of states and k is the number of LDA topics. In our experiments the "binarized" version of the LDA representation proved to be most effective.

Actually, only after including covariates to a binary LDA representation via the logistic regression we have obtained satisfactory results in our experiments. The covariates are defined in such a way that the observation distribution – at a given point in time/position in the case – depends not only on the current state but also on the actual form of the preceding document expressed using normalized `tf` based representation. Formula (9) makes it possible to model the patterns of dependency between documents neighboring within a case such that occurrence in the preceding document of the terms strongly represented in a given LDA topic increases or decreases the probability of this topic in the next document in the sequence.

The proposed solution is based on a rather simple extension of the classic HMM. An interesting and natural alternative seems to be the use of a discriminative model, such as, e.g., the conditional random field. However, it should be noted that the MTC task resembles rather a time-series prediction problem than a sequential supervised learning problem [6]. In particular, in general, we do not assume the availability of the training data comprising cases where each document is assigned to a class (a label). Such a labeling may be envisaged, e.g., assuming that a specific stage of a business process may be associated with each document but this leads to a different class of possible approaches referring to the concept of business processes mining which we do not consider here. Anyway, in our research agenda for the MTC problem we consider the use of the Hidden Conditional Random Fields [10] which do not require labeled training sequences.

4 Computational Experiments

We have verified the proposed algorithm using an enlarged version of the collection of documents we adopted and used in our previous works. A detailed description of the collection may be found in [17,18]. The starting point is the set of articles on computational linguistics available in the framework of the ACL Anthology Reference Corpus (ACL ARC) [3]; see also http://atmykitchen.info/datasets/acl_rd_tec/cleansed_text/index_cleansed_text.htm. We use a subset of 664 papers which are composed of sections. In order to group the documents into categories we cluster the whole set of 664 papers into 6 clusters (the number 6 has been chosen experimentally to obtain reasonably sized categories). Then, we treat each paper as a case composed of documents corresponding to the sections of this paper.

Thus, we obtain 664 cases comprising 6884 documents in total. The number of cases and a cut-off point in each of them are randomly chosen. All documents at the cut-off positions are treated as test data while the documents following them are deleted from the collection. In each experiment, for each category a number of test documents has been selected proportionally to the size of this category, 64 documents in total in each experiment, i.e., 10 % of cases are each time treated as on-going.

The results obtained, averaged over 10 runs and 6 categories, are the following: microaveraged and macroaveraged accuracy of classification equal 0.54 and 0.57, respectively. The results are encouraging though one can well imagine a number of ways the proposed algorithm may be tuned and there seem to be a real potential for improvement thanks to employing a more semantic oriented document representation and an explicit modeling of dependencies between the documents within cases. In our previous papers we reported the results for other approaches we proposed earlier, including also a recent technique developed for the topic tracking task in TDT. However, most of them concerned a smaller subset of the ACL ARC corpus and also a smaller number of cases are there assumed to be on-going. It should be noted that if a case is considered as a class the respective classification problem gets usually more difficult with the growing number of classes; cf., e.g., [5]. However, recently we have tested (and compared against its newly proposed modified version) the method introduced in [18] on the same, larger version of the ACL ARC corpus which is adopted in this paper. We have obtained comparable results but the current proposed solution attempts to grasp the logic behind the order of the documents in a case in a more explicit way and is thus more promising as a starting point for some further improvements.

All computations are carried out using the R platform [11] and the following packages: `tm` [7], `topicmodels` [9], `depmixS4` [14] and our own R scripts. The most important parameters of the methods involved are the following: for the LDA – the number of topics $k = 30$, the α parameter of the Dirichlet distribution $= 1.67$, i.e., $50/k$, the beta parameter is automatically estimated; for the HMM – the number of states $= 6$, observation distributions are binomial (actually, Bernoulli as 1 trial is assumed) with the logit link.

5 Concluding Remarks

We have proposed a novel hybrid approach to solving the new multiaspect text categorization (MTC) problem proposed in our previous works. In comparison to our earlier approaches it assumes as a point of departure a more sophisticated explicit model of the whole collection of documents and, in particular, of the sequences of documents forming cases. In the new hybrid approach proposed, our earlier solution proposal based on the HMM is combined in a synergistic way with the LDA modeling of the collection of documents which certainly opens new vistas on the capability of this modeling. In particular, the possibility to link the probability of occurrence of an LDA topic in a given document with the vocabulary of the preceding document seems to be particularly interesting and promising. This is a type of dependency modeling we are looking for, i.e., such which to some extent abstracts from the actual value of the features of the documents and makes it possible to discover more universal patterns typical for different cases belonging to the same category.

Acknowledgments. This work is supported by the National Science Centre under contracts no. UMO-2011/01/B/ST6/06908 and UMO-2012/05/B/ST6/03068.

References

1. Allan, J. (ed.): Topic Detection and Tracking: Event-based Information. Kluwer Academic Publishers, Norwell (2002)
2. Baeza-Yates, R., Ribeiro-Neto, B.: Modern Information Retrieval. ACM Press and Addison Wesley, New York (1999)
3. Bird, S., et al.: The ACL anthology reference corpus: A reference dataset for bibliographic research in computational linguistics. In: Proceedings of Language Resources and Evaluation Conference (LREC 08), pp. 1755–1759. Marrakesh, Morocco (2008)
4. Blei, D.M., Ng, A.Y., Jordan, M.I.: Latent Dirichlet Allocation. J. Mach. Learn. Res. **3**, 993–1022 (2003)
5. Bayou, L., Espes, D., Cuppens-Boulahia, N., Cuppens, F.: Security issue of WirelessHART based SCADA systems. In: Lambrinoudakis, C., et al. (eds.) CRiSIS 2015. LNCS, vol. 9572, pp. 225–241. Springer, Heidelberg (2016). doi:10.1007/978-3-319-31811-0_14
6. Dietterich, T.G.: Machine learning for sequential data: a review. In: Caelli, T.M., Amin, A., Duin, R.P.W., Kamel, M.S., de Ridder, D. (eds.) SPR 2002 and SSPR 2002. LNCS, vol. 2396, pp. 15–30. Springer, Heidelberg (2002)
7. Feinerer, I., Hornik, K., Meyer, D.: Text mining infrastructure in R. J. Stat. Softw. **25**(5), 1–54 (2008)
8. Gajewski, M., Kacprzyk, J., Zadrożny, S.: Topic detection and tracking: a focused survey and a new variant. Informatyka Stosowana **2014**(1), 133–147 (2014)
9. Grün, B., Hornik, K.: topicmodels: An R package for fitting topic models. J. Stat. Softw. **40**(13), 1–30 (2011). http://www.jstatsoft.org/v40/i13/
10. Quattoni, A., Wang, S.B., Morency, L., Collins, M., Darrell, T.: Hidden conditional random fields. IEEE Trans. Pattern Anal. Mach. Intell. **29**(10), 1848–1852 (2007). http://dx.org/10.1109/TPAMI.2007.1124

11. R Core Team: R: A Language and Environment for Statistical Computing. R Foundation for Statistical Computing, Vienna, Austria (2014). http://www.R-project.org
12. Rabiner, L.: A tutorial on HMM and selected applications in speech recognition. Proc. IEEE **77**(2), 257–286 (1989)
13. Sebastiani, F.: Machine learning in automated text categorization. ACM Comput. Surv. **34**(1), 1–47 (2002)
14. Visser, I., Speekenbrink, M.: depmixS4: An R package for Hidden Markov Models. J. Stat. Softw. **36**(7), 1–21 (2010)
15. Yang, Y., Zhang, J., Carbonell, J., Jin, C.: Topic-conditioned novelty detection. In: Proceedings of the Eighth ACM SIGKDD International Conference on Knowledge Discovery and Data Mining, pp. 688–693. ACM, New York (2002)
16. Zadrożny, S., Kacprzyk, J., Gajewski, M., Wysocki, M.: A novel text classification problem and its solution. Tech. Trans. **4–AC**, 7–16 (2013)
17. Zadrożny, S., Kacprzyk, J., Gajewski, M.: A new two-stage approach to the multiaspect text categorization. In: 2015 IEEE Symposium on Computational Intelligence for Human-like Intelligence, CIHLI 2015, Cape Town, South Africa, December 8–10, 2015, pp. 1484–1490. IEEE (2015)
18. Zadrożny, S., Kacprzyk, J., Gajewski, M.: A novel approach to sequence-of-documents focused text categorization using the concept of a degree of fuzzy set subsethood. In: Proceedings of the Annual Conference of the North American Fuzzy Information processing Society NAFIPS 2015 and 5th World Conference on Soft Computing 2015, Redmond, WA, USA, 17–19 August 2015 (2015)
19. Zadrożny, S., Kacprzyk, J., Gajewski, M.: On the detection of new cases in multiaspect text categorization: a comparison of approaches. In: Proceedings of the Congress on Information Technology, Computational and Experimental Physics, pp. 213–218. AGH University of Science and Technology (2015)
20. Zaki, M.J.: SPADE: an efficient algorithm for mining frequent sequences. Mach. Learn. **42**(1/2), 31–60 (2001)

How Much Is "About"? Fuzzy Interpretation of Approximate Numerical Expressions

Sébastien Lefort[1(✉)], Marie-Jeanne Lesot[1], Elisabetta Zibetti[2], Charles Tijus[2], and Marcin Detyniecki[1,3]

[1] Sorbonne Universités, UPMC Univ Paris 06, CNRS, LIP6 UMR 7606, 4 place Jussieu, 75005 Paris, France
{sebastien.lefort,marie-jeanne.lesot,marcin.detyniecki}@lip6.fr
[2] Laboratoire CHArt-LUTIN, EA 4004, Université Paris 8, 2 rue de la liberté, 93526 Saint-Denis - Cedex 02, France
{ezibetti,tijus}@univ-paris8.fr
[3] Polish Academy of Sciences, IBS PAN, Warsaw, Poland

Abstract. Approximate Numerical Expressions (ANEs) are linguistic expressions involving numbers and referring to imprecise ranges of values, such as "*about 100*". This paper proposes to interpret ANEs as fuzzy numbers. A model, taking into account the cognitive salience of numbers and based on critical points from Pareto frontiers, is proposed to characterise the support, the kernel and the 0.5-cut of the corresponding membership functions. An experimental study, based on real data, is performed to assess the quality of these estimated parameters.

Keywords: Approximate numerical expression · Fuzzy number · Pareto frontier · Empirical study · Number salience

1 Introduction

Approximate numerical expressions (ANEs) are vague linguistic expressions of the general form "*about* x" where x is a number. They are used in daily life to denote imprecise ranges of values, e.g., "Berlin is located at *about 900* km from Paris"; "The patient has had fever for *about one week*". In the field of Human-Computer Interfaces, ANEs raise the issues of their interpretation, i.e., the estimation of the range of values they designate and their representation in information systems, for instance as intervals of values or as fuzzy sets.

From a linguistic perspective, Lasersohn [10] proposes to formalise vagueness in a general context, beyond the case of numerical expressions, through the use of pragmatic halos, defined as the union of the entity that is explicitly referred to by a vague expression and entities of the same semantic type that are implicitly denoted. For instance, in the proposition "*there were about 100 participants at the meeting*", the pragmatic halo of the vague expression "*about 100*" corresponds to 100 exactly and a range of possible values around 100 (e.g., [90; 110]). Therefore, interpreting an ANE corresponds to estimating the range of values that satisfy it, i.e., the values that are included in its pragmatic halo.

J.P. Carvalho et al. (Eds.): IPMU 2016, Part I, CCIS 610, pp. 226–237, 2016.
DOI: 10.1007/978-3-319-40596-4_20

A natural approach to model the fuzziness in boundary values is to use fuzzy sets [14,15], that lead to represent ANEs as fuzzy numbers [16], defined by their membership functions. Fuzzy numbers are classically used to represent uncertainty or imprecision in numerical data [5]. However, to the best of our knowledge, no attempt has been made to empirically characterise the membership functions of fuzzy numbers related to ANEs in natural language.

The aim of this paper is to propose a model to characterise the support, the kernel and the 0.5-cut of fuzzy numbers corresponding to ANEs of the form "*about* x", for $x \in \mathbb{N}$. More specifically, the model is based on critical points from Pareto frontiers, as a compromise between the numbers cognitive salience and their distance to the reference value x. An empirical study is conducted to collect real data and to perform an experimental validation to highlight the quality of the estimations provided by the model.

The paper is structured as follows: Sect. 2 describes previous works and existing models. The proposed model is presented in Sect. 3. The data collection procedure is described in Sect. 4. Section 5 presents the experimental study and its results. Finally, conclusions and future works are discussed in Sect. 6.

2 Related Works

This section introduces the notations and definitions of dimensions and properties of ANEs used in this work. Two models from the literature, estimating the range of denoted values, are then presented: a scale-based model [8,13] and a regression model [4]. Finally, the fuzzy set approach to vagueness is discussed.

2.1 Definitions and Notations

The ANEs considered in this paper are of the form "*about* x", for $x \in \mathbb{N}$. In the decimal system, x can be written as $x = \sum_{i=0}^{q} a_i \cdot 10^i$, where $a_i \in [\![0, 9]\!]$. We propose four dimensions, formally defined in Table 1, to characterise x: granularity $Gran(x)$ is the power of ten x belongs to, relative magnitude $R_m(x)$ is the value of its last significant digit and precision $Prec(x)$ is the product of granularity and relative magnitude. These dimensions are expected to influence the interpretation of ANEs. For instance, precision is meant to reflect the expectation that the width of the interval corresponding to "*about 30.050*" is comparable to the one of "*about 150*", 50 being the common part.

From these dimensions, two classes of natural numbers can be distinguished. Round numbers are classically defined as multiples of 10 with a single significant digit (e.g., 50 or 8000). We propose to define pseudo-round numbers as multiples of 10 with at least two significant digits (e.g., 320 or 8150).

Beyond these arithmetical characteristics, we propose another one, taking into account a cognitive component. Indeed, it has been observed than some numbers occur more frequently than others in corpuses [2,7] and complexity $Cpx(x)$ aims at capturing this salience. It appears that, firstly, the more significant digits a number has, the lower its frequency. Secondly, numbers whose last

Table 1. Dimensions of a natural number $x = \sum_{i=0}^{q} a_i \cdot 10^i$, illustrated by $x = 4750$ in the last column. $B(x)$, used in the complexity definition, is defined in Eq. (1).

Dimension	Formal definition	Example $x = 4750$
Granularity	$Gran(x) = 10^{i^*}$ where $i^* = \min\{i \mid a_i \neq 0\}$	10
Relative magnitude	$R_m(x) = a_{i^*}$	5
Precision	$Prec(x) = a_{i^*} \cdot 10^{i^*}$	50
Number of significant digits	$NSD(x) = q - i^* + 1$	3
Complexity	$Cpx(x) = NSD(x) - B(x)$	2.5

significant digit is 5 or, to a lower extent, 2, occur more frequently. For symmetry reasons around multiples of 10, we propose to process numbers with $R_m(x) = 8$ (e.g., $18 = 20 - 2$) as numbers with $R_m(x) = 2$ (e.g., $22 = 20 + 2$). Thus, we propose to formalise the complexity of a number as its number of significant digits minus a bonus to capture these specific cases, if the number of significant digits is at least 2.

The bonus function thus distinguishes three categories, depending on the value of the last significant digit $R_m(x)$ and respecting the order of frequency of appearance: $B(x_1) > B(x_2) > B(x_3)$, for $x_1, x_2, x_3 \in \mathbb{N}$ such that $R_m(x_1) = 5$, $R_m(x_2) \in \{2, 8\}$ and $R_m(x_3) \notin \{2, 5, 8\}$. We arbitrarily propose to set these values at 0.5, 0.25 and 0. The bonus function is therefore formalised as:

$$B(x) = \begin{cases} 0.5 & \text{if } R_m(x) = 5 \text{ and } NSD(x) > 1 \\ 0.25 & \text{if } R_m(x) = 2 \text{ or } R_m(x) = 8 \text{ and } NSD(x) > 1 \\ 0 & \text{otherwise} \end{cases} \quad (1)$$

The plus signs on Fig. 1 illustrate the complexity $Cpx(x)$ for all integers x between 400 and 500.

2.2 Scale-Based Models (SBM)

The first approach in interpreting ANEs is proposed from a linguistic perspective and models the range of denoted values as an interval. Scale-based models (SBM) [8,12,13] rely on scale systems $S = \{s_1, \ldots, s_n\}$, where s_i are granularity levels such that $s_i < s_{i+1}$. As examples, one can mention the time scale-system, $S = \{1 \text{ min}, 5 \text{ min}, 15 \text{ min}, \ldots\}$, or the decimal one, $S = \{1, 10, 100, \ldots\}$.

The interpretation of a numerical expression can occur at any granularity level. For instance, in the decimal system, the numerical expression "*100*" can be interpreted at the 1, 10 or 100 levels. The finer the granularity, the narrower the interval. Speakers express the intended level through the use of approximators [12]: "*exactly*" refers to the finest granularity level the expression belongs to, while "*about*" refers to the coarsest one (e.g., the level of thousands for "*about 1000*"), formally defined as $Gran_C(x) = \sup(\{s_i \in S | x \mod s_i = 0\})$. If the scale-system S is the decimal system, $Gran_C(x) = Gran(x)$ (see Table 1).

SBM proposes that the values denoted by an ANE x are the ones closer to x than to any other number on $Gran_C(x)$. The interval is formally defined as:

$$I_{SBM}(x) = [x - Gran_C(x)/2; x + Gran_C(x)/2] \tag{2}$$

For instance, $I_{SBM}(300) = [250; 350]$; $I_{SBM}(8150) = [8145; 8155]$. This approach has the advantage of taking into account the ANE granularity; however, it does not address the issue of the relative magnitude: all ANEs at the same granularity level result in the same interval width, although, one may expect, for instance, that the interval of "*about 100*" would be narrower than the one of "*about 800*".

2.3 Regression Model (REGM)

Ferson et al. [4] propose an empirical approach using real data to test the relevance of predictors of the interval width. Semantically contextualised ANEs (e.g., "*Roughly 25 % of Canadians are Protestant.*") were presented to participants, who were asked to estimate the boundaries of the corresponding intervals. The proposed model then estimates the interval as:

$$\begin{aligned} I_{REGM}(x) &= \left[x - \frac{10^{L(x)}}{2}; x + \frac{10^{L(x)}}{2}\right] \\ \text{where } L(x) &= A + B \cdot O_m(x) + C \cdot R(x) + D \cdot f(x) \\ &+ E \cdot O_m(x) \cdot R(x) + F \cdot O_m(x) \cdot f(x) + G \cdot R(x) \cdot f(x) \\ &+ H \cdot O_m(x) \cdot R(x) \cdot f(x) \end{aligned} \tag{3}$$

where A to H are parameters empirically set by performing a regression on the data. $O_m(x)$ is the ANE order of magnitude ($O_m(x) = \log_{10}(x)$), $R(x)$ its roundness ($R(x) = i^* + 1$), and $f(x)$ its "fiveness", defined as $f(x) = 1$ if $a_{i^*} = 5$, $f(x) = 0$ otherwise. $O_m(x)$, $R(x)$, $f(x)$ and their combinations have been empirically selected as predictors for the interval width.

This model presents the advantage of allowing the adaptation to different contexts by learning parameters on a dataset. However, it can be noted that the semantic context is not controlled in the experimental setting although mixing different contexts may result in interactions between this factor and the ones related to the ANE reference number.

2.4 Fuzzy Representation of Vagueness

From a linguistic perspective, Lakoff [9] considers that every term in natural language is, to some extent, fuzzy: category membership is not a matter of all or nothing, but rather a matter of degrees. As supported by empirical evidence [6], fuzzy logic is therefore a relevant formalisation of the vagueness inherent to natural language: any term can be modeled by a membership function.

Among the natural language terms, numerical expressions can be represented as fuzzy numbers [16], defined as fuzzy sets on the universe $\mathbb{R}$. From this point of view, approximators are modifiers of the membership function of the fuzzy reference value [11]. For instance, the approximator *exactly* narrows the curve of the membership function whereas *approximately* widens it.

Interpreting an ANE x therefore consists in estimating its membership function, $f_{\tilde{x}}(y)$, where y are values that can be denoted by "*about* x". Among various methods to elicit such membership functions (see, e.g. [1]), the random set view interprets the membership degree of a candidate number (e.g., 95 for "*about 100*") as the cumulative frequency of participants thinking that it belongs to the interval denoted by the ANE. Thus, if half of the population think that 95 is included in "*about 100*", the truth value of 95 is 0.5. The median of the distribution is therefore a critical point for membership functions that corresponds to the 0.5 membership degree.

3 Proposed Model

This section describes the model we propose to estimate the support, the kernel and the 0.5-cut of fuzzy numbers corresponding to ANEs.

The Pareto Frontiers Model (PFM): The model we propose is based on the assumption that, when interpreting an ANE, human beings tend to make a compromise between the cognitive cost of boundary values, which can be measured by the complexity $Cpx(x)$, on one hand, and the range of denoted values, measured by the distance between the boundaries of the interval and the ANE x, on the other hand. It implies that, for a given range of denoted values, the cognitive cost is minimised; reciprocally, for a given cognitive cost, the range of denoted values is minimised. For instance, given the ANE "*about 500*", participants of the empirical study (see Sect. 4) tend to give answers such as [499; 501], [490; 510] or [450; 550]. The boundaries of these intervals are the closest to the ANE when $Cpx(x)$ is 3, 2 and 1.5. Therefore, the values that optimise the compromise are better candidates to be the boundaries than all other values.

As a consequence, the model we propose first consists in determining these good candidates by generating Pareto frontiers [3]: all possible candidate values v in $[1; x[$ for the lower boundary, and in $]x, +\infty[$ for the upper boundary of the ANE x are compared on two criteria (i) the absolute distance from the ANE: $d_x(v) = |v - x|$; (ii) the complexity $Cpx(v)$. The selected values, constituting the Pareto frontier, are those that are not dominated by any other value. For a given ANE, two Pareto frontiers are considered: $P^-(x) = [y_1^-, \ldots, y_{n^-}^-]$ relates to the lower boundary of the interval and $P^+(x) = [y_1^+, \ldots, y_{n^+}^+]$ to the upper one, ordered by increasing distance to x, $d_x(y_i)$. Figure 1 illustrates these Pareto frontiers for the ANE "*about 440*": $P^-(440) = [439, 438, 435, 430, 420, 400]$ and $P^+(440) = [441, 442, 445, 450, 500]$. One can notice that the model naturally captures the asymmetry observed in the data (see Sect. 4) due to salient numbers (e.g., 420, 450) in the reference number neighborhood.

The second step of the model we propose consists in using the values in the Pareto frontiers as candidates to be the boundaries of the support, kernel and 0.5-cut limits of fuzzy numbers corresponding to ANEs.

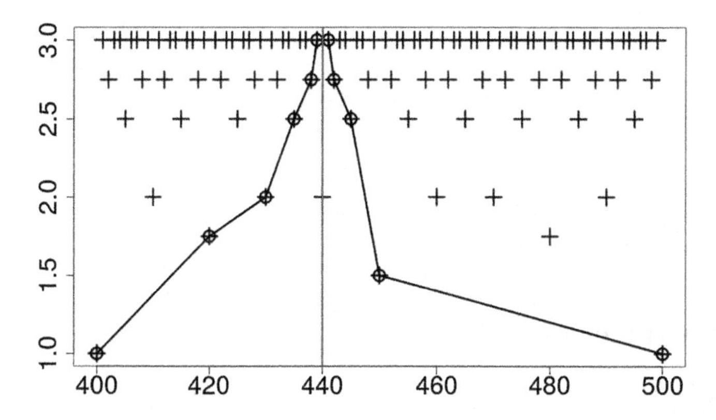

Fig. 1. Pareto frontiers (red lines) for lower (left from green line) and upper (right from green line) boundaries of the ANE "*about 440*". Black plus signs represent the complexity $Cpx(x)$ for each integer value in [400; 500] (see Table 1). (Color figure online)

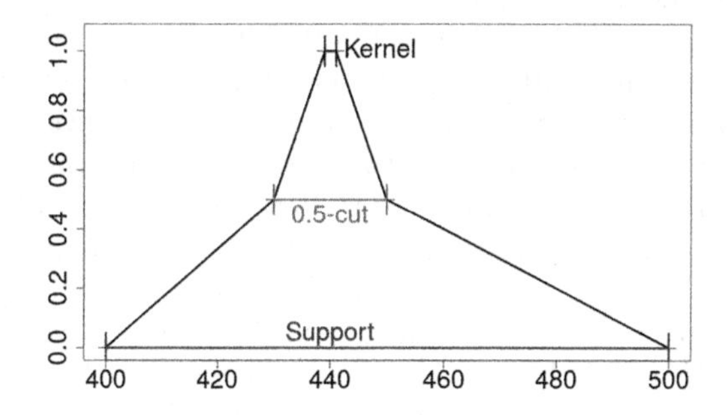

Fig. 2. Support, kernel and 0.5-cut of the membership function for ANE *about 440*, based on critical points from Pareto frontiers with piecewise linear interpolation.

Support, Kernel and 0.5-cut Estimations: Any value outside the support interval, noted $I_S(x)$, is considered as not referred by the ANE. We therefore propose to define the farthest values from x of the Pareto frontiers as boundaries of this interval, formally: $I_S(x) = [y^-_{n^-}; y^+_{n^+}]$.

Any value inside the kernel interval, noted $I_K(x)$, is considered as being fully denoted by the ANE. We propose to define the nearest values from x of the Pareto frontiers as boundaries of this interval, formally: $I_K(x) = [y^-_1; y^+_1]$.

The boundaries of the 0.5-cut interval, noted $I_M(x)$, are also selected according to their rank in $P^-(x)$ and $P^+(x)$. We propose to make the chosen rank dependent on the considered ANE x, so as to make the model more flexible. More, precisely, we propose that the rank of the boundary estimation depends on the number of significant digits $NSD(x)$ and the precision $Prec(x)$ of the ANE: an exhaustive analysis of empirical data (omitted in this paper for reasons of space) has validated them as factors influencing ANE interpretation. The rank is computed as:

$$r_P(x) = round\left(\log(Prec(x)) - 1 + \sum_{k=1}^{NSD(x)} k\right) \quad (4)$$

The estimation of the 0.5-cut interval is then $I_M(x) = [y^-_{r_P(x)}; y^+_{r_P(x)}]$.

For the example $x = 440$, as illustrated in Figs. 1 and 2, one obtains: $I_S(440) = [400; 500]$, $I_K(440) = [439; 441]$ and $I_M(440) = [430; 450]$.

4 Data Collection

We conducted an empirical study to collect real intervals corresponding to ANEs so as to experimentally validate our proposed model. This section presents the methods used to collect and process the data.

Material: An online questionnaire containing 24 uncontextualised ANEs, 15 round (20, 30, 40, 50, 80, 100, 200, 400, 500, 600, 800, 1000, 2000, 6000 and 8000) and 9 pseudo-round (110, 150, 440, 560, 1100, 1500, 4700, 4730 and 8150) was designed. These values have been selected in order to cover different combinations of dimensions, to avoid biases towards any specific one: several relative magnitudes at a granularity level (e.g., 20/40/80), several granularity levels at a relative magnitude (e.g., 80/800/8000), several numbers of significant digits at the same precision (e.g., 50/150/8150). ANEs are presented in a random order. The instructions, given in French, can be translated as "*In your opinion, what are the MINIMUM and MAXIMUM values associated with "about x"?*". This questionnaire meets the criteria proposed by [1] to elicit membership functions in a random set perspective. This method is also similar to the one used by [4].

146 participants have been recruited through an announcement diffused on mailing-lists: 102 women and 44 men, aged 20 to 70 ($M = 38.6$; $\sigma = 14.2$).

Data Preprocessing: The answer to ANE x given by participant p is noted $I_p(x) = [I^-_p(x); I^+_p(x)]$. It is considered as an outlier if: (i) it is inadequate (e.g., $[0; infinity]$), (ii) $I^-_p(x) > x$ or $I^+_p(x) < x$ (e.g., $I(800) = [700; 750]$ or $I(800) = [810; 850]$), or (iii) $I^-_p(x) < x/10$ or $I^+_p(x) > 10x$ (e.g., $I(100) = [10; 1100]$). In a second step, mean and standard deviation are computed for the remaining boundaries of each ANE. Any boundary value beyond three standard deviations of the mean is considered as an outlier. Finally, participants with more than 70 % missing values or outliers are considered as untrustworthy and all their answers are excluded. The analyses include 3177 (91 %) of the 3504 collected intervals.

Global Observations: In the collected data, not detailed here, it can be observed that participants tend not to agree on the intervals: on average, 15.4 different answers per boundary are obtained, ranging from 9 (for "*about 20*") to 22 (for "*about 8150*"). However, 84.4 % of the boundaries are located on the

Pareto frontiers as defined in Sect. 3, which validates the principle underlying the model we propose.

When examining whether the provided intervals are symmetric around the reference value, the collected data show that symmetry depends on the ANE: 74.2 % are symmetric with respect to the considered ANE, but intervals of some ANEs, such as 440 or 4730, are less often symmetric (63 % and 50 % respectively). This observation validates the definition of a flexible model allowing for non-symmetric observations.

5 Experimental Study

This section presents the experimental study we performed in order to assess the quality of the three estimated parameters of fuzzy numbers corresponding to ANEs: 0.5-cut, support and kernel. The used quality criteria and the results of each parameter are described in the next subsections.

5.1 Evaluation of the 0.5-cut Estimation

In the random set view of membership functions [1], 0.5-cuts correspond to the median of the intervals given by the participants. Thus, to evaluate the 0.5-cut estimation, we propose to compare it to this median interval.

As the models from the literature [4,13] are not fuzzy, they can be used to estimate either the support, the kernel or the 0.5-cut. We propose to use them to predict the 0.5-cut as it is a central indicator of the boundary distributions.

Quality Criteria: We note X the set of considered ANEs and $P(x)$ the set of participants whose intervals are not considered as outliers for $x \in X$. Moreover, we note the prediction of model m $[m^-(x); m^+(x)]$, $\Delta M_m^b(x) = |m^b(x) - x|$ its distance from x for $b \in \{-,+\}$, and $\Delta Med^b(x)$ the median of the distances $\Delta P_p^b(x) = |I_p^b(x) - x|$ over all participants p in $P(x)$.

To assess whether the estimations are correct, we first propose to use the accuracy score of the median prediction, i.e., the number of boundary values for which the relative distance to the observed median is lower than 10 %. The median accuracy, MA, to be maximised, can be formalised as:

$$MA(m) = \frac{1}{2 \cdot |X|} \sum_{x \in X} \left| \left\{ b \in \{-,+\} \,\middle|\, \frac{|\Delta M_m^b(x) - \Delta Med^b(x)|}{\Delta Med^b(x)} \leq 0.1 \right\} \right| \quad (5)$$

Secondly, to assess the degree of error, we propose to evaluate the balance between participants who are above and below the estimated 0.5-cut, formally defined as: $N_+ = |\{p \in P(x) | \Delta P_p^b(x) > \Delta M_m^b(x)\}|$ and $N_- = |\{p \in P(x) | \Delta P_p^b(x) < \Delta M_m^b(x)\}|$. A correct estimation of the median interval implies that the model m should be such that $N_+ = N_-$ for all x, b.

However, since interval boundaries given by the participants are distributed on few points, a perfect balance may not be possible. Therefore, the score takes into account the balance of the actual median, i.e., $N_+^* = |\{p \in P(x) | \Delta P_p^b(x) > \Delta Med^b\}|$ and $N_-^* = |\{p \in P(x) | \Delta P_p^b(x) < \Delta Med^b\}|$.

The score of the model then depends on the difference between N_+ and N_+^* and between N_- and N_-^*. Averaging over the two boundaries $b \in \{-,+\}$ and all considered ANEs, the median error, to be minimised, can be defined as:

$$MErr(m) = \frac{1}{2 \cdot |X|} \cdot \sum_{x \in X} \sum_{b \in \{-,+\}} (|N_+ - N_+^*| + |N_- - N_-^*|) \qquad (6)$$

Experimental Procedure: Using these quality criteria, we compare the performances of our proposed Pareto frontiers model PFM, the scale-based model SBM [8,13] with the decimal system (i.e., $S = \{1, 10, 100, \ldots\}$), and the regression model REGM [4]. The latter only provides the size of the intervals and no information about their location or symmetry around the ANE. We make the assumption that they are symmetric and centered on x.

A cross-validation procedure is performed on two benchmarks, (i) Participant (PB): REGM learning is performed on the intervals given by 75 % of the participants, the remaining 25 % constitute the test dataset. (ii) ANE (AB): REGM learning is performed on the intervals given by all participants on 17 (66.7 %) of the ANEs. The 7 remaining ANEs are used as test dataset. Each benchmark consists in 1000 random decompositions of the learning/test datasets, with the constraint that they must include a mix of round and pseudo-round ANEs.

In order to determine which model shows the best results in each benchmark, statistical analyses using ANOVA tests with model as factor, and Tukey's HSD post-hoc tests are performed. The significance threshold is set at $p = .01$.

Results: Table 2 shows the performances of the models. Results are similar in both the Participant and the ANE benchmarks.

It can firstly be observed that our proposed model PFM shows the best performances, both in median prediction accuracy (MA) and in median estimation error ($MErr$), providing an empirical validation.

Table 2. Means and standard deviations of the two criteria for each model on the Participant (PB, left) and the ANE benchmarks (AB, right). Bold scores are the statistically best ones according to the ANOVA and post-hoc tests.

Model	MA (%) - PB	$MErr$ - PB	MA (%) - AB	$MErr$ - AB
SBM	28.0 (6.9)	0.76 (0.08)	24.9 (14.5)	0.79 (0.18)
REGM	20.0 (7.2)	0.67 (0.18)	15.7 (13.2)	0.65 (0.14)
PFM	**58.3 (8.9)**	**0.35 (0.12)**	**63.8 (14.0)**	**0.27 (0.16)**

The behaviour of REGM (poor MA but an average $MErr$) can be due to the fact that it provides real-numbered boundary estimations while participants tend to give round or pseudo-round numbers, leading to erroneous predictions. However, the average $MErr$ indicates that these real-numbered estimations are close to the actual medians. On the contrary, SBM appears to perform better than REGM on prediction accuracy while the prediction errors are much more important.

5.2 Evaluation of the Support and Kernel Estimations

Quality Criterion: Assessing the quality of the support and the kernel estimations the same way as the 0.5-cut raises the issue of the outliers. Indeed, in the random set view, the support corresponds to the largest interval, and the kernel corresponds to the narrowest one. Therefore, the presence of a single extreme answer results in aberrant support or kernel values. Prediction accuracy or distance to actual values thus lack robustness with respect to extreme values.

To overcome this issue, we propose to build a basic piecewise linear membership function, $f^G_{\tilde{x}}(y)$, obtained by linking the generated points of support, 0.5-cut and kernel and to compare it to an elicited reference fuzzy set $f^E_{\tilde{x}}(y)$. We build the latter in a random set view [1], defining $f^E_{\tilde{x}}(y)$ as the cumulative relative frequency of participants including y in the interval corresponding to x.

We propose to compare $f^G_{\tilde{x}}(y)$ to $f^E_{\tilde{x}}(y)$ using the area of their difference, relatively to the area of the reference $f^E_{\tilde{x}}(y)$. This criterion, measuring the membership function quality, to be minimised, can be formalised as:

$$MFQ(x) = \frac{\int_y |f^G_{\tilde{x}}(y) - f^E_{\tilde{x}}(y)|}{\int_y f^E_{\tilde{x}}(y)} \tag{7}$$

Results: Figure 3 illustrates four examples of elicited and generated membership functions. The high steps observed in $f^E_{\tilde{x}}(y)$ are due to boundary values frequently given by participants.

The generated membership functions visually fit well the elicited ones of 150, 400 and 8150, corresponding to MFQ scores 0.211, 0.397 and 0.618 respectively. Moreover, the asymmetry of the $f^E_{\tilde{x}}(y)$ is captured, validating our PFM model.

The mean quality score is 0.502 ($\sigma = 0.175$), ranging from 0.211 ($x = 150$) to 0.950 ($x = 1100$). Setting a threshold at $MFQ = 0.6$ to consider a good estimation, 17 over 24 (70.1 %) generated membership functions are correct.

As expected, the presence of outliers (i.e., 7500 and 10000 for $x = 8150$; 100 and 600 for $x = 400$) lowers the score of some ANEs. In the particular case of $x = 1100$ (Fig. 3, top right), the poor obtained fitting and score ($MFQ = 0.950$) can be explained by the fact that the upper Pareto frontier ends at 2000, a value not given by participants.

When detailing the difference between round and pseudo-round ANEs, it appears that the mean scores obtained for round (0.488) and pseudo-round numbers (0.524) are similar. However, the standard deviation reveals a higher significantly variability for pseudo-round numbers (0.272) than for round numbers

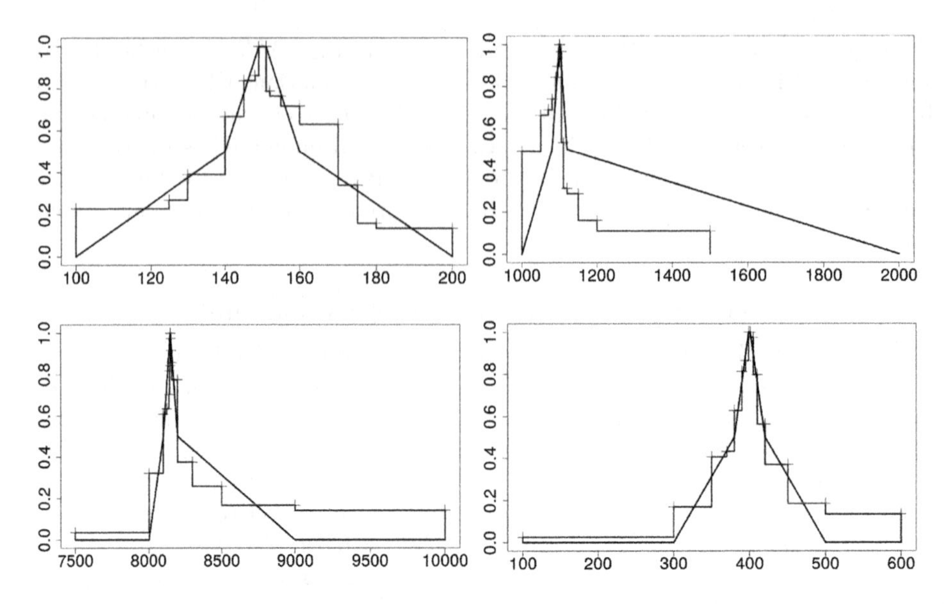

Fig. 3. Generated (red) and elicited (black) membership functions of four ANEs: $x = 150$ (top, left), $x = 1100$ (top, right), $x = 8150$ (down, left), and $x = 400$ (down, right). (Color figure online)

(0.087), indicating that some ANEs are well captured while some other are less. In particular, $x = 1100$ ($MFQ = 0.950$) and $x = 4730$ ($MFQ = 0.864$) result in scores far from the mean, compared to other ANEs.

6 Conclusion and Future Works

In this paper, we propose a model to interpret ANEs of the form "*about x*" as fuzzy numbers. More specifically, a computational model, based on critical points from Pareto frontiers and capturing the cognitive dimension of number salience, is proposed to characterise the support, the kernel and the 0.5-cut of the corresponding membership functions.

We conducted an experimental study on real data collected from an online questionnaire, which validates the proposed model: it shows that PFM performs better than the models from the literature in 0.5-cut estimation. Moreover, the piecewise linear membership functions generated from the estimations are close approximations of the elicited ones.

Future work will study the relevance of including other points from the Pareto frontiers as specific α-cuts to better fit the elicited membership functions. It will also focus on extension of the model to take into account the context of an ANE occurence as it has an effect on ANE interpretation [10,13]. Indeed, "*about 10.000 euros*", for instance, may not be interpreted the same way it is said by a seller or a buyer. Extensions of the model will focus on other linguistic approximators, such as "*at least*" or "*less than*".

Finally, the proposed model will be implemented in applications such as search engines to improve the relevance of answers provided to approximate queries.

Aknowledgments. This work was performed within the Labex SMART (ANR-11-LABX-65) supported by French state funds managed by the ANR within the Investissements d'Avenir programme under reference ANR-11-IDEX-0004-02.

References

1. Bilgiç, T., Türkşen, I.B.: Measurement of membership functions: Theoretical and empirical work. In: Fundamentals of Fuzzy Sets. The Handbooks of Fuzzy Sets Series, vol. 7, pp. 195–227. Springer, US (2000)
2. Dehaene, S., Mehler, J.: Cross-linguistic regularities in the frequency of number words. Cognition **43**(1), 1–29 (1992)
3. Ehrgott, M.: Multicriteria Optimization, vol. 491. Springer Science & Business Media, Berlin (2013)
4. Ferson, S., O'Rawe, J., Antonenko, A., Siegrist, J., Mickley, J., Luhmann, C.C., Sentz, K., Finkel, A.M.: Natural language of uncertainty: numeric hedge words. Int. J. Approximate Reasoning **57**, 19–39 (2015)
5. González, A., Pons, O., Vila, M.A.: Dealing with uncertainty and imprecision by means of fuzzy numbers. Int. J. Approximate Reasoning **21**(3), 233–256 (1999)
6. Hersh, H.M., Caramazza, A.: A fuzzy set approach to modifiers and vagueness in natural language. J. Exp. Psychol. Gen. **105**(3), 254–276 (1976)
7. Jansen, C.J.M., Pollmann, M.M.W.: On round numbers: Pragmatic aspects of numerical expressions. J. Quant. Linguist. **8**(3), 187–201 (2001)
8. Krifka, M.: Approximate interpretations of number words: A case for strategic communication. In: Cognitive Foundations of Interpretation, pp. 111–126, Amsterdam (2007)
9. Lakoff, G.: Hedges: A study in meaning criteria and the logic of fuzzy concepts. J. Philos. Logic **2**(4), 458–508 (1973)
10. Lasersohn, P.: Pragmatic halos. Language **75**(3), 522–551 (1999)
11. Prince, E.F., Frader, J., Bosk, C., Dipietro, R.J.: On hedging in physician-physician discourse. In: Linguistics and the professions. Ablex, Norwood, NJ (1982)
12. Sauerland, U., Stateva, P.: Scalar vs. epistemic vagueness: Evidence from approximators. Proceedings of SALT (1995). 228–245 (2007)
13. Solt, S.: An alternative theory of imprecision. In: Semantics and Linguistic Theory,vol. 24, pp. 514–533 (2014)
14. Zadeh, L.A.: Fuzzy sets. Inf. Control **8**(3), 338–353 (1965)
15. Zadeh, L.A.: The concept of a linguistic variable and its application to approximate reasoning. Inf. Sci. **8**(3), 199–249 (1975)
16. Zadeh, L.A.: A computational approach to fuzzy quantifiers in natural languages. Comput. Math. Appl. **9**(1), 149–184 (1983)

Towards a Non-oriented Approach for the Evaluation of Odor Quality

Massissilia Medjkoune[1,2](✉), Sébastien Harispe[1], Jacky Montmain[1], Stéphane Cariou[2], Jean-Louis Fanlo[2], and Nicolas Fiorini[1]

[1] Laboratory of Computer Science and Production Engineering (LGI2P), Ecole des Mines d'Alès, Parc Scientifique Georges Besse, 30035 Nîmes cedex 5, France
{massissilia.medjkoune,sebastien.harispe, jacky.montmain,nicolas.fiorini}@mines-ales.fr

[2] Laboratory of Engineering for Industrial Environment (LGEI), Ecole des Mines d'Alès, 6 avenue de Clavières, 30319 Alès cedex, France
{stephane.cariou,jean-louis.fanlo}@mines-ales.fr

Abstract. When evaluating an odor, non-specialists generally provide descriptions as bags of terms. Nevertheless, these evaluations cannot be processed by classical odor analysis methods that have been designed for trained evaluators having an excellent mastery of professional controlled vocabulary. Indeed, currently, mainly oriented approaches based on learning vocabularies are used. These approaches too restrictively limit the possible descriptors available for an uninitiated public and therefore require a costly learning phase of the vocabulary. The objective of this work is to merge the information expressed by these free descriptions (terms) into a set of non-ambiguous descriptors best characterizing the odor; this will make it possible to evaluate the odors based on non-specialist descriptions. This paper discusses a non-oriented approach based on Natural Language Processing and Knowledge Representation techniques - it does not require learning a lexical field and can therefore be used to evaluate odors with non-specialist evaluators.

Keywords: Sensorial analysis · Distributional semantics · Information fusion · Taxonomy · Odor quality · Non-oriented approach

1 Introduction, Problem and Objective

A Major Societal Concern. According to several surveys, people have become increasingly sensitive to issues related to pollution and environment - see for example ISAAC study [1] and Aphekom study [2]. In particular, in large cities and large industrial areas, people have become increasingly attentive to air pollution. Indeed, odors are considered to be the second reason for complaint after noise. They are often considered to be aggressions and are generally perceived as threats for individual health. In this context, numerous companies focus on finding solutions to improve their image and their relationship with neighboring populations of industrial sites. It is therefore essential for these companies to control and to measure the acceptability

J.P. Carvalho et al. (Eds.): IPMU 2016, Part I, CCIS 610, pp. 238–249, 2016.
DOI: 10.1007/978-3-319-40596-4_21

related to their air emissions – and therefore to have tools at their disposal to perform these analyses. Assessing the acceptability of an odor is to check that the odor is not associated to an unpleasant character. Studying the acceptability of an odor therefore relies on the ability to analyze the natural language descriptions qualifying the odor.

The evaluation of odor quality is also considered vital for companies that want to remain competitive. Indeed, the quality of an odor highly impacts product design since smell plays a central role in defining the identity of a product - the odor of a product may have both positive and negative effects on product perception – almost everybody will remind the often artificial and carefully designed sweet smell of his favorite candy. Odor quality has therefore a direct impact on sells, which explains that controlling this product characteristic is of major importance for brands.

Identifying the Smell. The smell is the image that the brain has made in the presence of odor molecules. An odor is commonly characterized using three notions: (i) its intensity, which depends on the concentration of the odorous substance of the inspired air; (ii) its hedonic tone, which assigns a pleasant or an unpleasant character to the smell, and (iii) its quality, i.e. to what the odor refers to, that can be translated by linguistic descriptors. The quality refers to the perceived odor and remains the most subjective propriety of olfactory perception - as it relies on the personal memories of individuals [3].

In this setting, the complexity of evaluating odor quality returns to the difficulty to analyze odor descriptions. Currently, industrials have long relied on well-defined evaluation procedures in the framework of costly sensorial analyses sessions that have been designed for carefully analyzing, controlling and selecting odors [13]. Such evaluations of odors quality are commonly made through controlled linguistic descriptors provided by trained specialists capable of distinguishing precise panels of odors. Among the approaches proposed in the literature, the wheel of odors and odor fields are often used to qualify odors [4, 5]. These methods use an oriented approach based on a common referential to qualify the odors, which facilitates their characterization. Indeed, forcing evaluators to use specific descriptors facilitates understanding, interpreting and processing the results. Nevertheless, a learning phase in which valid descriptors have to be learned is required to use such methods. This (i) prevents their use by non-specialists, (ii) implies additional training costs, and (iii) limits the number of evaluators and experiments that can be used to evaluate an odor. In this article, we propose an alternative to this costly and restrictive approach by defining an automatic approach enabling to evaluate odors based on non-specialist descriptions. The aim of this work is therefore to propose an approach for evaluating an odor by analyzing natural language descriptions provided by non-specialists. The proposed approach is non-oriented and requires neither prior training nor learning of a specific lexical field by the evaluators.

We consider that the terms used by non-specialists to describe the quality of an odor are free, *i.e.* it is here assumed that people provide their assessments using their own words and vocabulary. Therefore, contrary to descriptions provided by experts in oriented approaches, the terms we have to deal with are not part of an implicit controlled and standardized vocabulary defining the terms commonly used for characterizing (specific) odors. The purpose of our work is therefore to identify the descriptors that best describe the smell by analyzing natural language descriptions provided by non-specialists. In this paper, we focus on the special case where we want an odor to be described by a set of non-ambiguous descriptors. These descriptors–denoted concepts in the rest of the paper–are assumed to be partially ordered into taxonomy $\mathcal{O} = (\preccurlyeq, C)$, with C the set of concepts. The knowledge organization confers their semantic nature to the concepts. The aim of this work is therefore to summarize the quality of an odor into a conceptual annotation, *i.e.* set of concepts.

Assumptions and Materials. For a given odor evaluation, a description provided by non-expert is assumed to be a set of terms in which each term may be associated to an intensity (the intensity degree is defined to be between 1 and 5, the higher its value, the more important the related term is in the smell description). As an example, the description d provided by an evaluator testing the smell of yoghurt could be the following:

$$d = [(orange, 4); (lemon, 4); (butter, 3); (brownsugar, 1); (lemoncake, 2)]$$

The problem we face can therefore simply be formulated by the following question: how to fusion the information expressed by description d in order to formally characterize the odor of the evaluated yoghurt into a conceptual annotation (set of partially-ordered concepts)? Defining a model for answering this question requires characterizing the semantics, *i.e.* meaning, of the terms used in the description, *e.g.* intuitively, looking at description d, people understand that the terms "*orange*" and "*lemon*" refer to "*citrus fruits*" (putting aside the problem of ambiguity). This is because our knowledge about the world provides us a taxonomical organization we can use to derive conclusions on the basis of deductive reasoning. Indeed, we know that the concepts `Orange` and `Lemon` both refer to specific types of concept `Citrus-fruit`, *i.e.* formally `Orange` $\preccurlyeq$ `Citrus-fruit` and `Lemon` $\preccurlyeq$ `Citrus-fruit` which implies that the description d implicitly refers to the concept `Citrus-fruit`, a concept that could therefore be a good candidate to characterize the odor. Such a reasoning approach relies on a taxonomic knowledge organization partially ordering concepts. However, in other cases, such reasoning is also based on the consideration of the semantic proximity between concepts, *e.g.* `Lemon cake` contains `Lemon` which reinforces the fact that `Citrus-fruit` seems to be relevant to describe the odor of yoghurt, despite the fact that, strictly speaking, `Lemon cake` is not ordered to `Lemon` and `Citrus-fruit`.

For defining our approach, we intuitively propose to consider that a term evokes concepts (*e.g.* the term *lemon cake* evokes with some degrees the concepts `Lemon` and `Cake`). Automatically assessing the degree with which a term evokes a concept is a difficult task. First we have to consider the fact that no consensus could be obtained in numerous cases – the appraisal may vary a lot between two persons since it depends on subjective notions. Second, even if our will is to mimic only one person estimation, it would require formalizing too much complex and extensive knowledge to represent all the interactions between the concepts. To tackle this complex problem, we propose to take advantage of distributional semantics models for evaluating term proximity with regard to their usage (*i.e.* meaning). These models are based on the distributional hypothesis which, in linguistics, states that words that are used and occur in the same contexts tend to purport similar meanings [10]. The first step to obtain such a model is to analyze word co-occurrences in a large corpus of texts (*e.g.* Wikipedia). These co-occurrences are used to derive a model, usually a matrix, providing a vector representation of terms. Finally, two terms can be compared by analyzing their vector representations; a variety of measures have been proposed for that purpose [6], *e.g.* cosine measure. Since distributional semantics models enable us to compare terms with regards to their semantics, we next consider that measuring the similarity between a term and a concept returns to calculate the similarity between the term and the terms associated to this concept in the input taxonomy. As an example, the degree to which the term *lemon cake* evokes the concept `Lemon` will be estimated using the semantic proximity between the term *lemon cake* and the term related to the concept `Lemon` (*e.g.* *lemon, citrus medica*).

The following section presents the model we propose for deriving term vector representations. They will next be used by fusion information techniques to derive the description of the odor.

2 Modeling

The model is composed of two principal parts; the first consists in defining a correspondence between the terms and the concepts of the taxonomy, *i.e.* to represent the terms and the terms related to concepts as vectors on a given vocabulary T. By comparing the vector representations, we will then compute the degree to which a term evokes a concept. The second part consists to aggregate the terms of a description in order to obtain a synthetic set of concepts (conceptual annotation) which formally best characterizes the odors.

In order to ease the readability of this section, the various notations which will be used to introduce the model are listed below:

C: the set of concepts, $C = \{c_1, c_2, \ldots, c_n\}$, $|C| = n$

$\mathcal{O} = (\preccurlyeq, C)$: the taxonomy partially ordering the set of concepts

T: the set of terms that constitute the vocabulary

T_e: the set of terms of the description provided by the evaluator e to qualify the odor, $T_e = \left\{t_1^e, t_2^e, \ldots, t_{k_e}^e\right\}$, $T_e \subset T$;

T_c: the set of terms associated to concept c, with $T_c \subset T$.

2.1 Computing Conceptual Annotations from Terms

The vector representations of the terms (including concepts' related terms) are used to estimate to which degree a term of the description evokes a concept. Informally, the strength of evocation a term has with regard to a concept can be regarded as the semantic proximity between the term and the concept. The correspondence or measure the semantic proximity between terms and concepts cannot be established in a straightforward manner without considering the terms associated to the concepts. For this, we first define a correspondence between the terms of the vocabulary T and the terms that refer to concepts ($\bigcup_{c \in C} T_c$). This correspondence between terms and concepts is defined as the measure of proximity between terms of T and terms associated to concepts.

The proximity between terms and concepts will be defined by $\sigma_{TC} : T \times C \rightarrow [0,1]$, according to the measure σ_{TT} which assesses the proximity of two terms $\sigma_{TT} : T \times T \rightarrow [0,1]$. Numerous measures for comparing terms have already been proposed in the literature [6] - co-occurrences or pointwise mutual information are classical ones in texts analysis. These measures basically verify $\sigma_{TT}(t,t) = 1$ and $\sigma_{TT}(t,t') = \sigma_{TT}(t',t)$. Then, the semantic proximity between a term t and a concept c can be estimated, for example, as the maximum of the similarity values between the term t and the terms in T_c associated with the concept c: $\sigma_{TC}(t,c) = \max_{\tau \in T_c} \sigma_{TT}(t,\tau)$.

The objective of this step is to synthesize the information expressed by the terms of a description T_e to characterize the odor through a conceptual annotation. To this aim, it is required to aggregate the information conveyed by the vector representations of the terms composing T_e. This conceptual annotation, $annot(T_e)$, with $annot : 2^T \rightarrow 2^C$, can be computed as follows:

$$annot(T_e) = \bigcup\nolimits_{t \in T_e} \{c \in C | max_{c \in C} \sigma_{TC}(t,c)\} \tag{1}$$

This model considers a simple one-to-one correspondance between terms and concepts. When two terms in T_e evoke the same concept no redundancy will be considered. The main drawback of this strategy is that long descriptions may lead to large conceptual annotations. In those cases, a conceptual summary of $annot(T_e)$ is desired. The situation is similar when there are several evaluators who individually provide descriptions that are to be conceptually synthetized. This summary is not so obvious because it requires having in mind the way concepts are organized in the taxonomy to make simplistic but relevant factorizations, *i.e.*, eliminate redundancies of too similar concepts but retain concepts that evoke obviously different ideas related to the perception of the odor.

2.2 Summarizing Conceptual Annotations

Let $E = \{e_1, e_2, \ldots, e_s\}$ be the set of evaluators. We consider that each evaluator e_i provides a set of terms characterizing the quality of the same odor. Using the model introduced in Sect. 2.1, a set of concepts $annot(T_{e_i})$ can be associated to each individual bag of terms T_{e_i}.

In the following, we propose an algorithm to semantically summarize a set of semantic annotations. When the number of evaluators s is not too large (*e.g.*, sensorial analysis sessions are composed of 6 or 8 evaluators), we search for a conceptual summary $annot^*$ that synthesizes the individual conceptual annotations. We suppose the search space to be 2^{f_0} where $f_0 = \bigcup_{c \in \bigcup_{i=1}^{s} annot(T_{e_i})} Anc(c)$, and $Anc(c)$ is the set of inclusive ancestors of concept c in the sense of the taxonomic order of $\mathcal{O} = (\preccurlyeq, C)$ (*i.e.*, the set of concepts composed of c and the concepts that subsume c). The search for $annot^*$ can be expressed as an optimization problem as proposed in [8]. The objective function is associated to a consistency criterion:$annot^*$ must be as similar as possible to all the annotations it summarizes. The second criterion is a concision constraint: a summary is by definition synthetic.

The similarity between a summarizing annotation $annot \in 2^{f_0}$ and the annotations $annot(T_{e_i})$ can be modelled using a taxonomic semantic similarity measure used to compare groups of concepts (groupwise measure). The objective function is then:

$$g_1(annot) = \frac{1}{s}\sum\nolimits_{i=1}^{s} sim_g(annot, annot(T_{e_i}))) \tag{2}$$

with $sim_g : 2^{\mathrm{C}} \times 2^{\mathrm{C}} \rightarrow [0, 1]$ the groupwise semantic similarity measure.

Two strategies may be envisaged when comparing two groups of concepts, called direct and indirect groupwise semantic measures. The former considers the steps of features of both sets of concepts while the latter aggregates individual pairwise values. The Jaccard index for example may be applied to create a direct groupwise semantic measure [14]: say A, B are two groups of concepts and $A^+ = \bigcup_{c \in \mathrm{A}} anc(c)$, $B^+ = \bigcup_{c \in B} anc(c)$, where $anc(c)$ corresponds to c and all its ancestors in O, then the semantic similarity of A and B can be computed by $sim_{Jaccard}(A, B) = \frac{|A^+ \cap B^+|}{|A^+ \cup B^+|}$. However, because direct groupwise measures are all hampered by a higher computation time than for indirect ones, we choose in order to evaluate the closeness between two groups of concepts a composite average of pairwise similarities called Best Match Average (BMA) defined as follows [15]:

$$sim_{BMA}(A, B) = \frac{1}{2|B|}\sum\nolimits_{c \in B} sim_m(c, A) + \frac{1}{2|A|}\sum\nolimits_{c \in A} sim_m(c, B)$$

where $sim_m(c, X) = max_{c' \in X} sim(c, c')$ with $sim(c, c')$ a pairwise semantic similarity measure. The field of similarity measures (SM) is wide and has been subject to many contributions [7], pursuing the idea that computing similarities of pairs of concepts is crucial in order to mimic the human thinking. We choose a graph based SM since the taxonomy O is basically a directed graph. A more comprehensive work on this topic is available in [16].

The optimal solution $annot^*$ must be as consistent as possible under the concision constraint. The more concepts in $(e.g., annot = f_0 = \bigcup_{i=1}^{s} annot(T_{e_i}))$, the more precise the summary and the more likely $g_1(annot)$ value is high but in return not synthetic at all. The concision is defined as a penalty function with regard to the number of concepts in the annotation:

$$g_2(annot) = \mu|annot| \tag{3}$$

where $\mu \in [0, 1]$ is a parameter controlling the importance of the constraint.

Finally, the function to be maximized is:

$$\begin{aligned} g(annot) &= g_1(annot) - g_2(annot) \\ \text{and } annot^* &= argmax_{annot \in 2^{f_0}}\, g(annot) \end{aligned} \tag{4}$$

Finding an exact optimal solution according to the objective function is not feasible when $|f_0|$ becomes large. Fiorini *et al.* have proposed a local heuristic using a greedy algorithm to remove one after the other elements from f_0 until $annot^*$ is found - details on the parameterization and performances of this algorithm are provided in [8, 9] and are not discussed in the following practical case.

3 Practical Case

The distributional models used for computing the proximity of terms are of critical importance for summarizing the term descriptions. Various approaches for computing and comparing word-vector representations have been proposed in the literature. Even if some strategies have been proved to be better suited for some specific use cases, the selection of the best-suited strategy for practical cases is still an open question and therefore requires domain-specific analyses and parameter settings [6]. In this study, several models using different parameter settings have been tested to compute word-vector representations. Considering the different parameter settings, obtained models differ (i) on the way the set of terms T is computed, (ii) on the size of vector representations, and (iii) on the semantics of each dimension of the vectors that are finally considered. A brief discussion on the different models that have been tested in this study is proposed in this section. All tested models have been obtained from the analysis of a lemmatized version of Wikipedia – English version 2015 – and have been computed using open source code (that will be made available if the paper is accepted).

The set of terms T is of major importance and must be carefully built since it defines the terms provided by the evaluators that can be mapped into concepts – ideally we want $\bigcup_{i=1}^{s} T_{e_i} \subset T$, *i.e.* we don't want to lose any information provided by a term description. The set of terms T also defines the taxonomy concepts that can be associated to evaluator term descriptions. Indeed, it's important to stress that it will not be possible to process the terms provided by the evaluators that are not in T since it will not be possible to compare them to the concepts defined in the taxonomy. Similarly, it will be impossible to take advantage of the concept for which the terms are not in T - those terms could not be used to link a term description into the conceptual space, which will also hamper the performance of the treatment. In this study, two approaches have been evaluated to distinguish the set of terms T:

1. Considering a custom English dictionary built from the free dictionary Wiktionary ($m = 200k$);
2. Applying some restrictions on the grams and bi-grams that can be found into Wikipedia, *e.g.* by only considering the words that have at least been seen 10 or 50 times in the whole corpora ($m_{10} = 1.2M, m_{50} = 209k$).

In all cases, the vector representations of the terms composing T have first been computed into $\mathbb{R}^{|T|}$ only considering syntagmatic relationships between words, *i.e.* by analyzing co-occurrences of words in a specific window size (30 and 100 words have been used for the experiments). Using such a model a term $t \in T$ is represented as a vector $\vec{t} = [coocc(t, t_i)]_{i=0,|T|}$, with $coocc : T \times T \rightarrow \mathbb{N}$ the function used to compute the number of times two words co-occurred into the same term window. A model in which $\vec{t} = [pmi(t, t_i)]_{i=0,|T|}$ has also been tested where $pmi(x, y) = \frac{p(x/y)}{p(x)}$ is the pointwise mutual information of (x, y); and p holds for probability. Since the set T is large, specific reduction techniques have next been applied on the models in order to reduce the size of term vectors, *e.g.* by only considering dimensions that are associated to the most frequent words – reductions based on matrix factorization techniques, such as Single Value Decomposition, could have been used but have not been tested in this study. Reducing vector sizes is not only useful to reduce computational time but also has the benefit to remove some noises that will hamper vector comparison. Based on the tested models, the proximity between two terms, *i.e.* $\sigma_{TT} : T \times T \rightarrow [0, 1]$, is computed using cosine similarity between vector representations. Therefore the set of terms $T_e \subset T$ provided by an evaluator e is summarized by a conceptual annotation according to Eq. (1).

In the illustrative following example, a naïve evaluator has been invited to assess the quality of a set of honey items. It is assumed that we have a non-specialist evaluator, who is prompted to smell and then verbalize the odor by a set of terms in natural language. Let us suppose $T_e = \{cappuccino; grape\}$. The proximity between cappuccino (resp. grape) and its 40 closest terms in the sense of the proximity measure are provided in the Table 1. This result has been obtained with a model built from a collection of 10^6 texts of Wikipedia, the proximity measure is based on co-occurrences of terms in the corpus and the terms' co-occurrences have been computed with a symmetric 100 wide sliding window. A restriction $|T| = 3000$ has been applied in this practical case.

The results in Table 1 merely illustrate the *intuitive* notion of proximity: they are clearly debatable even if they suitably match with the intuitive terms we might commonly relate to cappuccino (resp. grape). The evaluation of the model would obviously require a test campaign which is out of the scope of this paper. Our aim here is rather to propose a general processing pipe from sets of terms in natural language to the synthetic conceptual annotation that summarizes the collective evaluation quality into notions that make sense for professional of specific domains. Each part of this pipe will be deeper analyzed in future works.

The set of terms $T_e = \{cappuccino; grape\}$ provided by an evaluator e is conceptualized according to Eq. (1). The taxonomy that is used to compute this conceptual annotation has been built from the sensory analysis applied to honey proposed in [11] where an odor and aroma wheel for honey sensory analysis is provided (Fig. 1). An "odor or aroma wheel" is a popular visual scheme for diagramming the range of smells that characterize a particular food or beverage. A well-known example is the wine aroma wheel, developed in 1984 by University of California-Davis chemist Ann Noble. It is laid out like a dartboard, with broad flavor categories (*e.g.*, "fruity") near the center and specific examples of that category (*e.g.*, "strawberry") on the outer ring. As a consequence, this abstraction hierarchy is interpreted in our framework as the partial order that organizes the concepts of our field of application. The taxonomy has been built with the free open-source taxonomy editor PROTEGE (see extract at the right side of Fig. 1) [12].

Table 1. *Grape* and *cappuccino* closest terms (scores are rounded up to 10^{-2})

Cappuccino			Grape		
0	1.0	Cappuccino	0	0.99	Grape
1	0.95	Latte	1	0.94	Vineyard
2	0.92	Espresso	2	0.93	Winemaking
3	0.91	Macchiato	3	0.92	Winery
4	0.89	Cortado	4	0.92	Varietal
5	0.88	Fluid ounce	5	0.92	Viticulture
6	0.88	Latte Macchiato	6	0.92	Winemaker
7	0.87	Portafilter	7	0.92	Chardonnay
8	0.87	Coffee bean	8	0.91	Vine
9	0.87	Coffee cup	9	0.91	Grape wine
10	0.86	Ristretto	10	0.90	Wine grape
11	0.85	Lait	11	0.90	Doc
12	0.85	Coffee maker	12	0.90	Blanc
13	0.85	Americano	13	0.90	SAUVIGNON
14	0.82	Quarter glass	14	0.90	blanc
15	0.81	Coffee liqueur	15	0.89	Phylloxera
16	0.81	Half-caf	16	0.89	Planting
17	0.81	Doppio	17	0.89	Terroir
18	0.81	Crema	18	0.89	Appellation
19	0.81	Nong	19	0.89	Syrah
20	0.80	Barista	20	0.87	Dessert wine
21	0.79	Frappe	21	0.87	Tasting
22	0.79	Coffee pot	22	0.86	AOC
23	0.78	Mocha	23	0.85	Mildew
24	0.77	Tastebud	24	0.85	Cabernet
25	0.77	Teaspoon	25	0.85	Sauvignon
26	0.77	Milk tea	26	0.85	Chianti
27	0.76	Hyperforeignism	27	0.84	Vintage
28	0.76	Sade	28	0.84	Merlot
29	0.76	Sugar spoon	29	0.83	Spoilage
30	0.74	Froth	30	0.83	Riesling
31	0.72	Demerara	31	0.83	Grapevine
32	0.70	Chaus	32	0.83	Winepress
33	0.70	Piloncillo	33	0.82	Vintner
34	0.69	Pharisee	34	0.82	Ice wine
35	0.69	Chicory	35	0.82	Pinot Noir
36	0.67	Ice cube	36	0.82	Wine bottle
37	0.65	Chocolate milk	37	0.82	Wine
38	0.64	Milk chocolate	38	0.82	Wine cellar
39	0.63	Red eye	39	0.82	Montrachet Riesling Vigneron

The Fig. 1 both provides the odor wheel of [11] and our free interpretation into a taxonomic order whose part is illustrated at the right part of the figure. The wheel provides both the concepts of the domain and their specificity levels that are interpreted

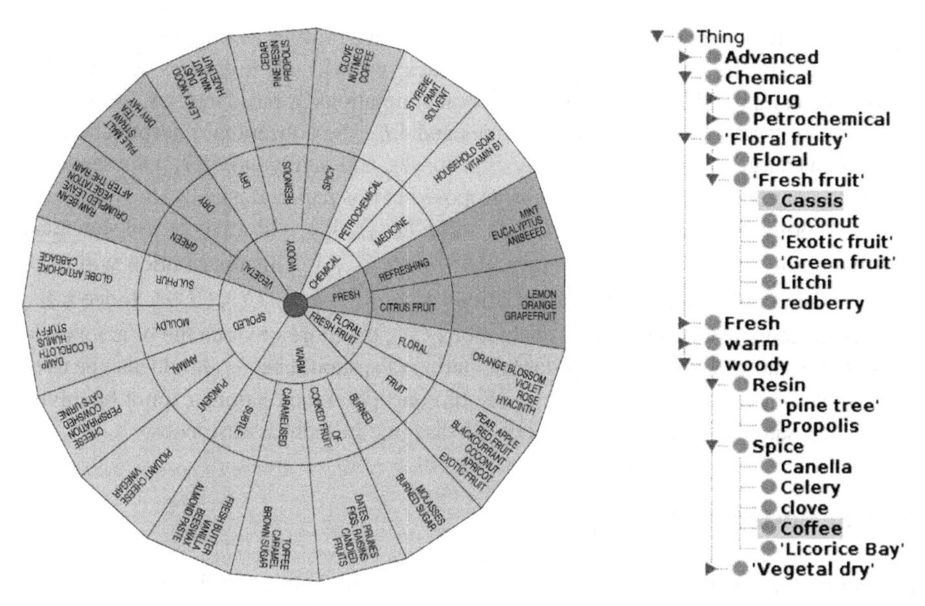

Fig. 1. (left) odor and aroma wheel for honey sensory analysis; (right) interpretation of the wheel as the partial order of a taxonomic relation

in the ontological model. This interpretation should be deeper discussed: the translation of the abstraction hierarchy of odors the wheel captures and the taxonomic order based on the relationships `rdfs:subClassOf` do not match so easily and should be more precisely analyzed in future works – since it may impact both the conceptual annotation calculus and in the summary of conceptual annotations. In our example, the conceptual annotation related to $T_e = \{cappuccino; grape\}$ is $annot(T_e) = \{coffee; blackcurrant\}$ where `Coffee` and `Blackcurrant`, two concepts of the aroma wheel, they then make sense for professionals in the honey sector (note that `Blackcurrant` and `Cassis` are equivalent).

The experiments concerning the way conceptual annotations characterizing the odor quality provided by a group of evaluators can be semantically summarized into a unique conceptual annotation are yet under progress. However, the interested reader can already consult another one of our publication in the application field of semantic indexing [9].

Our approach tries to mimic the analysis of naïve descriptions in natural language a professional sensorial analysis operator could perform. The vocabulary used to describe the smell is not limited and it does not require learning any specific vocabulary. The way the verbal descriptions are gathered and interpreted as well as the formal descriptors chosen is more or less arbitrary and often depends on the operator. For that purpose, our approach is supposed to automate the identification of the sets of concepts (descriptors) in order to increase both the reliability and the performance of the process. This automation should also allow expanding the assessment of the quality of odors to large groups of evaluators as it is the case in the analysis of olfactory nuisances around industrial sites.

To validate our model in future works, the sensory profiles evaluators provide will be compared to the ones our approach generates; we will use a groupwise similarity measure to compare the two groups conceptual annotations as basic metrics. Let us note T_{e_i} the set of terms the evaluator e_i provides and CI_{e_i}, the conceptual interpretation the operator in charge of the sensorial analysis proposes, and finally $annot(T_{e_i})$ the semantic annotation our approach relates to T_{e_i} as proposed in this paper. Basically, the relevancy of our algorithm is function of $sim_g(CI_{e_i}, annot(T_{e_i}))$. We will analyze the statistical distribution of the errors $\varepsilon_i = \left(1 - sim_g(CI_{e_i}, annot(T_{e_i}))\right)$. This distribution will have to be read in conjunction with the deviations $\delta_{ij} = \left(1 - sim_g(CI_{e_i}, CI_{e_j})\right)$. Indeed, the more homogeneous the interpreted evaluations CI_{e_i}, the more obvious the olfactory test seems to be and the smaller the standard deviation of ε_i should be. δ_{ij} should also be used to calibrate or normalize ε_i. The relevancy of conceptual annotations cannot be envisaged in the same way whereas even experts disagree in their assessments.

4 Conclusion and Perspectives

Evaluating the quality of odors is important for industrials to control the acceptability of odors related to its air emissions. The evaluation of odor quality also plays a very vital role in sensory analysis: pleasant smell of product increases the likelihood a product appeal to customers. Nowadays, industrial oriented approaches for assessing the quality of odors mostly require structured and controlled vocabularies that implies expensive and long training phases for experienced assessors. To address this issue, we propose an approach that automates the process of assessing the quality of an odor. Non-oriented, this approach is based on semantic analysis and does not require learning a lexical field - it that can therefore be used to evaluate odors with non-specialist natural language descriptions. Distributional models commonly used in natural language processing are introduced to capture the relationships between naïve terms and professional standardized descriptors. Finally, optimization and clustering techniques allow identifying the conceptual annotation that best summarizes the natural language descriptions and therefore the initially informal evaluations of odor quality.

In future works, the evaluation of the approach will require several test campaigns. A database on smells of candies is under construction. Expert and novice assessments are gathered for 6 candy types. Another database concerns flavor descriptions in red and white wines. The parameterization of the distributional model will be one of the major issues during this validation phase. We will also have to pay a particular attention on the effect the interpretation of the odor/aroma wheels into our taxonomic partial orders could have. In our current framework, the terms are represented as vectors on a given vocabulary T; it is also envisaged to project these representations onto the conceptual space C in order to take into account the similarity between concepts in the vector representation. Interestingly, this non-oriented approach for identifying the quality of odors appears to be an actual cognitive automation of the task entrusted to expert operators in sensorial analysis. It opens interesting perspectives for developing scalable sensorial analyses to large sets of evaluators when assessing olfactory nuisances around industrial site.

References

1. Asher, M.I., Montefort, S., Björkstén, B., Lai, C.K., Strachan, D.P., Weiland, S.K., Williams, H.: Worldwide time trends in the prevalence of symptoms of asthma, allergic rhinoconjunctivitis, and eczema in childhood: ISAAC phases one and three repeat multicountry cross-sectional surveys. Lancet **368**(9537), 733–743 (2006)
2. Declercq, C., Pascal, M., Chanel, O., Corso, M., Lefranc, A., Medina, S.: Impact sanitaire de la pollution atmosphérique dans neuf villes françaises. Resultats du projet aphekom. Rev. Epidemiol. Sante Publique **60**, S60–S61 (2012)
3. David, S., Dubois, D., Rouby, C., Schaal, B.: L'expression des odeurs en français: analyse lexicale et représentation cognitive. Intellectica **24**(1), 51–83 (1997)
4. Léger, C.: Smells supervision setting by AIR NORMAND, air pollution monitoring network Smells, straight perceptible nuisances. Pollut. Atmos. **47**(187), 373–384 (2005)
5. Suffet, I.H., Rosenfeld, P.: The anatomy of odour wheels for odours of drinking water, wastewater, compost and the urban environment. Water Sci. Technol. **55**(5), 335–344 (2007)
6. Harispe, S., Ranwez, S., Janaqi, S., Montmain, J.: Semantic similarity from natural language and ontology analysis. Synth. Lect. Hum. Lang. Technol. **8**(1), 1–254 (2015)
7. Harispe, S., Sánchez, D., Ranwez, S., Janaqi, S., Montmain, J.: A framework for unifying ontology-based semantic similarity measures: a study in the biomedical domain. J. Biomed. Inform. **48**, 38–53 (2014)
8. Fiorini, N.: Semantic similarities at the core of generic indexing and clustering ap-proaches. Ph.D. thesis of the Université de Montpellier (2015)
9. Fiorini, N., Ranwez, S., Harispe, S., Ranwez, V., Montmain, J.: USI at BioASQ 2015 : a semantic similarity-based approach for semantic indexing. In: Working Notes Conference on Labs of the Evaluation Forum (CLEF), Toulouse, France (2015)
10. Harris, Z.: Distributional structure. Word **10**(23), 146–162 (1954)
11. Piana, M., Oddo, L., Bentabol, A., Bruneau, E., Bogdanov, S., et al.: Sensory analysis applied to honey: state of the art. Apidologie **35**(Suppl. 1), S26–S37 (2004). doi:10.1051/apido:2004048, HAL Id: hal-00891892
12. Horridge, M., Knublauch, H., Rector, A., Stevens, R., Wroe, C.: A Practical Guide to Building OWL Ontologies Using the Protégé-OWL Plugin and CO-ODE Tools Edition 1.0. University of Manchester, Manchester (2004)
13. Standard, B.: Air quality–determination of odour concentration by dynamic olfactometry. BS EN **13725** (2003)
14. Gentleman, R.: Visualizing and Distances using GO (2010). https://www.bioconductor.org/packages/3.3/bioc/vignettes/GOstats/inst/doc/GOvis.pdf
15. Schlicker, A., Domingues, F.S., Rahnenführer, J., Legenhauer, T.: A new measure for functional similarity of gene products based on gene ontology. BMC Inform. **7**(1), 302 (2006)
16. Harispe, S., Ranwez, S., Janaqi, S., Montmain, J.: Semantic similarity from Natural Language and Ontology Analysis, vol. 8. Morgan & Claypool Publishers, San Rafael (2015)

References

Belief Function Theory and Its Applications

Joint Feature Transformation and Selection Based on Dempster-Shafer Theory

Chunfeng Lian[1,2(✉)], Su Ruan[2], and Thierry Denœux[1]

[1] Sorbonne Universités, Université de Technologie de Compiègne, CNRS, UMR 7253 Heudiasyc, 60205 Compiègne, France
{chunfeng.lian,thierry.denoeux}@utc.fr
[2] Université de Rouen, QuantIF - EA 4108 LITIS, 76000 Rouen, France
su.ruan@univ-rouen.fr

Abstract. In statistical pattern recognition, feature transformation attempts to change original feature space to a low-dimensional subspace, in which new created features are discriminative and non-redundant, thus improving the predictive power and generalization ability of subsequent classification models. Traditional transformation methods are not designed specifically for tackling data containing unreliable and noisy input features. To deal with these inputs, a new approach based on Dempster-Shafer Theory is proposed in this paper. A specific loss function is constructed to learn the transformation matrix, in which a sparsity term is included to realize joint feature selection during transformation, so as to limit the influence of unreliable input features on the output low-dimensional subspace. The proposed method has been evaluated by several synthetic and real datasets, showing good performance.

Keywords: Belief functions · Dempster-Shafer theory · Feature transformation · Feature selection · Pattern classification

1 Introduction

The performance of pattern classification methods depends crucially on the quality of input features: (1) with a small-sized training pool, a relatively high dimensional feature space increases the complexity of the learning algorithms, thus raising the risk of over-fitting on the training set; (2) it often happens that the input space contains features that are irrelevant, or even at odds with the class labels. These unreliable input features could decrease substantially the classification accuracy of the distance-based learning algorithms (e.g., the K-nearest neighbor rules).

Low-dimensional feature transformation is a feasible solution to the issues discussed above. It attempts to transform the original feature space to a discriminative subspace, in which new features are created for use in model construction. However, since traditional feature transformation methods, e.g., principal component analysis (PCA), neighborhood component analysis (NCA) [5] and large margin nearest neighbor method (LMNN) [18], were not designed specifically

J.P. Carvalho et al. (Eds.): IPMU 2016, Part I, CCIS 610, pp. 253–261, 2016.
DOI: 10.1007/978-3-319-40596-4_22

for tackling data that contains unreliable input features, their performance may severely decline with this kind of imperfect information.

The Dempster-Shafer Theory (DST) [15] is also known as the theory of belief functions or Evidence theory. As a powerful tool for modeling and reasoning with uncertain and/or imprecise information, it has shown remarkable applications in divers fields, such as unsupervised learning [3,13,20], supervised learning [4,6,8,10,11], information fusion [7,9,12,14,17], etc. These facts motivated us to design a new DST-based feature transformation method for data that contains unreliable and noisy features. To this end, a specific cost function consisting of two terms is constructed for learning a low-dimensional transformation matrix. The first term minimizes the imprecision regarding the class membership of each instance. The $\ell_{2,1}$-norm regularization of the transformation matrix acts as the second term. By means of feature selection, it aims to manage the influence of unreliable original features on the output transformation. The proposed cost function is minimized efficiently by a first order method (namely the Beck-Teboulle proximal gradient algorithm [1]). Finally, a low-dimensional transformation of the original feature space is realized to widely separate instances from different classes.

The rest of this paper is organized as follows. The background on DST is recalled in Sect. 2. The proposed method based on DST is then introduced in Sect. 3. In Sect. 4, the proposed method is tested on both synthetic and real-world datasets. Finally, we conclude paper in Sect. 5.

2 Background on Dempster-Shafer Theory

The necessary background on DST is briefly reviewed in this section. As a generalization of both probability theory and the set-membership approaches, DST has two main components, i.e., quantification of a piece of evidence and combination of different items of evidence.

2.1 Evidence Quantification

DST is a formal framework for reasoning under uncertainty based on the modeling of evidence [15]. Let ω be a variable taking values in a finite domain $\Omega = \{\omega_1, \cdots, \omega_c\}$, called the *frame of discernment*. An item of evidence regarding the actual value of ω can be represented by a *mass function* m on Ω, defined from the powerset 2^Ω to the interval $[0,1]$, such that

$$\sum_{A \subseteq \Omega} m(A) = 1. \tag{1}$$

Each number $m(A)$ denotes a *degree of belief* attached to the hypothesis that "$\omega \in A$". Function m is said to be normalized if $m(\emptyset) = 0$, which is assumed in this paper. Any subset A with $m(A) > 0$ is called a *focal element* of mass function m. If all focal elements are singletons, m is said to be *Bayesian*; it is then equivalent to a probability distribution. A mass function m with only one focal element is said to be *categorical* and is equivalent to a set.

Corresponding to a normalized mass function m, we can associate *belief* and *plausibility* functions from 2^Ω to $[0,1]$ defined as:

$$Bel(A) = \sum_{B \subseteq A} m(B); \quad Pl(A) = \sum_{B \cap A \neq \emptyset} m(B). \tag{2}$$

Quantity $Bel(A)$ (also known as *credibility*) can be interpreted as the degree to which the evidence supports A, while $Pl(A)$ can be interpreted as the degree to which the evidence is not contradictory to A. Functions Bel and Pl are linked by the relation $Pl(A) = 1 - Bel(\overline{A})$. They are in one-to-one correspondence with mass function m.

2.2 Evidence Combination

In DST, beliefs are elaborated by aggregating different items of evidence. *Dempster's rule of combination* [15], as well as its unnormalized version, i.e., the *conjunctive combination rule* defined in the Transferable Belief Model (TBM) [16], are basic mechanisms for evidence fusion. Let m_1 and m_2 be two mass functions derived from independent items of evidence. They can be fused via Dempster's rule to induce a new mass function $m_1 \oplus m_2$ defined as

$$(m_1 \oplus m_2)(A) = \frac{1}{1-Q} \sum_{B \cap C = A} m_1(B) m_2(C), \tag{3}$$

where $Q = \sum_{B \cap C = \emptyset} m_1(B) m_2(C)$ measures the *degree of conflict* between evidence m_1 and m_2.

3 Method

Let $\{(X_i, Y_i) | i = 1, \cdots, N\}$ be a collection of N training pairs, in which $X_i = [x_1, \cdots, x_V]^T$ is the ith instance with V input features, and Y_i is the corresponding class label taking values in a frame of discernment $\Omega = \{\omega_1, \cdots, \omega_c\}$ with an integer $c \geq 2$.

Feature Transformation. To realize a linear transformation of the input feature space, we need to learn a matrix $A \in \mathbf{R}^{v \times V}$, by which the squared distance between any two instances (e.g., X_i and X_j) is quantified as

$$d^2(X_i, X_j) = ||AX_i - AX_j||_2{}^2. \tag{4}$$

The size of the transformation matrix A should satisfy the constraint $v \ll V$, so as to output a low-dimensional transformation.

To learn such a matrix A, we successively set each X_i, $i \in \{1, \ldots, N\}$, as a query instance. Then, other samples in the training pool can be regarded as independent items of evidence that support different hypotheses concerning the class membership of X_i. The evidence offered by the training sample $(X_j, Y_j = \omega_q)$, $j \neq i$ and $q \in \{1, \ldots, c\}$, asserts that X_i is also originated from the class ω_q. However, this piece of evidence is partially reliable. It is inversely proportional to the dissimilarity between X_i and X_j, and can be quantified as a mass function

$$\begin{cases} m_{ij}(\{\omega_q\}) & = \exp\left(-d^2(X_i, X_j)\right) \\ m_{ij}(\Omega) & = 1 - \exp\left(-d^2(X_i, X_j)\right) \end{cases}, \tag{5}$$

where the distance, i.e., $d^2(X_i, X_j)$, is measured by (4). Let Γ_q $(q = 1, \ldots, c)$ be the set of training samples (except X_i) belonging to the same class ω_q. Since the corresponding mass functions point to the same hypothesis (i.e., $Y_i = \omega_q$), they can be combined via Dempster's rule (i.e., (3)) to deduce a global mass function for all training samples in Γ_q:

$$\begin{cases} m_i^{\Gamma_q}(\{\omega_q\}) & = 1 - \prod_{j \in \Gamma_q} \left[1 - \exp\{-d(X_i, X_j)\}\right] \\ m_i^{\Gamma_q}(\Omega) & = \prod_{j \in \Gamma_q} \left[1 - \exp\{-d(X_i, X_j)\}\right] \end{cases}. \tag{6}$$

The global mass function $m_i^{\Gamma_q}$ quantifies the evidence refined from the training pool that support the assertion $Y_i = \omega_q$. The mass of belief $m_i^{\Gamma_q}(\Omega)$ measures the imprecision of this hypothesis. If the actual value of Y_i is ω_q, this imprecision should then close to zero, i.e., $m_i^{\Gamma_q}(\Omega) \approx 0$; in contrast, imprecision pertaining to other hypotheses should close to one, i.e., $m_i^{\Gamma_r}(\Omega) \approx 1$, $\forall r \neq q$. According to this assumption, we propose to represent the prediction loss for training sample (X_i, Y_i) as a function of the matrix A, namely

$$loss_i(A) = \sum_{q=1}^{c} t_{i,q} \cdot \left\{1 - m_i^{\Gamma_q}(\{\omega_q\}) \cdot \prod_{r \neq q}^{c} m_i^{\Gamma_r}(\Omega)\right\}^2, \tag{7}$$

where $t_{i,q}$ is the qth element of a binary vector $t_i = [t_{i,1}, \ldots, t_{i,c}]$, with $t_{i,q} = 1$ iff $Y_i = \omega_q$. When $Y_i = \omega_q$ is true, minimizing $loss_i(A)$ can force both $m_i^{\Gamma_q}(\{\omega_q\}) = 1 - m_i^{\Gamma_q}(\Omega)$ and $\prod_{r \neq q}^{c} m_i^{\Gamma_r}(\Omega)$ to approach one as far as possible, thus achieving the goal to maximize the reliability of the right hypothesis $(Y_i = \omega_q)$ but minimize the reliability of other assertions. As the result, the learnt matrix A can lead X_i only close to samples from the same class in the transformed space.

Feature Selection. To control the influence of unreliable input features in the transformed feature subspace, the $l_{2,1}$-norm sparsity regularization of A, namely

$$||A||_{2,1} = \sum_{j=1}^{V} \left(\sum_{i=1}^{v} A_{i,j}^2\right)^{1/2}, \tag{8}$$

is adopted to realize the joint selection and transformation of input features. By forcing columns of A to be zero during the learning procedure, this sparsity term can only select the most reliable input features to calculate the low-dimensional transformation.

Finally, based on all training samples, the loss function to learn the matrix A is defined as

$$\arg\min_{A} \frac{1}{N}\sum_{i=1}^{N} loss_i(A) + \lambda||A||_{2,1}, \tag{9}$$

where λ is a hyper-parameter that controls the influence of the sparsity penalty.

Optimization. Considering that $loss_i$ (7) is differentiable concerning A, while $||A||_{2,1}$ (8) is partly smooth (it is non-smooth iff $A = 0$), the Beck-Teboulle proximal gradient algorithm [1], which belongs to the class of first-order optimization methods, is used in this paper to find the solution of (9).

4 Experimental Results

In this section, the proposed method was evaluated by a synthetic dataset and two real-world datasets. The Evidential K-nearest-neighbor (EK-NN) classification rule [2] was selected to classify the testing samples after feature transformation.

4.1 Evaluation by Synthetic Datasets

The synthetic dataset was generated using a process similar to the one described in [19]. It contains n_r relevant features uniformly and independently distributed between $[-1, 1]$. The output label of each instance is determined by

$$y = \begin{cases} \omega_1 & \text{if } \max_i(x_i) > 2^{1-\frac{1}{n_r}} - 1 \\ \omega_2 & \text{otherwise} \end{cases}, \tag{10}$$

where x_i is the ith relevant feature. Besides the relevant features, there are n_u irrelevant (noisy) features also uniformly distributed between $[-1, 1]$, without

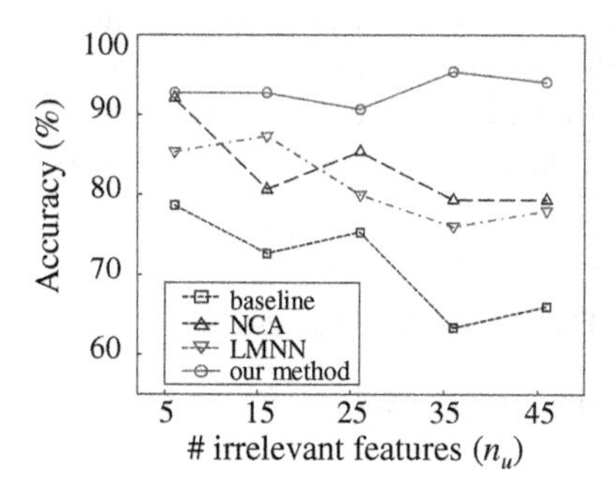

Fig. 1. Testing accuracy of the EK-NN classifier based on different feature transformation methods. Performance in the input feature space is presented as the baseline.

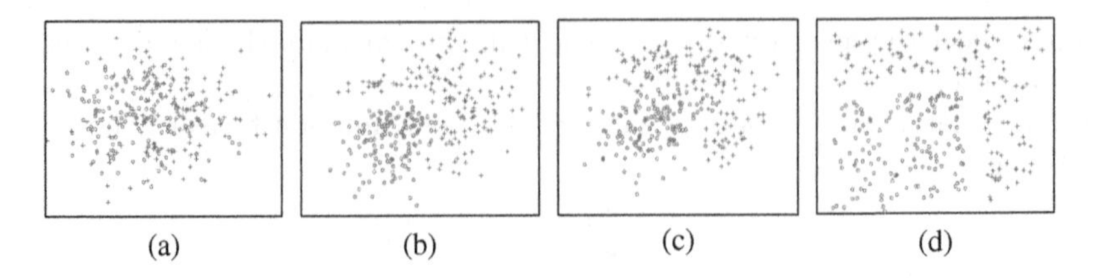

Fig. 2. Two-dimensional transformation results obtained by (a) PCA, (b) NCA, (c) LMNN and (d) our method, respectively.

any relation with the class label; and also n_i imprecise features copied as the cubic of the relevant features.

The numbers of relevant, irrelevant and imprecise features were set, respectively, as $n_r = 2$, $n_u \in \{6, 16, 26, 36, 46\}$ and $n_i = 2$ to simulate five different situations. Under each situation, 150 training instances and 150 testing instances were simulated. The proposed method was compared with PCA, NCA [5] and LMNN [18]. Each of the compared methods was used to learn a two-dimensional transformation (i.e., to learn a matrix $A \in \mathbf{R}^{n_r \times (n_r+n_u+n_i)}$) on the training dataset. After that, the EK-NN was used to classify the testing samples in the transformed subspace. The parameters used in the EK-NN classifier was determined by the method proposed in [21], and the number of nearest neighbors was set as $K = 3$.

Finally, the testing accuracy (in %) for different methods with respect to changing number of unreliable features are summarized in Fig. 1, in which the results obtained by the input features are also presented as the baselines for comparison. As can be seen, our method has higher testing accuracy than other methods under all the five different situations. It is also worth to note that the difference increases following the augment of unreliable input features, which reveals that the proposed method is stable and immune to severely deteriorated input information.

Apart from the classification performance, we also visualized a synthetic dataset (fifty input features with $n_r = 2$, $n_i = 2$ and $n_u = 46$) in the 2-D subspace, so as to evaluate whether the proposed method can effectively separate instances from different classes after feature transformation. The proposed method was still compared with PCA, NCA, and LMNN. As shown in Fig. 2, it outputs the largest margin between different classes as compared to the other three methods.

4.2 Evaluation by Real-World Datasets

The proposed method was further evaluated using two real-world datasets offered by oncologists[1]:

[1] They are with the Department of Nuclear Medicine, Centre Henri Becquerel, 76038 Rouen, France.

(1) *Lung Tumor Dataset:* This dataset contains twenty-five lung tumor patients (instances) treated with chemo-radiotherapy (CRT). For each patient, fifty-two intensity and texture features were extracted from the positron emission tomography (PET) images acquired before and during the treatment. The class label for each patient was *recurrence* or *no-recurrence*, which was clinically assessed at one year after the end of CRT.
(2) *Esophageal Tumor Dataset:* This dataset contains thirty-six esophageal tumor patients (instances) treated with chemo-radiotherapy. For each patient, twenty-nine features were extracted from the PET images and the clinical documents. The class label for each patient was *disease-free* or *disease-positive*, which was clinically assessed at one month after the end of CRT.

The two real-world datasets are briefly summarized in Table 1. Comparing to a limited number of instances (which is often encountered in the medical domain), a relatively large amount of input features were gathered for each clinical dataset. In addition, due to system noise and limited resolution of PET imaging, some features calculated from the PET images are unreliable, or even at odds with the class labels.

Since the datasets are small-sized, the leave-one-out cross-validation (LOOCV) was adopted to assess the performance. The proposed method was compared with the other three feature transformation methods, i.e., PCA, NCA and LMNN. For all the compared methods, the dimensionality of the output subspace was chosen between two to five. The best output dimension was determined according to the average testing accuracy. Finally, the average training accuracy and testing accuracy (more important) obtained by different methods

Table 1. Description of the real-world datasets.

Dataset	Classes	Instances	Features
Lung tumor	2	25	52
Esoph. tumor	2	36	29

Table 2. Comparing the performance (*ave* ± *std*) of our method with the other three feature transformation methods (PCA, NCA and LMNN). The results obtained by the EK-NN in the original feature space is served as the baseline for comparison.

Method	Lung Tumor Dataset		Esophageal Tumor Dataset	
	Training	Testing	Training	Testing
Original space	69.50±4.46	60.00±50.00	63.73±2.14	61.11±49.44
PCA	81.50±5.25	76.00±43.60	56.90±5.81	58.34±50.00
NCA	99.50±1.83	80.00±40.82	94.21±3.24	69.44±46.72
LMNN	100.00±0.00	68.00±47.61	85.48±4.50	80.56±40.14
Our method	100.00±0.00	**88.00±33.17**	97.46±1.64	**83.33±37.80**

are summarized in Table 2, in which results obtained by the input features are presented as baselines for comparison. As shown in Table 2, the proposed method leads to better testing accuracy than other methods on the studied real-world datasets.

5 Conclusion

An approach based on DST has been proposed to realize joint feature transformation and feature selection from the input space that contains unreliable features. To this end, a loss function consisting of two terms has been constructed, in which the first term attempts to minimize the imprecision regarding each training sample's class membership; while the second term, namely a sparsity regularization of the transformation matrix, serves to limit the influence of unreliable input features on the output feature transformation. The constructed loss function has been minimized by a proximal gradient algorithm to find a satisfactory transformation matrix. After that, the output matrix has been used to accomplish the low-dimensional transformation of the input space. Experimental results obtained on the synthetic dataset and the real-world datasets show that the proposed method can be used to improve the performance of classification methods (e.g., the EK-NN classifier) on low-quality data.

Acknowledgements. This work was partly supported by China Scholarship Council.

References

1. Beck, A., Teboulle, M.: A fast iterative shrinkage-thresholding algorithm for linear inverse problems. SIAM J. Imaging Sci. **2**(1), 183–202 (2009)
2. Denœux, T.: A K-nearest neighbor classification rule based on Dempster-Shafer theory. IEEE Trans. Syst. Man Cybern. **25**(5), 804–813 (1995)
3. Denœux, T., Kanjanatarakul, O., Sriboonchitta, S.: EK-NNclus: a clustering procedure based on the evidential k-nearest neighbor rule. Knowl.-Based Syst. **88**, 57–69 (2015)
4. Denœux, T., Smets, P.: Classification using belief functions: relationship between case-based and model-based approaches. IEEE Trans. Syst. Man Cybern. Part B Cybern. **36**(6), 1395–1406 (2006)
5. Goldberger, J., Roweis, S., Hinton, G., Salakhutdinov, R.: Neighbourhood components analysis. In: Advances in Neural Information Processing Systems, pp. 513–520 (2005)
6. Jiao, L., Pan, Q., Denoeux, T., Liang, Y., Feng, X.: Belief rule-based classification system: extension of FRBCS in belief functions framework. Inf. Sci. **309**, 26–49 (2015)
7. Lelandais, B., Ruan, S., Denœux, T., Vera, P., Gardin, I.: Fusion of multi-tracer PET images for dose painting. Med. Image Anal. **18**(7), 1247–1259 (2014)
8. Lian, C., Ruan, S., Denœux, T.: An evidential classifier based on feature selection and two-step classification strategy. Pattern Recogn. **48**(7), 2318–2327 (2015)

9. Lian, C., Ruan, S., Dencœux, T., Li, H., Vera, P.: Dempster-Shafer theory based feature selection with sparse constraint for outcome prediction in cancer therapy. In: Navab, N., Hornegger, J., Wells, W.M., Frangi, A.F. (eds.) MICCAI 2015. LNCS, vol. 9351, pp. 695–702. Springer, Heidelberg (2015)
10. Liu, Z., Pan, Q., Mercier, G., Dezert, J.: A new incomplete pattern classification method based on evidential reasoning. IEEE Trans. Cybern. **45**(4), 635–646 (2015)
11. Ma, L., Destercke, S., Wang, Y.: Online active learning of decision trees with evidential data. Pattern Recogn. **52**, 33–45 (2016)
12. Makni, N., Betrouni, N., Colot, O.: Introducing spatial neighbourhood in evidential C-means for segmentation of multi-source images: application to prostate multi-parametric MRI. Inf. Fusion **19**, 61–72 (2014)
13. Masson, M.H., Denœux, T.: ECM: an evidential version of the fuzzy C-means algorithm. Pattern Recogn. **41**(4), 1384–1397 (2008)
14. Nguyen, T., Boukezzoula, R., Coquin, D., Perrin, S.: Combination of sugeno fuzzy system and evidence theory for NAO robot in colors recognition. In: 2015 IEEE International Conference on Fuzzy Systems (FUZZ-IEEE), pp. 1–8 (2015)
15. Shafer, G.: A Mathematical Theory of Evidence. Princeton University Press, Princeton (1976)
16. Smets, P., Kennes, R.: The transferable belief model. Artif. Intell. **66**(2), 191–234 (1994)
17. Wang, F., Miron, A., Ainouz, S., Bensrhair, A.: Post-aggregation stereo matching method using Dempster-Shafer theory. In: 2014 IEEE International Conference on Image Processing (ICIP), pp. 3783–3787 (2014)
18. Weinberger, K.Q., Saul, L.K.: Distance metric learning for large margin nearest neighbor classification. J. Mach. Learn. Res. **10**, 207–244 (2009)
19. Weston, J., Elisseeff, A., Schölkopf, B., Tipping, M.: Use of the zero norm with linear models and kernel methods. J. Mach. Learn. Res. **3**, 1439–1461 (2003)
20. Zhou, K., Martin, A., Pan, Q., Liu, Z.-G.: Median evidential C-means algorithm and its application to community detection. Knowl.-Based Syst. **74**, 69–88 (2015)
21. Zouhal, L.M., Denœux, T.: An evidence-theoretic K-NN rule with parameter optimization. IEEE Trans. Syst. Man Cybern. Part C Appl. Rev. **28**(2), 263–271 (1998)

Recognition of Confusing Objects for NAO Robot

Thanh-Long Nguyen, Didier Coquin(✉), and Reda Boukezzoula

LISTIC Laboratory, Polytech Annecy-Chambery,
University of Savoie Mont-Blanc, 74940 Annecy-le-vieux, France
{thanh-long.nguyen,didier.coquin,reda.boukezzoula}@univ-smb.fr

Abstract. Visual processing is one of the most essential tasks in robotics systems. However, it may be affected by many unfavourable factors in the operating environment which lead to imprecisions and uncertainties. Under those circumstances, we propose a multi-camera fusing method applied in a scenario of object recognition for a NAO robot. The cameras capture the same scenes at the same time, then extract feature points from the scene and give their belief about the classes of the detected objects. Dempster's rule of combination is then used to fuse information from the cameras and provide a better decision. In order to take advantages of heterogeneous sensors fusion, we combine information from 2D and 3D cameras. The results of experiment prove the efficiency of the proposed approach.

Keywords: Object recognition · NAO robot · Uncertainty · Evidence theory · Camera fusion

1 Introduction

With the very fast development of high technologies, robotics is now more and more important to human life. Specifically, vision processing is one of the most focused areas, which helps a robot increase its ability to learn in explored environments. This work considers a scenario in which a NAO robot can recognize previously learned objects by fusing multi-camera to increase the quality of recognition and reduce uncertainties and imprecisions. We first have a look at how the other works have dealt with object recognition, then propose a solution for the considered case.

In fact, the problem of recognizing an object has been addressed for several decades. The number of methodologies is huge up to now; each of them tried to prove their strengths and overcame the weaknesses of the preceding solutions. For instances, Berg et al. [1] used Geometric Blur approach for feature descriptors and proposed an algorithm to calculate the correspondences between images. The query image was then classified according to its lowest cost of correspondence to the sample images. Besides that, Ling and Jacobs [2] introduced the term "inner-distance"as the length of the shortest path between

J.P. Carvalho et al. (Eds.): IPMU 2016, Part I, CCIS 610, pp. 262–273, 2016.
DOI: 10.1007/978-3-319-40596-4_23

landmark points within the shape silhouette. The inner-distance was used to build shape representations and they helped to obtain good matching results. For some texture-based approaches, [3] proposed a texture descriptor based on Random Sets and experimentally showed that it outperformed the co-occurrence matrix descriptor. Decision tree induction was used in that work to learn the classifier. Another example can be found in [4] where color and texture information were both used in an agricultural scenario to recognize fruits. On the other hand, some context-based methods like [5–7] considered contextual information surrounding the target objects. These information come from the interaction among objects in the scene and they help to disambiguate appearance inputs in recognition tasks. Similarly successful, the methods based on local feature description like SIFT [8] and SURF [9] have received many positive evaluations and have been widely applied [10–13]. SIFT extracts keypoints from object to build feature vectors. We then calculate the matching (using Euclidean distance) between an input object and the ones in database to find the best candidate class. After that, the agreement on the object and its location, scale, and orientation are determined by using a hash table implementation of the Generalized Hough Transform. In a different manner, SURF uses a blob detector based on the Hessian matrix to find interest points, then it calculates the descriptor by using the sum of Haar wavelet responses. Finally, by comparing the descriptors obtained from different images, the matching pairs can be found.

For the purpose of collecting spatial information about the detected objects, and avoiding imprecision of 2D images under non-ideal lighting conditions like outdoor environment, some works concentrated on 3D object recognition. In [14], an extended version of the Generalized Hough Transform was used in 3D scenes. Each point in the input cloud votes for a spatial position of the object's reference point and the accumulating bin with the maximum votes indicates an instance of the object in the scene. In [15,16], the 3D extensions of SIFT and SURF descriptor also gave positive recognition results. In addition, Zhong [17] introduced a new 3D shape descriptor called Intrinsic Shape Signature to characterize a local/semi-local region of a point cloud. This descriptor uses a view-independent representation of the 3D shape to match shape patches from different views directly, and a view-dependent transform encoding the viewing geometry to facilitate fast pose estimation. On the contrary, [18,19] considered the use of point pairs for the description and the feature matching is then done by implementing a hash table. Recently, the SHOT descriptor [20] has emerged as an efficient tool for 3D object recognition [21,22]. Indeed, the descriptor encodes histograms of basic first-order differential entities (i.e. the normals of the points within the support), which are more representative than plain 3D coordinates about the local structure of the surface. After defining an unique and robust 3D local reference frame, it is possible to enhance the discriminative power of the descriptor by concerning the location of the points within the support, from that describing a signature.

It is clear that all of the above mentioned approaches have experimentally shown good results in object recognition. Nevertheless, many of them did not

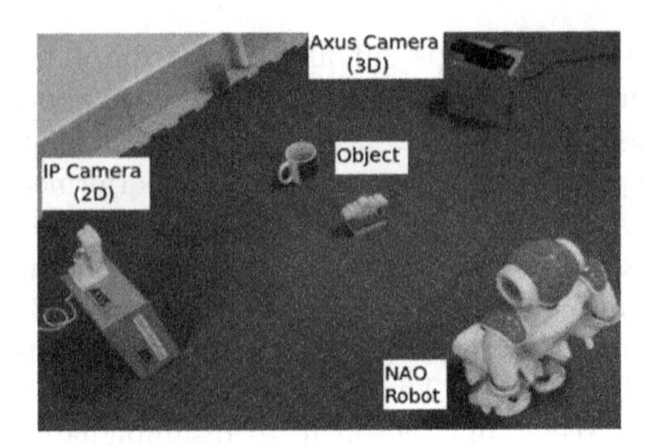

Fig. 1. Multi-camera helps NAO robot recognize objects.

focus on the problem of uncertainty and imprecision which might come from the quality of data and sensors, the lighting conditions, the viewing angles to the objects and particularly, *the similarity among confusing objects*. Therefore, in this work we propose to use multi-camera to recognize objects which have many similarities. The proposed method is implemented in a NAO robot due to our development in a robotics project, however it is not restricted to any other kind of vision-based platform. In order to take advantage of both 2D and 3D recognitions, we use not only a 2D camera of the NAO robot but also another 2D IP Axis camera and another 3D Axus camera; Fig. 1 shows the multi-camera environment where the robot is requested to recognize objects. The fusion of these three heterogeneous sensors brings additional advantages for each one because the NAO camera and the IP camera give characteristics about the 2D features of the detected objects whereas the Axus camera provides depth information. We propose an evidential classifier based on Dempster-Shafer theory (or Evidence theory) [23] for each camera, then we combine them in decision level in order to give more reasonable results of object recognition.

The outline of the paper is as follow. First, we describe our approach step-by-step in Sect. 2, then we give an illustrative example in Sect. 3. Section 4 shows our results of experiment to validate the approach, finally Sect. 5 gives the conclusion.

2 Our Recognition Approach

2.1 An Evidential Classifier for Each Camera

Processing Flow: Figure 2 shows the flow of classification by each camera. First, an input image in 2D or 3D form is captured based on the type of camera sensor. For the NAO camera and the IP camera (2D), the input data is 640×480 images; for the Axus camera (3D), the input images are in form of Point Cloud since we implement 3D processing by using the PCL library [24]. To focus on the classification, we use only one instance of object appearing in the captured scene.

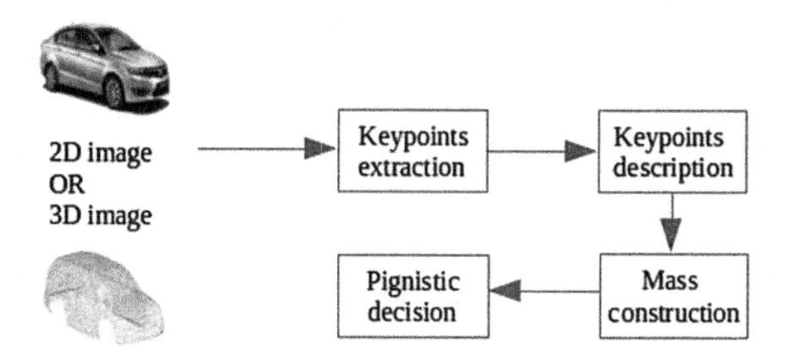

Fig. 2. Evidential classifier for each camera

First, interest points (or key points) of the object in the scene are extracted. In an image, an interest point can be described as a point that has rich information about local image structure around it, and these points characterize well the patterns in the image. After that, we use methods of descriptor to build a feature vector for each interest point. We use the word "feature points" for the interest points that have been described by the descriptor. The methods of descriptors used in this work are SURF [9] for 2D data and SHOT [20] for 3D data according to their strong properties as explained above. From the set of feature points acquired, we build a mass function which describes the camera's degree of belief about the classes of detected object. Thereafter, a decision is made by choosing the class with the maximum pignistic probability. The processing flow is described with more detail later.

Evidence Theory in the Scenario: Suppose the robot has to recognize an object that can be only in one of N classes, i.e. the space of discernment is:

$$\Omega = \{O_1, O_2, ..., O_N\} \tag{1}$$

Then we have the power set which contains the subsets of the space of discernment:

$$2^{\Omega} = \{\{\emptyset\},\ \{O_1\},\ \{O_2\},\ ...,\ \{O_N\},\ \{O_1 \cup O_2\},\ ...,\ \{O_1 \cup O_N\},\ ...,\ \{\Omega\}\} \tag{2}$$

In Evidence Theory, we have to determine a mass function which describes the degree of belief for all possible hypotheses in the power set. This function satisfies:

$$\begin{aligned} m : 2^{\Omega} \to [0, 1] \\ \sum_{H \in 2^{\Omega}} m(H) = 1 \end{aligned} \tag{3}$$

To illustrate the proposed approach, we consider a simple case in Fig. 3 where we suppose that there are three classes of object: A, B and C. For the sake of explanation, we assume that we have only one training image for each class. With an input image which contains a set X of feature points of object, our mission is to decide the appropriate class for X. The basic idea is that each

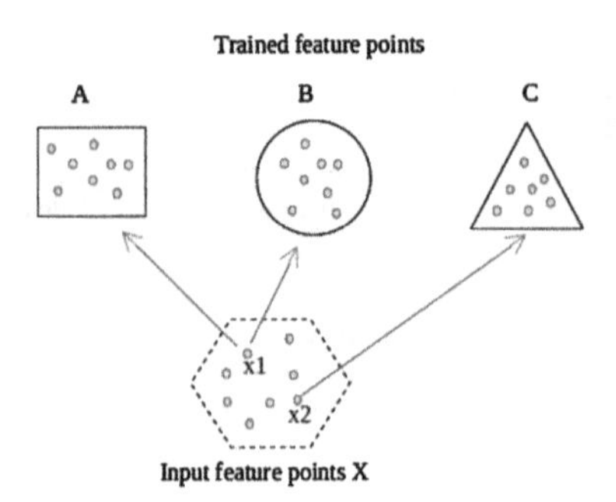

Fig. 3. Illustration of the idea. Each input feature point votes for a hypothesis.

feature point $x_i \in X$ will vote for a hypothesis $H \in 2^{\Omega}$ based on its matching to the training images. In Fig. 3, the feature point x_1 matches both images of class A and B, so we accumulate one vote for the hypothesis $H = \{A \cup B\}$. Similarly, the feature point x_2 votes for $H = \{C\}$. By doing the same principle for all the feature points of X, we can construct all elements of the mass function after doing a normalization step. Due to the need of clear explanation in a scientific work, the step of defining the matching and constructing mass function will be mathematically described thereafter.

Construction of Mass and Decision: First, let us denote $\Delta(p_i, p_j)$ the normalized distance between two feature points p_i and p_j; the shorter the distance is, the more similar the two feature points are.

$$\Delta(p_i, p_j) \in [0, 1] \tag{4}$$

In order to decide the matching between a feature point p_i^X of an input image X (X can also be understood as the set of feature points for the input image) and a training image M whose class is $O_j \in \Omega$, we use the idea in [25]. We will find the two nearest neighbours of p_i^X in M, called $p_{i_1}^M$ and $p_{i_2}^M$ (the feature points in M are previously extracted in the training phase). We suppose that $p_{i_1}^M$ is closer to p_i^X than $p_{i_2}^M$ i.e. $\Delta(p_i^X, p_{i_1}^M) \leq \Delta(p_i^X, p_{i_2}^M)$. After that, we define a matching function between the feature point p_i^X of an input image X and the model M :

$$\delta(p_i^X, M) = \begin{cases} 1, & \text{if } \Delta(p_i^X, p_{i_1}^M) \leq \alpha \text{ and } \frac{\Delta(p_i^X, p_{i_1}^M)}{\Delta(p_i^X, p_{i_2}^M)} \leq \beta \\ 0, & \text{otherwise} \end{cases} \tag{5}$$

where α and β are two user-defined parameters such that $0 \leq \alpha, \beta \leq 1$. The former guarantees that the distance between p_i^X and its most similar feature point found in M is small enough whereas the latter helps to avoid false matching. In this work, we choose $\beta = 0.8$ as suggested in [25], and we add $\alpha = 0.25$ in order to reduce noise. Indeed, these two parameters help us to find a strong and distinctive matching between the feature point p_i^X and its closest feature point in M. If $\delta(p_i^X, M) = 1$, we then say that p_i^X is matched to the training image M, i.e. matched to the class $O_j \in \Omega$ of M and vice versa. In the same way, we can find all the matches of the feature points in the input image X to the training image M.

For now, we define the *matching between* X *and the class* O_j by considering all the matches between feature points p_i^X in X and the class O_j. In the case that the class O_j has several training images M_k, we choose the training image M_{max} that has the maximum number of matches to X according to Eq. (5).

$$\delta^{max}(p_i^X, O_j) = \delta(p_i^X, M_{max}) \tag{6}$$

Table 1 shows an example illustrating the matches between input feature points and the output classes. A cell $c(p_i^X, O_j)$ implies the matching between the feature point p_i^X of X and the class O_j, $i = 1, 2, ...R_X$ - number of feature points in X, $j = 1, 2, ...N$ - number of classes. If the cell is red, it means that the feature point p_i^X matches the class O_j (i.e. $\delta^{max}(p_i^X, O_j) = 1$), otherwise not matched.

After we determine the matching between the input feature points and the output classes, we can construct the mass function as follow. Each feature point p_i^X will vote for a hypothesis in the power set such that the hypothesis is composed of the classes that match p_i^X. Mathematically, let's define a hypothesis-voted function that calculates the accumulated votes for each hypothesis:

$$accVote(X, H) = \sum_{p_i^X \in X} \phi(p_i^X, H), \quad H \in 2^\Omega \tag{7}$$

where $\phi(p_i^X, H)$ is a function indicating the matching between the feature point p_i^X and every element class in H:

$$\phi(p_i^X, H) = \begin{cases} 1, & \text{if } \sum_{O_j \in H} \delta^{max}(p_i^X, O_j) = |H| \\ 0, & \text{otherwise} \end{cases} \tag{8}$$

where $|H|$ be the cardinality of H and $\delta^{max}(p_i^X, O_j)$ was already explained above. Indeed, $\phi(p_i^X, H)$ indicates whether a feature point p_i^X matches every element class in the hypothesis H or not, and $accVote(X, H)$ calculates the number

Table 1. Matching between the feature points of input image X and the classes

	p_1^X	p_2^X	p_3^X	...	$p_{R_X}^X$
O_1	■			...	
O_2		■		...	■
O_3	■		■	...	■
...	...	...	...	...	...
O_N		■		■	

of feature points in X that matches every element class in H. After that, we calculate the mass function based on the hypothesis-voted function:

$$m^X(H) = \frac{accVote(X, H)}{G^X} \tag{9}$$

where G^X is the normalization factor that guaranties the condition in Eq. (3):

$$G^X = \sum_{H \in 2^{\Omega}, H \neq \emptyset} accVote(X, H) \tag{10}$$

It is worth noting that, in this work we assume that the class of object in the input image X is only in Ω, so we put $m^X(\emptyset) = 0$.

Once we have constructed the mass function, we can give decision about the class of the object. Since the maximum of belief is too pessimistic and the maximum of plausibility is too optimistic, we choose the class which has the maximum pignistic probability [26]:

$$BetP^X(O_j) = \frac{1}{1 - m^X(\emptyset)} \sum_{O_j \in H} \frac{m^X(H)}{|H|} \tag{11}$$

2.2 Fusion of Cameras

Base on the Evidence theory, each camera gives a decision about the classification of the detected object. In addition, by using Dempster's rule of combination [23], we can integrate information from multi-camera in order to give a better decision. Usually, the rule is defined for two sources, however it is enough to ensure a trivial extension to many sources due to its associativity and commutativity:

$$m_{comb}(\emptyset) = 0$$
$$m_{comb}(H) = \frac{\sum_{H_1 \cap H_2 \cap ... \cap H_S = H} m_1(H_1) m_2(H_2) ... m_S(H_S)}{1 - K}, H \in 2^{\Omega}, H \neq \emptyset \tag{12}$$

where S is the number of information source (i.e. number of cameras, 3 in this experiment) and:

$$K = \sum_{H_1 \cap H_2 \cap ... \cap H_S = \emptyset} m_1(H_1) m_2(H_2) ... m_S(H_S) \tag{13}$$

Finally, the decision about the class of the detected object can be made by using pignistic probability as in Eq. (11).

3 Illustrative Example

In this section, we provide an example to illustrate the proposed approach. Suppose that we want the robot to recognize an object in a captured scene with three classes in the space of discernment, that means:

$$\Omega = \{O_1, O_2, O_3\} \tag{14}$$

so there are 8 possible hypotheses in the power set:

$$2^{\Omega} = \{\{\emptyset\}, \{O_3\}, \{O_2\}, \{O_2 \cup O_3\}, \{O_1\}, \{O_1 \cup O_3\}, \{O_1 \cup O_2\}, \{\Omega\}\} \tag{15}$$

For simplicity, we suppose that for each class, we have only 1 training image. Assuming that the NAO camera captures the scene X and it found 10 feature points in the input image X_{NAO}. For each of those input feature points, we find two nearest neighbours feature points in each training image. After that, we use Eqs. (4), (5), and (6) to construct the matching between the input image and each class. Table 2 shows an example of the matching found. Each cell describes the matching between a feature point and a class; if $\delta^{max}(p_i^{X_{NAO}}, O_j) = 1$, the cell is red, otherwise white. The last row indicates the hypothesis voted by the associating feature point.

Table 2. Matching between the input image X^{NAO} and the classes

	p_1	p_2	p_3	p_4	p_5	p_6	p_7	p_8	p_9	p_{10}
O_1	■		■	■			■		■	
O_2		■			■	■			■	■
O_3				■			■	■		■
Vote for:	O_1	O_2	O_1	$O_1 \cup O_3$	O_2	O_2	$O_1 \cup O_3$	O_3	$O_1 \cup O_2$	$O_2 \cup O_3$

From Table 2, we have determined the strength of each hypothesis in the power set. Table 3 then shows the accumulated vote for each hypothesis which is calculated by Eqs. (7) and (8). Each cell in the table is the value of $\phi(p_i^{X_{NAO}}, H), H \in 2^{\Omega}$. Remind that if $\phi(p_i^{X_{NAO}}, H) = 1$, it means that the feature point $p_i^{X_{NAO}}$ votes for the hypothesis H. According to Eq. (10), we have $G^{X_{NAO}} = \sum accVote = 1 + 3 + 1 + 2 + 2 + 1 + 0 = 10$. From these information, we calculate the mass values as in the last column by using Eq. (9).

After that, we assume that we use not only the NAO camera but also another IP camera (2D) and another Axus camera (3D). By doing the same steps, we can obtain two mass vectors output from the two additional sensors. Table 4 shows example values of these mases. Additionally, we also calculate the combination of the masses using Dempster's rule (m_{comb}) and transform it to the pignistic probability ($BetP$) for each of singleton hypothesis. The last column is the final decision from the fusion of three cameras, which recognizes that the detected object belongs to the class O_1.

Table 3. Accumulated vote for each hypothesis

$H \in 2^{\Omega}$	p_1	p_2	p_3	p_4	p_5	p_6	p_7	p_8	p_9	p_{10}	accVote	Mass value
$\emptyset$	0	0	0	0	0	0	0	0	0	0	0	0.00
O_3	0	0	0	0	0	0	0	**1**	0	0	1	1/10
O_2	0	**1**	0	0	**1**	**1**	0	0	0	0	3	3/10
$O_2 \cup O_3$	0	0	0	0	0	0	0	0	0	**1**	1	1/10
O_1	**1**	0	**1**	0	0	0	0	0	0	0	2	2/10
$O_1 \cup O_3$	0	0	0	**1**	0	0	**1**	0	0	0	2	2/10
$O_1 \cup O_2$	0	0	0	0	0	0	0	0	**1**	0	1	1/10
Ω	0	0	0	0	0	0	0	0	0	0	0	0/10

Table 4. Mass values from there camera sensors

Hypothesis	m_{NAO}	m_{IP}	m_{Axus}	m_{comb}	$BetP$	Decision
$\emptyset$	0.00	0.00	0.00	0.00		
O_3	0.10	0.23	0.21	0.22	0.23	
O_2	0.30	0.17	0.12	0.26	0.27	
$O_2 \cup O_3$	0.10	0.08	0.00	0.00		
O_1	**0.20**	**0.32**	**0.09**	**0.49**	**0.50**	$\boldsymbol{O_1}$
$O_1 \cup O_3$	0.20	0.13	0.13	0.02		
$O_1 \cup O_2$	0.10	0.00	0.39	0.01		
Ω	0.00	0.07	0.06	0.00		

4 Experiments

As mentioned previously, the concentration of this work is how to resolve uncertainties and imprecisions during the object recognition process of the NAO robot. For that reason, we did three experiments, each of them contains a set of confusing objects as shown in Fig. 4. In the first set, there are 4 cups which can cause uncertainty in their spatial structures for the 3D camera to recognize. Conversely, the second experiment contains 4 boxes that have similar brand information on their surface, which may limit the recognition of the 2D cameras. Finally, we tested with 4 Lego bricks which are considered to have difficulties for both 2D and 3D cameras, in the third experiment.

For the training phase, we trained two images for each object with each camera in different view points. We then manually removed the background in these images in order to have only the model objects. For the test phase, NAO robot is requested to recognize an object appearing in front of it and say the result to human. The two cameras (IP and Axus) are on the two sides of the robot to help it improve the recognition. These three cameras capture the scene at the same time whenever the robot wants to recognize the object in the scene.

To focus on the work of recognition, the image region containing the object is restricted in order to avoid the noises in scene. For each of the three experiments, we did 32 recognition tests with different objects of 4 classes (so 8 tests for each object). The tested objects were turned around and put in different angles to the cameras in each test for the reason of challenging uncertainty.

Table 5 shows the results of experiment which is the comparison between the recognition rate of each camera (using the proposed classifier individually) and the fusion of three cameras. Remind that the rate for each camera cannot be high due to the confusing between similar objects and the objects are turned around each time of test. The fifth column is the result when we fuse the three cameras by using a simple voting based on majority: each camera gives its own recognition result based on the proposed classifier, then we choose the output class that is voted by the largest number of cameras. The last column shows the result of using Dempster-Shafer combination for the three cameras, which outperforms the majority voting to improve the recognition rate in average.

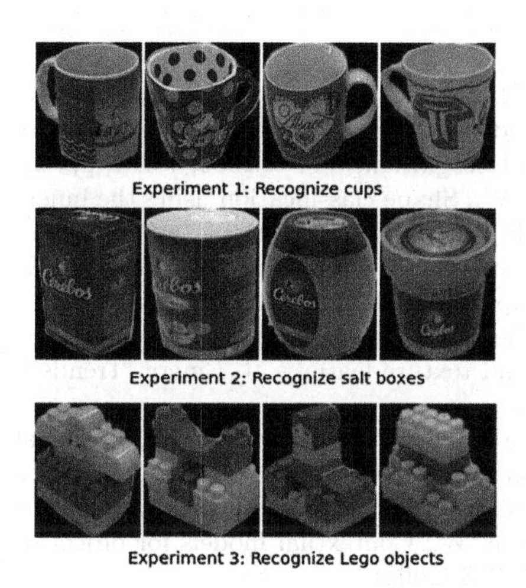

Fig. 4. Confusing objects used in the experiment

Table 5. Experiment result

Camera	NAO (2D)	IP (2D)	Axus (3D)	Majority voting fusion	Dempster-Shafer fusion
Experiment 1	78 %	88 %	75 %	100 %	**97 %**
Experiment 2	72 %	72 %	91 %	91 %	**97 %**
Experiment 3	59 %	59 %	69 %	72 %	**84 %**
Average	69.67 %	73 %	78 %	87.67 %	**92.67%**

5 Conclusion

The work in this paper focuses on how to resolve uncertainties and imprecisions in object recognition for a NAO robot. Since the robot may face difficulties during its visual operation due to lighting conditions, viewing angles and the quality of camera, we propose to add more cameras in order to improve the recognition rate. Each camera extracts feature points from the captured scene, then provides a mass function based on the matching between the input and the training images. After that, Dempster's rule of combination is used to fuse information from these cameras. As can be seen, the approach is generalized for both 2D and 3D cameras, and the experiment work gives positive results, which prove the advantage of the fusion. Our future works will consider a more complex scenario where the NAO robot can build a semantic map based on the recognition approach used in this work.

References

1. Berg, A.C., Berg, T.L., Malik, J.: Shape matching and object recognition using low distortion correspondences. In: IEEE Computer Society Conference on Computer Vision and Pattern Recognition, CVPR 2005, vol. 1, pp. 26–33. IEEE (2005)
2. Ling, H., Jacobs, D.W.: Shape classification using the inner-distance. IEEE Trans. Pattern Anal. Mach. Intell. **29**(2), 286–299 (2007)
3. Perner, P.: Cognitive aspects of object recognition-recognition of objects by texture. Procedia Comput. Sci. **60**, 391–402 (2015)
4. Arivazhagan, S., Shebiah, R.N., Nidhyanandhan, S.S., Ganesan, L.: Fruit recognition using color and texture features. J. Emerg. Trends Comput. Inf. Sci. **1**(2), 90–94 (2010)
5. Galleguillos, C., Rabinovich, A., Belongie, S.: Object categorization using co-occurrence, location and appearance. In: IEEE Conference on Computer Vision and Pattern Recognition, CVPR 2008, pp. 1–8. IEEE (2008)
6. Murphy, K., Freeman, W.: Contextual models for object detection using boosted random fields. In: NIPS (2004)
7. Wolf, L., Bileschi, S.: A critical view of context. Int. J. Comput. Vis. **69**(2), 251–261 (2006)
8. Lowe, D.G.: Object recognition from local scale-invariant features. In: The proceedings of the Seventh IEEE International Conference on Computer vision, 1999, vol. 2, pp. 1150–1157. IEEE (1999)
9. Tuytelaars, T., Van Gool, L., Bay, H., Ess, A.: Speeded-up robust features (surf). Comput. Vis. Image Underst. **110**(3), 346–359 (2008)
10. Abdel-Hakim, A.E., Farag, A. et al.: Csift: a sift descriptor with color invariant characteristics. In: 2006 IEEE Computer Society Conference on Computer Vision and Pattern Recognition, vol. 2, pp. 1978–1983. IEEE (2006)
11. Suga, A., Fukuda, K., Takiguchi, T., Ariki, Y.: Object recognition and segmentation using sift and graph cuts. In: 19th International Conference on Pattern Recognition, ICPR 2008, pp. 1–4. IEEE (2008)
12. Ruf, B., Kokiopoulou, E., Detyniecki, M.: Mobile museum guide based on fast SIFT recognition. In: Detyniecki, M., Leiner, U., Nürnberger, A. (eds.) AMR 2008. LNCS, vol. 5811, pp. 170–183. Springer, Heidelberg (2010)

13. Mehrotra, H., Majhi, B., Gupta, P.: Annular Iris recognition using SURF. In: Chaudhury, S., Mitra, S., Murthy, C.A., Sastry, P.S., Pal, S.K. (eds.) PReMI 2009. LNCS, vol. 5909, pp. 464–469. Springer, Heidelberg (2009)
14. Khoshelham, K.: Extending generalized hough transform to detect 3d objects in laserrange data. In: ISPRS Workshop on Laser Scanning and SilviLaser 2007, 12–14 September 2007, Espoo, Finland. International Society for Photogrammetry and Remote Sensing (2007)
15. Flitton, G.T., Breckon, T.P., Bouallagu, N.M.: Object recognition using 3d sift in complex ct volumes. In: BMVC, pp. 1–12 (2010)
16. Knopp, J., Prasad, M., Willems, G., Timofte, R., Van Gool, L.: Hough transform and 3D SURF for robust three dimensional classification. In: Daniilidis, K., Maragos, P., Paragios, N. (eds.) ECCV 2010, Part VI. LNCS, vol. 6316, pp. 589–602. Springer, Heidelberg (2010)
17. Zhong, Y.: Intrinsic shape signatures: a shape descriptor for 3d object recognition. In: 2009 IEEE 12th International Conference on Computer Vision Workshops (ICCV Workshops), pp. 689–696. IEEE (2009)
18. Drost, B., Ulrich, M., Navab, N., Ilic, S.: Model globally, match locally: efficient and robust 3d object recognition. In: 2010 IEEEConference on Computer Vision and Pattern Recognition (CVPR), pp. 998–1005. IEEE (2010)
19. Papazov, C., Burschka, D.: An efficient RANSAC for 3D object recognition in noisy and occluded scenes. In: Kimmel, R., Klette, R., Sugimoto, A. (eds.) ACCV 2010, Part I. LNCS, vol. 6492, pp. 135–148. Springer, Heidelberg (2011)
20. Tombari, F., Salti, S., Di Stefano, L.: Unique signatures of histograms for local surface description. In: Maragos, P., Paragios, N., Daniilidis, K. (eds.) ECCV 2010, Part III. LNCS, vol. 6313, pp. 356–369. Springer, Heidelberg (2010)
21. Tombari, F., Di Stefano, L.: Hough voting for 3d object recognition under occlusion and clutter. IPSJ Trans. Comput. Vis. Appl. **4**, 20–29 (2012)
22. Rodolà, E., Albarelli, A., Bergamasco, F., Torsello, A.: A scale independent selection process for 3d object recognition in cluttered scenes. Int. J. Comput. Vis. **102**(1–3), 129–145 (2013)
23. Shafer, G., et al.: A Mathematical Theory of Evidence, vol. 1. Princeton University Press, Princeton (1976)
24. Rusu, R.B., Cousins, S.: 3d is here: point cloud library (pcl). In: 2011 IEEE International Conference on Robotics and Automation (ICRA), pp. 1–4. IEEE (2011)
25. Lowe, D.G.: Distinctive image features from scale-invariant keypoints. Int. J. Comput. Vis. **60**(2), 91–110 (2004)
26. Smets, P.: Constructing the pignistic probability function in a context of uncertainty. In: UAI, vol. 89, pp. 29–40 (1989)

Evidential Missing Link Prediction in Uncertain Social Networks

Sabrine Mallek[1,2](✉), Imen Boukhris[1], Zied Elouedi[1], and Eric Lefevre[2]

[1] LARODEC, Institut Supérieur de Gestion de Tunis,
Université de Tunis, Tunis, Tunisia
sabrinemallek@yahoo.fr, imen.boukhris@hotmail.com, zied.elouedi@gmx.fr
[2] Univ. Lille Nord de France, UArtois EA 3926 LGI2A, Lille, France
eric.lefevre@univ-artois.fr

Abstract. Link prediction is the problem of determining future or missing associations between social entities. Most of the methods have focused on social networks under a certain framework neglecting some of the inherent properties of data from real applications. These latter are usually noisy, missing or partially observed. Therefore, uncertainty is an important feature to be taken into account. In this paper, proposals for handling the problem of missing link prediction while being attentive to uncertainty are presented along with a technique for uncertain social networks generation. Uncertainty is not only handled in the graph model but also in the method itself using the assets of the belief function theory as a general framework for reasoning under uncertainty. The approach combines sampling techniques and information fusion and returns good results in real-life settings.

Keywords: Social network analysis · Missing link prediction · Uncertain social network · Belief function theory · Information fusion · Graph sampling

1 Introduction

Social networks have been witnessing an inconceivable development in recent years, becoming possibly the main actor of the Web 2.0. Social Network Analysis (SNA) has provided a collection of specific models and methods designed for the investigation of social network data and extraction of knowledge from them. The main objective is to determine the conditions under which the patterning of social ties arise and uncover their consequences. From this perspective, link prediction of topology became the focus of many researchers from various domains. It is a task of link mining that aims at predicting new or existent links in the network. In fact, prediction of future links considers the dynamics of the social network. The task is to determine very likely but not yet existing associations based on the previous snapshots of the network. In contrast, prediction of missing links considers its static state rather than its evolution, where the current knowledge is incomplete [15]. In a word, the latter has no temporal aspect, the goal is to

J.P. Carvalho et al. (Eds.): IPMU 2016, Part I, CCIS 610, pp. 274–285, 2016.
DOI: 10.1007/978-3-319-40596-4_24

predict missing connections to get a more outright picture of the overall structure of the links from the data [27]. Here we are interested at the prediction of missing links under uncertainty in social networks.

As a matter of fact, missing link prediction is of theoretical and practical significance in modern science [26]. In many cases, links might exist at time t but not at t'. A possible reason is a change in privacy settings or when data are partially observed [13], e.g. a facebook user might decide to hide his friends between time t and t', which lead to missing links in the network. This has important ramifications as it may alter estimates of the network statistics [14]. Besides, inferring these missing links raises privacy matters in social networks since several algorithms can be applied to predict new and missing links [9].

On the other hand, data from real world applications are prone to observation errors. They are frequently missing, incomplete and noisy. As pointed out in [2], different degrees of uncertainty characterize several real-world networks especially the large-scale ones. Accordingly, it is an important feature that needs to be taken into account when dealing with social networks from real world data. To handle this uncertainty, the edges might be associated with weights describing their existence in the network. Most of the existing methods use probabilities [3,13], however, in our case, we propose to use the belief function theory [6,21] which is considered as a generalization of the probability theory. In fact, one of the practical uses of the belief functions is the representation and management of missing information. It provides convenient ways to handle real life missing data problems [23]. Furthermore, the belief function theory provides tools for combining of evidence induced from several pieces of information. More information about the interest of adopting the belief function theory to handle uncertainty in networks can be found in [5].

Additionally, we design a fruitful approach for missing link prediction that takes into account the uncertain aspects of the social network. It is completely different from methods from link prediction literature as it operates merely with the belief function tools. It uses popular structural measures based on local graph information to compute distances between the links. A fusion procedure is subsequently applied taking into account the reliability of the sources to predict missing links. Besides, a technique based on network sampling is operated for the creation of uncertain social networks to test the validity of our proposals.

This paper is organized as follows: in the next two sections, we examine related literature about link prediction and the belief function theory. In Sect. 4, we introduce our graph model for uncertain social networks. In Sect. 5, we design the approach for missing link prediction. In Sect. 6, we show the experiments we have carried out to test the performance of our approach. Finally, in Sect. 7, we draw our conclusions and sketch possible future works.

2 Missing Link Prediction

In recent years, topological link prediction in network evolution has gained the interest of many researchers from various fields. Its applications include exploration of protein-protein interactions, mining food relationships in biological and ecological networks, co-authorship retrieval in collaboration networks, mining frienships, uncovering hidden groups or investigating missing members in social networks.

The most straightforward assumption for link prediction is that two nodes that are similar tend to share a link. To this end, the main concern is how to compute the similarity between nodes accurately. As discussed in several works, methods and measures used in link prediction can be applied for both future and missing link prediction. For a review, see [16].

Typically, social networks are schematized as a graph $G = (V, E)$, where V is the set of social entities and E is the social ties linking them. On the basis of this graph formulation, the link prediction problem can be defined as follows: Let $T_l = (G_l, V_l)$ and $T_k = (G_k, V_k)$ be two states of a social network at times l and k. The link prediction task consists at using T_l to predict the social network structure G_k. We predict new links when $l < k$. In contrast, missing links are predicted when $l > k$ [8]. Most existing methods use the topological information of the networks, including the local or global similarity measures. The local methods consider indices based on neighborhoods in the network while the global methods use the ensemble of paths between the nodes.

2.1 Local Information Measures

These measures capture node similarity by considering their structural local properties. The most popular property is the set of neighbors $\tau(u)$ of a given node u. The most widely used index is the number of common neighbors [20], denoted by CN. The intuition is that two nodes u and v that share many common neighbors are more likely to form a link. It is defined as follows:

$$CN(u, v) = |\tau(u) \cap \tau(v)| \tag{1}$$

The Jaccard Coefficient (JC) uses all the the neighbors of the pair (u, v) as it considers the number of nodes that are adjacent to at least one of them. It is defined as follows:

$$JC(u, v) = \frac{|\tau(u) \cap \tau(v)|}{|\tau(u) \cup \tau(v)|} \tag{2}$$

On the other hand, the AdamicAdar measure [1], denoted by AA, penalizes high degree neighbors since a node with high degree is likely to be in the common neighborhood of other nodes anyway. The AA index is defined as follows:

$$AA(u, v) = \sum_{v_k \in (\tau(u) \cap \tau(v))} \frac{1}{log|\tau(v_k)|} \tag{3}$$

2.2 Global Information Measures

These measures derive nodes similarity between a pair of nodes (u, v) from paths based on the assumption: the closest two nodes are, the higher the chance for them to be connected. Global information measures include the shortest path,

SimRank [12], Hitting time, etc. For instance, the shortest path distance is simply the shortest distance between two nodes. The SimRank index assumes that two nodes tend to be connected if they are linked to similar nodes. The Hitting time consider random walks, it computes the expected number of steps required for a random walker to reach v from u.

2.3 Discussion

Both types of measures are simple and generic, they may be applied to networks from several domains. Yet, they have some shortcomings. The CN measure has proved its efficiency in several real networks and has shown the best performances in many comparisons with others measures based on local information [15,26]. However, it favors the nodes with large degrees. To solve this problem, variants such as the JC and AA have been proposed to clear up this tendency. On the other hand, path based metrics generally give accurate prediction however they suffer from two major drawbacks. Firstly, they are computationally expensive as they inquire for the global topological information of the network, and are usually impractical on large-scale networks. Secondly, the global topological information is frequently not available [16]. Besides, the additional complexity does not always enhance the prediction, since similar power can be obtained with local methods as well [15]. For that, our approach for missing link prediction uses local information measures.

3 Belief Function Theory

The belief function theory [6,21], is a suitable theory for the representation and management of imperfect knowledge. It allows to handle uncertainty and imprecision found in data, fuse evidence and make decisions. In fact, belief functions provide convenient solutions to deal with missing information problems, many real life examples are given in [23]. For these reasons, we have adopted this theory to address the missing link prediction problem.

Let Θ be the frame of discernment, an exhaustive and finite set of mutually exclusive events associated to a given problem, and let 2^Θ denote the set of all subsets of Θ. Knowledge in the belief function theory is represented by a basic belief assignment (*bba*), denoted by m, it is defined as follows:

$$m : 2^\Theta \rightarrow [0,1] \qquad \sum_{A \subseteq \Theta} m(A) = 1 \tag{4}$$

We call A a focal element if $m(A) > 0$.

Evidence induced from two reliable and distinct sources of information may be combined using the conjunctive rule of combination denoted by $\textcircled{\cap}$. It is defined as [22]:

$$m_1 \textcircled{\cap} m_2(A) = \sum_{B,C \subseteq \Theta : B \cap C = A} m_1(B) \cdot m_2(C) \tag{5}$$

On the other hand, to combine two masses m_1 and m_2 defined on two disjoint frames Θ and Ω, the vacuous extension is applied. For that, the *bba*'s have to be extended to the product space $\Theta \times \Omega$. The vacuous extension denoted by $\uparrow$ is defined by:

$$m^{\Theta \uparrow \Theta \times \Omega}(C) = \begin{cases} m^{\Theta}(A) & \text{if } C = A \times \Omega, A \subseteq \Theta, C \subseteq \Theta \times \Omega \\ 0, & \text{otherwise} \end{cases} \tag{6}$$

When combining evidence on Θ, it is important to take reliability of the sources into account. For that, a discounting operation can be applied [21]:

$$\begin{cases} {}^{\alpha}m(A) = (1-\alpha) \cdot m(A), \forall A \subset \Theta \\ {}^{\alpha}m(\Theta) = \alpha + (1-\alpha) \cdot m(\Theta) \end{cases} \tag{7}$$

where $\alpha \in [0, 1]$ is the discount rate.

In order to define the relation between two different frames of discernment Θ and Ω, one may use the multi-valued mapping [6]. In fact, a multi-valued mapping operation denoted by τ, joins the subsets $X_i \subseteq \Omega$ that can possibly correspond to $A_i \subseteq \Theta$:

$$m_{\tau}(A_i) = \sum_{\tau(X_i) = A_i} m(X_i) \tag{8}$$

The pignistic probability measure denoted by $BetP$ is usually used to make decisions under the belief function framework [24]:

$$BetP(A) = \sum_{B \subseteq \Theta} \frac{|A \cap B|}{|B|} \frac{m(B)}{(1 - m(\emptyset))}, \text{for all } A \in \Theta \tag{9}$$

4 Evidential Social Network

A graph $G = (V, E)$ is the most commonly used representation of social networks where V is the set of nodes representing the actors and E is the set of social links. Yet, binary relationships do not express uncertainty resulting from imperfect data and unreliability of the tools used when constructing the network.

Accordingly, we encapsulate the uncertainty degrees on the edges level using the belief function theory [17,18]. In fact, each edge uv is weighted by a *bba* denoted by m^{uv} defined on $\Theta^{uv} = \{E_{uv}, \neg E_{uv}\}$, where E_{uv} expresses the event exists and $\neg E_{uv}$ depicts the absence of the link. That is to say, m^{uv} encodes the degree of uncertainty regarding the existence of uv. In other terms, instead of assigning weights that can be either 1 or 0 to describe whether or not a link exists, a mass distribution with values in $[0, 1]$ is ascribed. It is important to notice that links uv with pignistic probability $BetP^{uv}(E_{uv}) < 0.5$ are considered not existing. In other words, the likelihood that uv exists is less than 50 %. It is therefore not schematized on the graph.

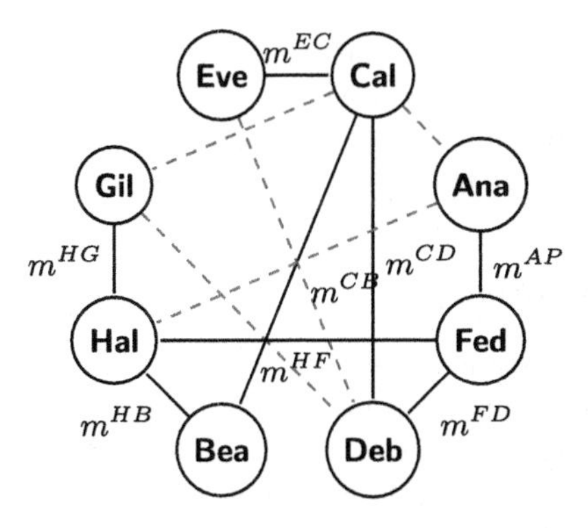

Fig. 1. An evidential social network with missing links and *bba*'s weighted edges

An example of such a graph structure is given in Fig. 1, where links are weighted by *bba*'s. The dashed links are the missing ones, they can match to a previous state of the graph or they are unobservable from the data due to noise. Assume that the graph in Fig. 1 is a social network of friendships, the nodes represent users and the links describe friend relationships. We can imagine the setting where the user "Gil" decided to hide his list of friends between time t and t', however, one of his friends, here "Hal", is showing his list of friends. Hence, we will have missing links i.e., *Gil Deb* and *Gil Cal* from the social network data at time t'. Since the hidden links do not have assigned masses, our task is to uncover them in order to decide whether the connections actually exist or not.

5 Evidential Missing Link Prediction

To properly deal with uncertainty, it is not enough to just handle links with mass functions attached, we have also to define a proper method for how to take the uncertainties into consideration when applying the link prediction task. As a matter of fact, as discussed in [25], sampling techniques and simulation-based approaches are promising methods to model and analyze social networks with uncertain data. Actually, sampling mechanisms are frequently applied when dealing with missing or partially observed data [4,13]. One of the reasons is that processing missing data and treatment of sampled data are much alike since what is not sampled can be considered as unobserved. In particular, various effective link tracing and link mining techniques use network sampling [7,11]. The authors in [10] discussed the connection between sampled and missing data in social networks.

To this end, our proposed method draws on n independent random samples of the social network graph generated from the data. We assume that the links have a priori *bba*'s, the task is to determine the missing ones. For a link to predict, a distance based on local information measures is computed with respect to the links in each graph G_i. The most similar link is considered as the most reliable source of evidence and the information is transferred to the frame of discernment of the analyzed. Finally, evidence gathered from all the graphs is combined to get an overall picture and make decision about the link existence. To this end, we propose a method fulfilling the task of inferring a missing link between a pair of nodes (u, v) based on the five steps presented below.

5.1 Similarity Measurement

At first, the Euclidean distance $D(uv, xy)$ between the link uv and each link xy included in each graph G_i is computed. We use structural similarity measures based on local information as features. CN (Eq. 1), JC (Eq. 2) and AA (Eq. 3) are employed since they are simple and they have proved their effectiveness in many scenarios [15,20,26]. The most similar link that has the smallest distance is considered. We divide the distance metric by its maximum value to get values in $[0, 1]$. It is computed as follows:

$$D(uv, xy) = \frac{\sqrt{\sum_{s=1}^{n}(sim_{uv}^{s} - sim_{xy}^{s})^2}}{D_{max}} \tag{10}$$

where s is the index of a structural similarity metric, sim_{uv} and sim_{xy} are respectively its values for uv and xy and D_{max} is the maximum value of the Euclidean distance.

5.2 Reliability Computation

Upon determining the most similar link, we quantify its degree of reliability using a discounting operation (Eq. 7). The value given by the distance measure is considered as a discount coefficient denoted by $\alpha = D(uv, xy)$. In fact, the more similar the two links are, the more reliable the similar link is, i.e., when the two links are totally similar $D(uv, xy)$ is equal to 0 thus xy is a totally reliable source of evidence ($\alpha = 0$). Hence, m^{xy} is discounted as follows:

$$\begin{cases} {}^{\alpha}m^{xy}(\{E_{xy}\}) = (1-\alpha) \cdot m^{xy}(\{E_{xy}\}) \\ {}^{\alpha}m^{xy}(\{\neg E_{xy}\}) = (1-\alpha) \cdot m^{xy}(\{\neg E_{xy}\}) \\ {}^{\alpha}m^{xy}(\Theta^{xy}) = \alpha + (1-\alpha) \cdot m^{xy}(\Theta^{xy}) \end{cases} \tag{11}$$

One should notice that when there are many similar links, i.e., equal smallest distances, the link with the highest mass on the event "exist" is chosen since the degree of certainty of its existence would be higher.

5.3 Information Mapping

The discounted *bba* of the most similar link xy defined on the frame of discernment Θ^{xy} has to be transfered to the frame Θ^{uv} of the link to predict. For that, a multi-valued mapping operation (Eq. 8) denoted by τ: $\Theta^{xy} \rightarrow 2^{\Theta^{uv}}$ is applied to match the elements as follows:

- The discounted mass ${}^{\alpha}m^{xy}(\{E_{xy}\})$ is transferred to $m^{uv}_{G_i}(\{E_{uv}\})$;
- The discounted mass ${}^{\alpha}m^{xy}(\{\neg E_{xy}\})$ is transferred to $m^{uv}_{G_i}(\{\neg E_{uv}\})$;
- The discounted mass ${}^{\alpha}m^{\Theta^{xy}}(\Theta^{xy})$ is transferred to $m^{uv}_{G_i}(\Theta^{uv})$.

Where $\alpha = D(uv, xy)$ and $m^{uv}_{G_i}$ denotes the *bba* of uv on Θ^{uv} given the most similar link, here xy in the graph G_i.

5.4 Global Fusion

Upon gathering information from the n sample graphs, the overall evidence is fused to get the final basic belief assignment denoted by m^{uv}_f. The masses $m^{uv}_{G_i}$ obtained from the n graphs are combined using the conjunctive rule of combination such that:

$$m^{uv}_f = m^{uv}_{G_1} \textcircled{\cap} m^{uv}_{G_2} \textcircled{\cap} \dots \textcircled{\cap} m^{uv}_{G_n} \tag{12}$$

At this step, the graphs are treated as independent sources of evidence, the combined information obtained from each most similar link in each graph is fused with the evidence collected from all the graphs.

5.5 Decision Process

At the final step, we make decision about whether or not the link is missing (existent). Fot that, we compute the pignistic probability $BetP^{uv}(E_{uv})$ (Eq. 9). Actually, if $BetP^{uv}(E_{uv}) > 0.5$ then the likelihood that a link between u and v exists has probability greater than 50 %, it would not be considered missing otherwise.

6 Experiments

In our experiments for testing the proposed evidential missing links prediction method, we generated samples of a real social network component of 1500 nodes and 20 K edges of facebook friendships obtained from [19]. A simulation phase is subsequently applied in order to transform the samples into evidential graphs. Mass functions are simulated randomly and attached to the edges. The link prediction task is then applied. We compared the predicted missing links with the actual existing ones in the initial graph to test the quality of the results.

6.1 Pre-processing

In the first part of our experiments, we generated 13 samples of the social network graph. A fraction of the existing links is removed and a number of false edges that do not exist either in the sample graph or the original graph is added randomly. Hence, the removed links are the missing ones that we aim to predict when applying the prediction task. Table 1 reports the percentage of false links added to the samples graphs.

Table 1. The percentage of false links

Graphs	G_1,G_2,G_3,G_4	G_5,G_6,G_7	G_8,G_9,G_{10}	G_{11},G_{12},G_{13}
False links %	10	15	20	25

At a second step, for all our dataset, we simulate mass functions according to the links' existence in each graph in order to get uncertain versions of the samples. For that, *bba*'s on the links that exist in the original graph corresponding to a pignistic probability that is greater than 0.5 on the event "exist" are randomly created. In contrast, the *bba*'s on the new added links in each sample are generated such that the pignistic probability on the event "not exist" is greater or equals 50 %.

6.2 Results

To test our proposals, six experiments E_1, E_2, E_3, E_4, E_5 and E_6 are performed. In E_1, E_2, E_3 and E_4, the missing link prediction approach is applied to respectively three graph samples with the same percentage of false added links, (G_1,G_2,G_3), (G_5,G_6,G_7), (G_8,G_9,G_{10}) and (G_{11},G_{12},G_{13}). To analyze the effect of the number of considered graphs, we used in the fifth and sixth experiments respectively two and four graphs samples with the same number of added false edges, (G_1,G_2) and (G_1,G_2,G_3,G_4). The predicted links are subsequently compared with the original graph. The performance is evaluated using two ppopular measures: precision and recall. The precision represents the ratio of the number of relevant predicted existing links n_c to the number of analyzed links n. It is defined as follows:

$$precision = \frac{n_c}{n} \tag{13}$$

The recall catches the correctly predicted existing links n_c versus the correctly and falsely predicted existing ones n_{cf}. It is defined as follows:

$$recall = \frac{n_c}{n_{cf}} \tag{14}$$

In each experiment, 50 % of the analyzed links correspond to true missing links that exist in the original graph. The other 50 % are false links that do not exist in both the original and sample graphs. The precision and recall results obtained in the experiments are shown in Figs. 2 and 3.

As it can be seen in Figs. 2 and 3, the prediction quality in terms of precision gives values higher than 60 % reaching a maximum performance of 71 % in E_6. Besides, the recall measure reaches 61 % in E_1 which means that 61 % of relevant existing links are predicted by the approach. In other words, the method is able to predict 61 % of the actual missing links. It clearly sticks out from these results that our method is applicable on uncertain social networks generated from real world data. That is, validity of our proposals is experimentally showed.

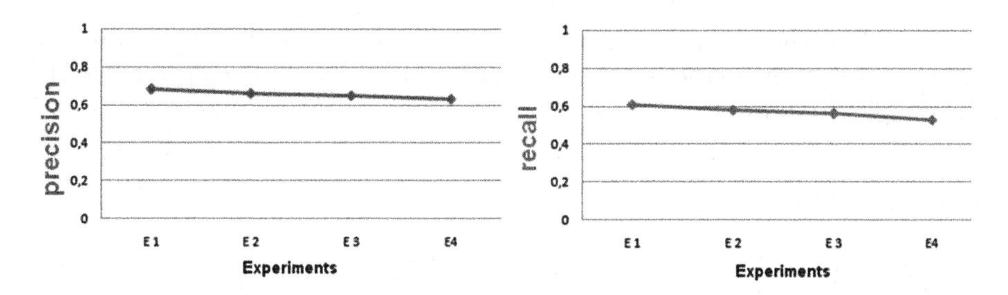

Fig. 2. Precision and recall values obtained in E_1, E_2, E_3 and E_4

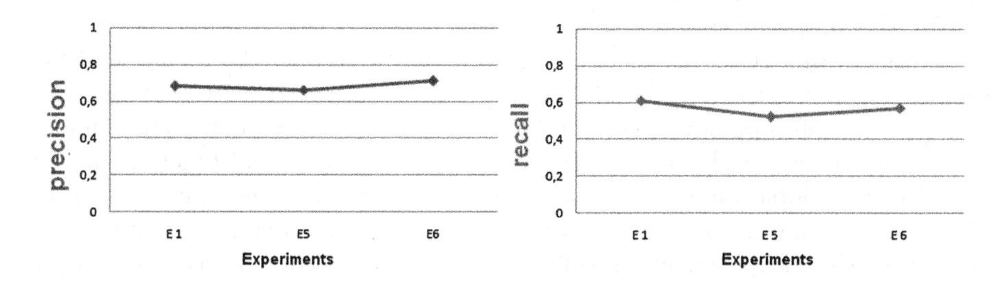

Fig. 3. Precision and recall values obtained in E_1, E_5 and E_6

In Fig. 2, we observe that prediction accuracy for the four experiments is above 60 % for precision and close to 60 % for recall. However, the precision decreases as the percentage of false edges increases, from 68 % in E_1 where the graphs have 10 % false edges to 62 % in E_4 where the graphs have 25 % false added edges. This can be due to the increase of the nodes' degrees. In fact, the proposed method is based on local information measures, these latter are sensitive to nodes neighborhoods. The more the nodes are connected and the more we get similar links when computing distances. The same applies to recall values, it decreases from 61 % in E_1 to 53 % in E_4. On the other hand, increasing the number of considered graphs enhance the prediction accuracy. As shown in Fig. 3, the precision and recall values increase respectively from 64 % and 52 % in E_5 (two considered graphs) to 71 % and 57 % in E_6 (four considered graphs). This can be related to the fact that further sources of evidence are considered and new information about the nodes becomes available. Accordingly, more evidence is investigated in the fusion procedure. We also note that in both Figs. 2 and 3, precision values are higher than the recall values which points out that the method predicts more incorrect missing links than incorrect non missing links. In other terms, the approach is omitting relevant missing links more than it is predicting false non existing ones. Although the results given by the recall measure are quite satisfactory i.e., 61 % in E_1, they are considerably smaller due to the large size of the dataset which challenges the algorithms.

Unfortunately, a comparative analysis cannot be accomplished at this point since, to the best of our knowledge, there is no existing approach that addresses missing link prediction under uncertainty. On the other hand, comparison with the state of the art methods is not engaging since they do not operate on the same graph structures.

7 Conclusion

Missing link prediction is a substantial problem in social networks as it helps analyze and understand social groups. It enables the implementation of efficient tools to discover hidden groups or to investigate missing members, etc. which are very crucial problems in security analysis and criminal investigation.

We have proposed a graph model for social networks that handles uncertainty degrees regarding the links existences using mass functions. A new method for the prediction of missing links have also been investigated. It uses local information of the graph topology to compute distances between the nodes. These information are subsequently transferred and fused using the belief function theory tools to get a global information and make decisions about the links' existence. Our proposals have been evaluated on a real world online social network of facebook friendship. Experimental results given by the precision and recall measures show that our method provides accurate prediction.

Interesting avenues for future research include prediction of jointly missing attributes of the nodes under uncertainty. In fact, several methods use additional information about the nodes and edges to predict future or missing links. Yet, these attributes are frequently missing from the data due to privacy or anonymization issues. Therefore, it would be interesting to study the problem of jointly missing links and attributes under an uncertain framework.

References

1. Adamic, L.A., Adar, E.: Friends and neighbors on the web. Soc. Netw. **25**(3), 211–230 (2003)
2. Adar, E., Ré, C.: Managing uncertainty in social networks. Data Eng. Bull. **30**(2), 23–31 (2007)
3. Ahmed, N.M., Chen, L.: An efficient algorithm for link prediction in temporal uncertain social networks. Inf. Sci. **331**, 120–136 (2016)
4. Clauset, A., Moore, C., Newman, M.E.J.: Hierarchical structure and the prediction of missing links in networks. Nature **453**, 98–101 (2008)
5. Dahlin, J., Svenson, P.: A method for community detection in uncertain networks. In: Proceedings of the 2011 European Intelligence and Security Informatics conference, pp. 155–162 (2011)
6. Dempster, A.P.: Upper and lower probabilities induced by a multivalued mapping. Ann. Math. Stat. **38**, 325–339 (1967)
7. Gile, K.J., Handcock, M.S.: Respondent-driven sampling: an assessment of current methodology. Sociol. Method. **40**(1), 285–327 (2010)
8. Gong, N.Z., Talwalkar, A., Mackey, L., Huang, L., Shin, E.C.R., Stefanov, E., Shi, E.R., Song, D.: Joint link prediction and attribute inference using a social-attribute network. ACM Trans. Intell. Syst. Technol. **5**(2), 27:1–27:20 (2014)

9. Gong, N.Z., Talwalkar, A., Mackey, L.W., Huang, L., Shin, E.C.R., Stefanov, E., Shi, E., Song, D.: Predicting links and inferring attributes using a social-attribute network (SAN). CoRR abs/1112.3265 (2011)
10. Handcock, M.S., Gile, K.: Modeling social networks with sampled data. Ann. Appl. Stat **4**(1), 5–25 (2010)
11. Heckathorn, D.D.: Comment: Snowball versus respondent-driven sampling. Sociol. Method. **41**(1), 355–366 (2011)
12. Jeh, G., Widom, J.: Simrank: a measure of structural-context similarity. In: Proceedings of the Eighth ACM SIGKDD International Conference on Knowledge Discovery and Data Mining, KDD 2002, pp. 538–543. ACM (2002)
13. Koskinen, J.H., Robins, G., Wang, P., Pattison, P.: Bayesian analysis for partially observed network data, missing ties, attributes and actors. Soc. Netw. **35**(4), 514–527 (2013)
14. Kossinets, G.: Effects of missing data in social networks. Soc. Netw. **28**, 247–268 (2003)
15. Liben-Nowell, D., Kleinberg, J.: The link prediction problem for social networks. J. Am. Soc. Inf. Sci. Technol. **58**(7), 1019–1031 (2007)
16. Lu, L., Zhou, T.: Link prediction in complex networks: a survey. Phys. A **390**(6), 1150–1170 (2011)
17. Mallek, S., Boukhris, I., Elouedi, Z., Lefevre, E.: Evidential link prediction based on group information. In: Prasath, R., Vuppala, A.K., Kathirvalavakumar, T. (eds.) MIKE 2015. LNCS, vol. 9468, pp. 482–492. Springer International Publishing, New York (2015)
18. Mallek, S., Boukhris, I., Elouedi, Z., Lefevre, E.: The link prediction problem under a belief function framework. In: Proceedings of the IEEE 27th International Conference on the Tools with Artificial Intelligence (ICTAI), pp. 1013–1020 (2015)
19. McAuley, J.J., Leskovec, J.: Learning to discover social circles in ego networks. In: Proceedings of the 26th Annual Conference on Neural Information Processing Systems 2012, pp. 548–556 (2012)
20. Newman, M.E.J.: Clustering and preferential attachment in growing networks. Phys. Rev. E **64**, 025102 (2001)
21. Shafer, G.R.: A Mathematical Theory of Evidence. Princeton University Press, Princeton (1976)
22. Smets, P.: Application of the transferable belief model to diagnostic problems. Int. J. Int. Syst. **13**(2–3), 127–157 (1998)
23. Smets, P.: Practical uses of belief functions. In: Proceedings of the Fifteenth Conference on Uncertainty in Artificial Intelligence, UAI 1999, pp. 612–621 (1999)
24. Smets, P., Kennes, R.: The transferable belief model. Artif. Intell. **66**(2), 191–234 (1994)
25. Svenson, P.: Social network analysis of uncertain networks. In: Proceedings of the 2nd Skövde workshop on information fusion topics (2008)
26. Zhou, T., Lü, L., Zhang, Y.: Predicting missing links via local information. Eu. Phys. J. B-Condens. Matter Complex Syst. **71**(4), 623–630 (2009)
27. Zhu, Y.X., Lü, L., Zhang, Q.M., Zhou, T.: Uncovering missing links with cold ends. Phys. A Stat. Mech. Appl. **391**(22), 5769–5778 (2012)

An Evidential Filter for Indoor Navigation of a Mobile Robot in Dynamic Environment

Quentin Labourey(✉), Olivier Aycard, Denis Pellerin, Michèle Rombaut, and Catherine Garbay

Univ. Grenoble Alpes, Grenoble, France
quentin.labourey@gmail.com

Abstract. Robots are destined to live with humans and perform tasks for them. In order to do that, an adapted representation of the world including human detection is required. Evidential grids enable the robot to handle partial information and ignorance, which can be useful in various situations. This paper deals with an audiovisual perception scheme of a robot in indoor environment (apartment, house..). As the robot moves, it must take into account its environment and the humans in presence. This article presents the key-stages of the multimodal fusion: an evidential grid is built from each modality using a modified Dempster combination, and a temporal fusion is made using an evidential filter based on an adapted version of the generalized bayesian theorem. This enables the robot to keep track of the state of its environment. A decision can then be made on the next move of the robot depending on the robot's mission and the extracted information. The system is tested on a simulated environment under realistic conditions.

Keywords: Active multimodal perception · Evidential filtering · Mobile robot

1 Introduction

Perceptually-driven robots have raised increased interest in different domains (autonomous vehicles [6], medecine, social robotics [3,9]). This work takes place in the context of companion robots, i.e. autonomous robot that monitor and estimate the needs of the persons in their environment and react accordingly by positioning themselves in a socially acceptable way. In order to do that, the humans in the scene must be detected and tracked in time, in an adequate world representation. In this paper, we consider the case of a small indoor mobile robot, equipped with video and depth (RGB-D) and audio sensors, that must keep track of humans in its surrounding in order to accomplish a mission. Each sensor is able to provide information on the possible presence of humans at different

Q. Labourey—This work has been partially supported by the LabEx PERSYVAL-Lab (ANR-11-LABX-0025-01) funded by the French program Investissement davenir.

J.P. Carvalho et al. (Eds.): IPMU 2016, Part I, CCIS 610, pp. 286–298, 2016.
DOI: 10.1007/978-3-319-40596-4_25

locations with respect to the robot, with a certain degree of uncertainty. The robot must then fuse at best the information at hand, while taking into account this uncertainty, in order to localize humans properly.

This requires the robot to be able to create and update a map of its environment and to localize itself in this map. The most common kind of map is the grid, where each cell of the grid represents a spatial portion of the world and contains information as to the content of this portion. Until recently, grids contained only information of occupancy to enable the robot to move around obstacles and detect moving objects [1,8].

In our case, we want the grid to contain more information in order to extract relevant information for the mission, i.e. human positions. Moreover uncertainty and doubt are crucial to the problem, as the perception algorithms and sensors are not perfect, and the spatial field of detection is only partial. A framework adapted to this representation of knowledge is the evidential framework [7]. Interesting works already exist on evidential grids [2,4], mainly in the field of autonomous cars: in both works, a fusion scheme is presented that takes sensor data (and a priori knowledge in the case of [2]) and fuse it into a time-evolutive model in order to produce an enhanced grid containing information relevant to vehicle navigation. However in both cases the only exteroceptive sensor considered is the LIDAR, which means that no multimodality is used. Moreover the perception is only used to increase the accuracy of navigation and does not take into account the mission of the robot.

We propose a fusion scheme based on extracted data from the sensors at each observation time. The information extracted by the sensors are directly integrated into evidential sensor grids that are fused together to produce a perception grid containing the information accross all sensors. To increase the robustness and add information about dynamicity, this grid is fused into an evolutive evidential model. This enables the robot to keep all the extracted data directly into an information grid from which decisions concerning its next motion can be taken depending on its mission.

2 World Model and Evidential Grids

An *occupancy grid* is a grid in which each cell contains information about the state of occupancy of this cell. The most common type of occupancy grid is an estimation for each cell of a probability distribution over the two states "occupied" or "free".

In our case, the robot has missions to fulfill in dynamic environment, in human presence (e.g. monitor the state of the persons in the environment). In order to accomplish its mission, the robot must navigate through its environment, i.e. know where it is allowed to go without collisions with its environment and what paths to consider to reach a goal, and at the same time extract information relevant to its mission, i.e. places where the humans might be present, places where objects might be, and the dynamics of the scene (mobility, staticity of its surrounding). The robot must also keep in memory the places where it does not know the content of the scene, as they might become the next important locations to visit.

We are thus trying to estimate the state of the spatial surrounding of the robot, in the state space $\Omega = \{H, O, F\}$, where:

- H: Presence of a human, mobile or static
- O: Presence of a non-human object, mobile or static
- F: Free navigable space

As the perceptual capabilities of the robot are not perfect (faulty sensors or perception algorithms), a deterministic approach can present a quantity of drawbacks. To have all the information relevant to navigation and the mission of the robot in one unified stochastic representation, we propose a fusion method based on evidential grids [2].

An *evidential* grid is a grid in which each cell contains information about the state of a portion of the robot's environment in the form of a belief function.

The outline of the method is presented in Fig. 1:

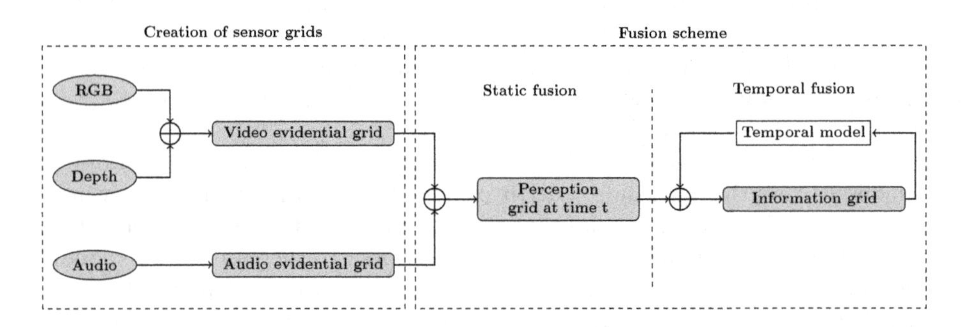

Fig. 1. Global fusion scheme

At each time step, information is extracted from sensors, and sensor grids are created accordingly. The creation of such grids is detailed in Sect. 3. The sensor grids are then fused together in order to obtain a perception grid. This perception grid is then fused in an evolutive fusion model, in order to extract information about the dynamics of the scene. The creation of the perception grid and the temporal fusion are detailed in Sect. 4.

Throughout the article the following notations are used: m^{Ω}, q^{Ω}, and bel^{Ω} represent respectively a basic belief assignment (bba), commonality function, and belief function over discernment space Ω.

$m^{\Omega}[S]$,$q^{\Omega}[S]$, and $bel^{\Omega}[S]$ represent respectively a conditional basic belief assignment, commonality function, and belief function over discernment space Ω, which represents the belief repartition knowing hypothesis H to be true.

3 Robot Perceptual System and Sensor Grids

3.1 Perceptual Capabilities of the Robot

It is assumed that an occupancy grid of the environment has been built in a previous patrol of the robot, thanks to a Simultaneous Localization and Mapping (SLAM) algorithm, and contain information about purely static objets (wall, furniture). This is not a necessity but enables us not to consider the loop-closing problem in mapping, and gives a predefined configuration of the environment of the robot (a static map of the residential place of the robot).

The field of perception for each sensor is different, as shown in Fig. 2. The field of vision, represented by the purple cone in Fig. 2(a) is narrower than the field of audition, represented by the purple zone in Fig. 2(b), which means that a person can be heard without being part of the field of vision.

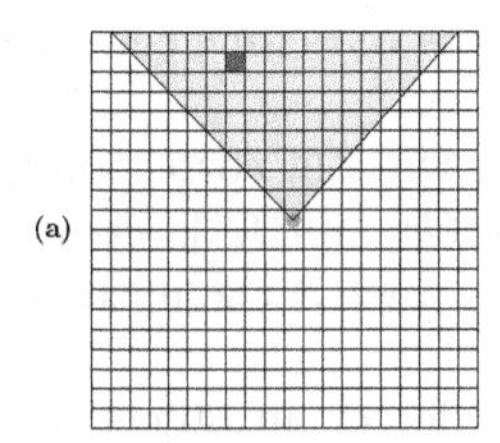

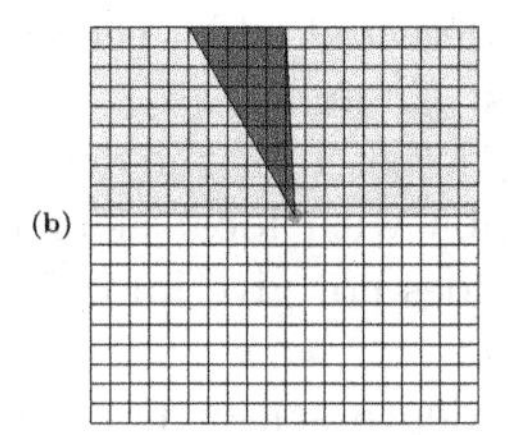

Fig. 2. Perception characteristics: In both maps the robot is represented by a green point and is assumed to be facing north. In (a), the field of vision is represented by the purple zone. When a human or an object detection happens (in red), it can directly be put at its precise location inside the grid. In (b), the field of audition is represented by the purple zone. A detection (in red) is less precise. The source can be anywhere in the red zone. (Color figure online)

We assume that human detection is performed on each frame of the video (e.g. face or skeleton detection). On each frame, if a human is detected, its position in the image is known. A video observation cannot directly be placed in a grid, as the depth information is missing. In order to be able to localize video observation inside sensor grids, a registration is performed between the information provided by the RGB sensor and the depth information given by the depth sensor. This registration enables the linking between each pixel of the video and corresponding cells inside the grid (i.e. correspondance between the image frame of reference and the world frame of reference) Fig. 2(a). In the same way, we assume that a human sound-source localization is performed on the audio in synchronization with the video (typicall thanks to a generalized cross-correlation method), so that each audio observation is made at the same time as the video observation. As the robot only has two microphones, the localization cannot yield a precise position of the source but a cone in which the source is likely to be, as shown in red in Fig. 2(b).

3.2 Sensor Grids

An evidential grid is created for each sensor. The aim of evidential sensor grids is to fill each cell of the perceived spatial zone with a belief function on discernment space $\Omega = \{H, O, F\}$, described in Sect. 2, matching the information extracted from the sensor for this particular cell.

Video Evidential Grid: The video itself can only deliver information on the possible presence of a human in the scene, while the depth sensor gives information on the presence of objects (human or not) and free space. This leaves four possibilities for a given cell:

- If a human is detected a bba is created in the matching cell with the following focal elements: H[1] and Ω
- If an object is detected by the depth sensor without corresponding to a human detection, a bba with focal elements $\{H, O\}$ and Ω is created. Indeed, the depth sensor does not give information as to what the nature of the detected object is.
- If neither the depth sensor nor the video detect an object, a bba with focal elements F and Ω is created.
- If a spatial zone is not part of the field of vision of the robot or is occluded because of an object, a vacuous belief function is created for each cell of this zone.

The way the belief are attributed reflects the trust in the detection algorithm.

Audio Evidential Grid: As shown in Fig. 2(b), sound-source localization with two microphones is not spatially precise. The sound source can be anywhere in the cone. Moreover, the sound can only give out information on the possible presence of a human in a given cell (not on the fact that a given cell is free space). Thus for the audio perception map, two cases are possible for a given cell:

- If a detection happens and the considered cell is part of the audio cone, a bba with focal elements H and Ω is created.
- If the considered cell is not part of the audio cone or if no detection is made (silence), a vacuous bba is created for the cell.

In case of a detection, all the cells in the spatial zone of detection receive the same bba. The way the beliefs are attributed reflects the trust in the detection algorithm.

[1] In order to simplify the notations, singletons of the discernment frame are noted without brackets, hence H instead of {H}.

4 Fusion Scheme

4.1 Perception Grid

In order to obtain a unified perception map accross all sensors, the audio and video perception map need to be fused at each observation time. The grids are fused cell by cell, thanks to an unnormalized conjunctive combination. However, the detections do not have the same spatial uncertainty, depending on the modality, which means that conflict can appear during fusion. Knowing the possible focal elements of the belief functions, only one combination can create conflict, as shown in Table 1: it is the case where a cell is part of the audio cone, which mean that there is a possible presence of a human in this cell, whereas the vision sensors detect the cell as being free.

Table 1. Intersections between possible focal elements of the audio belief function m_a^Ω and the video belief function m_v^Ω. The states are: F for free space, H for human presence, and O for non-human object

m_a^Ω \ m_v^Ω	H	$\{H, O\}$	F	Ω
H	H	H	$\emptyset$	H
Ω	H	$\{H, O\}$	F	Ω

In that case we chose to assign the conflict to hypothesis F, which means trusting the vision over the audio detection, as the vision is spatially more precise. The fused basic belief assignment m_P^Ω is then obtained through this modified fusion rule:

$$m_P^\Omega(S) = \left(m_a^\Omega \bigcirc\!\!\!\!\cap\, m_v^\Omega\right)(S), \quad \forall S \subseteq \Omega, S \neq F \tag{1}$$

$$m_P^\Omega(F) = \left(m_a^\Omega \bigcirc\!\!\!\!\cap\, m_v^\Omega\right)(F) + (m_a^\Omega \bigcirc\!\!\!\!\cap\, m_v^\Omega)(\emptyset) \tag{2}$$

After this fusion, we obtain the perception grid, representing the information from all sensors, in which each cell contains a basic belief assignement m_P^Ω, with four possible focal elements: $H, \{H, O\}, F, \{\Omega\}$.

4.2 Temporal Fusion

A type of information that can be useful to a companion robot is to differentiate between mobile and static objects, through temporal integration. In order to be able to differentiate between static and mobile objects, we introduce the frame of discernment $\Theta = \{H_s, H_m, O_s, O_m, F\}$ where the new states are:

- H_s: Presence of a static human, i.e. a human has been occupying this cell for an arbitrary period of time
- H_m: Presence of a mobile human, i.e. a human has filled the cell recently

- O_s: Presence of a static non-human object, i.e. an object has been occupying this cell for an arbitrary period of time
- O_m: Presence of a mobile non-human object, i.e. an object has filled the cell recently
- F: free space

To be able to work in this discernment space we define the following refining:

$$\begin{array}{rcl} \gamma: \Omega & \rightarrow & 2^{\Theta} \\ \{H\} & \mapsto & \{H_s, H_m\} \\ \{O\} & \mapsto & \{O_s, O_m\} \\ \{F\} & \mapsto & \{F\} \end{array} \tag{3}$$

$$\begin{array}{rcl} 2^{\Omega} & \rightarrow & 2^{\Theta} \\ m_P^{\Omega}(B) & \mapsto & m_P^{\Theta}(\gamma(B)), \quad \forall B \subseteq \Omega \end{array} \tag{4}$$

This refining is applied to each cell of the perception grid, which enables us to obtain a grid with discernment frame Θ. The sensor grid cannot add belief on dynamic hypotheses as sensors are not able to detect motion in the scene. This can be done by detecting changes in the scene between the time t-1 and time t. In order to do that, we propose the use of a credibilist hidden markov model [5].

Credibilist Hidden Markov Model: A credibilist HMM estimates a belief function at time t from the estimate of this belief function at t-1 and observations at time t.
To be able to perform the HMM, two elements are needed:

- $q_t^{\Theta}[S_i^{t-1}](S_j^t)$ $\forall S_j, S_i \subseteq \Theta$, is the conditional commonality of transition from credibilist state S_i at t-1 to the state S_j at time t. Thus the matrix q_t^{Θ} is a matrix of size $2^{|\Theta|} * 2^{|\Theta|}$
- $q_b^{\Theta_t}(P)$ $\forall P \subseteq \Theta$ which represents the commonality repartition on subsets of Θ given by observations at time t

With those elements, the estimation is a two-step process. First the commonality function at time t $\widehat{q_{\alpha}^{\Theta_t}}$ is predicted:

$$\widehat{q_{\alpha}^{\Theta_t}}(S_j^t) = \sum_{S_i^{t-1} \subseteq \Theta_{t-1}} m_{\alpha}^{\Theta_{t-1}}(S_i^{t-1}) \cdot q_t^{\Theta}[S_i^{t-1}](S_j^t), \quad \forall S_j^t \subseteq \Theta_t \tag{5}$$

where $m_{\alpha}^{\Theta_{t-1}}$ is the estimated basic belief assignment at time t-1.

The estimated commonality is then corrected with the observations at time t, by conjunctive combination:

$$q_{\alpha}^{\Theta_t}(S_j^t) = \widehat{q_{\alpha}^{\Theta_t}}(S_j^t) \cdot q_b^{\Theta}(S_j^t), \quad \forall S_j^t \subseteq \Theta_t \tag{6}$$

Prediction Step: In our case, no model of evolution of a cell is available, as predicting the way each cell of the scene is going to change is complicated. That is why we propose to perform a temporal discounting on each cell of the estimated grid at t-1 as a prediction. This enables the model to slowly decrease towards ignorance in case no further observation is made on the cell:

$$\widehat{m}_\alpha^{\Theta_t}(S_i) = m_\alpha^{\Theta_{t-1}}(S_i) \cdot e^{-\gamma} \quad \forall S_i \subseteq \Theta_{t-1} \tag{7}$$

$$\widehat{m}_\alpha^{\Theta_t}(\Theta) = 1 - \sum_{S_i \subseteq \Theta_{t-1}} \widehat{m}_\alpha^{\Theta_t}(S_i) \tag{8}$$

where γ is the decrease rate toward ignorance. This is equivalent to defining:

$$m_t^{\Theta}[S_i^{t-1}](S_j^t) = 0, \quad \forall i \neq j, S_j^t \neq \Theta \tag{9}$$

$$m_t^{\Theta}[S_i^{t-1}](S_i^t) = e^{-\gamma}, \quad S_i \neq \Theta \tag{10}$$

$$m_t^{\Theta}[S_i^{t-1}](\Theta) = 1 - e^{-\gamma}, \quad \forall S_i^{t-1} \neq \Theta \tag{11}$$

$$m_t^{\Theta}[\Theta](\Theta) = 1 \tag{12}$$

Correction Step: For each cell, the correction step is supposed to be a conjunctive combination between $\widehat{m}_\alpha^{\Theta_t}$ and $m_P^{\Theta_t}$. However, each time the fusion is performed, some conflict will appear if the cell's true state changes between t-1 and t, as the model does not predict any change. Information must be extracted from the conflict. Once again the only case where conflict appears is when a cell changes state from occupied to free, or conversely when it becomes occupied. In the former case, the only viable option is to trust the sensors and transfer the conflict on the hypothesis F (trusting the prediction step would be dangerous as it does not account for any possible state change in the cell). However, in the latter case, the cell filling with an object means that a mobile object just entered the cell, and we propose to transfer the conflict on the underlying mobile hypothesis of $m_P^{\Theta_t}$. This corresponds to the following rule of fusion:

$$m_\alpha^{\Theta_t}(A) = \left(\widehat{m}_\alpha^{\Theta_t} \bigcirc m_P^{\Theta_t}\right)(A) + \sum_{\substack{A=(B\cap\{H_m,O_m\}) \\ B\subseteq\Theta\backslash F}} \left(\widehat{m}_\alpha^{\Theta_t}(F) \cdot m_P^{\Theta_t}(B)\right), \quad \forall A \subseteq \{H_m, O_m\} \tag{13}$$

$$m_\alpha^{\Theta_t}(F) = \left(\widehat{m}_\alpha^{\Theta_t} \bigcirc m_P^{\Theta_t}\right)(F) + \sum_{A\subseteq\Theta\backslash F} \left(\widehat{m}_\alpha^{\Theta_t}(A) \cdot m_P^{\Theta_t}(F)\right) \tag{14}$$

$$m_\alpha^{\Theta_t}(A) = \left(\widehat{m}_\alpha^{\Theta_t} \bigcirc m_P^{\Theta_t}\right)(A), \quad \forall A \;\; else \tag{15}$$

Staticity Check: At that point, the system never adds belief on static hypotheses H_s and $\{H_s, O_s k\}$. It can also artificially increase the belief on mobile hypotheses in the case where a previously free cell is filled, e.g. by a human, and the object that fills it stops: in this case, as the sensor continues to detect the human (focal element $\{H_s, H_m\}$), when the temporal fusion happens, the mass will be transferred on H_m for as long as the person stays in the cell, because of the conjunctive combination. To answer this problem, we propose a staticity check, which is done after each temporal fusion. This test enables the transfer of masses on static hypotheses in case a cell has been occupied for a predetermined time. A cell is considered occupied if the belief in the hypothesis $\Theta \backslash F$ is larger than a predefined number, i.e. if the percentage of belief in occupation is higher than an arbitrary threshold.

The occupation is defined as follow:

$$Occ = bel(\{H_m, H_s, O_m, O_s\}) \tag{16}$$

if Occ is larger than a predefined δ for a time longer than Δt then all masses on mobile hypotheses are transferred on static hypotheses i.e.:

$$m(\{H_m\}) \mapsto m(\{H_s\}) \tag{17}$$

$$m(\{H_m, O_m\}) \mapsto m(\{H_s, O_s\}) \tag{18}$$

$$m(\{H_m, O_m, H_s, O_s\}) \mapsto m(\{H_s, O_s\}) \tag{19}$$

By doing this, if a mobile object suddenly stops, and stays detected, the belief will be naturally transferred on static hypotheses.

5 Results

5.1 Simulation Conditions

In the simulation, an occupancy grid is available, showing the true state of the scene, of size 50×50 cells. With cells of size 20×20 cm, this represents a place of $100\,\text{m}^2$. In order to properly illustrate the fusion scheme without overcomplicating the analysis, we propose a simplified environment for the robot, shown in Fig. 3. The environment is composed of free space and an obstacle, e.g. a wall.

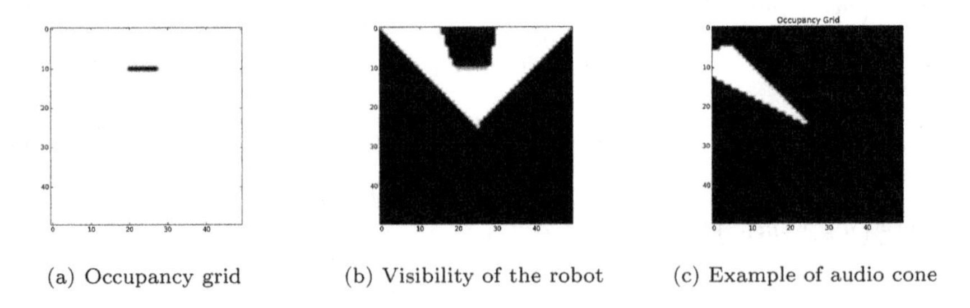

(a) Occupancy grid (b) Visibility of the robot (c) Example of audio cone

Fig. 3. Occupancy grid and perceptual capabilities of the robot. (a) shows the occupancy grid: white cells represent free space and black cells represent obstacles. (b) shows the visual perception space of the robot. In this case the robot is placed in the center of the grid, facing north. White cells represent places the robot can see, black cells that are not visible. (c) shows an audio detection by the robot (white cells).

The robot occupies one cell at a time. This is not restrictive as the robot is not part of its own perception. We assume that humans of the scene also occupy one cell at a time. This might cause some problems locally as humans are never truly motionless, but as it depends of the spatial resolution of the grids, we consider it a sufficient first approximation.

As for the perceptual capabilities of the robot, they are shown in Fig. 3b and c. We used a cone of visibility of angular size 45 degrees centered on the robot, and an audio cone of angular size 20 degrees centered on the observation to account for audio localization uncertainty. We assume that the robot is able to localize itself properly in the map, considering the performances of available state-of-the-art localization algorithms [8].

5.2 Scenario

In the scenario described in this section, the robot does not move. This is not restrictive as we assume that the robot localizes itself properly, as the state of each cell is estimated independently of the state of its neighbourhood. We consider a scenario containing one moving and talking person, e.g. on the phone. The person crosses the field of vision of the robot. His voice may continuously be heard. The steps are the following ones:

1. When the robot starts perceiving, the human is outside its field of vision, but inside its audio cone detection, talking
2. The human starts to move and enters the field of vision of the robot
3. The human continues walking in front of the robot and becomes silent
4. The human stops for a while and stand without moving, e.g. listening

The parameters used are:

- $\gamma = 0.4$: we use a rather slow decrease rate, in order to see the influence of temporal discounting
- $\Delta t = 6$: after 6 consecutive observations of occupancy, the staticity check will perform on the cell. This leaves a good amount of time before adding belief to staticity
- $\delta = 0.5$: if there is more belief that support the occupied state of the cell than the free state, we consider the cell to be occupied

In order to present the information in a compact way, only the subset of Θ with the highest assigned belief is shown for each cell. The three first step of the scenario can be observed in Fig. 4.

At t = 0 (Fig. 4a and d), the robot begins to perceive its environment. As there is no previous fused map, the sensor grid is the final map. The human is not inside the field of vision, and the audio cone is filled with belief functions on $\{H_s, H_m\}$. As expected, the obstacle in the field of vision receives belief functions with belief on $\{H_s, H_m, O_s, O_m\}$. The rest of the space is accurately separated between free space in the rest of the field of vision and ignorance on non-perceived zones.

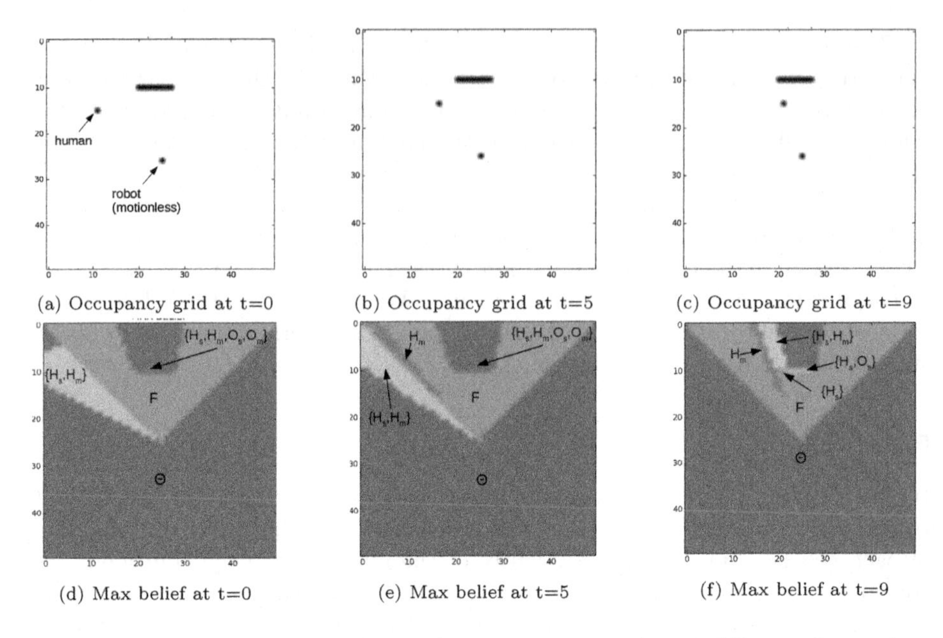

(a) Occupancy grid at t=0 (b) Occupancy grid at t=5 (c) Occupancy grid at t=9

(d) Max belief at t=0 (e) Max belief at t=5 (f) Max belief at t=9

Fig. 4. Scenario realization: on the top line, occupancy grids at 3 different observation time, and on the bottom line, corresponding max belief grid after fusion. On each zone, the subset of Θ with the maximum belief assigned is indicated.

At t = 5 (Fig. 4b and e), the human has entered the cone of vision, and is still talking. The temporal discounting leaves a part of previous audio cones on the left (hypothesis $\{H_s, H_m\}$). As the human is walking, both the audio cone and the video detection are changing at each observation, filling cells that were previously considered free, thus the presence of belief functions with maximal belief on H_m. As expected, the case of conflict between video putting belief on F and audio putting belief on $\{H_s, H_m\}$ is solved by trusting the video, hence the fact that the audio cone is not appearing once the human enter the field of vision.

At t = 9 (Fig. 4c and f), the human continues walking and talking and is now walking in front of the obstacle. As before, in the visually perceived zone, the hypothesis H_m is dominant in cells that are becoming occupied from previously free. This time the audio cone has two different kinds of impact: on the zone behind the obstacle, it is fused with ignorance, hence the dominance of hypothesis $\{H_s, H_m\}$, which seems fair, as the sound might in fact come from behind the obstacle with a non-visible person. The cone also has an influence on the obstacle itself, as the part of the obstacle that is inside the audio cone will receive a belief function on $\{H_s, H_m\}$ (due to the fact the audio is increasing the belief in the presence of a human and the video cannot contredict the information). However as those cells have been occupied for longer than 6 consecutive observations, the belief is transferred on $\{H_s\}$. This can be an advantage, as there are a lot of common obstacles on which a human can be (tables, chair, sofas, etc.).

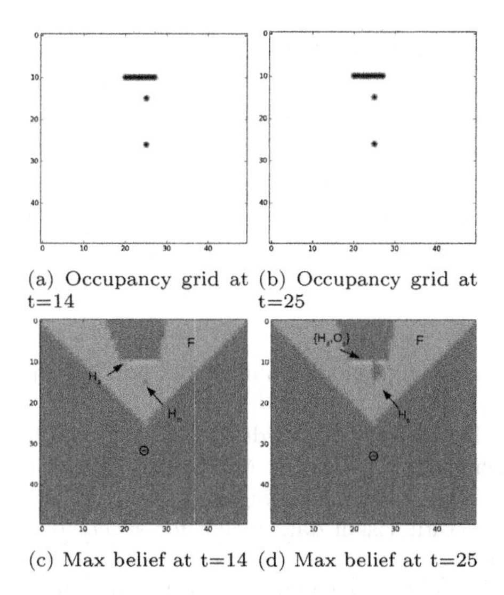

(a) Occupancy grid at t=14 (b) Occupancy grid at t=25

(c) Max belief at t=14 (d) Max belief at t=25

Fig. 5. End of the scenario

However, to obtain a more accurate state estimate, a good idea would be to include information from the occupancy grid (e.g. a cell containing a wall cannot receive belief on H). The staticity check also happens on the rest of the obstacle, which transfers the belief from $\{H_s, O_s, H_m, O_m\}$ to $\{H_s, O_s\}$.

The rest of the scenario can be seen in Fig. 5.

At t = 14, the person has stopped talking and stops moving. Not enough time has passed for the staticity check to enter into effect. Moreover the temporal discounting, linked with the video observation of the obstacle has not allowed yet to go back to the dominance of $\{H, O\}$ on the totality of the obstacle. Finally, at t = 25, the staticity has had effect on the human detection and the cell containing the person is accurately dominated by a belief on H_s. The obstacle is detected as such again, and the temporal discounting created a zone of ignorance behind the human. The filter describes the environment of the robot accurately.

6 Conclusion and Perspectives

This paper presented an evidential filter for indoor navigation of a mobile robot in dynamic environment. The filter is based on two main steps: the creation of evidential grids to represent sensor data, and the fusion scheme divided into two parts. First the sensor data is fused accross all sensors and then the information is integrated inside an evolutive model. The filter describes the environment of the robot accurately, based on information extracted from conflict between the sensors and the evolutive model.

The computational complexity aspect was not explored in this paper and can be problematic as the fusion process is performed at each time step on each step, however methods to increase efficiency in belief functions exist in litterature. One interesting addition could be the use of neighbourhood to estimate the state of a cell: this would enable the inclusion of spatial information in the model. Moreover it would help face the possible loss of detection from sensors.

References

1. Elfes, A.: Using occupancy grids for mobile robot perception and navigation. Computer **22**(6), 46–57 (1989)
2. Kurdej, M.: Map-aided fusion using evidential grids for mobileperception in urban environment. In: Denoeux, T., Masson, M.-H. (eds.) Belief Functions: Theory and Appl. AISC, vol. 164, pp. 343–350. Springer, Heidelberg (2012)
3. Labourey, Q., et al.: Audiovisual data fusion for successive speakers tracking. In: VISIGRApp (2014)
4. Moras, J., et al.: Credibilist occupancy grids for vehicle perception in dynamic environments. In: ICRA (2011)
5. Ramasso, E.: State sequence recognition based on the Transferable BeliefModel. Theses, Université Joseph-Fourier - Grenoble I (2007)
6. Siagian, C., et al.: Autonomous mobile robot localization and navigation using a hierarchical map representation primarily guided by vision. J. Field Robot. **31**, 408–440 (2014)
7. Smets, P.: Belief functions: The disjunctive rule of combination and the generalized bayesian theorem. In: Yager, R.R., Liu, L. (eds.) Classic Works of the Dempster-Shafer Theory of Belief Functions. STUDFUZZ, vol. 219, pp. 633–664. Springer, Heidelberg (2008)
8. Thrun, S., et al.: Probabilistic Robotics. The MIT Press, Cambridge (2005)
9. Zaraki, A., et al.: Designing and evaluating a social gaze-control system for a humanoid robot. Trans. Human-Mach. Syst. **44**(2), 157–168 (2014)

A Solution for the Learning Problem in Evidential (Partially) Hidden Markov Models Based on Conditional Belief Functions and EM

Emmanuel Ramasso(✉)

Department of Applied Mechanics, and Department of Automatic Control and Micro-Mechatronic Systems, FEMTO-ST Institute, UMR CNRS 6174 - UBFC/ENSMM/UTBM, 25000 Besançon, France
emmanuel.ramasso@femto-st.fr

Abstract. Evidential Hidden Markov Models (EvHMM) is a particular Evidential Temporal Graphical Model that aims at statistically representing the kynetics of a system by means of an Evidential Markov Chain and an observation model. Observation models are made of mixture of densities to represent the inherent variability of sensor measurements, whereas uncertainty on the latent structure, that is generally only partially known due to lack of knowledge, is managed by Dempster-Shafer's theory of belief functions. This paper is dedicated to the presentation of an Expectation-Maximization procedure to learn parameters in EvHMM. Results demonstrate the high potential of this method illustrated on complex datasets originating from turbofan engines where the aim is to provide early warnings of malfunction and failure.

Keywords: Evidential Temporal Graphical Model · Evidential latent variable · Markov chain · Belief functions · Parameter learning

1 Introduction

The statistical representation of multi-dimensional time-series originating from a dynamical system consists in finding a concise and meaningful mathematical model that can be easily interpreted and used to undertand the behavior of the system. Those models can then be used to enhance data-driven phenomenological physics model with better prediction capabilities in in-service applications. However, sources of uncertainty are numerous in real-world applications which accounts for systems' oversizing to ensure people safety and equipments' availability. Uncertainty quantification thus plays a critical role during both systems' design (upstream) and in-service monitoring (downstream).

Dempster-Shafer's theory of belief functions is a mathematical framework that allows to represent, quantify and propagate uncertainties. Its application in mechanical engineering has however been limited due to a lack of tools to handle temporal data. The additional temporal dimension compared to static

J.P. Carvalho et al. (Eds.): IPMU 2016, Part I, CCIS 610, pp. 299–310, 2016.
DOI: 10.1007/978-3-319-40596-4_26

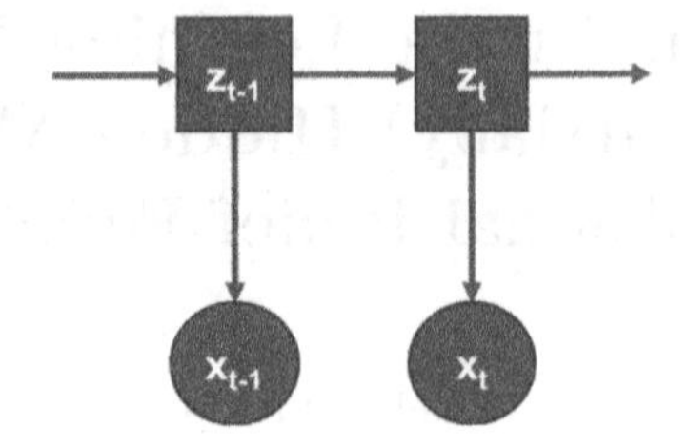

Fig. 1. Graphical representation of a hidden Markov model: $\mathbf{z}$ is discrete and hidden, $\mathbf{x}$ is continuous and observed, t is the time index.

data makes both the inference and learning problems more difficult than with probability theory.

This paper is focused on a simple – yet not tackled – problem that is the estimation of parameters in Hidden Markov Models when uncertainty is no more represented by probabilities but by belief functions. This model has been initially called Evidential Hidden Markov Models (EvHMM) and represents a particular statistical discrete-latent Markov model which is depicted in Fig. 1. This model assumes that the system is driven by a doubly stochastic process: A Markov chain on discrete hidden (not observed) variables called states, and an observation model that statistically represents the distribution of sensor measurements (recorded on the system) defined conditionally to the states.

Inference mechanisms in EvHMM has been proposed by the author in [16, 21] (not recalled here) to estimate the belief functions over hidden variables given both an observation model and data. Those mechanisms provide exact belief functions and enable one to compute the equivalent of a likelihood of a given model for some sequences of observations, as well as to estimate sequences of hidden states. Those procedures can then be used to explore relevant regions in the feature or parameter space.

Only the learning problem is considered subsequently, with some assumptions required to make the problem tractable [18]. Section 2 presents a criterion for learning parameters in evidential discrete-latent models, which is applied in Sect. 3 for EvHMM. Some results are presented in Sect. 4.

2 Parameter Learning in Evidential Discrete-Latent Models

2.1 The Criterion

The quality of a model such as depicted in Fig. 1 can be quantified by minimizing the *amount of conflict* between the model and the data. Time-series are denoted as $\mathbf{X} = [\mathbf{x}_1; \mathbf{x}_2; \dots \mathbf{x}_T]$ with length T in D dimensions with $\mathbf{x}_t = (x_1, \dots, x_D)'$ called feature vector. The latent states are represented by discrete random variables $z_1, z_2 \dots z_t$ taking values in a finite set $\Omega_z = \{s_1, s_2, \dots, s_K\}, s_i \cap s_j = \emptyset$.

Finding the parameters θ^* in a latent variable model, in which uncertainty is managed by belief functions, and made of one observed variable $\mathbf{x}_t$ and one discrete hidden variable $\mathbf{z}_t$ with $t = 1 \ldots T$, can be turned into the maximization of the potential support assigned to the subset $(\mathbf{x}_1, \Omega_z), (\mathbf{x}_2, \Omega_z) \ldots (\mathbf{x}_T, \Omega_z)$ after observing all data vectors:

$$\theta^* = \underset{\theta}{\text{argmax}}\ pl^{\mathbb{R}^T \times \Omega_z^T} \left((\mathbf{x}_1, \Omega_z), (\mathbf{x}_2, \Omega_z) \ldots (\mathbf{x}_T, \Omega_z) \mid \theta \right) \tag{1}$$

$\mathbb{R}$ is the domain of $\mathbf{x}_t$. For short, this criterion is rewritten as $\text{argmax}_{\theta}\, pl(\mathbf{X}, \boldsymbol{\Omega}_{\mathbf{z}} \mid \theta)$. By definition of the plausibility function, the criterion can be computed by summing belief masses assigned to all configurations of the hidden variables $\mathbf{S}$:

$$pl(\mathbf{X}, \boldsymbol{\Omega}_{\mathbf{z}} \mid \theta) = \sum_{\mathbf{S} \not\ni \emptyset} m(\mathbf{X}, \mathbf{S} \mid \theta) \tag{2}$$

Direct maximization of this criterion is untractable but by making use of the latent structure, it is possible to formulate the problem differently in order to use an EM procedure (Expectation-Maximization) [6]. In EM, the E-step can indeed be used to estimate the distribution over latent variables given both the data and the current values of the parameters $\theta^{(q)}$ (at iteration q) while the M-step can be used to find the parameters $\theta^{(q+1)}$ that allows to maximize an "auxiliary" function such that the criterion (that is not directly maximized) does not decrease. For that, we can rewrite the criterion as

$$pl(\mathbf{X}, \boldsymbol{\Omega}_{\mathbf{z}} \mid \theta) = \sum_{\mathbf{S} \not\ni \emptyset} R(\mathbf{S}) \frac{m(\mathbf{X}, \mathbf{S} \mid \theta)}{R(\mathbf{S})} \tag{3}$$

where $S \subseteq \Omega_z$ and R is a distribution such that $\sum_A R(A) = 1$ that allows Jensen's inequality to be applied [9, Eq. 5]:

$$\log pl(\mathbf{X}, \boldsymbol{\Omega}_{\mathbf{z}} \mid \theta) \geq \mathcal{Q}_{m,m}(\theta^{(q)}, \theta) - H_{m,m}(\theta^{(q)}, \theta^{(q)}) \tag{4a}$$

$$\mathcal{Q}_{m,m}(\theta^{(q)}, \theta) = \sum_{\mathbf{S} \not\ni \emptyset} R(\mathbf{S}, \theta^{(q)}) \log m(\mathbf{X}, \mathbf{S} \mid \theta) \tag{4b}$$

$$H_{m,m}(\theta^{(q)}, \theta^{(q)}) = \sum_{\mathbf{S} \not\ni \emptyset} R(\mathbf{S}, \theta^{(q)}) \log R(\mathbf{S}, \theta^{(q)}) \tag{4c}$$

$$s.t. \sum_{\mathbf{S}} R(\mathbf{S}, \theta^{(q)}) = 1 \tag{4d}$$

where $H_{m,m}$ depends only on previous estimates $\theta^{(q)}$ and allows to underline that when the function in the logarithm ideally evolves towards the target R (which can change at each iteration) then $\mathcal{Q}_{m,m} - \mathcal{H}_{m,m} \to 0$. Since $R(\mathbf{S})$ must sum up to 1, it follows that a rational choice for R is a BBA denoted as m_γ subsequently and $\mathcal{Q}_{m,m}$ is an expectation taken with respect to m_γ.

2.2 E/M-steps

An EM-like procedure can thus be applied. At iteration q, the E-step aims at maximizing the expectation 4b given fixed parameters $\theta^{(q)}$. We can cancel its derivative with repect to R using appropriate Lagrangian multipliers (integrating the aforementionned constraint on R) to get the maximizer $m_\gamma^{(q)}$:

$$\text{E-step: } \Rightarrow m_\gamma^{(q)} = \frac{m(\mathbf{X}, \mathbf{S} \mid \theta^{(q)})}{\sum_{\mathbf{S}' \not\ni \emptyset} m(\mathbf{X}, \mathbf{S}' \mid \theta^{(q)})} \equiv m(\mathbf{S} \mid \mathbf{X}, \theta^{(q)}) \tag{5}$$

$m_\gamma^{(q)}(\cdot \mid \mathbf{X})$ is the posterior BBA on states given observations. The posterior is then used in the M-step to find the best estimate $\theta^{(q+1)}$ for the next iteration so that it maximizes the expectation under $m_\gamma^{(q)}$:

$$\text{M-step: } \Rightarrow \theta^{(q+1)} = \underset{\theta}{\text{argmax}}\ \mathbb{E}_{m_\gamma^{(q)}} [\log m(\mathbf{X}, \mathbf{S} \mid \theta)] \tag{6}$$

The algorithm iterates likewise to standard EM until the relative increase of the support $pl(\mathbf{X}, \boldsymbol{\Omega}_{\mathbf{z}})$ between two consecutive iterations remains below a threshold.

Property 1. Since R is a BBA, then Jensen's inequality holds so that this algorithm is guaranteed to converge.

The proof follows the same line of reasoning as in standard EM.

Conjecture 1. Similarly to the auxiliary function in EM [1, Theorem 2.1], the maximization of the lower bound $\mathcal{Q}_{m,m}$ does not decrease the total support.

This conjecture was implicitly assumed in [23] for the Credal EM algorithm applied to Gaussian Mixture Model.

Remark 1. According to the model considered, it can be practically feasible to check whether the conjecture holds or not. It is the case for EvHMM [18] by using the evidential forward propagation [16].

2.3 Incorporating Evidential Prior to Adjust the Posterior BBA

The target BBA $m_\gamma^{(q)}$ computed in the E-step is of paramount interest to reestimate the parameters. In cases of model's misspecification (choice of $\mathbb{A}$ for instance) or biases induced by the data collection process, this BBA may eventually lead to wrong parameter estimates.

One solution was proposed in [23]. It considers that the prior knowledge on hidden variables are encoded by a set of T belief functions. For temporal data, the prior may be defined on Ω_z with a BBA $m_{\text{prior}(t)}^{\Omega_z}$. Note that it is likely to encounter situations where the prior can be defined on $\Omega_z \times \Omega_z$. If nothing is known about the hidden variables, then $\forall t, m_{\text{prior}(t)}^{\Omega_z}(\Omega_z) = 1$. Those priors can then be incorporated into the computation of the mathematical expectation (Eq. 4a) by Dempster's rule $\textcircled{\cap}$ as initially proposed in [23]:

$$m_{\gamma}^{(q)} \leftarrow m_{\gamma}^{(q)} \textcircled{\cap} m_{\text{prior}(t)} \tag{7}$$

see also [4,7]. The second solution relies on the Theory of Weighted Distributions (TWD) [11] that allows to incorporate prior knowledge on expectations computed in EM [3].

3 Learning Parameters in EvHMM

3.1 What Is an EvHMM?

An EvHMM is a particular evidential discrete-latent model enhanced by a Markov chain [16] in which the states can be partially observable with some degree of uncertainties. It is defined by two main sets of parameters:

- Transition matrix $\mathbb{A}$: An entry a_{ij} represents the belief mass of observing subset S_j at time t given that the system was in subset S_i at $t-1$.
- Observation model: Allows to generate the belief mass on subset S_j at t given observation $\mathbf{x}_t$. Observations are supposed to follow a multivariate Gaussian Mixture Model (GMM) for each state, characterized by parameters $\Phi = \{\boldsymbol{\mu}, \mathbf{c}, \boldsymbol{\Sigma}\}$ representing the means, covariances and mixing weights.

The symbol $\theta = \{\mathbb{A}, \boldsymbol{\mu}, \mathbf{c}, \boldsymbol{\Sigma}\}$ represents the set of parameters of an EvHMM.

In this model, the joint BBA (Eq. 6) located in the logarithm of the criterion can not be expressed using only products which makes the estimation of parameters untractable.

Assumption 1. *It is possible to decouple the estimation of the transitions parameters in the Markov chain from the parameters in the observation model.*

This decoupling appears *naturally* in standard HMM due to factorisation [2, Chap. 13]. The criterion can thus be rewritten as $\mathcal{Q}_{m,m} = \mathcal{Q}^a_{m,m} + \mathcal{Q}^b_{m,m}$ where $\mathcal{Q}^a_{m,m}$ is related to the transitions while $\mathcal{Q}^b_{m,m}$ to the observation model.

3.2 M-step for the Markov Chain

Suppose that the transition matrix is made of BBAs $m_a^{\Omega_z}(\cdot \mid S_{t-1}), S_{t-1} \subseteq \Omega_z$. A sequence $\mathbf{S} = (S_1, S_2, \ldots S_t \ldots S_T), S_t \subseteq \Omega_z$ starting at S_1 requires to considering that S_1 is true at $t=1$, S_2 is true at $t=2$ and so on.

Proposition 1. *The total support assigned to a sequence* $\mathbf{S} = (S_1, S_2, \ldots S_t \ldots S_T), S_t \subseteq \Omega_z$ *can be quantified by the plausibility on* $\Omega_z^T = \Omega_z \times \Omega_z \times \ldots \Omega_z$ *(*T *times) after conditioning on the sequence. Given a vacuous BBA on initial states, the total support is given by:*

$$pl^{\Omega_z^T}(\mathbf{S}) = \prod_{t=2}^{T} pl_a^{\Omega_z}(S_t \mid S_{t-1}) \tag{8}$$

It defines an Evidential Markov Chain (EMC).

Note that the solution is different if the prior is not vacuous. The solution is also different from the result proposed in [12, Definition 4.1]:

Definition 1 (proposed in [12]). *An EMC has been defined as*

$$m^{\Omega_z^T}(\mathbf{S}) = \prod_{t=2}^{T} m_a^{\Omega_z}(S_t \mid S_{t-1}) \tag{9}$$

This definition is of practical interest in the sequel since estimating the transition given plausibilities (Eq. 8), although exact, would lead to incoherences due to the presence of BBA in the suggested EM procedure (Sect. 2). The proposed criterion has thus the following form:

$$\mathcal{Q}^a_{m,m}(\mathbb{A}^{(q)}, \mathbb{A}) = \sum_{t=2}^{T} \sum_{S_j \subseteq \Omega_z} \sum_{S_i \subseteq \Omega_z} m_{\xi(t,t-1)}^{\Omega_z \times \Omega_z}(S_i, S_j \mid \mathbb{A}^{(q)}) \log m_a^{\Omega_z}(S_j \mid S_i, \mathbb{A}) \tag{10}$$

where $m_{\xi(t,t-1)}$ is a BBA defined on two consecutive time slices that represents the probability mass of observing two given subsets. The maximization of $\mathcal{Q}^a_{m,m}$ with respect to m_a at iteration (q) requires to take the derivative of $\mathcal{Q}^a_{m,m}$ and using appropriate Lagrangian multipliers (ensuring that $\sum_B m_a^{\Omega_z}(B \mid S_{t-1}) = 1, \forall S_{t-1} \subseteq \Omega_z$ yielding:

$$m_a^{(q+1)}(S_{j,t} \mid S_{i,t-1}) = \frac{\sum_{t=2}^{T} m_{\xi(t,t-1)}^{\Omega_z \times \Omega_z}(S_i, S_j \mid \mathbb{A}^{(q)})}{\sum_{t=2}^{T} \sum_{\emptyset \neq S_l \subseteq \Omega_z} m_{\xi(t,t-1)}^{\Omega_z \times \Omega_z}(S_i, S_l \mid \mathbb{A}^{(q)})} \tag{11}$$

By assuming that the BBAs defined conditionally to subsets are computed by the disjunctive rule of combination [22] using only on BBAs defined conditionally to singletons, it follows that Eq. 11 allows to estimate $|\Omega_z| \times 2^{|\Omega_z|}$ parameters.

3.3 M-step for the Observation Model

In [23], the authors suggested an approach (EM-like) to estimate the parameters in a GMM using belief functions to represent uncertainty on mixing (discrete latent) variables. The criterion relies on both BBA and plausibilities generating inconsistencies for reestimation formulas. We can thus aim at maximizing an approximation of the support similarly to the Markov chain given by:

$$Q^b_{m,m}(\theta^{(q)}, \theta) = \sum_{t=1}^{T} \sum_{S \subseteq \Omega_z} m_{\gamma,t}^{\Omega_z}(S \mid \mathbf{X}, \mathbb{A}^{(q)}, \Phi^{(q)}) \log m_b^{\Omega_z}(S \mid \mathbf{x}_t, \Phi) \tag{12}$$

where it is important to remark that m_γ is made dependent not only on the current parameters of the observation model ($\Phi^{(q)}$) but also on the EMC ($\mathbb{A}^{(q)}$). Indeed, the Evidential Forward-Backward algorithm proposed in [16] can compute this quantity, which is related to Eq. 11 by a marginal operation likewise to standard HMM [13, Eq. 38].

The Generalized Bayesian Theorem (GBT) [22] allows to deduce the BBA $m_b^{\Omega_z}(S \mid \mathbf{x}_t, \Phi)$ given plausibilities conditional to singleton $pl^{\Omega_z}(\mathbf{x}_t \mid S_t, \Phi)$ [5]:

$$m_b^{\Omega_z}(S \mid \mathbf{x}_t, \Phi) = \prod_{s_k \in S} pl^{\mathbb{R}}(\mathbf{x}_t \mid s_k, \Phi) \prod_{s_k \notin S} \left(1 - pl^{\mathbb{R}}(\mathbf{x}_t \mid s_k, \Phi)\right) \tag{13}$$

where $pl^{\mathbb{R}}(\mathbf{x}_t \mid s_k, \theta), \forall s_k \in \Omega_z$ is given by a GMM [13, Sect. 4A]. Making use of Eq. 13, the criterion $\mathcal{Q}^b_{m,m}$ can be rewritten as:

$$\begin{aligned}\mathcal{Q}^b_{m,m}(\theta^{(q)}, \theta) = \sum_{t=1}^{T} \sum_{s_l \in \Omega_z} & \Big\{ pl^{\Omega_z}_{\gamma,t}(s_l \mid \theta^{(q)}) \log pl(\mathbf{x}_t \mid s_l, \theta) \\ + bel_{\gamma,t}(\overline{s_l} \mid \theta^{(q)}) & \log\left(1 - pl(\mathbf{x}_t \mid s_l, \theta)\right) \Big\}\end{aligned} \tag{14}$$

where $bel_{\gamma,t}$ is the belief function.

Assumption 2. *The contribution of "$bel_{\gamma,t}(\overline{s_l} \mid \theta^{(q)}) \log\left(1 - pl(\mathbf{x}_t \mid s_l, \theta)\right)$" is negligible compared to $pl^{\Omega_z}_{\gamma,t}(s_l \mid \theta^{(q)}) \log pl(\mathbf{x}_t \mid s_l, \theta)$.*

This assumption does not narrow the expression down to a probabilistic formulation because the weight $pl^{\Omega_z}_{\gamma,t}(s_l \mid \theta^{(q)})$ makes use of the information held by all subsets that contain s_l.

For illustration purpose, we consider one Gaussian component for each singleton state. The criterion can thus be approximated as:

$$\mathcal{Q}^b_{m,m}(\theta^{(q)}, \theta) \approx \sum_{t=1}^{T} \sum_{s_l \in \Omega_z} pl^{\Omega_z}_{\gamma,t}(s_l \mid \theta^{(q)}) \log pl(\mathbf{x}_t \mid s_l, \theta) \tag{15}$$

The means $\mu_k, k = 1 \ldots K$ for the next iteration are obtained by

$$\mu_j^{(q+1)} = \frac{\sum_t pl^{(q)}_{\gamma,t}(s_j)\, \mathbf{x}_t}{\sum_t pl^{(q)}_{\gamma,t}(s_j)} \tag{16}$$

and the covariances by

$$\Sigma_j^{(q+1)} = \frac{\sum_t pl^{(q)}_{\gamma,t}(s_j) \cdot (\mathbf{x}_t - \mu_j)\,(\mathbf{x}_t - \mu_j)'}{\sum_t pl^{(q)}_{\gamma,t}(s_j)} \tag{17}$$

Due to component annealing observed in practice [10], the mixture weights were not considered.

3.4 E-step

$m_{\gamma,t}$ represents the knowledge on subsets of states after observing $\mathbf{X}$ which is obtained by the evidential forward-backward algorithm [16]. This algorithm can be written using commonality functions which allows point-wise multiplication and therefore with limited complexity.

4 Results

4.1 Turbofan Engine Datasets

The turbofan datasets were generated using the CMAPSS simulation environment that represents an engine model of the 90,000 lb thrust class [8,20]. The authors used a number of editable input parameters to specify operational profile, closed-loop controllers, environmental conditions (various altitudes and temperatures). Some efficiency parameters were modified to simulate various degradations in different sections of the engine system. Selected fault injection parameters were varied to simulate continuous degradation trends. The datasets generated possess unique characteristics that make them very useful and suitable for developing classification and prognostics algorithms [17]: Multi-dimensional response from a complex non-linear system, high levels of noise, effects of faults and operational conditions, and plenty of units simulated with high variability.

In the present paper, the 100 training instances of dataset #1 are considered to illustrate the EvHMM on a complex system. These instances were generated by considering one operating condition and one fault mode. The data were collected from various parts of the system to record effects of different degradation mechanisms on 21 sensor measurements. The time-series thus represent different degradation behaviors in multiple units. From sensor measurements in each instance of the training dataset #1, a health indicator is built as proposed in [14]. The health indicators (HI) are depicted in Fig. 2 where we can observe high variability in terms of noise and degradation level. For classification purpose, a ground truth of the state sequences corresponding to each time-series was proposed in [19] which is used subsequently, and available at https://fr.mathworks.com/matlabcentral/fileexchange/54808-segmentation-of-cmapss-trajectories-into-states. The comparison between the ground truth and the estimations provided by both HMM and EvHMM is made by the Adjusted Rand Index (ARI) [24] that tends to 1 if the sequence estimated and the ground truth are equal.

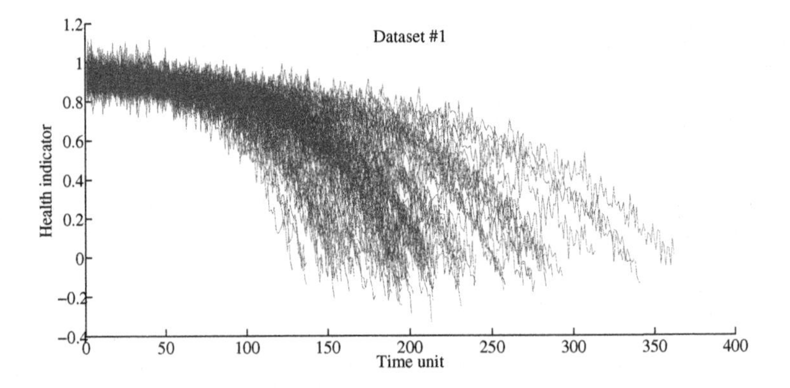

Fig. 2. Instances in training dataset #1.

4.2 Influence of a Possibly Wrong Ground Truth

The ground truth may be corrupted by errors due, for instance, to the lack of expertise on the degradation, to the noise on the HI or to some parameter tuning. To evaluate the influence of labeling errors, we proceed as proposed in [4,15] where at each time step t of a training instance, an error probability q_t is drawn randomly from a beta distribution with mean ρ and standard deviation 0.2. Then, with probability q_t, the state y_t is replaced by a completely random value $\tilde{y}_t$ with a uniform distribution over possible states. We thus obtain noisy labels corresponding to a crisp random labeling. Note that $\rho = 1$ corresponds to the unsupervised case (no labels). The use of prior in HMM was proposed in [15] with available code at https://fr.mathworks.com/matlabcentral/fileexchange/55172.

Training and testing sets were then generated to evaluate the EvHMM. 20 instances and the corresponding labels were randomly selected for training, the remaining 80 are kept for testing (without labels), and this process is repeated 8 times. For each run, the labels were corrupted 10 times with the random process explained previously. For each value of ρ, 6400 results were thus obtained which are represented by box plots in Fig. 3.

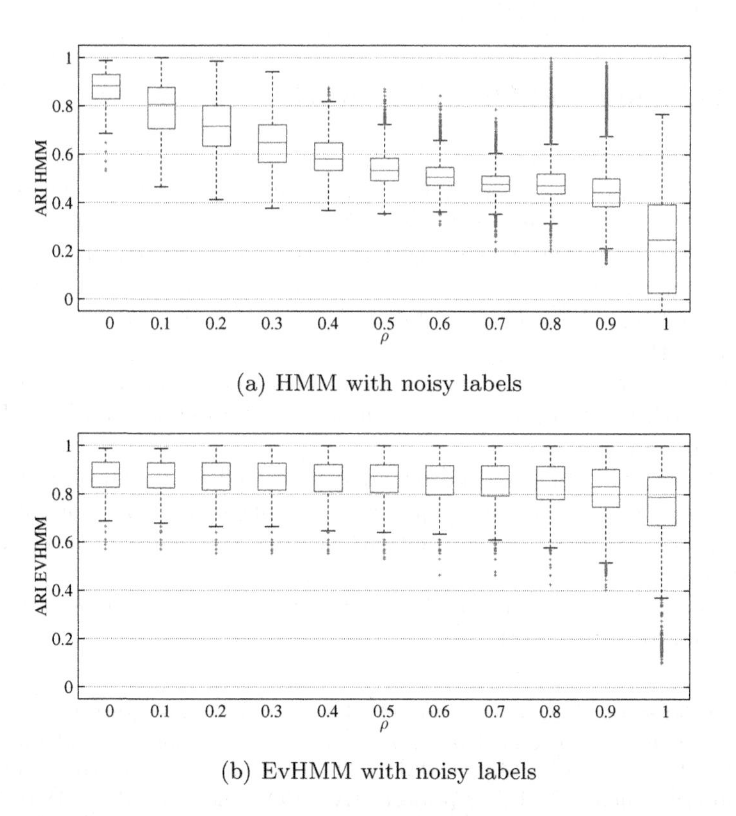

(a) HMM with noisy labels

(b) EvHMM with noisy labels

Fig. 3. Performance (ARI) of EvHMM and HMM for the classification of hidden states with respect to the quantity and quality of prior about states (controlled by ρ).

Figure 3(a) and (b) depict the evolution of the performance of both HMM and EvHMM when *noisy* labels are considered. It can be observed that the behavior of the EvHMM is highly different from standard HMM. Note that the low performance of the HMM in this particular case (noisy labels) was also underlined in [15]. The tests made with uncertain labels (not shown here) led to similar results for both models. The EvHMM appears more robust to label switching compared to standard HMM, with a stable performance around 90 % until high level of noise ($\rho = 0.8$), whereas the performance of the HMM highly decreases with report to the noise level ρ. For noisy labels, the EvHMM always outperformed the HMM, except for the supervised case with $\rho = 0$ yielding similar results. The 75-th percentiles of the EvHMM's performance are almost always close to 1 while the boxes for the HMM follows the decreasing trend of the median with report to the noise level. Therefore, with an appropriate initialization (made similarly for both models), the EvHMM may lead to a high recognition rate for this dataset. The difference between both models is partly attributed to the ability of the EvHMM to manage disjunctive sets in particular in presence of conflicting information between prior and estimates. This eventually allows belief masses to be assigned more gradually to particular singletons compared to the probabilistic case.

5 Conclusion and On-going Work

EvHMM (Evidential Hidden Markov Model) is a new method for time-series modelling based on Dempster-Shafer's theory of belief functions. The main difference with standard Hidden Markov Models is the consideration of random disjunctive sets in the Markov chain. The use of belief functions to quantify and propagate uncertainty and imprecision on subsets of random latent variables allows to represent the gradual evolutions of a state variables of a system which is monitored through sensors.

Some preliminary results are presented on complex datasets on which the proposed EvHMM depicts high performance compared to standard HMM in presence of noisy labels.

In [18], it is shown that the likelihood (termed as a plausibility) can be exactly computed so that the conjecture and approximations proposed in this learning procedure to make the solution tractable can be easily checked in real applications.

The application of the proposed procedure to various evidential latent models is considered for future work.

Acknowledgments. The author would like to express his gratitude to Michèle Rombaut, Denis Pellerin and Thierry Denoeux for discussions around inference in EvHMM and EM-based learning in HMM. This work has been carried out in the following projects: the CNRS-PEPS project "EVIPRO", the "SMART COMPOSITES" project (FRI2). It also got support from the Laboratory of Excellence "ACTION" (reference ANR-11-LABX-01-01).

References

1. Baum, L.E., Petrie, T., Soules, G., Weiss, N.: A maximization technique occurring in the statistical analysis of probabilistic functions of Markov chains. Ann. Math. Stat. **41**, 164–171 (1970)
2. Bishop, C.: Pattern Recognition and Machine Learning. Springer, New York (2006)
3. Juesas, P., Ramasso, E.: Ascertainment-adjusted parameter estimation approach to improve robustness against misspecification of health monitoring methods. In: Mechanical Systems and Signal Processing (2016). http://dx.doi.org/10.1016/j.ymssp.2016.03.022
4. Côme, E., Oukhellou, L., Denoeux, T., Aknin, P.: Learning from partially supervised data using mixture models and belief functions. Pattern Recogn. **42**(3), 334–348 (2009)
5. Delmotte, F., Smets, P.: Target identification based on the transferable belief model interpretation of Dempster-Shafer model. IEEE Trans. Syst. Man Cybern. Part A Syst. Hum. **34**(4), 457–471 (2004)
6. Dempster, A., Laird, N., Rubin, D.: Maximum likelihood from incomplete data via the EM algorithm. J. Roy. Stat. Soc. **39**(1), 1–38 (1977)
7. Denoeux, T.: Maximum likelihood estimation from uncertain data in the belief function framework. IEEE Trans. Knowl. Data Eng. **25**(1), 119–130 (2013)
8. Frederick, D., DeCastro, J., Litt, J.: User's guide for the commercial modular aero-propulsion system simulation (C-MAPSS). Technical report, National Aeronautics and Space Administration (NASA), Glenn Research Center, Cleveland, Ohio 44135, USA (2007)
9. Jensen, J.L.W.V.: Sur les fonctions convexes et les inégalités entre les valeurs moyennes. Acta Mathematica **30**(1), 175–193 (1906)
10. Naim, I., Gildea, D.: Convergence of the EM algorithm for gaussian mixtures with unbalanced mixing coefficients. In: Langford, J., Pineau, J. (eds.) Proceedings of the 29th International Conference on Machine Learning (ICML-12), pp. 1655–1662. ACM, New York (2012)
11. Patil, G.: Weighted Distributions, vol. 4. Wiley, Chichester (2002)
12. Pieczynski, W.: Multisensor triplet Markov chains and theory of evidence. Int. J. Approximate Reasoning **45**(1), 1–16 (2007)
13. Rabiner, L.: A tutorial on hidden Markov models and selected applications in speech recognition. Proc. IEEE **77**, 257–285 (1989)
14. Ramasso, E.: Investigating computational geometry for failure prognostics. Int. J. Prognostics Health Manage. **5**(5), 1–18 (2014)
15. Ramasso, E., Denoeux, T.: Making use of partial knowledge about hidden states in HMMs: an approach based on belief functions. IEEE Trans. Fuzzy Syst. **22**(2), 395–405 (2014)
16. Ramasso, E., Rombaut, M., Pellerin, D.: Forward-backward-viterbi procedures in the transferable belief model for state sequence analysis using belief functions. In: Mellouli, K. (ed.) ECSQARU 2007. LNCS (LNAI), vol. 4724, pp. 405–417. Springer, Heidelberg (2007)
17. Ramasso, E., Saxena, A.: Performance benchmarking and analysis of prognostic methods for CMAPSS datasets. Int. J. Prognostics Health Manage. **5**(2), 1–15 (2014)
18. Ramasso, E.: Inference and learning in evidential discrete-latent Markov models. IEEE Trans. Fuzzy Syst. (2016). submitted, ver. 2, 30 January 2016

19. Ramasso, E.: Segmentation of CMAPSS health indicators into discrete states for sequence-based classification and prediction purposes. Technical report 6839, FEMTO-ST institute, January 2016
20. Saxena, A., Goebel, K., Simon, D., Eklund, N.: Damage propagation modeling for aircraft engine run-to-failure simulation. In: IEEE Prognostics and Health Management (2008)
21. Serir, L., Ramasso, E., Zerhouni, N.: Time-sliced temporal evidential networks: the case of evidential HMM with application to dynamical system analysis. In: 2011 IEEE Conference on Prognostics and Health Management (PHM), pp. 1–10, June 2011
22. Smets, P.: Beliefs functions: the disjunctive rule of combination and the generalized Bayesian theorem. IJAR **9**, 1–35 (1993)
23. Vannoorenberghe, P., Smets, P.: Partially supervised learning by a **Cr**edal **EM** approach. In: Godo, L. (ed.) ECSQARU 2005. LNCS (LNAI), vol. 3571, pp. 956–967. Springer, Heidelberg (2005)
24. Vinh, N., Epps, J., Bailey, J.: Information theoretic measures for clustering comparison: is a correction for chance necessary? In: Proceedings of the 26th Annual International Conference on Machine Learning, pp. 1073–1080 (2009)

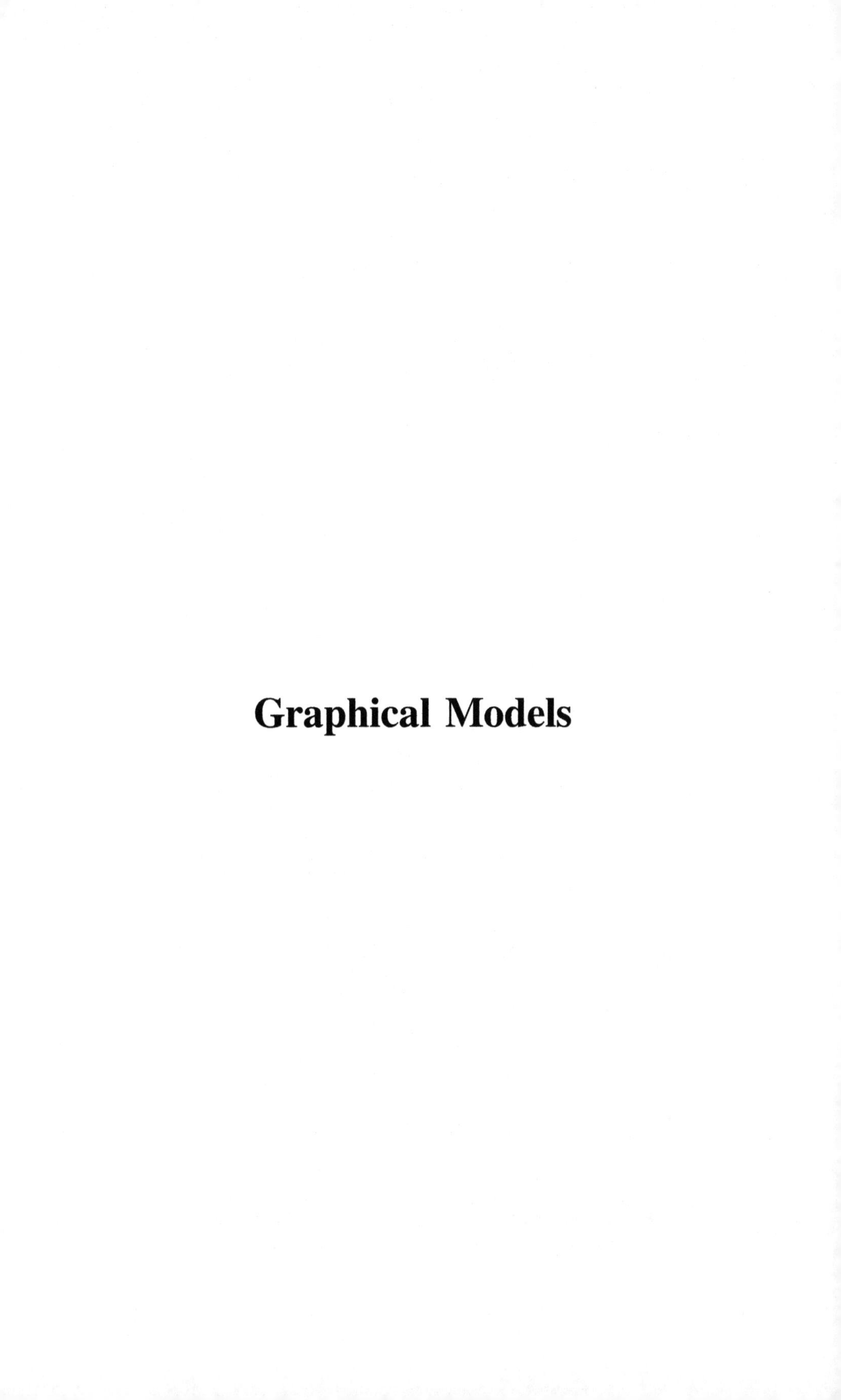

Graphical Models

Determination of Variables for a Bayesian Network and the Most Precious One

Esma Nur Cinicioglu(✉) and Taylan Yenilmez

Quantitative Methods Division, School of Business,
Istanbul University, 34322 Avcilar, Istanbul, Turkey
{esmanurc, taylan.yenilmez}@istanbul.edu.tr

Abstract. To ensure the quality of a learned Bayesian network out of limited data sets, evaluation and selection process of variables becomes necessary. With this purpose, two new variable selection criteria N_2S_j and N_3S_j are proposed in this research which show superior performance on limited data sets. These newly developed variable selection criteria with the existing ones from prior research are employed to create Bayesian networks from three different limited data sets. On each step of variable elimination, the performance of the resulting BNs are evaluated in terms of different network performance metrics. Furthermore, a new variable evaluation criteria, IH_j, is proposed which measures the impact of a variable to all the other variables in the network. IH_j serves as an indicator of the most important variables in the network which has a special importance for the use of BNs in social science research, where it is crucial to identify the most important factors in a setting.

Keywords: Bayesian networks · Variable selection in Bayesian networks · Importance hierarchy of variables in network · Variable evaluation scores · Limited data sets

1 Introduction

In the art of transforming data into information, statistical community has long dealt with the problem of variable selection and there is an immense literature for different models and approaches, which acted as footprints for the machine learning community in efforts of identifying the relevant features of a problem. These efforts are particularly praiseworthy in the world of today where the advancement of data collection tools and automatic capture of data is ever increasing with an enormous speed and thus the number of redundant variables without contribution.

Having more features does not always contribute to problem solving or the correct representation of the problem. [1] have shown that with their proposed algorithm the feature space may be reduced in many learning tasks while also improving the accuracy. [2] states that for interpretation of the results only a small representative subset of the original feature space of the data may be sufficient. The reasons for selecting only a subset of variables, instead of including all, may be listed as concerns for loss in prediction accuracy, loss in speed of the predictor, cost considerations for observation of extra features [3], model overfitting [4].

J.P. Carvalho et al. (Eds.): IPMU 2016, Part I, CCIS 610, pp. 313–325, 2016.
DOI: 10.1007/978-3-319-40596-4_27

Maybe the bliss is in ignoring and that has long been identified by the researchers; the only debate is about on how to decide what to ignore. However, when it comes to the topic of variable selection in probabilistic graphical networks, particularly Bayesian networks (BN), the same level of prosperity of research cannot be observed and that qualifies the subject as an open and unsaturated research field. The previous work on variable selection in BNs mostly focused on selection of the variables for BNs where BNs are used as a classifier. [5] have studied the feature-selection behavior of the minimum description length for feature selection function for learning BN classifiers from data. Many researchers have argued that the knowledge of the values of variables of Markov blanket should be sufficient for classification problems [6, 7] and suggested to use the Markov blanket information for selecting the important variables. There exist considerable research where a BN is generated from a data set and then the Markov blanket of the class variable is used to the feature subset selection task for the classification problem [8, 9].

Though there exist considerable research for variable selection in BNs with the purpose of creating a better classifier, considering the capabilities of BNs, it can be well appreciated that the role and importance of a BN is not limited to act just as a classifier. BNs are used in many different domains ranging from biology to document classification, targeted advertising, legal services and the variety of these fields is increasing every day. A natural consequence of this is that depending on the field of application the expectations from a BN may differ and hence the expectations from its performance may vary [10]. Additionally, though there exist much research for structure learning algorithms in BN, most of the research assume to have access to large data sets available which is not always the case for many real-world applications [11]. All these point out the need for further research for variable selection in BN aiming for improved BN representations of the considered problem and hence better performing Bayesian networks out of limited data sets.

For the purpose of creation of better performing BNs [12] proposed a heuristic procedure for selecting the variables to be used in the final BN. In their approach, first an initial BN is learned using all of the variables in data. Then using the information contained in the conditional probability tables (cpt) of the variables the S_j score is calculated for each variable j in the network, which acts as a stepwise criteria for selecting the variables to be used in the final BN. The proposed heuristic is applied using a market basket data set and its performance is evaluated in terms of the logscore of the final BN. Later [10] have argued that though the S_j score demonstrates a sound performance on prediction capacity, its formula leads to the problem that the variables without parents or children in the network are punished and that in turn affects the overall performance of the heuristic. To that end they proposed a modification on the score and tested the performance of the revised score, NS_j using a credit data set, again in terms of the logscore of the resulting BN. Though both of these works obtain successful results, there is need for more data sets to evaluate the process and for different criteria to evaluate the performance of the resulting BN.

In this work we develop two new variable selection scores N_2S_j and N_3S_j. Additionally we develop a new variable evaluation score, IH_j, which identifies the variable in the network with the most contribution to other variables. Identifying the most important variable in the network is especially important in terms of the social science research, where it is crucial to identify the most important factors in a setting. We test the performance of all the variable selection scores d_j, S_j, NS_j, N_2S_j, N_3S_j, IH_j using three data sets. The performance of the scores are evaluated in terms of the network performance measures AIC, BIC, loglikelihood and K2.

The outline of the remainder of paper is as follows: In Sect. 2 we review the existing scores d_j, S_j, NS_j, introduce the new ones IH, N_2S_j, N_3S_j and explain the application and evaluation process of the proposed scores. Section 3 gives details about the data sets used for this study. In Sect. 4, we present the results of all the scores and discuss its implications. In Sect. 5 we summarize and conclude.

2 Variable Evaluation for Bayesian Networks: Prior Work, New Scores and the Application of the Heuristic

2.1 Prior Work: d_j, S_j, NS_j

The prior work in variable evaluation scores for creating better performing BNs, [10, 12], are both based on the idea that once an initial BN is learned from the data set, the cpts of the variables in the network can be used to identify the variables which have strong association with each other. With that in mind, [12] applied the distance measure to the cpt of each variable, in order to measure the degree of change of the conditional probabilities of a child node depending on the states of its parents. Here a high average distance is desired as an indication of a strong relationship. The average distance of each variable j in the network, where N is the set of variables in the network, may be calculated as follows: Here d_j represents the average distance of the variable of interest j with its parent variables. p_{jk} and q_{jk} stand for the conditional probabilities of this variable j for the different states of its parents, k stands for the different states of the variable and n stands for the number of states of the set of parent nodes.

$$d_j = \sum (p_{jk} - q_{jk})^2 / \binom{n}{2}, \forall p \wedge q \quad for\ j \in N \tag{1}$$

Basing the evaluation of the variables just on their level of association with their parents might us, overlook the ones which have a strong association with their children. This fact and the idea that the same average distance shows the degree of association of a child node jointly with the child's other parents is considered by the development of the S_j score. In this formula the S_j score of a variable j in the network, for $j \in N$, is the sum of the average distance of the same variable d_j and the average of the average distances of its children. Here ij denotes the child variable i of the variable j and c_j denotes the number of j's children. Hence according the proposed heuristic for each variable in the network first their average distance and then S_j will be calculated, where the variables with the lowest performance on the S_j score will be eliminated from the network and then a new BN will be learned where again the information contained in the new cpts will be used for the calculation of the variable evaluation scores of the variables.

$$S_j = \begin{cases} d_j + \left(\sum_i \frac{d_{ij}}{C_j}\right), & for\ c_j > 0 \\ d_j, & for\ c_j = 0 \end{cases} \tag{2}$$

As presented in Eq. 2, S_j diminishes to d_j when the variable does not possess any children and equals to the latter part (average of the average distances of its children) if the variable does not have any parents. For that reason, later [10] criticized the S_j score suggesting that its formula leads to the problem that the variables without parents or children in the network are punished which in turn affects the overall performance of the heuristic. As a modification to the S_j score they offered the NS_j score for the evaluation of variables. NS_j considers three cases, the case of a variable without children, without parents and the case of a variable with both parents and children. The ones without any parents or children will be eliminated from the network. Accordingly, NS_j, for $j \in N$, is computed as follows:

$$NS_j = \begin{cases} d_j, & for\ c_j = 0 \\ \sum_i \frac{d_{ij}}{C_j}, & for\ d_j = 0 \\ \frac{d_j + \left(\sum_i \frac{d_{ij}}{C_j}\right)}{2}, & for\ c_j \wedge d_j > 0 \end{cases} \tag{3}$$

2.2 New Scores for Variable Evaluation in BN: IH_j, N_2S_j, N_3S_j

In this section we present three new variable evaluation scores IH_j, N_2S_j and N_3S_j. IH_j is a score defined for identifying the variables in the network which contribute most to other variables in the network and may hence be called as the importance hierarchy indicator for variables present in the network. The calculation of IH_j is based on the results of the sensitivity analysis calculated as the entropy reduction. Entropy reduction, I, is calculated as

$$I = H(Q) - H(Q|F) = \sum_q \sum_f \frac{P(q,f) log_2[p(q,f)]}{P(q)P(f)} \tag{4}$$

where $H(Q)$ is the entropy of Q before any findings, $H(Q|F)$ is the entropy of Q after new findings from variable F, and Q is measured in information bits [13]. For the calculation of IH_j, for each variable j in the network, where N is the set of variables, sensitivity analysis based on entropy reduction are performed. Accordingly, IH_j, is calculated as the sum of entropy reduction that the variable j has with the other variables, i, in the network. This sum is normalized over the sum of the sensitivity analysis conducted for each variable in the network. In terms of contribution to the other variables, IH_j, can identify the most important variables in the network, hence the precious one of the BN. This has a special importance for the use of BNs in social science research, where it is crucial to identify the most important factors in a setting. Previously, for the same purpose [14] used the Borda score based on the entropy reduction of the variables in the network. However, that approach considers only the top four results and overlooks the effect of all the variables in the network.

$$IH_j = \frac{\sum_{i \in N} I_{ji}}{\sum_{j \in N} \sum_{i \in N} I_{ji}}, for\, j \in N \tag{5}$$

As a follow up on the previous work, N_2S_j is defined as the maximum between the average distance of a variable j and the average of the average distances of its children. With this modification, similar to NS_j score, N_2S_j prevents the punishment of the variables without parents or children whereas at the same time also induces the selection process towards the variables with a strong association with either parents and or children. On the other hand, different than the NS_j score which uses the average of the scores for variables with both parents and children, in the formulation of N_2S_j the Max function is used. This change, leverages the selection of the variables to the network which have both parents and children but the association level with one of them (with either its parents or children) is weak whereas with the other one, the variable has a strong association.

$$N_2S_j = \begin{cases} Max\left(d_j, \frac{\sum_i d_{ij}}{c_j}\right), & for\ c_j > 0 \\ d_j, & for\ c_j = 0 \end{cases} \tag{6}$$

N_3S_j, on the other hand, aims to also consider the effect of importance a variable plays towards the other variables in the network and not just considers the level of association with parents and or children. Hence, N_3S_j is defined as the product of N_2S_j with IH_j, for $j \in N$.

$$N_3S_j = \begin{cases} Max\left(d_j, \frac{\sum_i d_{ij}}{c_j}\right) * IH_j, & for\ c_j > 0 \\ d_j * IH_j, & for\ c_j = 0 \end{cases} \tag{7}$$

2.3 Application and Evaluation Process of the Variable Evaluation Scores

For the application of the heuristic and the comparison of the variable selection scores, first a BN is learned using all of the variables in data set. Afterwards, using the cpts of the variables out of the learned networks, the corresponding scores, d_j, S_j, NS_j, N_2S_j, N_3S_j, IH_j are calculated and a predefined number of variables with the lowest scores are eliminated from the network. Later, for each score considered and according its result in variable selection, a new BN is created with the remaining variables and new cpts are learned which will be again used for the calculation of the variable evaluation scores and hence the selection of variables. This procedure is repeated until the desired number of variables is reached. After each step of the variable elimination, the performance of the resulting networks are evaluated in terms of four network performance measures Akaike information criterion (AIC) [15], Schwarz Bayesian information criterion (BIC) [16], loglikelihood [17] and logarithm of the K2 score [18]. A nice review of the performance metrics for Bayesian networks are given by [13]. The measures used in this study are defined as follows:

$$AIC = \log\left(p\left(D|\hat{\theta}, G\right)\right) - n_p \tag{8}$$

$$BIC = \log\left(p\left(D|\hat{\theta}, G\right)\right) - \frac{n_p}{2}\log(N) \tag{9}$$

where $p\left(D|\hat{\theta}, G\right)$ is the likelihood of the data D according the estimated parameters $\hat{\theta}$ and structure G of the created BN, N is the sample size of the data set and n_p is the number of parameters.

The log-likelihood L is defined as

$$L = \sum\nolimits_i n_{i-test} * log\frac{n_i}{w_i * N} \tag{10}$$

where n_i is the number of instances in bin i, $n_{i\text{-test}}$ the number of instances of the test set that fall into this bin, w_i the bin width, and N the total number of training instances. L measures the performance of a learned BN structure with its estimated parameters on a given test set.

The logarithm of the K2 score is given as:

$$\log(K2(X_i)) = \sum\nolimits_{j=1}^{q_i} ((\ln\left(\frac{(r_i - 1)!}{(N_{ij} + r_i - 1)!}\right) + \sum\nolimits_{k=1}^{r_i} \ln(N_{ijk}!) \tag{11}$$

where N_{ijk} represents the number of cases in the database in which the variable X_i took its kth value, where k = 1, 2,…, r_i and j represents the unique combination of values of its set of parents (j = 1, 2,…, q_i). With respect to BIC, AIC measure penalizes less harshly for the inclusion of additional edges and K2 score can be considered as an intermediate one between AIC and BIC in terms of the penalization of network complexity [19]. All of the networks are learned using the Bayesian search algorithm in Genie [20], the sensitivity analysis are conducted in Netica [21] and for the evaluation of the networks bnlearn [22] package in R is employed.

3 Data Sets: Credit, WEF, Trade

Three different data sets are used for this research which will be referred to as Credit, WEF and Trade data throughout the paper. The selection of these data sets in particular, is in line with the research idea of this paper which is to develop efficient variable evaluation scores for BNs. It is aimed that, with the application of these variable selection scores better performing BNs will be obtained and hence the hidden information contained in limited data sets may be revealed. With that in mind, the data sets used in this study are limited in terms of the representative cases, the corresponding variables and the number of states and in terms of that each shows different characteristics.

The first data set is a free data set, called the German credit data, provided by the UCI Center for Machine Learning and Repository Systems. For the analysis the cleaned and discretized version of this dataset, previously used by [10] is employed, which constitutes of 21 columns and 1000 lines, referring the number of variable and cases consequently.

The second data set employed is the data released by WEF [23] for the 2009–2012 period. The data set used to learn the BN constitute the data of 21 variables considered for all the countries[1] participating in the WEF study over the four years indicated. The data is first transformed into a form where each variable is discretized with seven states, each having equal width of range. Accordingly, the final form of the data set used has 21 discrete variables with 539 cases.

As the third data set the BACI database by [25] is used. It is a cleaned version of COMTRADE Database. The number of cases contained in the data set is 143. In this paper, trade data is aggregated to indicate exports of each country in each HS-two digit product categories which refers to 69 variables where each variable is discretized to have four states with equal width. Trade data sets can be used for different levels of aggregation of products. As the data is more disaggregated, the product groups have of a more detailed definition. HS-6 digit data includes over five thousand product groups and the construction of a BN using all, would significantly increase the computational cost. On the other hand, by eliminating irrelevant variables, we can both work with disaggregated trade data, identify the effect of the important variables and also reduce the computational cost.

4 Score Performance Results, Evaluation and Discussion

The performance of the resulting BNs using each variable selection criteria on three data sets are presented in Figs. 1, 2 and 3, which are reported in terms of the network scores AIC, BIC, loglikelihood and K2. The variable elimination process is repeated for five steps in Credit and WEF data, and for six[2] steps in Trade data. The number of variables eliminated on each step is two on Credit and WEF data, and six on Trade data which is decided upon the inherent number of variables of the data set. Step 0 indicates the BN created using all of the variables in the data set. Since on some steps the same set of variables are selected by different criteria, the same results are obtained, hence in figures the bars of different variable selection criteria do have the same height.

In addition to the variable selection criteria, d_j, S_j, NS_j, N_2S_j, N_3S_j, IH_j, considered in this research, the results given in Figs. 1, 2 and 3 also includes the results of the BNs created by eliminating the variables with the highest d_j scores, called top d_j. The purpose for the inclusion of these results is to demonstrate that the considered variable elimination scores, indeed select the correct variables to be eliminated from the network and the improvement obtained in the BNs using these, is not due to overfit in data

[1] Oil exporting coutries are excluded from the data set [24].

[2] Since all of the criteria selected the same set of variables on the first step an additional step was added.

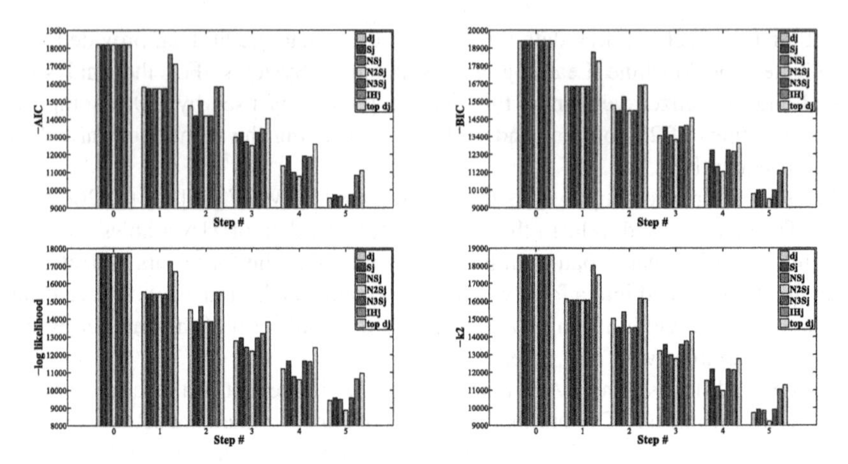

Fig. 1. Credit data: performance of the resulting BNs in different network performance metrics (Color figure online)

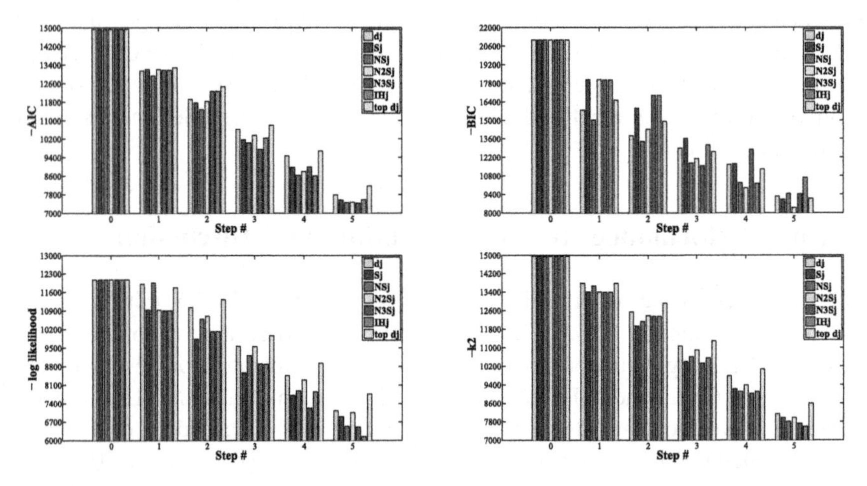

Fig. 2. WEF data: performance of the resulting BNs in different network performance metrics (Color figure online)

because of the reduction of the data set. Consequently, as it is demonstrated in Figs. 1, 2 and 3, the results of the BNs created by eliminating the variables with top d_j scores from the network, are worse than all of the variable selection criteria in almost every step of the three data sets. This result is also apparent given in Table 1 below. In Table 1, a pairwise comparison is made between all the variable selection scores plus the top d_j, in terms of difference in mean values of AIC, BIC, loglike and k2 of the resulting BNs. The variable selection criteria with the better performance is reported with its corresponding significance level. Accordingly, comparing the results obtained using the considered variable elimination criteria to the results with top d_j we see that in

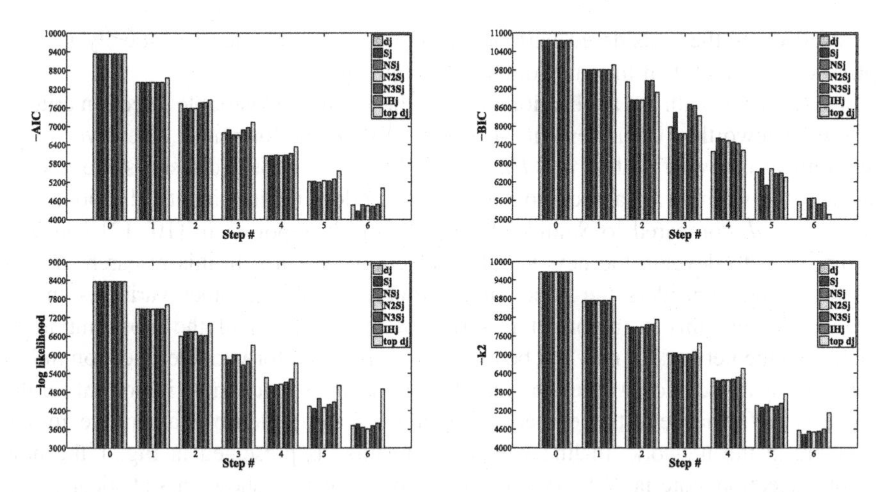

Fig. 3. Trade data: performance of the resulting BNs in different network performance metrics (Color figure online)

Table 1. Pairwise comparison of criteria in terms of the mean (***, **, * represent consecutively 1 %, 5 % and 10 % level of significance for difference in mean) values

	AIC	BIC	loglike	k2
dj vs S_j	S_j	d_j^{**}	S_j^{**}	S_j^{*}
dj vs NS_j	NS_j^{***}	NS_j^{*}	NS_j^{**}	NS_j^{**}
dj vs N_2S_j	$N_2S_j^{***}$	N_2S_j	$N_2S_j^{***}$	$N_2S_j^{***}$
d_j vs N_3S_j	N_3S_j	d_j^{**}	$N_3S_j^{**}$	$N_3S_j^{*}$
d_j vs IH_j	d_j^{*}	d_j^{***}	d_j	d_j
S_j vs NSj	NS_j	NS_j^{**}	S_j	NS_j
S_j vs N_2Sj	$N_2S_j^{**}$	$N_2S_j^{***}$	S_j	N_2S_j
S_j vs N_3Sj	S_j	S_j	N_3S_j	S_j
S_j vs IH_j	S_j^{**}	S_j^{**}	S_j^{**}	S_j^{**}
NS_j vs N_2S_j	N_2S_j	N_2S_j	N_2S_j	N_2S_j
NS_j vs N_3S_j	NS_j	NS_j^{**}	N_3S_j	NS_j
NS_j vs IH_j	NS_j^{***}	NS_j^{***}	NS_j	NS_j^{**}
N_2S_jvs N_3S_j	$N_2S_j^{**}$	$N_2S_j^{**}$	N_3S_j	N_2S_j
N_2S_j vs IH_j	$N_2S_j^{***}$	$N_2S_j^{***}$	$N_2S_j^{*}$	$N_2S_j^{**}$
N_3S_j vs IH_j	$N_3S_j^{**}$	N_3S_j	$N_3S_j^{**}$	$N_3S_j^{**}$
top d_j vs d_j	d_j^{***}	d_j^{**}	d_j^{***}	d_j^{***}
top d_j vs S_j	S_j^{***}	S_j	S_j^{***}	S_j^{***}
top d_j vs NS_j	NS_j^{***}	NS_j^{***}	NS_j^{***}	NS_j^{***}
top d_j vs N_2S_j	$N_2S_j^{***}$	$N_2S_j^{**}$	$N_2S_j^{***}$	$N_2S_j^{***}$
top d_j vs N_3S_j	$N_3S_j^{***}$	–	$N_3S_j^{***}$	$N_3S_j^{***}$
top d_j vs IH_j	IH_j^{***}	IH_j^{*}	IH_j^{***}	IH_j^{***}

almost all cases the considered variable selection score obtained statistically significantly better results than the ones obtained using 'top d_j'.

In all data sets with the application of all of the considered variable selection criteria we see noteworthy improvement in the network performance. However, this improvement is lowest with d_j and IH_j scores. This is an expected result since d_j only considers the strength of association of a child node with its parent. The poor performance of d_j compared to S_j and NS_j are previously reported in [10, 12] and it is measure used to develop the new variable selection measures in this research. IH_j on the other hand, considers the effect of a variable on all the other variables in the network. Though this is important in terms of identification of the most important variables in the network, it may not be the correct approach for variable selection in BN since a strong association with either parents or children is much more important for the prediction performance of the learned BN, than the wide influence of a variable among the others in the network. In line with these results, as presented in Fig. 1 the new variable selection criteria N_2S_j, developed in this research, shows a statistically significant superior performance in Credit data set compared to all the other scores. N_2S_j, according to its formulation using the max function, favors the variables in the network with a strong association with either its parents or children. The Credit data in this research represent a richer data set compared to the other two and hence the selection of variables can be made according to the max function.

On the other hand, though WEF data includes the same number of variables as in Credit, it has a smaller number of cases but a greater number of states. Accordingly, as presented in Fig. 2 below, on WEF data, N_3S_j shows a superior performance to all the other criteria. It can be observed that, though also NS_j and N_2S_j got noteworthy results, the newly suggested criteria N_3S_j, differentiates itself among the others. This difference is statistically significant in loglike compared to NS_j and S_j. With the involvement of IH_j term in its formulation N_3S_j not only values the strength of association of a variable with its parents and children but also the variable's impact to the whole network. In that matter, restricting the selection process to the strength of association to parents and children only, deteriorates the findings in limited data sets with a big number of states but a small number of cases. However, when the success of the network is evaluated in terms of the BIC measure which punishes harshly for the inclusion of more edges we observe that N_2S_j obtains better results compared to S_j and N_3S_j, the criteria which favor the well connectivity among the network.

The variable selection process for BNs requires the consideration of different dimensions inherent in the data set. In order to assure to end up with a high quality network, the weight of these dimensions should differ according the data characteristics. In that instance the trade data set represents a very sparse form with its big number of variables and very limited number of cases. Accordingly, results given in Fig. 3 below show that the S_j score, which considers the total contribution of a variable has both with its parents and children and thus punishes the ones without parents or children, shows a sound performance compared to others. BNs created from such limited data sets have the problem of well connectivity among the network since the big number of variables are represented through a few number of cases. In that matter, having a selection criteria which praises the well connectivity of a variable with its prodecessor and successor is crucial for the rightful selection of the variables which will be eliminated from the network.

5 Summary and Conclusions

Though there exist much research for structure learning algorithms in BNs, the topic of variable selection with the purpose of creating better performing BNs is a neglected area of research. The existing work on this topic focuses on the case where BNs are used as a classifier. However, the use and thus the extensiveness of BNs is increasing every day which creates the need for efficient variable selection criteria capable of creating better performing BNs out of limited data sets.

In this research, two new variable selection criteria N_2S_j and N_3S_j are proposed. Using these variable selection criteria in addition to the existing ones in previous work, BNs out of three different limited data sets are created. On each step of variable elimination, the performance of the resulting BNs using six different variable selection criteria d_j, S_j, NS_j, N_2S_j, N_3S_j, IH_j are evaluated in terms of four different network performance metrics. The new variable evaluation criteria, IH_j, measures the impact of a variable to all the other variables in the network. IH_j serves as an indicator of the identification of the most important variables in the network. This is especially important in terms of the social science research, where it is crucial to identify the most important factors in a setting.

Our findings indicate that the BNs created using the newly suggested criteria N_2S_j and its predecessor NS_j show a superior performance on limited data sets. N_2S_j, according to its formulation favors the variables in the network with a strong association with either its parents or children. The newly suggested N_3S_j score on the other hand differentiates itself from its previous alternatives through the involvement of IH_j term. N_3S_j not only values the strength of association of a variable with its parents and children but also the variable's impact to the whole network. In that matter, restricting the selection process to the strength of association to parents and children only, deteriorates the findings in limited data sets with a big number of states but a small number of cases. We conclude that the variable selection process for BNs requires the consideration of different characteristics of the data sets. Hence to end up with a high quality network, the rightful choice of criteria should be based on those characteristics inherent in data set.

References

1. Koller, D., Sahami, M.: Toward optimal feature selection. In: 13th International Conference on Machine Learning, pp. 284–292 (1996)
2. Fernández, A., Gómez, Á., Lecumberry, F., Pardo, Á., Ramírez, I.: Pattern recognition in latin America in the "big data" era. Pattern Recogn. **48**(4), 1185–1196 (2015)
3. Reunanen, J.: Overfitting in making comparisons between variable selection methods. J. Mach. Learn. Res. **3**, 1371–1382 (2003)

4. Sawalha, Z., Sayed, T.: Traffic accident modeling: some statistical issues. Can. J. Civ. Eng. **33**(9), 1115–1124 (2006)
5. Drugan, M.M., Wiering, M.A.: Feature selection for Bayesian network classifiers using the MDL-FS score. Int. J. Approximate Reasoning **51**(6), 695–717 (2010)
6. Cooper, G.F., Aliferis, C.F., Ambrosino, R., Aronis, J., Buchanan, B.G., Caruana, R., Janosky, J.E.: An evaluation of machine-learning methods for predicting pneumonia mortality. Artif. Intell. Med. **9**(2), 107–138 (1997)
7. Cheng, J., Greiner, R.: Comparing Bayesian network classifiers. In: Proceedings of the Fifteenth Conference on Uncertainty in Artificial Intelligence, pp. 101–108. Morgan Kaufmann Publishers Inc. (1999)
8. Hruschka Jr., E.R., Hruschka, E.R., Ebecken, N.F.: Feature selection by Bayesian networks. In: Tawfik, A.Y., Goodwin, S.D. (eds.) Canadian AI 2004. LNCS (LNAI), vol. 3060, pp. 370–379. Springer, Heidelberg (2004)
9. Castro, P.A., Von Zuben, F.J.: Learning Bayesian networks to perform feature selection. In: International Joint Conference on Neural Networks, IJCNN 2009, pp. 467–473. IEEE Press (2009)
10. Cinicioglu, E.N., Büyükuğur, G.: How to create better performing Bayesian networks: a heuristic approach for variable selection. In: Laurent, A., Strauss, O., Bouchon-Meunier, B., Yager, R.R. (eds.) IPMU 2014, Part I. CCIS, vol. 442, pp. 527–535. Springer, Heidelberg (2014)
11. Tonda, A.P., Lutton, E., Reuillon, R., Squillero, G., Wuillemin, P.-H.: Bayesian network structure learning from limited datasets through graph evolution. In: Moraglio, A., Silva, S., Krawiec, K., Machado, P., Cotta, C. (eds.) EuroGP 2012. LNCS, vol. 7244, pp. 254–265. Springer, Heidelberg (2012)
12. Cinicioglu, E.N., Shenoy, P.P.: A new heuristic for learning Bayesian networks from limited datasets: a real-time recommendation system application with RFID systems in grocery stores. Ann. Oper. Res. 1–21 (2012). doi:10.1007/s10479-012-1171-9
13. Marcot, B.G., Steventon, J.D., Sutherland, G.D., McCann, R.K.: Guidelines for developing and updating Bayesian belief networks applied to ecological modeling and conservation. Can. J. For. Res. **36**(12), 3063–3074 (2006)
14. Cinicioglu, E.N., Önsel Ekici, Ş., Ülengin, F.: Bayes Ağ Yapısının Oluşturulmasında Farklı Yaklaşımlar: Nedensel Bayes Ağları ve Veriden Ağ Öğrenme. In: Halil Sarıaslan Armağan Kitap, pp. 267–286. Siyasal Kitabevi, Ankara (2015)
15. Akaike, H.: Information theory and an extension of the maximum likelihood principle. In: Petrov, B.N, Csaki, F. (eds.) Second International Symposium on Information Theory, pp. 267–281. Akademiai Kiado, Budapest (1973)
16. Schwarz, G.: Estimating the dimension of a model. Ann. Stat. **6**(2), 461–464 (1978)
17. Witten, I.H., Frank, E.: Data Mining: Practical Machine Learning Tools and Techniques. Morgan Kaufmann, Burlington (2005)
18. Cooper, G.F., Herskovits, E.: A Bayesian method for the induction of probabilistic networks from data. Mach. Learn. **9**(4), 309–347 (1992)
19. Su, C., Andrew, A., Karagas, M.R., Borsuk, M.E.: Using Bayesian networks to discover relations between genes, environment, and disease. BioData Min. **6**(1), 1 (2013)
20. Druzdzel, M.J.: SMILE: structural modeling, inference, and learning engine and GeNIe: a development environment for graphical decision-theoretic models. In: AAAI/IAAI, pp. 902–903 (1999)
21. Netica (2012). http://www.norsys.com/dl/Netica_Win.exe
22. Scutari, M.: Learning Bayesian networks with the bnlearn R package. J Stat. Soft **35**, 1–22 (2010)

23. Schwab, K.: The global competitiveness report 2009–2010. World Economic Forum, 2009, 2010, 2011, 2012. World Economic Forum, Geneva
24. Cinicioglu, E.N., Ulusoy, G., Önsel, S., Ülengin, F., Ülengin, B.: The basic competitiveness factors shaping the innovation performance of countries. In: Proceedings of the International Conference of Institute of Industrial Engineers, Istanbul, pp. 1–5 (2013)
25. Gaulier, G., Zignago, S.: BACI: international trade database at the product-level. The 1994–2007 Version. Working Papers 2010–23, CEPII Research Center (2010)

Incremental Junction Tree Inference

Hamza Agli[1], Philippe Bonnard[1], Christophe Gonzales[2(✉)], and Pierre-Henri Wuillemin[2]

[1] IBM France Lab, Gentilly, France
{hamza.agli,philippe.bonnard}@fr.ibm.com
[2] Sorbonne Universités, UPMC Univ Paris 6, CNRS, UMR 7606 LIP6, Paris, France
{christophe.gonzales,pierre-henri.wuillemin}@lip6.fr

Abstract. Performing probabilistic inference in multi-target dynamic systems is a challenging task. When the system, its evidence and/or its targets evolve, most of the inference algorithms either recompute everything from scratch, even though incremental changes do not invalidate all the previous computations, or do not fully exploit incrementality to minimize computations. This incurs strong unnecessary overheads when the system under study is large. To alleviate this problem, we propose in this paper a new junction tree-based message-passing inference algorithm that, given a new query, minimizes computations by identifying precisely the set of messages that differ from the preceding computations. Experimental results highlight the efficiency of our approach.

Keywords: Bayesian networks · Incremental inference · Junction tree

1 Introduction

Bayesian networks (BN) [10,17] are one of the most popular framework for reasoning with uncertainty in expert systems. They are used in a wide range of real-world applications, including medical diagnosis, risk management and clinical decision support. A *BN* is a compact graphical representation of a joint probability distribution. It can be considered as a probabilistic knowledge base, in which the process of querying/requesting is called inference. Different queries exist, including the computation of most probable explanations or that of the posterior marginal distributions of some random variables (hereafter called *targets*). In this paper, we focus on the latter. It is known to be NP-hard in general [2,3] but many exact and approximate inference algorithms have been proposed in the literature [12,15,20]. Extensions to handle very large systems [11,18,21] and temporal features have also been proposed [6,16,19]. Their increased complexity requires even more efficient inference algorithms.

Rule-based systems, which originated our research, are nowadays a very popular tool for automating decision making. To quantify uncertainties in the domain, they most often use heuristic models, e.g., *certainty factors* [1], which have theoretical and practical limitations [8] that could be overcome by exploiting probabilities. In this context, BNs could prove to be useful. In addition,

J.P. Carvalho et al. (Eds.): IPMU 2016, Part I, CCIS 610, pp. 326–337, 2016.
DOI: 10.1007/978-3-319-40596-4_28

their efficient inference engines, notably cluster-based and junction tree-based algorithms [9,15,20], seem to be good candidates to speed-up the rules inference process. But, by essence, rule-based systems are incremental multi-target environments, so BN inference shall also be performed incrementally. Some algorithms exploit partially this feature (see [15]) but they are far from optimal when the set of targets is smaller than the set of all the random variables or when it changes. The problem is even worse when the structure of the junction tree (JT) evolves over time. This can become an issue in rule-based systems in which changes in the *BN* structure, the evidence and the *targets*, occur frequently.

In [4], 4 incrementality criteria relevant to probabilistic inference were introduced: incrementality w.r.t resources, queries, evidence and representation. In this paper, we are interested in all these criteria, especially in the last three. Surprisingly, very few inference algorithms address all these aspects. In [5], for instance, the query point of view is taken into account by reconfiguring dynamically some join trees when queries change but the *BN* structure is assumed to remain static, which may not necessarily be the case in rule-based systems. In [14], the authors exploit relevance-based reasoning to identify the parts of the network that are relevant for computations and, then, update several subnetworks whose union covers the original one. Unfortunately, this algorithm does not take into account computations performed previously. In [15], an incremental JT-based inference algorithm has been proposed that exploits independences induced by incremental evidence updates. But the JT structure never evolves and it is assumed that all the nodes are targets, which is not optimal in our context. On the opposite, the incremental JT structure is addressed in [7] but not the queries incrementality nor the exploitation of previous probabilistic computations. Along similar lines, Li *et al.* argue that compiling the original *BN* into a conjunctive normal form coupled with caching techniques improves inference when the network structure is updated [13]. But this does not take optimally into account evidence and queries. In this paper, we investigate a new approach to overcome the above shortcomings. This approach aims at improving the efficiency of inference for very large and dynamic systems. The key idea of our algorithm, called Incremental Junction Tree Inference (*IJTI*), consists of restricting the computations only to parts of the JT that are relevant to *targets* and that have been invalidated by incremental changes. As a consequence, *IJTI* minimizes the probabilistic computations.

The paper is organized as follows. In the next section, we introduce the necessary background. In Sect. 3, we present our approach and justify its correctness. Then we highlight the efficiency of our contribution with a set of experiments. Finally, some conclusion and future works are provided in Sect. 5. All the proofs are given in an appendix.

2 Preliminaries and Notations

A BN is a pair $(\mathcal{G}, \Theta)$, where $\mathcal{G} = (\mathcal{V}, \mathcal{A})$ is a directed acyclic graph (DAG). $\mathcal{V}$ is a set of nodes representing random variables. $\mathcal{A}$ is a set of arcs and

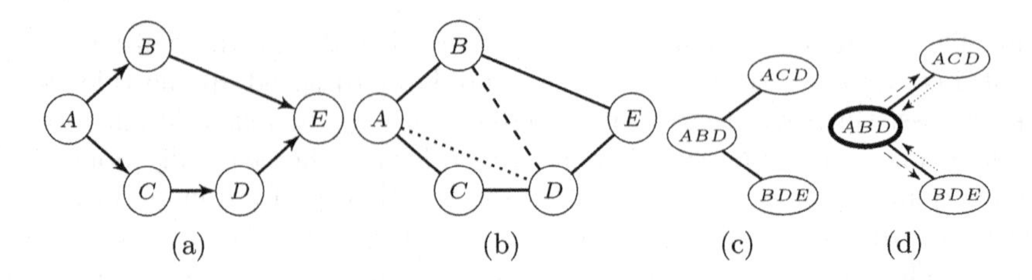

Fig. 1. A JT construction

$\Theta = \{P(X|\text{Pa}(X)) : X \in \mathcal{V}\}$ is the set of the conditional probability tables (CPT) of the variables in $\mathcal{V}$ given their parents in $\mathcal{G}$. The BN encodes the joint probability over $\mathcal{V}$ as the product of these CPTs. In this paper, probabilistic inference is based on a message-passing algorithm within a JT. Constructing the latter consists of, first, converting DAG $\mathcal{G}$ into an undirected graph by adding, for each node in $\mathcal{V}$, edges between all of its parents (*moralization*) and removing the orientations of the remaining arcs, and, then, by adding an edge between a pair of non-adjacent nodes in every cycle of at least four nodes (*triangulation*). The nodes of the JT correspond to complete maximal subgraphs (cliques) of the resulting graph. These nodes are linked by edges in such a way that *i*) the JT contains no loop; and *ii*) any pair of cliques with a nonempty intersection are linked by a path on which all cliques contain this intersection. Figure 1a shows an example of a DAG, its moralized and triangulated graph are given in Fig. 1b, where dashed and dotted edges represent those added during *moralization* and *triangulation* respectively. Finally Fig. 1c depicts a corresponding JT. Note that a JT can be a forest, e.g., when DAG $\mathcal{G}$ is not connected. In our approach, dealing with a forest is equivalent to iterate the same process on its connected components. Hence, without loss of generality, we will consider in the sequel that the JT on which we will perform inference is a tree $\mathcal{T}$. Hereafter, for any JT $\mathcal{T}$, we will denote by $\mathcal{V}(\mathcal{T})$ and $\mathcal{E}(\mathcal{T})$ its set of cliques and edges respectively.

The message-passing algorithm consists of performing a *collect* and a *distribution* from a predetermined root $r \in \mathcal{V}(\mathcal{T})$. During the *collect*, messages are sent along edges from leaves toward r and, during the *distribution*, they are sent in the opposite direction. To guarantee the correctness of computations, for any edge $(i, j) \in \mathcal{E}(\mathcal{T})$, the message sent from i to j, denoted by $\psi_{i\to j}$, is computed only when clique i has received messages from all its neighbors except j. Figure 1d shows an example of message-passing with $r = ABD$ (thick clique) and dotted and dashed arcs representing the *collect* and *distribution* messages respectively. The computation of these messages is beyond the scope of this paper but can be found in [15]. In an incremental environment, not all the messages need be recomputed each time a modification occurs because, in practice, many will remain the same. As we shall see, using the following definitions, we can characterize precisely those that need some update given new set of evidence, structural changes or/and targets.

Definition 1 (Path). *Let $\mathcal{T} = (\mathcal{V}(\mathcal{T}), \mathcal{E}(\mathcal{T}))$ be a JT. $i_1, \ldots, i_{n+1}$ is said to be a path in $\mathcal{T}$ if $(i_\alpha, i_{\alpha+1}) \in \mathcal{E}(\mathcal{T})$ for all $\alpha \in \{1, \ldots, n\}$. For simplicity, this path is denoted by $i_1 - i_{n+1}$ and its length by $len(i_1 - i_{n+1})$ (which is equal to n).*

Definition 2 (Adjacency). *Let $i, j \in \mathcal{V}(\mathcal{T})$, $i \neq j$. i and j are adjacent in $\mathcal{T}$ iff $(i, j) \in \mathcal{E}(\mathcal{T})$. The set of cliques adjacent to i is denoted by $\mathrm{Adj}(i)$, i.e., $\mathrm{Adj}(i) := \{k \in \mathcal{V}(\mathcal{T}) : (i, k) \in \mathcal{E}(\mathcal{T})\}$. Let $r \in \mathcal{V}(\mathcal{T})$, $r \neq i$, then $Adj_r(i)$ denotes the singleton set containing the clique adjacent to i that is on the path between i and r, i.e., $Adj_r(i) := \{k \in \mathrm{Adj}(i) : k \in i - r\}$. We also define $Adj_r(r) := \emptyset$. Finally, let $\mathrm{Adj}_{\text{-}j}(i) := \mathrm{Adj}(i) \setminus \{j\}$.*

For instance, in Fig. 2a, $Adj_r(i) = \{k_3\}$ and $Adj_r(k_3) = \{r\}$. Finally, let $\mathcal{V}_{\text{-}j}(i)$ stands for the set of nodes of the maximal subtree in $\mathcal{T}$ that contains i and not $Adj_j(i)$, and let $\mathcal{V}_j(i) = \mathcal{V}_{\text{-}j}(i) \cup \{j\}$ (see the shadowed area in Fig. 2a).

A message $\psi_{i \to j}$ sent within $\mathcal{T}$ is directed by nature. It propagates toward j (and, by induction, toward $\mathcal{V}_{\text{-}i}(j)$) all the relevant information coming from the cliques in $\mathcal{V}_{\text{-}j}(i)$, notably all the evidence they received (by abuse, we say that a clique received evidence when at least one of its random variables received evidence). As a consequence, if $\psi_{i \to j}$ has already been computed previously and no new evidence has been received nor structural changes occured in $\mathcal{V}_{\text{-}j}(i)$, there is no need to recompute it. But even if $\mathcal{V}_{\text{-}j}(i)$ received evidence, $\psi_{i \to j}$ needs not be computed/updated if $\mathcal{V}_{\text{-}i}(j)$ contains no target. In this case, $\psi_{i \to j}$'s state becomes "invalid" since the content of $\psi_{i \to j}$ is now incorrect. This is not an issue for the current inference but, for future ones, we have to take this state into account to recompute $\psi_{i \to j}$ if it is to be used. Let $\mathcal{A}(\mathcal{T})$ be the set of all arcs induced from $\mathcal{E}(\mathcal{T})$, taking into account orientations, i.e., $\mathcal{A}(\mathcal{T}) := \bigcup_{(i,j) \in \mathcal{E}(\mathcal{T})} \{(i, j)\} \cup \{(j, i)\}$. To formalize the above conditions, we begin with characterizing the information that is "local" to i and j by:

Definition 3 (Local label-message λ). *$\lambda : \mathcal{A}(\mathcal{T}) \mapsto 2^{\{\epsilon, T\}}$ is a function s.t.*

$$(i, j) \longmapsto \lambda_{i \to j} := \begin{cases} \{\epsilon\} & \text{if } \psi_{i \to j} \text{ is in "invalid state" or "new evidence or} \\ & \text{structural changes" have affected } i \text{ (1)} \\ \{T\} & \text{if } i \text{ contains targets (2)} \\ \{T, \epsilon\} & \text{if (1) and (2)} \\ \emptyset & \text{otherwise} \end{cases}$$

To simplify the notation, hereafter, we will remove braces and denote $\{T, \epsilon\}$ by $T\epsilon$. Then, the idea of our algorithm consists of marking every arc (i, j) in $\mathcal{A}(\mathcal{T})$ by labels $\mu_{i \to j}$ expressing all the "local" information that $\mathcal{V}_{\text{-}j}(i)$ contains.

Definition 4 (Label-message μ). *For $(i, j) \in \mathcal{A}(\mathcal{T})$, the label-message sent from i to j is a function $\mu : \mathcal{A}(\mathcal{T}) \mapsto 2^{\{\epsilon, T\}}$ such that $\mu_{i \to j} := \bigcup_{\substack{k' \in \mathcal{V}_{\text{-}j}(i) \\ \{k\} = Adj_j(k')}} \lambda_{k' \to k}$.*

As an example of the previous discussion, imagine that a first incremental update impacts the initial DAG and consequently the initial $\mathcal{T}$ of Fig. 2a. This consists of the removal of k_4, the insertion of an evidence on r and a new target

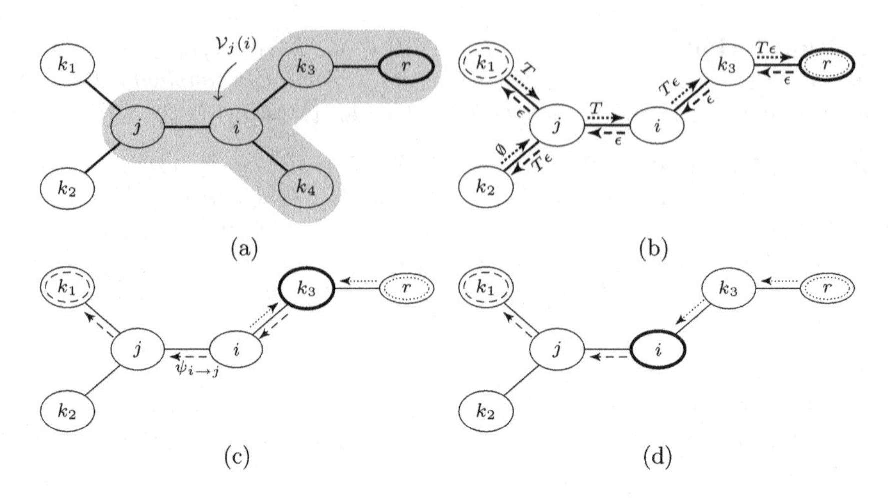

Fig. 2. Message passing within a JT $\mathcal{T}$.

on k_1. Figure 2b depicts the μ-messaging within $\mathcal{T}$ after this update, where dashed and dotted ellipses stand for the cliques containing targets and evidence respectively. One can easily see that, for instance, $\mu_{i \to j} = \epsilon$, $\mu_{j \to i} = T$ and $\mu_{j \to k_2} = T\epsilon$. The following proposition allows to recursively construct the μ-messages:

Proposition 1 (μ construction). *Let $(i, j) \in \mathcal{A}(\mathcal{T})$, then we have:* $\mu_{i \to j} = \lambda_{i \to j} \cup \bigcup_{k \in \text{Adj}_{\text{-}j}(i)} \mu_{k \to i}$.

3 Inference Optimization: *IJTI*

Let us recall that $\psi_{i \to j}$ denotes the message exchanged between cliques i and j during an inference computation. It shall not be confused with the label message $\mu_{i \to j}$ of Definition 4.

3.1 Optimal Roots

Usually, the number of computations performed by a JT-based message-passing algorithm does not depend on the root clique selected for collect/distribution because the $\psi_{i \to j}$ messages are sent on both directions on *all* the edges of the JT. For IJTI, this is not the case, since this algorithm computes and sends *only* the $\psi_{i \to j}$ messages necessary for the computation of the posterior distributions of its target nodes. On some edges, IJTI will therefore not compute some $\psi_{i \to j}$ messages because they are irrelevant w.r.t. the targets posterior distributions. As a consequence, in IJTI, the number of computations performed is sensitive to the selection of the root: for instance, in the JT of Fig. 2a, if clique i received evidence and the only target is j, only message $\psi_{i \to j}$ from i to j is necessary, which is precisely what is sent if clique i is selected as root (here, only a distribution is necessary). But if clique k_4 is selected instead, message $\psi_{i\to k_4}$ needs to be sent during the collect and messages $\psi_{k_4\to i}$ and $\psi_{i\to j}$ need to be sent during the distribution, which is clearly not optimal. To determine the optimal roots, let us define $\delta_{i\to j}(r)$ as an indicator of whether message $\psi_{i\to j}$ is recomputed (in this case, $\delta_{i\to j}(r) = 1$) or not ($\delta_{i\to j}(r) = 0$) when r is selected as a root. In IJTI, we therefore seek to minimize $\delta(r) = \sum_{(k',k)\in\mathcal{A}(\mathcal{T})} \delta_{k'\to k}(r)$, which corresponds to the total number of messages recomputed and sent. Based on the discussion of the preceding section, we can write:

$$\delta_{i\to j}(r) = \begin{cases} 1 \text{ if } (\epsilon \in \mu_{i\to j} \text{ and } \{j\} = Adj_r(i)) \text{ or} \\ \qquad (\epsilon \in \mu_{i\to j} \text{ and } \{i\} = Adj_r(j) \text{ and } T \in \mu_{j\to i}) \\ 0 \text{ otherwise} \end{cases} \tag{1}$$

The first line of Eq. (1) concerns *collect* messages ($\{j\} = Adj_r(i)$). It asserts that *collect* message $\psi_{i\to j}$ needs to be recomputed only if it is currently in an invalid state or if new evidence or structural changes have occurred in $\mathcal{V}_{\text{-}j}(i)$ ($\epsilon \in \mu_{i\to j}$). When this is not the case, clearly, this message is up to date and does not need recomputation. The second line of Eq. (1) concerns *distribution* messages ($\{i\} = Adj_r(j)$). It asserts that $\psi_{i\to j}$ needs to be recomputed only if there exists a target farther toward the leaves of the JT ($T \in \mu_{j\to i}$) and if some evidence has been received on $\mathcal{V}_{\text{-}j}(i)$ or some message coming from $\mathcal{V}_{\text{-}j}(i)$ has been updated ($\epsilon \in \mu_{i\to j}$). Equation (1) can be rewritten more compactly as:

$$\delta_{i\to j}(r) = \begin{cases} 1 \text{ if } \epsilon \in \mu_{i\to j} \text{ and} \\ \qquad (\{j\} = Adj_r(i) \text{ or } (\{i\} = Adj_r(j) \text{ and } T \in \mu_{j\to i})) \\ 0 \text{ otherwise} \end{cases} \tag{2}$$

Figure 2c and d illustrate that $\delta(k_3) = 5$ and $\delta(i) = 4$ respectively. In this case, it is better to select i as a root rather than k_3 since this avoids the unnecessary computation of one message. The following theorem states the existence of some optimal roots and characterize them:

Theorem 1 (Optimal roots). *Suppose we computed the μ-messages within $\mathcal{T}$. Then there exists $r \in \mathcal{V}(\mathcal{T})$ fulfilling one of the following mutually exclusive and exhaustive properties:*

(a) $(\mathcal{V}(\mathcal{T}), \mathcal{E}(\mathcal{T})) = (\{r\}, \emptyset)$
(b) $\exists r' \in \mathcal{V}(\mathcal{T}) : \mu_{r'\to r} = \mu_{r\to r'} = T\epsilon$
(c) $\forall k \in Adj(r) : \mu_{k\to r} \in \{T, \epsilon, \emptyset\}$

In addition, $r \in Argmin_{k\in\mathcal{V}(\mathcal{T})}\delta(k)$, i.e., r is an optimal root w.r.t. inference computations.

3.2 A New Incremental Inference

In this section, we propose a new algorithm designed to deal with incremental inference. We assume that a first inference has been performed by message-passing within $\mathcal{T}$, using for instance a collect-distribute algorithm in a Lazy

Algorithm 1. IJTI

```
input  : modified T, Q targets cliques
output : posteriors on targets
// set the number of neighbors
   visited during the collect
1  for i ∈ V(T) do
2      i.nbVN ← 0
3  Compute the μ-labels in T
4  Find r using Theorem 1
5  L ← the set of leaves of T
   // collect phase
6  foreach clique i ∈ L do
7      p ← Adj_r(i)
8      if δ_{i→p}(r) = 1 then
9          Compute ψ_{i→p}
10     p.nbVN ← p.nbVN + 1
11     if p ≠ r and p.nbVN = |Adj(p)| − 1
         then L ← L ∪ {p}
12     L ← L \ {i}
   // distribution phase
13 L ← {r}
14 foreach clique i ∈ L do
15     foreach j ∈ Adj(i) \ Adj_r(i) do
16         if δ_{i→j}(r) = 1 then
17             Compute ψ_{i→j}
18             L ← L ∪ {j}
19 foreach clique t ∈ Q do
20     Compute the posterior distributions
       of the target nodes in clique t
21 return posterior distributions
```

Propagation-like architecture. Afterwards, incremental changes occur. Then *IJTI* is called to optimize the inference process. We recall that we use a target-driven approach, hence, we recompute only invalidated *collect* messages and we only *distribute* messages up to the *targets.* Under these assumptions, the proposed algorithm is described in Algorithm 1. It runs a revised message-passing algorithm to compute $\psi_{i\to j}$ only when $\delta_{i\to j}(r)=1$ for all i, j in the modified junction tree $\mathcal{T}$. In line 5, a leaf clique i is such that $|\mathrm{Adj}(i)|=1$. We emphasize that computing messages is performed similarly to a classic JT-based inference algorithm. The correctness of IJTI is guaranteed by the following proposition:

Proposition 2. *The IJTI algorithm is sound, i.e., computing only messages $\psi_{i\to j}$ such that $\delta_{i\to j}(r)=1$, for all $(i,j)\in\mathcal{A}(\mathcal{T})$, results in the correct computation of the posterior distributions of the target variables.*

4 Experiments

In this section, we highlight the effectiveness of our algorithm by comparing the gain of using it instead of any non-incremental JT-based inference algorithm. This gain is equal to $1-\delta(r)/(2|\mathcal{E}(\mathcal{T})|)$, i.e., this is the percentage of unnecessary messages that IJTI avoids to compute compared to the messages sent by classical inference algorithms on both directions on all the edges.

For this purpose, we performed tests using the aGrUM library[1] on 9 real-world BNs of different complexities as well as on randomly generated BNs. The

[1] http://agrum.lip6.fr.

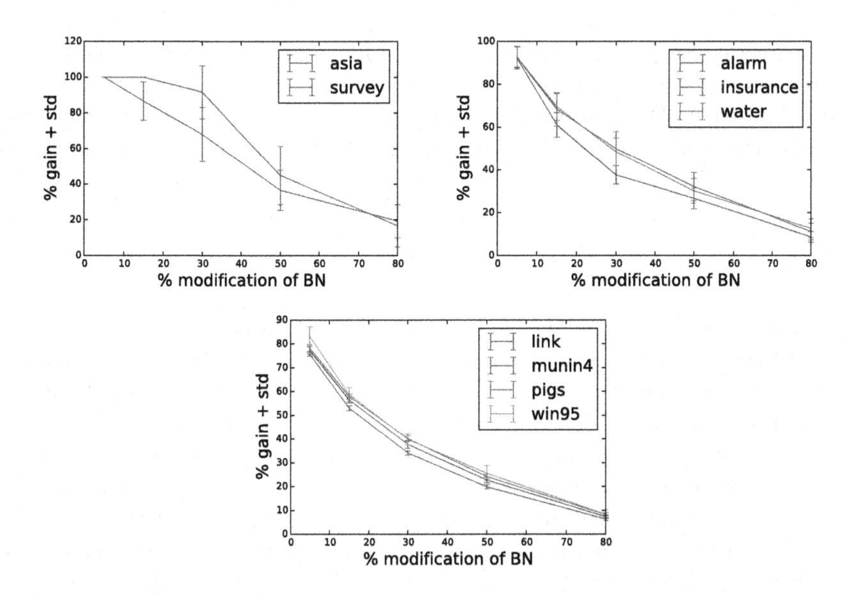

Fig. 3. *IJTI* gain for real BNs (Color figure online)

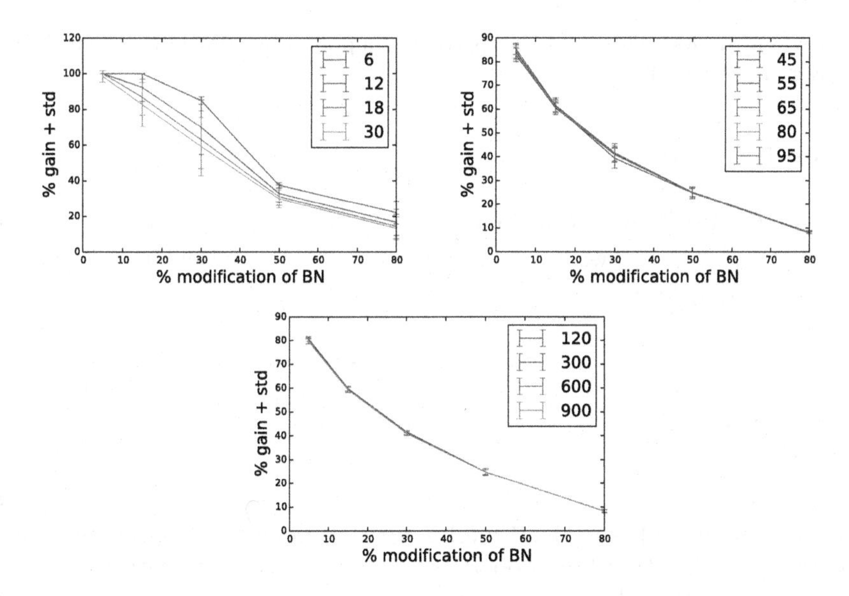

Fig. 4. *IJTI* gain for artificial BNs (Color figure online)

latter contained $nbNodes$ Boolean random variables, ($6 \leq nbNodes \leq 900$, see Fig. 4) and, for each value of $nbNodes$, 3 BNs were generated with $nbArcs$ arcs, $nbArcs$ being chosen randomly in the interval $[nbNodes - 1, 4/3 * nbNodes - 1]$.

We simulated the incrementality by randomly choosing for each inference a set of targets and modified cliques. This induced invalid messages in $\mathcal{T}$. Figures 3 and 4 show the average resulting gains and their standard deviations (error bars[2]) over 20 incremental inference queries. Note that the behavior of the algorithm is the same for real-world BNs and for randomly generated ones. As could be expected, the smaller the modifications, the bigger the gain. Note also that the gain is not too sensitive to the size of the BN.

5 Discussion and Future Work

In this paper, we introduced *IJTI*, a new incremental junction-tree-based inference algorithm for multi-target dynamic systems. Assuming that a first complete inference has been performed, it extracts an optimal root and optimizes the inference accordingly. The correctness of these two optimizations is proved and experiments highlight that our approach allows for important savings compared to classical ones. For future works, we plan to improve our algorithm, notably by taking into account caching for the determination of the roots. We also plan to apply IJTI in Probabilistic Relational Models in order to speed-up their inference. Finally, we aim at coupling our approach with rule-based expert systems to improve their probabilistic reasoning.

Acknowledgments. This work was partially supported by IBM France Lab/ANRT CIFRE grant #2014/421.

Appendix: Proofs

Proof of Proposition 1: Note that $\mathcal{V}_{\text{-}j}(i) = \{i\} \cup \bigcup_{k \in \text{Adj}_{\text{-}j}(i)} \mathcal{V}_{\text{-}i}(k)$ and, for $k \in \text{Adj}_{\text{-}j}(i)$, $l' \in \mathcal{V}_{\text{-}i}(k)$, we have $Adj_j(l') = Adj_i(l')$. Using Definition 4, one can thus rewrite $\mu_{i \to j}$ into:

$$\mu_{i \to j} = \bigcup_{\substack{l' \in \mathcal{V}_{\text{-}j}(i) \\ \{l\} = Adj_j(l')}} \lambda_{l' \to l} = \lambda_{i \to j} \cup \bigcup_{k \in \text{Adj}_{\text{-}j}(i)} \overbrace{\bigcup_{\substack{l' \in \mathcal{V}_{\text{-}i}(k) \\ \{l\} = Adj_j(l')}} \lambda_{l' \to l}}^{\mu_{k \to i}} = \lambda_{i \to j} \cup \bigcup_{k \in \text{Adj}_{\text{-}j}(i)} \mu_{k \to i}$$

■

Proof of Theorem 1 – mutual exclusivity: if property (a) is satisfied, then $\mathcal{T}$ contains no edge, therefore properties (b) and (c) cannot be satisfied.

Now, assume that there exist r_1, r'_1 such that $\mu_{r'_1 \to r_1} = \mu_{r_1 \to r'_1} = \mathcal{T}\epsilon$ (property b). Let r_2 be any clique in $\mathcal{V}(\mathcal{T})$. Without loss of generality, assume that r_1 lies on the path $i_1 = r_2, i_2, \ldots, i_p = r'_1$ between r_2 and r'_1. Then, by Proposition 1, $\mu_{i_2 \to r_2} \supseteq \mu_{i_3 \to i_2} \supseteq \cdots \supseteq \mu_{r'_1 \to r_1} = \mathcal{T}\epsilon$. Therefore, properties (b) and (c) cannot hold simultaneously. ■

[2] In Fig. 3a, due to the small number of nodes and arcs in the BNs, percentages of modifications lower than 10 % imply no modification at all, hence the lack of error bars.

Proof of Theorem 1 – r's existence: if $\mathcal{A}(\mathcal{T}) = \emptyset$, then property (a) holds and r is the unique node of $\mathcal{T}$. Now, assume that $\mathcal{A}(\mathcal{T}) \neq \emptyset$. If there exists an edge $(i,j) \in \mathcal{E}(\mathcal{T})$ such that $\mu_{i \to j} = \mu_{j \to i} = T\epsilon$, then $r = i$ satifies property (b). Otherwise, neither properties (a) nor (b) hold. Assume that property (c) neither holds. Then, for all edges (i,j), exactly one of $\mu_{i \to j}$ or $\mu_{j \to i}$ is equal to $T\epsilon$ and the other one belongs to $\{\emptyset, \epsilon, T\}$. Let (i_0, j_0) be such that $\mu_{i_0 \to j_0} = T\epsilon$ and $\mu_{j_0 \to i_0} \neq T\epsilon$. Then, if $|\mathrm{Adj}(i_0)| = 1$, clique i_0 satisfies property (c), a contradiction. As we assume that property (b) neither holds, there exists $i_1 \in \mathrm{Adj}(i_0)$ such that $\mu_{i_1 \to i_0} = T\epsilon$ and $\mu_{i_0 \to i_1} \neq T\epsilon$. The same reasoning holds for i_1, hence either i_1 is a leaf, which contradicts property (c) or i_1 has another neighbor i_2 such that $\mu_{i_2 \to i_1} = T\epsilon$ and $\mu_{i_1 \to i_2} \neq T\epsilon$. By induction, we create a path $i_1, \ldots, i_n$ of maximal size. This path is necessarily finite since $\mathcal{T}$ is a finite tree, hence clique i_n is a leaf which, therefore, satisfies property (c), a contradiction. Consequently, when properties (a) and (b) do not hold, property (c) holds. ■

One can now prove separately the optimality for each property of Theorem 1, since these properties are mutually exclusive:

Proof of Theorem 1 – property a's optimality: r is the only node in $\mathcal{T}$. Choosing it as a root is therefore optimal. ■

Lemma 1. *Let $i, j \in \mathcal{V}(\mathcal{T})$ be such that $\epsilon \in \mu_{j \to i}$ and $\mu_{i \to j} = \emptyset$, then $\forall l \in \mathcal{V}_{\text{-}j}(i) : \delta(l) = \delta(j) + len(l - j)$.*

Proof. Note that when $\epsilon \notin \mu_{j \to i}$, $\mathcal{T}$ is up-to-date in the current inference and there is no need to perform any computation. The proof is achieved by induction on $n = len(l - j)$. For $n = 1$, we have $l = i$, so by Eq. (2) and the fact that $\epsilon \in \mu_{j \to i}$ and $i \in Adj_i(j)$, we get $\delta_{j \to i}(i) = 1$. As a consequence, $\delta(i) = \sum_{(k',k) \in \mathcal{A}(\mathcal{T}) \setminus \{(j,i)\}} \delta_{k' \to k}(i) + 1$. Yet, as $T \notin \mu_{i \to j}$ we have $\delta_{j \to i}(j) = 0$; so $\delta(j) = \sum_{(k',k) \in \mathcal{A}(\mathcal{T})) \setminus \{(j,i)\}} \delta_{k' \to k}(j)$. Since $\epsilon \notin \mu_{i \to j}$, $\delta_{i \to j}(i) = \delta_{i \to j}(j) = 0$. For $(k',k) \neq (i,j), (j,i)$, we have $Adj_i(k) = Adj_j(k)$ and $Adj_i(k') = Adj_j(k')$. In this case, it follows that $\delta_{k' \to k}(i) = \delta_{k' \to k}(j)$. We conclude that $\delta(i) = \delta(j) + 1$.

Now suppose this property is satisfied for $n - 1 > 1$, let us prove that it remains true for n. Let l be such that $len(l - j) = n - 1$. Let $\{p\} = Adj_i(l)$. Then $\delta(l) = 1 + \sum_{(k',k) \in \mathcal{A}(\mathcal{T}) \setminus \{(p,l)\}} \delta_{k' \to k}(l)$ because $\delta_{p \to l}(l) = 1$ (since $\epsilon \in \mu_{p \to l}$ and $\{l\} = Adj_l(p)$). Knowing that $T \notin \mu_{l \to p}$, we get $\delta_{p \to l}(p) = 0$, it follows that $\delta(p) = \sum_{(k',k) \in \mathcal{A}(\mathcal{T}) \setminus \{(p,l)\}} \delta_{k' \to k}(p)$. Now using the same reasoning as in the case $n = 1$ and by remarking $\delta_{l \to p}(p) = \delta_{l \to p}(l) = 0$ because $\epsilon \notin \mu_{l \to p}$, we conclude that $\delta(l) = 1 + \sum_{(k',k) \in \mathcal{A}(\mathcal{T}) \setminus \{(p,l)\}} \delta_{k' \to k}(l) = 1 + \sum_{(k',k) \in \mathcal{A}(\mathcal{T}) \setminus \{(p,l)\}} \delta_{k' \to k}(p) = 1 + \delta(p)$. By applying the induction hypothesis on l, where $len(l - j) = n - 1$, we obtain: $\delta(l) = 1 + \delta(p) = 1 + n - 1 + \delta(j) = \delta(j) + n$. ■

Lemma 2. *Let $\mathcal{V}_1 = \{r \in \mathcal{V}(\mathcal{T}) : \exists k \in \mathrm{Adj}(r), \mu_{r \to k} = \mu_{k \to r} = T\epsilon\}$, then for any r, r' in $\mathcal{V}_1$ we have $\delta(r) = \delta(r')$.*

Proof. Assume that $|\mathcal{V}_1| > 1$. By Proposition 1, the nodes in $\mathcal{V}_1$ form a connected subgraph. Let $r, r' \in \mathcal{V}_1$ be such that $(r, r') \in \mathcal{E}(\mathcal{T})$. Finally, let $(k', k) \in \mathcal{A}(\mathcal{T}) \setminus \{(r, r'), (r', r)\}$. If $k' \notin \{r, r'\}$, then either $k = r$, $k = r'$

or $k \notin \{r, r'\}$ and in all these cases we have: $Adj_r(k') = Adj_{r'}(k')$, hence $\delta_{k' \to k}(r) = \delta_{k' \to k}(r')$. Otherwise, let $k' = r'$ then $k \neq r$ and we have also[3] $Adj_r(k) = Adj_{r'}(k)$ and again $\delta_{k' \to k}(r) = \delta_{k' \to k}(r')$. As a consequence: $\sum_{(k',k) \in \mathcal{A}(\mathcal{T}) \setminus \{(r,r'),(r',r)\}} \delta_{k' \to k}(r) = \sum_{(k',k) \in \mathcal{A}(\mathcal{T}) \setminus \{(r,r'),(r',r)\}} \delta_{k' \to k}(r')$. By Eq. (2), we get: $\delta_{r \to r'}(r) + \delta_{r' \to r}(r) = \delta_{r \to r'}(r') + \delta_{r' \to r}(r') = 2$. We conclude that $\delta(r) - \sum_{(k',k) \in \mathcal{A}(\mathcal{T}) \setminus \{(r,r'),(r',r)\}} \delta_{k' \to k}(r) = \delta(r') - \sum_{(k',k) \in \mathcal{A}(\mathcal{T}) \setminus \{(r,r'),(r',r)\}} \delta_{k' \to k}(r')$. Hence $\delta(r) = \delta(r')$. ■

Proof of Theorem 1 – property b's optimality: Under the notations of property b), it is sufficient to prove that for any i not in $\mathcal{V}_1$, $\delta(r) \leq \delta(i)$[4]. Without loss of generality, assume that $i \in \mathcal{V}_{\text{-}r'}(r)$. Let $(k, k') \in \mathcal{A}(i - r)$, where $\mathcal{A}(i - r)$ is the set of arcs induced from $i - r$. We either have $\{k'\} = Adj_r(k)$ or $\{k\} = Adj_r(k')$. Assume for instance that $\{k'\} = Adj_r(k)$, $k \neq r$, the second case should be treated similarly. Then $\mu_{k' \to k} = T\epsilon$ and by applying Eq. 2, we summarize the results on the following table:

$\mu_{k \to k'}$	$\delta_{k \to k'}(i) + \delta_{k' \to k}(i)$	$\delta_{k \to k'}(r) + \delta_{k' \to k}(r)$
$\emptyset$	1	0
T	1	1
ϵ	2	1

we conclude that $\sum_{(k',k) \in \mathcal{A}(i - r)} \delta_{k' \to k}(r) \leq \sum_{(k',k) \in \mathcal{A}(i - r)} \delta_{k' \to k}(i)$. (1)

Now for $(k, k') \notin \mathcal{A}(i - r)$ it is easy to see that $\delta_{k \to k'}(i) = \delta_{k \to k'}(r)$ and hence: $\sum_{(k,k') \in \mathcal{A}(\mathcal{T}) \setminus \mathcal{A}(i - r)} \delta_{k' \to k}(r) = \sum_{(k,k') \in \mathcal{A}(\mathcal{T}) \setminus \mathcal{A}(i - r)} \delta_{k' \to k}(i)$. (2).

By comparing (1) and (2) we get that $\delta(r) \leq \delta(i)$ for $i \notin \mathcal{V}_1$. So far, we obtain, by Lemma 2, for any i in $\mathcal{V}_1$, $\delta(r) = \delta(i)$ and for any i not in $\mathcal{V}_1, \delta(r) \leq \delta(i)$, therefore we have $r \in Argmin_{i \in \mathcal{V}(\mathcal{T})} \delta(i)$. ■

Proof of Theorem 1 – property c's optimality: Let i in $\mathcal{V}(\mathcal{T})$ s.t. $i \neq r$.

first case: $\mu_{Adj_i(r) \to r} = \emptyset$. Assume that $T, \epsilon \in \mathcal{V}_{\text{-}i}(r)$, because otherwise there is no need to perform any computation, as either there is no query or no modification in $\mathcal{T}$; so by Lemma 1 we have $\delta(i) = \delta(r) + len(i - r)$ because $i \in \mathcal{V}_{\text{-}r}(Adj_i(r))$. Hence $\delta(r) < \delta(i)$.

second case: we omit the case $\mu_{Adj_i(r) \to r} \in \{T, \epsilon\}$, but one should use the same methodology as in property b)'s proof and the fact that for any k, k' in $i - r$ s.t $\{k'\} = Adj_r(k) : \mu_{k \to k'} = \mu_{i \to Adj_r(i)}$ and examine $\delta_{k' \to k}(r)$ and $\delta_{k' \to k}(i)$. ■

Proof of Proposition 2: Given a root r, $\delta_{i \to j}(r)$ corresponds, by construction, to the fact that $\psi_{i \to j}$ is necessary during the current inference and was invalidated in the previous one. As a consequence, the current inference needs to recompute only such a message for any i, j in $\mathcal{V}(\mathcal{T})$. ■

[3] If $k' = r$ then $k \neq r'$ and the equality also verified.

[4] All the nodes are computationally equivalent if $\forall i \in \mathcal{V}(\mathcal{T}), i \in \mathcal{V}_1$ since $\mathcal{V}(\mathcal{T}) = \mathcal{V}_1$.

References

1. Buchanan, B.G., Shortliffe, E.H.: Rule Based Expert Systems: The Mycin Experiments of the Stanford Heuristic Programming Project. Addison-Wesley, Reading (1984)
2. Cooper, G.F.: The computational complexity of probabilistic inference using Bayesian belief networks. Artif. Intell. **42**(2–3), 393–405 (1990)
3. Dagum, P., Luby, M.: Approximating probabilistic inference in Bayesian belief networks is NP-hard. Artif. Intell. **60**(1), 141–153 (1993)
4. D'Ambrosio, B.: Incremental probabilistic inference. In: Proceedings of the 9th Conference on Uncertainty in Artificial Intelligence (UAI), pp. 301–308 (1993)
5. Darwiche, A.: Dynamic join trees. In: Proceedings of the Fourteenth Conference on Uncertainty in Artificial Intelligence (UAI), pp. 97–104 (1998)
6. Dean, T., Kanazawa, K.: A model for reasoning about persistence and causation. Comput. Intell. **5**(2), 142–150 (1989)
7. Flores, M.J., Gámez, J.A., Olesen, K.G.: Incremental compilation of Bayesian networks. In: Proceedings of the Nineteenth Conference on Uncertainty in Artificial Intelligence (UAI), pp. 233–240 (2003)
8. Heckerman, D.E., Shortliffe, E.H.: From certainty factors to belief networks. Artif. Intell. Med. **4**(1), 35–52 (1992)
9. Jensen, F., Lauritzen, S., Olesen, K.: Bayesian updating in causal probabilistic networks by local computations. Comput. Stat. Q. **4**, 269–282 (1990)
10. Koller, D., Friedman, N.: Probabilistic Graphical Models: Principles and Techniques. MIT Press, Cambridge (2009)
11. Koller, D., Pfeffer, A.: Probabilistic frame-based systems. In: Proceedings of the 15th National Conference on Artificial Intelligence (AAAI), pp. 580–587 (1998)
12. Lauritzen, S., Spiegelhalter, D.J.: Local computations with probabilities on graphical structures and their applications to expert systems. J. Roy. Stat. Soc. **50**(2), 157–224 (1988)
13. Li, W., van Beek, P., Poupart, P.: Performing incremental Bayesian inference by dynamic model counting. In: Proceedings of the National Conference on Artificial Intelligence (AAAI), pp. 1173–1179 (2006)
14. Lin, Y., Druzdzel, M.J.: Relevance-based sequential evidence processing in Bayesian networks. In: Proceedings of the Eleventh International Florida Artificial Intelligence Research Society Conference (FLAIRS), pp. 446–450 (1998)
15. Madsen, A.L., Jensen, F.V.: Lazy propagation: a junction tree inference algorithm based on lazy evaluation. Artif. Intell. **113**(12), 203–245 (1999)
16. Murphy, K.P.: Dynamic Bayesian networks: representation, inference and learning. Ph.D. thesis, UC Berkeley (2002)
17. Pearl, J.: Probabilistic Reasoning in Intelligent Systems: Networks of Plausible Inference. Morgan Kaufmann, San Mateo (1988)
18. Pfeffer, A.J.: Probabilistic reasoning for complex systems. Ph.D. thesis, Stanford University (2000)
19. Robinson, J., Hartemink, A.: Non-stationary dynamic Bayesian networks, pp. 1369–1376 (2009)
20. Shenoy, P., Shafer, G.: Axioms for probability and belief-function propagation. In: Proceedings of the Conference Uncertainty in Artificial Intelligence, vol. 4, pp. 169–198 (1990)
21. Torti, L., Gonzales, C., Wuillemin, P.H.: Speeding-up structured probabilistic inference using pattern mining. Int. J. Approximate Reasoning **54**(7), 900–918 (2013)

Real Time Learning of Non-stationary Processes with Dynamic Bayesian Networks

Matthieu Hourbracq[1,2](✉), Pierre-Henri Wuillemin[1], Christophe Gonzales[1], and Philippe Baumard[2]

[1] Sorbonne Universités, UPMC Univ Paris 6, CNRS, UMR 7606 LIP6, Paris, France
{matthieu.hourbracq,pierre-henri.wuillemin,christophe.gonzales}@lip6.fr
[2] Akheros, Paris, France
{matthieu.hourbracq,philippe.baumard}@akheros.com

Abstract. Dynamic Bayesian Networks (DBNs) provide a principled scheme for modeling and learning conditional dependencies from complex multivariate time-series data and have been used in a wide scope. However, in most cases, the underlying generative Markov model is assumed to be homogeneous, meaning that neither its topology nor its parameters evolve over time. Therefore, learning a DBN to model a non-stationary process under this assumption will amount to poor predictions capabilities. To account for non-stationary processes, we build on a framework to identify, in a streamed manner, transition times between underlying models and a framework to learn them in real time, without assumptions about their evolution. We show the method performances on simulated datasets. The goal of the system is to model and predict incongruities for an Intrusion Dectection System (IDS) in near real-time, so great care is attached to the ability to correctly identify transitions times. Our preliminary results reveal the precision of our algorithm in the choice of transitions and consequently the quality of the discovered networks. We finally suggest future works.

Keywords: DBN · ns-DBN · tv-DBN · Non-stationnary · Learning · Real time · Change point

1 Introduction

In many fields, particularly in information systems and biology modeling, observed processes evolve over time on many scales. Their system states change with time, describing complex trajectories. Some events or entities may influence others at any given time, but those correlations do not necessarily hold forever. Which entity influences another may therefore vary, and any model wishing to capture such a process, without observing the mechanism responsible for such changes, cannot be stationary, that is its structure and/or parameters need to evolve with time too. Otherwise, only one behavior is seen, *averaging* all observations. Since we wish to model the behavior of information systems within a network of computers - in real time - it seems reasonable to assume

J.P. Carvalho et al. (Eds.): IPMU 2016, Part I, CCIS 610, pp. 338–350, 2016.
DOI: 10.1007/978-3-319-40596-4_29

non-stationarity of the observed processes. Indeed, any program or application can accept a wide range of inputs, communicate with other programs, and paths chosen by the process (where it goes and what it does) are often at least input dependent. Thereby, we need a framework for learning non-stationary processes, in real time, and set our focus on non-stationary dynamic Bayesian networks.

Dynamic Bayesian Networks [5,13] are a probabilistic graphical formalism describing, through conditional dependencies, complex dynamical systems under uncertainty. Yet, the use of *dynamic* in DBN refers to the system evolving over time, not the dynamics of the network structure or its parameters. Once determined on a subset of observations, conditional dependencies and parameters are never revisited. In many applications, even more so when data are not produced in a controlled manner, assuming homogeneity of the underlying model(s) describing which state the system is in seems too strong an assumption. This issue has received attention in the last years from the academic field giving rise to non-stationary dynamic Bayesian Networks (ns-DBN) [6–8,17,18] or time-varying dynamic Bayesian Networks (TV-DBN) [20] with applications for system biology [9]. Since processes have many execution paths and a huge input space, it seems unwise to assume that one homogeneous model could accurately capture a process evolution. Two different invocations could result in two completely different traces. Thus we build our system on ns-DBNs.

In this paper, we propose a new algorithm to model non-stationary processes using non-stationary dynamic Bayesian networks. It is organized as follows. We start with (d)BNs and non-stationary dynamic Bayesian Networks. We then build on a non-stationary learning algorithm and present our framework before evaluating its performances on a number of simulated cases to reveal strengths and weaknesses. We finally conclude and extend on our future work.

2 (Non-stationary) Dynamic Bayesian Networks

DBNs are classical Bayesian networks [16] in which nodes $\{X_i(t), i = 1 \dots n\}$, representing (discrete) random variables, are indexed by time t. They provide a factored representation of the joint probability distribution P on a finite time interval $[1, \tau]$ defined as follows:

$$P(\mathbf{X}(1) \dots, \mathbf{X}(\tau)) = \prod_{i=1}^{n} \prod_{t=1}^{\tau} P\left(X_i\left(t\right) \mid \mathbf{U}_i\left(t\right)\right) \tag{1}$$

where $\mathbf{U}_i(.)$ denotes the set of parent nodes of $X_i(.)$ and $P\left(X_i\left(t\right) \mid \mathbf{U}_i\left(t\right)\right)$ denotes the conditional probability function associated with random variable $X_i(t)$ given $\mathbf{U}_i(t)$. $\mathbf{X}(t) = \{X_1(t), \dots, X_n(t)\}$, is called a "slice" and represents the set of all variables indexed by the same time t. This joint probability $P(\mathbf{X}(1), \dots, \mathbf{X}(\tau))$ represents the beliefs about possible trajectories of the dynamic process $\mathbf{X}(t)$.

DBNs assume the *first-order Markov property* which means that the parents of a variable in time slice t must occur in either slice $t-1$ or t:

$$\mathbf{U}_i(t) \subseteq \mathbf{X}(t-1) \cup \mathbf{X}(t) \backslash X_i(t) \tag{2}$$

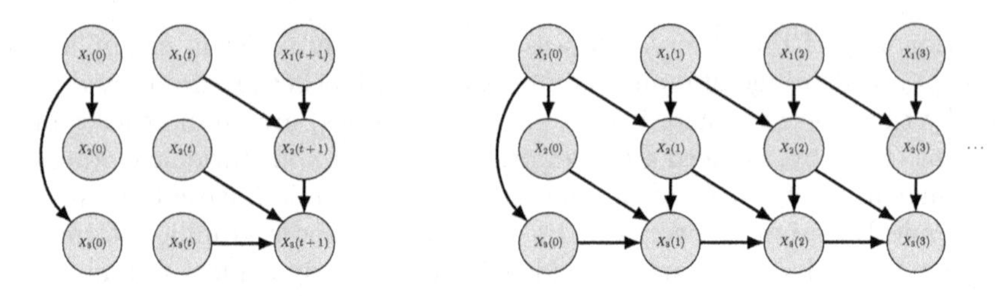

Fig. 1. A 2-Time-Slice BN (2TBN) and the (unrolled) dynamic Bayesian network.

Moreover, the conditional probabilities are time-invariant (*first-order homogeneous Markov property*):

$$P\left(X_i\left(t\right) \mid \mathbf{U}_i\left(t\right)\right) = P(X_i(2)|\mathbf{U}_i(2))), \forall t \in [2, \tau] \tag{3}$$

Hence, to specify a DBN, we only need to define the intra-slice topology (within a time slice), the inter-slice topology (between two time slices), as well as the parameters (*i.e* conditional probabilities, see Eq. 3) for the first two time slices. We obtain a 2TBN such as in Fig. 1.

In this paper, we consider that $X_i(t)$ are all discrete variables and let P^t_{ijk} be the probability that $X_i(t) = k$, given that its parents have instantiation j, *i.e.*

$$P^t_{ijk} = P\left(X_i(t) = k \mid \mathbf{U}_i(t) = j\right), \begin{array}{l} i = 1, \ldots, n \\ j = 1, \ldots, c_i \\ k = 1, \ldots, r_i \end{array} \tag{4}$$

where r_i is the number of values that node $X_i(t)$ can take and c_i is the number of distinct configurations of $\mathbf{U}_i(t)$.

DBNs have been applied in a variety of domains such as speech recognition [12], fault detection [11], medical diagnosis [4] or system biology [19] but their applications on Intrusion Detection Systems are rare [1]. However, (Hidden) Markov Models have been extensively proposed to model system call traces and shell commands [23,24] or network data flow [15]. Bayesian Networks are mainly used in this field in a static manner and often for classification purposes, as a deciding mechanism aggregating smaller models outputs that offer a summary of input data [10,14]. A variant of dynamic Bayesian Networks, called Continuous Time Bayesian Networks (CTBNs), has been used to model network traffic [21]. CTBNs leverage continuous time to solve the issue of time-granularity when using DBNs, which require a time-slice width, thus making them computationally inefficient when dealing with long period of "inactivity" or irregularly spaced observations. The main drawbacks of the framework are that two variables cannot change states simultaneously and a parameter needs to be chosen to scale timing correlations. The work in [21,22] is close to ours in approach; the use of a hidden variable allows to model the machine unknown state - the structure is manually specified and does not evolve. After training, they use a sliding window and selection by likelihood threshold to flag anomalous behavior. However,

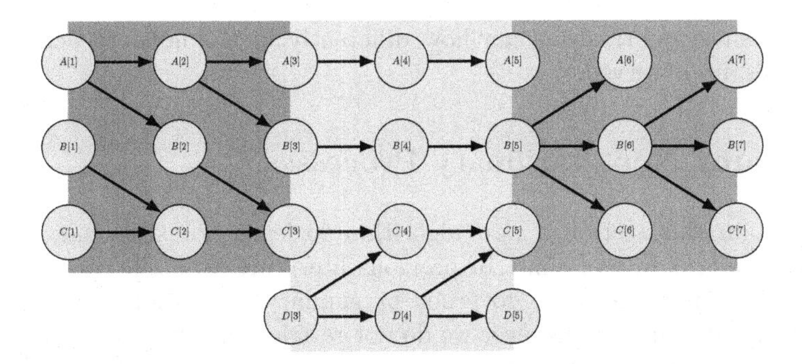

Fig. 2. Non-Stationary dynamic Bayesian network (ns-DBN) with 3 different epochs. Note that DBNs in different epochs may have different parameters, structures and even variables. For simplicity's sake, priors BNs as in Fig. 1 are not represented.

the number of states of the hidden variable needs to be known in advance (two in this case) and new models cannot be discovered on the fly. There is also no mechanism for windows overlapping events from different models.

ns-DBNs are dynamic bayesians networks $\mathcal{B} : (\Theta, \mathcal{G})$ organized by epochs of varying size or transition times $\mathcal{T} : \{(\mathcal{B}_m : (\Theta, \mathcal{G})_m, \mathcal{T}_m)\}$. There is no framework yet to model the behavior of the transition times for ns-DBNs. As a consequence, ns-DBNs represent non-stationary processes assuming piece-wise stationarity over epochs, which seems a reasonable assumption. They inherit all advantages and inconvenients of DBNs.

Although different epochs in ns-DBN may alter parameters, structures and even set of variables for the DBNs (see Fig. 2), [7] focuses on parameters evolution with fixed structure. In [18], the focus is set on structural evolution. For this later papers, the number of variables and their domains remain constant over time (even if some are not observed during whole epochs). [6] is close to our approach allowing structure, parameters as well as variables and their domains to evolve over time. However we use different criteria and a mechanism for windows overlapping events from different models to refine transition times.

ns-DBNs learning algorithms consist in identifying the different epochs and the DBNs associated with each. Current learning algorithms focus on either structure or parameter evolution (mainly to cope with the size of the search space). Sadly they pretty much all require the availability of the whole database and cannot be used in our online framework except. Offline learning of ns-DBNs is usually achieved through the use of an updated traditional DBNs scoring function used to account for the sufficient statistics that need to be specified by epoch to find the best time transitions and an updated structural move set for learning the structure which also need to be specified by epoch. However, we need simpler schemes to achieve real time performance while streaming observations. [6] proposes an interesting avenue: to take into account arcs strength using the mutual information of a node and its parent and to use previous parameters as priors for nodes that do not change from one model to another. Further inquiry is required since information can flow differently in the network according to which nodes are observed.

3 Learning Non-stationary Processes

We present in this section a new algorithm to learn non-stationary dynamic Bayesian networks in real time. In previous literature [7,8,18], the assumption that two adjacent models are governed by similar distributions and/or similar structures is often made. However we do not restrict ourselves to smooth evolutions from model to model.

Real-time data are streamed in a continuous manner using a sliding window. For each new window of data, our algorithm has to choose between using an already known model or creating a new one. This choice is based on the likelihood of the windowed data. Indeed it is expected that the likelihood of a model will decrease if the underlying behavior changes (see Fig. 3). The algorithm begins with a burn-in resulting in a first network that serves as a starting point.

More formally, at any current time τ, the algorithm confronts a collection of M DBN models $\{\mathcal{B}_m : (\boldsymbol{\Theta}, \mathcal{G})_m\}_M$ with the windowed data $\mathbf{w}[\tau, \tau + r]$.

3.1 Learning with Fixed Variables and Static Window Size

To evaluate how a model m is able to explain the data $\mathbf{w}[\tau, \tau+r]$, we use a simple criterion based on the likelihood on $\mathbf{w}$. In a stationary DBN, the log-likelihood of the data against a network with structure $\mathcal{G}$ and parameters $\boldsymbol{\Theta}$ is:

$$LL(\mathbf{w} : \boldsymbol{\Theta}, \mathcal{G}) \propto \sum_{t=\tau}^{\tau+r} \sum_{i,j,k} N_{ijk} \log(\theta_{ijk}) \tag{5}$$

with $\theta_{ijk} = P(X_i(t) = k \mid \mathbf{U}_i(t) = j)$ and N_{ijk} the number of cases where $X_i(t) = k$ and $\mathbf{U}_i(t) = j$ in $\mathbf{w}$.

For each DBN $(\boldsymbol{\Theta}, \mathcal{G})_m$, we then compute $LL(\mathbf{w} : \boldsymbol{\Theta}_m, \mathcal{G}_m)$. However the best matching model cannot be selected only by maximizing LL since the algorithm may also discover new models on the fly. For this purpose, one can note that the distribution of the log-likelihood of a window $\mathbf{w}$ is approximately normally distributed (as a sum of $r + 1$ i.i.d random variables using the central limit theorem). We then design a statistical hypothesis test in order to find the log-likelihood p-value LL_{tr} such that 99 % of matches occur with greater or equal log-likelihood (see Fig. 3). To produce a first estimate of LL_{tr} after discovering a new network, we use Gibbs sampling [3]: trajectories are sampled from the network, as many observations as needed to fit several windows, before computing their likelihood by sliding the window. For each model m, we compute $\frac{LL_{tr}(\boldsymbol{\Theta}_m, \mathcal{G}_m)}{LL(\mathbf{w}:\boldsymbol{\Theta}, \mathcal{G})}$. Our selection rule becomes:

$$\arg\max_m \left\{ \frac{LL_{tr}(\boldsymbol{\Theta}_m, \mathcal{G}_m)}{LL(\mathbf{w} : \boldsymbol{\Theta}_m, \mathcal{G}_m)} \geq 0.97 \right\} \tag{6}$$

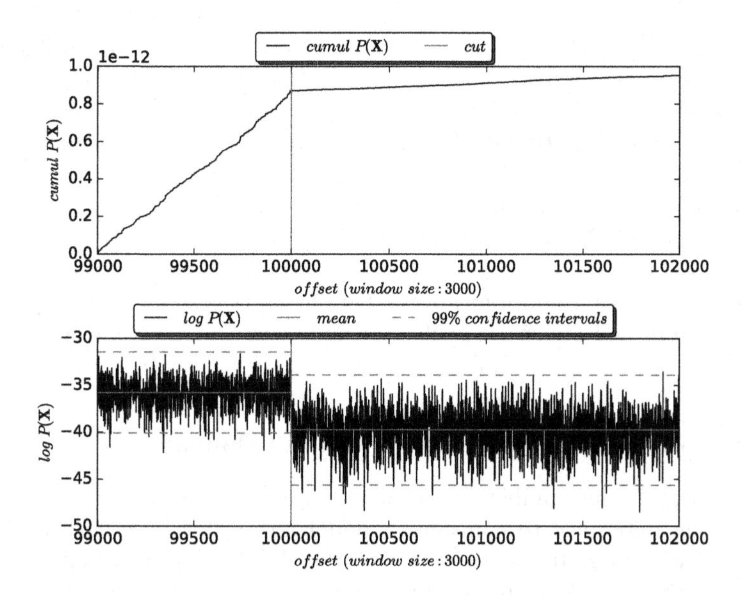

Fig. 3. Cumulative $P(\mathbf{X})$ and $log\ P(\mathbf{X})$ for a window covering two different behaviors. The vertical red line is a transition between models, and horizontal red lines are Gaussian with 99 % confidence interval. (Color figure online)

We use 0.97 as the threshold on the likelihood ratio instead of 1 since the first learning are quite inaccurate - a few events to learn a lot of parameters - and we allow some divergence to occur. A time varying threshold on the likelihood ratio could be designed to take into account parameters and structure convergence. The higher the threshold the more specific the discovered and learned networks will be (as a side-effect we will have more networks, for a given database, than with a lower threshold).

If the set of models in Eq. 6 is empty, the algorithm will learn a new DBN from the window and select it for the current window. If an existing model m is selected, there is still a learning phase in order to update the parameters and eventually the structure with the new data. Indeed, as observations increase for the models, their structures will need to be reevaluated: at each order of magnitude, we then re-estimate the network structures.

In experiments, Fig. 6 shows how a badly sized window can mislead the learning. We therefore propose to adapt the window.

3.2 Dynamic Adaptation of the Window

When the algorithm predicts for the current window a model m_t different from model m_{t-1} of the last window (i.e. m_t would be a newly created DBN or an already existing DBN), the prediction is not only about the change of behaviors of the dynamic process but also about the time τ of this change (the change point). In this section, we propose to investigate more exactly the value of this point by looking at the distribution of the likelihood within $\mathbf{w}[\tau - r, \tau + r]$.

Figure 3 shows the cumulative value of $P(\mathbf{X})$ and $log\ P(\mathbf{X})$ for such a window (with change point $c = 100\,000$). In order to identify a correct value for c, one could rely on the change of slope in the cumulative $P(\mathbf{X})$. To be more accurate, we maximize over c the likelihood of a model where c separates two different Gaussian processes. We then update m_{t-1} on $\mathbf{w}[\tau - r, \tau - r + c^*]$ and update or learn m_t on $\mathbf{w}[\tau - r + c^*, \tau + r]$ with the optimized change point c^*. If m_t is a new model, we use a non-informative Dirichlet prior, making the assumption that parameters and structure evolve without correlations from one model to another.

3.3 Learning with Incompatible Variables Domain

As seen in Fig. 2, the number of variables may change during the process. In this case, one may have to confront a model and a database with a different numbers of variables. If the variables of the database form a sub-set of the variables of the model, with variable X_e in $\mathcal{G}_m$ but not in the database, we use inference to estimate $P(X_i \mid \mathbf{U}_i \setminus X_e)$ and then compute the likelihoods. On the other hand, if the variables of the model form a sub-set of the variables of the database, those informations in the database are not exploitable for this model and then are simply discarded. Such a model will not be selected for the current window. If variables domains Ω_{X_i} differs, we add the missing states

Algorithm 1. Main loop

Data: previous model id m^*_{t-1}, observations $\mathbf{w}[\tau - r, \tau]$, $\mathbf{w}[\tau, \tau + r]$
Data: $\mathbf{D} = \{\mathcal{B}_m\}$, $\mathcal{B}_{m^*_{t-1}}$

1 **begin**
2 $\quad \Phi \leftarrow find\_match(\mathbf{D}, \mathbf{w})$
3 $\quad$ **if** $\Phi \neq \{\}$ **then**
4 $\quad\quad m^*_t \leftarrow \arg\max_m \left\{ \frac{LL_{tr}(\Theta, \mathcal{G})}{LL(\mathbf{w}:\Theta, \mathcal{G})} : (LL, LL_{tr}, m) \in \Phi \right\}$
5 $\quad\quad$ **if** $m^*_t \neq m^*_{t-1}$ **then**
6 $\quad\quad\quad$ find the change point c on $\mathbf{w}[\tau - r, \tau + r]$
7 $\quad\quad\quad$ update and validate previous model $\mathcal{B}_{m^*_{t-1}}$ on $\mathbf{w}[\tau - r, \tau - r + c]$
8 $\quad\quad$ **if** *new observations* $\geq$ *10 * previous observations* **then**
9 $\quad\quad\quad$ update $\mathcal{B}_{m^*_t}$ structure and parameters with $\mathbf{w}[\tau - r + c, \tau + r]$
10 $\quad\quad\quad$ *previous observations* $\leftarrow$ *previous observations* + *new observations*
11 $\quad\quad$ **else**
12 $\quad\quad\quad$ update $\mathcal{B}_{m^*_t}$ parameters with $\mathbf{w}[\tau - r + c, \tau + r]$
13 $\quad\quad$ *return*
14 $\quad$ find the change point c on $\mathbf{w}[\tau - r, \tau + r]$
15 $\quad$ update and validate previous model $\mathcal{B}_{m^*_{t-1}}$ on $\mathbf{w}[\tau - r, \tau - r + c]$
16 $\quad$ learn new model on $\mathbf{w}[\tau - r + c, \tau + r]$

Algorithm 2. Find match

Data: observations $\mathbf{w}[\tau, \tau + r]$
Data: $\mathbf{D} = \{\mathcal{B}_m\}$, $Dir(\{\alpha_{ijk}\})$-(*Dirichlet parameters*)

1 **begin**
2 $\quad \Phi \leftarrow \{\}$
3 $\quad$ **for** $\mathcal{B}_m = (\boldsymbol{\Theta}, \mathcal{G})_m \in \mathbf{D}$ **do**
4 $\quad\quad$ **while** $\exists X_e \in \mathcal{B}_m, X_e \notin \mathbf{w}$ **do**
5 $\quad\quad\quad$ $\forall j \in [\![1, c_e]\!]$, eliminate X_e using inference :
6 $\quad\quad\quad$ $P(X_i \mid (\mathbf{U}_i \setminus X_e) = j)$
7 $\quad\quad$ **while** $\exists X_e \in \mathbf{w}, X_e \notin \mathcal{B}_m,$ **do**
8 $\quad\quad\quad$ discard X_e
9 $\quad\quad$ **while** $\exists X_i \in \mathbf{w}, X_i = k$ **and** $X_i \in \mathcal{B}_m, k \notin \Omega_{X_i}$ **do**
10 $\quad\quad\quad$ $\Omega_{X_i} \leftarrow \Omega_{X_i} \cup k$
11 $\quad\quad\quad$ $\theta_{ijk} \leftarrow \frac{\alpha_{ijk}}{N_{ij}+\alpha_{ij}}$
12 $\quad\quad\quad$ $\theta_{ij\{o \neq k\}} \leftarrow \frac{N_{ijo}+\alpha_{ijo}}{N_{ij}+\alpha_{ij}}$
13 $\quad\quad\quad$ **for** $X_l \in \mathcal{B}_m : X_i \in \mathbf{U}_l$ **do**
14 $\quad\quad\quad\quad$ $\forall j \in [\![1, c_i]\!]$, compute using inference :
15 $\quad\quad\quad\quad$ $P(X_l \mid (\mathbf{U}_l \setminus X_i) = j, X_i = k) \leftarrow P(X_l \mid (\mathbf{U}_l \setminus X_i) = j)$
16 $\quad\quad$ **if** $\frac{LL_{tr}(\boldsymbol{\Theta}, \mathcal{G})}{LL(\mathbf{w}:\boldsymbol{\Theta}, \mathcal{G})} \geq 0.97$ **then**
17 $\quad\quad\quad$ $\Phi \leftarrow \Phi \cup (LL, LL_{tr}, m)$
18 $\quad$ *return* Φ

using the (non-informative) Dirichlet priors α_{ijk} parameters and then compute the likelihoods.

Algorithms 1 and 2 describe our framework for online learning of non-stationary processes with ns-DBN. While the next section will investigate our experiments, it is noteworthy that the complexity of our algorithm does not depend of the size of the database but only of the size of the window and the number of known models which is an important quality for online learning.

4 Experiments and Results

Our experiment consists in modeling simulated non-stationary processes. Using the aGrUM library (http://agrum.lip6.fr), we generated a DBN of 10 nodes by time-step of average domain size 7 ($[\![3, 10]\!]$) and average node degree 3. We then perturbed the structure and parameters of the model using the hellinger distance [2] between the two models as stopping criterion. Multiple thresholds were used to see how far apart two networks need to be for them to be recognized as two independent models. Hellinger distances greater than 0.8 always involve changes in parameters for all nodes and sometimes structure for a few set of nodes. Hellinger distances under this threshold involve parameter changes for one or two nodes, with small degrees, and sometimes an arc is added, adding very little information.

The databases were then sampled from each model before being combined to form a unique dataset consisting of 600.000 events. Different epoch sizes were used in order to see the impact of sample size against network distance as well as different resolutions of the sliding window to see how the system performs when overlapping datasets from two distinct models (i.e. the epoch is not a multiple of the window size). We ran each settings with and without the dynamic window scheme. It is important to note that our algorithm have no prior information about the number of networks, their variables and variables domains or the number of transitions.

The fictive Fig. 4 explains how to read experiments' figures and tables, where *FN* stands for transitions false negatives (percentage of missed transitions over all true transitions), *FP* stands for transitions false positives (percentage of false transitions over all discovered transitions) and *TP* stands for transitions true positives (percentage of true transitions over all discovered transitions). Also, tp is the number of (true) events learned by correct networks, fp the number of (false) events learned by incorrect networks and fn the number of (true) missed events by networks that are learned by others. Adaptive windows can be seen with curves being extended either on the left (for the current matching model moving the window) or the right (non matching models that do not move the window). In experiments' tables, cuts average, minimal and maximal errors are shown, with standard deviation. Finally, $precision$ $tp/(tp+fp)$ and $recall$ $tp/(tp+fn)$ for events are also shown, that is average $precision$ and $recall$ over discovered networks. *Recall* amounts to the percentage of correct events found for all correct events that should have been found. *Precision* is inversely proportional to noise (events generated from another model used to update the current model). Due to pages restriction, results were averaged for all thresholds of hellinger distance. We focus on the cases with the epoch not being a multiple of the window size - and show a best case (Fig. 5) and worst case (Figs. 6, 7, 8 and 9) scenario with and without the adaptive window. The results for static and adaptive windows are presented in Tables 1 and 2, respectively.

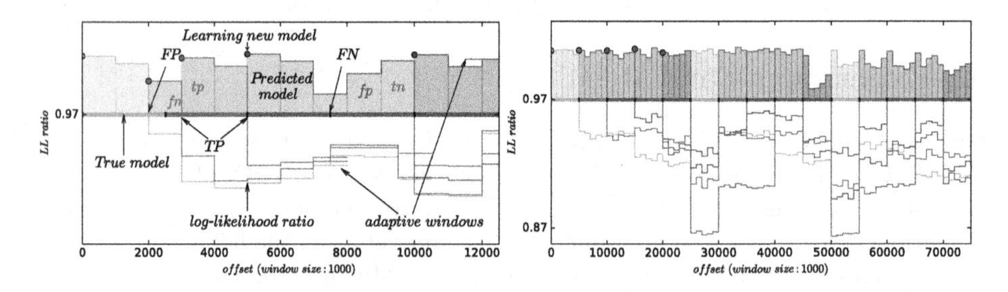

Fig. 4. How to read figures.

Fig. 5. Results for epochs of $5K$ observations, *hellinger* < 0.8, fixed window size

4.1 Static Windows

Figure 5 is an example of a successful run: the epoch is a multiple of the window size, consequently the sliding window always contains observations from one model at a time. In such settings, correct transition times and models are always identified, with and without adaptive window. However, errors arise when using arbitrary window sizes without dynamical windows as shown in Figs. 6 and 7.

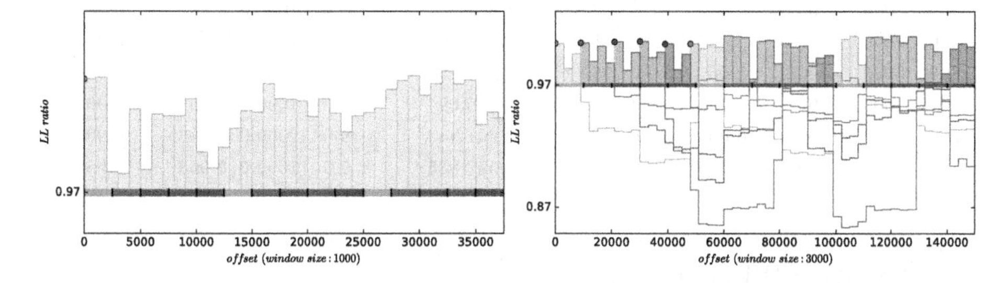

Fig. 6. Results for epochs of $2K5$ observations, *hellinger* < 0.8, fixed window size

Fig. 7. Results for epochs of $10K$ observations, *hellinger* < 0.8, fixed window size

In the static case as in Table 1, two issues explain the poor *precision* and *recall* for some experiments. The first issue arises when we have discovered fewer networks than we should, mainly with lower hellinger thresholds, in which case transitions were missed and some models are averaging several true models, increasing noise and making further transitions harder to detect, hence increasing *FN* of transitions and decreasing *precision* and *recall* over events. Such a case is highlighted by Fig. 6 and by the first two rows of Table 1, with the first row and Fig. 6 showing results for close true networks and the second row results for distinct true networks. The second issue arises when we have discovered more networks than we should, mainly with higher hellinger thresholds. When it happens, most networks in excess were made when the window overlaps events from two true networks, thus modeling the transition itself (the next window matches or creates another model, the true one), such as in Fig. 7 (the brown network). Hence, we have two transitions instead of one, increasing *FP* for transitions. *Precision* and *recall* are less affected by those *FP* since only a few transitions give rise to very specific models, slightly reducing the *recall* of other discovered (true) networks, but increasing their *precision* (reducing noise).

Results in the static case could be worse: in our setting, one epoch is not a period of the window size but a multiple of the epoch can be a multiple of the window size, in which case the window ends or starts at a true transition, therefore "increasing" our probability of correctly identifying a transition or model. Thus, *FN* and cuts errors could be higher whereas *precision* and *recall* could be lower.

Table 1. Results for static windows, showing missed transitions over all true transitions (false negatives *FN*), false positive transitions (*FP*) and true positive transitions (*TP*) over all discovered transitions. For cuts, minimal, average and maximal error in events, with variance. For discovered networks, *precision* and *recall* over events.

epoch	*window size*	*FN*	*FP*	*TP*	*avg. error*	*std. deviation*	*min*	*max*	*precision*	*recall*
2500	1000	1.0	0.0	0.0	*NA*	*NA*	*NA*	*NA*	0.2	1.0
2500	1000	0.0	0.0	1.0	251.046	249.998	0.0	500.0	0.829	0.875
2500	2000	0.602	0.0	1.0	521.052	361.644	0.0	1000.0	0.5	0.952
5000	1500	0.101	0.035	0.965	351.216	258.546	0.0	1000.0	0.833	0.935
5000	3000	0.0	0.06	0.94	752.1	579.236	0.0	2000.0	0.856	0.854
10000	1500	0.0	0.131	0.869	381.356	295.694	0.0	1000.0	0.961	0.96
10000	3000	0.1017	0.0083	0.992	772.81	601.843	0.0	2000.0	0.833	0.929
15000	2000	0.0	0.204	0.795	512.82	499.835	0.0	1000.0	0.963	0.958

4.2 Adaptive Windows

The results for adaptive windows, shown in Table 2, reveal that the size of the window has little impact on the correct identification of transitions and models, and it should hold as long as the window size is lower than the epoch. Surprisingly, results are not worse for small epochs given the domain size of the network. With adaptive windows, both previous issues are solved by looking for a change point, as in Figs. 8 and 9: in the first case, we do not learn from overlapping windows which reduces noise, making future transitions easier to discover. In the second case, looking for a change point itself avoids the creation of a network to represent the transition alone.

The ability of the algorithm to add modalities to known variables avoids the creation of unnecessary networks in both settings, thus reducing *FP* for transitions, and is of crucial importance for outliers that happen every now and then.

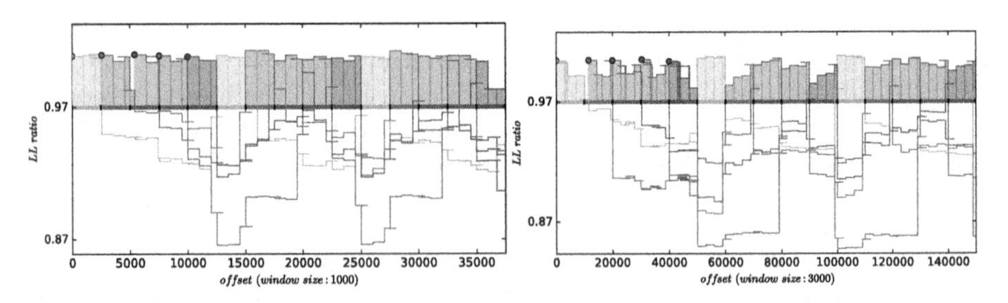

Fig. 8. Results for epochs of $2K5$ observations, *hellinger* < 0.8, dynamic window size

Fig. 9. Results for epochs of $10K$ observations, *hellinger* < 0.8, dynamic window size

Table 2. Results for adaptive windows, with columns as in Table 1.

epoch	*window size*	*FN*	*FP*	*TP*	*average error*	*std. deviation*	*min*	*max*	*precision*	*recall*
2500	1000	0.0	0.0	1.0	4.399	18.045	0.0	254.0	0.998	0.998
2500	2000	0.0	0.0	1.0	2.435	4.683	0.0	38.5	0.999	0.999
5000	1500	0.0	0.0	1.0	3.0966	7.784	0.0	62.5	0.999	0.999
5000	3000	0.0	0.0	1.0	8.702	43.383	0.0	390.5	0.998	0.998
10000	1500	0.0	0.0	1.0	32.923	80.526	0.0	311.5	0.997	0.996
10000	3000	0.0	0.0	1.0	19.559	103.199	0.0	778.0	0.998	0.998
15000	2000	0.0	0.0	1.0	8.551	32.423	0.0	202.5	0.999	0.999

5 Conclusions and Future Work

We built a framework around Dynamic Bayesian Networks to learn non-stationary processes in a continuous manner, designed to be fast and accurate. However, several enhancements comes to mind: we mentioned a dynamical threshold on the log-likelihood ratio to take into account convergence, as well as the need for merging and deleting schemes, since we expect results to be poorer the closer the original networks are from each other. While a naive deleting scheme could consists of using a parameter for each network, decreasing over time when not matching, merging networks require to compare their joint probability distributions which involves heavy computations. The problem of looking for a cut could also be investigated further, since if the cumulative likelihood is a stepping function, we should cut at the first step, which we do not (the algorithm is maximizing the likelihood over both Gaussians and the cut is most often than not in between the steps). A most important enhancement would be to model transitions from behavior to behavior, and predict to some extent the next behavior for a given time or identify critical events (weak signals) that would allow such predictions. Finally, we will apply this work to detect anomalies in a host and network intrusion detection system.

Acknowledgments. This work was supported by Akheros S.A.S./ANRT CIFRE grant #2014/0268, and the European project SCISSOR H2020-ICT-2014-1 #644425.

References

1. An, X., Jutla, D., Cercone, N.: Privacy intrusion detection using dynamic Bayesian networks. In: ACM International Conference Proceeding Series, vol. 156, pp. 208–215 (2006)
2. Beran, R.: Minimum hellinger distance estimates for parametric models. Ann. Stat. **5**, 445–463 (1977)
3. Casella, G., George, E.I.: Explaining the Gibbs sampler. Am. Stat. **46**(3), 167–174 (1992)
4. Charitos, T., Van Der Gaag, L.C., Visscher, S., Schurink, K.A., Lucas, P.J.: A dynamic Bayesian network for diagnosing ventilator-associated pneumonia in ICU patients. Expert Syst. Appl. **36**(2), 1249–1258 (2009)

5. Dean, T., Kanazawa, K.: A model for reasoning about persistence and causation. Comput. Intell. **5**(2), 142–150 (1989)
6. Gonzales, C., Dubuisson, S., Manfredotti, C.: A new algorithm for learning non-stationary dynamic Bayesian networks with application to event detection. In: The Twenty-Eighth International Flairs Conference (2015)
7. Grzegorczyk, M., Husmeier, D.: Non-stationary continuous dynamic Bayesian networks. In: Advances in Neural Information Processing Systems, pp. 682–690 (2009)
8. Grzegorczyk, M., Husmeier, D.: Non-homogeneous dynamic Bayesian networks for continuous data. Mach. Learn. **83**(3), 355–419 (2011)
9. Grzegorczyk, M., Husmeier, D., Edwards, K.D., Ghazal, P., Millar, A.J.: Modelling non-stationary gene regulatory processes with a non-homogeneous Bayesian network and the allocation sampler. Bioinformatics **24**(18), 2071–2078 (2008)
10. Kruegel, C., Mutz, D., Robertson, W., Valeur, F.: Bayesian event classification for intrusion detection. In: 2003 Proceedings of the 19th Annual Computer Security Applications Conference, pp. 14–23. IEEE (2003)
11. Lerner, U., Parr, R., Koller, D., Biswas, G., et al.: Bayesian fault detection and diagnosis in dynamic systems. In: AAAI/IAAI, pp. 531–537 (2000)
12. Mitra, V., Nam, H., Espy-Wilson, C.Y., Saltzman, E., Goldstein, L.: Gesture-based dynamic Bayesian network for noise robust speech recognition. In: 2011 IEEE International Conference on Acoustics, Speech and Signal Processing (ICASSP), pp. 5172–5175. IEEE (2011)
13. Murphy, K.P.: Dynamic Bayesian networks: representation, inference and learning. Ph.D. thesis, University of California, Berkeley (2002)
14. Mutz, D., Valeur, F., Vigna, G., Kruegel, C.: Anomalous system call detection. ACM Trans. Inf. Syst. Secur. (TISSEC) **9**(1), 61–93 (2006)
15. Ourston, D., Matzner, S., Stump, W., Hopkins, B.: Applications of hidden Markov models to detecting multi-stage network attacks. In: 2003 Proceedings of the 36th Annual Hawaii International Conference on System Sciences, 10 p. IEEE (2003)
16. Pearl, J.: Probabilistic Reasoning in Intelligent Systems: Networks of Plausible Inference. Morgan Kaufmann, San Mateo (2014)
17. Robinson, J.W., Hartemink, A.J.: Non-stationary dynamic Bayesian networks. In: Advances in Neural Information Processing Systems, pp. 1369–1376 (2009)
18. Robinson, J.W., Hartemink, A.J.: Learning non-stationary dynamic Bayesian networks. J. Mach. Learn. Res. **11**, 3647–3680 (2010)
19. Sicard, M., Baudrit, C., Leclerc-Perlat, M., Wuillemin, P.H., Perrot, N.: Expert knowledge integration to model complex food processes. Application on the camembert cheese ripening process. Expert Syst. Appl. **38**(9), 11804–11812 (2011)
20. Song, L., Kolar, M., Xing, E.P.: Time-varying dynamic Bayesian networks. In: Advances in Neural Information Processing Systems, pp. 1732–1740 (2009)
21. Xu, J., Shelton, C.R.: Continuous time Bayesian networks for host level network intrusion detection. In: Daelemans, W., Goethals, B., Morik, K. (eds.) ECML PKDD 2008, Part II. LNCS (LNAI), vol. 5212, pp. 613–627. Springer, Heidelberg (2008)
22. Xu, J., Shelton, C.R.: Intrusion detection using continuous time Bayesian networks. J. Artif. Intell. Res. **39**, 745–774 (2010)
23. Yeung, D.Y., Ding, Y.: Host-based intrusion detection using dynamic and static behavioral models. Pattern Recogn. **36**(1), 229–243 (2003)
24. Zanero, S., Serazzi, G.: Unsupervised learning algorithms for intrusion detection. In: 2008. IEEE Network Operations and Management Symposium, NOMS 2008, pp. 1043–1048. IEEE (2008)

Fuzzy Implication Functions

About the Use of Admissible Order for Defining Implication Operators

Maria Jose Asiain[1], Humberto Bustince[2,3](✉), Benjamin Bedregal[4], Zdenko Takáč[5], Michal Baczyński[6], Daniel Paternain[2], and Graçaliz Dimuro[7]

[1] Dept. de Matemáticas, Universidad Pública de Navarra, Campus Arrosadia, s/n, 31.006 Pamplona, Spain
asiain@unavarra.es

[2] Dept. de Automática y Computación, Universidad Pública de Navarra, Campus Arrosadia s/n, 31.006 Pamplona, Spain
{bustince,daniel.paternain}@unavarra.es

[3] Institute of Smart Cities, Universidad Pública de Navarra, Campus Arrosadia, s/n, 31.006 Pamplona, Spain

[4] Departamento de Informática e Matemática Aplicada, Universidade Federal do Rio Grande do Norte, Campus Universitario, s/n, Lagoa Nova, Natal 59078-970, Brazil
bedregal@dimap.ufrn.br

[5] Institute of Information Engineering, Automation and Mathematics, Slovak University of Technology in Bratislava, Radlinskeho 9, Bratislava, Slovakia
zdenko.takac@stuba.sk

[6] Institute of Mathematics, University of Silesia, ul. Bankowa 14, 40-007 Katowice, Poland
michal.baczynski@us.edu.pl

[7] Centro de Ciências Computacionais, Universidade Federal do Rio Grande, Av. Itália, km 08, Campus Carreiros, Rio Grande 96201-900, Brazil
gracaliz@furg.br

Abstract. Implication functions are crucial operators for many fuzzy logic applications. In this work, we consider the definition of implication functions in the interval-valued setting using admissible orders and we use this interval-valued implications for building comparison measures.

Keywords: Interval-valued implication operator · Admissible order · Similarity measure

1 Introduction

Implication operators are crucial for many applications of fuzzy logic, including approximate reasoning or image processing. Many works have been devoted to the analysis of these operators, both in the case of fuzzy sets [1,2,14,15] and in the case of extensions [3–5,13,16]. A key problem in order to define these operators is that of monotonicity. When implication operators are extended to fuzzy extensions, this problem is not trivial, since for most of the fuzzy extensions do not exist a linear order, whereas for some applications, as it is the case of

J.P. Carvalho et al. (Eds.): IPMU 2016, Part I, CCIS 610, pp. 353–362, 2016.
DOI: 10.1007/978-3-319-40596-4_30

fuzzy rules-based classification systems, it is necessary to have the possibility of comparing any two elements [12].

In this work, we propose the definition of implication operators in the interval-valued setting defining its monotonicity in terms of the so-called admissible orders [11]. This is a class of linear orders which extends the usual order between intervals and which include the most widely used examples of linear orders between intervals, as lexicographical and Xu and Yager ones.

As a first step in a deeper study of these interval-valued implications with admissible orders, we show how implications which are defined in terms of admissible orders can be used to build comparison measures which are of interest from the point of view of applications.

The structure of the present work is as follows. In Sect. 2 we present some preliminary definitions and results. In Sect. 3 we present the definition of interval-valued implication function with respect to an admissible order. Section 4 is devoted to obtaining equivalence and restricted equivalence functions with respect to linear orders. In Sect. 5 we use our previous results to build comparison measures. We finish with some conclusions and references.

2 Preliminaries

In this section we introduce several well known notions and results which will be useful for our subsequent developments.

We are going to work with closed subintervals of the unit interval. For this reason, we define:

$$L([0,1]) = \{[\underline{X}, \overline{X}] \mid 0 \leq \underline{X} \leq \overline{X} \leq 1\}\,.$$

By $\leq_L$ we denote an arbitrary order relation on $L([0,1])$ with $0_L = [0,0]$ as its minimal element and $1_L = [1,1]$ as maximal element. This order relation can be partial or total. If we must consider an arbitrary total order, we will denote it by $\leq_{TL}$.

Example 1. The partial order relation on $L([0,1])$ induced by the usual partial order in $\mathbb{R}^2$ is:

$$[\underline{X}, \overline{X}] \precsim_L [\underline{Y}, \overline{Y}] \text{ if } \underline{X} \leq \underline{Y} \text{ and } \overline{X} \leq \overline{Y}. \tag{1}$$

As an example of total order in $L([0,1])$ we have Xu and Yager's order (see [17]):

$$[\underline{X}, \overline{X}] \leq_{XY} [\underline{Y}, \overline{Y}] \text{ if } \begin{cases} \underline{X} + \overline{X} < \underline{Y} + \overline{Y} \text{ or} \\ \underline{X} + \overline{X} = \underline{Y} + \overline{Y} \text{ and } \overline{X} - \underline{X} \leq \overline{Y} - \underline{Y}. \end{cases} \tag{2}$$

Definition 1. *An admissible order in $L([0,1])$ is a total order $\leq_{TL}$ which extends the partial order $\precsim_L$.*

In the following, whenever we speak of a total order we assume it is an admissible order.

Definition 2. *Let $\leq_L$ be an order relation in $L([0,1])$. A function $N\colon L([0,1]) \to L([0,1])$ is an interval-valued negation function (IV negation) if it is a decreasing function with respect to the order $\leq_L$ such that $N(0_L) = 1_L$ and $N(1_L) = 0_L$. A negation N is called strong negation if $N(N(X)) = X$ for every $X \in L([0,1])$. A negation N is called non-filling if $N(X) = 1_L$ iff $X = 0_L$, while N is called non-vanishing if $N(X) = 0_L$ iff $X = 1_L$.*

We recall now the definition of interval-valued aggregation function.

Definition 3. *Let $n \geq 2$. An (n-dimensional) interval-valued (IV) aggregation function in $(L([0,1]), \leq_L, 0_L, 1_L)$ is a mapping $M : (L([0,1]))^n \to L([0,1])$ which verifies:*

(i) $M(0_L, \cdots, 0_L) = 0_L$.
(ii) $M(1_L, \cdots, 1_L) = 1_L$.
(iii) M *is an increasing function with respect to* $\leq_L$.

Example 2. Fix $\alpha \in [0,1]$. With the order $\leq_{XY}$, the function

$$M_\alpha\colon L([0,1])^2 \to L([0,1])$$

defined by

$$M_\alpha([\underline{X}, \overline{X}], [\underline{Y}, \overline{Y}]) = [\alpha \underline{X} + (1-\alpha)\underline{Y}, \alpha \overline{X} + (1-\alpha)\overline{Y}]$$

is an IV aggregation function.

3 Interval-Valued Implication Functions

Definition 4 (cf. [2,5]). *An interval-valued (IV) implication function in $(L([0,1]), \leq_L, 0_L, 1_L)$ is a function $I\colon (L([0,1]))^2 \to L([0,1])$ which verifies the following properties:*

(i) I *is a decreasing function in the first component and an increasing function in the second component with respect to the order* $\leq_L$.
(ii) $I(0_L, 0_L) = I(0_L, 1_L) = I(1_L, 1_L) = 1_L$.
(iii) $I(1_L, 0_L) = 0_L$.

Some properties that can be demanded to an IV implication function are the following [10]:

$I4$: $I(X,Y) = 0_L \Leftrightarrow X = 1_L$ and $Y = 0_L$.
$I5$: $I(X,Y) = 1_L \Leftrightarrow X = 0_L$ or $Y = 1_L$.
NP: $I(1_L, Y) = Y$ for all $Y \in L([0,1])$.
EP: $I(X, I(Y,Z)) = I(Y, I(X,Z))$ for all $X, Y, Z \in L([0,1])$.
OP: $I(X,Y) = 1_L \Leftrightarrow X \leq_L Y$.
SN: $N(X) = I(X, 0_L)$ is a strong IV negation.
$I10$: $I(X,Y) \geq_L Y$ for all $X, Y \in L([0,1])$.

IP: $I(X,X) = 1_L$ for all $X \in L([0,1])$.
CP: $I(X,Y) = I(N(Y),N(X))$ for all $X,Y \in L([0,1])$, where N is an IV negation.
*I*14: $I(X,N(X)) = N(X)$ for all $X \in L([0,1])$, where N is an IV negation.

We can obtain IV implication functions from IV aggregation functions as follows.

Proposition 1. *Let M be an IV aggregation function such that*

$$M(1_L,0_L) = M(0_L,1_L) = 0_L$$

and let N be an IV negation in $L([0,1])$, both with respect to the same order $\leq_L$.
Then the function $I_M\colon L([0,1])^2 \to L([0,1])$ given by

$$I_M(X,Y) = N(M(X,N(Y)))$$

is an IV implication function.

Proof. It follows from a straight calculation. □

However, in this work we are going to focus on a different construction method for IV implication functions.

Proposition 2. *Let $\leq_{TL}$ be a total order in $L([0,1])$, and let N be an IV negation function with respect to that order. The function $I\colon L([0,1])^2 \to L([0,1])$ defined by*

$$I(X,Y) = \begin{cases} 1_L, & \text{if } X \leq_{TL} Y, \\ \vee(N(X),Y), & \text{if } X >_{TL} Y. \end{cases}$$

is an IV implication function.

Proof. It is clear that the function I is an increasing function in the second component and a decreasing function in the first component. Moreover

$$I(0_L,0_L) = I(0_L,1_L) = I(1_L,1_L) = 1_L$$

and $I(1_L,0_L) = 0_L$. □

This result can be further generalized as follows [15]:

Proposition 3. *Let $\leq_{TL}$ be a total order in $L([0,1])$, and let N be an IV negation function with respect to that order. If $M\colon L([0,1])^2 \to L([0,1])$ is an IV aggregation function, then the function $I\colon L([0,1])^2 \to L([0,1])$ defined by*

$$I(X,Y) = \begin{cases} 1_L, & \text{if } X \leq_{TL} Y, \\ M(N(X),Y), & \text{if } X >_{TL} Y, \end{cases}$$

is an IV implication function.

4 Equivalence and Restricted Equivalence Functions in $L([0,1])$ with Respect to a Total Order

Along this section only total orders are considered.

The equivalence functions [6–8] are a fundamental tool in order to build measures of similarity between fuzzy sets. In this section we construct interval-valued equivalence functions from IV aggregation and negation functions.

Definition 5. *A map $F: L([0,1])^2 \to L([0,1])$ is called an interval-valued (IV) equivalence function in $(L([0,1]), \leq_{TL})$ if F verifies:*

(1) $F(X,Y) = F(Y,X)$ for every $X, Y \in L([0,1])$.
(2) $F(0_L, 1_L) = F(1_L, 0_L) = 0_L$.
(3) $F(X,X) = 1_L$ for all $X \in L([0,1])$.
(4) If $X \leq_{TL} X' \leq_{TL} Y' \leq_{TL} Y$, then $F(X,Y) \leq_{TL} F(X', Y')$.

Theorem 1. *Let $M_1: L([0,1])^2 \to L([0,1])$ be an IV aggregation function such that $M_1(X,Y) = M_1(Y,X)$ for every $X, Y \in L([0,1])$, $M_1(X,Y) = 1_L$ if and only if $X = Y = 1_L$ and $M_1(X,Y) = 0_L$ if and only if $X = 0_L$ or $Y = 0_L$. Let $M_2: L([0,1])^2 \to L([0,1])$ be an IV aggregation function such that $M_2(X,Y) = 1_L$ if and only if $X = 1_L$ or $Y = 1_L$ and $M_2(X,Y) = 0_L$ if and only if $X = Y = 0_L$. Then the function $F: L([0,1])^2 \to L([0,1])$ defined by*

$$F(X,Y) = M_1(I(X,Y), I(Y,X)),$$

with I the IV implication function defined in the Proposition 3 taking $M = M_2$, is an IV equivalence function.

Proof. Since

$$F(X,Y) = \begin{cases} 1_L, & \text{if } X = Y, \\ M_1(M_2(N(Y), X), 1_L), & \text{if } X <_{TL} Y, \\ M_1(M_2(N(X), Y), 1_L), & \text{if } Y <_{TL} X, \end{cases}$$

then F verifies the four properties in Definition 5. □

In [8] the definition of equivalence function (in the real case) was modified in order to define the so-called restricted equivalence function. Now we develop a similar study for the case of IV equivalence functions.

Definition 6. *Let N be an IV negation. A map $F: L([0,1])^2 \to L([0,1])$ is called an interval valued (IV) restricted equivalence function (in $(L([0,1]), \leq_{TL})$) if F verifies the following properties:*

1. *$F(X,Y) = F(Y,X)$ for all $X, Y \in L([0,1])$.*
2. *$F(X,Y) = 1_L$ if and only if $X = Y$.*
3. *$F(X,Y) = 0_L$ if and only if $X = 0_L$ and $Y = 1_L$, or, $X = 1_L$ and $Y = 0_L$.*
4. *$F(X,Y) = F(N(X), N(Y))$ for all $X, Y \in L([0,1])$.*
5. *If $X \leq_{TL} Y \leq_{TL} Z$, then $F(X,Z) \leq_{TL} F(X,Y)$ and $F(X,Z) \leq_{TL} F(Y,Z)$.*

Theorem 2. *Let N be an IV negation function. Let $M_1\colon L([0,1])^2 \to L([0,1])$ be an IV aggregation function such that $M_1(X,Y) = M_1(Y,X)$ for every $X, Y \in L([0,1])$, $M_1(X,Y) = 1_L$ if and only if $X = Y = 1_L$ and $M_1(X,Y) = 0_L$ if and only if $X = 0_L$ or $Y = 0_L$. Let $M_2\colon L([0,1])^2 \to L([0,1])$ be an IV aggregation function such that $M_2(X,Y) = 1_L$ if and only if $X = 1_L$ or $Y = 1_L$ and $M_2(X,Y) = 0_L$ if and only if $X = Y = 0_L$. Then the function $F\colon L([0,1])^2 \to L([0,1])$ defined by*

$$F(X,Y) = M_1(I(X,Y), I(Y,X))$$

with I an IV implication function defined by

$$I(X,Y) = \begin{cases} 1_L & \text{if } X \leq_{TL} Y \\ M_2(N(X), Y) & \text{otherwise,} \end{cases}$$

verifies the properties (1) and (5) of Definition 6. Moreover, it satisfies property (2) if N is non-filling and property (3) if N is non-vanishing.

Proof. Since

$$F(X,Y) = \begin{cases} 1_L, & \text{if } X = Y \\ M_1(M_2(N(Y), X), 1_L), & \text{if } X <_{TL} Y \\ M_1(M_2(N(X), Y), 1_L), & \text{if } Y <_{TL} X \end{cases}$$

then F verifies:

(1) $F(X,Y) = F(Y,X)$ trivially.
(5) If $X \leq_{TL} Y \leq_{TL} Z$, then $N(Z) \leq_{TL} N(Y) \leq_{TL} N(X)$. Since M_1 is an increasing function then $F(X,Z) \leq_{TL} F(X,Y)$ and $F(X,Z) \leq_{TL} F(Y,Z)$.

Since $M_1(X,Y) = 1_L$ if and only if $X = Y = 1_L$, then, if N is non-filling, $F(X,Y) = 1_L$ if and only if $X = Y$ because

$$\begin{cases} M_2(N(Y), X) \neq 1_L, & \text{if } X <_{TL} Y \\ M_2(N(X), Y) \neq 1_L, & \text{if } X >_{TL} Y. \end{cases}$$

Moreover, $F(X,Y) = 0_L$ if and only if $X >_{TL} Y$ and $M_2(N(X), Y) = 0_L$ or $X <_{TL} Y$ and $M_2(N(Y), X) = 0_L$. Therefore, as N is non-vanishing, $F(X,Y) = 0_L$ if and only if

$$\begin{cases} X = 0_L \text{ or } Y = 1_L \text{ or} \\ Y = 0_L \text{ or } X = 1_L. \end{cases}$$

with $X \neq Y$.

□

5 Similarity Measures, Distances and Entropy Measures in $L([0, 1])$ with Respect to a Total Order

Our constructions in the previous section can be used to build comparison measures between interval-valued fuzzy sets, and, more specifically, to obtain similarity measures, distances in the sense of Fang and entropy measures. Along this section, we only deal with a total order $\leq_{TL}$.

To start, let us consider a finite referential set of n elements, $U = \{u_1, \ldots, u_n\}$. We denote by $IVFS(U)$ the set of all interval-valued fuzzy sets over U. Recall that an interval-valued fuzzy set A over U is a mapping $A : U \rightarrow L([0, 1])$ [9]. Note that the order $\leq_{TL}$ induces a partial order $\leq_{TL}$ in $IVFS(U)$ given, for $A, B \in IVFS(U)$, by

$$A \leq_{TL} B \text{ if } A(u_i) \leq_{TL} B(u_i) \text{ for every } u_i \in U \ .$$

First of all, we show how we can build a similarity between interval-valued fuzzy sets defined over the same referential U. We start recalling the definition.

Definition 7 [8]. *An interval-valued (IV) similarity measure on $IVFS(U)$ is a mapping $SM : IVFS(U) \times IVFS(U) \rightarrow L([0, 1])$ such that, for every $A, B, A', B' \in IVFS(U)$,*

(SM1) SM is symmetric.
(SM2) $SM(A, B) = 1_L$ if and only if $A = B$.
(SM3) $SM(A, B) = 0_L$ if and only if $\{A(u_i), B(u_i)\} = \{0_L, 1_L\}$ for every $u_i \in U$.
(SM4) If $A \leq_{TL} A' \leq_{TL} B' \leq_{TL} B$, then $SM(A, B) \leq_{TL} SM(A', B')$.

Then we have the following result.

Theorem 3. *Let $M : L([0, 1])^n \rightarrow L([0, 1])$ be an IV aggregation function with respect to the total order $\leq_{TL}$ and such that $M(X_1, \ldots, X_n) = 1_L$ if and only if $X_1 = \cdots = X_n = 1_L$ and $M(X_1, \ldots, X_n) = 0_L$ if and only if $X_1 = \cdots = X_n = 0_L$. Then, the function $SM : IVFS(U) \times IVFS(U) \rightarrow L([0, 1])$ given by*

$$SM(A, B) = M(F(A(u_1), B(u_1)), \ldots, F(A(u_n), B(u_n)))$$

where F is defined as in Theorem 2 with non-filling and non-vanishing negation, is an IV similarity measure.

Proof. It follows from a straightforward calculation. □

We can make use of this construction method to recover both distances and entropy measures. First of all, let's recall the definition of both concepts.

Definition 8 [6]. *A function $D : IVFS(U) \times IVFS(U) \rightarrow L([0, 1])$ is called an IV distance measure on $IVFS(U)$ if, for every $A, B, A', B' \in IVFS(U)$, D satisfies the following properties:*

(D1) $D(A,B) = D(B,A)$;
(D2) $D(A,B) = 0_L$ *if and only if* $A = B$;
(D3) $D(A,B) = 1_L$ *if and only if* A *and* B *are complementary crisp sets;*
(D4) *If* $A \leq_{TL} A' \leq_{TL} B' \leq_{TL} B$, *then* $D(A,B) \geq_{TL} D(A',B')$.

Definition 9 [6]. *A function* $E : IVFS(U) \to L([0,1])$ *is called an entropy on* $IVFS(U)$ *with respect to a strong IV negation* N *(with respect to* $\leq_{TL}$ *such that there exists* $\varepsilon \in L([0,1])$ *with* $N(\varepsilon) = \varepsilon$ *if* E *has the following properties:*

(E1) $E(A) = 0_L$ *if and only if* A *is crisp;*
(E2) $E(A) = 1_L$ *if and only if* $A = \{(u_i, A(u_i) = \varepsilon) | u_i \in U\}$;
(E3) $E(A) \leq_{TL} E(B)$ *if* A *refines* B*; that is,* $A(u_i) \leq_{TL} B(u_i) \leq_{TL} \varepsilon$ *or* $A(u_i) \geq_{TL} B(u_i) \geq_{TL} \varepsilon$;
(E4) $E(A) = E(N(A))$.

Then the following two results are straight from Theorem 3.

Corollary 1. *Let* $M : L([0,1])^n \to L([0,1])$ *be an IV aggregation function with respect to the total order* $\leq_{TL}$ *such that* $M(X_1, \ldots, X_n) = 1_L$ *if and only if* $X_1 = \cdots = X_n = 1_L$ *and* $M(X_1, \ldots, X_n) = 0_L$ *if and only if* $X_1 = \cdots = X_n = 0_L$ *and let* N *be an IV negation with respect to the order* $\leq_{TL}$ *which is non filling and non-vanishing. Then, the function* $D : IVFS(U) \times IVFS(U) \to L([0,1])$ *given by*

$$D(A,B) = N(M(F(A(u_1), B(u_1)), \ldots, F(A(u_n), B(u_n))))$$

where F *is defined as in Theorem 2, is an IV distance measure.*

Proof. It is straight from Theorem 3, since a similarity measure defines a distance in a straightforward way. □

Theorem 4. *Let* N *be a strong IV negation (with respect to* $\leq_{TL}$*) and such that there exists* $\varepsilon \in L([0,1])$ *with* $N(\varepsilon) = \varepsilon$. *Let* $M : L([0,1])^n \to L([0,1])$ *be an IV aggregation function with respect to the total order* $\leq_{TL}$ *and such that* $M(X_1, \ldots, X_n) = 1_L$ *if and only if* $X_1 = \cdots = X_n = 1_L$ *and* $M(X_1, \ldots, X_n) = 0_L$ *if and only if* $X_1 = \cdots = X_n = 0_L$. *Then, the function* $E : IVFS(U) \to L([0,1])$ *given by*

$$E(A) = M(F(A(u_1), N(A(u_1))), \ldots, F(A(u_n), N(A(u_n))))$$

where F *is defined as in Theorem 2 with non-filling and non-vanishing negation, is an IV entropy measure.*

Proof. It follows from the well known fact that, for a given IV similarity SM, the function $E(A) = SM(A, N(A))$ is an IV entropy measure [6]. □

6 Conclusions

In this paper we have considered the problem of defining interval-valued implications when the order relation is a total order. In particular, we have considered the case of admissible orders. We have also studied the construction of interval-valued equivalence and similarity functions constructed with appropriate interval-valued implication functions. Finally we have shown how our constructions can be used to get IV similarity measures, distances and entropy measures with respect to total orders. In future works we will consider the use of these functions in different image processing, classification or decision making problems.

Acknowledgements. H. Bustince was supported by Project TIN2013-40765-P of the Spanish Government. Z. Takáč was supported by Project VEGA 1/0420/15. B. Bedregal and G. Dimuro were supported by Brazilian funding agency CNPQ under Processes 481283/2013-7, 306970/2013-9, 232827/2014-1 and 307681/2012-2.

References

1. Baczyński, M., Beliakov, G., Bustince, H., Pradera, A.: Advances in Fuzzy Implication Functions, Advances in Fuzzy Implication Functions, Studies in Fuzziness and Soft Computing, 300. Springer, Berlin (2013)
2. Baczyński, M., Jayaram, B.: Fuzzy Implications, Studies in Fuzziness and Soft Computing, vol. 231. Springer, Berlin (2008)
3. Bedregal, B., Dimuro, G., Santiago, R., Reiser, R.: On interval fuzzy S-implications. Inf. Sci. **180**(8), 1373–1389 (2010)
4. Burillo, P., Bustince, H.: Construction theorems for intuitionistic fuzzy sets. Fuzzy Sets Syst. **84**, 271–281 (1996)
5. Bustince, H., Barrenechea, E., Mohedano, V.: Intuitionistic fuzzy implication operators. an expression and main properties. Int. J. Uncertainty Fuzziness Knowl.-Based Syst. **12**(3), 387–406 (2004)
6. Bustince, H., Barrenechea, E., Pagola, M.: Relationship between restricted dissimilarity functions, restricted equivalence functions and normal E_N-functions: Image thresholding invariant. Pattern Recogn. Lett. **29**(4), 525–536 (2008)
7. Bustince, H., Barrenechea, E., Pagola, M.: Image thresholding using restricted equivalence functions and maximizing the measure of similarity. Fuzzy Sets Syst. **128**(5), 496–516 (2007)
8. Bustince, H., Barrenechea, E., Pagola, M.: Restricted equivalence functions. Fuzzy Sets Syst. **157**(17), 2333–2346 (2006)
9. Bustince, H., Barrenechea, E., Pagola, M., Fernandez, J., Xu, Z., Bedregal, B., Montero, J., Hagras, H., Herrera, F., De Baets, B.: A historical account of types of fuzzy sets and their relationship. IEEE Trans. Fuzzy Syst. **24**(1), 179–194 (2016)
10. Bustince, H., Burillo, P., Soria, F.: Automorphisms, negations and implication operators. Fuzzy Sets Syst. **134**, 209–229 (2003)
11. Bustince, H., Fernández, J., Kolesárová, A., Mesiar, R.: Generation of linear orders for intervals by means of aggregation functions. Fuzzy Sets Syst. **220**, 69–77 (2013)

12. Bustince, H., Galar, M., Bedregal, B., Kolesárová, A., Mesiar, R.: A new approach to interval-valued Choquet integrals and the problem of ordering in interval-valued fuzzy set applications. IEEE Trans. Fuzzy Syst. **21**(6), 1150–1162 (2013)
13. Cornelis, C., Deschrijver, G., Kerre, E.E.: Implication in intuitionistic fuzzy and interval-valued fuzzy set theory: construction, classification, application. Int. J. Approximate Reason. **35**(1), 55–95 (2004)
14. Massanet, S., Mayor, G., Mesiar, R., Torrens, J.: On fuzzy implication: an axiomatic approach. Int. J. Approximate Reason. **54**, 1471–1482 (2013)
15. Pradera, A., Beliakov, G., Bustince, H., De Baets, B.: A review of the relationship between implication, negation and aggregation functions from the point of view of material implication. Inf. Sci. **329**, 357–380 (2016)
16. Riera, J.V., Torrens, J.: Residual implications on the set of discrete fuzzy numbers. Inf. Sci. **247**, 131–143 (2013)
17. Xu, Z., Yager, R.R.: Some geometric aggregation operators based on intuitionistic fuzzy sets. Int. J. Gen. Syst. **35**, 417–433 (2006)

Generalized Sugeno Integrals

Didier Dubois[1(✉)], Henri Prade[1], Agnès Rico[2], and Bruno Teheux[3]

[1] IRIT, Université Paul Sabatier, 31062 Toulouse Cedex 9, France
dubois@irit.fr
[2] ERIC, Université Claude Bernard Lyon 1, 69100 Villeurbanne, France
[3] Mathematics Research Unit, FSTC, University of Luxembourg,
1359 Luxembourg, Luxembourg

Abstract. Sugeno integrals are aggregation functions defined on a qualitative scale where only minimum, maximum and order-reversing maps are allowed. Recently, variants of Sugeno integrals based on Gödel implication and its contraposition were defined and axiomatized in the setting of bounded chain with an involutive negation. This paper proposes a more general approach. We consider totally ordered scales, multivalued conjunction operations not necessarily commutative, and implication operations induced from them by means of an involutive negation. In such a context, different Sugeno-like integrals are defined and axiomatized.

Keywords: Sugeno integral · Conjunctions · Implications · Multifactorial evaluation

1 Introduction and Prerequisites

In a recent paper [4], we introduced variants of Sugeno integrals based on Gödel implication and its contraposition using an involutive negation. It models qualitative aggregation methods that extend min and max, based on the idea of tolerance threshold beyond which a criterion is considered satisfied. These new aggregation operations have been axiomatized in [5] in the setting of a complete bounded chain with an involutive negation. In the present paper, we try to cast this approach in a more general totally ordered algebraic setting, using multivalued conjunction operations that are not necessarily commutative, and implication operations induced from them by means of an involutive negation.

We adopt the terminology and notations usual in multi-criteria decision making, where some alternatives are evaluated according to a common set $\mathcal{C} = \{1, \ldots, n\} = [n]$ of criteria. A common evaluation scale L is assumed to provide ratings according to the criteria: each alternative is thus identified with a function $f \in L^{\mathcal{C}}$ which maps every criterion i of $\mathcal{C}$ to the local rating f_i of the alternative with regard to this criterion. We assume that L is a totally ordered set with 1 and 0 as top and bottom, respectively (L may be the real unit interval $[0, 1]$ for instance). For any $a \in L$, we denote by $\mathbf{a}_{\mathcal{C}}$ the constant alternative equals to a on $\mathcal{C}$. In addition, we assume that L is equipped with a unary order reversing involutive operation $t \mapsto 1 - t$, that we call *negation*.

J.P. Carvalho et al. (Eds.): IPMU 2016, Part I, CCIS 610, pp. 363–374, 2016.
DOI: 10.1007/978-3-319-40596-4_31

We denote by $\wedge$ and $\vee$ the minimum and maximum operation on L. These two aggregation schemes can be slightly generalised by means of importance levels or priorities $\pi_i \in L$, on the criteria $i \in [n]$. Suppose π_i is increasing with the importance of i. A fully important criterion has importance weight $\pi_i = 1$. In the following, we assume $\pi_i > 0$ for every $i \in [n]$, *i.e.*, there is no useless criterion. In this section, we also assume $\pi_i = 1$, for some criterion i (the most important one). It is a kind a normalization assumption that ensures that the whole scale L is useful, and that is typical of possibility theory. These importance levels can interact with each local evaluation f_i in different manners. Usually, a weight π_i acts as a saturation threshold that blocks the global score under or above a certain value dependent on the importance level of criterion i. Such weights truncate the evaluation scale from above or from below. The rating f_i is taken into acount in the form of either $(1-\pi_i)\vee f_i \in [1-\pi_i, 1]$, or $\pi_i \wedge f_i \in [0, \pi_i]$. A fully important criterion can alone bring the whole global score to 1 or to 0. The weighted minimum and maximum operations then take the following forms:

$$MIN_\pi(f) = \bigwedge_{i=1}^{n} ((1-\pi_i) \vee f_i); \quad MAX_\pi(f) = \bigvee_{i=1}^{n} (\pi_i \wedge f_i). \tag{1}$$

It is well-known that if the evaluation scale L is reduced to $\{0, 1\}$ (Boolean criteria) then letting $A_f = \{i : f_i = 1\}$ be the set of criteria satisfied by alternative f, the function $\Pi\colon A_f \mapsto MAX_\pi(f) = \bigvee\{\pi_i : i \in A_f\}$ is a possibility measure on $\mathcal{C}$ [12] (*i.e*, a set function Π that satisfies $\Pi(A \cup B) = \Pi(A) \vee \Pi(B)$ for every $A, B \subseteq \mathcal{C}$), and $N\colon f \mapsto MIN_\pi(f) = \bigwedge\{1-\pi_i : i \notin A_f\}$ is a necessity measure [3] (*i.e.*, a set function N that satisfies $N(A \cap B) = N(A) \wedge N(B)$ for every $A, B \subseteq \mathcal{C}$). Note that the well-known duality property $\Pi(A) = 1 - N(\overline{A})$, where $\overline{A}$ denotes the set complement of A in $\mathcal{C}$, immediately generalizes to the scale L in the following way:

$$MAX_\pi(f) = 1 - MIN_\pi(1-f). \tag{2}$$

There are two possible lines of action to extend the definition of the aggregation operations in (1):

- Replacing possibility and necessity measures by more general monotonic set functions that attach weights to groups of criteria.
- Extending the rating modification schemes using more general conjunctions and implications.

Sugeno Integral. The first extension leads to modeling relative weights of the sets of criteria via a *capacity*, which is an order-preserving map $\gamma\colon 2^{\mathcal{C}} \to L$ that satisfies $\gamma(\varnothing) = 0$ and $\gamma(\mathcal{C}) = 1$. The *conjugate capacity* γ^c of γ is defined by $\gamma^c(A) = 1 - \gamma(\overline{A})$ for every $A \subseteq \mathcal{C}$. The Sugeno integral [11], of an alternative f can be defined by means of several expressions, among which the two following normal forms [9]:

$$\int_\gamma f = \bigvee_{A \subseteq \mathcal{C}} (\gamma(A) \wedge \bigwedge_{i \in A} f_i) = \bigwedge_{A \subseteq \mathcal{C}} (1 - \gamma^c(A)) \vee \bigvee_{i \in A} f_i. \tag{3}$$

These expressions, which generalise the conjunctive and disjunctive normal forms in logic, can be simplified as follows:

$$\int_{\gamma} f = \bigvee_{a \in L} \gamma(\{i : f_i \geq a\}) \wedge a = \bigwedge_{a \in L} \gamma(\{i : f_i > a\})) \vee a. \quad (4)$$

Moreover, for the necessity measure N associated with a possibility distribution π, we have $\int_N(f) = MIN_{\pi}(f)$; and for the possibility measure Π associated with π, we have $\int_{\Pi}(f) = MAX_{\pi}(f)$.

There is a duality relation between Sugeno integrals with respect to conjugate capacities, extending (2):

$$\int_{\gamma} f = 1 - \int_{\gamma^c} (1 - f). \quad (5)$$

Two alternatives $f, g \in L^C$ are said to be *comonotone* if for every $i, j \in [n]$, if $f(i) < f(j)$ then $g(i) \leq g(j)$ and if $g(i) < g(j)$ then $f(i) \leq f(j)$. By means of this notion, Sugeno integral can be characterized as follows:

Theorem 1 [1]. *Let $I : L^C \to L$. There is a capacity γ such that $I(f) = \int_{\gamma} f$ for every $f \in L^C$ if and only if the following properties are satisfied*

1. *$I(f \vee g) = I(f) \vee I(g)$, for any comonotone $f, g \in L^C$.*
2. *$I(a \wedge f) = a \wedge I(f)$, for every $a \in L$ and $f \in L^C$.*
3. *$I(\mathbf{1}_C) = 1$.*

Equivalently, conditions (1-3) can be replaced by conditions (1'-3') below.

1'. *$I(f \wedge g) = I(f) \wedge I(g)$, for any comonotone $f, g \in L^C$.*
2'. *$I(b \vee f) = b \vee I(f)$, for every $a \in L$ and $f \in L^C$.*
3'. *$I(\mathbf{0}_C) = 0$.*

The existence of these two equivalent characterisations is due to the possibility of writing Sugeno integral in conjunctive and disjunctive forms (3) equivalently.

Generalized Rating Modification. The second extension yields weighted min and max operations of the form

$$MIN_{\pi}^{\to}(f) = \bigwedge_{i=1}^{n} \pi_i \to f_i; \quad MAX_{\pi}^{\otimes}(f) = \bigvee_{i=1}^{n} \pi_i \otimes f_i, \quad (6)$$

where $\to$ is an implication connective, and $\otimes$ a conjunction, understood as multi-valued connectives that coincide with Boolean implication and conjunction when restricted to $\{0, 1\}$. In order to preserve the duality property (2), these operations must be related by a property that we call *semi-duality*, defined by the equation $a \to b = 1 - (a \otimes (1 - b))$, or equivalently $a \otimes b = 1 - (a \to (1 - b))$.

One may then consider both generalizations together and define, given a pair of semi-dual implication $\to$ and conjunction $\otimes$, the integrals $\int^{\otimes}$ and $\int^{\to}$ by

$$\int_{\gamma}^{\otimes} f = \bigvee_{A \subseteq C} (\gamma(A) \otimes \bigwedge_{i \in A} f_i); \quad \int_{\gamma}^{\to} f = \bigwedge_{A \subseteq C} (\gamma^c(A) \to \bigvee_{i \in A} f_i), \quad (7)$$

for every capacity γ and every $f \in L^{\mathcal{C}}$. In what follows, we refer to expressions of the form $\int_\gamma^{\rightarrow}$ as *co-integrals*. The assumption of semi-duality ensures that the duality Eq. (5) holds between integrals and co-integrals.

Sugeno integral is a particular instance of (7), since the minimum $\wedge$ and the Kleene-Dienes implication $\rightarrow_K$ defined as $a \rightarrow_K b := (1-a) \vee b$ exchange by semi-duality. So, Eq. (3) actually states that

$$\int_\gamma f = \int_\gamma^{\wedge} f = \int_\gamma^{\rightarrow_K} f$$

for every capacity γ and every $f \in L^{\mathcal{C}}$. It means that integrals and co-integrals defined by means of the operation $\wedge$ and $\rightarrow_K$, respectively, coincide. As we shall see in the sequel, this is not generally the case.

2 Variants of Sugeno Integrals: An Example

Let us recall previous results [4] in the qualitative setting of a *complete bounded totally ordered set* $L = (L, \wedge, \rightarrow_G, 0, 1)$ where $\rightarrow_G$ is the Gödel implication defined by residuation of $\wedge$:

$$a \rightarrow_G b := \sup\{x : a \wedge x \leq b\} = \begin{cases} 1 \text{ if } a \leq b \\ b \text{ otherwise.} \end{cases} \tag{8}$$

As previously, L is equipped with an involutive operation $1 - \cdot$. The following (non-commutative) conjunction, introduced in [2] is defined by semi-duality:

$$a \otimes_G b := 1 - (a \rightarrow_G (1-b)) = \begin{cases} b \text{ if } a > 1-b, \\ 0 \text{ otherwise.} \end{cases} \tag{9}$$

The qualitative integral $\int_\gamma^{\otimes_G}$ and co-integral $\int_\gamma^{\rightarrow_G}$ have simplified expressions that extend those of Sugeno integrals, assuming $f_1 \leq \cdots \leq f_n$:

$$\int_\gamma^{\otimes} f = \bigvee_{i=1}^{n} \gamma(\{i, \cdots, n\}) \otimes f_{(i)} = \bigvee_{a \in L} \gamma(\{f \geq a\}) \otimes a \tag{10}$$

$$\int_\gamma^{\rightarrow} (f) = \bigwedge_{i=1}^{n} \gamma^c(\{1, \cdots, i\}) \rightarrow f_{(i)} = \bigwedge_{a \in L} \gamma^c(\{f \leq a\}) \rightarrow a. \tag{11}$$

Note also that if N is a necessity measure and Π is a possibility measure, then $\int_N^{\rightarrow_G} = MIN_\pi^{\rightarrow_G}$ and $\int_\Pi^{\otimes_G} = MAX_\pi^{\otimes_G}$. However we cannot exchange N and Π in those results.

As $\otimes_G$ is not commutative, there is an alternative definition for those aggregation operations, replacing $\otimes_G$ by the operation $\otimes_{GC}$ defined by $a \otimes_{GC} b := b \otimes_G a$, and the operation $\rightarrow_G$ by the implication $\rightarrow_{GC}$ associated with $\otimes_{GC}$ by semi-duality (*i.e.*, the operation $\rightarrow_{GC}$ is the contrapositive version of $\rightarrow_G$):

$$a \rightarrow_{GC} b := 1 - (a \otimes_{GC} (1-b)) = (1-b) \rightarrow_G (1-a) = \begin{cases} 1 \text{ if } a \leq b, \\ 1-a \text{ otherwise.} \end{cases}$$

Properties (11)-(10) hold for $\int_\gamma^{\to_{GC}} f$ and $\int_\gamma^{\otimes_{GC}} f$ as well as for their reductions to a form of weighted min and max for necessity and possibility measures.

Noticeably, the integral and co-integral based on Gödel implications and their associated semi-dual conjunctions do not coincide. We have proved [4] that

$$\int_\gamma^{\otimes_G} f \geq \int_\gamma^{\to_G} f \text{ and } \int_\gamma^{\otimes_{GC}} f \geq \int_\gamma^{\to_{GC}} f, \qquad (12)$$

but the inequalities may be strict. For instance, $\int_\gamma^{\to_G}(f) = 1$ if for all $A \subseteq \mathcal{C}$, there is some $i \in A$ such that $f_i \geq \gamma_c(A)$, and $\int_\gamma^{\otimes_G}(f) = 1$ if there is some subset $A \subseteq \mathcal{C}$ such that $\gamma(A) > 0$ and $f_i = 1$ for every $i \in A$.

Some characterization theorems for these variants of Sugeno integrals have been obtained [5]:

Theorem 2. *Let $I\colon L^{\mathcal{C}} \to L$ be a mapping. There is a capacity γ such that $I(f) = \int_\gamma^{\otimes_G} f$ for every $f \in L^{\mathcal{C}}$ if and only if*

1. *$I(f \vee g) = I(f) \vee I(g)$, for any comonotone $f, g \in L^{\mathcal{C}}$.*
2. *There is a capacity $\lambda\colon 2^{\mathcal{C}} \to L$ such that $I(\mathbf{1}_A \otimes_G a) = \lambda(A) \otimes_G a$ for every $a \in L$ and every $A \subseteq \mathcal{C}$.*

In that case, we have $\gamma = \lambda$.

Theorem 3. *Let $I\colon L^{\mathcal{C}} \to L$ be a mapping. There is a capacity γ such that $I(f) = \int_\gamma^{\to_G} f$ for every $f \in L^{\mathcal{C}}$ if and only if*

1. *$I(f \wedge g) = I(f) \wedge I(g)$, for any comonotone $f, g \in L^{\mathcal{C}}$.*
2. *There is a capacity $\rho\colon 2^{\mathcal{C}} \to L$ such that $I(\mathbf{1}_A \to_G a) = \rho(A) \to_G a$ for every $a \in L$.*

If these conditions are satisfied then $\gamma = \rho^c$.

Similar theorems hold [5] for $\int_\gamma^{\otimes_{GC}}(f)$ and $\int_\gamma^{\to_{GC}}(f)$. The above results suggest that it is possible to find a more general algebraic structure to define generalized Sugeno integrals, while keeping the same properties.

Note that the three implications and conjunctions in the above setting are related in the following way. We consider the three following transformations that can be applied to any operation $\star$ on a bounded totally ordered set with involutive negation $L = (L, \vee, \wedge, 1 - \cdot, 0, 1)$:

- Residuation: $aRes(\star)b := \bigvee\{a : a \star b \leq c\}$ if this supremum exists,
- Semi-duality: $a\mathcal{S}(\star)b := 1 - a \star (1 - b)$,
- Contraposition: $a\mathcal{C}(\star)b := (1 - b) \star (1 - a)$.
- Argument exchange: $a\mathcal{A}(\star)b := b \star a$

Note that semi-duality and contraposition are involutive transformations. Moreover the diagram in Fig. 1 commutes [2].

In the sequel, we focus on generalized Sugeno integrals on a finite total order equipped with a conjunction that is not necessarily commutative, and the co-integral obtained by semi-duality. For simplicity, the word "q-integral" is used here in the sense of generalized Sugeno integrals on a qualitative scale.

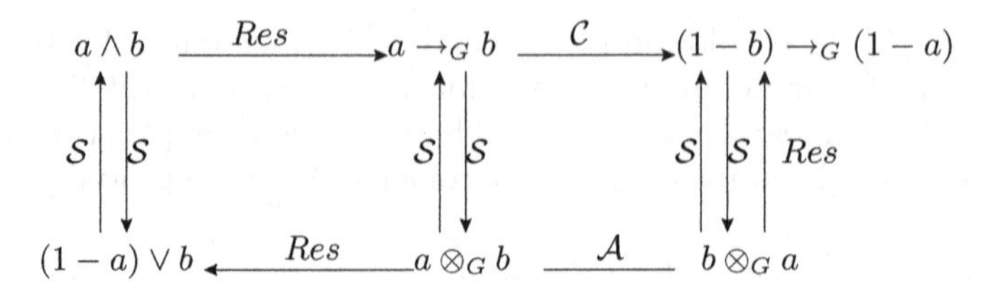

Fig. 1. Connectives induced by the minimum on a finite chain

3 Sugeno-Like q-Integrals Based on Left-Conjunctions

We consider a bounded complete totally ordered value scale $(L, 0, 1, \leq)$, equipped with an operation $\otimes$ called *left-conjunction*, which has the following properties:

- the top element 1 is a left-identity: $1 \otimes x = x$,
- the bottom element 0 is a left-anihilator $0 \otimes x = 0$,
- the maps $x \mapsto a \otimes x$, $x \mapsto x \otimes a$ are order-preserving for every $a \in L$.

It follows that $a \otimes 0 = 0$ for every $a \in L$ (0 is an anihilator on both sides), and so, a left-conjunction coincides with a Boolean conjunction on $\{0, 1\}$; but we assume neither associativity nor commutativity. The following operations are examples of left-conjunctions.

- T-norms on $[0, 1]$, in particular $\wedge$, the product t-norm, the Łukasiewicz t-norm and the nilpotent minimum $\overline{\wedge}$ defined by $a \overline{\wedge} b = 0$ if $a + b \leq 1$ and $a \overline{\wedge} b = a \wedge b$ otherwise.
- Weak t-norms [8], i.e., left conjunctions such that $a \otimes 1 \leq a$.
- The non-commutative Gödel conjunction $\otimes_G$ previously introduced, and the non-commutative conjunction $\otimes_{rTC}$ defined by $a \otimes_{rTC} b = 0$ if $a = 0$, and $a \otimes_{rTC} b = b$ if $a \neq 0$ (see [5]).
- Pseudo-multiplications used by Klement et al. [10] in the definition of universal integrals. A pseudo-multiplication has genuine identity 1 and anihilator 0 (on both sides).

Definition 1. *Let $\otimes$ be a left-conjunction on L and $\gamma\colon 2^{\mathcal{C}} \to L$ be a capacity. The q-integral $\int_\gamma^\otimes$ is the mapping $\int_\gamma^\otimes\colon L^{\mathcal{C}} \to L$ defined by*

$$\int_\gamma^\otimes f = \bigvee_{A \subseteq C} \left(\gamma(A) \otimes \bigwedge_{i \in A} f_i\right), \ \textit{for all}\ f \in L^{\mathcal{C}}.$$

We show that q-integrals can be characterized similarly as in Theorem 2. In the following when we consider $f \in L^{\mathcal{C}}$, $(\cdot)$ denotes a permutation on the set of criteria such that $f_{(1)} \leq \cdots \leq f_{(n)}$ and we let $A_{(i)} = \{(i), \cdots, (n)\}$ with the convention $A_{(n+1)} = \varnothing$.

Lemma 1. *If $f \in L^{\mathcal{C}}$ then $f = \bigvee_{i=1}^{n} \mathbf{1}_{A_{(i)}} \otimes f_{(i)}$.*

Proof. For any $i, k \in [n]$, $f_{(i)} \leq f_k$ if $k \in A_{(i)}$. It follows that $\mathbf{1}_{A_{(i)}}(k) \otimes f_{(i)} = 0$ if $f_k < f_{(i)}$ and $f_{(i)}$ otherwise; hence $\bigvee_{i=1}^{n} \mathbf{1}_{A_{(i)}}(k) \otimes f_{(i)} = \bigvee\{f_{(i)} \mid f_{(i)} \leq f_k\} = f_k$.

Proposition 1. $\int_{\gamma}^{\otimes} f = \bigvee_{i=1}^{n} \gamma(A_{(i)}) \otimes f_{(i)}$.

Proof. $\int_{\gamma}^{\otimes} f = \bigvee_{A \subseteq C} \gamma(A) \otimes \bigwedge_{i \in A} f_i$. Let us denote $\bigwedge_{i \in A} f_i$ by f_{i_A}. It follows that $A \subseteq A_{(i_A)}$ which entails $\gamma(A) \leq \gamma(A_{(i_A)})$ and $\gamma(A) \otimes f_{i_A} \leq \gamma(A_{(i_A)}) \otimes f_{i_A}$.

Lemma 2. *For every capacity* γ, *the map* $\int_{\gamma}^{\otimes} : L^C \to L$ *is order-preserving.*

Proof. Directly from the assumption that the map $x \mapsto a \otimes x$ is order-preserving.

Lemma 3. $\int_{\gamma}^{\otimes} f = \bigvee_{a \in L} \gamma(\{f \geq a\}) \otimes a$.

Proof. We use Proposition 1. Let $a \in L \setminus \{f_1, \ldots, f_n\}$.
If $a > f_{(n)}$ then $\gamma(\{f \geq a\}) \otimes a = 0 \otimes a = 0$.
If $a < f_{(1)}$ then $\gamma(\{f \geq a\}) \otimes a = \gamma(\{f \geq f_{(1)}\}) \otimes a \leq \gamma(\{f \geq f_{(1)}\}) \otimes f_{(1)}$.
If $f_{(i-1)} < a < f_{(i)}$ then $\gamma(\{f \geq a\}) \otimes a = \gamma(\{f \geq f_{(i)}\}) \otimes a \leq \gamma(\{f \geq f_{(i)}\}) \otimes f_{(i)}$.

Lemma 4. *For any comonotone* $f, g \in L^C$, *we have* $\int_{\gamma}^{\otimes}(f \vee g) = \int_{\gamma}^{\otimes} f \vee \int_{\gamma}^{\otimes} g$.

Proof. The inequality $\int_{\gamma}^{\otimes}(f \vee g) \geq \int_{\gamma}^{\otimes} f \vee \int_{\gamma}^{\otimes} g$ follows from Lemma 2. Let us prove the other inequality. Let $a \in L$. For any two comonotone functions $f, g \in L^C$ we have either $\{f \geq a\} \subseteq \{g \geq a\}$ or $\{g \geq a\} \subseteq \{f \geq a\}$.
If $\{f \geq a\} \subseteq \{g \geq a\}$ then $\{f \vee g \geq a\} = \{f \geq a\} \cup \{g \geq a\} = \{g \geq a\}$ and $\gamma(\{f \vee g \geq a\}) \otimes a = \gamma(\{g \geq a\}) \otimes a \leq \big(\gamma(\{g \geq a\}) \otimes a\big) \vee \big(\gamma(\{f \geq a\}) \otimes a\big)$. By symmetry, the inequality is also true when $\{g \geq a\} \subseteq \{f \geq a\}$; hence $\int_{\gamma}^{\otimes}(f \vee g) \leq \bigvee_{a \in L} \Big(\big(\gamma(\{g \geq a\}) \otimes a\big) \vee \big(\gamma(\{f \geq a\}) \otimes a\big)\Big) = \int_{\gamma}^{\otimes} f \vee \int_{\gamma}^{\otimes} g$.

Lemma 5. *For every* $f \in L^C$ *and every* $\ell \in \{1, \ldots, n-1\}$, *the maps* $\mathbf{1}_{A_{(\ell)}} \otimes f_{(\ell)}$ *and* $\bigvee_{i=\ell+1}^{n} \mathbf{1}_{A_{(i)}} \otimes f_{(i)}$ *are comonotone.*

Proof. We represent both maps as vectors of components ordered according to $(1), \ldots, (n)$, so that $A_{(\ell)} = \{(\ell), \ldots, (n)\}$. In consequence, $\mathbf{1}_{A_{(\ell)}} \otimes f_{(\ell)}(i) = \bigvee_{i=\ell+1}^{n} \mathbf{1}_{A_{(i)}} \otimes f_{(i)}(i) = 0$ if $i \leq \ell$ while $\mathbf{1}_{A_{(\ell)}} \otimes f_{(\ell)}(i) = f_{(\ell)}$ and $\bigvee_{i=\ell+1}^{n} \mathbf{1}_{A_{(i)}} \otimes f_{(i)}(i) = f_{(i)}$ if $i > \ell$. Hence it is easy to check that the two maps are comonotone.

Lemma 6. *For any capacity* γ, *any* $B \subseteq C$ *and any* $a \in L$ *we have* $\int_{\gamma}^{\otimes}(\mathbf{1}_B \otimes a) = \gamma(B) \otimes a$. *In particular* $\int_{\gamma}^{\otimes} \mathbf{1}_C = 1$.

Proof. $\int_{\gamma}^{\otimes} f = \bigvee_{i=1}^{n} \gamma(A_{(i)}) \otimes f_{(i)}$, where $f = \mathbf{1}_B \otimes a$. Note that $\mathbf{1}_B(i) \otimes a = a$ if $i \in B$ and $\mathbf{1}_B(i) \otimes a = 0$ otherwise. So, there is j such that $B = A_{(j)} = \{(j), \ldots, (n)\}$. So we get $\int_{\gamma}^{\otimes}(\mathbf{1}_B \otimes a) = \bigvee_{i \geq j}^{n} \gamma(A_{(i)}) \otimes a$, and the maximum is attained for $i = j$. Further, $\int_{\gamma}^{\otimes} \mathbf{1}_C = \int_{\gamma}^{\otimes} \mathbf{1}_C \otimes 1 = \gamma(C) \otimes 1 = 1$.

We can now prove our first characterization result.

Theorem 4. *Let $I: L^{\mathcal{C}} \to L$ be a mapping and $\otimes$ a left-conjunction. There is a capacity γ such that $I(f) = \int_{\gamma}^{\otimes} f$ for every $f \in L^{\mathcal{C}}$ if and only if*

1. *$I(f \vee g) = I(f) \vee I(g)$, for any comonotone $f, g \in L^{\mathcal{C}}$.*
2. *There is a capacity $\lambda: 2^{\mathcal{C}} \to L$ such that $I(\mathbf{1}_A \otimes a) = \lambda(A) \otimes a$ for every $a \in L$ and every $A \subseteq \mathcal{C}$.*

In that case, we have $\gamma = \lambda$.

Proof. Necessity is obtained by previous Lemmas. For sufficiency, assume that I is a mapping that satisfies conditions 1 and 2 and let $f \in L^{\mathcal{C}}$. We have $I(f) = I(\bigvee_{i=1}^{n} \mathbf{1}_{A_{(i)}} \otimes f_{(i)}) = \bigvee_{i=1}^{n} I(\mathbf{1}_{A_{(i)}} \otimes f_{(i)}) = \bigvee_{i=1}^{n} \lambda(A_{(i)}) \otimes f_{(i)} = \int_{\lambda}^{\otimes} f$.

Note that we have used all properties of left-conjunctions in our proof of the previous result. Moreover, contrary to universal integrals, $\int_{\gamma}^{\otimes} \mathbf{1}_A = \gamma(A) \otimes 1 \neq \gamma(A)$, generally, since 1 is not an identity on the right. To get the property $\int_{\gamma}^{\otimes} \mathbf{1}_A = \gamma(A)$, it is enough to assume the left-conjunction $\otimes$ is commutative. The set function $\hat{\gamma}(A) = \gamma(A) \otimes 1$ generally differs from γ. The following counter-example shows that we may have $\hat{\gamma}(A) \neq \gamma(A)$ and $\int_{\gamma}^{\otimes} \neq \int_{\hat{\gamma}}^{\otimes}$.

Example 1. $\mathcal{C} = \{1, 2\}$, $L = [0, 1]$ and $\otimes_G$ is the Gödel conjunction. We have $a \otimes_G 1 = 0$ if $a = 0$ and 1 otherwise so $a \otimes_G 1 \neq a$ if $a < 1$. Let us consider γ such that $\gamma(\{1\}) = 0$ and $\gamma(\{2\}) = 0.1$. If f is defined by $f_1 = 0$ and $f_2 = 0.8$ we obtain $\int_{\gamma}^{\otimes_G} f = (1 \otimes_G 0) \vee (0.1 \otimes 0.8) = 0.8$ and $\int_{\hat{\gamma}}^{\otimes_G} f = 1$ since $\hat{\gamma}(\{2\}) = 1$.

In the case when the functional I is maxitive, we prove the following:

Theorem 5. *Assume that $\otimes$ is a left conjunction that is right-cancellative, that is, for every $a, b, c \in L$, if $a \otimes c = b \otimes c$ then $a = b$. Let $I: L^{\mathcal{C}} \to L$ be a mapping. There is a possibility measure Π such that $I(f) = \int_{\Pi}^{\otimes} f$ for every $f \in L^{\mathcal{C}}$ if and only if I satisfies the following properties:*

1. *$I(f \vee g) = I(f) \vee I(g)$, for any $f, g \in L^{\mathcal{C}}$.*
2. *There is a capacity $\lambda: 2^{\mathcal{C}} \to L$ such that $I(\mathbf{1}_A \otimes a) = \lambda(A) \otimes a$ for every $a \in L$ and every $A \subseteq \mathcal{C}$.*

In that case, we have $\Pi = \lambda$.

Proof. A q-integral with respect to a possibility measure satisfies the requested properties. Let us prove the converse. According to the previous theorem, there exists a capacity λ such that $I(\mathbf{1}_A \otimes a) = \lambda(A) \otimes a$ for all a. For all $a \neq 0$, $(\mathbf{1}_A \otimes a) \vee (\mathbf{1}_B \otimes a) = (\mathbf{1}_{A \cup B} \otimes a)$ by the order-preservingness property; so $(\lambda(A) \otimes a) \vee (\lambda(B) \otimes a) = \lambda(A \cup B) \otimes a$. Again by order-preservingness, $(\lambda(A) \otimes a) \vee (\lambda(B) \otimes a) = (\lambda(A) \vee \lambda(B)) \otimes a$ hence the cancellativeness property allows us to conclude $\lambda(A \cup B) = \lambda(A) \vee \lambda(B)$.

The above result also holds for commutative conjunctions (so, for triangular norms), and also pseudo-multiplications, since in that case $I(\mathbf{1}_A) = I(\mathbf{1}_A \otimes \mathbf{1}_{\mathcal{C}}) = \Pi(A) \otimes 1 = \Pi(A)$. But it does not hold for the Gödel conjunction, nor the other non-commutative conjunction mentioned above.

4 Integrals Defined with a Right-Conjunction

We consider a binary operation $\otimes_C$ defined by $a \otimes_C b := b \otimes a$ where $\otimes$ is a left-conjunction. Clearly, $a \otimes_C 1 = a$, $a \otimes_C 0 = 0$, $0 \otimes_C a = 0$, and the maps $x \mapsto a \otimes_C x$ and $x \mapsto x \otimes_C a$ are order-preserving. We call $\otimes_C$ a *right-conjunction.* The associated q-integral is

$$\int_{\gamma}^{\otimes_C} f = \bigvee_{A \subseteq C} (\bigwedge_{i \in A} f_i \otimes \gamma(A)).$$

It generally differs from $\int_{\gamma}^{\otimes} f$ [4]. Using the results presented in Sect. 3 it is easy to prove that if $f \in L^{\mathcal{C}}$, then $f = \bigvee_{i=1}^{n} f_{(i)} \otimes_C \mathbf{1}_{A_{(i)}}$ where for every $\ell \in \{1, \ldots, n-1\}$, the maps $f_{(\ell)} \otimes_C \mathbf{1}_{A_{(\ell)}}$ and $\bigvee_{i=\ell+1}^{n} f_{(i)} \otimes_C \mathbf{1}_{A_{(i)}}$ are comonotone.

Since $x \mapsto x \otimes_C a$ is increasing, $\int_{\gamma}^{\otimes_c} f = \bigvee_{i=1}^{n} \gamma(A_{(i)}) \otimes_C f_{(i)}$, and since $x \mapsto a \otimes_C x$ is increasing, the map $\int_{\gamma}^{\otimes_C} : L^{\mathcal{C}} \to L$ is order-preserving for every capacity γ. Moreover we have $\int_{\gamma}^{\otimes_C} f = \bigvee_{a \in L} (a \otimes \gamma(\{f \geq a\}))$ and for any comonotone $f, g \in L^{\mathcal{C}}$ we have $\int_{\gamma}^{\otimes_C} (f \vee g) = \int_{\gamma}^{\otimes_C} (f) \vee \int_{\gamma}^{\otimes_C} (g)$.

For every capacity γ, every $A \subseteq C$ and every $a \in L$, it holds

$$\int_{\gamma}^{\otimes_C} (a \otimes_C \mathbf{1}_A) = \bigvee_{B \subseteq C} \gamma(B) \otimes_C \bigwedge_{i \in B} (a \otimes_C \mathbf{1}_A(i)) = \bigvee_{B \subseteq C} \bigwedge_{i \in B} (\mathbf{1}_A(i) \otimes a) \otimes \gamma(B).$$

We have $\bigwedge_{i \in B} (\mathbf{1}_A(i) \otimes a) = a$ if $B \subseteq A$ and $\bigwedge_{i \in B} (\mathbf{1}_A(i) \otimes a) = 0$ otherwise, so

$$\int_{\gamma}^{\otimes_C} (a \otimes_C \mathbf{1}_A) = \bigvee_{B \subseteq A} a \otimes \gamma(B) = a \otimes \gamma(A).$$

In particular, as $1 \otimes a = a$, we have $\int_{\gamma}^{\otimes_C} \mathbf{1}_A = \gamma(A)$, and so $\int_{\gamma}^{\otimes_C} (a \otimes_C \mathbf{1}_A) = a \otimes \int_{\gamma}^{\otimes_C} \mathbf{1}_A$. We are ready to prove the following characterization result.

Theorem 6. *Let $I \colon L^{\mathcal{C}} \to L$ be a mapping and $\otimes_C$ a right-conjunction. There is a capacity γ such that $I(f) = \int_{\gamma}^{\otimes_C} f$ for every $f \in L^{\mathcal{C}}$ if and only if*

1. *$I(f \vee g) = I(f) \vee I(g)$, for any comonotone $f, g \in L^{\mathcal{C}}$.*
2. *For every $A \subseteq C$ and every $a \in L$ we have $I(\mathbf{1}_A \otimes_C a) = I(\mathbf{1}_A) \otimes_C a$.*
3. *$I(\mathbf{1}_C) = 1$.*

In that case γ is defined by $\gamma(A) = I(\mathbf{1}_A)$ for every $A \subseteq C$.

Proof. The proof that $I(f) = \bigvee_{i=1}^{n} I(\mathbf{1}_{A_{(i)}}) \otimes_C f_{(i)}$ is similar to that of Theorem 4. Then we must prove that the set function $\lambda \colon A \mapsto I(\mathbf{1}_A)$ is a capacity. We do have that $\lambda(\mathcal{C}) = 1$, and $\lambda(\varnothing) = I(0) = I(0 \otimes_C \mathbf{1}_C) = 0 \otimes \lambda(\mathcal{C}) = 0$. Finally, for every $A \subseteq B$, as $1 \otimes_C \mathbf{1}_A \leq 1 \otimes_C \mathbf{1}_B$,we get by conditions 1 and 2 that $\lambda(A) = \lambda(A) \otimes 1 = I(1 \otimes_C \mathbf{1}_A) \leq I(1 \otimes_C \mathbf{1}_B) = 1 \otimes \lambda(B) = \lambda(B)$.

Note that if the maxitivity condition 1 is extended to any pair of mappings f, g, then $I(\mathbf{1}_B) = \Pi(B)$ and $I(f) = MAX_{\pi}^{\otimes_C}(f) = \bigvee_{i=1}^{n} f_i \otimes \pi_i$. The contraposed Gödel conjunction $\otimes_{GC}$ and the right conjunction associated with the conjunction $\otimes_{rTC}$ introduced above are examples of right-conjunctions.

5 Q-cointegrals Defined from Left-Conjunctions

As L is equipped with an involutive negation $t \mapsto 1-t$, we can define an implication $\rightarrow$ from $\otimes$ by semi-duality: $a \rightarrow b := 1-(a \otimes (1-b))$. This implication satisfies the following very usual properties:

- $a \rightarrow 1 = 1$, $0 \rightarrow b = 1$ and $1 \rightarrow b = b$,
- $\rightarrow$ is decreasing according to its first argument,
- $\rightarrow$ is increasing according to the second one.

Under property $1 \rightarrow b = b$, implication $\rightarrow$ is called a *border implication*. The implication $\rightarrow$ and the conjunction $\otimes$ exchange via semi-duality.

The following implication operations satisfy the properties presented above.

- S-implications obtained from triangular norms by semi-duality, among them, the Kleene-Dienes implication, the Łukasiewicz implication $a \rightarrow_{\text{Ł}} b = \max(1-a+b, 0)$, the nilpotent implication induced by the nilpotent minimum $a \rightarrow b = 1$ if $a \geq b$ and $(1-a) \vee b$ otherwise, Reichenbach implication (induced by product).
- Implications obtained from triangular norms by residuation Res, among which the Gödel and Łukasiewicz implications, the nilpotent implication, Goguen implication (induced by product). Operations of the form $\otimes = \mathcal{S}(Res(\star))$, where $\star$ is a left-continuous t-norm, are (generally non-commutative) left-conjunctions instrumental in the construction of Sect. 4.
- $a \rightarrow_{XC} b = 1$ if $a = 0$, and b if $a \neq 0$ [5].

Definition 2. *Let $\rightarrow$ be a border implication as defined above on L and $\gamma\colon 2^{\mathcal{C}} \rightarrow L$ be a capacity. The q-cointegral $\int_{\gamma}^{\rightarrow}\colon L^{\mathcal{C}} \rightarrow L$ is the mapping*

$$\int_{\gamma}^{\rightarrow} f = \bigwedge_{A \subseteq C} (\gamma^c(A) \rightarrow \bigvee_{i \in A} f_i), \text{ for all } f \in L^{\mathcal{C}}.$$

Using semi-duality, q-cointegrals can be expressed in terms of q-integrals.

Proposition 2. $\int_{\gamma}^{\rightarrow}(f) = 1 - \int_{\gamma^c}^{\otimes}(1-f)$.

As in [5], using semi-duality, we derive the following results from Sect. 4.

For any $f \in L^{\mathcal{C}}$, we have $f = \bigwedge_{i=1}^{n} \mathbf{1}_{\overline{A_{(i+1)}}} \rightarrow f_{(i)}$ where for every $\ell \in \{1, \ldots, n-1\}$, the maps $\mathbf{1}_{\overline{A_{(\ell+1)}}} \rightarrow f_{(\ell)}$ and $\bigwedge_{i=\ell+1}^{n} \mathbf{1}_{\overline{A_{(i+1)}}} \rightarrow f_{(i)}$ are comonotone. Also, $\int_{\gamma}^{\rightarrow}(f) = \bigwedge_{i=1}^{n} \gamma^c(\overline{A_{(i+1)}}) \rightarrow f_{(i)} = \bigwedge_{a \in L} \gamma^c(\{f \leq a\}) \rightarrow a$.

Moreover we have the following characterisation result.

Theorem 7. *Let $I\colon L^{\mathcal{C}} \rightarrow L$ be a mapping. There is a capacity γ such that $I(f) = \int_{\gamma}^{\rightarrow} f$ for every $f \in L^{\mathcal{C}}$ if and only if the following properties are satisfied.*

1. *$I(f \wedge g) = I(f) \wedge I(g)$, for any comonotone $f, g \in L^{\mathcal{C}}$.*
2. *There is a capacity $\rho\colon 2^{\mathcal{C}} \rightarrow L$ such that $I(\mathbf{1}_A \rightarrow a) = \rho^c(A) \rightarrow a, \forall a \in L$.*

In that case $\rho = \gamma$.

We can run a similar study for q-cointegrals induced by a right-conjunction. They use implications that differ from the above ones by the property $a \to 0 = 1 - a$ that replaces $1 \to b = b$, i.e. these implications reconstruct the involutive negation.

Remark 1. The equality (3) between the q-cointegral and the q-integral, in the case of conjunction $\wedge$ and its semi-dual $(1-a) \vee b$ does not extend to conjunction-based q-integrals. For example only the inequality (12): $\int_\gamma^{\otimes} f \geq \int_\gamma^{\to} f$ holds for $(\otimes, \to) \in \{(\otimes_G, \to_G), (\otimes_{GC}, \to_{GC})\}$ [4]. This inequality cannot even be generalised to other conjunctions. For example, for the nilpotent minimum and the nilpotent maximum ($a \underline{\vee} b = a \vee b$, if $a \leq 1 - b$ and 1 otherwise) the relations $a \overline{\wedge} b \leq a \wedge b$ and $a \vee b \leq a \underline{\vee} b$ imply the opposite inequality $\int_\gamma^{\overline{\wedge}} f \leq \int_\gamma^{\Rightarrow} f$, where $\Rightarrow$ is $\mathcal{S}(\overline{\wedge})$, but $\int_\gamma^{\overline{\wedge}} f \neq \int_\gamma^{\Rightarrow} f$.

Example 2. Consider $\mathcal{C} = \{1, 2\}$, $L = [0, 1]$, a capacity γ defined by $\gamma(\{1\}) = \gamma(\{2\}) = 0$ and f such that $f_1 = 0.5$ and $f_2 = 0.5$; then $\int_\gamma^{\overline{\wedge}} f = 0.5 < \int_\gamma^{\Rightarrow} f = 1$.

Remark 2. The diagram on Fig. 1 holds for transforms of many more operations $\star$ than $\wedge$. Fodor [8] has shown that the existence of $Res(\star)$ is a necessary and sufficient condition for the square on the left-hand side to commute. The whole diagram commutes if and only if moreover $\star$ is commutative. $Res(\star)$ exists for left and right conjunctions. So from any q-integral based on a left-conjunction $\otimes$, one can generate two q-cointegrals (based on $Res(\otimes)$ and $\mathcal{S}(\otimes)$), and another q-integral based on $Res \circ \mathcal{S}(\otimes) = \mathcal{S} \circ Res(\otimes)$. We can do likewise for the associated right-conjunction $\mathcal{A}(\otimes)$.

Remark 3. The q-cointegral-like expression defined on a complete residuated lattice in [6] is based on an anticapacity ν i.e., a set function such that $\nu(\emptyset) = 1$, $\nu(C) = 0$ and $A \subseteq B$ implies $\nu(A) \geq \nu(B)$. For all $f \in L^{\mathcal{C}}$, it takes the form: $\oint_\nu^{\to} f = \bigwedge_{A \subseteq C} \bigvee_{i \in A} (f_i \to \nu(A))$, for all $f \in L^{\mathcal{C}}$. It is what we call a desintegral in [4] as it is decreasing with f_i.

6 Conclusion

In this paper, we have proposed a very general setting for generalized forms of Sugeno integrals where the inside operation is either a not-necessarily commutative multivalued conjunction or a multivalued implication. The properties in the algebraic setting were chosen to be minimal in order to preserve representation theorems by means of comonotonic minitive or maxitive functionals: integrals are maxitive, while cointegrals are minitive and differ from each other, in contrast with the case of standard Sugeno integrals. One remaining open problem is to find necessary and sufficient conditions for a conjunction $\otimes$ to ensure the equality between integrals and their semi-dual cointegrals.

Acknowledgements. This work is partially supported by ANR-11-LABX-0040-CIMI (Centre International de Mathématiques et d'Informatique) within the program ANR-11-IDEX-0002-02, project ISIPA.

References

1. Chateauneuf, A., Grabisch, M., Rico, A.: Modeling attitudes toward uncertainty through the use of the Sugeno integral. J. Math. Econ. **44**, 1084–1099 (2008)
2. Dubois, D., Prade, H.: A theorem on implication functions defined from triangular norms. Stochastica **8**, 267–279 (1984)
3. Dubois, D., Prade, H.: Possibility Theory. Plenum Press, New-York (1988)
4. Dubois, D., Prade, H., Rico, A.: Residuated variants of Sugeno integrals. Inf. Sci. **329**, 765–781 (2016)
5. Dubois, D., Rico, A., Teheux, B., Prade, H.: Characterizing variants of qualitative Sugeno integrals in a totally ordered Heyting algebra. In: Proceedings of 9th Conference of the European Society for Fuzzy Logic and Technology (Eusflat), Gijon, pp. 865–872 (2015)
6. Dvořák, A., Holčapek, M.: Fuzzy integrals over complete residuated lattices. In: Proceedings of the IFSA-EUSFLAT Conference, Lisbon, pp. 357–362 (2009)
7. Dvořák, A., Holčapek, M.: Fuzzy measures, integrals defined on algebras of fuzzy subsets over complete residuated lattices. Inf. Sci. **185**, 205–229 (2012)
8. Fodor, J.: On fuzzy implication operators. Fuzzy Sets Syst. **42**, 293–300 (1991)
9. Marichal, J.-L.: On Sugeno integrals as an aggregation function. Fuzzy Sets Syst. **114**, 347–365 (2000)
10. Klement, E., Mesiar, R., Pap, E.: A universal integral as common frame for Choquet and Sugeno integral. IEEE Trans. Fuzzy Syst. **18**, 178–187 (2010)
11. Sugeno, M.: Fuzzy measures, fuzzy integrals: A survey. In: Gupta, M.M., et al. (eds.) Fuzzy Automata and Decision Processes, pp. 89–102. North-Holland, Amsterdam (1977)
12. Zadeh, L.A.: Fuzzy sets as a basis for a theory of possibility. Fuzzy Sets Syst. **1**, 3–28 (1978)

A New Look on Fuzzy Implication Functions: FNI-implications

Isabel Aguiló, Jaume Suñer, and Joan Torrens(✉)

Department of Mathematics and Computer Science, University of the Balearic Islands, Crta. de Valldemossa, km. 7,5, 07122 Palma, Spain
{isabel.aguilo,jaume.sunyer,jts224}@uib.es

Abstract. Fuzzy implication functions are used to model fuzzy conditional and consequently they are essential in fuzzy logic and approximate reasoning. From the theoretical point of view, the study of how to construct new implication functions from old ones is one of the most important topics in this field. In this paper a construction method of implication functions from a t-conorm S (or any disjunctive aggregation function F), a fuzzy negation N and an implication function I is studied. Some general properties are analyzed and many illustrative examples are given. In particular, this method shows how to obtain new implications from old ones with additional properties not satisfied by the initial implication function.

Keywords: Fuzzy implication function · t-conorm · Disjunctive aggregation function · Construction methods · Natural negation

1 Introduction

Fuzzy implication functions play a fundamental role in fuzzy logic and approximate reasoning since they are used not only to model fuzzy conditionals but also in the inference processes. Moreover, they have a lot of applications in many different fields that vary from fuzzy control and fuzzy subsethood measures to fuzzy mathematical morphology and image processing. For this reason, the interest of implication functions from a pure theoretical point of view has been growing in last decades, see for instance the survey [15] and the monographs [4,6] entirely devoted to this kind of operators.

Due to this great quantity of applications, it was pointed out for instance in [25] the necessity of having a lot of different classes of fuzzy implication functions. In this sense, one of the main topics in the theoretic study of implications is the research of new construction methods of fuzzy implication functions. The most important methods to construct fuzzy implication functions are the following:

(i) From different kinds of aggregation functions, especially t-norms, t-conorms and uninorms, leading to the classes of R, (S, N), QL and D-implications derived from t-norms and t-conorms (see mainly [6]), as well as their counterparts derived from uninorms (see [2,7,11,14,21,22]). Moreover, some of these results were collected and completed later in the survey [20].

J.P. Carvalho et al. (Eds.): IPMU 2016, Part I, CCIS 610, pp. 375–386, 2016.
DOI: 10.1007/978-3-319-40596-4_32

(ii) From different kinds of generators leading mainly to the Yager's f and g generated implications that are constructed from generators of continuous Archimedean t-norms and t-conorms (see [5,8,28]), the so-called h-implications constructed from generators of representable uninorms (see [17]), and many generalizations that can be found in the recent survey [13].

(iii) From other implication functions like for instance the minimum and the maximum of two implications, any convex linear combination, the φ-conjugate or the N-reciprocal of an implication, and so on [6]. More recent constructions like the threshold and vertical threshold generation methods [18,19], some algebraic operations between implications [26,27] and some others, see the recent survey [20].

Some new construction methods of implication functions have recently appeared for instance in [3,10] or [23]. The method presented in [23] and later recalled and completed in [24] allows to construct fuzzy implication functions from a fuzzy negation N. In that paper, a more general method from a t-conorm S, a fuzzy negation N and a fuzzy implication function I was posed as an open question.

The idea of this paper is to develop such method and to study the properties that it preserves, generalizing the t-conorm S to a disjunctive aggregation function F. Since it depends on three different operators, F, N and I, there are many directions to investigate. For instance, one can fix two of the three operators and study the implications obtained varying the third one, or one can fix one of them varying the other two, and so on. Thus, we are dealing with a field of study with many possibilities and we highlight some of them in the current paper as a first approximation. In particular, it is shown how to obtain new implications from old ones with additional properties not satisfied by the initial implication function.

The paper is organized as follows. After this introduction, Sect. 2 is devoted to some preliminaries in order to make the paper as self-contained as possible. Section 3 presents the main results of the paper. In such section the construction method is investigated, some general results are shown, many examples are presented and some possible lines of study are pointed out. Finally, the paper ends with Sect. 4 devoted to some conclusions and future work.

2 Preliminaries

We will suppose the reader to be familiar with the theory of t-norms, t-conorms and fuzzy negations (all necessary results and notations can be found in [12]). We recall here only some facts on aggregation functions and fuzzy implications in order to establish the necessary notation that we will use along the paper and to make it as self-contained as possible.

Definition 1. *A binary operator* $F : [0,1] \times [0,1] \to [0,1]$ *is said to be an* aggregation function *if it is increasing in each variable and it satisfies* $F(0,0) = 0$ *and* $F(1,1) = 1$.

Definition 2. *An aggregation function* $F : [0,1] \times [0,1] \to [0,1]$ *is said to be* disjunctive *when* $F(x,y) \geq \max(x,y)$ *for all* $x, y \in [0,1]$.

For more details on aggregation functions, their classes and their properties see for instance [9].

Definition 3. *A binary operator* $I : [0,1] \times [0,1] \to [0,1]$ *is said to be a fuzzy* implication function, *or an* implication, *if it satisfies:*

(I1) $I(x,z) \geq I(y,z)$ *when* $x \leq y$, *for all* $z \in [0,1]$.
(I2) $I(x,y) \leq I(x,z)$ *when* $y \leq z$, *for all* $x \in [0,1]$.
(I3) $I(0,0) = I(1,1) = 1$ *and* $I(1,0) = 0$.

Note that, from the definition, it follows that $I(0,x) = 1$ and $I(x,1) = 1$ for all $x \in [0,1]$ whereas the symmetrical values $I(x,0)$ and $I(1,x)$ are not derived from the definition.

Definition 4. *Given a fuzzy implication function* I, *the function* $N_I(x) = I(x,0)$ *for all* $x \in [0,1]$ *is always a fuzzy negation, known as the* natural negation *of* I.

Among many other properties usually required for fuzzy implications we recall here some of the most important ones.

– The (Left) Neutrality Property:

$$I(1,y) = y \text{ for all } y \in [0,1]. \tag{NP}$$

– The Consequent Boundary:

$$I(x,y) \geq y \text{ for all } y \in [0,1]. \tag{CB}$$

– The Ordering Property:

$$I(x,y) = 1 \iff x \leq y \text{ for all } x, y \in [0,1]. \tag{OP}$$

– The Identity Principle:

$$I(x,x) = 1 \text{ for all } x \in [0,1]. \tag{IP}$$

– The Strong Negation Principle:

$$I(x,0) \text{ is a strong negation for all } x \in [0,1]. \tag{SNP}$$

– The Continuity condition:

$$I \text{ is a continuous mapping} \tag{CO}$$

- The Laws of Contraposition with respect to a fuzzy negation N:

$$I(x,y) = I(N(y), N(x)) \text{ for all } x,y \in [0,1]. \qquad CP(N)$$

$$I(x, N(y)) = I(y, N(x)) \text{ for all } x,y \in [0,1]. \qquad R-CP(N)$$

$$I(N(x), y) = I(N(y), x) \text{ for all } x,y \in [0,1]. \qquad L-CP(N)$$

  Recall that the three contrapositions are equivalent when the negation considered is strong.
- The Exchange Principle:

$$I(x, I(y,z)) = I(y, I(x,z)) \text{ for all } x,y,z \in [0,1]. \qquad (EP)$$

In [24] the following method to construct implication functions from fuzzy negations was presented.

Proposition 5. *Let N be a fuzzy negation. The function F^N given by:*

$$F^N(x,y) = \begin{cases} 1 & \text{if } x \leq y \\ \frac{(1-N(x))y}{x} + N(x) & \text{if } x > y \end{cases}$$

is always an implication function.

All the above mentioned properties of implication functions were studied in detail and they were characterized depending on the fuzzy negation N used in the construction. However, the authors pointed out that the previous construction is a particular case of a more general one, where implication functions are obtained from a t-conorm S, a fuzzy negation N and an implication function I through the formula:

$$I_{SNI}(x,y) = S(N(x), I(x,y)) \quad \text{for all } x,y \in [0,1]. \qquad (1)$$

3 FNI-implications

In this section we want to investigate those implication functions given by the previous equation. However, we will generalize the method since function S needs not be a t-conorm in order to generate fuzzy implications. In fact any disjunctive aggregation function F is enough for that purpose as it is posed in the following proposition.

Proposition 6. *Let F be a disjunctive aggregation function, N a fuzzy negation, and I an implication function. The function I_{FNI} given by:*

$$I_{FNI}(x,y) = F(N(x), I(x,y)) \quad \text{for all } x,y \in [0,1], \qquad (2)$$

is always an implication function.

Definition 7. *A fuzzy implication function I_{FNI} constructed through Eq. (2) will be called an FNI-implication. Whenever the aggregation function F is in fact a t-conorm S, I_{FNI} will be called a SNI-implication.*

According to Proposition 6, Eq. (2) gives a new method to construct implication functions. It is clear that such methods are especially interesting depending on which usual properties of implications they preserve. Thus, one of the main objectives of this paper is to investigate which properties are preserved by the construction given by Eq. (2).

Remark 8. *Note that in fact the previous proposition remains true if the aggregation function F satisfies only the condition $F(0,1) = 1$. However, we require that F must be disjunctive because it will be essential in the preservation of some properties of fuzzy implication functions.*

Let us first give some illustrative examples of the new construction.

Examples 9

(i) As a first example of course we can take F the probabilistic sum t-conorm, N a fuzzy negation and I the Goguen implication given by

$$I_{\mathbf{GG}}(x,y) = \begin{cases} 1 & \text{if } x \le y \\ \frac{y}{x} & \text{if } x > y, \end{cases}$$

and we obtain implications I^N given by Shi et al. in [23, 24] (see also Proposition 5 in the preliminaries).

(ii) Taking a t-conorm S, N a fuzzy negation and I the implication function given by

$$I(x,y) = \begin{cases} 1 & \text{if } x = 0 \\ y & \text{if } x > 0, \end{cases}$$

then we clearly retrieve the class of (S,N)-implications.

(iii) Take a t-conorm S, N the greatest fuzzy negation given by

$$N_{\mathbf{D_2}}(x) = \begin{cases} 1 & \text{if } x < 1 \\ 0 & \text{if } x = 1, \end{cases}$$

and I any implication function. The corresponding SNI-implication is then given by

$$I_{SNI}(x,y) = \begin{cases} 1 & \text{if } x < 1 \text{ and } y \in [0,1] \\ I(1,y) & \text{if } x = 1 \text{ and } y \in [0,1]. \end{cases}$$

(iv) Similarly, taking F a t-conorm S, N a fuzzy negation and I the Gödel implication given by

$$I_{\mathbf{GD}}(x,y) = \begin{cases} 1 & \text{if } x \le y \\ y & \text{if } x > y, \end{cases}$$

we obtain the implications:

$$I_{SNI}(x,y) = \begin{cases} 1 & \text{if } x \le y \\ S(N(x),y) & \text{if } x > y. \end{cases}$$

For instance, if we take the classical negation $N(x) = 1 - x$, we obtain the Fodor implication from the the maximum t-conorm.

(v) *Taking $I = I_{\mathbf{RC}}$ the Reichenbach implication, that is $I_{\mathbf{RC}}(x,y) = 1 - x + xy$, and the classical negation we obtain*

$$I_{SNI}(x,y) = S(1-x, 1-x+xy) \quad \text{for all } x,y \in [0,1].$$

Thus, for instance, from the maximum t-conorm we retrieve the proper Reichenbach implication, whereas considering the probabilistic sum t-conorm we obtain the polynomial implication (see [16]):

$$I(x,y) = 1 - x^2 + x^2 y \quad \text{for all } x,y \in [0,1],$$

which is also the f-generated Yager implication with generator $f(x) = \sqrt{1-x}$.

(vi) *Consider F the dual representable aggregation function (see [1]) with generating pair (f, N), where $f(x) = -\ln x$ and $N(x) = 1 - x$, that is,*

$$F(x,y) = \min\left(1, \frac{\max(x,y)}{1 - \min(x,y)}\right) \quad \text{for all } x,y \in [0,1]$$

Then if we take $I = I_{\mathbf{RC}}$ the Reichenbach implication and the classical negation, we obtain

$$I_{FNI}(x,y) = \begin{cases} 1 & \text{if } y \ge \dfrac{2x-1}{x} \\ \dfrac{1-x+xy}{x} & \text{if } y < \dfrac{2x-1}{x} \end{cases}$$

The structure of this implication function can be viewed in Fig. 1.

Let us begin our discussion by proving that all fuzzy implication functions are in fact SNI-implications.

Proposition 10. *Let F be a disjunctive aggregation function F with neutral element 0 and $N = N_{\mathbf{GD}}$ the Gödel negation given by*

$$N_{\mathbf{GD}}(x) = \begin{cases} 1 & \text{if } x = 0 \\ 0 & \text{if } x > 0. \end{cases}$$

Let I be a fuzzy implication function and I_{FNI} the corresponding FNI-implication. Then $I_{FNI} = I$.

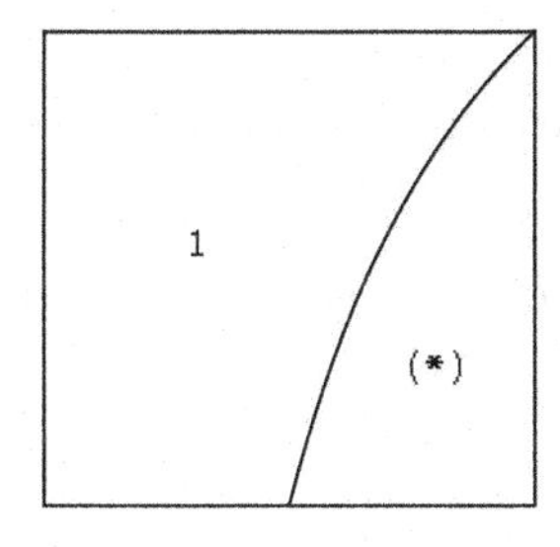

Fig. 1. Structure of the FNI-implication of Example 9(vi), where $(*)$ stands for $\frac{1-x+xy}{x}$.

Consequently we have that any fuzzy implication can be seen as an FNI-implication. Thus both the class $\mathcal{FNI}$ of all FNI-implications and the class $\mathcal{FI}$ of all fuzzy implication functions are exactly the same. That is $\mathcal{FNI} = \mathcal{FI}$. In spite of this fact, FNI-implications allow us to construct new implications from other already known, as we have seen in the examples before. Now, we want to deal with the properties that are preserved by FNI-implications.

Proposition 11. *Let F be a disjunctive aggregation function, N a fuzzy negation, I an implication function, and I_{FNI} the corresponding FNI-implication. The following items hold:*

(i) If I satisfies (IP) then I_{FNI} also satisfies it.
(ii) If I satisfies (CB) then I_{FNI} also satisfies it.

Moreover,

(iii) If F has neutral element 0 then I satisfies (NP) if and only if I_{FNI} also satisfies it.
(iv) If F and N are continuous then I_{FNI} preserves the continuity condition (CO).

It is easy to see that, without the conditions stated in the previous proposition, items (iii) and (iv) do not longer hold. Next we want to study the preservation of the ordering property. First, let us recall that an aggregation function F is said to have *trivial* 1*-region* whenever it satisfies $F(x, y) = 1$ if and only if $\max(x, y) = 1$. Similarly, a negation N is said to be *non-filling* whenever it satisfies $N(x) = 1$ if and only if $x = 0$.

Proposition 12. *Let F be a disjunctive aggregation function, N a fuzzy negation, I an implication function, and I_{FNI} the corresponding FNI-implication. If F has trivial 1-region and N is non-filling then I_{FNI} preserves (OP).*

Again it is clear that without the conditions stated in the proposition above the (OP) is not preserved and again Example 9-(iii) gives a counterexample. Let us now deal with the natural negation of an FNI-implication.

Let F be a disjunctive aggregation function, N a fuzzy negation, I an implication function, and I_{FNI} the corresponding FNI-implication. Then the natural negation of I_{FNI} is given by

$$N_{I_{FNI}}(x) = F(N(x), N_I(x)) \quad \text{for all } x \in [0,1].$$

Before to deal with the other usual properties of fuzzy implications, we want to highlight an interesting property of this new construction method. Namely, it is possible to consider adequate aggregation functions and fuzzy negations in such a way that the corresponding FNI-implication, I_{FNI} obtained from I, satisfies additional properties that the initial implication I does not satisfy. The idea comes from Proposition 11, since most of the preserved properties do not satisfy the converse. That is, the FNI-implication I_{FNI} can satisfy (IP), (CB) or (CO) even when the initial implication function I does not satisfy it. For instance, we have the following result concerning (IP).

Proposition 13. *Let S be a t-conorm and N a fuzzy negation with $S(N(x), x) = 1$ for all $x \in [0,1]$. Let I be an implication function and I_{SNI} the corresponding SNI-implication. If I satisfies (CB) then I_{SNI} satisfies (IP).*

Example 14. *Take $N = N_c$ the classical negation and $S = S_{\mathbf{LK}}$ the Łukasiewicz t-conorm. Consider $I = I_{\mathbf{RC}}$ the Reichenbach implication that satisfies (CB) but not the (IP). Then the corresponding I_{SNI}-implication is given by*

$$I_{SNI}(x,y) = S_{\mathbf{LK}}(1-x, I_{\mathbf{RC}}(x,y)) = \min(2-2x+xy, 1) \text{ for all } x, y \in [0,1].$$

The structure of this implication can be viewed in Fig. 2.

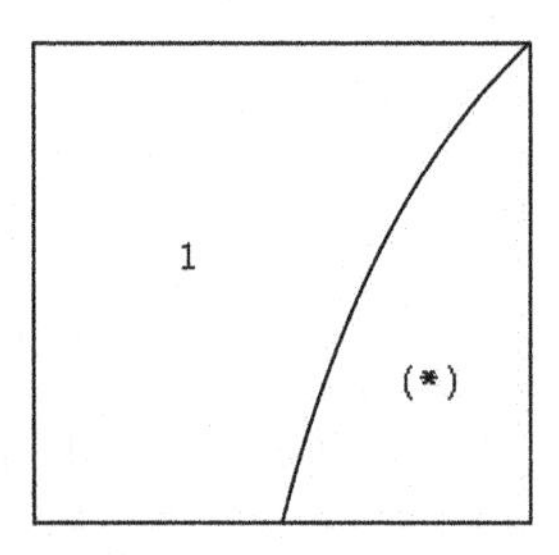

Fig. 2. Structure of the FNI-implication of Example 14, where $(*)$ stands for $2-2x+xy$.

Similarly one can construct FNI-implications that are continuous and that satsfy (SNP) even from implication functions that do not satisfy such properties, like in the following proposition.

Proposition 15. *Let F be a continuous disjunctive aggregation function and N a continuous fuzzy negation. Let I be an implication function which is continuous except at point $(0,0)$.*

(i) Then I_{FNI} satisfies (CO).
(ii) Moreover, if F has 0 as neutral element, N is strong, and I has $N_{\mathbf{D_1}}$ as natural negation, then I_{FNI} satisfies (SNP).

Examples 16. *There are many fuzzy implication functions satisfying the conditions given in the previous proposition. For instance, all residual implications derived from strict t-norms as well as all Yager implications f-generated with generator f such that $f(0) = +\infty$.*

1. *Consider $N = N_c$ the classical negation, $S = \max$ and $I = I_{\mathbf{GG}}$ the Goguen implication, which is the residual implication derived from the product t-norm. The corresponding SNI-implication is given by*

$$I_{SNI}(x,y) = \begin{cases} 1 & \text{if } x \leq y \\ \frac{y}{x} & \text{if } x > y \geq x - x^2 \\ 1 - x & \text{otherwise,} \end{cases}$$

   and is a continuous implication with strong natural negation. The structure of this implication can be viewed in Fig. 3.
2. *Consider $N = N_c$ the classical negation, $S = \max$ and $I = I_{\mathbf{YG}}$ the Yager implcation, which is a Yager f-generated implication with generator $f(x) = -\ln(x)$. The corresponding SNI-implication is given by*

$$I_{SNI}(x,y) = \begin{cases} 1 & \text{if } x = y = 0 \\ \max(1 - x, y^x) & \text{otherwise,} \end{cases}$$

   and is a continuous implication with strong natural negation. The structure of this implication can be viewed in Fig. 4.

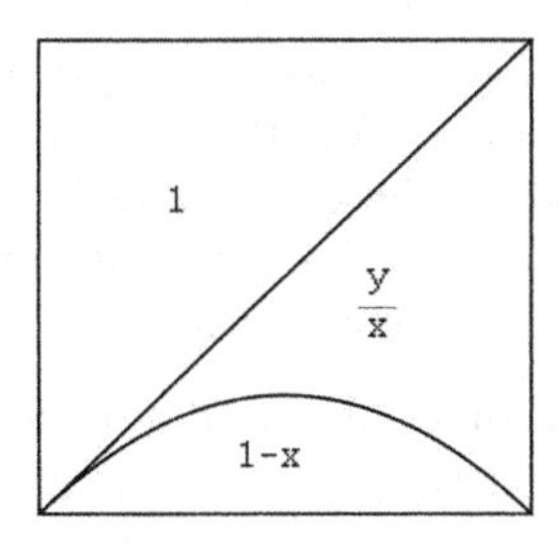

Fig. 3. Structure of the FNI-implication of Example 16(1).

Finally, it is clear that the FNI-construction does not preserve the (EP) as it is usually the case in almost all known construction methods. However, with respect to the contraposition properties, we want to finish this section with a new result involving the $R - CP(N)$ and its preservation.

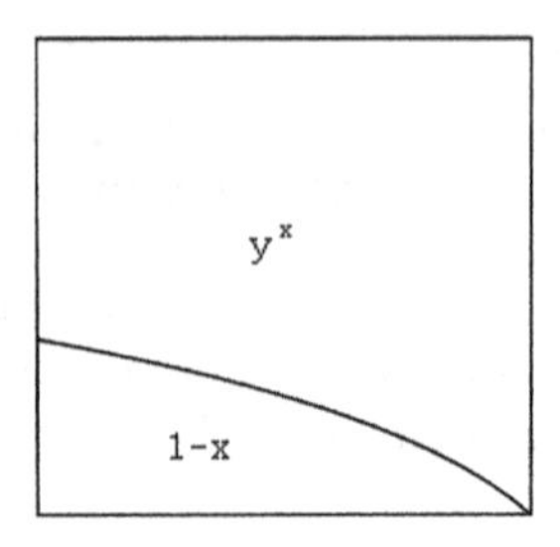

Fig. 4. Structure of the FNI-implication of Example 16(2).

Proposition 17. *Let $S = \max$, N a continuous fuzzy negation and I an implication function satisfying (CB) and $R - CP(N)$. Let I_{SNI} the corresponding SNI-implication, then $I_{SNI} = I$.*

Of course, there are many other cases for which the corresponding SNI-implication coincides with the initial implication I. See for instance Proposition 10. Thus, a natural question arises: In which cases a fuzzy implication function I remains invariant through the FNI-construction?

4 Conclusions and Future Work

In this paper the FNI-construction method of fuzzy implication functions based on a disjunctive aggregation function F, a fuzzy negation N and an implication function I is presented and studied. Many examples of this method are given to illustrate the situation, and as a consequence it is proved that any fuzzy implication function can be obtained through this method. It is investigated which of the usual properties of fuzzy implication functions are preserved through this method and moreover, it is proved that the obtained FNI-implication can satisfy additional properties that the initial implication does not satisfy. Thus, the FNI-construction can be viewed also as a method of modifying an implication function not satisfying a concrete property, in order to obtain a new one that satisfies it. The FNI-method becomes specially interesting because it depends on three different operators and consequently it can be investigated in many different directions. Let us detail some of them.

1. One can fix an implication function I and a disjunctive aggregation function F obtaining implications constructed just from a fuzzy negation N. Therefore one can investigate which properties satisfy these new implications depending on the properties of N. This is the case of the paper [24] which is the real origin of the current work. However, there are many different options. For instance, take the probabilistic sum t-conorm and the Łukasiewicz implication. Then the obtained FNI-implications are given by

$$I_{SNI}(x,y) = \begin{cases} 1 & \text{if } x \le y \\ 1-(x-y)(1-N(x)) & \text{if } x > y, \end{cases}$$

and one can investigate the properties of these implications depending on those properties of the corresponding negation N.

2. Similarly, one can fix I and N and then obtain implications just from a disjunctive aggregation function F (or from a t-conorm S). For instance, taking the classical negation and the Reichenbach implication we obtain the FNI-implications given by

$$I_{FNI}(x, y) = F(1 - x, 1 - x + xy) \quad \text{for all } x, y \in [0, 1].$$

3. Finally, by fixing N and F one obtain implication functions from a fixed one. For instance, taking the classical negation and the probabilistic sum t-conorm we obtain

$$I_{SNI}(x, y) = 1 - x + xI(x, y) \quad \text{for all } x, y \in [0, 1].$$

In all the previous cases one obtain a new class of implication functions which properties can be studied in detail.

Acknowledgments. This paper has been partially supported by the Spanish grant TIN2013-42795-P.

References

1. Aguiló, I., Carbonell, M., Suñer, J., Torrens, J.: Dual representable aggregation functions and their derived S-implications. In: Hüllermeier, E., Kruse, R., Hoffmann, F. (eds.) IPMU 2010. LNCS, vol. 6178, pp. 408–417. Springer, Heidelberg (2010)
2. Aguiló, I., Suñer, J., Torrens, J.: A characterization of residual implications derived from left-continuous uninorms. Inf. Sci. **180**(20), 3992–4005 (2010)
3. Aguiló, I., Suñer, J., Torrens, J.: New types of contrapositivisation of fuzzy implications with respect to fuzzy negations. Inf. Sci. **322**, 223–226 (2015)
4. Baczyński, M., Beliakov, G., Bustince Sola, H., Pradera, A. (eds.): Advances in Fuzzy Implication Functions. STUDFUZZ, vol. 300. Springer, Heidelberg (2013)
5. Baczyński, M., Jayaram, B.: Yager's classes of fuzzy implications: Some properties and intersections. Kybernetika **43**, 157–182 (2007)
6. Baczyński, M., Jayaram, B.: Fuzzy Implications. STUDFUZZ, vol. 231. Springer, Heidelberg (2008)
7. Baczyński, M., Jayaram, B.: (U, N)-implications and their characterizations. Fuzzy Sets Syst. **160**, 2049–2062 (2009)
8. Balasubramaniam, J.: Yager's new class of implications J_f and some classical tautologies. Inf. Sci. **177**, 930–946 (2007)
9. Beliakov, G., Pradera, A., Calvo, T.: Aggregation Functions: A Guide for Practitioners. STUDFUZZ, vol. 221. Springer, Heidelberg (2007)
10. Jayaram, B., Mesiar, R.: On special fuzzy implications. Fuzzy Sets Syst. **160**(14), 2063–2085 (2009)
11. De Baets, B., Fodor, J.C.: Residual operators of uninorms. Soft. Comput. **3**, 89–100 (1999)

12. Klement, E.P., Mesiar, R., Pap, E.: Triangular Norms. Kluwer Academic Publishers, Dordrecht (2000)
13. Hlinená, D., Kalina, M., Kral, P.: Implications functions generated using functions of one variable. In: [4], pp. 125–153 (2013)
14. Mas, M., Monserrat, M., Torrens, J.: Two types of implications derived from uninorms. Fuzzy Sets Syst. **158**, 2612–2626 (2007)
15. Mas, M., Monserrat, M., Torrens, J., Trillas, E.: A survey on fuzzy implication functions. IEEE Trans. Fuzzy Syst. **15**(6), 1107–1121 (2007)
16. Massanet, S., Riera, J.V., Ruiz-Aguilera, D.: On Fuzzy Polynomial Implications. In: Laurent, A., Strauss, O., Bouchon-Meunier, B., Yager, R.R. (eds.) IPMU 2014, Part I. CCIS, vol. 442, pp. 138–147. Springer, Heidelberg (2014)
17. Massanet, S., Torrens, J.: On a new class of fuzzy implications: h-Implications and generalizations. Inf. Sci. **181**, 2111–2127 (2011)
18. Massanet, S., Torrens, J.: Threshold generation method of construction of a new implication from two given ones. Fuzzy Sets Syst. **205**, 50–75 (2012)
19. Massanet, S., Torrens, J.: On the vertical threshold generation method of fuzzy implication and its properties. Fuzzy Sets Syst. **226**, 232–252 (2013)
20. Massanet, S., Torrens, J.: An overview of construction methods of fuzzy implications. In: [4], pp. 1–30 (2013)
21. Ruiz, D., Torrens, J.: Residual implications and co-implications from idempotent uninorms. Kybernetika **40**, 21–38 (2004)
22. Ruiz-Aguilera, D., Torrens, J.: S- and R-implications from uninorms continuous in $]0,1[^2$ and their distributivity over uninorms. Fuzzy Sets Syst. **160**, 832–852 (2009)
23. Shi, Y., Van Gasse, B., Ruan, D., Kerre, E.: On a new class of implications in fuzzy logic. In: Hüllermeier, E., Kruse, R., Hoffmann, F. (eds.) IPMU 2010. CCIS, vol. 80, pp. 525–534. Springer, Heidelberg (2010)
24. Shi, Y., Van Hasse, B., Ruan, D., Kerre, E.: Fuzzy implications: Classification and a new class. In: [4], pp. 31–53 (2013)
25. Trillas, E., Mas, M., Monserrat, M., Torrens, J.: On the representation of fuzzy rules. Int. J. Approximate Reasoning **48**, 583–597 (2008)
26. Vemuri, N.R., Jayaram, B.: Representations through a monoid on the set of fuzzy implications. Fuzzy Sets Syst. **247**, 51–67 (2014)
27. Vemuri, N.R., Jayaram, B.: The $\star$-composition of fuzzy implications: Closures with respect to properties, powers and families. Fuzzy Sets Syst. **275**, 58–87 (2015)
28. Yager, R.R.: On some new classes of implication operators and their role in approximate reasoning. Inf. Sci. **167**, 193–216 (2004)

On a Generalization of the Modus Ponens: U-conditionality

Margalida Mas, Miquel Monserrat, Daniel Ruiz-Aguilera, and Joan Torrens(✉)

Department of Mathematics and Computer Science, University of the Balearic Islands, Crta. de Valldemossa, km. 7,5, 07122 Palma, Spain
{mmg448,mma112,daniel.ruiz,jts224}@uib.es

Abstract. In fuzzy logic, the Modus Ponens property for fuzzy implication functions is usually considered with respect to a continuous t-norm T and for this reason this property is also known under the name of T-conditionality. In this paper, the t-norm T is substituted by a uninorm U leading to the property of U-conditionality. The new property is studied in detail and it is shown that usual implications derived from t-norms and t-conorms do not satisfy it, but many solutions appear among those implications derived from uninorms. In particular, the case of residual implications derived from uninorms or RU-implications is investigated in detail for some classes of uninorms.

Keywords: Fuzzy implication function · Modus ponens · t-norm · Uninorm · Natural negation

1 Introduction

In the framework of fuzzy logic and approximate reasoning, fuzzy implication functions are used not only in modelling fuzzy conditionals but also in the inference process. To manage forward inferences the Modus Ponens becomes essential in the process, especially when the Zadeh's compositional rule of inference is used. In these cases, the Modus Ponens is guaranteed when the fuzzy operators used, that is, the conjunction and the fuzzy conditional, satisfy the following inequality:

$$T(x, I(x,y)) \leq y \quad \text{for all} \quad x, y \in [0,1], \tag{1}$$

where T is usually considered a (continuous) t-norm and I a fuzzy implication function. For this reason the previous inequality is known as the Modus Ponens property, and also as T-conditionality.

Due to its importance in the inference process, those t-norms T and fuzzy implication functions I that satisfy Eq. (1) have been investigated by many researchers for decades (see for instance, [2,3,17,19,26–29]). The main studies are related to implications derived from t-norms and t-conorms. Thus, residual implications and (S,N)-implications were investigated in detail in [2,26,27], and QL and D-implications in [28]. Moreover, these results were collected and completed later in [3] (see Sect. 7.4).

J.P. Carvalho et al. (Eds.): IPMU 2016, Part I, CCIS 610, pp. 387–398, 2016.
DOI: 10.1007/978-3-319-40596-4_33

However, there are many other kinds of implication functions to be considered and even different methods of constructing implication functions from others (see [21]). Among them, a kind of implication functions extensively studied are those derived from more general aggregation functions than t-norms and t-conorms, specially those derived from uninorms (see for instance [1,4,7,23,24]). Recently, T-conditionality was already studied for these two kinds of implication functions, that is, for the so-called RU-implications and (U, N)-implications (see [14,15]).

The idea of this paper is to extend such study to the property of conditionality with respect to a uninorm U instead of a t-norm T. This is done for a uninorm U and a fuzzy implication function I in general. However, some initial results prove that the considered uninorm must be conjunctive and the implication function must satisfy some properties that lead us to center our investigation in the case of residual implications derived from a uninorm U_0, that is, the case of RU-implications. We extend our study to two of the most usual classes of uninorms, that is, uninorms in $\mathcal{U}_{\min}$ and idempotent uninorms.

The paper is organized as follows. After this introduction, Sect. 2 is devoted to some preliminaries in order to make the paper as self-contained as possible. Section 3 deals with the Modus Ponens with respect to a uninorm U, including some general results for any kind of implication functions as well as some particular ones for the case of RU-implications. This last part is divided in two subsections, one for each class of uninorms, uninorms in $\mathcal{U}_{\min}$ and idempotent uninorms. Finally, the paper ends with Sect. 4 devoted to some conclusions and future work.

2 Preliminaries

We will suppose the reader to be familiar with the theory of t-norms, t-conorms and fuzzy negations (all necessary results and notations can be found in [10]). We also suppose that some basic facts on uninorms are known (see for instance [9]) as well as their most usual classes, that is, uninorms in $\mathcal{U}_{\min}$ and $\mathcal{U}_{\max}$ ([9]) and idempotent uninorms ([6,12,25]). See also the recent survey in [13]. We recall here only some facts on implications and uninorms in order to stablish the necessary notation that we will use along the paper.

Definition 1. *A binary operator $I : [0,1] \times [0,1] \to [0,1]$ is said to be a fuzzy* implication function, *or an* implication, *if it satisfies:*

(I1) $I(x, z) \geq I(y, z)$ *when* $x \leq y$, *for all* $z \in [0,1]$.
(I2) $I(x, y) \leq I(x, z)$ *when* $y \leq z$, *for all* $x \in [0,1]$.
(I3) $I(0,0) = I(1,1) = 1$ *and* $I(1,0) = 0$.

Note that, from the definition, it follows that $I(0, x) = 1$ and $I(x, 1) = 1$ for all $x \in [0,1]$ whereas the symmetrical values $I(x, 0)$ and $I(1, x)$ are not derived from the definition.

Definition 2. *Given a fuzzy implication function I, the function $N_I(x) = I(x,0)$ for all $x \in [0,1]$ is always a fuzzy negation, known as the* natural negation *of I.*

Definition 3. *A* uninorm *is a two-place function $U : [0,1]^2 \longrightarrow [0,1]$ which is associative, commutative, increasing in each place and such that there exists some element $e \in [0,1]$, called* neutral element, *such that $U(e,x) = x$ for all $x \in [0,1]$.*

Evidently, a uninorm with neutral element $e = 1$ is a t-norm and a uninorm with neutral element $e = 0$ is a t-conorm. For any other value $e \in]0,1[$ the operation works as a t-norm in the $[0,e]^2$ square, as a t-conorm in $[e,1]^2$ and its values are between minimum and maximum in the set of points $A(e)$ given by

$$A(e) = [0,e[\times]e,1] \cup]e,1] \times [0,e[.$$

We will usually denote a uninorm with neutral element e and underlying t-norm and t-conorm, T and S, by $U \equiv \langle T,e,S \rangle$. For any uninorm it is satisfied that $U(0,1) \in \{0,1\}$ and a uninorm U is called *conjunctive* if $U(1,0) = 0$ and *disjunctive* when $U(1,0) = 1$. On the other hand, let us recall two of the most studied classes of uninorms in the literature.

Theorem 4 ([9]). *Let $U : [0,1]^2 \to [0,1]$ be a uninorm with neutral element $e \in]0,1[$.*

(a) If $U(0,1) = 0$, then the section $x \mapsto U(x,1)$ is continuous except in $x = e$ if and only if U is given by

$$U(x,y) = \begin{cases} eT\left(\frac{x}{e},\frac{y}{e}\right) & \text{if } (x,y) \in [0,e]^2, \\ e + (1-e)S\left(\frac{x-e}{1-e},\frac{y-e}{1-e}\right) & \text{if } (x,y) \in [e,1]^2, \\ \min(x,y) & \text{if } (x,y) \in A(e), \end{cases}$$

where T is a t-norm, and S is a t-conorm.

(b) If $U(0,1) = 1$, then the section $x \mapsto U(x,0)$ is continuous except in $x = e$ if and only if U is given by the same structure as above, changing minimum by maximum in $A(e)$.

The set of uninorms as in case (a) will be denoted by $\mathcal{U}_{\min}$ and the set of uninorms as in case (b) by $\mathcal{U}_{\max}$. We will denote a uninorm in $\mathcal{U}_{\min}$ with underlying t-norm T, underlying t-conorm S and neutral element e as $U \equiv \langle T,e,S \rangle_{\min}$ and in a similar way, a uninorm in $\mathcal{U}_{\max}$ as $U \equiv \langle T,e,S \rangle_{\max}$.

Idempotent uninorms were analysed first in [5] and they were characterized in [6] for those with a lateral continuity and in [12] for the general case. An improvement of this last result was done in [25] as follows.

Theorem 5 ([25]). *U is an idempotent uninorm with neutral element $e \in [0,1]$ if and only if there exists a non increasing function $g : [0,1] \to [0,1]$, symmetric with respect to the identity function, with $g(e) = e$, such that*

$$U(x,y) = \begin{cases} \min(x,y) & \textit{if } y < g(x) \textit{ or } (y = g(x) \textit{ and } x < g^2(x)), \\ \max(x,y) & \textit{if } y > g(x) \textit{ or } (y = g(x) \textit{ and } x > g^2(x)), \\ x \textit{ or } y & \textit{if } y = g(x) \textit{ and } x = g^2(x), \end{cases}$$

being commutative in the points (x,y) such that $y = g(x)$ with $x = g^2(x)$.

Any idempotent uninorm U with neutral element e and associated function g, will be denoted by $U \equiv \langle g, e \rangle_{\text{ide}}$ and the class of idempotent uninorms will be denoted by $\mathcal{U}_{\text{ide}}$. Obviously, for any of these uninorms the underlying t-norm T is the minimum and the underlying t-conorm S is the maximum.

On the other hand, different classes of implications derived from uninorms have been studied. We recall here *RU*-implications.

Definition 6. *Let U be a uninorm. The* residual operation *derived from U is the binary operation given by*

$$I_U(x,y) = \sup\{z \in [0,1] \mid U(x,z) \leq y\} \textit{ for all } x,y \in [0,1].$$

Proposition 7 *([7]). Let U be a uninorm and I_U its residual operation. Then I_U is an implication if and only if the following condition holds*

$$U(x,0) = 0 \quad \textit{for all} \quad x < 1. \tag{2}$$

In this case I_U is called an RU-implication.

This includes all conjunctive uninorms but also many disjunctive ones, like for instance disjunctive idempotent uninorms $U \equiv \langle g, e \rangle_{\text{ide}}$ with $g(0) = 1$ (see [22]). Many other disjunctive uninorms are also included in other classes of uninorms like the representable uninorms (see [7]) or uninorms continuous in the unit open square (see [24]). However, when we deal with left-continuous uninorms U we clearly have that U satisfies condition (2) if and only if it is conjunctive.

Some properties of *RU*-implications have been studied involving the main classes of uninorms: those previously stated, uninorms in $\mathcal{U}_{\min}$ and idempotent uninorms, and also involving many other classes like representable uninorms, uninorms continuous in the open unit square and even uninorms with continuous underlying operators (for more details see [1,3,7,8,11,16,18,23,24]). Recently, the Modus Ponens property with respect to a t-norm T has been studied in detail also for implications derived from uninorms (not only for *RU*, but also for (U,N)-implications) in [14,15].

3 U-conditionality

In this section we want to generalize the definition of Modus Ponens with respect to a t-norm T by introducing the so-called U-conditionality:

Definition 8. *Let I be an implication function and U a uninorm. It is said that I satisfies the* Modus Ponens property *with respect to U, or that I is an* U-conditional *if*

$$U(x, I(x, y)) \leq y \quad \textit{for all}\, x, y \in [0, 1]. \tag{3}$$

The purpose of this paper is to study which conditions must satisfy a fuzzy implication I and a uninorm U in order to be I an U-conditional. Since we take a uninorm that generalizes a conjunction, it could be natural to take a conjunctive uninorm in the definition before. However, this is not necessary because it can be deduced from the definition as follows.

Lemma 9. *Let I be an implication function and U a uninorm. If I is an* U-conditional *then U must be conjunctive.*

Proof. Just taking $x = 1$ in Eq. (3) we obtain $U(1, 0) = 0$ and U must be conjunctive. □

Let us now give some properties that must satisfy an implication function to be an U-conditional.

Proposition 10. *Let U be a conjunctive uninorm with neutral element $e \in]0, 1[$ and let I be an U-conditional. The following items hold:*

1. *$I(e, y) \leq y$ for all $y \in [0, 1]$.*
2. *The natural negation N_I must satisfy*

$$N_I(x) = 0 \quad \textit{for all} \;\; x \geq e, \quad \textit{and} \quad N_I(x) < e \quad \textit{for all} \;\; 0 < x < e.$$

   In particular, N_I can not be continuous.
3. *It must be $U(x, N_I(x)) = 0$ for all $x \in [0, 1]$.*
4. *$I(x, y) < e$ for all $x > y \geq e$. In particular, $I(1, y) < e$ for all $y < 1$.*
5. *$U(1, I(1, y)) \leq y$ for all $y \in [0, 1]$.*

The previous proposition gives some necessary conditions on the uninorm U as well as on the implication I in order they satisfy U-conditionality. For those conditions involving the uninorm U, note that condition 3 above can be trivial or easily satisfied depending on how is the negation N_I, and condition 5 is also trivially satisfied for instance for any uninorm in $\mathcal{U}_{\min}$.

On the other hand, from the proposition above it is clear that the usual classes of fuzzy implication functions, that is, R, (S, N), QL and D-implications derived from t-norms and t-conorms, as well as f and g-generated Yager's implications, can not be U-conditionals (note that all of them satisfy $I(1, y) = y$ for all $y \in [0, 1]$ which is incompatible with property 4 in the previous proposition). However, this is not the case of RU and (U, N)-implications derived from uninorms.

For instance, RU-implications (see Definition 6) satisfy $I_U(e, y) = y$ for all $y \in [0, 1]$ and so they are good candidates to be U-conditionals. In what follows we will deal with U-conditionality for RU-implications derived from uninorms.

First of all recall that when we take a left-continuous uninorm U, then the corresponding RU-implication always satisfies U-conditionality with respect to the proper uninorm U (see [7]). From this fact we directly have the following result.

Proposition 11. *Let U, U_0 be two uninorms with neutral elements $e, e_0 \in]0,1[$ respectively, such that one of them is left-continuous and let I_{U_0} be the residual implication derived from U_0. If $U \leq U_0$ then I_{U_0} is an U-conditional.*

Example 12. *Let U be a left-continuous uninorm with neutral element $e \in]0,1[$ and U_0 the least uninorm with neutral element e, that is,*

$$U_0(x,y) = \begin{cases} 0 & if\, x, y < e \\ \max(x,y) & if\, x, y \geq e \\ \min(x,y) & otherwise. \end{cases}$$

Then I_U is an U_0-conditional.

Another general result for RU-implications is given in the following proposition.

Proposition 13. *Let U, U_0 be two uninorms with neutral elements $e, e_0 \in]0,1[$ respectively and let I_{U_0} be the residual implication derived from U_0. If I_{U_0} is an U-conditional then it must be $e_0 \leq e$.*

From now on, let us deal with the cases when U_0 is a uninorm in one of the classes recalled in the preliminaries, that is, when U_0 is in $\mathcal{U}_{\min}$ or an idempotent uninorm. We will divide our study in two subsections, one for each class of uninorms.

3.1 Case When U_0 Is in $\mathcal{U}_{\min}$

In this section we want to deal with residual implications derived from uninorms in $\mathcal{U}_{\min}$. Let us recall first how are this kind of implications, that can be found for instance in [7] (see also [23] for the version recalled in the next proposition).

Proposition 14. *Let $U_0 \equiv \langle T_0, e_0, S_0 \rangle_{\min}$ be a uninorm in $\mathcal{U}_{\min}$ with neutral element $e_0 \in]0,1[$. Then its residual operator I_{U_0} is always an implication function and it is given by*

$$I_U(x,y) = \begin{cases} 1 & if\, 0 \leq x < e_0\, and\, x \leq y \\ e_0 I_{T_0}\left(\frac{x}{e_0}, \frac{y}{e_0}\right) & if\, 0 \leq x < e_0\, and\, x > y \\ y & if\, y \leq e_0 \leq x \\ e_0 & if\, e_0 \leq y < x \\ e_0 + (1-e_0) R_{S_0}\left(\frac{x-e_0}{1-e_0}, \frac{y-e_0}{1-e_0}\right) & if\, e_0 \leq x \leq y, \end{cases}$$

where R_{S_0} denotes the residual operator associated to the t-conorm S_0, that is, $R_{S_0}(x,y) = \sup\{z \in [0,1] \mid S(x,z) \leq y\}$ for all $x, y \in [0,1]$.

Thus, for this kind of RU-implications we have the following results.

Proposition 15. *Let U be a uninorm with neutral element $e \in]0,1[$ and $U_0 \equiv \langle T_0, e_0, S_0\rangle_{\min}$ a uninorm in $\mathcal{U}_{\min}$ with $e_0 \le e$. If the RU-implication I_{U_0} is an U-conditional then it must be $e_0 < e$ and $U(x,y) = \min(x,y)$ for all $x, y \in R_0$ where the region R_0 is given by*

$$R_0 = [0, e_0] \times [e, 1] \cup [e, 1] \times [0, e_0]. \tag{4}$$

The region R_0 in the proposition above, where the uninorm U must be given by the minimum is depicted in Fig. 1.

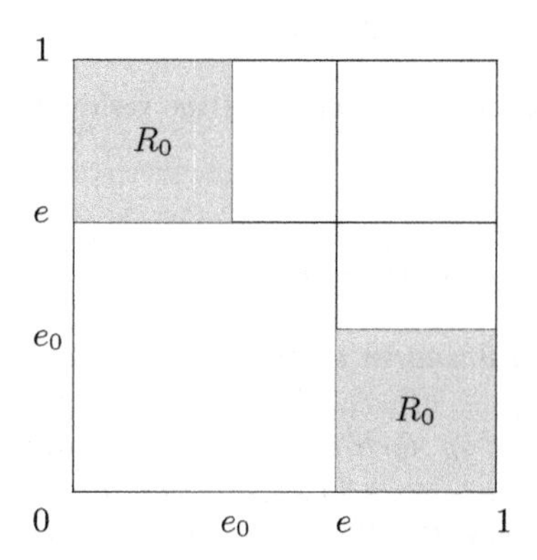

Fig. 1. Region R_0 given in Eq. (4) in the previous proposition.

Proposition 16. *Let U be a uninorm with neutral element $e \in]0,1[$ and underlying operators T_U, S_U, and let $U_0 \equiv \langle T_0, e_0, S_0\rangle_{\min}$ be a uninorm in $\mathcal{U}_{\min}$ with $e_0 < e$. Suppose that $T_0 = \min$ and $S_U = \max$. Then the RU-implication I_{U_0} is an U-conditional if and only if $U(x,y) = \min(x,y)$ for all $x, y \in R_0$ where R_0 is the region given in (4).*

Example 17. *Let $U_0 \equiv \langle T_0, e_0, S_0\rangle_{\min}$ a uninorm in $\mathcal{U}_{\min}$ with $e_0 \in]0,1[$, $T_0 = \min$, that is, U_0 is given by*

$$U_0(x,y) = \begin{cases} e_0 + (1-e_0) S_0\left(\frac{x-e_0}{1-e_0}, \frac{y-e_0}{1-e_0}\right) & \text{if } x, y \ge e_0 \\ \min(x,y) & \text{otherwise}, \end{cases}$$

and let $U \equiv \langle T_U, e, S_U\rangle_{\min}$ be a uninorm in $\mathcal{U}_{\min}$ with $e > e_0$ and $S_U = \max$, that is U is given by

$$U(x,y) = \begin{cases} e T_U\left(\frac{x}{e}, \frac{y}{e}\right) & \text{if } x, y \le e \\ \max(x,y) & \text{if } x, y \ge e \\ \min(x,y) & \text{otherwise}. \end{cases}$$

In this case we have that I_{U_0} is always an U-conditional. The structures of the uninorm U and the residuated implication I_{U_0} can be viewed in Fig. 2.

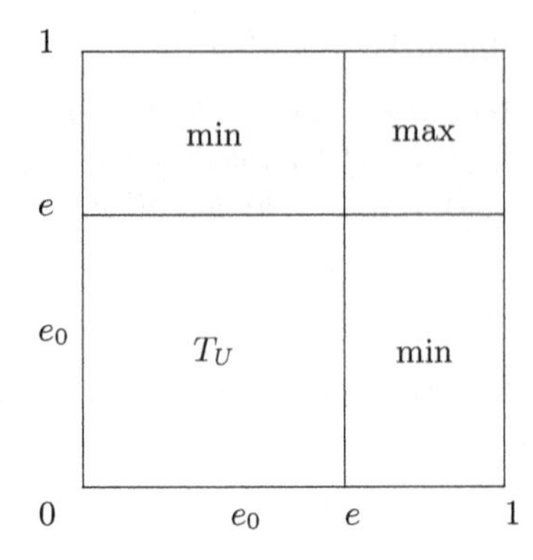

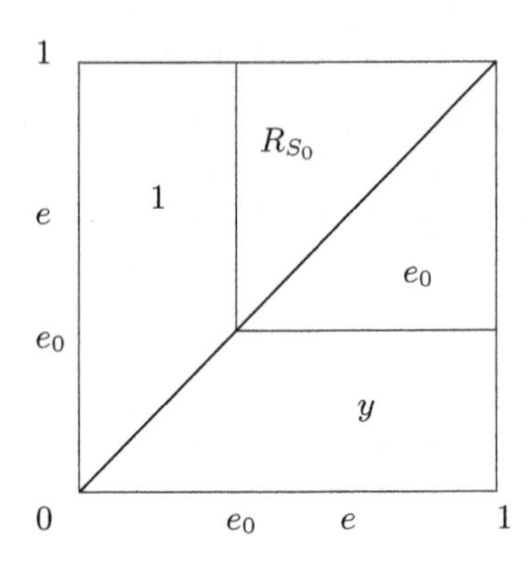

Fig. 2. Structures of the uninorm U (left) and the residuated implication I_{U_0} (right) that is an U-conditional, given in Example 17.

Now, we can give the general result for uninorms U in $\mathcal{U}_{\min}$.

Theorem 18. *Let U be a uninorm with neutral element $e \in]0,1[$ and underlying operators T_U, S_U and $U_0 \equiv \langle T_0, e_0, S_0 \rangle_{\min}$ a uninorm in $\mathcal{U}_{\min}$ with $e_0 < e$. Suppose that $U(e_0, e_0) = e_0$ in such a way that U is given by a t-norm T_1 in the square $[0, e_0]^2$ and that $U_0(e, e) = e$ in such a way that U_0 is given by a t-conorm S_1 in the square $[e, 1]^2$. Then the RU-implication I_{U_0} is an U-conditional if and only if the following items hold:*

i) $U(x, y) = \min(x, y)$ for all $x, y \in R_0$ where R_0 is the region given in (4).
ii) The residual implication derived from the t-norm T_0, I_{T_0}, is a T_1-conditional.
iii) $S_U(x, R_{S_1}(x, y)) \leq y$ for all $x \leq y$.

3.2 Case When U_0 Is an Idempotent Uninorm

Let us deal in this section with residual implications derived from idempotent uninorms and let us recall first how are this kind of implications. First, recall that the residual operator derived from a uninorm U_0 is an implication function if and only if $U_0(x, 0) = 0$ for all $x < 1$ (see [7], or Proposition 7 in the preliminaries). Consequently, in our case when U_0 is idempotent, say $U_0 \equiv \langle g_0, e_0 \rangle_{\text{ide}}$ with $0 < e_0 < 1$, then it is necessary to have $g_0(0) = 1$ and so this condition will be always assumed from now on. Let us recall now how are the residual implications derived from idempotent uninorms (see for instance [22]).

Proposition 19. *Let $U_0 \equiv \langle g_0, e_0 \rangle_{\text{ide}}$ be an idempotent uninorm with neutral element $e_0 \in]0,1[$ and such that $g_0(0) = 1$. Then its residual operator I_{U_0} is an implication function and is given by*

$$I_{U_0}(x, y) = \begin{cases} \max(g_0(x), y) & \text{if } x \leq y \\ \min(g_0(x), y) & \text{if } y < x. \end{cases}$$

We already know from the general result given in Proposition 13 that it must be $e_0 \leq e$ in order I_{U_0} be an U-conditional. Moreover, in this case we can give also a necessary condition on the uninorm U.

Proposition 20. *Let U be a conjunctive uninorm with neutral element $e \in]0,1[$, $U_0 \equiv \langle g_0, e_0 \rangle_{\text{ide}}$ an idempotent uninorm with neutral element $0 < e_0 \leq e$ and I_{U_0} the residual implication derived from U_0. If I_{U_0} is an U-conditional then the underlying t-conorm of U must be $S_U = \max$.*

There are some cases when the necessary condition before is also sufficient as it is shown in the following result.

Proposition 21. *Let U be a conjunctive uninorm with neutral element $e \in]0,1[$, $U_0 \equiv \langle g_0, e_0 \rangle_{\text{ide}}$ an idempotent uninorm with neutral element $0 < e_0 < e$ and such that $g_0(e) = 0$. Let I_{U_0} be the residual implication derived from U_0. Then I_{U_0} is an U-conditional if and only if the underlying t-conorm of U is $S_U = \max$.*

Example 22. *Let us consider a fixed element $e \in]0,1[$ and U any conjunctive uninorm with neutral element e and underlying t-conorm $S_U = \max$. Let g_0 be the Id-symmetrical decreasing function given by*

$$g_0(x) = \begin{cases} 1 & \text{if } x = 0 \\ e - x & \text{if } x \leq e \\ 0 & \text{otherwise,} \end{cases}$$

and let U_0 be any idempotent uninorm with associated function g_0. With these conditions, it is clear that the neutral element of U_0 is $e_0 = \frac{e}{2}$ and the corresponding residual implication is given by

$$I_{U_0}(x,y) = \begin{cases} \max(e - x, y) & \text{if } x \leq y \\ \min(e - x, y) & \text{if } y < x < e \\ 0 & \text{otherwise.} \end{cases}$$

From the previous proposition, this residual implication I_{U_0} is always an U-conditional. The structures of the uninorm U_0 and its residual implication I_{U_0} can be viewed in Fig. 3.

On the other hand, in the case when $g_0(e) > 0$ an additional condition must be required to the uninorm U to ensure U-conditionality of I_{U_0}. Specifically, we have the following result.

Proposition 23. *Let U be a conjunctive uninorm with neutral element $e \in]0,1[$, $U_0 \equiv \langle g_0, e_0 \rangle_{\text{ide}}$ an idempotent uninorm with neutral element $0 < e_0 \leq e$ and such that $g_0(e) > 0$. Let I_{U_0} be the residual implication derived from U_0. Then I_{U_0} is an U-conditional if and only if the following two items hold:*

1. *The underlying t-conorm of U is $S_U = \max$.*
2. *$U(x, g_0(x)) = g_0(x)$ for all $x \in [e, 1]$.*
3. *$U(1, y) = y$ for all $y < g_0(1)$*

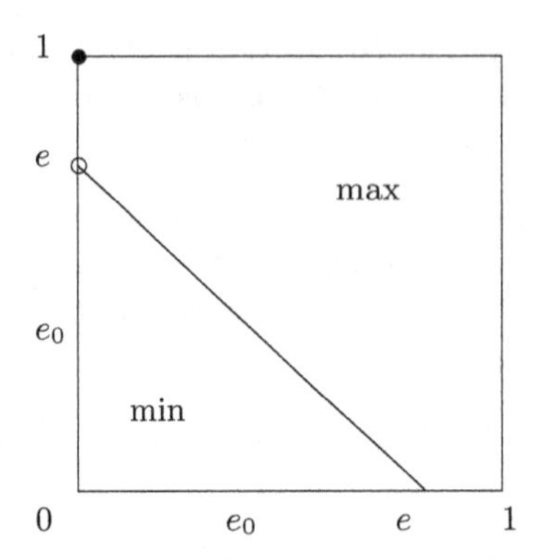

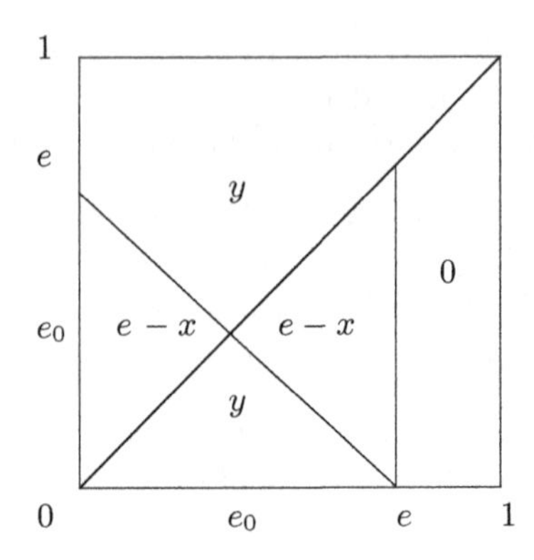

Fig. 3. Structures of the uninorm U_0 (left) and its residual implication I_{U_0} (right) given in Example 22.

Example 24. *Note that any uninorm* U *in* $\mathcal{U}_{\min}$ *with underlying t-conorm* $S_U = \max$ *satisfies the conditions in the proposition above and consequently, the residual implication* I_{U_0} *of any idempotent uninorm* $U_0 \equiv \langle g_0, e_0 \rangle_{\text{ide}}$ *with* $e_0 < e$ *and* $g_0(0) = 1$ *is an* U*-conditional.*

4 Conclusions and Future Work

Forward inference schemes in approximate reasoning are based on the *Modus Ponens* property, also called T-conditionality. Thus, fuzzy implication functions used in the inference process of any fuzzy rule based system are required to satisfy this property, which becomes essential in approximate reasoning and fuzzy control. In this paper we have extended such property to the so-called U-conditionality, by substituting the t-norm T by a uninorm U. Fixed a uninorm U, we studied in this paper which fuzzy implication functions satisfy U-conditionality leading to the fact that the most appropriate implications in this case are those derived from uninorms. We have given a detailed study of the case of *RU*-implications when the uninorm used to derive the residual implication is a uninorm in $\mathcal{U}_{\min}$ or an idempotent uninorm.

As a future work, we want to extend this study to other kinds of uninorms like representable uninorms, uninorms continuous in the open unit square, compensatory uninorms and so on. Moreover, we want to deal also with other kinds of implications like (U, N)-implications derived from disjunctive uninorms (see [3]) or h and (h, e)-implications recently introduced in [20]. Finally, a similar generalization through uninorms of the Modus Tollens property would be also worth of study.

Acknowledgments. This paper has been partially supported by the Spanish grant TIN2013-42795-P.

References

1. Aguiló, I., Suñer, J., Torrens, J.: A characterization of residual implications derived from left-continuous uninorms. Inf. Sci. **180**, 3992–4005 (2010)
2. Alsina, C., Trillas, E.: When (S, N)-implications are (T, T_1)-conditional functions? Fuzzy Sets Syst. **134**, 305–310 (2003)
3. Baczyński, M., Jayaram, B.: Fuzzy Implications. STUDFUZZ, vol. 231. Springer, Heidelberg (2008)
4. Baczyński, M., Jayaram, B.: (U, N)-implications and their characterizations. Fuzzy Sets Syst. **160**, 2049–2062 (2009)
5. Czogala, E., Drewniak, J.: Associative monotonic operations in fuzzy set theory. Fuzzy Sets Syst. **12**, 249–269 (1984)
6. De Baets, B.: Idempotent uninorms. Eur. J. Oper. Res. **118**, 631–642 (1999)
7. De Baets, B., Fodor, J.C.: Residual operators of uninorms. Soft Comput. **3**, 89–100 (1999)
8. Fodor, J., De Baets, B.: A single-point characterization of representable uninorms. Fuzzy Sets Syst. **202**, 89–99 (2011)
9. Fodor, J.C., Yager, R.R., Rybalov, A.: Structure of uninorms. Int. J. Uncertainty Fuzziness Knowl. Based Syst. **5**, 411–427 (1997)
10. Klement, E.P., Mesiar, R., Pap, E.: Triangular Norms. Kluwer Academic Publishers, Dordrecht (2000)
11. Li, G., Liu, H.W., Fodor, J.: Single-point characterization of uninorms with nilpotent underlying t-norm and t-conorm. Int. J. Uncertainty Fuzziness Knowl. Based Syst. **22**, 591–604 (2014)
12. Martín, J., Mayor, G., Torrens, J.: On locally internal monotonic operators. Fuzzy Sets Syst. **137**, 27–42 (2003)
13. Mas, M., Massanet, S., Ruiz-Aguilera, D., Torrens, J.: A survey on the existing classes of uninorms. J. Intell. Fuzzy Syst. **29**, 1021–1037 (2015)
14. Mas, M., Monserrat, M., Ruiz-Aguilera, D., Torrens, J.: Residual implications derived from uninorms satisfying the Modus Ponens. In: IFSA-EUSFLAT-2015, pp. 233–240. Atlantis Press, Gijón (2015)
15. Mas, M., Monserrat, M., Ruiz-Aguilera, D., Torrens, J.: *RU* and (U, N)-implications satisfying Modus Ponens. International Journal of Approximate Reasoning. In press. doi:10.1016/j.ijar.2016.01.003
16. Mas, M., Monserrat, M., Torrens, J.: Two types of implications derived from uninorms. Fuzzy Sets Syst. **158**, 2612–2626 (2007)
17. Mas, M., Monserrat, M., Torrens, J.: Modus Ponens and Modus Tollens in discrete implications. Int. J. Approximate Reason. **49**, 422–435 (2008)
18. Mas, M., Monserrat, M., Torrens, J.: A characterization of (U, N), RU, QL and D-implications derived from uninorms satisfying the law of importation. Fuzzy Sets Syst. **161**, 1369–1387 (2010)
19. Mas, M., Monserrat, M., Torrens, J., Trillas, E.: A survey on fuzzy implication functions. IEEE Trans. Fuzzy Syst. **15**(6), 1107–1121 (2007)
20. Massanet, S., Torrens, J.: On a new class of fuzzy implications: h-implications and generalizations. Inf. Sci. **181**, 2111–2127 (2011)
21. Massanet, S., Torrens, J.: An overview of construction methods of fuzzy implications. In: Baczynski, M., Beliakov, G., Bustince, H., Pradera, A. (eds.) Adv. in Fuzzy Implication Functions. STUDFUZZ, vol. 300, pp. 1–30. Springer, Heidelberg (2013)

22. Ruiz, D., Torrens, J.: Residual implications and co-implications from idempotent uninorms. Kybernetika **40**, 21–38 (2004)
23. Ruiz-Aguilera, D., Torrens, J.: Distributivity of residual implications over conjunctive and disjunctive uninorms. Fuzzy Sets Syst. **158**, 23–37 (2007)
24. Ruiz-Aguilera, D., Torrens, J.: S- and R-implications from uninorms continuous in $]0,1[^2$ and their distributivity over uninorms. Fuzzy Sets Syst. **160**, 832–852 (2009)
25. Ruiz-Aguilera, D., Torrens, J., De Baets, B., Fodor, J.: Some remarks on the characterization of idempotent uninorms. In: Hüllermeier, E., Kruse, R., Hoffmann, F. (eds.) IPMU 2010. LNCS, vol. 6178, pp. 425–434. Springer, Heidelberg (2010)
26. Trillas, E., Alsina, C., Pradera, A.: On MPT-implication functions for fuzzy logic. Rev. Real Acad. Cienc. Ser. A Matemáticas (RACSAM) **98**(1), 259–271 (2004)
27. Trillas, E., Alsina, C., Renedo, E., Pradera, A.: On contra-symmetry and MPT-conditionality in fuzzy logic. Int. J. Intell. Syst. **20**, 313–326 (2005)
28. Trillas, E., Campo, C., Cubillo, S.: When QM-operators are implication functions and conditional fuzzy relations. Int. J. Intell. Syst. **15**, 647–655 (2000)
29. Trillas, E., Valverde, L.: On Modus Ponens in fuzzy logic. In: 15th International Symposium on Multiple-Valued Logic, pp. 294–301. Kingston, Canada (1985)

A New Look on the Ordinal Sum of Fuzzy Implication Functions

Sebastia Massanet(✉), Juan Vicente Riera, and Joan Torrens

Dept. Mathematics and Computer Science, University of the Balearic Islands, Crta. Valldemossa km 7,5, 07122 Palma, Spain
{s.massanet,jvicente.riera,jts224}@uib.es

Abstract. Fuzzy implication functions are logical connectives commonly used to model fuzzy conditional and consequently they are essential in fuzzy logic and approximate reasoning. From the theoretical point of view, the study of how to construct new implication functions from old ones is one of the most important topics in this field. In this paper new ordinal sum construction methods of implication functions based on fuzzy negations N are presented. Some general properties are analysed and particular cases when the considered fuzzy negation is the classical one or any strong negation are highlighted.

Keywords: Ordinal sum · Fuzzy implication function · Fuzzy negation · t-norm · t-conorm · (S,N)-implication

1 Introduction

From a theoretical point of view fuzzy logical connectives play an essential role in the theory of fuzzy sets and fuzzy logic. Usually these connectives are modelled in a functionally expressible framework and then, they are modelled through operations defined on the unit interval $[0,1]$. Thus, conjunctions and disjunctions are commonly modelled by t-norms and t-conorms, complements by fuzzy negations and fuzzy conditionals by fuzzy implication functions (see [3,10]). Along the theoretic study of fuzzy sets, these logical operations have been deeply investigated and many generalizations have appeared, leading to the use of not only t-norms and t-conorms, but also uninorms and many other classes of conjunctive and disjunctive aggregation functions (see [5,9]).

The necessity of having many different kinds of logical connectives have led to the study of construction methods of some of these operations from given ones. One of the most important of these construction methods is the ordinal sum construction of t-norms and t-conorms and some generalizations (see [10,19]). In fact, this construction methods comes from a more general framework which is the ordinal sum of (commutative) semigroups (see [6,11]).

Among these logical connectives, fuzzy implication functions play a fundamental role in approximate reasoning and fuzzy control because they are used

J.P. Carvalho et al. (Eds.): IPMU 2016, Part I, CCIS 610, pp. 399–410, 2016.
DOI: 10.1007/978-3-319-40596-4_34

not only to model fuzzy conditionals, but also in the inference processes. Moreover, they have a lot of applications in many different fields that vary from fuzzy control and fuzzy subsethood measures, to fuzzy mathematical morphology and image processing. For this reason, the interest of implication functions from a pure theoretical point of view has grown in last decades, see for instance the surveys [4,13] and the monographs [2,3] entirely devoted to this kind of operators.

Due to this great quantity of applications, it was pointed out for instance in [22] the necessity of having a lot of different classes of fuzzy implication functions. In this sense, one of the main topics in the theoretic study of implications is the research of new construction methods of fuzzy implication functions. There are many different methods to construct fuzzy implication functions and most of them can be found for instance in [2,3]. However, the research in this direction is currently very active and new constructions are constantly being proposed. For instance, there are some of them constructed from fuzzy negations [1,20], from unary functions [12,14], from other fuzzy implication functions [15–17,23,24], and so on. The interest of these construction methods lies in the potential preservation of the usual properties of fuzzy implications. That is, when the resulting implication function constructed from an initial one keeps the properties that satisfies this initial implication function. For this reason, the study of which properties are preserved by any new construction method which is presented, is the first matter of study for such method.

Curiously, unlike in the aggregation functions framework, the ordinal sum construction of fuzzy implication functions has not been as studied as other construction methods. In [7], the structure of residual implications of ordinal sums of continuous Archimedean t-norms was studied and the definition of ordinal sums of residual implications was presented. Then, in [18] some few notes on residual implications derived from ordinal sums of left-continuous t-subnorms were given and the structure of the related residual implications was shown to be an ordinal sum of residual implications linked to the corresponding left-continuous t-subnorms. Nevertheless, a general study on this construction method was missing until the publication of the very recent paper [21] where an ordinal sum of fuzzy implications functions (not necessarily residual implications) is presented leading to fuzzy implication functions whose structure resembles the one based on residual implications.

However, there are other possibilities in the study of ordinal sums of fuzzy implication functions and this paper is focused precisely in this question. The idea is based not on residual implications derived from ordinal sums of t-norms, but on the (S, N)-implications derived from ordinal sums of t-conorms. From this starting point, we present many different ordinal sum constructions, one for each possible fuzzy negation N, leading to the so-called N-ordinal sums.

The paper is organized as follows. After this introduction, Sect. 2 is devoted to some preliminaries in order to make the paper as self-contained as possible. Section 3 presents the main results of the paper. In such section the ordinal construction based on a fuzzy negation N is investigated, some examples are given and it is investigated which of the most usual properties of implication functions are preserved under the N-ordinal sum construction. Finally, the paper ends with Sect. 4 devoted to some conclusions and future work.

2 Preliminaries

We will suppose the reader to be familiar with the theory of t-norms and t-conorms. For more details in this particular topic, we refer the reader to [10]. To make this work self-contained, we recall here some of the concepts and results used in the rest of the paper. First of all, the definition of fuzzy negation is given.

Definition 1 ([3, Definition 1.4.2]). *A decreasing function $N : [0,1] \to [0,1]$ is called a fuzzy negation if $N(0) = 1$ and $N(1) = 0$. A fuzzy negation N is called*

(i) strict, if it is strictly decreasing and continuous,
(ii) strong, if it is an involution, i.e., $N(N(x)) = x$ for all $x \in [0,1]$,

Important examples of fuzzy negations are *the classical (standard) negation* given by $N_C(x) = 1 - x$ for all $x \in [0,1]$ and *the least fuzzy negation* given by

$$N_{D_1}(x) = \begin{cases} 1 \text{ if } x = 0, \\ 0 \text{ if } x \in (0,1]. \end{cases}$$

Now, we recall the definition of fuzzy implication function.

Definition 2 ([8, Definition 1.15]). *A binary operator $I : [0,1]^2 \to [0,1]$ is said to be a fuzzy implication function if it satisfies:*

(I1) *$I(x,z) \geq I(y,z)$ when $x \leq y$, for all $z \in [0,1]$.*
(I2) *$I(x,y) \leq I(x,z)$ when $y \leq z$, for all $x \in [0,1]$.*
(I3) *$I(0,0) = I(1,1) = 1$ and $I(1,0) = 0$.*

Note that, from the definition, it follows that $I(0,x) = 1$ and $I(x,1) = 1$ for all $x \in [0,1]$ whereas the symmetrical values $I(x,0)$ and $I(1,x)$ are not derived from the definition.

Given a fuzzy implication function I, its 0-horizontal section defines always a fuzzy negation.

Definition 3 ([3, Definition 1.4.15]). *Let I be a fuzzy implication function. The function N_I defined by $N_I(x) = I(x,0)$ for all $x \in [0,1]$, is called the* natural negation *of I.*

Fuzzy implication functions may satisfy some additional properties which come from tautologies in crisp logic. Some interesting ones which will be studied in this paper are the following:

- The *exchange principle*,

$$I(x, I(y,z)) = I(y, I(x,z)), \quad x,y,z \in [0,1]. \qquad \textbf{(EP)}$$

- The *consequent boundary*,

$$I(1,y) \geq y, \quad y \in [0,1]. \tag{CB}$$

- The *left neutrality principle*,

$$I(1,y) = y, \quad y \in [0,1]. \tag{NP}$$

- The *ordering property*,

$$x \leq y \Longleftrightarrow I(x,y) = 1, \quad x,y \in [0,1]. \tag{OP}$$

- The *identity principle*,

$$I(x,x) = 1, \quad x \in [0,1]. \tag{IP}$$

- The *contrapositive symmetry* with respect to a fuzzy negation N,

$$I(x,y) = I(N(y),N(x)), \quad x,y \in [0,1]. \tag{CP(N)}$$

- The *law of right contraposition* with respect to a fuzzy negation N,

$$I(x,N(y)) = I(y,N(x)), \quad x,y \in [0,1]. \tag{R − CP(N)}$$

Finally, let us recall the well-known family of (S,N)-implications. Given a t-conorm S and a fuzzy negation N, an (S,N)-implication is defined as

$$I_{S,N}(x,y) = S(N(x),y), \qquad x,y \in [0,1].$$

3 The Opposite Ordinal Sum Construction of Fuzzy Implication Functions

Before giving the definition of an ordinal sum of fuzzy implication functions, let us motivate this new look to this construction method of fuzzy implication functions from some given ones. As we have already commented in the introduction, the idea comes from the structure of an (S,N)-implication generated by an ordinal sum t-conorm S. Next example illustrates this structure.

Example 1. Let us consider the ordinal sum t-conorm S given by $S = (\langle 0, 0.25, S_{LK}\rangle, \langle 0.75, 1, S_P\rangle)$ where S_{LK} stands for the Łukasiewicz t-conorm and S_P for the probabilistic sum t-conorm. Concretely, S is given by

$$S(x,y) = \begin{cases} \min\{x+y, 0.25\} & \text{if } x,y \in [0,0.25], \\ -4xy + 4x + 4y - 3 & \text{if } x,y \in [0.75,1], \\ \max\{x,y\} & \text{otherwise.} \end{cases} \tag{1}$$

Now, taking the strict fuzzy negation $N(x) = 1 - x^2$ for all $x \in [0,1]$, we can consider the corresponding (S,N)-implication given by

$$I_{S,N}(x,y) = \begin{cases} \min\{1 - x^2 + y, 0.25\} & \text{if } x \in [\sqrt{0.75}, 1], y \in [0,0.25], \\ 1 - 4x^2 + 4x^2 y & \text{if } x \in [0, \sqrt{0.25}], y \in [0.75,1], \\ \max\{1-x^2, y\} & \text{otherwise.} \end{cases}$$

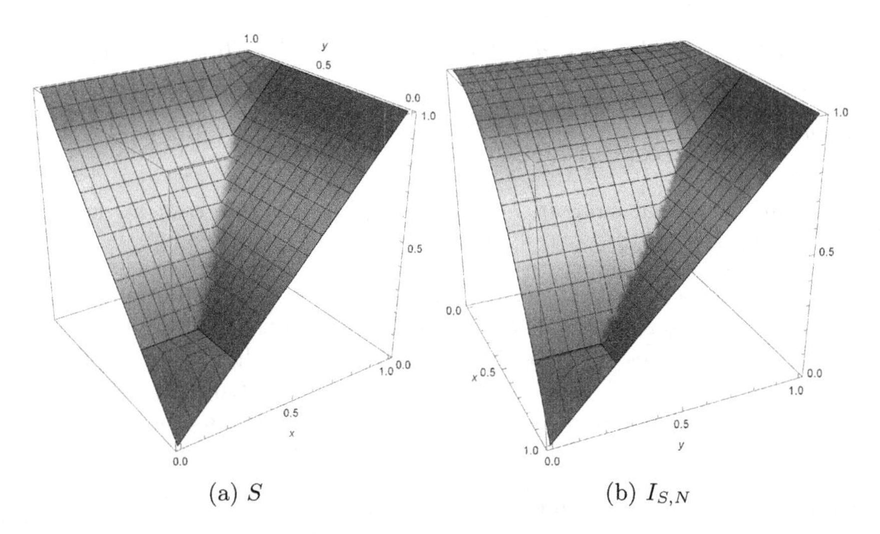

(a) S (b) $I_{S,N}$

Fig. 1. The t-conorm S and the (S, N)-implication considered in Example 1.

In Fig. 1, the graphical representations of S and $I_{S,N}$ are given. However, a careful look to this expression enables to establish a relationship between $I_{S,N}$ and the implication functions I_{S_{LK},N_1} and I_{S_P,N_2} where

$$N_1(x) = \frac{1-(\sqrt{0.75}+(1-\sqrt{0.75})x)^2}{0.25}, \quad N_2(x) = 1 - x^2, \quad \text{for all } x \in [0,1].$$

Indeed,

$$I_{S,N}(x,y) = \begin{cases} 0.25 \cdot I_{S_{LK},N_1}\left(\frac{x-\sqrt{0.75}}{1-\sqrt{0.75}}, \frac{y}{0.25}\right) & \text{if } x \in [\sqrt{0.75}, 1], y \in [0, 0.25], \\ 0.75 + 0.25 \cdot I_{S_P,N_2}\left(\frac{x}{\sqrt{0.25}}, \frac{y-0.75}{0.25}\right) & \text{if } x \in [0, \sqrt{0.25}], y \in [0.75, 1], \\ \max\{1-x^2, y\} & \text{otherwise.} \end{cases}$$

Thus, $I_{S,N}$ is in fact constructed from some (S, N)-implications generated by the summands (S_{LK} and S_P) of the ordinal sum t-conorm S.

The underlying structure of an (S, N)-implication where S is an ordinal sum t-conorm is what we want to explain and generalize with the following novel definition of an ordinal sum of fuzzy implication functions.

Definition 4. *Let J be a finite or countably infinite index set, $\{(a_j, b_j)\}_{j \in J}$ be a family of pairwise disjoint open intervals of $[0,1]$, N be a fuzzy negation and let $\{(c_j, d_j)\}_{j\in J}$ be such that $c_j = N(b_j)$ and $d_j = N(a_j)$. Let $\{I_j\}_{j\in J}$ be a family of fuzzy implication functions satisfying*

$$N_{I_j}(x) = \frac{N(a_j + (b_j - a_j)x) - c_j}{d_j - c_j}, \quad \textit{for all } x \in [0,1], \tag{2}$$

for every $j \in J$ *such that* $c_j < d_j$. *A function* $I : [0,1]^2 \to [0,1]$ *given by*

$$I(x,y) = \begin{cases} c_j + (d_j - c_j) \cdot I_j\left(\frac{x-a_j}{b_j-a_j}, \frac{y-c_j}{d_j-c_j}\right) & \text{if } x \in [a_j, b_j] \text{ and } y \in [c_j, d_j], \\ \max\{N(x), y\} & \text{otherwise,} \end{cases}$$

is called the N-ordinal sum *of the family* $\{I_j\}_{j\in J}$. *In this case we will write* $I = \langle N, (a_j, b_j, I_j)_{j\in J}\rangle$.

Remark 1. When a non-strict negation N is considered in Definition 4, it may occur that $c_j = d_j$ for some $j \in J$. In this case, the corresponding fuzzy implication function I_j does not play any role in the N-ordinal sum implication. In particular, its natural negation does not need to satisfy Eq. (2) and the N-ordinal sum implication I is understood that it satisfies $I(x, c_j) = c_j$ for all $x \in [a_j, b_j]$.

Remark 2. Definition 4 includes the fuzzy implication function studied in Example 1. In particular, taking the family of open intervals $\{(0, \sqrt{0.25}), (\sqrt{0.75}, 1)\}$, $N(x) = 1 - x^2$ for all $x \in [0,1]$, $I_1 = I_{S_P,N_1}$ and $I_2 = I_{S_P,N_2}$, we have that $I_{S,N}$ where S is the t-conorm given by Eq. (1) is in fact

$$I = \langle N, (0, \sqrt{0.25}, I_1), (\sqrt{0.75}, 1, I_2)\rangle.$$

The first question that arises is whether the N-ordinal sum of a family of fuzzy implications $\{I_j\}$ is always a fuzzy implication function in the sense of Definition 2. Next example shows that this is not true in general.

Example 2. Let us consider $N = N_C$ and $I = \langle N, (0, 0.5, I_{RS})\rangle$ where I_{RS} is the Rescher implication. In this case, I is given by

$$I(x,y) = \begin{cases} 1 & \text{if } x \in [0,0.5], y \in [0.5,1], x + 0.5 \leq y, \\ 0.5 & \text{if } x \in [0,0.5], y \in [0.5,1], x + 0.5 > y, \\ \max\{1-x, y\} & \text{otherwise.} \end{cases}$$

However, I is not a fuzzy implication function since $I(0.5, 0.75) = 0.5 < 0.75 = I(0.75, 0.75)$ and therefore, I is not decreasing in the first variable.

Thus, we have to characterize which fuzzy implication functions are suitable to generate a fuzzy implication function through the N-ordinal sum construction method. Next theorem shows that the consequent boundary property is the key property to ensure that this construction method provides a fuzzy implication function in the sense of Definition 2.

Theorem 1. *Let* J *be a finite or countably infinite index set,* $\{(a_j, b_j)\}_{j\in J}$ *be a family of pairwise disjoint open intervals of* $[0,1]$, N *a fuzzy negation,* $\{I_j\}_{j\in J}$ *a family of implication functions with associated negations satisfying Eq. (2), and* $I = \langle N, (a_j, b_j, I_j)_{j\in J}\rangle$ *its* N*-ordinal sum. Then the following statements are equivalent:*

(i) I *is a fuzzy implication function.*

(ii) $\{I_j\}_{j\in J}$ *and* $\{(a_j,b_j)\}_{j\in J}$ *satisfy:*

$$I_j \text{ satisfies (CB) for all } j \in J \text{ such that } b_j < 1. \tag{3}$$

Remark 3. It would be possible to generalize Definition 4 in the following way:

$$I(x,y) = \begin{cases} c_j + (d_j - c_j) \cdot I_j\left(\frac{x-a_j}{b_j-a_j}, \frac{y-c_j}{d_j-c_j}\right) & \text{if } x \in [a_j,b_j] \text{ and } y \in [c_j,d_j], \\ \max\{N'(x), \varphi(y)\} & \text{otherwise,} \end{cases}$$

where $N' \leq N$ and $\varphi : [0,1] \to [0,1]$ is a nondecreasing function such that $\varphi \geq id$. This definition would change all the results presented in this paper in the sense that they would be more complex involving φ and N'. For the sake of simplicity and in order to focus on the relationship with (S,N)-implications, we have decided to restrict the ordinal sum to Definition 4.

At a first glance, the N-ordinal sum construction method given in Definition 4 would force the researcher to fix a fuzzy negation N and then to choose some specific family of fuzzy implication functions $\{I_j\}_{j\in J}$ satisfying Eq. (2). This would be a quite restrictive construction method that would not allow to construct an ordinal sum of any given family of fuzzy implication functions satisfying Property (3) with the chosen family of open intervals. However, this method allows the researcher to construct an ad-hoc fuzzy negation N for any family $\{I_j\}_{j\in J}$ in order to obtain a fuzzy implication function I through the N-ordinal sum construction method, as it is detailed in the following remark.

Remark 4. Given a family of fuzzy implication functions $\{I_j\}_{j\in J}$ satisfying Property (3) with a family of pairwise disjoint open intervals of $[0,1]$ $\{(a_j,b_j)\}_{j\in J}$, we can always construct a fuzzy negation N in such a way that the natural negation N_{I_j} of each I_j satisfies Eq. (2). Specifically, we have several cases to consider:

1. If $\{(a_j,b_j)\}_{j\in J}$ gives a partition of $[0,1]$, we can construct the fuzzy negation N given by:
$$N(x) = c_j + (d_j - c_j) \cdot N_{I_j}\left(\frac{x-a_j}{b_j-a_j}\right) \tag{4}$$
for all $x \in [a_j,b_j]$ and $j \in J$ where $\{(c_j,d_j)\}_{j\in J}$, stated in the decreasing order, is a family of pairwise disjoint open intervals of $[0,1]$. It is immediate to check that in this case Eq. (2) holds and we can construct the N-ordinal sum of $\{I_j\}_{j\in J}$.
2. When J is a finite set and $\{(a_j,b_j)\}_{j\in J}$ is not a partition of $[0,1]$, N can be constructed as in Eq. (4) for all values in $[a_j,b_j]$ and $j \in J$, but then it is not uniquely defined in $[0,1] \setminus \bigcup_{j\in J}[a_j,b_j]$. Indeed, for all $x \in [b_j,a_{j+1}]$ with $J \in J$, N needs only to be a non-increasing function satisfying $N(b_j) = c_j$ and $N(a_{j+1}) = d_{j+1}$.
3. When J is an infinite set and $\{(a_j,b_j)\}_{j\in J}$, it is also possible to define a fuzzy negation N but the construction method of N is more complex and due to the lack of space, we have omitted it.

Let us illustrate some examples of the N-construction method.

Example 3. The following two examples deal with on the one hand, a case when we have a partition of $[0,1]$ and on the other hand, a case when the ad-hoc fuzzy negation N can be chosen in an infinite number of ways.

(i) Let us consider the family of fuzzy implication functions $\{I_{LK}, I_{RC}, I_{FD}\}$ where I_{LK}, I_{RC} and I_{FD} denote the Łukasiewicz, Reichenbach and Fodor implications and the partition of $[0,1]$ given by $\{(0,0.25),(0.25,0.75),(0.75,1)\}$. These three fuzzy implication functions satisfy (CB) and their natural negation is N_C. In this case, through Eq. (4) in Remark 4, the ad-hoc construction of the fuzzy negation N retrieves $N(x) = N_C(x)$ for all $x \in [0,1]$. Thus, the N-ordinal sum

$$I_1 = \langle N_C, (0, 0.25, I_{LK}), (0.25, 0.75, I_{RC}), (0.75, 1, I_{FD})\rangle$$

is given by

$$I_1(x,y) = \begin{cases} \min\{1, 0.25 - x + y\} & \text{if } x \in [0,0.25], y \in [0.75,1], \\ 1.125 - 1.5x - 0.5y + 2xy & \text{if } x, y \in [0.25, 0.75], \\ 0.25 & \text{if } x \in [0.75,1], y \in [0,0.25], 4x - 3 \leq 4y, \\ \max\{1-x, y\} & \text{otherwise.} \end{cases}$$

(ii) Let us consider now the family of fuzzy implication functions $\{I_{GD}, I_{GG}\}$ where I_{GD} and I_{GG} stand for the Gödel and the Goguen implications, which also satisfy (CB) and they have N_{D_1} as natural negation, and the family of open intervals $\{(0,0.25),(0.75,1)\}$. In this case, using Eq. (4), we have that

$$N(x) = \begin{cases} 1 & \text{if } x = 0, \\ 0.75 & \text{if } 0 < x \leq 0.25, \\ 0.25 & \text{if } x = 0.75, \\ 0 & \text{if } 0.75 < x \leq 1. \end{cases}$$

Note that N is not defined in $(0.25, 0.75)$ and we can consider an infinite number of extensions of N to the whole interval $[0,1]$. In particular, we can consider the following two ones:

$$N_2(x) = \begin{cases} 1 & \text{if } x = 0, \\ 0.75 & \text{if } 0 < x \leq 0.25, \\ 1-x & \text{if } 0.25 < x < 0.75, \\ 0.25 & \text{if } x = 0.75, \\ 0 & \text{if } 0.75 < x \leq 1, \end{cases} \qquad N_3(x) = \begin{cases} 1 & \text{if } x = 0, \\ 0.75 & \text{if } 0 < x \leq 0.25, \\ 0.25 & \text{if } 0.25 < x \leq 0.75, \\ 0 & \text{if } 0.75 < x \leq 1. \end{cases}$$

Using these two fuzzy negations, we can construct the following two N-ordinal sums given by

$$I_2 = \langle N_2, (0, 0.25, I_{GD}), (0.75, 1, I_{GG})\rangle,$$
$$I_3 = \langle N_3, (0, 0.25, I_{GD}), (0.75, 1, I_{GG})\rangle,$$

and for $i = 2, 3$, we have

$$I_i(x, y) = \begin{cases} 1 & \text{if } x \in [0, 0.25], y \in [0.75, 1], 4x \leq 4y - 3, \\ y & \text{if } x \in [0, 0.25], y \in [0.75, 1], 4x > 4y - 3, \\ 0.25 & \text{if } x \in [0.75, 1], y \in [0, 0.25], 4x - 3 \leq 4y, \\ \frac{y}{4x-3} & \text{if } x \in [0.75, 1], y \in [0, 0.25], 4x - 3 > 4y. \\ \max\{N_i(x), y\} & \text{otherwise.} \end{cases}$$

All these fuzzy implication functions are depicted in Fig. 2.

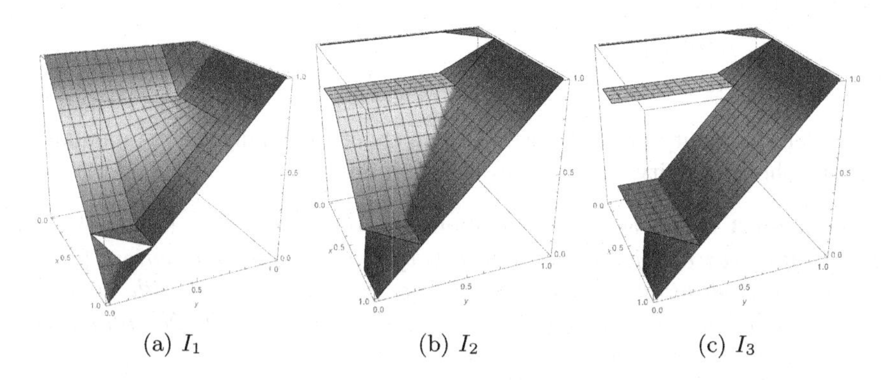

(a) I_1 (b) I_2 (c) I_3

Fig. 2. Some fuzzy implication functions constructed via the N-ordinal sum construction method in Example 3.

From now on, we will study some of the most interesting additional properties which can be satisfied by a fuzzy implication function. Since the N-ordinal sum implication is constructed from a family of fuzzy implication functions, its properties will heavily rely on the properties satisfied by the members of the family. Let us start studying the natural negation of the N-ordinal sum.

Proposition 1. *Let J be a finite or countably infinite index set, $\{(a_j, b_j)\}_{j \in J}$ be a family of pairwise disjoint open intervals of $[0, 1]$, N a fuzzy negation, $\{I_j\}_{j \in J}$ a family of implication functions satisfying Property (3) and with associated negations satisfying Eq. (2). Let $I = \langle N, (a_j, b_j, I_j)_{j \in J} \rangle$ be its N-ordinal sum. Then the natural negation of I is equal to N, that is, $N_I = N$.*

Remark 5. If we want the N-ordinal sum implication with (strict) continuous natural negation N of a given family of fuzzy implication functions $\{I_j\}_{j \in J}$ satisfying Property (3), then it is necessary that the natural negations of the initial fuzzy implication functions I_J are (strict) continuous for all $j \in J$. This condition is also sufficient when the family of intervals is a partition of $[0, 1]$. Otherwise, recall that N is not uniquely defined in $[0, 1] \setminus \bigcup_{j \in J} [a_j, b_j]$ and to obtain a (strict) continuous natural negation N, we must choose continuous (strictly) decreasing functions there.

The fulfilment of the left neutrality principle by the N-ordinal sum implication relies only on the left neutrality principle of only one of the fuzzy implication functions of the family. In fact, when $b_j < 1$ for all $j \in J$, it always satisfies the left neutrality principle. The same analysis is also valid for the consequent boundary property.

Proposition 2. *Let J be a finite or countably infinite index set, $\{(a_j, b_j)\}_{j\in J}$ be a family of pairwise disjoint open intervals of $[0,1]$, N a fuzzy negation, $\{I_j\}_{j\in J}$ a family of implication functions satisfying Property (3) and with associated negations satisfying Eq. (2). Let $I = \langle N, (a_j, b_j, I_j)_{j\in J}\rangle$ be its N-ordinal sum. Then the following statements are equivalent:*

(i) I satisfies (NP) ((CB)).
(ii) If there exists $j \in J$ such that $b_j = 1$ then I_j must satisfy (NP) ((CB)).

The left neutrality principle of the members of the family $\{I_j\}$ is mandatory to ensure the continuity of the N-ordinal sum implication.

Proposition 3. *Let J be a finite or countably infinite index set, $\{(a_j, b_j)\}_{j\in J}$ be a family of pairwise disjoint open intervals of $[0,1]$, N a fuzzy negation, $\{I_j\}_{j\in J}$ a family of implication functions satisfying Property (3) and with associated negations satisfying Eq. (2). Let $I = \langle N, (a_j, b_j, I_j)_{j\in J}\rangle$ be its N-ordinal sum. Then the following statements are equivalent:*

(i) I is continuous.
(ii) The two following properties hold:
  (a) For all $j \in J$, I_j is continuous.
  (b) For all $j \in J$ such that $b_j < 1$, I_j satisfies (NP).

Now let us study the (right) contrapositive symmetry of the N-ordinal sum implications. Next result shows that these two properties are preserved by this construction method.

Proposition 4. *Let J be a finite or countably infinite index set, $\{(a_j, b_j)\}_{j\in J}$ be a family of pairwise disjoint open intervals of $[0,1]$, N a fuzzy negation, $\{I_j\}_{j\in J}$ a family of implication functions satisfying Property (3) and with associated negations satisfying Eq. (2). Let $I = \langle N, (a_j, b_j, I_j)_{j\in J}\rangle$ be its N-ordinal sum. Then the following items hold:*

(i) I satisfies $R{-}CP(N)$ with respect to N if, and only if, I_j satisfies $R{-}CP(N)$ with respect to N_{I_j} for all $j \in J$.
(ii) If N is a strong negation, then I satisfies $CP(N)$ with respect to N if, and only if, I_j satisfies $CP(N)$ with respect to N_{I_j} for all $j \in J$.

Now, let us study the exchange principle for this class of implications. Next result shows that the exchange principle for the N-ordinal sum implication is related not only to the exchange principle for the summands of the ordinal sum but also to the right contrapositive symmetry.

Proposition 5. *Let J be a finite or countably infinite index set, $\{(a_j, b_j)\}_{j \in J}$ be a family of pairwise disjoint open intervals of $[0,1]$, N a fuzzy negation, $\{I_j\}_{j \in J}$ a family of implication functions satisfying Property (3) and with associated negations satisfying Eq. (2). Let $I = \langle N, (a_j, b_j, I_j)_{j \in J} \rangle$ be its N-ordinal sum. Then I satisfies (EP) if, and only if, the following items hold:*

(i) I_j satisfies (EP) for all $j \in J$.
(ii) I_j satisfies $R - CP(N)$ with respect to N_{I_j} for all $j \in J$ such that $N(b_j) = c_j > 0$.

4 Conclusions and Future Work

The ordinal sum construction method is a well-known construction method of some aggregation functions, such as t-norms and t-conorms. Until recently with the publication of [21], this construction method had not been proposed for fuzzy implication functions. However, the method proposed in [21] is mainly based on residual residual implications since the structure of the ordinal sum defined there resembles heavily to the ordinal construction method of t-norms. Therefore, in this paper, a totally different ordinal sum construction method for fuzzy implication functions has been presented with the aim of generalizing the structure of the (S, N)-implications generated by an ordinal sum t-conorm S. This construction method allows to define a new fuzzy implication function from a given family of fuzzy implication functions satisfying a property related to the consequent boundary. Some examples have been presented and some important additional properties such as the exchange principle, the left-neutrality principle, the continuity or the contrapositive symmetry, among others, have been analysed.

As future work, we plan to study more additional properties of the N-ordinal sum implications such as the law of importation, the ordering property, the identity principle or the distributivity properties. In addition, we want to fully characterize the intersections of the family of N-ordinal sum implications with respect to the most well-known families of fuzzy implications.

Acknowledgement. This paper has been partially supported by the Spanish Grant TIN2013-42795-P.

References

1. Aguiló, I., Suñer, J., Torrens, J.: New types of contrapositivisation of fuzzy implications with respect to fuzzy negations. Inf. Sci. **322**, 223–226 (2015)
2. Baczyński, M., Beliakov, G., Bustince Sola, H., Pradera, A. (eds.): Advances in Fuzzy Implication Functions. Studies in Fuzziness and Soft Computing, vol. 300. Springer, Heidelberg (2013)
3. Baczyński, M., Jayaram, B.: Fuzzy Implications. Studies in Fuzziness and Soft Computing, vol. 231. Springer, Heidelberg (2008)

4. Baczyński, M., Jayaram, B., Massanet, S., Torrens, J.: Fuzzy implications: past, present, and future. In: Kacprzyk, J., Pedrycz, W. (eds.) Springer Handbook of Computational Intelligence, pp. 183–202. Springer, Heidelberg (2015)
5. Beliakov, G., Pradera, A., Calvo, T.: Aggregation Functions: A Guide for Practitioners. Studies in Fuzziness and Soft Computing, vol. 221. Springer, Heidelberg (2007)
6. Clifford, A.H.: Naturally totally ordered commutative semigroups. Am. J. Math. **76**(3), 631–646 (1954)
7. De Baets, B., Mesiar, R.: Residual implicators of continuous t-norms. In: Proceedings of EUFIT 1996, pp. 37–41 (1996)
8. Fodor, J.C., Roubens, M.: Fuzzy Preference Modelling and Multicriteria Decision Support. Kluwer Academic Publishers, Dordrecht (1994)
9. Grabisch, M., Marichal, J.L., Mesiar, R., Pap, E.: Aggregation Functions. Encyclopedia of Mathematics and Its Applications, vol. 127. Cambridge University Press, New York (2009)
10. Klement, E.P., Mesiar, R., Pap, E.: Triangular Norms. Kluwer Academic Publishers, Dordrecht (2000)
11. Klement, E.P., Mesiar, R., Pap, E.: Triangular norms as ordinal sums of semigroups in the sense of A.H. Clifford. Semigroup Forum **65**, 71–82 (2002)
12. Hlinená, D., Kalina, M., Kral, P.: Implications functions generated using functions of one variable. In: [2], pp. 125–153 (2013)
13. Mas, M., Monserrat, M., Torrens, J., Trillas, E.: A survey on fuzzy implication functions. IEEE Trans. Fuzzy Syst. **15**(6), 1107–1121 (2007)
14. Massanet, S., Torrens, J.: On a new class of fuzzy implications: h-Implications and generalizations. Inf. Sci. **181**, 2111–2127 (2011)
15. Massanet, S., Torrens, J.: Threshold generation method of construction of a new implication from two given ones. Fuzzy Sets Syst. **205**, 50–75 (2012)
16. Massanet, S., Torrens, J.: On the vertical threshold generation method of fuzzy implication and its properties. Fuzzy Sets Syst. **226**, 232–252 (2013)
17. Massanet, S., Torrens, J.: An overview of construction methods of fuzzy implications. In: [2], pp. 1–30 (2013)
18. Mesiar, R., Mesiarová, A.: Residual implications and left-continuous t-norms which are ordinal sums of semigroups. Fuzzy Sets Syst. **143**, 47–57 (2004)
19. Mesiarová, A.: Continuous triangular subnorms. Fuzzy Sets Syst. **142**, 75–83 (2004)
20. Shi, Y., Van Hasse, B., Ruan, D., Kerre, E.: Fuzzy implications: classification and a new class. In: [2], pp. 31–53 (2013)
21. Su, Y., Xie, A., Liu, H.: On ordinal sum implications. Inf. Sci. **293**, 251–262 (2015)
22. Trillas, E., Mas, M., Monserrat, M., Torrens, J.: On the representation of fuzzy rules. Int. J. Approximate Reason. **48**, 583–597 (2008)
23. Vemuri, N.R., Jayaram, B.: Representations through a monoid on the set of fuzzy implications. Fuzzy Sets Syst. **247**, 51–67 (2014)
24. Vemuri, N.R., Jayaram, B.: The composition of fuzzy implications: closures with respect to properties, powers and families. Fuzzy Sets Syst. **275**, 58–87 (2015)

Distributivity of Implication Functions over Decomposable Uninorms Generated from Representable Uninorms in Interval-Valued Fuzzy Sets Theory

Michał Baczyński(✉) and Wanda Niemyska

Institute of Mathematics, University of Silesia, Ul. Bankowa 14, 40-007 Katowice, Poland
{michal.baczynski,wniemyska}@us.edu.pl

Abstract. In this work we investigate two distributivity equations $\mathcal{I}(x, \mathcal{U}_1(y,z)) = \mathcal{U}_2(\mathcal{I}(x,y), \mathcal{I}(x,z))$, $\mathcal{I}(\mathcal{U}_1(x,y), z) = \mathcal{U}_2(\mathcal{I}(x,z), \mathcal{I}(y,z))$ for implication operations and uninorms in interval-valued fuzzy sets theory. We consider decomposable (t-representable) uninorms generated from two conjunctive or disjunctive representable uninorms. Our method reduces to solve the following functional equation $f(u_1 + v_1, u_2 + v_2) = f(u_1, u_2) + f(v_1, v_2)$, thus we present new solutions for this equation.

Keywords: Aggregation operators · Uninorms · Interval-valued fuzzy sets · Distributivity equations · Functional equations

1 Introduction

The distributivity of (classical) fuzzy implications over different fuzzy logical connectives, like t-norms, t-conorms or uninorms has been studied in the recent past by many authors (see [1–3,7,9,13,15,30,31,33]). Distributivity equations have a very important role to play in efficient inferencing in approximate reasoning, especially in fuzzy control systems. Given an input "$\widetilde{x}$ is A'", the role of an inference mechanism is to obtain a fuzzy output B' that satisfies some desirable properties. The most important inference schemas are fuzzy relational inference and similarity based reasoning. In the first case, the inferred output B' is obtained either as

(i) $\sup -T$ composition, where T is a t-norm, as in the compositional rule of inference (CRI) of Zadeh (see [35]), or
(ii) $\inf -I$ composition, where I is a fuzzy implication, as in the Bandler-Kohout subproduct (BKS) (see [16]),

of A' and given rules. Since all the rules of an inference engine are exercised during every inference cycle, the number of rules directly affects the computational duration of the overall application.

J.P. Carvalho et al. (Eds.): IPMU 2016, Part I, CCIS 610, pp. 411–422, 2016.
DOI: 10.1007/978-3-319-40596-4_35

To reduce the complexity of fuzzy "IF-THEN" rules, Combs and Andrews [18] proposed an equivalent transformation of the CRI to mitigate the computational cost. In fact, they demanded the following classical tautology

$$(p \wedge q) \to r = (p \to r) \vee (q \to r),$$

written in fuzzy logic language, i.e., using t-norms, t-conorms and fuzzy implications. Subsequently, there were many discussions (see [17,21,29]), most of them pointed out the need for a theoretical investigation required for employing such equations. Later, the similar method but for similarity based reasoning was presented by Jayaram [27]. For an overview of the most important these methods see [12, Chapter 8].

In [4–6] (for the full article see [10]), [8,11] we discussed the following distributivity equations

$$\mathcal{I}(x, \mathcal{T}_1(y,z)) = \mathcal{T}_2(\mathcal{I}(x,y), \mathcal{I}(x,z)),$$
$$\mathcal{I}(\mathcal{S}(x,y), z) = \mathcal{T}(\mathcal{I}(x,z), \mathcal{I}(y,z)),$$

for t-representable (decomposable) t-norms and t-conorms (in interval-valued fuzzy sets theory) generated from continuous Archimedean operations. In these articles we obtained the solutions for each of the following functional equations, respectively:

$$f(u_1 + v_1, u_2 + v_2) = f(u_1, u_2) + f(v_1, v_2), \tag{A}$$
$$g(\min(u_1 + v_1, a), \min(u_2 + v_2, a)) = g(u_1, u_2) + g(v_1, v_2), \tag{B}$$
$$h(\min(u_1 + v_1, a), \min(u_2 + v_2, a)) = \min(h(u_1, u_2) + h(v_1, v_2), b), \tag{C}$$
$$k(u_1 + v_1, u_2 + v_2) = \min(k(u_1, u_2) + k(v_1, v_2), b), \tag{D}$$

where $a, b > 0$ are fixed real numbers, $f\colon L^\infty \to [0,\infty]$, $g\colon L^a \to [0,\infty]$, $h\colon L^a \to [0,b]$, $k\colon L^\infty \to [0,b]$ are unknown functions and

$$L^\infty = \{(u_1, u_2) \in [0,\infty]^2 \mid u_1 \geq u_2\},$$
$$L^a = \{(u_1, u_2) \in [0,a]^2 \mid u_1 \geq u_2\}.$$

More precisely, the solutions of Eq. (A) are presented in [4, Proposition 3.2], the solutions of Eq. (B) are presented in [5, Proposition 4.2], the solutions of Eq. (C) are presented in [10, Proposition 5.2] and the solutions of Eq. (D) are presented in [8, Proposition 3.2].

These investigations have been extended to the following functional equation

$$\mathcal{I}(x, \mathcal{U}_1(y,z)) = \mathcal{U}_2(\mathcal{I}(x,y), \mathcal{I}(x,z)), \tag{D-UU1}$$

when $\mathcal{U}_1, \mathcal{U}_2$ are decomposable uninorms on $\mathcal{L}^I$ generated from two conjunctive representable uninorms and $\mathcal{I}$ is an unknown function (see [14]). In that article we presented the solutions of the following functional equation

$$f(u_1 + v_1, u_2 + v_2) = f(u_1, u_2) + f(v_1, v_2), \qquad (u_1, u_2), (v_1, v_2) \in L^{\overline{\infty}}, \tag{F}$$

where $L^{\overline{\infty}} = \{(x_1, x_2) \in [-\infty, \infty]^2 \mid x_1 \leq x_2\}$, with the assumption $(-\infty)+\infty = \infty + (-\infty) = -\infty$ in both sets of domain (formally in both projections) and codomain of a function f.

In this article we continue this research for two distributivity laws: Eq. (D-UU1) and the following equation

$$\mathcal{I}(\mathcal{U}_1(x, y), z) = \mathcal{U}_2(\mathcal{I}(x, z), \mathcal{I}(y, z)), \qquad x, y, z \in L^I, \tag{D-UU2}$$

where $\mathcal{U}_1$, $\mathcal{U}_2$ are given decomposable uninorms, and function $\mathcal{I}$ is unknown, in particular an implication. We show that in this case solving both Eqs. (D-UU1) and (D-UU2) reduces to finding solutions of Eq. (F) for some combinations of the following assumptions

$$(-\infty) + \infty = \infty + (-\infty) = -\infty, \tag{$A-$}$$

$$(-\infty) + \infty = \infty + (-\infty) = \infty. \tag{$A+$}$$

in both sets of domain (formally in both projections) and codomain of a function f.

2 Interval-Valued Fuzzy Sets, Implications and Uninorms

One possible extension of fuzzy sets theory is interval-valued fuzzy sets theory introduced by Sambuc [32] (see also [26,36]), in which to each element of the universe a closed subinterval of the unit interval is assigned – it can be used as an approximation of the unknown membership degree. Let us define

$$L^I = \{(x_1, x_2) \in [0, 1]^2 \mid x_1 \leq x_2\},$$
$$(x_1, x_2) \leq_{L^I} (y_1, y_2) \Longleftrightarrow x_1 \leq y_1 \text{ and } x_2 \leq y_2.$$

In the sequel, if $x \in L^I$, then we denote it by $x = [x_1, x_2]$. In fact, $\mathcal{L}^I = (L^I, \leq_{L^I})$ is a complete lattice with units $0_{\mathcal{L}^I} = [0, 0]$ and $1_{\mathcal{L}^I} = [1, 1]$.

Definition 2.1. *An interval-valued fuzzy set on X is a mapping $A\colon X \to L^I$.*

We assume that the reader is familiar with the classical results concerning basic fuzzy logic connectives, but we briefly mention some of the results employed in the rest of the work.

One possible definition of an implication on $\mathcal{L}^I$ is the following one (cf. [12,20,24]).

Definition 2.2. *Let $\mathcal{L} = (L, \leq_L)$ be a complete lattice. A function $\mathcal{I}\colon L^2 \to L$ is called a fuzzy implication on $\mathcal{L}$ if it is decreasing with respect to the first variable, increasing with respect to the second variable and fulfills the following conditions:*

$$\mathcal{I}(0_{\mathcal{L}}, 0_{\mathcal{L}}) = \mathcal{I}(1_{\mathcal{L}}, 1_{\mathcal{L}}) = \mathcal{I}(0_{\mathcal{L}}, 1_{\mathcal{L}}) = 1_{\mathcal{L}}, \qquad \mathcal{I}(1_{\mathcal{L}}, 0_{\mathcal{L}}) = 0_{\mathcal{L}}. \tag{1}$$

Uninorms (in the unit interval) were introduced by Yager and Rybalov in 1996 (see [34]) as a generalization of triangular norms and conorms. For the recent overview of this family of operations see [23,28].

Definition 2.3. *Let $\mathcal{L} = (L, \leq_L)$ be a complete lattice. An associative, commutative and increasing operation $\mathcal{U}: L^2 \to L$ is called a uninorm on $\mathcal{L}$, if there exists $e \in L$ such that $\mathcal{U}(e,x) = \mathcal{U}(x,e) = x$, for all $x \in L$.*

Remark 2.4

(i) The neutral element e corresponding to a uninorm $\mathcal{U}$ is unique. Moreover, if $e = 0_\mathcal{L}$, then $\mathcal{U}$ is a t-conorm and if $e = 1_\mathcal{L}$, then $\mathcal{U}$ is a t-norm.
(ii) For a uninorm $\mathcal{U}$ on any $\mathcal{L}$ we get $\mathcal{U}(0_\mathcal{L}, 0_\mathcal{L}) = 0_\mathcal{L}$ and $\mathcal{U}(1_\mathcal{L}, 1_\mathcal{L}) = 1_\mathcal{L}$.
(iii) For a uninorm U on $([0,1], \leq)$ we get $U(0,1) \in \{0,1\}$.
(vi) For a uninorm $\mathcal{U}$ on $\mathcal{L}^I$ with the neural element $e \in L^I \setminus \{0_{\mathcal{L}^I}, 1_{\mathcal{L}^I}\}$ we get $\mathcal{U}(0_{\mathcal{L}^I}, 1_{\mathcal{L}^I}) \in \{0_{\mathcal{L}^I}, 1_{\mathcal{L}^I}\}$ or $\mathcal{U}(0_{\mathcal{L}^I}, 1_{\mathcal{L}^I}) \| e$, i.e., $\mathcal{U}(0_{\mathcal{L}^I}, 1_{\mathcal{L}^I})$ is not comparable with e (cf. [19,22]).
(v) In general, for any lattice $\mathcal{L}$, if $\mathcal{U}(0_\mathcal{L}, 1_\mathcal{L}) = 0_\mathcal{L}$, then it is called conjunctive and if $\mathcal{U}(0_\mathcal{L}, 1_\mathcal{L}) = 1_\mathcal{L}$, then it is called disjunctive.

In the literature one can find several classes of uninorms (see [25,28]). Uninorms that can be represented as in point (ii) of Theorem 2.5 are called representable uninorms.

Theorem 2.5 ([25, **Theorem 3**]). *For a function $U: [0,1]^2 \to [0,1]$ the following statements are equivalent:*

(i) U is a strictly increasing and continuous on $]0,1[^2$ uninorm with the neutral element $e \in]0,1[$ such that U is self-dual, except in points $(0,1)$ and $(1,0)$, with respect to a strong negation N with the fixed point e, i.e.,

$$U(x,y) = N(U(N(x), N(y))), \qquad x, y \in [0,1]^2 \setminus \{(0,1),(1,0)\}.$$

(ii) U has a continuous additive generator, i.e., there exists a continuous and strictly increasing function $h: [0,1] \to [-\infty, \infty]$, such that $h(0) = -\infty$, $h(e) = 0$ for $e \in]0,1[$ and $h(1) = \infty$, which is uniquely determined up to a positive multiplicative constant, such that for all $x, y \in [0,1]$ either

$$U(x,y) = \begin{cases} 0 & \text{if } (x,y) \in \{(0,1),(1,0)\}, \\ h^{-1}(h(x) + h(y)), & \text{otherwise,} \end{cases}$$

when U is conjunctive, or

$$U(x,y) = \begin{cases} 1 & \text{if } (x,y) \in \{(0,1),(1,0)\}, \\ h^{-1}(h(x) + h(y)), & \text{otherwise,} \end{cases}$$

when U is disjunctive.

Remark 2.6 (cf. [3]). If a representable uninorm U is conjunctive, then $U(x,y) = h^{-1}(h(x) + h(y))$ holds for all $x, y \in [0,1]$ with the assumption

$$(-\infty) + \infty = \infty + (-\infty) = -\infty. \tag{A-}$$

If a representable uninorm U is disjunctive, then $U(x,y) = h^{-1}(h(x) + h(y))$ holds for all $x, y \in [0,1]$ with the assumption

$$(-\infty) + \infty = \infty + (-\infty) = \infty. \tag{A+}$$

Now we shall consider the following special class of uninorms on $\mathcal{L}^I$.

Definition 2.7 (see [19,22]). *A uninorm $\mathcal{U}$ on $\mathcal{L}^I$ is called decomposable (or t-representable) if there exist uninorms U_1, U_2 on $([0,1], \leq)$ such that*

$$\mathcal{U}([x_1,x_2],[y_1,y_2]) = [U_1(x_1,y_1), U_2(x_2,y_2)], \qquad [x_1,x_2],[y_1,y_2] \in L^I,$$

and $U_1 \leq U_2$. In this case we will write $\mathcal{U} = (U_1, U_2)$.

It should be noted that not all uninorms on $\mathcal{L}^I$ are decomposable (see [22]).

Lemma 2.8 ([22, Lemma 8]). *If $\mathcal{U}$ on $\mathcal{L}^I$ is a decomposable uninorm, then $\mathcal{U}(0_{\mathcal{L}^I}, 1_{\mathcal{L}^I}) = 0_{\mathcal{L}^I}$ or $\mathcal{U}(0_{\mathcal{L}^I}, 1_{\mathcal{L}^I}) = 1_{\mathcal{L}^I}$ or $\mathcal{U}(0_{\mathcal{L}^I}, 1_{\mathcal{L}^I}) = [0,1]$.*

Therefore it is not possible that for decomposable uninorm $\mathcal{U} = (U_1, U_2)$ on $\mathcal{L}^I$ we have that U_1 is disjunctive and U_2 is conjunctive.

Lemma 2.9 (cf. [22, Theorems 5 and 6]). *If $\mathcal{U} = (U_1, U_2)$ on $\mathcal{L}^I$ is a decomposable uninorm with the neutral element $e = [e_1, e_2]$, then $e_1 = e_2$ is the neutral element of U_1 and U_2.*

Lemma 2.10 ([14, Lemma 3.9]). *Let a function $\mathcal{I}\colon (\mathcal{L}^I)^2 \to \mathcal{L}^I$ satisfy* (1) *and Eq.* (D-UU1) *with some uninorms $\mathcal{U}_1$, $\mathcal{U}_2$ defined on $\mathcal{L}^I$. Then $\mathcal{U}_1$ is conjunctive if and only if $\mathcal{U}_2$ is conjunctive.*

Lemma 2.11 *Let a function $\mathcal{I}\colon (\mathcal{L}^I)^2 \to \mathcal{L}^I$ satisfy* (1) *and Eq.* (D-UU2) *with some uninorms $\mathcal{U}_1$, $\mathcal{U}_2$ defined on $\mathcal{L}^I$. Then $\mathcal{U}_1$ is conjunctive if and only if $\mathcal{U}_2$ is disjunctive and $\mathcal{U}_1$ is disjunctive if and only if $\mathcal{U}_2$ is conjunctive.*

Proof. Putting $x = z = 0_{\mathcal{L}^I}$ and $y = 1_{\mathcal{L}^I}$ in (D-UU2) we have

$$\mathcal{I}(\mathcal{U}_1(0_{\mathcal{L}^I}, 1_{\mathcal{L}^I}), 0_{\mathcal{L}^I}) = \mathcal{U}_2(\mathcal{I}(0_{\mathcal{L}^I}, 0_{\mathcal{L}^I}), \mathcal{I}(1_{\mathcal{L}^I}, 0_{\mathcal{L}^I})).$$

If $\mathcal{U}_1$ is conjunctive, then $\mathcal{U}_1(1_{\mathcal{L}^I}, 0_{\mathcal{L}^I}) = 0_{\mathcal{L}^I}$ and from (1) we obtain $1_{\mathcal{L}^I} = \mathcal{U}_2(1_{\mathcal{L}^I}, 0_{\mathcal{L}^I})$, thus $\mathcal{U}_2$ is disjunctive. If on the other hand $\mathcal{U}_1$ is disjunctive, then $\mathcal{U}_1(1_{\mathcal{L}^I}, 0_{\mathcal{L}^I}) = 1_{\mathcal{L}^I}$ and by (1) we have $0_{\mathcal{L}^I} = \mathcal{U}_2(1_{\mathcal{L}^I}, 0_{\mathcal{L}^I})$, so $\mathcal{U}_2$ is conjunctive. □

The above results allow us to investigate Eqs. (D-UU1) and (D-UU1) only for some decomposable uninorms.

3 Method for Solving Distributivity Eqs. (D-UU1) and (D-UU2) for Decomposable Uninorms

In this section we derive the Eq. (F) from distributivity Eqs. (D-UU1) and (D-UU2). Let $\mathcal{U}_1 = (U_1, U_2)$, $\mathcal{U}_2 = (U_3, U_4)$ be decomposable uninorms on $\mathcal{L}^I$. Assume that the projection mappings on $\mathcal{L}^I$ are defined as the following:

$$pr_1([x_1, x_2]) = x_1, \qquad pr_2([x_1, x_2]) = x_2, \qquad \text{for } [x_1, x_2] \in L^I.$$

Eqs. (D-UU1) and (D-UU2) have the following form:

$$\begin{aligned}\mathcal{I}([x_1, x_2],&[U_1(y_1, z_1), U_2(y_2, z_2)]) \\ =&[U_3(pr_1(\mathcal{I}([x_1, x_2], [y_1, y_2])), pr_1(\mathcal{I}([x_1, x_2], [z_1, z_2]))), \\ &U_4(pr_2(\mathcal{I}([x_1, x_2], [y_1, y_2])), pr_2(\mathcal{I}([x_1, x_2], [z_1, z_2])))],\end{aligned}$$

$$\begin{aligned}\mathcal{I}([U_1(x_1, y_1),&U_2(x_2, y_2)], [z_1, z_2]) \\ =&[U_3(pr_1(\mathcal{I}([x_1, x_2], [z_1, z_2])), pr_1(\mathcal{I}([y_1, y_2], [z_1, z_2]))), \\ &U_4(pr_2(\mathcal{I}([x_1, x_2], [z_1, z_2])), pr_2(\mathcal{I}([y_1, y_2], [z_1, z_2])))],\end{aligned}$$

for $[x_1, x_2], [y_1, y_2], [z_1, z_2] \in L^I$. As a consequence we obtain the following four equations, which are satisfied for all $[x_1, x_2], [y_1, y_2], [z_1, z_2] \in L^I$,

$$\begin{aligned}pr_1(\mathcal{I}([x_1, x_2],&[U_1(y_1, z_1), U_2(y_2, z_2)])) \\ &= U_3(pr_1(\mathcal{I}([x_1, x_2], [y_1, y_2])), pr_1(\mathcal{I}([x_1, x_2], [z_1, z_2]))), \\ pr_2(\mathcal{I}([x_1, x_2],&[U_1(y_1, z_1), U_2(y_2, z_2)])) \\ &= U_4(pr_2(\mathcal{I}([x_1, x_2], [y_1, y_2])), pr_2(\mathcal{I}([x_1, x_2], [z_1, z_2]))),\end{aligned}$$

$$\begin{aligned}pr_1(\mathcal{I}([U_1(x_1, y_1),&U_2(x_2, y_2)], [z_1, z_2])) \\ &= U_3(pr_1(\mathcal{I}([x_1, x_2], [z_1, z_2])), pr_1(\mathcal{I}([y_1, y_2], [z_1, z_2]))), \\ pr_2(\mathcal{I}([U_1(x_1, y_1),&U_2(x_2, y_2)], [z_1, z_2])) \\ &= U_4(pr_2(\mathcal{I}([x_1, x_2], [z_1, z_2])), pr_2(\mathcal{I}([y_1, y_2], [z_1, z_2]))).\end{aligned}$$

Next, let us fix arbitrarily $[x_1, x_2], [z_1, z_2] \in L^I$ and define four functions $k^1_{[x_1,x_2]}, k^2_{[x_1,x_2]}, l_1^{[z_1,z_2]}, l_2^{[z_1,z_2]} : \mathcal{L}^I \to \mathcal{L}^I$ by

- $k^1_{[x_1,x_2]}(\cdot) := pr_1 \circ \mathcal{I}([x_1, x_2], \cdot)$,
- $k^2_{[x_1,x_2]}(\cdot) := pr_2 \circ \mathcal{I}([x_1, x_2], \cdot)$,
- $l_1^{[z_1,z_2]}(\cdot) := pr_1 \circ \mathcal{I}(\cdot, [z_1, z_2])$,
- $l_2^{[z_1,z_2]}(\cdot) := pr_2 \circ \mathcal{I}(\cdot, [z_1, z_2])$,

where $\circ$ denotes the standard composition of functions. Thus we have shown that if $\mathcal{U}_1$ and $\mathcal{U}_2$ on $\mathcal{L}^I$ are decomposable, then Eqs. (D-UU1) and (D-UU2) are equivalent, respectively, to the following systems of equations:

$$\begin{aligned}k^1_{[x_1,x_2]}([U_1(y_1, z_1), U_2(y_2, z_2)]) &= U_3(k^1_{[x_1,x_2]}([y_1, y_2]), k^1_{[x_1,x_2]}([z_1, z_2])), \\ k^2_{[x_1,x_2]}([U_1(y_1, z_1), U_2(y_2, z_2)]) &= U_4(k^2_{[x_1,x_2]}([y_1, y_2]), k^2_{[x_1,x_2]}([z_1, z_2])),\end{aligned} \tag{DUU-1'}$$

$$l_1^{[z_1,z_2]}([U_1(x_1,y_1),U_2(x_2,y_2)]) = U_3(l_1^{[z_1,z_2]}([x_1,x_2]),l_1^{[z_1,z_2]}([y_1,y_2])),$$
$$l_2^{[z_1,z_2]}([U_1(x_1,y_1),U_2(x_2,y_2)]) = U_4(l_2^{[z_1,z_2]}([x_1,x_2]),l_2^{[z_1,z_2]}([y_1,y_2])). \tag{DUU-2'}$$

Let us look closer to Eq. (DUU-1'). Assume that $U_1 = U_2$ and $U_3 = U_4$ are representable uninorms generated from h_1 and h_3, respectively. Next, by Lemma 2.10, let us assume that both U_1, U_3 are conjunctive or disjunctive. From Remark 2.6, if both uninorms are conjunctive, then we assume the assumption $(A-)$ on the codomains of h_1 and h_3, while if both uninorms are disjunctive, then we the assumption $(A+)$ on the codomains of h_1 and h_3.

Using the representation for representable uninorms i.e., Theorem 2.5, we can transform our problem to the following equation (for a simplicity we deal only with k^1 now)

$$\begin{aligned} &k^1_{[x_1,x_2]}([h_1^{-1}(h_1(y_1)+h_1(z_1)),h_1^{-1}(h_1(y_2)+h_1(z_2))]) \\ &\quad = h_3^{-1}(h_3(k^1_{[x_1,x_2]}([y_1,y_2]))+h_3(k^1_{[x_1,x_2]}([z_1,z_2]))), \end{aligned}$$

where $[x_1,x_2],[y_1,y_2],[z_1,z_2] \in L^I$. Let us put $h_1(y_1) = u_1$, $h_1(y_2) = u_2$, $h_1(z_1) = v_1$ and $h_1(z_2) = v_2$. It is obvious that $u_1,u_2,v_1,v_2 \in [-\infty,\infty]$ and $u_1 \le u_2$, $v_1 \le v_2$, since $y_1 \le y_2$, $z_1 \le z_2$, and generator h_1 is strictly increasing. If we define

$$f^1_{[x_1,x_2]}(u,v) := h_3 \circ k^1_{[x_1,x_2]}([h_1^{-1}(u),h_1^{-1}(v)]), \qquad u,v \in [-\infty,\infty],\ u \le v,$$

then we get the following functional equation

$$f^1_{[x_1,x_2]}(u_1+v_1,u_2+v_2) = f^1_{[x_1,x_2]}(u_1,u_2)+f^1_{[x_1,x_2]}(v_1,v_2), \tag{2}$$

where $(u_1,u_2),(v_1,v_2) \in L^{\overline{\infty}}$ and $f^1_{[x_1,x_2]}\colon L^{\overline{\infty}} \to [-\infty,\infty]$ is an unknown function. By $L^{\overline{\infty}}$ we denoted the set $\{(x_1,x_2) \in [-\infty,\infty]^2 : x_1 \le x_2\}$.

Repeating all of the above calculations for the function k^2, we get analogous functional equation:

$$f^2_{[x_1,x_2]}(u_1+v_1,u_2+v_2) = f^2_{[x_1,x_2]}(u_1,u_2)+f^2_{[x_1,x_2]}(v_1,v_2), \tag{3}$$

where $f^2_{[x_1,x_2]}\colon L^{\overline{\infty}} \to [-\infty,\infty]$ is an unknown function defined by

$$f^2_{[x_1,x_2]}(u,v) := h_3 \circ k^2_{[x_1,x_2]}([h_1^{-1}(u),h_1^{-1}(v)]), \qquad (u,v) \in L^{\overline{\infty}}.$$

Observe that Eqs. (2) and (3) are exactly the same functional Eq. (F), i.e.,

$$f(u_1+v_1,u_2+v_2) = f(u_1,u_2)+f(v_1,v_2),$$

where $f\colon L^{\overline{\infty}} \to [-\infty,\infty]$ is an unknown function.

As a summary of this case we see that conjunctive representable uninorms U_1, U_3 leads us to Eq. (F) with the assumption $(A-)$ on the domain and codomain of a function f, while the case of disjunctive representable uninorms U_1, U_3 leads us to Eq. (F) with the assumption $(A+)$ on the domain and codomain of function f.

Now let us return to Eq. (DUU-2'). As before, let $U_1 = U_2$ and $U_3 = U_4$ will be representable uninorms generated by h_1 and h_3, respectively. By Lemma 2.11 we know that it is enough to consider again only two cases: when U_1 is conjunctive and U_3 disjunctive, or vice versa - when the U_1 is an disjunctive, and U_3 conjunctive. We still assume $(A-)$ on the codomains of generators of conjunctive uninorms and $(A+)$ on the codomains of generators of disjunctive uninorms. For fixed $[z_1, z_2] \in L^I$ let us define

$$g_1^{[z_1,z_2]}(u,v) := h_3 \circ l_1^{[z_1,z_2]}([h_1^{-1}(u), h_1^{-1}(v)]), \qquad (u,v) \in L^{\overline{\infty}},$$
$$g_2^{[z_1,z_2]}(u,v) := h_3 \circ l_2^{[z_1,z_2]}([h_1^{-1}(u), h_1^{-1}(v)]), \qquad (u,v) \in L^{\overline{\infty}}.$$

Repeating, for functions l_1, l_2, all the calculations which we carried out earlier for functions k_1 and k_2, we obtain that also functions $g_1^{[z_1,z_2]}$ and $g_2^{[z_1,z_2]}$ satisfy the functional Eq. (F). This time the case of conjunctive uninorm U_1 and disjunctive uninorm U_3 leads to the Eq. (F) with the assumption $(A-)$ on the domain of f and $(A+)$ on the codomain of f, while the case of disjunctive uninorm U_1 and conjunctive uninorm U_3 lead to the Eq. (F) with the assumption $(A+)$ on the domain of f and $(A-)$ on the codomain of f.

4 Some New Results Pertaining to Functional Equations

In [3] we solved the additive Cauchy functional equation:

$$f(x+y) = f(x) + f(y), \qquad x, y \in [-\infty, \infty],$$

for an unknown function $f\colon [-\infty,\infty] \to [-\infty,\infty]$. It should be noted that the main problem in this context was with the adequate definition of the additions $\infty+(-\infty)$ and $(-\infty)+\infty$. Recently, in [14] we presented solutions of the Eq. (F) for all $(u_1,u_2),(v_1,v_2) \in L^{\overline{\infty}}$, with the assumption (A-), i.e., $(-\infty)+\infty = \infty+(-\infty) = -\infty$ in both sets of domain (formally in both projections) and codomain.

In this article we present new theorem which shows all solutions of Eq. (F) with the assumption $(A+)$ on the domain (formally in both projections) and codomain of function f.

Theorem 4.1. *Let $L^{\overline{\infty}} = \{(u_1,u_2) \in [-\infty,\infty]^2 \mid u_1 \le u_2\}$. For a function $f\colon L^{\overline{\infty}} \to [-\infty,\infty]$ the following statements are equivalent:*

(i) f satisfies functional Eq. (F) for $(u_1,u_2),(v_1,v_2) \in L^{\overline{\infty}}$, with the assumption $(A+)$, i.e., $(-\infty)+\infty = \infty+(-\infty) = \infty$, in both sets of domain (formally in both projections) and codomain of f.
(ii) Either $f = -\infty$, or $f = 0$, or $f = \infty$ or

$$f(u,v) = \begin{cases} -\infty, & u = \infty, \\ 0, & u < \infty, \end{cases} \qquad or \qquad f(u,v) = \begin{cases} -\infty, & v = \infty, \\ 0, & v < \infty, \end{cases}$$

or

$$f(u,v) = \begin{cases} \infty, & u = \infty, \\ 0, & u < \infty, \end{cases} \quad \text{or} \quad f(u,v) = \begin{cases} \infty, & v = \infty, \\ 0, & v < \infty, \end{cases}$$

or

$$f(u,v) = \begin{cases} -\infty, & u < \infty, \\ \infty, & u = \infty, \end{cases} \quad \text{or} \quad f(u,v) = \begin{cases} -\infty, & v < \infty, \\ \infty, & v = \infty, \end{cases}$$

or

$$f(u,v) = \begin{cases} -\infty, & v \in \mathbb{R}, \\ \infty, & v \in \{-\infty, \infty\}, \end{cases} \quad \text{or} \quad f(u,v) = \begin{cases} -\infty, & u \in \mathbb{R}, \\ \infty, & u \in \{-\infty, \infty\}, \end{cases}$$

or

$$f(u,v) = \begin{cases} -\infty, & u, v \in \mathbb{R}, \\ \infty, & u = -\infty \text{ or } v = \infty, \end{cases}$$

or

$$f(u,v) = \begin{cases} -\infty, & u < \infty \text{ and } v = \infty, \\ 0, & v < \infty, \\ \infty, & u = \infty, \end{cases}$$

or there exists a unique additive function $c\colon \mathbb{R} \to \mathbb{R}$ *such that*

$$f(u,v) = \begin{cases} -\infty, & u \in \{-\infty, \infty\}, \\ c(u), & u \in \mathbb{R}, \end{cases} \quad \text{or} \quad f(u,v) = \begin{cases} -\infty, & v \in \{-\infty, \infty\}, \\ c(v), & v \in \mathbb{R}, \end{cases}$$

or

$$f(u,v) = \begin{cases} \infty, & u \in \{-\infty, \infty\}, \\ c(u), & u \in \mathbb{R}, \end{cases} \quad \text{or} \quad f(u,v) = \begin{cases} \infty, & v \in \{-\infty, \infty\}, \\ c(v), & v \in \mathbb{R}, \end{cases}$$

or

$$f(u,v) = \begin{cases} -\infty, & u = -\infty, \\ c(u), & u \in \mathbb{R}, \\ \infty, & u = \infty, \end{cases} \quad \text{or} \quad f(u,v) = \begin{cases} -\infty, & v = -\infty, \\ c(v), & v \in \mathbb{R}, \\ \infty, & v = \infty, \end{cases}$$

or

$$f(u,v) = \begin{cases} -\infty, & (u < \infty \text{ and } v = \infty) \text{ or } v = -\infty, \\ c(v), & v \in \mathbb{R}, \\ \infty, & u = \infty, \end{cases}$$

or there exist unique additive functions $c_1, c_2\colon \mathbb{R} \to \mathbb{R}$ *such that*

$$f(u,v) = \begin{cases} -\infty, & u = -\infty \text{ or } v = \infty, \\ c_1(u) + c_2(v), & u, v \in \mathbb{R}, \end{cases}$$

or

$$f(u,v) = \begin{cases} \infty, & u = -\infty \text{ or } v = \infty, \\ c_1(u) + c_2(v), & u, v \in \mathbb{R}, \end{cases}$$

or

$$f(u,v) = \begin{cases} -\infty, & u = -\infty \text{ and } v < \infty, \\ c_1(u) + c_2(v), & u, v \in \mathbb{R}, \\ \infty, & v = \infty, \end{cases}$$

or

$$f(u,v) = \begin{cases} -\infty, & (u < \infty \text{ and } v = \infty) \text{ or } u = -\infty, \\ c_1(u) + c_2(v), & u, v \in \mathbb{R}, \\ \infty, & u = \infty, \end{cases}$$

or

$$f(u,v) = \begin{cases} -\infty, & u \in \mathbb{R} \text{ and } v = \infty, \\ c_1(u) + c_2(v), & u, v \in \mathbb{R}, \\ \infty, & u \in \{-\infty, \infty\}, \end{cases}$$

or

$$f(u,v) = \begin{cases} -\infty, & u = -\infty \text{ and } v \in \mathbb{R}, \\ c_1(u) + c_2(v), & u, v \in \mathbb{R}, \\ \infty, & v \in \{-\infty, \infty\}, \end{cases}$$

for all $(u,v) \in L^{\overline{\infty}}$.

5 Conclusions

In this article we presented method for reducing Eqs. (D-UU1) and (D-UU2) to Eq. (F) for implication operations and decomposable uninorms (generated from two conjunctive or disjunctive representable uninorms) in interval-valued fuzzy sets theory. We showed that with this assumption it is enough to solve Eq. (F) for some combinations of the assumptions ($A-$) and/or ($A+$) in both sets of domain (formally in both projections) and codomain of a function f.

Theorem 4.1 solves the considered functional equation with the assumption ($A+$) in both the domain (in fact in both projections) and the codomain of function f. We would like to underline that cases combining ($A-$) in the domain and ($A+$) in the codomain (and vice versa) were also analyzed by us and we will present them soon.

Now, using Theorem 4.1, we are able to solve Eq. (2) and (3), i.e., we can obtain the description of the two projections of the vertical section $\mathcal{I}([x_1, x_2], \cdot)$, for fixed $[x_1, x_2] \in L^{\overline{\infty}}$, of the solutions of our main distributivity Eq. (D-UU1) for decomposable uninorms generated from disjunctive representable uninorms. In our future work we will consider these problems in details.

References

1. Baczyński, M.: On a class of distributive fuzzy implications. Internat. J. Uncertain. Fuzziness Knowl. Based Syst. **9**, 229–238 (2001)
2. Baczyński, M.: On the distributivity of fuzzy implications over continuous and Archimedean triangular conorms. Fuzzy Sets Syst. **161**, 1406–1419 (2010)
3. Baczyński, M.: On the distributivity of fuzzy implications over representable uninorms. Fuzzy Sets Syst. **161**, 2256–2275 (2010)
4. Baczyński, M.: On the distributivity of implication operations over t-representable t-norms generated from strict t-norms in interval-valued fuzzy sets theory. In: Hüllermeier, E., Kruse, R., Hoffmann, F. (eds.) IPMU 2010. CCIS, vol. 80, pp. 637–646. Springer, Heidelberg (2010)
5. Baczyński, M.: On the distributive equation for t-representable t-norms generated from nilpotent and strict t-norms. In: Galichet, S., et al. (eds.) Proceedings of EUSFLAT-LFA 2011, pp. 540–546. Atlantis Press, Amsterdam (2011)
6. Baczyński, M.: Distributivity of Implication Operations over t-Representable T-Norms Generated from Nilpotent T-Norms. In: Fanelli, A.M., Pedrycz, W., Petrosino, A. (eds.) WILF 2011. LNCS, vol. 6857, pp. 25–32. Springer, Heidelberg (2011)
7. Baczyński, M.: A note on the distributivity of fuzzy implications over representable uninorms. In: Greco, S., Bouchon-Meunier, B., Coletti, G., Fedrizzi, M., Matarazzo, B., Yager, R.R. (eds.) IPMU 2012, Part II. CCIS, vol. 298, pp. 375–384. Springer, Heidelberg (2012)
8. Baczyński, M.: Distributivity of implication operations over t-representable t-norms generated from continuous and Archimedean t-norms. In: Greco, S., Bouchon-Meunier, B., Coletti, G., Fedrizzi, M., Matarazzo, B., Yager, R.R. (eds.) IPMU 2012, Part II. CCIS, vol. 298, pp. 501–510. Springer, Heidelberg (2012)
9. Baczyński, M.: On two distributivity equations for fuzzy implications and continuous, Archimedean t-norms and t-conorms. Fuzzy Sets Syst. **211**, 34–54 (2013)
10. Baczyński, M.: Distributivity of implication operations over t-representable t-norms in interval-valued fuzzy set theory: the case of nilpotent t-norms. Inform. Sci. **257**, 388–399 (2014)
11. Baczyński, M.: The equation $\mathcal{I}(\mathcal{S}(x,y),z) = \mathcal{T}(\mathcal{I}(x,z),I(y,z))$ for t-representable t-conorms and t-norms generated from continuous, Archimedean operations. In: Masulli, F. (ed.) WILF 2013. LNCS, vol. 8256, pp. 131–138. Springer, Heidelberg (2013)
12. Baczyński, M., Jayaram, B.: Fuzzy Implications. STUDFUZZ, vol. 231. Springer, Heidelberg (2008)
13. Baczyński, M., Jayaram, B.: On the distributivity of fuzzy implications over nilpotent or strict triangular conorms. IEEE Trans. Fuzzy Syst. **17**(3), 590–603 (2009)
14. Baczyński, M., Niemyska, W.: On the Distributivity Equation $\mathcal{I}(x,\mathcal{U}_1(y,z)) = \mathcal{U}_2(\mathcal{I}(x,y),\mathcal{I}(x,z))$ for Decomposable Uninorms (in Interval-Valued Fuzzy Sets Theory) Generated from Conjunctive Representable Uninorms. In: Torra, V., Narukawa, Y., Endo, Y. (eds.) MDAI 2014. LNCS, vol. 8825, pp. 26–37. Springer, Heidelberg (2014)
15. Balasubramaniam, J., Rao, C.J.M.: On the distributivity of implication operators over t and s-norms. IEEE Trans. Fuzzy Syst. **12**, 194–198 (2004)
16. Bandler, W., Kohout, L.J.: Semantics of implication operators and fuzzy relational products. Internat. J. Man Mach. Stud. **12**, 89–116 (1980)
17. Combs, W.E.: Author's reply. IEEE Trans. Fuzzy Syst. **7**, 371–373, 477–478 (1999)

18. Combs, W.E., Andrews, J.E.: Combinatorial rule explosion eliminated by a fuzzy rule configuration. IEEE Trans. Fuzzy Syst. **6**, 1–11 (1998)
19. Deschrijver, G., Kerre, E.E.: Uninorms in L^*-fuzzy set theory. Fuzzy Sets Syst. **148**, 243–262 (2004)
20. Deschrijver, G., Cornelis, C., Kerre, E.E.: Implication in intuitionistic and interval-valued fuzzy set theory: construction, classification and application. Internat. J. Approx. Reason. **35**, 55–95 (2004)
21. Dick, S., Kandel, A.: Comments on "Combinatorial rule explosion eliminated by a fuzzy rule configuration". IEEE Trans. Fuzzy Syst. **7**, 475–477 (1999)
22. Drygaś, P.: On a class of operations on interval-valued fuzzy sets. In: Atanassov, K.T., et al. (eds.) New Trends in Fuzzy Sets, Intuitionistic Fuzzy Sets, Generalized Nets and Related Topics. Foundations, vol. 1, pp. 67–83. SRI PAS/IBS PAN, Warsaw (2013)
23. Fodor, J., De Baets, B.: Uninorm basics. In: Wang, P.P., et al. (eds.) Fuzzy Logic: A Spectrum of Theoretical and Practical Issues. STUDFUZZ, vol. 215, pp. 49–64. Springer, Heidelberg (2007)
24. Fodor, J., Roubens, M.: Fuzzy Preference Modelling and Multicriteria Decision Support. Kluwer Academic Publishers, Dordrecht (1994)
25. Fodor, J.C., Yager, R.R., Rybalov, A.: Structure of uninorms. Int. J. Uncertainty Fuzziness Knowl. Based Syst. **5**, 411–427 (1997)
26. Grattan-Guinness, I.: Fuzzy membership mapped onto interval and many-valued quantities. Z. Math. Logik. Grundladen Math. **22**, 149–160 (1975)
27. Jayaram, B.: Rule reduction for efficient inferencing in similarity based reasoning. Int. J. Approx. Reason. **48**, 156–173 (2008)
28. Mas, M., Massanet, S., Ruiz-Aguilera, D., Torrens, J.: A survey on the existing classes of uninorms. J. Intell. Fuzzy Syst. **29**, 1021–1037 (2015)
29. Mendel, J.M., Liang, Q.: Comments on "Combinatorial rule explosion eliminated by a fuzzy rule configuration". IEEE Trans. Fuzzy Syst. **7**, 369–371 (1999)
30. Ruiz-Aguilera, D., Torrens, J.: Distributivity of strong implications over conjunctive and disjunctive uninorms. Kybernetika **42**, 319–336 (2006)
31. Ruiz-Aguilera, D., Torrens, J.: Distributivity of residual implications over conjunctive and disjunctive uninorms. Fuzzy Sets Syst. **158**, 23–37 (2007)
32. Sambuc, R.: Fonctions Φ-floues. Application á l'aide au diagnostic en pathologie thyroidienne, Ph.D. thesis, Univ. Marseille, France (1975)
33. Trillas, E., Alsina, C.: On the law $[(p \wedge q) \rightarrow r] = [(p \rightarrow r) \vee (q \rightarrow r)]$ in fuzzy logic. IEEE Trans. Fuzzy Syst. **10**, 84–88 (2002)
34. Yager, R.R., Rybalov, A.: Uninorm aggregation operators. Fuzzy Sets Syst. **80**, 111–120 (1996)
35. Zadeh, L.A.: Outline of a new approach to the analysis of complex systems and decision processes. IEEE Trans. Syst. Man Cyber. **3**, 28–44 (1973)
36. Zadeh, L.A.: The concept of a linguistic variable and its application to approximate reasoning-I. Inform. Sci. **8**, 199–249 (1975)

On Functions Derived from Fuzzy Implications

Przemysław Grzegorzewski[1,2(✉)]

[1] Systems Research Institute, Polish Academy of Sciences, Newelska 6, 01-447 Warsaw, Poland
[2] Faculty of Mathematics and Computer Science, Warsaw University of Technology, Koszykowa 75, 00-662 Warsaw, Poland
pgrzeg@ibspan.waw.pl

Abstract. Recently, fuzzy implications based on copulas, i.e. probabilistic implications and probabilistic S-implications, were introduced and their properties were explored. However, the reverse problem of copulas derived from fuzzy implications, suggested by Massanet et al. [11,12], is also of interest. In the paper we consider geometric properties of those fuzzy implications that generate copulas. Moreover, we consider the reverse problem for some generalizations of copulas like quasi-copulas and semi-copulas.

Keywords: Copula · Fuzzy implication · Probabilistic implication · Spearman's rho · Diagonal section · Quasi-copula · Semi-copula

1 Introduction

Fuzzy implications still arouse curiosity of many researchers both because of their interesting theoretical properties and various applications in approximate reasoning, fuzzy control and so on. In the literature one can find several methods for constructing fuzzy implications, like S-implications, R-implications etc. (see [4]). Recently Grzegorzewski introduced new families of fuzzy implications based on copulas, i.e. probabilistic implications, probabilistic S-implications [7,8,10], survival implications and survival S-implications [9]. Another family of fuzzy implications based on copulas was proposed by Dolati et al. [5]. The common feature of all these contributions is to deliver such fuzzy implication that combine both imprecision modeled by fuzzy theory and randomness described by probability theory. And this is a justification for taking a copula to lay the foundations of such desired fuzzy implication. Actually, by the Sklar Theorem a copula expresses the dependence between random variables. Hence it can also form a link between a premise and a consequent.

The properties of implications based on copulas were examined in several papers (e.g. [2,3,7,8,10]). It is known, for instance, that for any copula the probabilistic S-implication is a fuzzy implication. However, Massanet et al. [11,12] set an interesting reverse problem: Can we say that using reverse reasoning each fuzzy implication leads to a copula? In general the answer is negative. But Massanet et al. [11,12] show some conditions that a fuzzy implication has to satisfy to generate a copula.

J.P. Carvalho et al. (Eds.): IPMU 2016, Part I, CCIS 610, pp. 423–434, 2016.
DOI: 10.1007/978-3-319-40596-4_36

In the present paper we try to exploit more thoroughly the geometric properties of those fuzzy implications which produce copulas. Moreover, since each copula is a particular case of a quasi-copula or semi-copula one may consider fuzzy implications based on those more general objects (see [1]). Thus, it seems quite natural to consider the reverse problem of fuzzy implications leading to quasi- or semi-copulas as well.

The paper is organized as follows. In Sect. 2 we recall some preliminaries on fuzzy implications, copulas and probabilistic implications. In Sect. 3 we present the reverse problem mentioned above and show some of its solutions. Then, in Sect. 4, we consider different aspects of the geometry of fuzzy implications in the context of their possible links with copulas. Section 5 is devoted to the reverse problem in relation to quasi-copulas, semi-copulas and binary aggregation operators. We end the paper with conclusions and suggestions for further research.

2 Preliminaries

Fuzzy implication functions are generalizations of the classical implication to fuzzy logic. According to the well-established fact that fuzzy concepts have to generalize adequately the corresponding crisp concepts, the most commonly accepted definitions of fuzzy connectives are the following.

Definition 2.1 ([4]). *A function $I\colon [0,1]^2 \to [0,1]$ is called a* ***fuzzy implication*** *if it satisfies the following conditions*

(I1) $I(x_1, y) \geq I(x_2, y)$ *if* $x_1 \leq x_2$, *for all* $y \in [0,1]$,
(I2) $I(x, y_1) \leq I(x, y_2)$ *if* $y_1 \leq y_2$, *for all* $x \in [0,1]$,
(I3) $I(0,0) = I(1,1) = 1$ *and* $I(1,0) = 0$.

Definition 2.2 ([13]). *A function $C\colon [0,1]^2 \to [0,1]$ is called a* **copula** *(specifically, a 2-copula) if it satisfies the following conditions*

(C1) $C(x,0) = C(0,y) = 0$, *for all* $x, y \in [0,1]$,
(C2) $C(x,1) = x$, *for all* $x \in [0,1]$,
(C3) $C(1,y) = y$, *for all* $y \in [0,1]$,
(C4) C *is 2-increasing, i.e.* $C(x_2, y_2) - C(x_2, y_1) - C(x_1, y_2) + C(x_1, y_1) \geq 0$ *for all* $x_1, x_2, y_1, y_2 \in [0,1]$ *such that* $x_1 \leq x_2, y_1 \leq y_2$.

It can be shown that every copula is bounded by the so-called Fréchet-Hoeffding bounds, i.e., for any copula C and for all $x, y \in [0,1]$ the following inequalities hold

$$W(x,y) \leq C(x,y) \leq M(x,y), \tag{1}$$

where $M(x,y) = \min\{x,y\}$ and $W(x,y) = \max\{x+y-1, 0\}$ are also copulas.

The notion of a copula was applied by Grzegorzewski for defining probabilistic implications and probabilistic S-implications.

Definition 2.3 ([7,10]). *Let C be a copula. A function $I_C\colon [0,1]^2 \to [0,1]$ given by*

$$I_C(x,y) = \begin{cases} 1, & x = 0 \\ \frac{C(x,y)}{x}, & x > 0 \end{cases}, \qquad x, y \in [0,1], \tag{2}$$

*is called a **probabilistic implication** (based on a copula C). The set of all probabilistic implications will be denoted by $\mathbb{I}^{\mathbb{C}}_{prob}$.*

It is worth noting that a probabilistic implication is not necessarily a fuzzy implication. To guarantee that a probabilistic implication is also a fuzzy implication we need to add condition **(I1)** that I_C is antitone with respect to the first variable (other conditions in Definition 2.1 are satisfied by any probabilistic implication, see [10]).

Definition 2.4 ([10]). *Let C be a copula. A function $\tilde{I}_C\colon [0,1]^2 \to [0,1]$ given by*

$$\tilde{I}_C(x,y) = C(x,y) - x + 1, \qquad x, y \in [0,1], \tag{3}$$

*is called a **probabilistic S-implication** (based on a copula C). The set of all probabilistic S-implications will be denoted by $\mathbb{I}^{\mathbb{C}}_{probS}$.*

It should be stressed that any probabilistic S-implication - contrary to probabilistic implication - is a fuzzy implication. It can be shown (see [2]) that the family of all probabilistic implications and the family of all probabilistic S-implications are disjoint, i.e., $\mathbb{I}^{\mathbb{C}}_{prob} \cap \mathbb{I}^{\mathbb{C}}_{probS} = \emptyset$.

Grzegorzewski [9] introduced also two families of implications based on survival copulas, i.e. survival implications and survival S-implications.

3 PSI-functions

As it was mentioned in the previous section each probabilistic implication and probabilistic S-implication, similarly as their survival counterparts, is defined by a copula. Massanet et al. [11,12] considered the reverse problem, i.e. how to construct copulas from implication functions. Since for any copula the corresponding probabilistic S-implication is a fuzzy implication, they discussed methods reverse to those used in construction of probabilistic S-implications. They have started by introducing the following notion.

Definition 3.1 ([11]). *Let I be a fuzzy implication function. A function C_I^{PSI} defined for all $x, y \in [0,1]$ as*

$$C_I^{PSI}(x,y) = I(x,y) + x - 1, \tag{4}$$

*is called a **probabilistic S-implication function (PSI-function** for short) derived from the fuzzy implication function I. Moreover, if C_I^{PSI} is a copula it will be called a **PSI-copula**.*

It is clear by the definition $C^{PSI}_{\tilde{I}_C} = C$ and $\tilde{I}_{C^{PSI}_I} = I$ for any copula C and fuzzy implication function I. However, a PSI-function C^{PSI}_I is not always a copula. For example, it is obvious that the implication function I must be continuous in order to derive a copula. Massanet et al. gave in [11] the necessary and sufficient conditions on a fuzzy implication function I to obtain a PSI-copula. Before we cite them let us recall two concepts important in fuzzy implication theory.

Firstly, we say that a fuzzy implication I satisfies the **left-neutrality principle** (NP) if $I(1, y) = y$ for all $y \in [0, 1]$. Secondly, a function defined for a given fuzzy implication I as follows $N_I(x) = I(x, 0)$, where $x \in [0, 1]$, is called the **natural negation** of I.

Theorem 3.2 ([11]). *Let I be a fuzzy implication function. A function C^{PSI}_I given by (4) is a copula if and only if the following conditions hold:*

(i) The natural negation of I is $N_I(x) = 1 - x$.
(ii) I satisfies the left-neutrality principle.
(iii) I is 2-increasing.

Massanet et al. [11,12] explored also some broad families of fuzzy implications like (S,N)-implications or R-implications in this context and proved some criteria dedicated especially for those families.

Theorem 3.3 ([11]). *Let S be a t-conorm, I_S the (S,N)-implication given by $I_S(x, y) = S(1 - x, y)$ for all $x, y \in [0, 1]$, and $C^{PSI}_{I_S}$ is the PSI-function derived from I_S. Then the following conditions are equivalent:*

(i) $C^{PSI}_{I_S}$ is a copula.
(ii) S is 2-decreasing, i.e. $S(x_2, y_2) - S(x_2, y_1) - S(x_1, y_2) + S(x_1, y_1) \leq 0$ for all $x_1, x_2, y_1, y_2 \in [0, 1]$ such that $x_1 \leq x_2, y_1 \leq y_2$.
(iii) S satisfies the 1-Lipschitz property, i.e. $S(x_2, y) - S(x_1, y) \leq x_2 - x_1$ for all $x_1, x_2, y \in [0, 1]$ such that $x_1 \leq x_2$.

From Theorem 3.3 one can conclude (see [11]) that a PSI-function $C^{PSI}_{I_S}$ derived from the (S,N)-implication I is a copula if and only if $N(x) = 1 - x$ and S is one of the following t-conorms: the maximum, an Archimedean t-conorm with convex additive generator or an ordinal sum of Archimedean t-conorms with convex additive generators.

Theorem 3.4 ([11]). *Let T be a left-continuous t-norm, I_T its R-implication. Then the PSI-function $C^{PSI}_{I_T}$ derived from I_T is a copula if and only if T is the Lukasiewicz t-norm.*

By Theorem 3.4 it is clear that $C^{PSI}_{I_T} = M$, where M is the upper Fréchet-Hoeffding bound (1).

Massanet et al. discussed also in [11,12] the similar reverse problem starting from survival S-implications.

4 Geometric Conditions for PSI-copulas

4.1 Introductory Example

To verify whether given PSI-function is a copula one has to check three conditions in Theorem 3.2. While the left-neutrality principle and the shape of the natural negation of I could be verified just by a glance on the plot of I, to check if I is 2-increasing is usually not so immediate. Let us consider the following example.

Example 4.1. Let us have a look on the plots of some well-known fuzzy implications: the Goguen implication (Fig. 1), the reciprocal Yager implication (Fig. 2) and the Baczynski implication (Fig. 3). For the expressions of those fuzzy implications we refer the reader to [4]. None of the PSI-functions derived from these three fuzzy implications are copulas. Actually, it is so because the natural negation of the Goguen implication is not a classical negation, the reciprocal Yager implication does not satisfy the neutrality principle while in the case of the Baczynski implication both conditions (i) and (ii) in Theorem 3.2 fail.

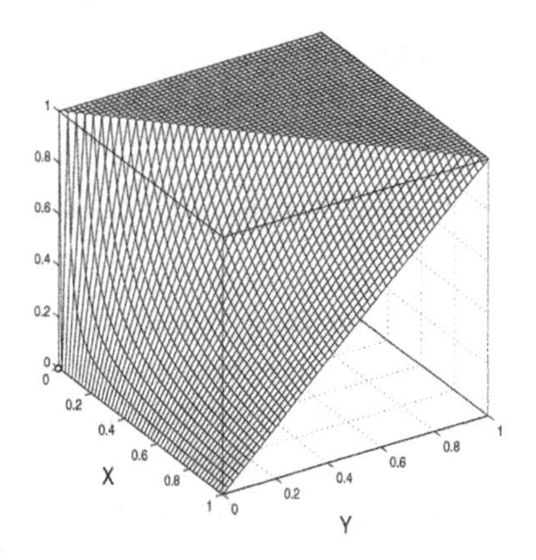

Fig. 1. The Goguen implication.

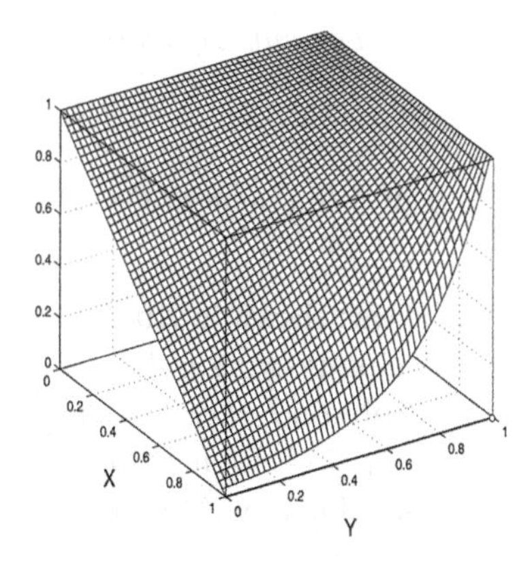

Fig. 2. The reciprocal Yager implication.

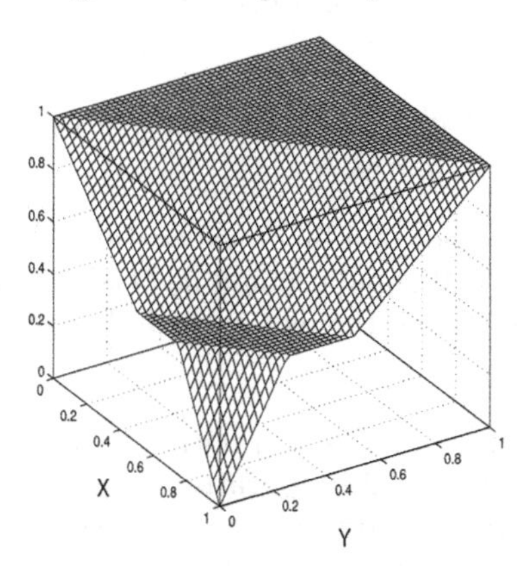

Fig. 3. The Baczynski implication.

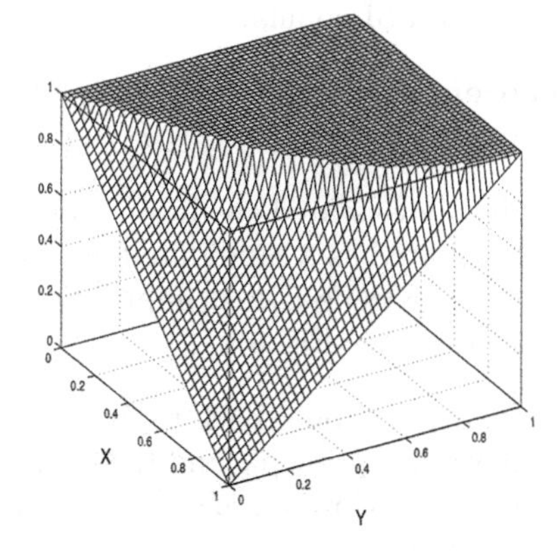

Fig. 4. I_{PC} implication.

Let us also consider the following fuzzy implication (see [4], p. 96):

$$I_{PC} = 1 - \left(\max\{x(x + xy^2 - 2y), 0\}\right)^{\frac{1}{2}} \tag{5}$$

depicted in Fig. 4. It is easily seen that $I_{PC}(1, y) = y$ and $I_{PC}(x, 0) = 1 - x$, so both conditions (*i*) and (*ii*) in Theorem 3.2 are satisfied. Does it mean that the PSI-function derived from I_{PC} is a PSI-copula? Can we answer this question without checking if I_{PC} is 2-increasing? □

It would be interesting to discover some simple criteria to eliminate those fuzzy implications which cannot generate a copula. In other words, we wish to find some simple necessary conditions for a PSI-function to be a PSI-copula. In fact, we may suggest several approaches that could be helpful in practice.

4.2 Bounds-Based Condition

Let us start from the following lemma.

Lemma 4.2. *For any copula C the following condition holds*

$$I_{KD} \leq \tilde{I}_C \leq I_{LK}, \tag{6}$$

where I_{KD} and I_{LK} denote the Kleene-Dienes implication and the Łukasiewicz implication, respectively.

Proof. As it was shown in [10] for the lower Fréchet-Hoeffding bound W we get $\widetilde{I}_W(x, y) = \max\{1-x, y\} = I_{KD}(x, y)$. Similarly, for the upper Fréchet-Hoeffding bound M we obtain $\widetilde{I}_M(x, y) = \min\{1, 1 - x + y\} = I_{LK}(u, v)$. Moreover, for any two copulas $C_1 \leq C_2$ we have $\widetilde{I}_{C_1}(x, y) = C_1(x, y) - x + 1 \leq C_2(x, y) - x + 1 = \widetilde{I}_{C_2}(x, y)$. Therefore, by (1) we conclude that $I_{KD} \leq \tilde{I}_C \leq I_{LK}$ for any copula C. ■

This way we obtain a natural necessary condition for a fuzzy implication to generate a PSI-copula.

Theorem 4.3. *Let I be a fuzzy implication such that C_I^{PSI} is a PSI-copula. Then $I_{KD}(x, y) \leq I(x, y) \leq I_{LK}(x, y)$ for any $x, y \in [0, 1]$.*

The above stated condition means that the plot of a fuzzy implication which can be treated as a candidate for generating a PSI-copula should lie between the plot of the Kleene-Dienes implication (see Fig. 5) and the plot of the Łukasiewicz implication (see Fig. 6).

Example 4.4. Going back to fuzzy implication (5), as we can notice, it satisfies the neutrality principle and the natural negation of this implication is the classical negation. However, $I_{PC}(0.95, 0.9) > I_{LK}(0.95, 0.9)$, so I_{PC} cannot generate a PSI-copula. □

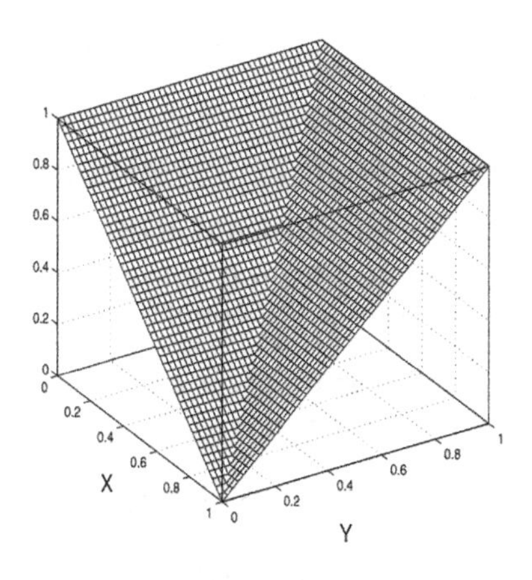

Fig. 5. The Kleene-Dienes implication.

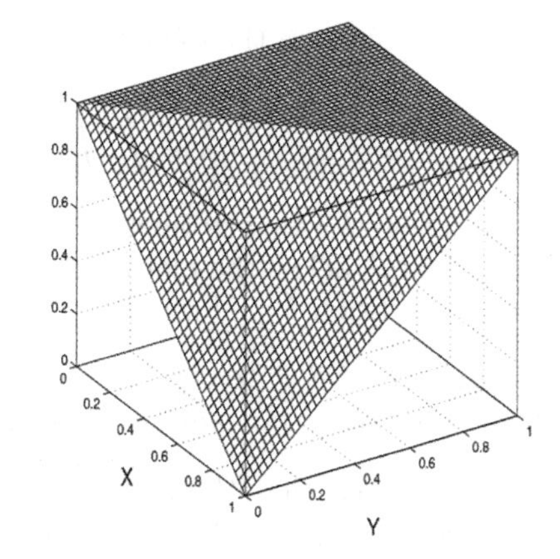

Fig. 6. The Łukasiewicz implication.

It is also easily seen that

Lemma 4.5. *If* $I_{KD} \leq I \leq I_{LK}$ *then for any* $x, y \in [0, 1]$

(a) $I(1, y) = y$*,*
(b) $I(x, 0) = 1 - x$*.*

Since property (a) in Lemma 4.5 means that I satisfies the left-neutrality property (NP), while (b) means that the natural negation of I is $N_I(x) = 1 - x$, the necessary and sufficient conditions for a PSI function to be a copula might be expressed as follows:

Theorem 4.6. *Let* I *be a fuzzy implication function. A function* C_I^{PSI} *given by* (4) *is a copula if and only if the following conditions hold:*

1. $I_{KD} \leq I \leq I_{LK}$*,*
2. I *is 2-increasing.*

4.3 Area-Based Conditions

Suppose X and Y are continuous random variables whose copula is C. The association between these two variables can be measured by Spearman's rho defined as follows

$$\rho_C = \rho(X, Y) = 12 \int_0^1 \int_0^1 (C(x, y) - xy)\, dxdy = 12 \int_0^1 \int_0^1 C(x, y) dxdy - 3. \quad (7)$$

As it is known $-1 \leq \rho_C \leq 1$ for any copula C. Therefore, assuming that C is a PSI-copula we may substitute it by $I(x, y) + x - 1$ in (7) and then we get

$$-1 \leq 12 \int_0^1 \int_0^1 (I(x, y) + x - 1)\, dxdy - 3 \leq 1,$$

which is equivalent to the following inequalities

$$\frac{2}{3} \leq \int_0^1 \int_0^1 I(x,y)dxdy \leq \frac{5}{6} \tag{8}$$

having a straightforward geometric interpretation. Indeed, condition (8) gives as the desired area below the plot of a fuzzy implication. This way we get another necessary condition for a fuzzy implication to generate a PSI-copula.

Theorem 4.7. *Let I be a fuzzy implication such that C_I^{PSI} is a PSI-copula. Then the area below the plot of I should take value in the interval $\left[\frac{2}{3}, \frac{5}{6}\right]$.*

It can be seen that the borders of this interval given in Theorem 4.7 are obtained for the Kleene-Dienes implication and the Łukasiewicz implication, respectively, i.e. $\int_0^1 \int_0^1 I_{KD}(x,y)dxdy = \frac{2}{3}$ and $\int_0^1 \int_0^1 I_{LK}(x,y)dxdy = \frac{5}{6}$.

Besides Spearman's rho one may consider some other measure of association between random variables defined using copulas. Let us recall the so-called *concordance function* $\Theta = \Theta(C_1, C_2)$ defined for any two copulas C_1 and C_2 as follows (see [13]):

$$\Theta(C_1, C_2) = 4 \int_0^1 \int_0^1 C_2(x,y)dC_1(x,y) - 1. \tag{9}$$

It is worth noting that Spearman's rho can be also defined by the concordance function, i.e. $\rho_C = 3\Theta(C, \Pi)$, where Π is the product copula. Because of the distinguished role of the Fréchet-Hoeffding bounds, the concordance function between given copula C and these bounds is also of interest, i.e. $\Theta(C, M)$ and $\Theta(C, W)$. It can be shown (see [13]) that $0 \leq \Theta(C, M) \leq 1$ and $-1 \leq \Theta(C, W) \leq 0$ for any copula C. Therefore, assuming $C = C_I^{PSI}(x,y) = I(x,y) + x - 1$, simple calculations lead us to the following conclusion

$$\begin{aligned}\Theta(C, M) &= 4 \int_0^1 \int_0^1 C(x,y)dM(x,y) - 1 = 4 \int_0^1 C(x,x)dx - 1 \\ &= 4 \int_0^1 (I(x,x) + x - 1)dx - 1 = 4 \int_0^1 I(x,x)dx - 3.\end{aligned}$$

Thus, finally, if I is a fuzzy implication such that C_I^{PSI} is a PSI-copula, then I satisfies the following inequalities

$$\frac{3}{4} \leq \int_0^1 I(x,x)dx \leq 1. \tag{10}$$

In the same way we can show that

$$\begin{aligned}\Theta(C, W) &= 4 \int_0^1 \int_0^1 C(x,y)dW(x,y) - 1 = 4 \int_0^1 C(x, 1-x)dx - 1 \\ &= 4 \int_0^1 (I(x, 1-x) + x - 1)dx - 1 = 4 \int_0^1 I(x, 1-x)dx - 3\end{aligned}$$

and therefore, we may conclude that if I is a fuzzy implication such that C_I^{PSI} is a PSI-copula, then the following inequalities are satisfied

$$\frac{1}{2} \leq \int_0^1 I(x, 1-x)dx \leq \frac{3}{4}. \tag{11}$$

4.4 Diagonal-Based Conditions

One of the crucial concepts connected with a copula C is its *diagonal section*, i.e. a function $\delta_C : [0,1] \to [0,1]$ defined as $\delta_C(x) = C(x,x)$. Sometimes the so-called *secondary diagonal section* given by $\delta_C^*(x) = C(x, 1-x)$ is also of interest. It seems that the similar functions can be useful in the framework of fuzzy implications. Let us introduce the following two functions.

Definition 4.8. *The* ***main diagonal*** *of a fuzzy implication I is a function $\delta_I : [0,1] \to [0,1]$ defined by*

$$\sigma_I(x) = I(1-x, x). \tag{12}$$

The ***second diagonal*** *of a fuzzy implication I is a function $\sigma_I^* : [0,1] \to [0,1]$ given by*

$$\sigma_I^*(x) = I(x, x). \tag{13}$$

One may easily seen that for any fuzzy implication I its main diagonal is nondecreasing and such that $\sigma_I(0) = 0$ and $\sigma_I(1) = 1$. On the other hand for any second diagonal we have $\sigma_I^*(0) = \sigma_I^*(1) = 1$. We can also say that a fuzzy implication satisfies the *identity principle* if and only if its second diagonal is constant and $\sigma_I^*(x) = 1$ for any $x \in [0,1]$.

We can utilize both above defined diagonals to specify other necessary conditions for a fuzzy implication to generate a copula.

Theorem 4.9. *Let I be a fuzzy implication such that C_I^{PSI} is a PSI-copula. Then its main diagonal is bounded as follows: $x \leq \sigma_I(x) \leq \min\{1, 2x\}$.*

Proof. By Lemma 4.2 each fuzzy implication that generates a copula is bounded from below by the Kleene-Dienes implication and from above by the Łukasiewicz implication. Hence the same relation holds for their main sections. Then we can easily calculate that $\sigma_{I_{KD}}(x) = \max\{1-(1-x), x\} = x$ and $\sigma_{I_{LK}}(x) = \min\{1, 1-(1-x)+x\} = \min\{1, 2x\}$. ■

In a similar way we prove the following theorem.

Theorem 4.10. *If C_I^{PSI} is a PSI-copula derived from a fuzzy implication I then its second diagonal is bounded as follows:* $\max\{1-x, x\} \leq \sigma_I^*(x) \leq 1$.

One can also notice that using diagonals necessary conditions (10) and (11) for a fuzzy implication to generate a PSI-function can be expressed as follows.

Theorem 4.11. *If C_I^{PSI} is a PSI-copula derived from a fuzzy implication I then $\frac{1}{2} \leq \int_0^1 \sigma_I(x)dx \leq \frac{3}{4}$.*

Theorem 4.12. *If C_I^{PSI} is a PSI-copula derived from a fuzzy implication I then $\frac{3}{4} \leq \int_0^1 \sigma_I^*(x)dx \leq 1$.*

One can also ask about fuzzy implications which generate copulas with some specific diagonal sections. The following lemmas can be easily proved.

Lemma 4.13. *If C_I^{PSI} is an Archimedean PSI-copula derived from a fuzzy implication I then $\sigma_I^*(x) \leq 1$.*

Lemma 4.14. *C_I^{PSI} is a PSI-copula such that $\delta_{C_I^{PSI}}(x) = x$ for $x \in [0,1]$ if and only if C_I^{PSI} is derived from the Łukasiewicz implication.*

Lemma 4.15. *C_I^{PSI} is a PSI-copula such that $\delta_{C_I^{PSI}}(x) = \max\{1-x, x\}$ for $x \in [0,1]$ if and only if C_I^{PSI} is derived from the Kleene-Dienes implication.*

5 PSI-quasi-copulas and Other Generalizations

As we know a PSI-function may not be a copula. Thus a natural question arises about the relation between PSI-functions and some generalizations of the copulas like quasi-copulas, etc. Let us start this section by recalling some definitions.

Definition 5.1 ([13]). *A function $Q\colon [0,1]^2 \to [0,1]$ is called a* ***quasi-copula*** *if it satisfies conditions (**C1**)-(**C3**) and*

(C4') *for all $x_1, x_2, y_1, y_2 \in [0,1]$ such that $x_1 \leq x_2, y_1 \leq y_2$ it holds*

$$Q(x_2, y_2) - Q(x_2, y_1) - Q(x_1, y_2) + Q(x_1, y_1) \geq 0,$$

where at least one of $x_1, x_2, y_1, y_2 \in \{0,1\}$.

It is worth noting that condition (**C4'**) is equivalent to requiring that quasi-copulas are nondecreasing in each variable, i.e. for all $x_1, x_2, y_1, y_2 \in [0,1]$ such that $x_1 \leq x_2, y_1 \leq y_2$

$$Q(x_1, y_1) \leq Q(x_2, y_2), \tag{ND}$$

and satisfy the 1-Lipschitz property, i.e. for all $x_1, x_2, y_1, y_2 \in [0,1]$

$$|Q(x_1, y_1) - Q(x_2, y_2)| \leq |x_1 - x_2| + |y_1 - y_2|. \tag{Lip}$$

Another interesting family of functions is given by the semicopulas.

Definition 5.2 ([6]). *A function $B\colon [0,1]^2 \to [0,1]$ is called a* ***semicopula*** *if it satisfies conditions (**C2**)-(**C3**) and* (ND).

By Definition 5.2 we have $0 \leq B(x,0) \leq B(1,0) = 0$ and $0 \leq B(0,y) \leq B(0,1) = 0$ which shows that each semicopula satisfies condition (**C1**).

It is worth noting that the notion of semicopula generalizes some concepts mentioned above. In particular, a semicopula C which is 2-increasing, i.e. satisfying condition (**C4**), is a copula. Moreover, a semicopula Q which satisfies the 1-Lipschitz property (Lip) is a quasi-copula.

It can be shown that a PSI-function automatically is neither a quasi-copula nor a semi-copula. However, similarly as in the case of a PSI-copula we may specify some requirements to be satisfied by a fuzzy implication generating quasi-copulas or a semi-copulas.

Theorem 5.3. *If I is a fuzzy implication such that*

(i) the natural negation of I is $N_I(x) = 1 - x$,
(ii) I satisfies the left-neutrality principle,
(iii) I satisfies the 1-Lipschitz property (Lip),

then the function C_I^{PSI} derived from I is a quasi-copula.

Theorem 5.4. *If I is a fuzzy implication such that*

(i) the natural negation of I is $N_I(x) = 1 - x$,
(ii) I satisfies the left-neutrality principle,
(iii) I satisfies the 1-Lipschitz property with respect to the first argument, i.e. $|I(x_1, y) - I(x_2, y)| \leq |x_1 - x_2|$ for all $x_1, x_2, y \in [0, 1]$,

then the function C_I^{PSI} derived from I is a semi-copula.

Please note, that both copulas and quasi-copulas are special cases of the 1-Lipschitz binary aggregation operators defined as follows.

Definition 5.5. *A (binary)* ***aggregation operator*** *is a function $A\colon [0,1]^2 \to [0,1]$ which is nondecreasing in each component and satisfies $A(0,0) = 0$ and $A(1,1) = 1$. Moreover, an aggregation operator A satisfying the Lipschitz condition with constant 1* (Lip) *is called a 1-Lipschitz aggregation*

The following theorems can be proved.

Theorem 5.6. *If I is a 1-Lipschitz fuzzy implication with respect to the first argument then the function C_I^{PSI} derived from I is a binary aggregation operator.*

Theorem 5.7. *If I is a 1-Lipschitz fuzzy implication then the function C_I^{PSI} derived from I is a 1-Lipschitz binary aggregation operator.*

6 Conclusions

In the paper we have tried to give more light in understanding the nature and geometric properties of those fuzzy implications that may generate copulas. We have discussed this reverse problem with respect to probabilistic S-implications only. First steps in the reverse problem with respect to survival S-implications were done by Massanet et al. [11,12]. We have abandoned this topic because of the limited size of this contribution. Quite a new challenge, not discussed yet, is the reverse problem with respect to usual probabilistic implications and survival implications. Moreover, many interesting topics for further research are connected with fuzzy implications required for deriving copulas with predefined properties (like given sections) or for generating required quasi-copulas or semi-copulas.

References

1. Baczyński, M., Grzegorzewski, P., Mesiar, R.: Fuzzy implications based on semicopulas. In: Alonso, J.M., Bustince, H., Reformat, M. (eds.) Proceedings of the 2015 IFSA and EUSFLAT Conference, pp. 792–798. Atlantis Press (2015)
2. Baczyński, M., Grzegorzewski, P., Helbin, P., Niemyska, W.: Properties of the probabilistic implications and S-implications. Inf. Sci. **331**, 2–14 (2016)
3. Baczyński, M., Grzegorzewski, P., Niemyska, W.: Laws of contraposition and law of importation for probabilistic implications and probabilistic s-implications. In: Laurent, A., Strauss, O., Bouchon-Meunier, B., Yager, R.R. (eds.) IPMU 2014, Part I. CCIS, vol. 442, pp. 158–167. Springer, Heidelberg (2014)
4. Baczyński, M., Jayaram, B.: Fuzzy Implications. Springer, Heidelberg (2008)
5. Dolati, A., Sánchez, J.F., Úbeda-Flores, M.: A copula-based family of fuzzy implication operators. Fuzzy Sets Syst. **211**, 55–61 (2013)
6. Durante, F., Sempi, C.: Semicopulae. Kybernetika **41**, 315–328 (2005)
7. Grzegorzewski, P.: Probabilistic implications. In: Proceedings of the EUSFLAT-2011 and LFA-2011 Conference, pp. 254–258. Atlantis Press (2011)
8. Grzegorzewski, P.: On the properties of probabilistic implications. In: Melo-Pinto, P., Couto, P., Serôdio, C., Fodor, J., De Baets, B. (eds.) Eurofuse 2011. AISC, vol. 107, pp. 67–78. Springer, Heidelberg (2011)
9. Grzegorzewski, P.: Survival implications. In: Greco, S., Bouchon-Meunier, B., Coletti, G., Fedrizzi, M., Matarazzo, B., Yager, R.R. (eds.) IPMU 2012, Part II. CCIS, vol. 298, pp. 335–344. Springer, Heidelberg (2012)
10. Grzegorzewski, P.: Probabilistic implications. Fuzzy Sets Syst. **226**, 53–66 (2013)
11. Massanet, S., Ruiz-Aguilera, D., Torrens, J.: Defining copulas from fuzzy implication functions. In: Baczynski, M., De Baets, B., Mesiar, R. (eds.) Proceedings of the AGOP 2015 Conference, pp. 181–186 (2015)
12. Massanet, S., Ruiz-Aguilera, D., Torrens, J.: On two construction methods of copulas from fuzzy implication functions. Prog. Artif. Intell. **5**, 1–14 (2016)
13. Nelsen, R.B.: An Introduction to Copulas, 2nd edn. Springer, Heidelberg (2006)

Applications in Medicine and Bioinformatics

Non-commutative Quantales for Many-Valuedness in Applications

Patrik Eklund[1(✉)], Ulrich Höhle[2], and Jari Kortelainen[3]

[1] Department of Computing Science, Umeå University, Umeå, Sweden
peklund@cs.umu.se
[2] FB C Bergische Universität Wuppertal, Wuppertal, Germany
uhoehle@uni-wuppertal.de
[3] Department of Electrical Engineering and Information Technology, Mikkeli University of Applied Sciences, Mikkeli, Finland
jari.kortelainen@mamk.fi

Abstract. In this paper we show how the diversity of properties for quantales is well suited for describing multivalence in many-valued logic. Tensor products of quantales will play an important role in showing how more simple valuation scales can be tensored together to provide more complex valuation scales. In health care applications, this is typically seen for disorders and functioning. Classification of disorder is typically quite bivalent, whereas scales used in functioning classifications are multivalent. The role 'not specified' or 'missing' is shown to be of importance.

Keywords: Assessment · Logic · Quantale

1 Introduction

The diversity of properties for quantales make them well suited for describing multivalence in many-valued logic, and as involving different carriers of uncertain information. This information is frequently subjected to various algebraic operations, which raises expectations of the logical machinery to deliver desired properties with a proper logical and mathematical foundation, and in particular to meet the requirement of richness needed in real-world applications. Non-commutativity of operation is a typically important consideration from application point of view. The notion of logic as a structure embraces signatures and constructed terms and sentences *latively* constructed as based on these terms. Similarly, sentence and conglomerates of sentences are fundamental for entailments, models and satisfactions, in turn latively to become part of axioms, theories and proof calculi. This lativity is always produced and maintained by functors and monads, and as acting over underlying categories in form of monoidal categories. Category theory is thus a suitable metalanguage for logic, in particular when applications and typing of information must be considered. Uncertainty may reside in generalized powerset functors, and may be internalized in underlying categories. In both cases, suitable algebras must motor this

J.P. Carvalho et al. (Eds.): IPMU 2016, Part I, CCIS 610, pp. 437–449, 2016.
DOI: 10.1007/978-3-319-40596-4_37

uncertainty representation, and quantales are very suitable in this context. In Sects. 2 and 3, we provide background and prerequisite for quantales and tensors [11–14]. In Sect. 4 we present an example from health and social care, and how multivalent classification is supported by tensored quantales.

2 Quantales

Let `Preord` be the category of preordered sets and `Sup` be the category of complete lattices and join preserving maps. A pair $(X, *)$ is a *prequantale* if X is a complete lattice and $X \times X \xrightarrow{*} X$ is a bimorphism of `Sup`. A prequantale $(X, *)$ is *unital* iff there exists an element $e \in X$ with $x * e = x = e * x$ for all $x \in X$, and $(X, *)$ is a *quantale* if $*$ is associative. A quantale $(X, *)$ is *semi-integral* if the relation $x_1 * \top * x_2 \leq x_1 * x_2$ holds for all $x_1, x_2 \in X$, and it is *semiunital* if the relations $x \leq x * \top$ and $x \leq \top * x$ hold for all $x \in X$. An element $x \in X$ is called *idempotent* if $x * x = x$, and the quantale is *idempotent* if every element of X is idempotent. An element $x \in X$ is called *left-sided* (resp. *right-sided*) if $\top * x \leq x$ (resp. $x * \top \leq x$). An element $x \in X$ is *two-sided* if it is left-sided and right-sided. The quantale is *left-sided* (resp. *right-sided*) if every element of X is left-sided (resp. right-sided), and *two-sided* if every element is two-sided.

An element $p \in X$ is called *prime* in $(X, *)$ if $p \neq \top$ and the implication $x * y \leq p \implies x * \top \leq p$ or $\top * y \leq p$ holds for all $x, y \in X$.

In the following we introduce our main quantale example, which will be used in our application description. Let C_3 be the chain consisting of three elements $\bot$, a and $\top$ where $\bot$ and $\top$ are the universal bounds. On C_3 we consider two non-commutative and associative multiplications:

$$
\begin{array}{lllll}
\top *_\ell \top = \top, & a *_\ell a = a, & \bot *_\ell \bot = \bot, & \top *_\ell a = a, & a *_\ell \top = \top \\
\top *_\ell \bot = \bot, & a *_\ell \bot = \bot, & \bot *_\ell a = \bot, & \bot *_\ell \top = \bot, & \\
\top *_r \top = \top, & a *_r a = a, & \bot *_r \bot = \bot, & \top *_r a = \top, & a *_r \top = a, \\
\top *_r \bot = \bot, & a *_r \bot = \bot, & \bot *_r a = \bot, & \bot *_r \top = \bot, &
\end{array}
$$

Then $C_3^\ell = (C_3, *_\ell)$ is a left-sided and idempotent quantale, while $C_3^r = (C_3, *_r)$ is right-sided and idempotent. The prime elements of C_3^ℓ and C_3^r are $\bot$ and a.

From application point of view, we will interpret $x_1 *_r x_2$ as x_1 "juncted with" x_2, or x_1 "juncted with, as followed by," x_2. We will further discuss the view that a corresponds to a valuation 'not (yet) specified'. Since $a *_r \top = a$, we have the interpretation

$$\text{'not (yet) specified'} *_r \text{"given and true"} = \text{'not (yet) specified'}.$$

Interpretations of the multiplication $*_\ell$ and $*_r$ on C_3 shows e.g. that $a *_r \top$ is 'not (yet) specified' and then 'true' = 'not (yet) specified'. Further $\top *_r a$is 'true' and then 'not (yet) specified'= 'true', $\top *_\ell a$ is 'true' and then 'not (yet) specified'= 'not (yet) specified', and $a *_\ell \top$ is 'not (yet) specified' and then 'true'= 'true'. In this sense we have to do with two different types of 'not (yet) specified', one being right-sided and the other one being left-sided. Thus our interpretation refers directly to the interpretation of 'right-sidedness' and 'left-sidedness'. Boole [1] never distinguished this non-commutativity and sidedness, so in his formulations, left-sidedness would be the same as right-sidedness, i.e., two-sidedness which leads to two values only, namely $\top$ as 'all beings' and $\bot$ as 'no beings'.

3 Tensors

If $\mathsf{C} = (\mathsf{C}_0, \otimes, \mathbb{1}, a, \ell, r)$ is a monoidal category, then the bifunctor $\otimes$ is called the *tensor product* of C and $\mathbb{1}$ the *unit object* of C. Tensor products exist for a wide variety of algebraic structures. For complete lattices X and Y, the pair (B, β) is called a *tensor product* of X and Y, if B is a complete lattice and $X \times Y \xrightarrow{\beta} B$ is a bimorphism such that, for every bimorphism $X \times Y \xrightarrow{b} Z$ there exists a unique join preserving map $B \xrightarrow{h_b} Z$ with $b = h_b \circ \beta$.

Tensors can be constructed using the set $\mathcal{G}(X,Y)$ of all join reversing maps $X \xrightarrow{f} Y$ provided with the following partial order:

$$f_1 \le f_2 \iff f_1(x) \le f_2(x) \quad \text{for all } x \in X.$$

Obviously, $(\mathcal{G}(X,Y), \le)$ is a complete lattice in which meets (but in general not joins) are computed pointwise. In particular, the universal upper bound $\underline{\top}$ and the universal lower bound $\underline{\bot}$ in $\mathcal{G}(X,Y)$ have the following form:

$$\underline{\top}(x) = \top \quad \text{and} \quad \underline{\bot}(x) = \begin{cases} \top, & x = \bot, \\ \bot, & x \neq \bot, \end{cases} \qquad x \in X.$$

Now we introduce a map $X \times Y \xrightarrow{\beta} \mathcal{G}(X,Y)$ as follows:

$$[\beta(x,y)](x') = \begin{cases} \top, & \text{if } x' = \bot, \\ y, & \text{if } \bot \neq x' \le x, \\ \bot, & \text{if } x' \not\le x, \end{cases} \qquad x, x' \in X,\, y \in Y.$$

Firstly, $\beta(x,y)$ is an element of $\mathcal{G}(X,Y)$, and the relation $\beta(x,\bot) = \beta(\bot,y) = \underline{\bot}$ holds for all $x \in X$ and $y \in Y$, and, secondly, β is a bimorphism that fulfills the following important properties

$$\beta(\top,\top) = \underline{\top} \tag{1}$$

$$\bigwedge_{i\in I} \beta(x_i, y_i) = \beta(\bigwedge_{i\in I} x_i, \bigwedge_{i\in I} y_i) \tag{2}$$

$$f = \bigvee_{x\in X} \beta(x, f(x)) \quad \text{for any } f \in \mathcal{G}(X,Y). \tag{3}$$

The properties (1) and (2) mean that β is *meet preserving*, while the property (3) expresses the fact that $\{\beta(x,y) \mid x \in X, y \in Y\}$ forms a *join basis* of $\mathcal{G}(X,Y)$.

Based on the example in the previous section, we compute the tensor product $C_3^{\ell} \otimes C_3^{r} = (C_3 \otimes C_3, \star)$ and observe that $C_3 \otimes C_3$ consists of six elements

$$\top = \top \otimes \top, \quad \alpha = (a \otimes \top) \vee (\top \otimes a), \quad \lambda = a \otimes \top, \quad \varrho = \top \otimes a, \quad \beta = a \otimes a,$$
$$\bot = \bot \otimes \bot = \bot \otimes a = a \otimes \bot.$$

Note also that $\bot \otimes \top = \top \otimes \bot = \bot$.

The order-theoretic structure of $C_3 \otimes C_3$ can be visualized by the following Hasse diagram:

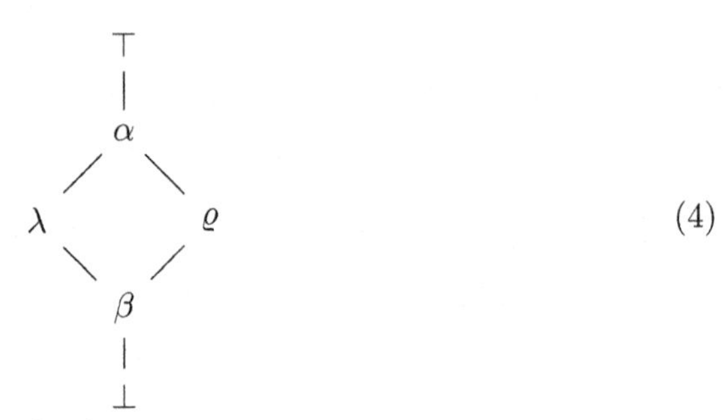

and the multiplication table has the form

$x\backslash y$	$\bot$	β	λ	ϱ	α	$\top$
$\bot$	$\bot$	$\bot$	$\bot$	$\bot$	$\bot$	$\bot$
β	$\bot$	β	β	ϱ	ϱ	ϱ
λ	$\bot$	λ	λ	$\top$	$\top$	$\top$
ϱ	$\bot$	β	β	ϱ	ϱ	ϱ
α	$\bot$	λ	λ	$\top$	$\top$	$\top$
$\top$	$\bot$	λ	λ	$\top$	$\top$	$\top$

The unique non-elementary tensor α is the unique non-idempotent element in $(C_3 \otimes C_3, \star)$. Further, the elementary tensor $\beta = a \otimes a$ is neither left-sided nor right-sided, while λ is left-sided, but not right-sided and ϱ is right-sided, but not left-sided. Moreover, the tensor product $(C_3 \otimes C_3, \star)$ is semi-integral.

The prime elements of $(C_3 \otimes C_3, \star)$ are α, λ, ϱ and $\bot$.

Further, note that $C_3 \otimes C_3$ is the (complete) lattice of all join reversing self-maps of C_3, and is therefore order isomorphic to the lattice of all join preserving self-maps C_3. The identity id_{C_3} of C_3 corresponds to the tensor (= join reversing map) $(a \otimes \top) \vee (\top \otimes a)$, where a is the element of C_3 being strictly between the $\top$ and $\bot$ elements of C_3.

This order isomorphism reveals that behind $\star$ there is hidden the composition. Hence the interpretation of $\star$ as "and then" makes sense.

As a further clarification of the algebraic structure of $C_3 \otimes C_3$, we should also recall the following. Firstly, $\star$ is the tensor product of the multiplication of the left-sided, idempotent, non-commutative chain C_3 with the multiplication of the right-sided, idempotent, non-commutative chain C_3. This is a general mathematical construction and happens in various places of algebra. Secondly, let L be a complete lattice provided with an order reversing involution. Then the unital quantale $Q(L)$ of all join preserving maps with the multiplication given by the composition of maps goes back to Mulvey, Pelletier and Rosický ([16,17]). An element $f \in Q(L)$ is left-sided iff there exists $a \in L$ such that f has the following form:

$$f(x) = \begin{cases} \top, x \not\leq a \\ \bot, x \leq a, \end{cases}, \qquad x \in L.$$

An element $g \in Q(L)$ is right-sided iff there exists $a \in L$ such that g has the following form:

$$g(x) = \begin{cases} a, x \neq \bot \\ \bot, x = \bot, \end{cases}, \qquad x \in L.$$

Further, the semi-integral regularization of $(Q(L), \circ)$ leads to a multiplication $*$ on $Q(L)$ given by $g * f = g \circ \top \circ f$. This quantale is no longer unital (the semi-integral regularization destroys the unit). The semi-integral regularization of a unital quantale means that the product, i.e., result of the multiplication, is determined by left-sided and right-sided elements.

In the special case $L = C_3$ the semi-integral regularization of $Q(C_3)$ is isomorphic to the tensor product $C_3^{\ell} \otimes C_3^{r}$. In particular this means that the multiplication in $C_3 \otimes C_3$ has the form $f \star g = f \circ \top \circ g$ in $Q(C_3)$. Here $\top$ is then the universal upper bound in of $Q(C_3)$, $\star$ is the tensor product of the respective multiplications in the left-sided, idempotent C_3 with the multiplication in the right-sided, idempotent C_3, and, $\circ$ is the composition of maps. So when we interpret $\star$ as "and then", we obviously hide the composition of maps.

From application point of view, the question is where in $(C_3 \otimes C_3, \star)$ we have the element for 'not (yet) specified'. Viewing $\otimes$ similarly "juncted with, as followed by,", we have $a \otimes \top = \lambda$, so in this interpretation, λ is the candidate for 'not (yet) specified' in $(C_3 \otimes C_3, \star)$. In applications the intuition of $\star$ is more application specific, but derives its meaning from the theoretical observations. Thus we are able to remain precise in our formal treatment, and at the same time enable the transition from the theory to practice. In general the approach is indeed the other way around. We start from applications and try to identify impact of the underlying theory. Often the application side needs to pay attention to the outcome of this theoretical change.

Once we use the tensor product $C_3^{\ell} \otimes C_3^{r}$, we indeed have the possibility to treat the concept 'right-sidedness' and 'left-sidedness' at the same time. The point of departure is the unique idempotent element $\beta = a \otimes a$ with is neither left-sided nor right-sided. Then we construct the unique left-sided element λ, being strictly between top and bottom, and the unique right-sided element ϱ, being strictly between top and bottom, as $\lambda = \top \star \beta (= a \otimes \top)$ and $\varrho = \beta * \top (= \top \otimes a)$. In this algebraic context, without still too much anticipating what may happen in a particular application context, we may interpret β as 'not know'. Then 'true' and then 'not know' = 'not (yet) specified' in the left-sided sense, and 'not know' and then 'true' = 'not (yet) specified' in the right-sided sense. We therefore have two sides of 'not knowing', so that we can polarize 'not knowing' to become attributed both within and outside the scale of factual truth values.

Known but not know for fact is different from not yet known. This distinction is subtle in particular when interpreting it with an application context.

Note finally that $(\{\bot, \lambda, \top\}, \star)$ is isomorphic to C_3^ℓ, and $(\{\bot, \varrho, \top\}, \star)$ is isomorphic to C_3^r. This observation underlines the fact that $C_3^\ell \otimes C_3^r$ is the smallest, non-idempotent quantale which covers the three-valued phenomenon of 'left-sidedness' and 'right-sidedness'.

4 Applications

In this section we suggest how juncted qualification of human functioning can be viewed as based on underlying tensored qualification of respective qualifications for capacity and mood. This makes sense in a multivalent setting, and, as we shall see, even based on the three-valued C_3. A key feature in C_3 is the availability of both the left-sided and the right-sided binary operation. This enables to have non-commutativity, when desired. In a bivalent situation, such an effort would make no sense, since non-trivial non-commutative junction does not exist if we always expect $\top \star \top = \top$ and $\bot \star \bot = \bot$. Here the only two remaining cases for non-commutativity are either $x_1 \star x_2 = x_1$ or $x_1 \star x_2 = x_2$. In both these cases we have situations where one of the evaluations totally ignore the other.

In the three-valued situation of C_3, the element a is the most interesting one as far as interpretations are concerned. Obviously, there are a number of candidates for the interpretation $\top$ and $\bot$, but given that we expect $\top \star \top = \top$ and $\bot \star \bot = \bot$, there is only little room for interpretation. We may view $\top$ and $\bot$ e.g. as 'good' and 'bad', 'right' and 'wrong', or (logically) as 'true' and 'false'.

In the case of 'true' and 'false', we typically say "what is known [about something] is true — what is known [about something] is false". In a propositional logic situation, that 'something' is never described, but indeed remains modelled by propositional constants. In the case of 'good' and 'bad', we similarly have "[this] is good — [this] is bad", and we would need at least signatures and terms [8,9] in order properly to model what we precisely mean by 'this'. However, in the case of 'good' and 'bad', there is an epistemic-dialogic aspect often overlooked in purely logical treatments. "Knowing what is good" and "knowing when doing good" is to be distinguished not just within the actions of one individual, but in particular when considering individuals "in dialogue", and how information is *processed in dialogue*. Dialogic or dialogism was also treated by Mikhail Bakhtin (1895–1975) in his literary theory. Bakhtin goes further than Buber, as Bakhtin includes the Other "for Me" aspect, i.e., kind of as an opposite arrow for the corresponding [dia]logic morphism representing Bakhtin's "I for the Other". Heteroglossia has unfortunately not been subjected to formal logic considerations, but doing might open up interesting avenues for logical investigations involving *logic in dialogue*. Even in Shakespeares theory of drama [2] it is unclear if Shakespeare actually was concerned with dramatic theory and even logic. Clearly, Shakespeare is understood as a poet, but he was obviously also much concerned with theatrical considerations. His literary part is then more of "information, knowledge and logic", where the theatrical part is more about "process and logic" [7].

The distinction between information and process is indeed important, and a distinction must also be made between "information in [a] process" and "information as processed". One could also say that "information processing" and "processed information" is not the same thing but rather two sides of the same "information and process coin". Here 'right' and 'wrong' seems more of a process-oriented issue as we tend to distinguish between "doing [things] right" and "doing [things] wrong".

Developing logical notions in these respects requires predicativity, and much more logical structure, so for the purpose of this paper we restrict to focus on the algebraic aspect of valuations. Now leaving $\top$ and $\bot$ mostly out of the discussion, let us turn our focus on a in C_3. Again there are a number of intuitive interpretation, but here we will look at it as an 'not specified' or 'missing'. In numerics we compute $2 + 2 = 4$ and if we must extend our computations with *missing* it will mostly be like $2 + missing = missing$. With *missing* as the a, we can logically compute with it. As 'missing' is a more technical matter, 'unspecificed' is broader and e.g. a 'not (yet) given', whether as a 'cannot say' or a '(still) not known'. Obviously, this a can also be viewed as the grade between $\top$ and $\bot$, in a linear view of C_3 as a set of truth values.

Non-commutativity now comes into play, and we illuminate this situation with capacity and mood being described by dementia and depression. Dementia is about cognitive decline and depression is about gradation of mood.

Discussions on dementia versus depression is known to lead to the question about which one coming first [18,19], and their is a vast amount of literature generally about the topic. Cognitive impairment may be seen as a component of late-life depression, so depression is in some sense seen as coming before dementia or accelerating the decline of cognitive capacity. This view says that depression increases the risk of dementia in an individual otherwise cognitively capable. On the other hand, depressive symptoms may be seen as commonly appearing because of neurodegeneration, so dementia is a cause, in some sense, of depression. The question of which comes first or which one should primarily be in focus in diagnostics or prevention.

There are several studies specifically about these two options. On the one hand, cognitive decline "juncted with" depression, or cognitive decline "juncted with, as followed by," depression [15], can be seen as valuated by $dementia *_r depression$. On the other hand, depression "juncted with, as followed by," cognitive decline [3], i.e., depression as a risk factor for dementia, means treating or preventing depression supports prevention of dementia. The valuation in this case is closer to $dementia *_\ell depression$.

In comparison, a similar but different duality and non-commutatitivity appears with falls and depression. Here its less intertwined as depression leading to higher fall risk is different from depression appearing months after a fall injury. In the latter case, depression is due to functional decline after a fall injury, rather that due to the fall injury itself. Other situations can be found, but in the subsequent steps we focus on $dementia * depression$.

To start with, let us look at dementia "juncted with" depression, and suppose *Dementia* $= \top$ and *depression* $= a$. Then *dementia* $*_r$ *depression* $= \top$, so an 'not specified' depression would not weaken *dementia* $*$ *depression* to 'not specified'. This amounts to saying that truth of *dementia* would preserve the truth of the statement *dementia "juncted with, as followed by," (still) not known depression.* On the other hand, we will have *depression* $*_r$ *dementia* $= a$, so not knowing *depression* would mean not knowing *depression "juncted with, as followed by," (still) not known dementia.* Since $x_1 *_\ell x_2 = x_2 *_r x_1$, the left-sided and right-sided are in this case just dual situations. However, non-commutativity is here a key factor, and this non-commutativity must not be confused with causality e.g. as appearing in computations related to conditional probabilities.

At this point it is appropriate to point out a fundamental difference between, on the one hand, sampling and hypothesis testing in statistics, with "how many" and related mean values as ingredients in statistical evidence. Logical evidence states "how" and is related to truth values. Bridging the gap between "how many" and "how" is the first step in a successful bridging of the gap between analytics (statistics) and guidelines (logic).

Dementia and depression are within the realm of disorders, as classified by WHO's (World Health Organization) ICD (International Classification of Diseases). Even if disorders in many meanings can have magnitude and severity, a diagnosis code is nevertheless bivalent in the sense of representing a disease being diagnosed or not, i.e., no gradation is enabled within ICD. WHO's ICF (International Classification of Functioning, Disability and Health), on the other hand, comes with a generic 5-scale for valuation of the severity of specific functioning. The so called *ICF Core Sets* are potentially many-valued subsets of ICF codes, even if the existing Core Sets do not explicitly recognize that many-valuedness in a more strict logical sense.

ICF as a classification for Functioning, Disability and Health distinguishes between *health domains* and *health-related domains*, where the domains are described in two basic lists, respectively, for *Body Functions and Structures*, and *Activities and Participation.* Functioning is seen as an umbrella term *encompassing all body functions, activities and participation*, whereas Disability is an umbrella term for *impairments, activity limitations or participation restrictions.* In WHOs international classifications, health conditions (diseases, disorders, injuries, etc.) are classified primarily in ICD-10, which provides the etiological framework as complementary to ICF. Functioning and disability associated with health conditions are indeed classified in ICF.

ICF has moved away from being a "consequences of disease" classification (1980 version) to become a "components of health" classification. Components of health are then seen as constituents of health, whereas "consequences" focuses on the impacts of diseases or other health conditions that may follow as a result. Disorder and functioning seen as intertwined does obviously not exclude causality and consequence. On the contrary, consequence becomes even a more complicated matter.

Components of functioning and disability are interpreted by means *constructs*, and these constructs are operationalized by using *qualifiers*. For the Activities and Participation component, two constructs are available, nemaly, *capacity* and *performance*. Consequences within or between constructs are never considered, and ICF indeed describes interactions between its components only informally, and on a very general level.

ICF uses the same generic scale, respectively, for Body Functions and Structures, Activities, and Participation, and as related to health conditions as disorder or disease (ICD).

All these components are quantified using the same generic scale. The scale is not logically explained, but leans on statistics. A typical statement in ICF is that "for this quantification to be used in a universal manner, assessment procedures need to be developed through research". One of the qualifiers, `MODERATE problem`, is defined as up to half of the time or half the scale of total difficulty. Here "scale" is intentionally left as undefined, but relates to percentage scales and how they appear in statistics.

There is indeed no logical explanations, and no indications whatsoever on how to junct qualifiers in an algebraic and logical setting. Qualifiers like `MODERATE problem` are more seen as quantifiers by percentages as to be calibrated in different domains with reference to population standards as percentiles. They are never explained or seen as logical qualifiers. Therefore, the no specific algebra for the generic is ever defined, nor is it explained or assumed how algebraic properties might be preserved e.g. when transforming from Body Function to Activity.

The ICF datatypes and its generic scale of quantifiers correspond to elements in $C_3^\ell \otimes C_3^r$ as follows:

`xxx.0 NO problem`	$\top$	*full capacity*
`xxx.1 MILD problem`	α	*sufficient capacity*
`xxx.2 MODERATE problem`	ϱ	*capacity in transition*
`xxx.3 SEVERE problem`	β	*capacity almost lost*
`xxx.4 COMPLETE problem`	$\bot$	*no capacity*
`xxx.8 not specified`	λ	*capacity not (yet) known*

In ICF's generic scale, there is also a `xxx.9` for 'not applicable', but this should not be seen as part of the valuation, but is rather a lack of typing.

In the table above we have provided our example focus on capacity, and in our brief algebraic view we focus e.g. on *in transition*. Clearly, several annotations and contexts are possible, and we do claim that there is a variety of possible annotations and contexts, all requiring specific properties as far as algebraic structures are concerned.

Before going deeper into how ICF's generic scale can be explained algebraically in a variety of ways using quantales and tensors of quantales, it is first important to note how $\otimes$ comes from a monoidal closed category, so objects are tensored, i.e., C_3 is to be viewed as a category. This is clearly quite challenging to connect to medical knowledge and preparations for sampling as basis for a clinical trial. Informally treated information in trials and studies of various kind is identified and used by medical professionals within the realm of evidence-based medicine, where statistics is the only method of computation, and logic is usually comprehended in quite rudimentary forms.

As we shall see, there are many candidates for representing ICF's generic scale as a quantale. An interpretation of ICF's constructs in relation to diseases (ICD) could suggest viewing ICF's generic scale as a quantale in form of

$$ICF_d = ICD^\ell \otimes ICD^r$$

reflecting the situation that a valuation of a multi-morbidity medical condition-condition interaction of ICD codes corresponds to the way valuation of functioning is done with respect to ICF codes. Similar relations and structure can be provided as involving several other classifications [6].

Specific chains of values in $C_3^\ell \otimes C_3^r$ now come into play, and reflect valuations appearing in care pathways. We use the ICF core set for rheumatoid arthritis (RA) to illuminate this situation. RA is an autoimmune, chronic inflammatory, disorder that affects the joints, and shows a relatively high prevalence of comorbidities [4], with depression as the most commonly observed comorbidity. Typical functioning affected included the following ICF codes:

```
b280 Sensation of pain
b455 Exercise tolerance functions
b710 Mobility of joint functions
```

Suppose we now have an overall functioning valuation for RA at $\bot$, and at some point in time and in a certain position in the care pathway. A complete problem ($\bot$) related to depression has been treated to become a severe problem (β). The valuation for RA then jumps to β. The sensation of pain, as caused by RA, has been brought to level α, so $\beta \star \alpha$ produces a new RA level at ϱ. Severity β of pulmonary disease may again bring the RA level back to $\varrho \star \beta = \beta$.

Concerning ICF's generic scale, we have the situation 'full capacity' and then 'capacity almost lost' = 'capacity not (yet) known', and the situation 'capacity almost lost' and then 'full capacity' = 'capacity in transition'. It therefore seems that the binary operation 'and then' gives an orientation! Moreover, we have 'full capacity' and then 'capacity not (yet) known' ='capacity not (yet) known', 'capacity not (yet) known' and then 'full capacity' = 'full capacity', 'capacity in transition' and 'full capacity' = 'capacity in transition', and 'full capacity' and then 'capacity in transition' = 'full capacity'. In this way we never reach $\alpha = (a \otimes \top) \vee (\top a \otimes a)$= 'sufficient capacity'. It is also remarkable to observe that

'sufficient capacity' and then 'sufficient capacity' = 'full capacity',

because α is the unique non-idempotent element, a phenomenon which is also beyond Boole's [1] idempotent view of conjunction.

Obviously, this is a quite general example, and without detail concerning mechanisms involved in this multimorbitiy and multifunctioning view.

The example nevertheless serves as a demonstration about how conditions of various kind interact in a non-commutative manner. Clearly, C_3 and $C_3^\ell \otimes C_3^r$ are not the only options even in the case of ICF_d, nor do we claim that there are canonic all-purpose scales. However, ICF has adopted its 5-scale with the 'not specified' not clearly related to elements in the 5-scale. Note also in the tensor how $\top$ dominates the quantale, in some sense reflecting the bivalence of disorder qualification. Looking at the algebra of the generic scale e.g. for the capacity construct, ICF_c, the Hasse diagram of the scale remains, but the quantale must be different and must be identified without such domination of the $\top$ qualifier.

5 Conclusions and Future Work

The use of quantales enables a non-commutative setting. Our medical example illuminates a situation where non-commutativity must be considered, and in health care there are many similar situations where multivalence invites to considering non-commutativity. Quantales as such or as tensored from other quantales clearly provide application oriented uncertainty modelling in particular in situations where classifications and nomenclatures play an important role. The role of 'not specified' in various forms is important. In health and social care this is clearly relevant for WHO's classifications, and in particular concerning their interrelations. This is expected to have bearing on considerations for SNOMED and HL7 as well.

For appropriate typing of items and objects within applications, valuations cannot simply be of the form $v : X \to C_3^\ell \otimes C_3^r$, where X would be seen as unstructured and based on no terminologies or classifications. From application point of view we do not rule out the possibility that we may have *sidedness per type*. This also raises the question whether or not one should aim at finding a universally valid quantale for all types within one application context, or is it more reasonable to think in terms of having different quantales per type. In a junction $x_1 \otimes x_2$ we need to understand this beyond just the algebraic machinery in particular when x_1 and x_2 are terms of different type [9].

From enrichment of logic language point of view, it is important to note how quantales can be arranged to appear in underlying Goguen categories [8], and when terms over a signature [9] are constructed over such Goguen categories, we enable multivalent annotations for expressions as terms, which then in turn enable multivalence annotations in sentences like, e.g., clauses in logic programming [10].

Several real-world application domains can make use of these structures and techniques. Logic and many-valuedness as proposed in this paper will be considered also e.g. when showing how to enrich the language used in the manufacturing industry regarding information structure and its representation for products and production processes. This underlines the importance of introducing a structure for functioning classification in order to complement and interact with the traditional view of faults and failures in product and production subsystems.

The potential use of a many-valued logic enriched classification of functioning in machines and manufacturing (MCFu) should then be related to enriched classification of faults (MCFa). As compared to the ICF tensor, for the manufacturing industry and design structures in engineering we have a similar

$$MCFu = MCFa^{\ell} \otimes MCFa^{r},$$

which is expected to be important in ontology considerations for various systems-of-systems. This clearly reflects the situation that a valuation of a multiple fault system-of-systems fault-fault interaction of MCFa encoding corresponds to the way valuation of MCFu based functioning is done with respect to MCFu encoding. More generally, encoding in this manner will need to be integrated in modelling standards like UML, SysML and BPMN [5].

References

1. Boole, G.: The Laws of Thought. Walton and Maberly, London (1853)
2. Cunningham, J.V.: Logic and lyric. Mod. Philol. **51**, 33–41 (1953)
3. Diniz, M.A., Butters, S.M., Albert, M.A.D., Dew, M.A., Reynolds, C.F.: Late-life depression and risk of vascular dementia and Alzheimers disease: systematic review and meta-analysis of community-based cohort studies. Brit. J. Psychiatry **202**, 329–335 (2013)
4. Dougados, M., et al.: Prevalence of comorbidities in rheumatoid arthritis, evaluation of their monitoring: results of an international, cross-sectional study (COMORA). Ann. Rheum. Dis. **73**, 62 (2013)
5. Eklund, P.: Lative Logic Modelling Languages and Notation, in preparation
6. Eklund, P.: Lative logic accomodating the WHO family of international classifications. In: Cruz-Cunha, M.M., Miranda, I. (eds.) Encyclopedia of E-Health and Telemedicine. IGI Global (in print)
7. Eklund, H.-T., Eklund, P.: Logic and process drama. In: Thorkelsdóttir, R.B., Ragnardóttir Fliss, Á.H., Háskólaprent, Reykjavik, (eds.) Drama Boreale, Earth-Air-Water-Fire, pp. 103–115 (2013)
8. Eklund, P., Galán, M.A., Helgesson, R., Kortelainen, J.: Fuzzy terms. Fuzzy Sets Syst. **256**, 211–235 (2014)
9. Eklund, P., Höhle, U., Kortelainen, J.: A survey on the categorical term construction with applications. Fuzzy Sets Syst. (2015)
10. Eklund, P., Galán, M.Á., Helgesson, R., Kortelainen, J., Moreno, G., Vázquez, C.: Towards categorical fuzzy logic programming. In: Masulli, F. (ed.) WILF 2013. LNCS, vol. 8256, pp. 109–121. Springer, Heidelberg (2013)
11. Gutiérrez García, J., Höhle, U., Kubiak, T.: Tensor products of complete lattices and their application in constructing quantales, preprint
12. Höhle, U.: Commutative, residuated ℓ-monoids. In: Höhle, U., Klement, E.P. (eds.) Non-classical Logics and their Applications to Fuzzy Subsets, pp. 53–106. Kluwer Academic Publishers, Dordrecht (1995)
13. Höhle, U.: Topological representation of right-sided and idempotent quantales. Semigroup Forum **90**, 648–659 (2015)
14. Höhle, U.: Prime elements of non-integral quantales and their applications. Order **32**, 329–334 (2015)
15. Lyketsos, C.G., Steele, C., Baker, L., Galik, E., Kopunek, S., Steinberg, M., Warren, A.: Major and minor depression in Alzheimer's disease: prevalence and impact. J. Neuropsychiatry Clin. Neurosci. **9**, 556–561 (1997)

16. Mulvey, C.J., Pelletier, J.W.: On the quantisation of the calculus of relations. In: CMS Proceedings, vol. 13, pp. 345–360. American Mathematical Society, Providence RI (1992)
17. Pelletier, J.W., Rosický, J.: Simple involutive quantales. J. Algebra **195**, 367–386 (1997)
18. Thorpe, L.: Depression vs. Dementia: how do we assess? The Canadian Review of Alzheimers Disease and Other Dementias, pp. 17–21, September 2009
19. Wilson, R.S., Capuano, A.W., Boyle, P.A., Hoganson, G.M., Hizel, L.P., Shah, R.C., Nag, S., Schneider, J.A., Arnold, S.E., Bennett, D.A.: Clinical-pathologic study of depressive symptoms and cognitive decline in old age. Neurology **83**, 702–709 (2014)

Evaluating Tests in Medical Diagnosis: Combining Machine Learning with Game-Theoretical Concepts

Karlson Pfannschmidt[1(✉)], Eyke Hüllermeier[1], Susanne Held[2], and Reto Neiger[2]

[1] Department of Computer Science, Paderborn University, Paderborn, Germany
kiudee@mail.upb.de

[2] Small Animal Clinic, Justus-Liebig University Gießen, Giessen, Germany

Abstract. In medical diagnosis, information about the health state of a patient can often be obtained through different tests, which may perhaps be combined into an overall decision rule. Practically, this leads to several important questions. For example, which test or which subset of tests should be selected, taking into account the effectiveness of individual tests, synergies and redundancies between them, as well as their cost. How to produce an optimal decision rule on the basis of the data given, which typically consists of test results for patients with or without confirmed health condition. To address questions of this kind, we develop an approach that combines (semi-supervised) machine learning methodology with concepts from (cooperative) game theory. Roughly speaking, while the former is responsible for optimally combining single tests into decision rules, the latter is used to judge the influence and importance of individual tests as well as the interaction between them. Our approach is motivated and illustrated by a concrete case study in veterinary medicine, namely the diagnosis of a disease in cats called *feline infectious peritonitis*.

1 Introduction

Different types of tests, such as measuring serum antibody concentrations, are commonly used in medical diagnostics in order to reveal the health condition of an individual. The effectiveness of a single test is typically determined by correlating the test outcome with the true condition. Moreover, classical statistical hypothesis testing can be used to compare different test procedures in terms of their effectiveness.

In this paper, we tackle the problem of evaluating or selecting a test procedure from a slightly different perspective using methods of (semi-)supervised machine learning. Roughly speaking, the idea is that, by learning a model in which various candidate tests play the role of predictor variables, information about the usefulness of individual tests as well as their combination is provided by properties of that model. An approach of that kind has at least two important advantages:

J.P. Carvalho et al. (Eds.): IPMU 2016, Part I, CCIS 610, pp. 450–461, 2016.
DOI: 10.1007/978-3-319-40596-4_38

- First, it not only allows for judging the usefulness of single tests but also of *combined tests*, i.e., the combination of different tests into one overall (diagnostic) decision rule. Thus, it informs about possible synergies (as well as redundancies) between individual tests and the potential to improve diagnostic accuracy thanks to a suitable combination of these tests.
- Second, going beyond the standard setting of supervised learning, a machine learning approach suggests various ways of improving the selection of tests by taking advantage of additional sources of information. An important special case is the use of *semi-supervised* learning to exploit "unlabeled" data coming from individuals for which tests have been made but the true health condition is unknown. This situation is highly relevant in medical practice, because tests can often be conducted quite easily, whereas determining the true health condition is very difficult or expensive.

Our approach is motivated by a concrete case study in veterinary medicine, namely the diagnosis of a disease in cats called *feline infectious peritonitis* (FIP). Complete certainty about whether or not a cat is FIP-positive, and eventually will die from the disease, requires a necropsy [1,10]; unfortunately, no test performed in a cat while still alive has a 100 % sensitivity or 100 % specificity. Consequently, while different tests can be applied to cats quite easily, "labeling" a cat in the sense of supervised learning is expensive, difficult and time-consuming.

In addition to the use of (semi-supervised) machine learning methodology in medical diagnosis, we propose a game-theoretical approach for measuring the usefulness of individual tests as well as model-based combinations of such tests. Roughly speaking, the idea is to consider a combination of tests as a "coalition" in the sense of cooperative game theory, and the "payoff" of the coalition as the diagnostic accuracy achieved by the test combination. This approach will be detailed in the next section, prior to elaborating more closely on our case study in Sect. 3, presenting experimental results in Sect. 4 and concluding the paper in Sect. 5.

2 Evaluating Single and Combined Tests

Suppose a set of tests $X_1, \ldots, X_K$ to be available. We consider the outcome of each test as a random variable $X_k : \Omega \longrightarrow \mathbb{R}$, where Ω is the population of individuals to which the test can be applied. Jointly, the K tests thus define a random vector

$$X = (X_1, \ldots, X_K) \in \mathcal{X} = \mathbb{R}^K .$$

The health state is a dichotomous variable $Y \in \mathcal{Y} = \{-1, +1\}$. Typically, each test is a positive indicator in the sense that $\mathbf{P}(Y = +1 \,|\, X_k)$ increases with X_k, i.e., the larger X_k, the larger the probability of the positive class. Using machine learning terminology, each test corresponds to a *feature* or predictor variable. Moreover, $\mathcal{X}$ is the *instance space*, each $X \in \mathcal{X}$ is an instance, and Y is the (binary) output or *response* variable.

2.1 Combined Tests

If a diagnostic decision $\hat{y} \in \{-1, +1\}$ is not necessarily based on a single test X_k alone, but possibly uses a combination of several tests, a first question concerns the way in which such a combination is realized. From a machine learning point of view, this question is related to the choice of an underlying models class (hypothesis space)

$$\mathcal{H} \subset \bigcup_{J=1}^{K} \mathcal{H}_J = \bigcup_{J=1}^{K} \mathcal{Y}^{\mathbb{R}^J},$$

where $J \leq K$ is the number of tests included in the decision rule. Formally, we specify a combined test in terms of the subset $A \subseteq [K] = \{1, \ldots, K\}$ of indices, i.e., test X_k is included if $k \in A$.

The model class $\mathcal{H}$ could be defined, for example, as the class of linear threshold functions of the form

$$h : \left(x_{\sigma(1)}, \ldots, x_{\sigma(J)}\right) \mapsto \left[\!\left[\sum_{j=1}^{J} w_j \cdot x_{\sigma(j)} > t \right]\!\right], \tag{1}$$

where $w_1, \ldots, w_J, t \in \mathbb{R}_+$ and $[\![\cdot]\!]$ maps true predicates to $+1$ and false predicates to -1; moreover, $\sigma(j)$ is the j-th test included in the combination, i.e., $\sigma(j) = k$ if $\sum_{i=1}^{k} [\![i \in A]\!] = j$.

2.2 Optimal Decision Rules

Let $L : \{-1, +1\}^2 \longrightarrow \mathbb{R}$ be a loss function, such that $L(y, \hat{y})$ denotes the penalty for making the diagnostic decision $\hat{y}$ if the true health state is y. For each combined test, specified by a subset $A \subseteq [K]$, there is an optimal decision rule

$$h_A^* \in \arg\min_{h \in \mathcal{H}} \int L\big(y, h(\boldsymbol{x})\big)\, d\,\mathbf{P}(\boldsymbol{x}, y),$$

i.e., a decision rule that minimizes the loss in expectation. We denote the expected loss of this model, which corresponds to the Bayes predictors in $\mathcal{H}_{|A|}$, by

$$e^*(A) = \int L\big(y, h_A^*(\boldsymbol{x})\big)\, d\,\mathbf{P}(\boldsymbol{x}, y). \tag{2}$$

2.3 Estimating Generalization Performance

In practice, of course, neither the Bayes predictor h_A^* nor the ideal generalization performance $e^*(A)$ are known. Instead, we only assume a data set $\mathcal{D} = \mathcal{D}_L \cup \mathcal{D}_U$ to be given, which consists of a set of labeled instances

$$\mathcal{D}_L = \big\{(\boldsymbol{x}_i, y_i)\big\}_{i=1}^{L} \subset \mathcal{X} \times \mathcal{Y}$$

and possibly another set of unlabeled instances (test results without ground truth) $\mathcal{D}_U = \{\boldsymbol{x}_j\}_{j=1}^{U} \subset \mathcal{X}$. From a machine learning point of view, it is then natural to estimate the generalization performance on the basis of $\mathcal{D}$ for each $A \subseteq [K]$. To this end, models (1) can be fitted and their generalization performance can be estimated, for example, using cross-validation techniques or the bootstrap. More specifically, what can be estimated in this way is the generalization performance of a model that is trained on a combination A and data in the form of L labeled and U unlabeled examples. Therefore, we shall denote a corresponding estimate by $\hat{e}(A, L, U)$ or simply $\hat{e}(A)$ (assuming the underlying data to be given).

Needless to say, the estimates $\hat{e}(A)$ thus obtained are not necessarily monotone in the sense that $\hat{e}(B) \leq \hat{e}(A)$ for $A \subseteq B$. In fact, while $e^*(A)$ is the generalization performance of the Bayes predictor, i.e., the model that is obtained in the limit of an infinite sample size (provided the underlying learner is consistent), the estimates $\hat{e}(A)$ are obtained from models trained on a finite (and possibly small) data set. Therefore, practical problems such as overfitting become an issue, i.e., including additional tests may deteriorate instead of improve generalization performance.

2.4 Correcting Generalization Performance

How can the ideal generalization performances

$$\{e^*(A) \mid A \in [K]\} \tag{3}$$

be estimated? Starting with the finite-sample estimates

$$\{\hat{e}(A) \mid A \subseteq [K]\}, \tag{4}$$

our proposal is to correct these estimates so as to assure monotonicity. In fact, monotonicity is the main difference between the ideal and finite-sample scores. Apart from that, the ideal scores (3) should not differ too much from the estimates (4), i.e., $e^*(A) \approx \hat{e}(A)$, at least if the training data is not too small.

These considerations suggest the following estimation principle: Find a set of values (3) that satisfy monotonicity while remaining as close as possible to the corresponding scores (4). This principle can be formalized as an optimization problem of the following kind:

$$\begin{aligned} \text{minimize} \quad & \sum_{A \subseteq [K]} \left| \hat{e}(A) - e^*(A) \right| \\ \text{s.t.} \quad & \\ & e^*(B) \leq e^*(A) \text{ for all } A \subseteq B \subseteq [K] \\ & 0 \leq e^*(A) \leq 1 \text{ for all } A \subseteq [K] \end{aligned}$$

The above problem can be tackled by means of methods for *isotonic regression*. More specifically, since the inclusion relation on subsets induces a partial order on $2^{[K]}$, methods for isotonic regression on partially ordered structures are needed [3,14].

2.5 Measuring the Usefulness of Tests

Consider the set function $\nu' : 2^{[K]} \longrightarrow [0,1]$ defined by $\nu'(A) = 1 - e^*(A)$. Obviously, ν' is a monotone measure (of the usefulness of combined tests). Moreover, this measure can be normalized by setting

$$\nu^*(A) = \frac{\nu'(A) - \nu'(\emptyset)}{\nu'([K]) - \nu'(\emptyset)},$$

where $\nu'(\emptyset)$ is the performance of the best (default) decision rule that does not use any test, i.e., which either always predicts $\hat{y} = +1$ or always $\hat{y} = -1$. The measure $\nu^*(\cdot)$ thus defined satisfies the following properties:

- $\nu^*(\emptyset) = 0$, $\nu^*([K]) = 1$,
- $\nu^*(A) \leq \nu^*(B)$ for all $A \subseteq B \subseteq [K]$.

Thus, ν^* is a normalized, monotone (but not necessarily additive) set function, referred to as *fuzzy measure* or *capacity* in the literature [5]. For each combined test A, $\nu^*(A)$ is a reasonable measure of the usefulness of this test.

In a similar way, a measure $v^\bullet$ can be defined on the basis of the finite-sample scores (4), that is, by normalizing $\nu'(A) = 1 - \hat{e}(A)$:

$$v^\bullet(A) = \frac{\nu'(A) - \nu'_{min}}{\nu'_{max} - \nu'_{min}},$$

where $\nu'_{min} = 1 - \max_{B \subseteq [K]} \hat{e}(B)$ and $\nu'_{max} = 1 - \min_{B \subseteq [K]} \hat{e}(B)$. Note, however, that this measure is not necessarily monotone.

Which of the two measures is more meaningful, ν^* or $\nu^\bullet$? The answer to this question depends on practical considerations and what the measure is actually supposed to capture. When being interested in the *potential* asymptotic usefulness of a test combination, then ν^* is the right measure. Otherwise, if a model induced from a concrete set of training data is supposed to be put into (medical) practice, $\nu^\bullet$ is arguably more relevant.

2.6 Shapley Value and Interaction Index

From the point of view of (cooperative) game theory, each (test) combination $A \subseteq [K]$ can be seen as a *coalition* and $\nu \in \{\nu^*, \nu^\bullet\}$ as the *characteristic function*, i.e., $v(A)$ is the *payoff* achieved by the coalition A. Thanks to this view, we can take advantage of various established game-theoretical concepts for analyzing the importance of individual players, which correspond to tests in our case, as well as the interaction between them. In particular, the *Shapley value*, also called importance index, is defined as follows [17]:

$$\varphi(k) = \sum_{A \subseteq [K] \setminus \{k\}} \frac{1}{K \binom{K-1}{|A|}} \Big(\nu(A \cup \{k\}) - \nu(A)\Big). \tag{5}$$

The Shapley value of ν is the vector $\varphi(\nu) = (\varphi(1), \ldots, \varphi(K))$. For monotone measures (such as $\nu = \nu^*$), one can show that $0 \leq \varphi(k) \leq 1$ and $\sum_{k=1}^{K} \varphi(k) = 1$; thus, $\varphi(k)$ is a measure of the *relative* importance of the test X_k.

The *interaction index*, as proposed by [13], is defined as follows:

$$I(i,j) = \sum_{A \subseteq [K] \setminus \{i,j\}} \frac{\Big(\nu(A \cup \{i,j\}) - \nu(A \cup \{i\}) - \nu(A \cup \{j\}) + \nu(A)\Big)}{(K-1)\binom{K-2}{|A|}} .$$

This index ranges between -1 and $+1$ and indicates a positive (negative) interaction between the tests X_i and X_j if $I_{i,j} > 0$ ($I_{i,j} < 0$).

It is worth mentioning that the approach put forward in this section is quite in line with the idea of *Shapley value regression* [11], which makes use of the Shapley value in order to quantify the contribution of predictor variables in (linear) regression analysis (quantifying the value of a set of variables in terms of the R^2 measure on the training data).

3 Feline Infectious Peritonitis in Cats

Feline infectious peritonitis (FIP) is a disease with an affinity to young cats, a predisposition to involve cats living in larger groups. As it exhibits typical physical examination and clinical laboratory findings, it appears to be easy to diagnose. However, while a presumptive diagnosis is quickly established, a definite diagnosis is difficult to impossible to obtain without gross and histopathological evaluation including immunohistochemistry [1,10].

The seroprevalence is high, especially in catteries where up to 90 % of the cats are positive [2], but also up to 50 % of cats living in single-cat households have coronavirus-specific antibodies [4]. Of these, 5–10 % will develop the deadly form of FIP. A characteristic symptom of FIP is body cavity effusion, which also appears in other diseases [8]. Several treatment options exist for some of these diseases while FIP is deadly and no reliable effective therapy is known so far [16]. Therefore, it is important to diagnose the correct disease early.

Several diagnostic tests are available that diagnose FIP, for which sensitivity, specificity, positive and negative predictive value vary between different studies, presumably because different forms of FIP (effusive and dry) were investigated and because various clinical signs, geographic locations, years of investigation, prevalence and combination of tests were used [4,6,7,9,15,18]. In studies so far, no cat had all available tests performed.

The data underlying our study includes the following diagnostic tests:

- Albumin to Globulin ratio, plasma (X_1) and effusion (X_2)
- Rivalta test (X_3)
- Presence of antibodies against feline coronavirus (FCoV, X_4)
- Reverse transcriptase nested polymerase chain reaction (RT-nPCR) to detect FCoV-RNA in EDTA-blood (X_5) and in the effusion (X_6)

- Immunofluorescence staining (IFA) of FCoV antigen in macrophages in the effusion (X_7)

4 Empirical Study

Our dataset consists of 100 cats in total. For 29 of these cats, a necropsy was performed to establish the gold standard diagnosis; 11 of the 29 cats were diagnosed with feline infectious peritonitis (FIP). Additionally, the above 7 diagnostic tests were performed on all cats (i.e., $K = 7$, $L = 29$ and $U = 71$).

To estimate the generalization accuracy (in terms of the simple 0/1 loss function) of each of the $2^7 = 128$ combined diagnostic tests, we employ a semi-supervised classification technique called maximum contrastive pessimistic likelihood estimation (MCPL) [12]. Logistic regression with L_2 penalization is used as the base learner in MCPL, i.e., individual tests are combined using a linear model of the form (1).

Estimates $\hat{e}(A)$ of the (finite-sample) classification errors are obtained as follows: We resample the set of 29 labelled cats and split the resulting sample into 16 training and 13 test examples. The remaining 71 cats without label information are added to the training set. This procedure is repeated 501 times for each of the 128 combinations of tests, and the results are averaged. To obtain estimates $e^*(A)$ of the ideal generalization performances, the finite-sample estimates are subsequently corrected using isotonic regression [3,14] as described in Sect. 2.4.

4.1 Test Importance for Finite-Sample Performance Estimates

Figure 1 shows the Shapley values calculated for each test on the basis of the finite-sample performances $\hat{e}(A)$, i.e., the measure $\nu^\bullet$. Note that, since this measure is not necessarily monotone, negative Shapley values are possible (as is the case for the Rivalta test). The highest Shapley values are obtained for the two RT-nPCR tests.

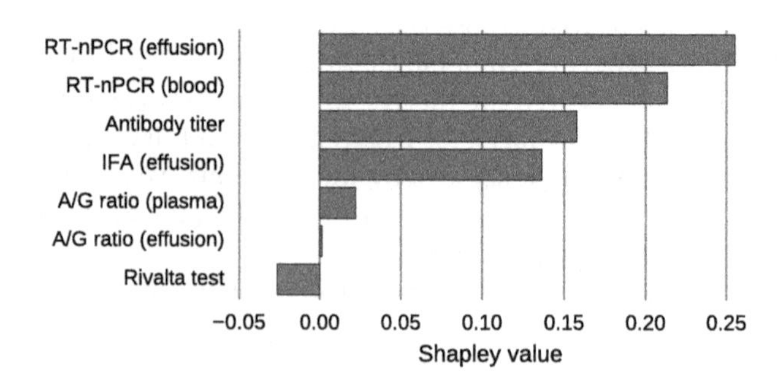

Fig. 1. Shapley values calculated for the finite-sample measure $\nu^\bullet$.

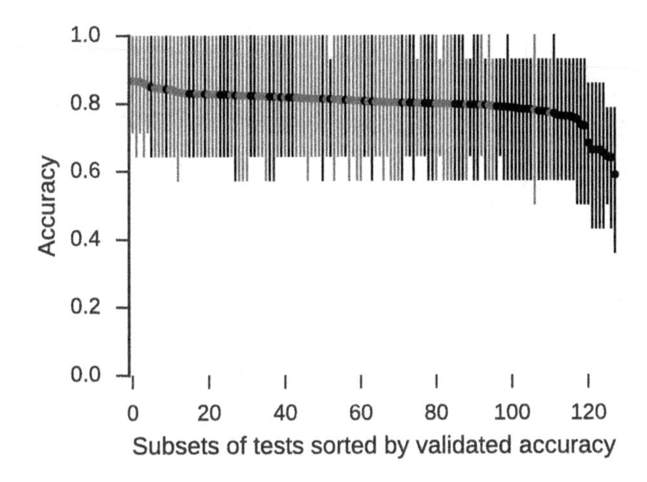

Fig. 2. Classification accuracy for all 128 test combinations (sorted by mean accuracy). The vertical lines show the 80 % empirical percentiles of the bootstrap estimates. The results for subsets including RT-nPCR (blood) are highlighted in blue. (Color figure online)

To further illustrate the importance of the diagnostic test RT-nPCR, Fig. 2 shows the mean validated classification accuracy for all 128 test combinations. The 80 % empirical percentiles are indicated by the vertical lines, and the subsets are sorted in decreasing order of their mean validated accuracy. Moreover, the results for those subsets including RT-nPCR (measured in blood) are highlighted in blue. Evidently, the concentration of subsets containing RT-nPCR (blood) is systematically higher to the left of the plot, which confirms that the inclusion of the test improves diagnostic accuracy.

4.2 Test Importance for Ideal Performance Estimates

The effect of isotonic regression on the finite-sample estimates is shown in Fig. 3. Here, each blue dot corresponds to an estimate $\hat{e}(A)$ for a particular subset A of diagnostic tests. Since partial monotonicity, which is assured by isotonic regression, cannot be visualized in a two-dimensional plot, the data points are sorted by their corrected classification accuracy (and ties are broken at random). The green line shows the isotonic regression fit.

The corrected performance estimates $\nu^*(A)$ can subsequently be used to calculate the Shapley values for each diagnostic test. The results are shown in Fig. 4. Due to the monotonicity of ν^*, all values are now positive. Again, the RT-nPCR tests achieve the highest Shapley values, but FCoV antibody titer and IFA (effusion) obtain values > 0.15, too. Note that the relative order of the RT-nPCR tests changed from the one in Fig. 1, probably due to their accuracy being very similar and the random nature of the bootstrap validation.

Figure 5 shows the accuracy estimates for all subsets. The dots indicate the corrected accuracies $\nu^*(A)$ and are used to sort subsets in decreasing order, while

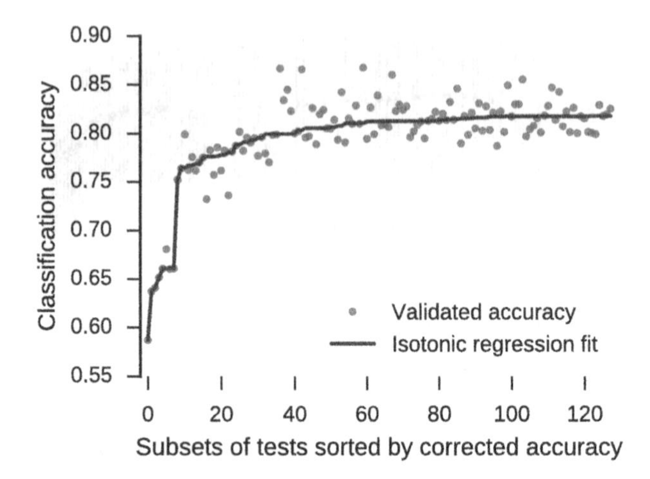

Fig. 3. Isotonic regression correction (green line) applied to the bootstrap validated classification accuracies (blue dots). (Color figure online)

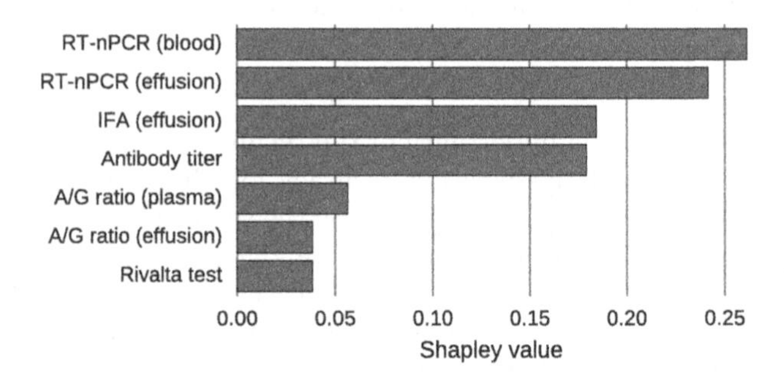

Fig. 4. Shapley values calculated using the corrected validation accuracies.

the vertical lines show the 80 % percentiles of the original bootstrap estimates. Again, the results are highlighted in blue if RT-nPCR (blood) is included in A. Like in the case of $\nu^{\bullet}$ (cf. Fig. 2), the subsets containing RT-nPCR (blood) can mostly be found on the left side of the plot; this trend is now even more pronounced.

4.3 Balancing Accuracy and Cost

An important question for a veterinary physician is which combination A of tests to perform, taking into account both diagnostic accuracy and effort. Figure 6 shows the corrected accuracies $\nu^*(A)$ (green dots) of all subsets of tests and their combined monetary cost in Euro. The Pareto set, consisting of those combinations that are not outperformed by any other combination in terms of both accuracy and cost at the same time, is indicated as a blue line. From a practical point of view, the result suggests to use a single diagnostic test, namely

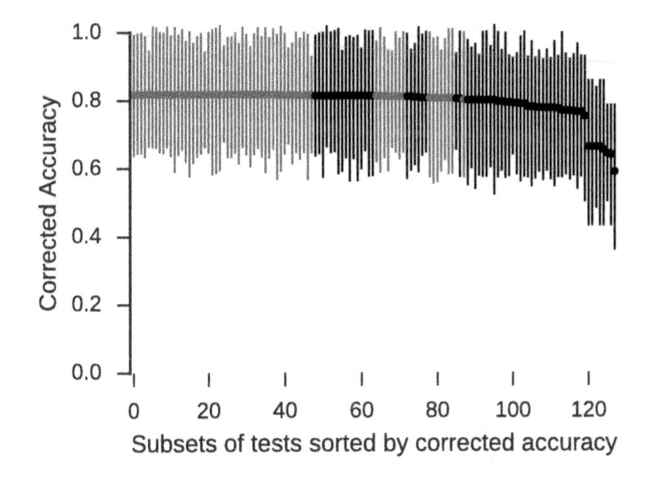

Fig. 5. Corrected accuracy $\nu^*(A)$ for all 128 subsets (sorted by mean accuracy). The vertical lines show the 80 % empirical percentiles of the original bootstrap estimates. Subsets including RT-nPCR (blood) are shown in blue. (Color figure online)

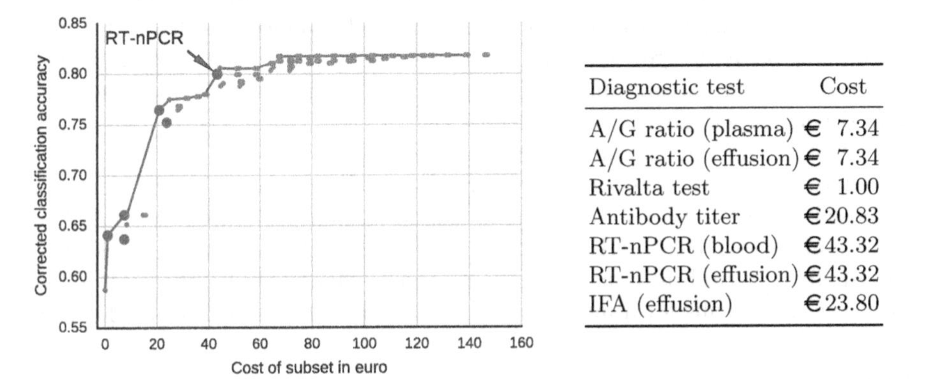

Diagnostic test	Cost
A/G ratio (plasma)	€ 7.34
A/G ratio (effusion)	€ 7.34
Rivalta test	€ 1.00
Antibody titer	€20.83
RT-nPCR (blood)	€43.32
RT-nPCR (effusion)	€43.32
IFA (effusion)	€23.80

Fig. 6. Scatter plot of the monetary costs of the subsets in Euro in relation to the corrected accuracies $\nu^*(A)$ shown as green dots. The blue line shows the Pareto front. The red points highlight the subsets which contain exactly one test. The costs for each individual test are shown in the table on the right. (Color figure online)

RT-nPCR (blood or effusion), because the inclusion of more tests yields only minor improvements. This is confirmed by the pairwise interaction indices shown for both measures $\nu^\bullet$ and ν^* in Table 1. All these measures are negative, suggesting that the tests are more redundant than complementary.

Note that, once a decision in favor of using a single test is made, the Shapley value, as a measure of average improvement achieved by adding a test, is no longer the best indicator of the usefulness of a test. Instead, a selection should be made based on the tests' individual performance. With a validated accuracy of 87 %, RT-nPCR (effusion) appears to be the best choice in this regard.

Table 1. Pairwise interaction indices for $\nu^{\bullet}$ (left) and ν^{*} (right).

	X_1	X_2		X_3		X_4		X_5		X_6		X_7	
X_1	0.0	−0.02	−0.03	−0.05	−0.05	0.00	−0.02	−0.20	−.21	−0.15	−.21	−0.17	−.19
X_2		0.0		−0.05	−0.05	−0.02	−0.03	−0.07	−0.05	−0.03	−0.04	−0.04	−0.04
X_3				0.0		−0.02	−0.04	−0.10	−0.06	−0.04	−0.06	−0.07	−0.06
X_4						0.0		−0.08	−0.05	−0.01	−0.04	−0.04	−0.03
X_5								0.0		−0.30	−.33	−0.24	−.32
X_6										0.0		−0.20	−.22

5 Summary and Conclusion

In this paper, we proposed a method for measuring the importance and usefulness of predictor variables in (semi-/supervised) machine learning, which makes use of concepts from cooperative game theory: subsets of variables are considered as coalitions, and their predictive performance plays the role of the payoff. Although our approach is motivated by a concrete application in veterinary medicine, namely the diagnosis of *feline infectious peritonitis* in cats, it is completely general and can obviously be used for other learning problems as well.

For the case study just mentioned, our method produces results that appear to be plausible and agree with the medical experts' experience. Roughly speaking, there are two strong diagnostic tests that are significantly more accurate than others; practically, it suffices to use one of them, since a combination with other tests yields only minor improvements.

There are several directions for future work. For example, the principle we proposed in Sect. 2.4 for inducing ideal generalization performances $e^*(A)$ from finite-sample estimates $\hat{e}(A)$ is clearly plausible and, moreover, seems to be indeed able to calibrate the original estimates thanks to an ensemble effect. Nevertheless, it calls for a more thorough analysis and theoretical justification.

References

1. Addie, D.D., Paltrinieri, S., Pedersen, N.C.: Recommendations from workshops of the second international feline coronavirus/feline infectious peritonitis symposium. J. Feline Med. Surg. **6**(2), 125–130 (2004)
2. Benetka, V., Kübber-Heiss, A., Kolodziejek, J., Nowotny, N., Hofmann-Parisot, M., Möstl, K.: Prevalence of feline coronavirus types I and II in cats with histopathologically verified feline infectious peritonitis. Vet. Microbiol. **99**(1), 31–42 (2004)
3. Block, H., Qian, S., Sampson, A.: Structure algorithms for partially ordered isotonic regression. J. Comput. Graph. Stat. **3**(3), 285–300 (1994)
4. Giori, L., Giordano, A., Giudice, C., Grieco, V., Paltrinieri, S.: Performances of different diagnostic tests for feline infectious peritonitis in challenging clinical cases. J. Small Anim. Pract. **52**(3), 152–157 (2011)
5. Grabisch, M., Nguyen, H., Walker, E.: Fundamentals of Uncertainty Calculi with Applications to Fuzzy Inference. Kluwer Academic Publishers, Dordrecht (1995)
6. Hartmann, K., Binder, C., Hirschberger, J., Cole, D., Reinacher, M., Schroo, S., Frost, J., Egberink, H., Lutz, H., Hermanns, W.: Comparison of different tests to diagnose feline infectious peritonitis. J. Vet. Intern. Med. **17**(6), 781–790 (2003)

7. Hirschberger, J., Hartmann, K., Wilhelm, N., Frost, J., Kraft, W.: Using direct immunofluorescence to detect coronaviruses in peritoneal in peritoneal and pleural effusions. Tierärztliche Praxis **23**, 92–99 (1995)
8. Hirschberger, J., DeNicola, D.B., Hermanns, W., Kraft, W.: Sensitivity and specificity of cytologic evaluation in the diagnosis of neoplasia in body fluids from dogs and cats. Vet. Clin. Pathol. **28**(4), 142–146 (1999)
9. Jeffery, U., Deitz, K., Hostetter, S.: Positive predictive value of albumin: globulin ratio for feline infectious peritonitis in a mid-western referral hospital population. J. Feline Med. Surg. **14**(12), 903–905 (2012)
10. Kipar, A., Köhler, K., Leukert, W., Reinacher, M.: A comparison of lymphatic tissues from cats with spontaneous feline infectious peritonitis (FIP), cats with FIP virus infection but no FIP, and cats with no infection. J. Comp. Pathol. **125**(2), 182–191 (2001)
11. Lipovetsky, S., Conklin, M.: Analysis of regression in game theory approach. Appl. Stochast. Models Bus. Ind. **17**(4), 319–330 (2001)
12. Loog, M.: Contrastive pessimistic likelihood estimation for semi-supervised classification. IEEE Trans. Pattern Anal. Mach. Intell. **38**(3), 462–475 (2016)
13. Murofushi, T., Soneda, S.: Techniques for reading fuzzy measures (iii): interaction index. In: 9th Fuzzy System Symposium, Sapporo, Japan, pp. 693–696 (1993)
14. Pardalos, P., Xue, G.: Algorithms for a class of isotonic regression problems. Algorithmica **23**(3), 211–222 (1999)
15. Parodi, M.C., Cammarata, G., Paltrinieri, S., Lavazza, A., Ape, F.: Using direct immunofluorescence to detect coronaviruses in peritoneal and pleural effusions. J. Small Anim. Pract. **34**(12), 609–613 (1993)
16. Ritz, S., Egberink, H., Hartmann, K.: Effect of feline interferon-omega on the survival time and quality of life of cats with feline infectious peritonitis. J. Vet. Intern. Med. **21**(6), 1193–1197 (2007)
17. Shapley, L.: A value for n-person games. Ann. Math. Stud. **28**, 307–317 (1953)
18. Soma, T., Wada, M., Taharaguchi, S., Tajima, T.: Detection of ascitic feline coronavirus RNA from cats with clinically suspected feline infectious peritonitis. J. Vet. Med. Sci. **75**(10), 1389–1392 (2013)

Fuzzy Modeling for Vitamin B12 Deficiency

Anna Wilbik[1(✉)], Saskia van Loon[2], Arjen-Kars Boer[2], Uzay Kaymak[1], and Volkher Scharnhorst[2,3]

[1] Information Systems, School of Industrial Engineering, Eindhoven University of Technology, P.O. Box 513, 5600 MB Eindhoven, The Netherlands
a.m.wilbik@tue.nl

[2] Catharina Hospital, Clinical Chemistry, Michelangelaan 2, 5623 EJ Eindhoven, The Netherlands

[3] Department of Biomedical Engineering, Eindhoven University of Technology, P.O. Box 513, 5600 MB Eindhoven, The Netherlands

Abstract. Blood vitamin B12 levels are not representative for actual vitamin B12 status in tissue. Instead plasma methylmalonic acid (MMA) levels can be measured because MMA concentrations increase relatively early in the course of vitamin B12 deficiency. However, MMA levels in plasma may also be increased due to renal failure. In this paper we estimate the influence of the kidney function on MMA levels in plasma by using fuzzy inference systems. Using this method diagnosing vitamin B12 deficiencies could be improved when kidney failure is present.

Keywords: Vitamin B12 deficiency · Kidney function · Takagi-Sugono fuzzy inference system

1 Introduction

A deficiency in vitamin B12 is a common disorder which can result in various hematological and neurological disorders. Until a significant vitamin B12 deficiency is developed in the tissue, patients often remain asymptomatic. In the detection of vitamin B12 deficiency, measuring vitamin B12 levels in serum is not a proper representation of actual vitamin B12 levels in tissue [9,11].

As an indirect but functional measure of tissue vitamin B12 status, methylmalonic acid (MMA) levels in plasma can be measured as MMA concentrations increase relatively early in the course of vitamin B12 deficiency [2,5]. Vitamin B12 serves as a cofactor in the enzymatic reaction where the coenzyme A-linked form of MMA, i.e. methylmalonyl-CoA, is converted into succinyl-CoA by methylmalonyl-CoA mutase, an important step in the extraction of energy from proteins and fats. When there is a vitamin B12 deficiency, methylmalonyl-CoA is not converted by the enzyme and accumulates. Upon release of methylmalonyl-CoA in the plasma, coenzyme A is released and MMA levels rise. As a consequence, the vitamin B12 supply to the tissues is reflected by the level of MMA in plasma, showing an inverse correlation.

J.P. Carvalho et al. (Eds.): IPMU 2016, Part I, CCIS 610, pp. 462–471, 2016.
DOI: 10.1007/978-3-319-40596-4_39

However, MMA plasma levels may also be increased due to renal failure as a result of impaired filtration and secretion [2,7]. Therefore, increased MMA levels in plasma are not specific to vitamin B12 deficiency. To determine if a vitamin B12 deficiency is present based on MMA plasma level measurements, it is preferred to include measurements assessing kidney function. It is known that there is an influence of kidney function on MMA plasma levels but to the authors' best knowledge the extend of this influence is unknown.

In this paper we estimate the influence of the kidney function on MMA plasma levels using data analysis techniques, in particular fuzzy inference systems (FIS). In order to achieve this goal, we first model the relationship between the MMA, vitamin B12 and kidney function with a FIS. Next, we use the model to estimate the influence of kidney function on MMA plasma levels.

This paper is structured as follows: in Sect. 2 we describe the collected data. Section 3 describes briefly the FIS and how it was created from the data. Section 4 presents the analysis of the influence of the kidney function. The paper is finished with the concluding remarks in Sect. 5.

2 Data Description

In order to model the relationship between MMA plasma levels, vitamin B12 and kidney function, we collected laboratory and physiological data from all patients at the Catharina Hospital Eindhoven in the Netherlands, where vitamin B12 measurements were performed in the period from July 2010 until April 2015. In this way we obtained 64331 records.

For vitamin B12, the exact value can be determined up to the value of 1400 pmol/L. If it is higher it is noted in the hospital information system as ">1400". The values of vitamin B12 below 90 pmol/L means deficiency, while values above 300 pmol/L are considered as indication for no deficiency.

Methylmalonic acid (MMA) levels in the plasma are measured in range from 100 (sometimes 140 depending on the equipment) till 1500 nmol/L. Concentrations lower and higher than the threshold values are indicated respectively as "<100" and ">1500". Levels of plasma MMA below 300 nmol/L are considered normal in the Catharina Hospital laboratory [9], while levels above 430 nmol/L are considered to indicate vitamin B12 deficiency.

Kidney functions can be described by creatinine values or glomerular filtration rate (GFR). In this study we use the GFR values calculated according to the CKD EPI formula, taking into account creatinine levels, gender and age. CKD EPI above 90 mL/min/1.73 m^2 indicate very good kidney function, values between 60 mL/min/1.73 m^2 and 90 mL/min/1.73 m^2 indicate good kidney function. Values between 30 mL/min/1.73 m^2 and 60 mL/min/1.73 m^2 are indication of moderate kidney function, while values below 30 mL/min/1.73 m^2 means poor kidney function (renal failure). For this study we may assume that the threshold of 60 mL/min/1.73 m^2 is the most important one.

Catharina Hospital is a leading center for bariatric surgery, and so many vitamin B12 measurements included in the dataset were performed for the bariatric patient population. For those patients, MMA measurements are performed only when the vitamin B12 results meet certain pre-determined criteria, i.e. so called reflex testing. As a result, this population is overrepresented in the dataset and therefore we decided to include only the data of non-bariatric patients, i.e. patients with a BMI between 18 and 25. We also excluded patients, that we suspect for being supplemented with vitamin B12 (vitamin B12 serum level above 900 pmol/L). Further we filtered out observations with missing values or non-numerical values of vitamin B12, MMA or CKD EPI. As a result we obtained 380 data points. The 3D scatter plot showing the data is shown in Fig. 1.

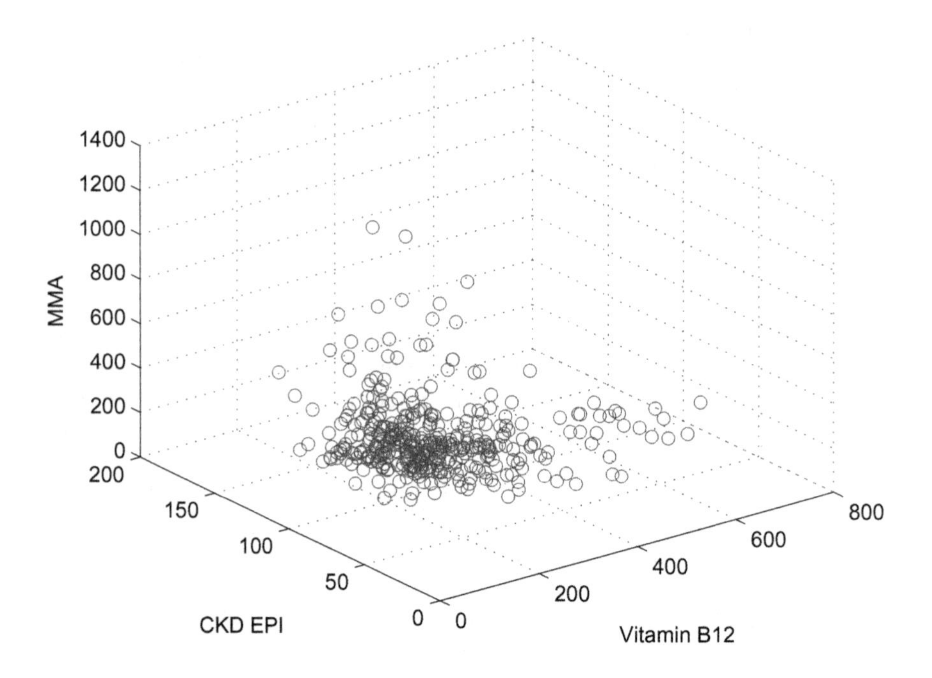

Fig. 1. Scatter plot MMA and vitamin B12 and CKD EPI

3 Fuzzy Model

We model the relationship between MMA, vitamin B12 and kidney function (CKD EPI) using first order Takagi-Sugeno Fuzzy Inference System. The fuzzy model is obtained by following a clustering-based methodology similar to the one in [10]. Fuzzy c-means clustering is used to determine the antecedents of the initial fuzzy model, after which the model parameters are optimized. The rule antecedents are distributed according to a scatter partition [4], which is the natural partitioning approach in the cluster-based approaches.

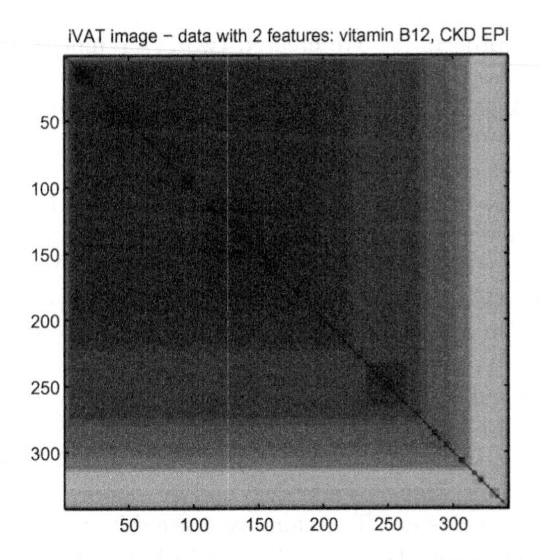

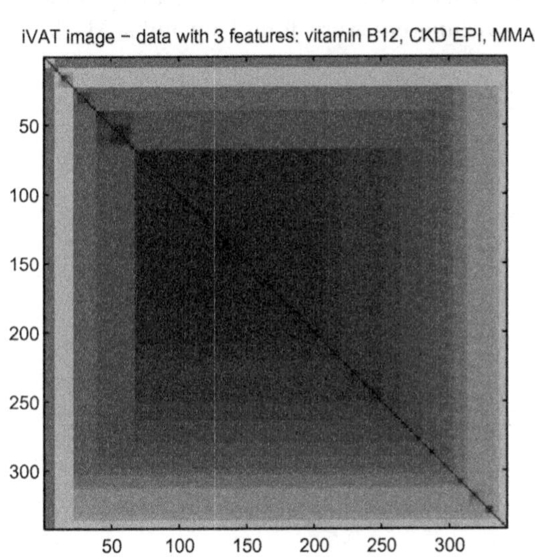

Fig. 2. iVAT images of dissimilarity of data points with two features: vitamin B12 and CKD EPI (above) and three features: vitamin B12, CKD EPI and MMA (below).

For testing purposes, we held out 10 % of the data as the test set. The selection of the test set is stratified based on the MMA values binned into five categories of equal sizes. The remaining data are used to build the model.

First we investigate the number of clusters and rules that the FIS should have. For this purpose we use a few cluster tendency assessment methods, such as iVAT [1] and cluster correlation validity indices [8]. The iVAT method is a visual method for determining the possible number of clusters in, or the cluster tendency of a set of objects [1]. We created two iVAT images for the data, as shown in Fig. 2, one for features Vitamin B12 and CKD EPI, and the other including also MMA as a feature. Those images does not give clear indication about number of clusters visible in this data set. However it seems that between two and seven clusters can be distinguished.

We also used Pearson's and Spearman's cluster correlation validity (CCV) indices [8] to determine number of clusters. If the clustering space contained only two features, namely vitamin B12 and CKD EPI, both CCV indices suggested creating six clusters. Once we added MMA as additional feature, then Pearson's CCV suggested creating five clusters, while Spearman's CCV suggested eight clusters.

We also used the validation set to find the best number of rules for the FIS. We used a repeated 10-fold stratified cross-validation approach. The quality of the FIS was evaluated with mean absolute error MAE. In this case choosing two clusters was indisputable winner. Therefore, and for sake of simplicity, we chose to model with only two clusters. Therefore we created first order Takagi-Sugeno Fuzzy Inference System with two rules. We used fuzzy c-means clustering. The optimal consequent parameters were found by anfis Matlab function [6]. The

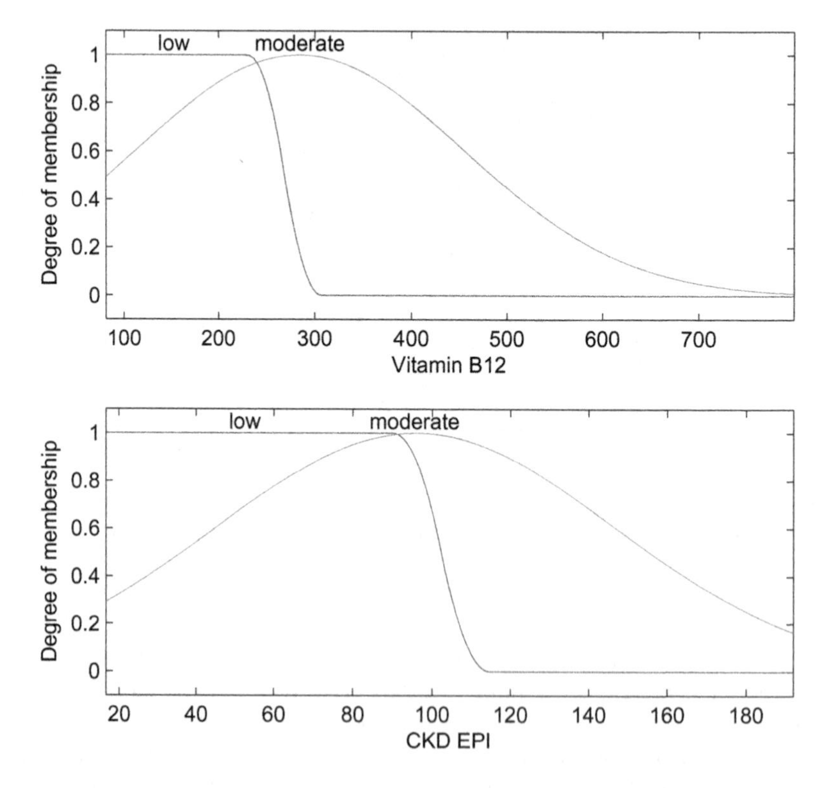

Fig. 3. Memberships of the linguistic labels *low* and *moderate* for Vitamin B12 and CKD EPI.

anfis function uses a network representation of a Takagi-Sugeno fuzzy system to determine the optimal model parameter through gradient-based learning. We used the hybrid learning approach in which the antecedent parameters are determined by back-propagation and the consequent parameters are learnt by least squares optimization in an iterative learning process [3].

However, before we trained the FIS with the data, we investigated the memberships of the antecedents. For both variables one membership function was targeting a small subset of the domain, which could be described as *low*, while the other was embracing the rest of the space. Because the boundaries were not covered by any of the membership functions, we decided to change the membership functions corresponding to the linguistic labels *low* to left shouldered membership functions. Only afterwards the output parameters were estimated using the hybrid learning algorithm.

The membership functions for both vitamin B12 and CKD EPI are shown in Fig. 3.

The following two rules were obtained after training:

- If (Vitamin B12 is moderate) and (CKD EPI is moderate) then MMA = -0,41 * Vitamin B12 -1,24 * CKD EPI + 473,46
- If (Vitamin B12 is low) and (CKD EPI is low) then MMA = -3.17* Vitamin B12 - 3,35 * CKD EPI + 1311,77

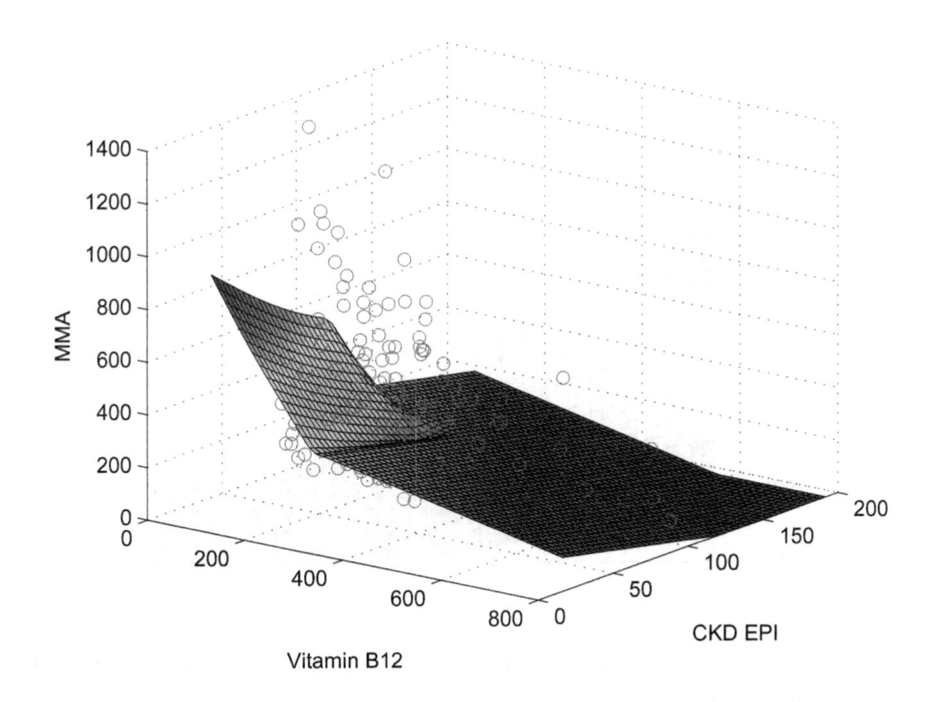

Fig. 4. The response surface generated by the fuzzy model together with the data points.

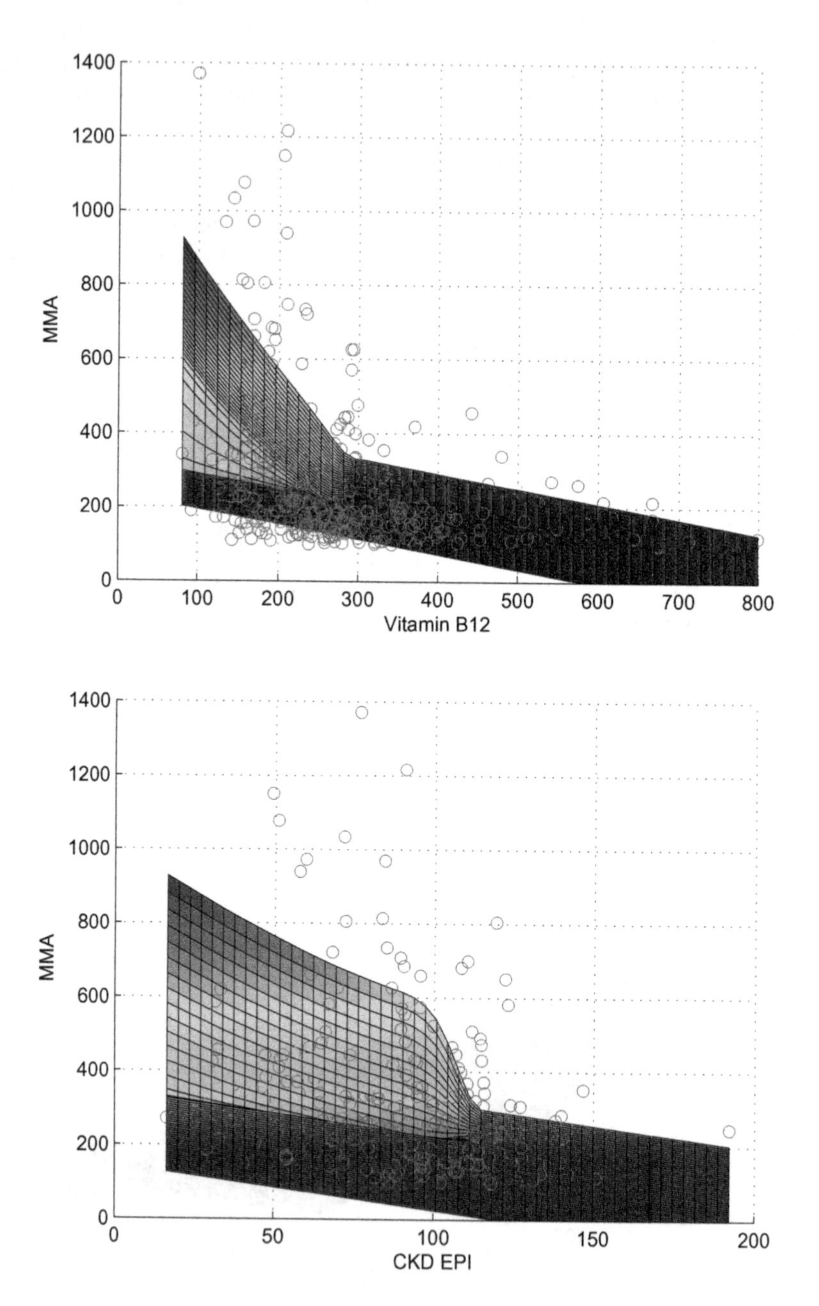

Fig. 5. Projections of the response surface of the fuzzy model to Vitamin B12 and CKD EPI dimensions.

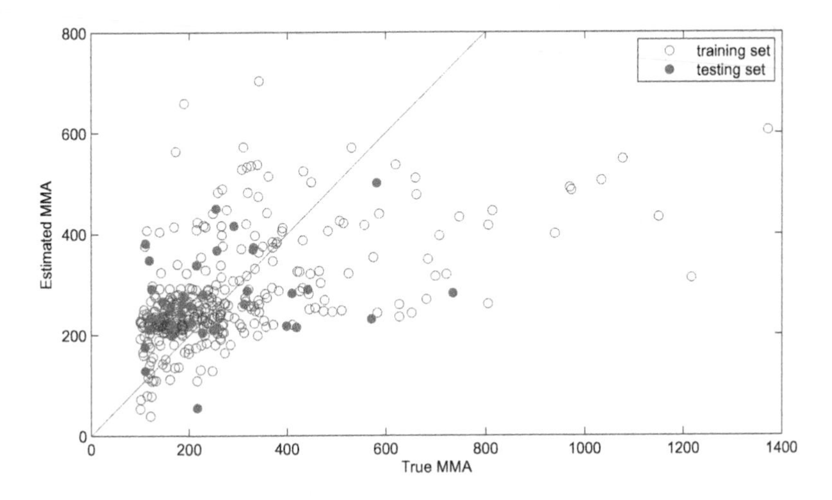

Fig. 6. Scatter plot of true vs estimated values of MMA. (Colour figure online)

The mean squared error (MSE) on the training set was 25759.3 and the MAE was 106.4. The MSE on the testing set was 21268.7 and the MAE was 113.6.

Figure 4 shows the surface plot of the fuzzy model together with the data points. Figure 5 are the projections of this surface to Vitamin B12 and CKD EPI dimensions, respectively. Figure 6 shows a scatter plot of the true values of plasma MMA level against the estimated ones. The observations of training and testing sets are marked with different colors.

4 Estimation of the Influence of the Kidney Functions

The model created may next be used to estimate the influence of the kidney function on the MMA levels in the plasma. Let us consider a patient with CKD EPI on the level of $30\,\mathrm{mL/min/1.73\,m^2}$ and vitamin B12 on the level of 90 pmol/L and MMA of 600 nmol/L. According to the model such patients have MMA of 829 nmol/L. A patient with same vitamin B12 level, but CKD EPI of $60\,\mathrm{mL/min/1.73\,m^2}$ 691. Hence 138 pmol/L of MMA can be explained by the poor kidney function. Therefore we can expect that for this patient MMA value should be corrected by 138 pmol/L, meaning that if his kidneys had been working normally, the patient should have had MMA value of 562 nmol/L.

In order to decide whether the patient has Vitamin B12 deficiency, the doctors prefer simple criterion with a single threshold, for instance MMA above 430 nmol/L. Therefore for the patients with decreased kidney function, based on the model we can estimate the increase of the level of MMA, that is caused by the poor kidney function compared to the patient with CKD EPI of $60\,\mathrm{mL/min/1.73\,m^2}$. Next we can adjust the observed MMA value with the correction value, and then this updated value can be compared with the threshold.

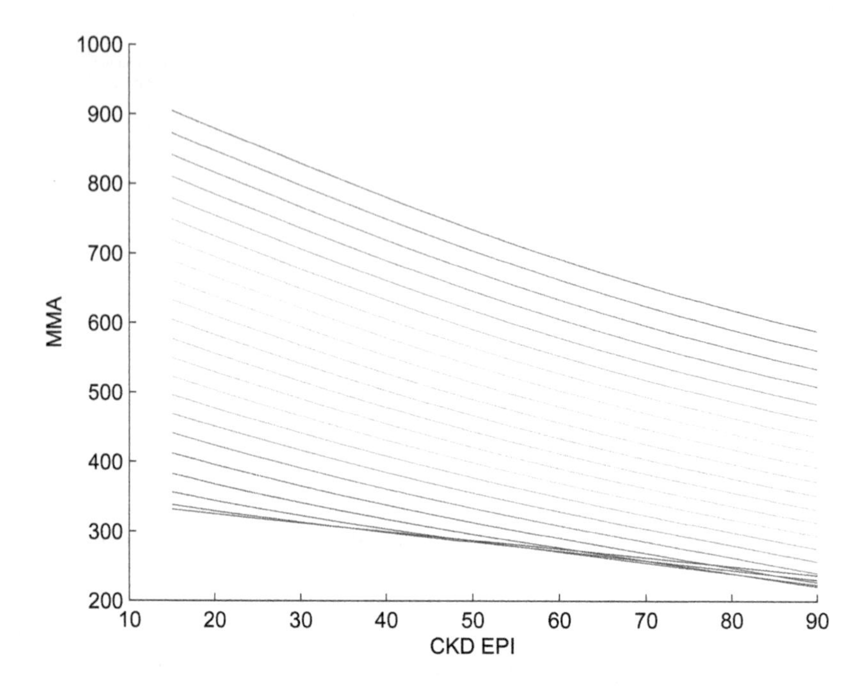

Fig. 7. MMA levels against CKD EPI for fixed vitamin B12 levels (from 90 pmol/L (dark red color) till 300 pmol/L (dark blue), every 10 pmol/L). (Color figure online)

Figure 7 plots the lines of MMA against CKD EPI for fixed vitamin B12 levels, every 10 pmol/L ranging from 90 pmol/L (dark red color) till 300 pmol/L (dark blue). We focus only on this area, because patients in this area are suspected to have vitamin B12 deficiency.

Detailed analysis of those values showed that the difference in MMA on each unit of CKD EPI increases as CKD EPI decreases. It means that the poorer is the kidney function, the higher are values of MMA. Similar relationship can be found between vitamin B12 and MMA, i.e. the difference in MMA on each unit of vitamin B12 increases with decrease of vitamin B12 levels. It means that greater vitamin B12 deficiency, the higher are values of MMA. Change of MMA for each 5 mL/min/1.73 m^2 of CKD EPI ranges from 6 till 25 pmol/L, with an average change of 16 pmol/L.

By using our fuzzy model, the clinicians can interpret the measured values of MMA more accurately by correcting it for the kidney function of the patient.

5 Concluding Remarks

We considered the problem of vitamin B12 deficiency. Vitamin B12 levels in serum is not a proper representative of actual vitamin B12 levels in tissue. Methylmalonic acid (MMA) levels in the plasma can be used as indication of vitamin B12 deficiency in tissue, because MMA concentrations increase relatively early in such situations.

However, MMA levels in the plasma may also be increased due to renal failure, complicating the relationship between B12 levels and the MMA.

To estimate the influence of the kidney function on MMA levels in the plasma we modelled the relationship between vitamin B12, MMA and kidney function using a first order Takagi-Sugeno fuzzy inference system.

By using this model it is possible to estimate the expected change of values of MMA for different levels of kidney functions and for the same levels of vitamin B12.

Our model gives the way for adaptive interpretation of MMA measurements based on the patient's kidney function. In this way, it becomes possible to detect B12 deficiency more accurately when kidney failure is present or when a patient's kidney function is reduced.

Despite our promising results, the model can still be improved further. Fore example, Fig. 6 shows that the model systematically underestimates true MMA values when they are high. In the future we will investigate the cause of this. We will also include additional features that may influence MMA levels in order to improve the predictive accuracy of the model.

Acknowledgements. This work is partially supported by TU/e Data Science Flagship.

References

1. Havens, T., Bezdek, J.: An efficient formulation of the improved visual assessment of cluster tendency (iVAT) algorithm. IEEE Trans. Knowl. Data Eng. **24**(5), 813–822 (2012)
2. Iqbal, N., Azar, D., Yun, Y.M., Ghausi, O., Ix, J., Fitzgerald, R.L.: Serum methylmalonic acid and holotranscobalamin-ii as markers for vitamin B12 deficiency in end-stage renal disease patients. Ann. Clin. Lab. Sci. **3**(43), 243–249 (2013)
3. Jang, J.S.R.: ANFIS: adaptive-network-based fuzzy inference system. IEEE Trans. Syst. Man Cybern. **3**(23), 665–685 (1993)
4. Jang, J.S.R., Sun, C.T., Mizutani, E.: Neuro-Fuzzy and Soft Computing. Prentice Hall, Englewood Cliffs (1997)
5. Klee, G.G.: Cobalamin and folate evaluation: Measurement of methylmalonic acid and homocysteine vs vitamin B12 and folate. Clin. Chem. **8**(46), 1277–1283 (2000)
6. MathWorks: Afis function documentation. http://nl.mathworks.com/help/fuzzy/anfis.html
7. Moelby, L., Rasmussen, K., Ring, T., Nielsen, G.: Relationship between methylmalonic acid and cobalamin in uremia. Kidney Int. **1**(57), 265–273 (2000)
8. Popescu, M., Bezdek, J., Havens, T., Keller, J.: A cluster validity framework based on induced partition dissimilarity. IEEE Trans. Cybern. **43**(1), 308–320 (2013)
9. Smelt, H., Smulders, J., Said, M., Nienhuijs, S., Boer, A.: Improving bariatric patient aftercare outcome by improved detection of a functional vitamin B12 deficiency. Obes. Surg. **11**, 1–5 (2015). doi:10.1007/s11695-015-1952-8
10. Sousa, J.M.C., Kaymak, U.: Fuzzy Decision Making in Modeling and Control, World Scientific Series in Robotics and Intelligent Systems, vol. 27. World Scientific (2002)
11. Wiersinga, W., de Rooij, S., Huijmans, J., Fischer, J., Hoekstra, J.: De diagnostiek van vitamine-B12-deficintie herzien. Ned. Tijdschr. Geneeskd. **149**, 2789–2794 (2005)

Real-World Applications

Using Geographic Information Systems and Smartphone-Based Vibration Data to Support Decision Making on Pavement Rehabilitation

Chun-Hsing Ho[1(✉)], Chieh-Ping Lai[2], and Anas Almonnieay[1]

[1] Department of Civil Engineering, Construction Management, and Environmental Engineering, Northern Arizona University, PO Box 15600, Flagstaff, AZ, USA
{chun-hsing.ho,Asa283}@nau.edu
[2] President of Civil Sensing Systems Inc., 511 Casey Lane, Rockville, MD 20850, USA
bmwlai@gmail.com

Abstract. This paper presents a data collecting process using smartphone-based accelerometer in association with geographic information systems (GIS) software to better manage pavement condition data and facilitate with decision making for maintenance and rehabilitation. The smartphone is equipped with an accelerometer (a mobile apps) could record 50 vibration data points per second in three direction (X, Y, and Z). The type of Traditional pavement survey is time-consuming and requires experienced technicians to travel along highway to visualize pavement conditions and record any failures. Combining vibration intensity data with a GIS platform can help public agencies with a strategic plan to prioritize maintenance schedules for both bike trails and highway roads. The objective of this paper is to (1) discuss the processes of vibration data analysis using a smartphone based accelerometer and to (2) demonstrate how to relate vibration intensity data to locate priority areas for immediate.

Keywords: Vibration · Smartphones · Mobile apps · Accelerometers · Geographic information systems

1 Background and Problem Statement

Pavement condition surveys involve data acquisition, interpretation, and documentation. These activities characterize surface condition, such as surface cracking, deformation, and other surface defects for both flexible and rigid pavements. Currently, there are three key major pavement distress detecting techniques; (1) manual inspection, (2) imaging process detection, and (3) vibration-based detection. Manual inspection perhaps is most popular and affordable method among highway agencies for pavement condition surveys. However, manual inspection remains labor intensive and time consuming procedure. It is subject to personal experience and bias, making the survey

J.P. Carvalho et al. (Eds.): IPMU 2016, Part I, CCIS 610, pp. 475–485, 2016.
DOI: 10.1007/978-3-319-40596-4_40

results inconsistent and different between distress raters [1, 2]. More recently, three-dimensional automated imaging and sensor technology have been widely used to survey pavement surface conditions and capture distress issues [3–5]. When traveling on highways, a vehicle equipped with sensors on board should integrate the dynamic vibration effect associated with sensors taking into account for a whole system that consists with (i) a vibration model with three dimensional components (x/y/z) along with varying vehicle speeds and vehicle weights that collects vibration responses based on the road roughness condition generated by an exogenous dynamical contact between tires of the vehicle and the road surface, and (ii) a signal processing model that can effectively transfer and analyze dynamic vibration data, and extract signatures of the pavement condition after the adaptive filtering and machine-learning process. More recently, bikes attached with mobile apps (sensors and accelerometers) have been used by local governments to perform condition surveys along their respective bike trails. One of issues with post- data processing is how to effectively display the relation between the survey results and the actual locations so that engineers would be able to prioritize areas the need to be repaired immediately. This paper presents a data collecting process using smartphone-based accelerometer in association with geographic information systems (GIS) software to better manage pavement condition data and facilitate with decision making for maintenance and rehabilitation.

2 Data Acquisition

2.1 Dynamic Vibration and Signal Processing

When a vehicle or bike travels on a road, vibration response is recorded based on the interaction between the pavement surface and tires of a vehicle or a bike. A dynamic vibration coupled model using motion equation ca be used to simulate the scenario as shown in Eq. 1.

$$[M]\{\ddot{u}\} + [C]\{\dot{u}\} + [K]\{u\} = \{F(t)\} \tag{1}$$

where $[M], [C]$ and $[K]$ represent the total mass matrix, the total damping matrix, and the total stiffness matrix of a vehicle; $\{\ddot{u}\}, \{\dot{u}\}, and\ \{u\}$, are acceleration, speed, and displacement vector of nodes in the three directions (x, y, and z) for a given vehicle/bike.

The traffic load matrix $\{F(t)\}$ is related to vehicle/bike structure, vehicle/bike weight, pavement irregularity, and driving speed and can be expressed as:

$$\{F(t)\} = \left(P_0 + m_0 \alpha \omega^2\right) |sin(\omega t)| \tag{2}$$

where P_0 = vehicle weight that caries from vehicle types; m_0 = spring coefficient related to the mass of vehicle and vehicle structure; α = geometric irregularity vector height that needs to be determined by the proposal; ω = vibration circular frequency, taking $\omega = 2\pi v/l$ in which v = vehicle speed and l = length of a vehicle/bike.

For a pavement system, Eq. 1 can be also applied with a different dynamic load as shown below:

$$\{F(t)\} = qsin^2\left(\frac{\pi}{2} + \frac{\pi t}{d}\right) \tag{3}$$

where d is the duration of load and the load intensity is q.

When the pavement receives moving load, the dynamic response can be expressed as:

$$\{R(t)\} = \int_{-d/2}^{0} R(t)\frac{dF}{dt}dt \tag{4}$$

The dynamic response equation can be rearranged to:

$$\{R(t)\} = \frac{q\pi^2}{2}\sum_{i=1}^{n} c_i \frac{1 + exp(-d/2t_i)}{\pi^2 + (d/2t_i)^2} \tag{5}$$

Sensor time-series signal from accelerometer used Z-axis as an example can be written as

$$S_Z(\mathrm{t}) = \mathrm{g} + V_Z(\mathrm{t}) + A_Z(\mathrm{t}) + \mathrm{n(t)} \tag{6}$$

where

g: Earth gravity-constant

V_Z(t) = vibration at z axis: Assume car on flatness road

A_Z(t) = Acceleration projection to Z axis. Assume car is accelerating (+ or −) when drive on the slopes.

n(t) = noise. It can come from system or thermal noise.

2.2 Data Acquisition: Smartphone Accelerometer and Data Transfer

The data were collected using a smartphone accelerometer, a mobile apps called My Vibrometer. This mobile apps allows the user to collect vibration data due to cell phone movements. The myVibrometer mobile app exports a .csv file format, which is easy to import into GIS software. The application collects data points at a rate of 50 points per second on a real time scale including date and time. The vibration data were reported in units of gravity. The vibration data collected were in X, Y and Z coordinates. The data collected from myVibrometer also directly provides the longitude and latitude within the same .csv file. Each data set can be quickly imported into the mapping software and would be easy to scale up the data sourcing and provides easy data management. Based on output files, the myVibrometer app records and exports the following data: vibration in X, Y and Z coordinates (g), heading (degrees), course (degrees), speed (m/s), altitude (m), latitude (degrees), longitude (degrees), and date and time. Recorded coordinates are associated with the smartphone's GPS system to help tracking the moving path of a vehicle/bike while traveling on roads. The above features make the smartphone accelerometer an ideal candidate for data collection and data analysis.

3 Preliminary Road Test: Vehicle Equipped with Mobile App

3.1 Vibration Accelerometer Setting

Before applying the vibration model, mobile apps, signal processing, and GIS platform

Fig. 1. Road surface with pothole and cracking (L) and smooth condition (R)

in roadway condition assessments, a preliminary road test was conduct on Northern Arizona University's campus to validate the applicability of the mobile sensing technology and mapping in real world application. A road displaying rough and smooth surface (Fig. 1) was selected.

For a vehicle, the entire vibration data collecting system consists of a smartphone equipped with vibration sensors that will capture vibration responses from the road

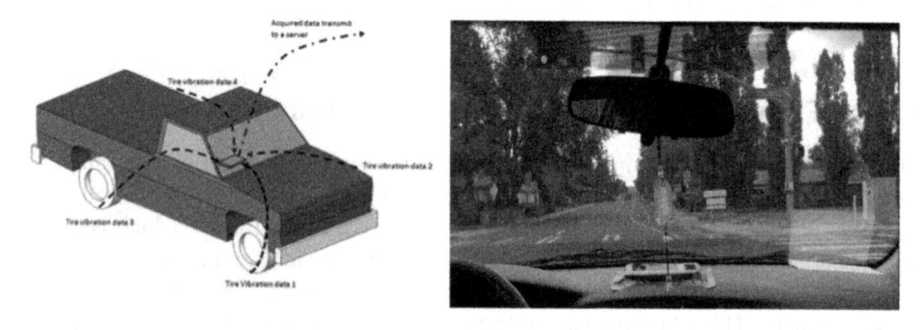

Fig. 2. Schematic of vibration data and placement of a smartphone

surface. When a vehicle travels on a roadway, the vibration sensors will gather vibration data based on the surface conditions as shown in Fig. 2. The collected data together with coordinates of points will be temporarily stored in the memory of the phone and then transmitted to the team's server.

3.2 Data Analysis and Result Discussions

During testing, the vehicle's speed was controlled at 35 km/h (20 mph) to keep all data to be consistent. Three road tests traveling on the same road with the same direction were implemented. Visual inspection of the data shows at least two potholes in Fig. 1 (Left). The presence of the potholes was manifested in the three data points, X, Y, and Z. It is also noted that the data for the first pothole (1st peak) is different than the second pothole (2nd peak). In particular, the X-and Z-axis samples show significantly wider impulse response around 2nd peak. We interpret this to reflect that the pothole depth is not deep and has wider uneven surface. We also analyzed the data using the variance, average, and mean square analysis. The objective of this analysis is to understand if the road condition is reflected on these numbers. By using signal processing method, data analysis result is shown in Fig. 3. We found that the variance for the smooth road (X = 0.07, Y = 0.05, Z = 0.05 g) is significantly smaller than the data for the medium road (X = 0.21, Y = 0.1, Z = 0.5 g), which is in turn smaller than the rough road (X = 1.03, Y = 0.76, Z = 0.6 g). Obviously, device sensors can successfully detect the signatures of road conditions and driving behaviors. The preliminary road test result was

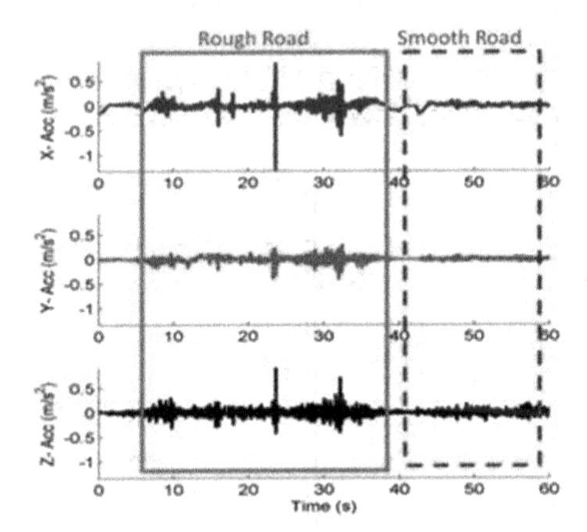

Fig. 3. Comparison of vibration magnitudes between rough and smooth road

in support of mobile app setting, dynamic vibration modeling, as well vibration data analysis and signal processing. A real project was therefore selected in cooperation with the Department of Facility Services on the university's bike trails using the same approach and technology to prioritize pavement areas for maintenance and rehabilitation. The reason of this project is to evaluate if the methodology presented in highways by a vehicle can be applied in urban/campus trails by a bike.

4 Road Test: Pavement Condition Assessment of Bike Trails

4.1 Application of Geographic Information Systems in Roadway Condition Surveys

The Geographic and Information System (GIS) is a software which can be used in managing and analyzing the geographical information. GIS has many applications in the fields of engineering and sciences. One of the fields which utilizes the GIS is the field of pavement management. The system can be used in maintaining, evaluating, and managing pavement systems. GIS can be used in pavement condition assessments of roadways and bike trails. The system develops maps and tables which can be used to identify the locations where the infrastructure needs maintenance. The roughness of the bike trails across Northern Arizona University (NAU) campus has long been a severe issue to maintenance technicians in the Department of Facility Services. NAU has different bike routes connecting between the North and South campus. About 22 % of NAU students use bikes for their trips, and about 30 % of NAU students use bikes for their school commutes. It is important to maintain bike trails in a good condition to provide better serviceability to all faculty, staff, and students who bike to and from their home. It is the priority that NAU's Facility Services' strategic planning to survey all bike trails, identify different levels of pavement failure mode, and prioritize the areas/locations for prompt rehabilitation. This project demonstrates the effectiveness of

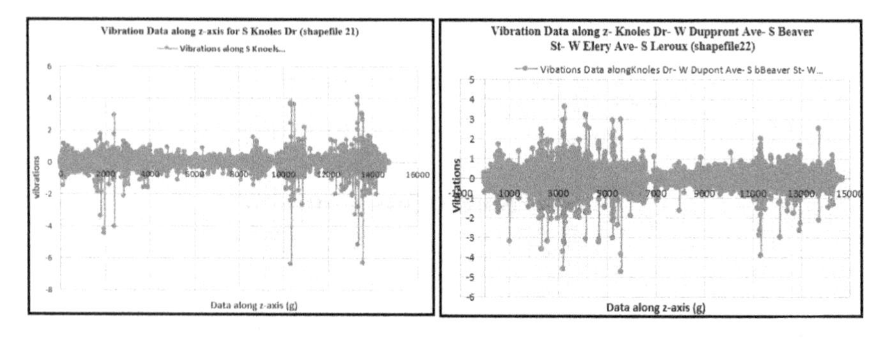

Fig. 4. Vibration data along S Knoles Dr

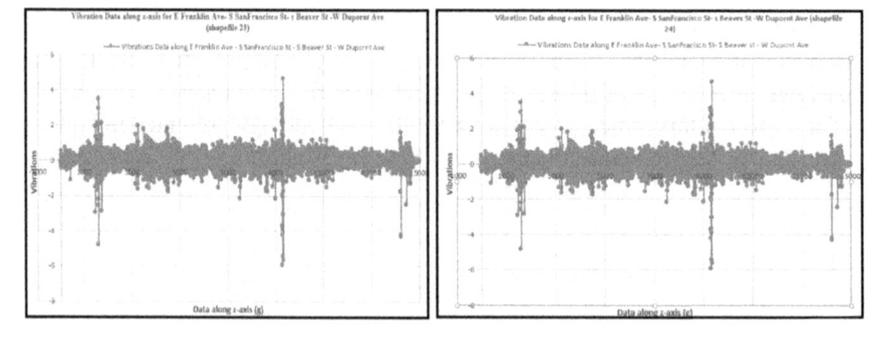

Fig. 5. Vibration data along S San Francisco Ave.

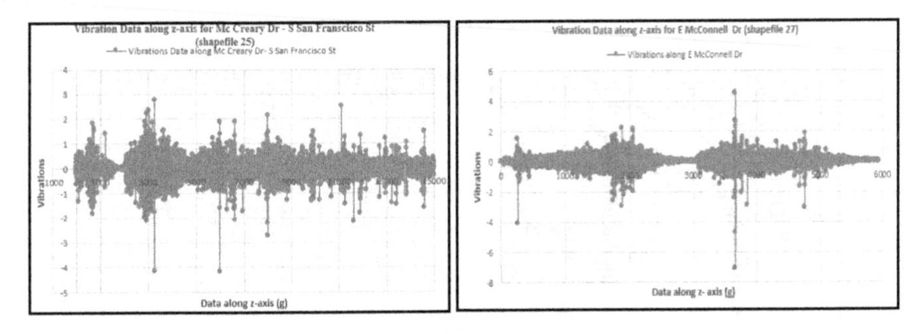

Fig. 6. Vibration data along *McCreary Dr. and McConnell Dr.*

combining mobile sensing technology, data analysis and GIS platform in bike trail condition management.

4.2 Data Collection

The roughness of the bike trail was evaluated using the accelerometer installed on a smartphone. The vibration data obtained from the application were used to evaluate the road roughness. After completion of biking along the campus trails, all vibration data were imported in excel to allow the team and NAU technicians to preview the vibration patterns before exporting all the data into the GIS platform. Figures 4, 5, 6 shows an example of raw vibration data sets collected from different bike trails.

The vibration data set obtained from the app were subsequently presented on an ArcGIS map. The bike trails were presented on the map with the cracks shown on the map. Pictures of the cracks were added to the ArcGIS map. Based on the above raw data, it is extremely difficult to identify where the pavement failures are even though many peak vertical magnitudes were noticed in the displayed excel graphs. Cracks were divided into two types which are large cracks and small cracks using 0.5 g value as a baseline to breakdown these two levels of pavement failure. It should be noted, as of present, there is no standard that specifies how to identify pavement failure modes by using the vibration intensity data. The baseline definition was made based on the previous preliminary road test. GPS data were recorded while biking along trails. All moving path were tracked by the mobile app and were displayed in a digital map using a GIS platform (Fig. 7). It was estimated that approximately a million vibration points were recorded in the GIS map. This moving path in Fig. 7 was a helpful tool that provides the team members with all tracks and trends the team members travelled. The reviewers in the office were able to observe the entire biking activities and make an adjustment if needed.

4.3 Data Analysis and Discussions

The data obtained from the accelerator application were imported to GIS software in order to present it on the map. It is feasible that using GIS software can be better used to manage numerous vibration data points, particularly when dealing with an

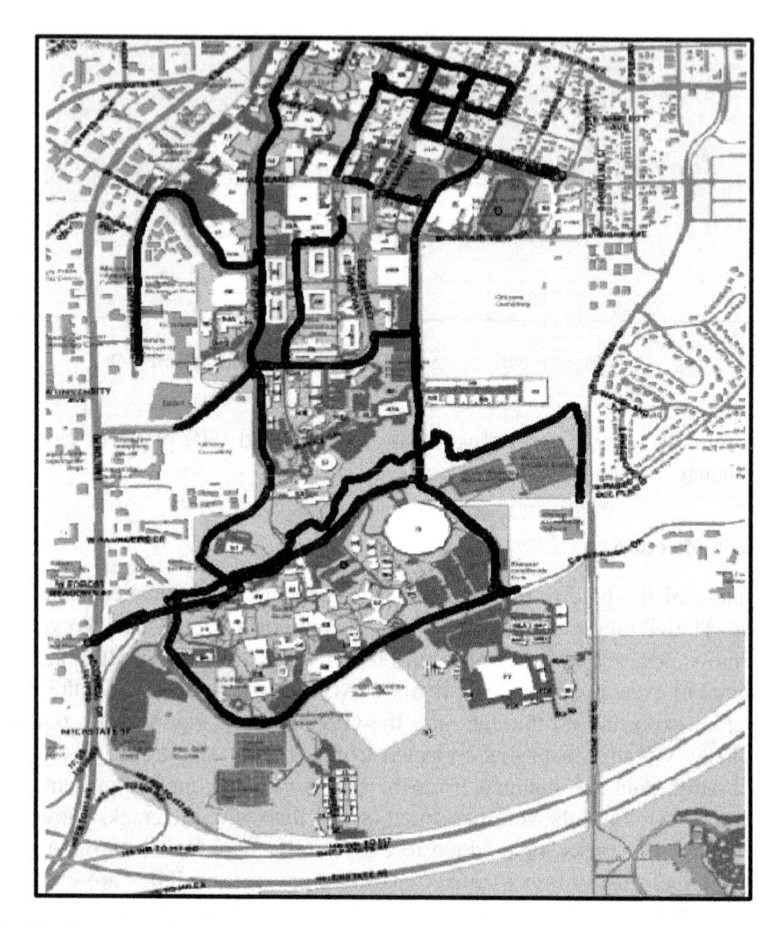

Fig. 7. Moving path of biking shown in a GIS platform (approximately a million data points)

approximately a million of data points. After the data was brought to the GIS map, it was converted to event themes where the x and y fields were set to longitude and latitude, respectively. All information of vibration data were stored in an attribute table for further analysis. The category function of GIS software was used to breakdown two major pavement failure mode: severe cracks and moderate/slight cracks. Given the discussion with NAU technicians, only severe cracking areas (large cracks in maps) will be considered for priority for rehabilitation. Based on the meeting discussion, all vibration data were analyzed and stored in the attribute table of the platform. A GIS map was produced to represent the analysis results (Fig. 8).

As shown in Fig. 8, all bike trails were displayed in a GIS map using five color polylines. The blue polyline was used to represent the bike lanes on road. The red polyline was used to represent bike trails where it's required to ride with caution. The green polyline was used to represent bike routes. The purple polyline was used to represent mixed pedestrian and bike path, and the orange polyline was used to represent the Flagstaff urban trail. The results show that the bike trails on the NAU north campus

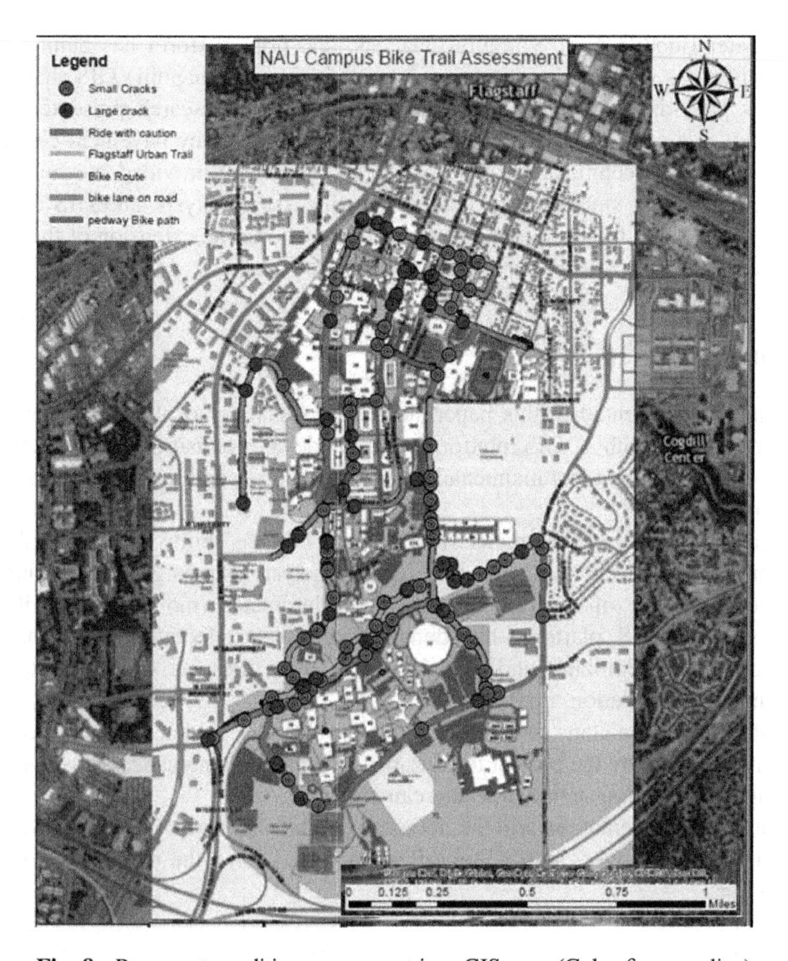

Fig. 8. Pavement condition assessment in a GIS map (Color figure online)

have the experienced a large number of cracks as shown in Fig. 8. During analysis, it was found that there are major types of cracking on the bike trails across the NAU campus, which are thermal cracking, block cracking, and alligator cracking, respectively. Based on visual inspection while biking, most of the cracks along the campus are thermal cracking, a dominant pavement failure mode in the Northern Arizona region.

After completion of data analysis and GIS digitizing and mapping, the research team and NAU technicians were able to sit together to review all pavement failures and their locations. In order to improve the infrastructure of the bike trails system at NAU, the severe cracks along all trails have to be fixed. A periodic maintenance consisting of crack seals and chip seals has been scheduled based on the data analysis and GIS map provided by the research. NAU technicians have been able to implement strategic planning for pavement rehabilitation and or preservation activities following the recommendation from the project results. The combination of smartphone based accelerometer (mobile app), vibration analysis, and GIS platform has demonstrated their effectiveness in analyzing vibration data, digitizing and mapping GIS shape files, as well as prioritizing pavement rehabilitation areas. The research presented in the paper provides a cost-effective method for technicians to survey pavement conditions in a regular basis. Technicians have enjoyed the advantage provided by the modern technology including smartphone, signal processing, data analysis, and GIS as useful knowledge databases to improve the pavement inspection. The extension of the project will be focused on infrastructure system for the near future.

5 Conclusions

1. The method presented in this paper shows the feasibility of utilizing smartphone accelerometers with a GIS platform to provide a promising and cost-effective pavement assessment and maintenance system being applied in both highway roads and bike trails.
2. Smartphone based accelerometer (mobile app) has the ability to record vibration responses and transfer to a server for further analysis using signal processing.
3. The combination of smartphone based accelerometer (mobile app), vibration analysis, and GIS platform has demonstrated their effectiveness in analyzing vibration data, digitizing and mapping GIS shape files, as well as prioritizing pavement rehabilitation areas.
4. Maintenance technicians have been able to follow the guidelines and recommendations provided by the project to prioritize pavement maintenance areas and better manage pavement system across the campus.
5. A future implementation will be focused on the improvement of infrastructure system (pipes, bridges, high volume freeways, etc.) using the method provided by the paper.

Acknowledgements. The authors would like to express their gratitude to Ms. Amal Abdelaziz, Ms. Noor Alsadi, and Ms. Shahad Aloqaili for helping collect vibration data, analyzing data, and providing GIS maps. Their contributions to the research is much appreciated.

References

1. Copp, R.: Field test of three video distress recognition systems. In: The Proceeding of Automated Pavement Distress Data Collection Seminar, Iowa, pp. 12–15 (1990)
2. Bursanescu, L., Hamdi, M.: Three-dimensional laser ranging image reconstruction using three-line laser sensors and fuzzy methods. In: The Proceeding of SPIE, vol. 3835, pp. 106–117 (1999)
3. Caroff, G., Joubert, P., Prudhommee, F., Soussain, G.: Classification of pavement distresses by image processing. In: Proceedings of the 1st International Conference Application of Advanced Technologies in Transportation Engineering, ASCE, San Diego, California, pp. 46–51 (1989)

4. Mohajeri, M.H., Manning, P.: ARIA: an operating system of pavement distress diagnosis. J. Transp. Res. Board. No. 1311, pp. 120–130 (1991)
5. Yu, B., Yu, X.: Vibration-based system for pavement condition evaluation. In: Applications of Advanced Technology in Transportation, pp. 183–189 (2006)

Automatic Synthesis of Fuzzy Inference Systems for Classification

Jorge Paredes, Ricardo Tanscheit(✉), Marley Vellasco, and Adriano Koshiyama

Department of Electrical Engineering,
Pontifical Catholic University of Rio de Janeiro, Rio de Janeiro, Brazil
{jparedes,ricardo,marley}@ele.puc-rio.br, as.koshiyama@gmail.com

Abstract. This work introduces AutoFIS-Class, a methodology for automatic synthesis of Fuzzy Inference Systems for classification problems. It is a data-driven approach, which can be described in five steps: (i) mapping of each pattern to a membership degree to fuzzy sets; (ii) generation of a set of fuzzy rule premises, inspired on a search tree, and application of quality criteria to reduce the exponential growth; (iii) association of a given premise to a suitable consequent term; (iv) aggregation of fuzzy rules to a same class and (v) decision on which consequent class is most compatible with a given pattern. The performance of AutoFIS-Class has been compared to those of other four rule-based systems for 21 datasets. Results show that AutoFIS-Class is competitive with respect to those systems, most of them evolutionary ones.

Keywords: Fuzzy inference system · Automatic synthesis · Classification

1 Introduction

Nowadays, a major part of accumulated knowledge is stored as datasets and many classification algorithms have been developed to extract that knowledge [1,2]. The question of an adequate knowledge representation becomes relevant in applications where a "black box" model [3] does not suffice. This model may provide high accuracy, but not a logical, functional or descriptive way of how results are obtained. In other words, in many cases not only accuracy but also linguistic interpretability is an important factor. Fuzzy Inference Systems [4] are well known for their capability of representing knowledge in a comprehensible way through inference rules and constitute therefore an appropriate modeling for merging interpretability and accuracy [5,6]. The assessment of how interpretable a fuzzy system is usually takes into account, for example, the model structure (Mamdani, Takagi-Sugeno), number of rules, number of antecedents and the membership functions format [6,7].

In order to build models that consider both accuracy and interpretability, most studies have used (i) Evolutionary Algorithms to elaborate fuzzy rules with or without membership function tuning [5,8,9] and (ii) Evolving Fuzzy

J.P. Carvalho et al. (Eds.): IPMU 2016, Part I, CCIS 610, pp. 486–497, 2016.
DOI: 10.1007/978-3-319-40596-4_41

Inference Systems [3,10–12] that create and adapt a fuzzy rule base by gathering new observations. A negative aspect in approach (i) is its computational cost and the number of parameters – due to the use of a Evolutionary Algorithm, in some cases multi-objective. As for approach (ii), several works succeed in obtaining low-lenght rules and compact rule bases [13–16], which are important aspects in the interpretability context [17]. However, the approach proposed here is a simpler one, in the sense that it assumes an initial and fixed allocation for membership functions and does not need post-processing.

This paper presents an off-line approach that generates a Fuzzy Inference System automatically, focused on classification and seeking accuracy but also favoring linguistic interpretability. The proposed AutoFIS-Class model has the following main characteristics: (i) generates premises to ensure minimum quality criteria, (ii) associates each premise to the more compatible consequent term and (iii) aggregates the degrees of activation of rules related to a same class by using aggregation operators that weigh rules on the basis of their impact on the classification.

The case studies take as a basis the work of Alcal-Fdez et al. [18] and consider 21 datasets. The performance of AutoFIS-Class is compared to those of other four Evolutionary Fuzzy Systems that generate rules.

The next section presents the AutoFIS-Class model while the third section describes the experiments and presents results and discussions. The fourth section concludes the paper and suggests future works.

2 AutoFIS-Class Model

In this section the stages of the AutoFIS-CLASS model are described, starting with the mapping of crisp values into membership degrees (Fuzzification). The inference process comprises definition of rules premises (Formulation), definition of the most appropriate consequent (Association) and the union of the activations of each rule (Aggregation). The winning class for a given pattern is defined in the Decision stage. The outline of the model is shown in Fig. 1. The inner annulus symbolizes the Formulation process, which contains all the premises; the yellow region corresponds to those that meet the minimum quality criteria to get through to the Association stage. Here premises are placed in different classes (hence the different colours) and some of them are not useful to any class (black region). Finally, in the Aggregation stage, rules are weighed and premises with small influence (black region) are excluded from the classification process.

2.1 Fuzzification

In classification, the main information consists of n patterns $\mathbf{x}_i = [x_{i1}, x_{i2}, ..., x_{iJ}]$ of J attributes X_j present in the database ($i = 1, ..., n$ and $j = 1, ..., J$). A number of L fuzzy sets $A_{jl} = \{(x_{ij}, \mu_{A_{jl}}(x_{ij}))|x_{ij} \in X_j\}$ is associated to each j−th attribute, where $\mu_{A_{jl}} : X_j \rightarrow [0, 1]$ is a membership function that associates to each observation x_{ij} a membership degree $\mu_{A_{jl}}(x_{ij})$ to the fuzzy

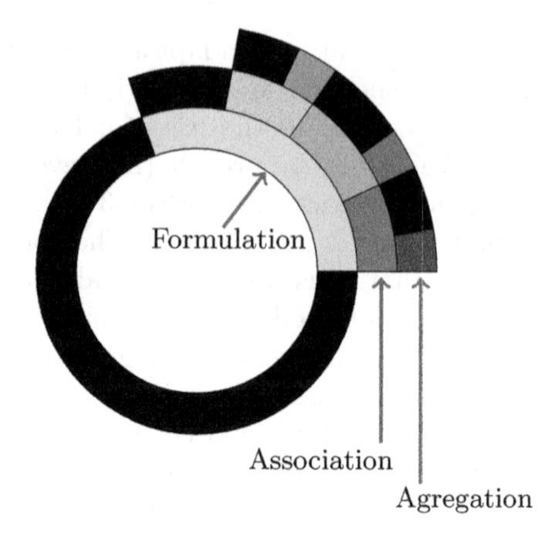

Fig. 1. AutoFIS-Class: inference stage. (Color figure online)

set A_{lj}. Each pattern is associated to a class C_i out of K possible ones, that is $C_i \in \{1, 2, ...k, ..., K\}$.

Three aspects are taken into account in the Fuzzification stage: membership functions format, the support of each membership function $\mu_{A_{jl}}(x_{ij})$ and the appropriate linguistic label, which qualifies the subspace defined by the membership function with a suitable term. In theory, this could be done by an expert, but in practice – due to the difficulty of finding such an expert – a strong partition is usually employed [8,19,20], as shown in Fig. 2a.

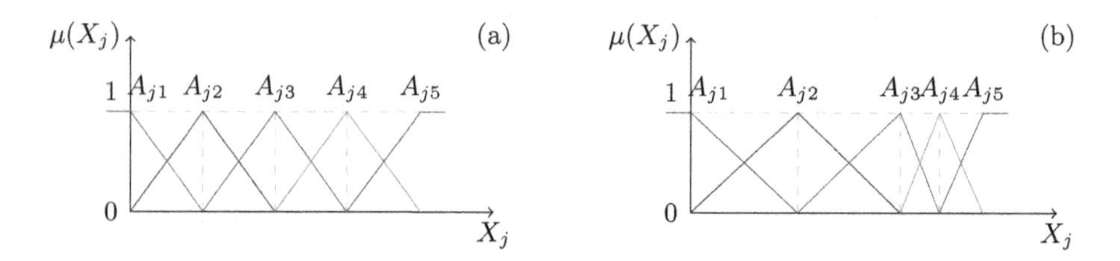

Fig. 2. Triangular membership functions: (a) Strongly partitioned and (b) Tukey.

Another approach considers the information from the quartiles: 0th quartile (minimum value), 1st quartile, 2nd quartile (median), 3rd quartile and 4th quartile (maximum value). This type of partitioning is called Tukey and is shown in Fig. 2b. It should be noted that in this work categorical attributes and the classes are binarized.

2.2 Fuzzy Inference

Formulation. This stage consists of building rule premises. A premise is usually defined as: "*If* X_1 *is* A_{1l} *and* ...*and* X_j *is* A_{jl} *and* ...*and* X_J *is* A_{Jl}", or, in mathematical terms:

$$\mu_{A_d}(\mathbf{x}_i) = \mu_{A_{1l}}(x_{i1}) * \ldots * \mu_{A_{jl}}(x_{ij}) * \ldots * \mu_{A_{Jl}}(x_{iJ}) \quad (1)$$

where $\mu_{A_d}(\mathbf{x}_i)$ is the joint membership degree of pattern i in premise d, $(d = 1, \ldots, D)$, computed by using a t-norm $*$ that combines every $\mu_{A_{jl}}(x_{ij})$. Thus, a premise can be built from a combination of the $\mu_{A_{jl}}(x_{ij})$ through the use of t-norms, t-conorms, negation operators and linguistic hedges.

In this work, the product t-norm is the only one employed. Besides, the negation operator can act upon each element in a premise:

"*If* X_1 *is not* A_{1l} *and* ...*and* X_j *is not* A_{jl} *and* ...*and* X_J *is not* A_{Jl}".

This procedure, as opposed to negating the result as a whole, that is No ("*If* X_1 *is* A_{1l} *and* ...*and* X_j *is* A_{jl} *and* ...*and* X_J *is* A_{Jl}"), doubles the number of possibilities of premises, enlarging therefore the search space.

The number of possible premises that will form a rule base grows exponentially with the quantity of attributes. As an example, consider a case with three attributes (X_1, X_2, X_3) where each of them is associated to two linguistic terms. Premises of size (number of antecedent elements) 1 are: "If X_1 is A_{1L}", "If X_1 is A_{1H}", ..., and "If X_3 is A_{3H}"; of size 2: "If X_1 is A_{1L} and X_2 is A_{2L}", ..., and "If X_3 is A_{3H} and X_1 is A_{1H}"; and of size 3: "If X_1 is A_{1L} and X_2 is A_{2L} and X_3 is A_{3L}", ..., and "If X_1 is A_{1H} and X_2 is A_{2H} and X_3 is A_{3H}". In total there will be six premises of size 1, twelve of size 2 and eight of size 3. When the number of attributes and of linguistic terms increase, a large quantity of premises is generated. However, not all of them will be used to form the rule base: many will either be redundant or will generate conflicts in the decision of the most appropriate class.

Regarding interpretability, it is desirable to have few rules with few antecedent elements. Thus, the following procedure is proposed.

1. Limit on the maximum size of premises;
2. Evaluation of a premise viability through a set of filters: support, similarity and conflict in classification;
3. Organized generation of premises: initialization with size-1 premises, creation of size-2 premises from the size-1 viable ones, generation of size-3 premises from viable size-2 premises and so on.

With respect to the first item above, the maximum size of premises is a user-defined parameter; its highest value is the number of attributes in the database. As for the third item, the objective is to build premises with few antecedent elements (more interpretable) and do not incur in computational overload by using premises that will not be adequate for a fuzzy rule (as a result of item 2). The filtering process is described below.

Support Filter: aims at building premises that cover a large number of patterns in the database. The *Relative Support* of a premise $\mu_{A_d}(\mathbf{x}_i)$ is given by:

$$Sup_d = \frac{\sum_{i=1}^{n} \mu_{A_d}(\mathbf{x}_i)}{n} \tag{2}$$

Given a user-defined tolerance ε_{Sup}, a premise is deemed viable if $Sup_d > \varepsilon_{sup}$. If $\mu_{A_*}(\mathbf{x}_i) = \mu_{A_{1L}}(x_{i1})$ does not present a Relative Support greater than ε_{sup}, any combination of $\mu_{A_*}(\mathbf{x}_i)$ with other membership functions will generate premises that are not viable. This may be verified through the degeneration property of most t-norms [21].

Similarity Filter: its objective is to reduce the occurrence of similar or identical premises. Given two premises $\mu_{A_d}(\mathbf{x}_i)$ and $\mu_{A_v}(\mathbf{x}_i)$, the similarity between them can be computed by:

$$Sim_{d,v} = \frac{\sum_{i=1}^{n} \min\{\mu_{A_d}(\mathbf{x}_i), \mu_{A_v}(\mathbf{x}_i)\}}{\sum_{i=1}^{n} \max\{\mu_{A_d}(\mathbf{x}_i), \mu_{A_v}(\mathbf{x}_i)\}} \tag{3}$$

Given a user-defined tolerance ε_{sim}, two premises are similar if $Sim_{d,v} > \varepsilon_{sim}$. If the similarity is identified, the premise with the lower Relative Support is removed. It should be noted that if a premise is excluded when $Sim_{d,v} > \varepsilon_{sim}$, its combination with another membership function (that is, a larger premise derived from it) would not necessarily be excluded. However, this exclusion is performed in order to generate the smallest possible number of premises that will get through to the next stages.

PCD Flter: this aims at reducing the occurrence of similar or conflicting rules by computing the Penalized Confidence Degree (PCD_k) [22]:

$$PCD_k = max\left\{\frac{\sum_{i\epsilon k} \mu_{A_d}(\mathbf{x}_i) - \sum_{i\notin k} \mu_{A_d}(\mathbf{x}_i)}{\sum_{i=1}^{n} \mu_{A_d}(\mathbf{x}_i)}, 0\right\} \tag{4}$$

The objective is to compute the activation degree of the remaining classes and discount them from CD_k. This should result in rules more specific to a given class and not in general ones. A premise is not viable if $PCD_k = 0$ for all K classes.

After generating premises $\mu_{A_1}(\mathbf{x}_i)$, $\mu_{A_2}(\mathbf{x}_i)$ ..., $\mu_{A_D}(\mathbf{x}_i)$ that satisfy size requirements and that have gone through all filters, the next step is the association of each premise to a consequent class.

Association. This determines the consequent class that is most compatible to a given premise $\mu_{A_d}(\mathbf{x}_i)$. Premise d associated to class k (i.e. a fuzzy rule) is denoted by $\mu_{A_{d(k)}}(\mathbf{x}_i)$, which describes, in linguistic terms:

"*If* X_1 *is* A_{1l} *and...and* X_J *is* A_{Jl}, *then* $\mathbf{x}_i$ *is Class* k"

Here, the Confidence Degree (CD_k) [19,23] is used for identifying the compatibility between premises and classes:

$$CD_k = \frac{\sum_{i \in k} \mu_{A_d}(\mathbf{x}_i)}{\sum_{i=1}^{n} \mu_{A_d}(\mathbf{x}_i)} \in [0,1] \tag{5}$$

where $\sum_{i \in k} \mu_{A_d}(\mathbf{x}_i)$ is the sum of the compatibility degree between the premise and class k and $\sum_{i=1}^{n} \mu_{A_d}(\mathbf{x}_i)$ is the total compatibility. The class to be assigned to a $\mu_{A_d}(\mathbf{x}_i)$, that is, $\mu_{A_{d(k)}}(\mathbf{x}_i)$, is given by the k-th class that maximizes CD_k. Premises for which $CD_k = 0$ will not be assigned to any consequent class.

Aggregation. In a Fuzzy Inference System a new pattern $\mathbf{x}_i^*$ may be compatible with several rules, pertinent to one or more classes. The aggregation stage combines the activation degree $\mathbf{x}_i^*$ of rules related to a same class, so that a consensual value is generated for discriminating the target class.

Consider $D^{(k)}$ as the number of rules related to class k. Given an aggregation operator [24], $g : [0,1]^{D^{(k)}} \to [0,1]$, the membership degree of $\mathbf{x}_i^*$ in each of the K classes ($\hat{\mu}_{C_i \in k}(\mathbf{x}_i^*)$) will be:

$$\hat{\mu}_{C_i \in 1}(\mathbf{x}_i^*) = g[\mu_{A_{1(1)}}(\mathbf{x}_i^*), ..., \mu_{A_{D(1)}}(\mathbf{x}_i^*)] \tag{6}$$

$$\hat{\mu}_{C_i \in 2}(\mathbf{x}_i^*) = g[\mu_{A_{1(2)}}(\mathbf{x}_i^*), ..., \mu_{A_{D(2)}}(\mathbf{x}_i^*)] \tag{7}$$

$$\dots$$

$$\hat{\mu}_{C_i \in K}(\mathbf{x}_i^*) = g[\mu_{A_{1(K)}}(\mathbf{x}_i^*), ..., \mu_{A_{D(K)}}(\mathbf{x}_i^*)] \tag{8}$$

In aggregation, the most used t-conorm is the maximum. Alternatively, the Weighted Average may be used similarly to the procedure of AsMQR already described [24]. In this work the use of Weighted Average estimated through Restricted Least Squares (RLS) is proposed:

$$\hat{\mu}_{C_i \in 1}(\mathbf{x}_i^*) = \sum_{d^{(1)}=1}^{D^{(1)}} w_{d^{(1)}} \mu_{A_{d(1)}}(\mathbf{x}_i) \tag{9}$$

$$\dots$$

$$\hat{\mu}_{C_i \in K}(\mathbf{x}_i^*) = \sum_{d^{(K)}=1}^{D^{(K)}} w_{d^{(K)}} \mu_{A_{d(K)}}(\mathbf{x}_i) \tag{10}$$

where $w_{d^{(k)}}$ is the weight, or the degree of influence, of $\mu_{A_{d(k)}}(\mathbf{x}_i)$ in the discrimination of patterns related to class k. The process described in the Association stage is used for finding the weights.

Once the membership degree $\mathbf{x}_i^*$ has been computed for each of the K classes, the information given by $\hat{\mu}_{C_i \in 1}(\mathbf{x}_i^*)$,...,$\hat{\mu}_{C_i \in K}(\mathbf{x}_i^*)$ helps decide which class the pattern $\mathbf{x}_i^*$ belongs to.

Decision. Given a new pattern $\mathbf{x}_i^*$, the decision on its membership to class k is computed by:

$$\hat{C}_i = \arg_k \max\{\hat{\mu}_{C_i \in 1}(\mathbf{x}_i^*), ..., \hat{\mu}_{C_i \in k}(\mathbf{x}_i^*), ..., \hat{\mu}_{C_i \in K}(\mathbf{x}_i^*)\} \tag{11}$$

where $\hat{C}_i$ is the predicted class: result of the k-th argument that minimizes (11). Therefore, this method associates pattern $\mathbf{x}_i^*$ to the most pertinent class according to the existing rule base. In the case of a tie, either a heuristic may be applied (for example: impose on $\mathbf{x}_i^*$ the class that has more patterns in the data base) or no class is defined for $\mathbf{x}_i^*$.

The information provided by $\hat{C}_i$ for the whole pattern makes it possible to evaluate the performance of AutoFIS-Class. A simple approach consists of computing the Average Classification Error (AVC):

$$AVC = \frac{\sum_{i=1}^{n} |C_i - \hat{C}_i|}{n} \tag{12}$$

where $|C_i - \hat{C}_i| = 0$ se $C_i = \hat{C}_i$, and 1 otherwise.

3 Case Studies

3.1 Experiments

The databases used in the experiments – 21 of the 26 employed by Alcal-Fdez et al. [18] – were taken from the KEEL site [25] and are shown in Table 1.

Some databases are small (Iris, Appendicitis and Wine) and others present high scalability and dimensionality(MSpambase, Texture, Satimage, etc.). Eleven of them are of a binary type and the remaining ten are multiple-class databases.

The work carried out in [18] shows results for four models: 2SLAVE, FH-GBML, SGERD e FARC-HD, alll of them Evolutionary Fuzzy Systems aiming at generating concise fuzzy rule bases by using an evolutionary metaheuristics. The model FARC-HD (Fuzzy Association Rule-Based Classification Model for High-Dimensional) proposed in [18] employs an evolutionary algorithm that selects a subset of rules initially generated through a procedure similar to that of AutoFIS-Class but without employing some of the filters.

Experiments with AutoFIS-Class have followed the same procedure used in [18]: (*i*) for each database a 10-fold cross-validation is performed; (*ii*) AutoFIS-Class is run for each cross-validation fold and accuracy and number of rules in the test stage are recorded; and (*iii*) the average accuracy for the 10 folds is reported as a result.

Table 2 shows the settings used in the experiments, both for binary an multiple classes databases. Those settings have been obtained from preliminary tests performed to evaluate (superficially) the sensitivity of AutoFIS-Class to the various parameters. It can be seen that the main differences between settings for binary and multiple classes databases reside on disabling the PCD filter and on

Table 1. Main characteristics of the databases.

Database	J	n	K
Iris	4	150	3
Cleveland	13	297	5
Phoneme	5	5404	2
Vowel	13	990	11
Monk-2	6	432	2
Crx	15	653	2
Appendicitis	7	106	2
Pen-based	16	10992	10
Pima	8	768	2
Two-norm	20	7400	2
Glass	9	214	6
Wdbc	30	569	2
Page-blocks	10	5472	5
Satimage	36	6435	6
Magic	10	19020	2
Texture	40	5500	11
Wine	13	178	3
German	20	1000	2
Heart	13	270	2
Spambase	57	4597	2
Ring	20	7400	2

the increase in the number of membership functions for multiple classes databases. This is due to the fact that rules were not generated for all classes, and the few that were generated for some classes were excluded by the PCD filter. Statistical analyses followed the recommendations of [26,27]; a significance level of 5 % was adopted.

3.2 Results

The main results - in terms of accuracy and number of rules – for each model in the test phase are shown in Table 3. The FARC-HD model presents the best average results in terms of accuracy.

In order to check whether the difference in accuracy is significant, Table 4 presents the results provided by the Friedman and Iman Davenport test and by the Holm method [27]. The FARC-HD model presented the lowest rank (1.3333): its accuracy is such that it almost always figures in first place, AutoFIS-Class occupies second place, with a rank of 2.6190. The Friedman and Iman-Davenport test show that some algorithms present significantly lower or higher ranks than

Table 2. AutoFIS-Class settings for binary and multiple classes databases.

Settings	Binary	Multiple classes
Membership functions format	Tukey	Tukey
Number of membership functions	3	5
t-norm	prod	prod
Negation	yes	yes
Premise size	2	2
ε_{sup}	0,05	0,05
ε_{sim}	0,95	0,95
PCD Filter	yes	no
Association	CD	CD
Aggregation	RLS	RLS

Table 3. Accuracy in test phase and no. of rules (# R) for each model and database.

Database	AutoFIS-Class		2SLAVE		FH-GBML		SGERD		FARC-HD	
	Tst. (%)	# R	Tst. (%)	# R	Tst. (%)	# R	Tst. (%)	# R	Tst. (%)	# R
Appendicitis	**89,82**	5,6	82,91	4,4	*86,00*	13,8	84,48	2,5	84,18	6,8
Crx	**86,77**	19,1	74,06	2,4	86,60	11,6	85,03	2,1	*86,03*	25,4
German	71,40	26,8	70,53	6,5	**87,01**	5,1	67,97	3,4	*72,8*	85,7
Heart	*83,33*	23,6	71,36	4,3	75,93	12,7	73,21	2,7	**84,44**	27,0
Magic	79,91	7,7	73,96	4,1	*81,23*	9,9	72,06	3,1	**84,51**	43,3
Monk-2	97,57	9,7	97,26	3,0	*98,18*	14,7	80,65	2,2	**99,77**	14,2
Phoneme	71,82	5,8	76,41	11,5	*79,66*	17,4	75,55	3,6	**82,14**	17,8
Pima	73,97	7,4	73,71	7,8	*75,26*	10,6	73,37	3,1	**75,66**	22,7
Ring	50,62	13,8	79,63	4,6	*86,92*	6,9	72,63	6,8	**94,08**	24,0
Spambase	*88,10*	21,7	70,14	7,9	77,22	3,9	72,98	3,7	**91,93**	29,8
Two-norm	*93,47*	31,4	86,99	26,5	85,97	12	73,98	3,1	**95,28**	60,9
Wdbc	91,91	6,6	*92,33*	5,2	92,26	7,2	90,68	3,7	**95,25**	10,4
Cleveland	**58,61**	56,4	48,82	11,9	53,51	6,9	51,59	6,4	*55,24*	61,3
Glass	*67,07*	46,7	58,05	15,1	57,99	9,4	58,49	6,9	**70,24**	22,7
Iris	*95,33*	10,9	94,44	4,0	94,00	14,9	94,89	3,4	**96,00**	4,0
Page-blocks	*94,68*	11,8	91,39	7,5	94,21	7,4	90,72	6,5	**95,01**	19,1
Penbased	*83,94*	111,9	81,16	40,0	50,45	18,4	67,93	15,9	**96,04**	152,8
Satimage	81,55	61,8	*81,69*	57,9	74,72	16,5	77,10	12,2	**87,32**	76,1
Texture	*87,24*	74,8	81,57	34,9	70,15	14,6	71,66	18,6	**92,89**	54,5
Vowel	63,84	113,6	*71,11*	63,1	67,07	9,2	65,83	18	**71,82**	72,3
Wine	*93,76*	16,8	89,47	5,5	92,61	9,2	91,88	4,2	**94,35**	8,7
Average	*81,18*	32,57	78,43	15,62	79,38	11,06	75,84	6,29	**85,95**	39,98
Std. Dev	13,15	33,22	11,85	18,19	13,49	4,10	11,01	5,25	11,32	35,68

Table 4. Friedman test and Holm method results.

i	Model	Rank		
4	SGERD	4.1905		
3	2SLAVE	3.7619		
2	FH-GBML	3.0952		
1	AutoFIS-Class	2.6190		
0	FARC-HD	**1.3333**		
Test	p-value			
Friedman	< 0.0001			
Iman Davenport	< 0.0001			
Model	$z = (R_0 - R_i)/SE$	p-value	Holm	Reject?
SGERD	5.8551	<0.0001	0.0125	yes
2SLAVE	4.9770	<0.0001	0.0167	yes
FH-GBML	3.6108	0.0003	0.0250	yes
AutoFIS-Class	2.6349	0.008415	0.0500	yes

their counterparts. As FARC-HD has the lowest rank, it has been selected to be compared pairwise with the other models. The FARC-HD model has presented the lowest rank (p-value $< 0,05$), which means that it outperforms all the other models in the test phase.

With respect to the rule base, AutoFIS-Class generated fewer rules than FARC-HD for 76,19 % of the databases. The main objective of obtaining a compact rule base without affecting accuracy in a significant way has been fully attained. It must also be stressed that, as opposed to the other models, AutoFIS-Class does not employ any evolutionary algorithm routine for rule base simplification or improvement.

4 Conclusion

This paper has presented AutoFIS-Class, a new methodology for automatic generation of Fuzzy Inference Systems for Classification. Its performance was evaluated for 21 databases and compared to those of four other models (that employ evolutionary approaches). Although AutoFIS-Class does not make use of any evolutionary algorithm, it has performed very well in terms of accuracy, falling behind FARC-HD only. On the other hand, AutoFIS-Class has generated a more compact rule base than FARC-HD has, reaching therefore a good compromise between accuracy an interpretability.

Future works will consider a hybridization with Wang & Mendel's algorithm [28] for the generation of longer premises, the use of nonlinear aggregation methods, new forms of association and new filters. Given the ease of execution of the proposed methodology, practical applications will also be considered.

References

1. Baitharu, T.R., Pani, S.K.: A survey on application of machine learning algorithms on data mining. Int. J. Innovative Technol. Explor. Eng. **3**(7), 17–20 (2013)
2. Phyu, T.N.: Survey of classification techniques in data mining. Int. Multiconference Eng. Comput. Sci. **1**, 18–20 (2009)
3. Lughofer, E.: On-line assurance of interpretability criteria in evolving fuzzy systems-achievements, new concepts and open issues. Inf. Sci. **251**, 22–46 (2013)
4. Alonso, J.M., Magdalena, L., González-Rodríguez, G.: Looking for a good fuzzy system interpretability index: an experimental approach. Int. J. Approximate Reasoning **51**(1), 115–134 (2009)
5. Herrera, F.: Genetic fuzzy systems: taxonomy, current research trends and prospects. Evol. Intell. **1**(1), 27–46 (2008)
6. Cordón, O.: A historical review of evolutionary learning methods for mamdani-type fuzzy rule-based systems: designing interpretable genetic fuzzy systems. Int. J. Approximate Reasoning **52**(6), 894–913 (2011)
7. Gacto, M.J., Alcalá, R., Herrera, F.: Interpretability of linguistic fuzzy rule-based systems: an overview of interpretability measures. Inf. Sci. **181**(20), 4340–4360 (2011)
8. Ishibuchi, H., Yamane, M., Nojima, Y.: Rule weight update in parallel distributed fuzzy genetics-based machine learning with data rotation. In: International Conference on Fuzzy Systems, pp. 1–8. IEEE (2013)
9. Fazzolari, M., Alcala, R., Nojima, Y., Ishibuchi, H., Herrera, F.: A review of the application of multiobjective evolutionary fuzzy systems: current status and further directions. IEEE Trans. Fuzzy Syst. **21**(1), 45–65 (2013)
10. Leite, D., Ballini, R., Costa, P., Gomide, F.: Evolving fuzzy granular modeling from nonstationary fuzzy data streams. Evol. Syst. **3**(2), 65–79 (2012)
11. Lughofer, E.: Evolving Fuzzy Systems-Methodologies, Advanced Concepts and Applications, vol. 53. Springer, Heidelberg (2011)
12. Angelov, P.P., Zhou, X.: Evolving fuzzy-rule-based classifiers from data streams. IEEE Trans. Fuzzy Syst. **16**(6), 1462–1475 (2008)
13. Lughofer, E., Cernuda, C., Kindermann, S., Pratama, M.: Generalized smart evolving fuzzy systems. Evol. Syst. **6**(4), 269–292 (2015)
14. Angelov, P.: Evolving takagi-sugeno fuzzy systems from streaming data. Evol. Intell. Syst. Method. Appl. **12**, 21 (2010)
15. Rosini, M.D.: Applications. In: Rosini, M.D. (ed.) Macroscopic Models for Vehicular Flows and Crowd Dynamics: Theory and Applications. UCS, vol. 12, pp. 217–226. Springer, Heidelberg (2013)
16. Leng, G., Zeng, X.J., Keane, J.A.: An improved approach of self-organising fuzzy neural network based on similarity measures. Evol. Syst. **3**(1), 19–30 (2012)
17. Angelov, P., Lughofer, E., Zhou, X.: Evolving fuzzy classifiers using different model architectures. Fuzzy Sets Syst. **159**(23), 3160–3182 (2008)
18. Alcalá-Fdez, J., Alcalá, R., Herrera, F.: A fuzzy association rule-based classification model for high-dimensional problems with genetic rule selection and lateral tuning. IEEE Trans. Fuzzy Syst. **19**(5), 857–872 (2011)
19. Berlanga, F., Rivera, A., del Jesus, M., Herrera, F.: Gp-coach: genetic programming-based learning of compact and accurate fuzzy rule-based classification systems for high-dimensional problems. Inf. Sci. **180**(8), 1183–1200 (2010)
20. Koshiyama, A.S., Vellasco, M.M., Tanscheit, R.: GPFIS-CLASS: a genetic fuzzy system based on genetic programming for classification problems. Appl. Soft Comput. **37**, 561–571 (2015)

21. Klement, E.P., Mesiar, R., Pap, E.: Triangular norms. Kluwer Academic Publishers, Dordrecht (2000)
22. Fernández, A., Calderón, M., Barrenechea, E., Bustince, H., Herrera, F.: Solving multi-class problems with linguistic fuzzy rule based classification systems based on pairwise learning and preference relations. Fuzzy Sets Syst. **161**(23), 3064–3080 (2010)
23. Ishibuchi, H., Yamamoto, T.: Rule weight specification in fuzzy rule-based classification systems. IEEE Trans. Fuzzy Syst. **13**(4), 428–435 (2005)
24. Calvo, T., Kolesárová, A., Komorníková, M., Mesiar, R.: Aggregation operators: properties, classes and construction methods. In: Calvo, T., Mayor, G., Mesiar, R. (eds.) Aggregation Operators: New Trends and Applications. Studies in Fuzziness and Soft Computing, vol. 97, pp. 3–104. Physica, Heidelberg (2002)
25. Alcalá-Fdez, J., Fernandez, A., Luengo, J., Derrac, J., García, S., Sánchez, L., Herrera, F.: Keel data-mining software tool: data set repository, integration of algorithms and experimental analysis framework. J. Multiple Valued Logic Soft Comput. **17**(2), 255–287 (2011)
26. Conover, W.J.: Practical Nonparametric Statistics, 2nd edn. John Wiley, New York (1980)
27. Derrac, J., García, S., Molina, D., Herrera, F.: A practical tutorial on the use of nonparametric statistical tests as a methodology for comparing evolutionary and swarm intelligence algorithms. Swarm Evol. Comput. **1**(1), 3–18 (2011)
28. Wang, L.X., Mendel, J.M.: Generating fuzzy rules by learning from examples. IEEE Trans. Syst. Man Cybern. **22**(6), 1414–1427 (1992)

A Proposal for Modelling Agrifood Chains as Multi Agent Systems

Madalina Croitoru[1](✉), Patrice Buche[2], Brigitte Charnomordic[2], Jerome Fortin[1], Hazael Jones[2], Pascal Neveu[2], Danai Symeonidou[2], and Rallou Thomopoulos[2]

[1] University of Montpellier, Montpellier, France
{croitoru,fortin}@lirmm.fr
[2] INRA, Montpellier, France
{buche,charnomordic,jones,neveu,symeonidou,thomopoulos}@supagro.inra.fr

Abstract. Viewing the modelling of agrifood chains (AFC) from a multi agent systems (MAS) point of view opens up numerous avenues for research while building upon existing advancements in the state of the art. This paper explores different aspects in MAS research areas in consensus and cooperation (argumentation, negotiation, normative systems, multi agent resource allocation and social affects) and provides insights into how viewing classical AFC problems from this perspective can bring new perspectives and research avenues.

Keywords: Multi agents systems · Agri-food chain modelling · Argumentation · Alignment · Coalition formation · Normative reasoning

1 Introduction

Understanding and controlling agri-food processes is of major importance when trying to ensure sustainability with respect to growing complexity and consumer expectation. Methodologies and tools from various sub-fields of Artificial Intelligence have showed their potential for advancing the state of the art.

Here we solely focus on the problem of dealing with the uncertain knowledge (elicitation, representation and reasoning) involved at different levels of the food chain. Such chains often model complex processes relying on numerous criteria, using various granularity of knowledge, most often inconsistent (due to the fact that complementary points of view can be expressed). The main aspect that characterises such knowledge is uncertainty that could be either regarded from a logical point of view or a provenance point of view.

Many approaches in the literature investigate the added value of a logical based representation to deal with the above mentioned problems. Such approaches [10] (mainly using ontologies and Linked Open Data) bring real added value within each step of the transformation but they have difficulty addressing overall chain transformations [9,23,27]. Approaches based on reasoning in presence of inconsistency (such as argumentation based approaches of

J.P. Carvalho et al. (Eds.): IPMU 2016, Part I, CCIS 610, pp. 498–509, 2016.
DOI: 10.1007/978-3-319-40596-4_42

[16,31,34,35]) look at integrating various steps of the food chain but they do not address multi objective optimisation problems common within food chains. Furthermore, recent advances in Linked Open Data and its potential for interoperability meant that more and more ontologies are developed by various actors of the food chain. Methods for integrating these ontologies in a principled manner are still to be developed within each food chain.

This paper investigates the following research question: "*What are the salient points of addressing knowledge representation and reasoning (KRR) in Agri-Food Chain (AFC) as a consensus and cooperation problem in multi-agent systems (MAS)?*" Our claim is that viewing the problem of KRR from a multi agent system point of view opens up numerous avenues for research while building upon existing advancements in the state of the art. We will explore different aspects in MAS research areas in consensus and cooperation (argumentation, negotiation, normative systems, multi agent resource allocation and social affects) and show how viewing classical AFC problems from this perspective can bring new perspectives and research avenues. We claim that agent technology can optimise food supply chain operation by employing intelligent agent applications (as shown in supply chain management case) but also facilitate reasoning with incomplete, inconsistent and missing knowledge - a key aspect of KRR management in AFC. Agents enhance the flexibility and efficiency of supply chain management while providing an unifying framework for various key problems in AFC. The main contribution of the paper resides in this unifying aspects: by modelling AFC problems as MAS problems we benefit from a unifying setting that encompasses a plethora of related research questions.

The paper is structures as follows. After a quick introduction on multi-agent systems in Sect. 2 we investigate the use of argumentation (Sect. 3.1), multi-agent resource allocation (Sect. 4.1), normative systems (Sect. 4.2), for AFC research. Section 5 concludes the paper.

2 Consensus and Cooperation in Multi Agent Systems

In agrifood chains, the products traditionally go through the intermediate stages of processing, storage, transport, packaging and reach the consumer (the demand) from the producer (the supply). More recently, due to an increase in quality constraints, several parties are involved in production process, such as consumers, industrials, health and sanitary authorities, etc. expressing their requirements on the final product as different point of views which could be conflicting. Such complex systems require to be addressed both at each individual transformation level as well as in its globality (from the genome to the final product).

Autonomous agents and multi agent systems represent a way of analysing, designing and implementing complex software systems. A multi agent system can be seen as a loosely coupled network of problem solvers that work together to solve problems beyond individual capabilities of each one of them. In multi agent systems each agent has its own incomplete information, the data is decentralised and computation is asynchronous. Such systems have the advantage of distributed and concurrent problem solving with a plethora of interactions possible [20,25]. Common types of interactions include: cooperation, coordination, negotiations, planning, norm compliance, blame assignment, etc.

When representing and reasoning about an agent's mind, one can distinguish between:

1. Cognitive models of rational action (representing the attitudes of agents, their beliefs, intentions etc.) and
2. Modelling of the strategic structure of the systems (how can agents accomplish their intentions either alone or in cooperation).

Regarding the first aspect (rational cognitive states) one can identify different attitudes such as information attitudes (beliefs), pro attitudes (desires, intentions, goals) and normative attitudes (obligations, permissions and authorisation). We will address these problems in Sect. 3.1 by explaining how we can model agent's beliefs in AFC and how the different agents can "defend" and justify their beliefs in the argumentation process.

Regarding the second above mentioned aspect, in multi agent systems, cooperation can be interpreted as giving consent to provide one's state and following a common protocol that serves the group objective [36]. We need to distinguish between unconstrained and constrained cooperation problems. Unconstrained cooperation is, for example, an alignment between two agents with the purpose of speaking the same logical language. Constrained cooperation refers, for example, to respective normative systems that impose a certain group behaviour. The strategic structure of a system has also been logically represented using coalition logic, temporal logic etc. In Sect. 4.1 we will explain how we can make use of multi agent resource allocation problems in order to model different cooperation problems that could arise in AFC. We will also investigate how normative reasoning can be used for AFC in Sect. 4.2.

3 Rational Cognitive Modelling

In this section we will focus first on the modelisation of the agent knowledge.

To clarify the notions we will propose, in Fig. 1 we show the multi agent system modelisation of agri-food chains. In the top part of the picture the agro-food chain is depicted, stemming from genomics all the way to the consumer's plate. The food, at every step, undergoes several transformations. For instance, the grape can be selected based on genomics to have desirable farming properties (such as draught resistance, disease resistance etc.). At the next step different technological itineraries are compared in order to select the best way of growing the plant according to different criteria (yield, pesticide treatment etc.). The product may undergo several transformations at this step depending on its final form (for example, the durum wheat may be transformed in flour or in couscous etc.). At the various next steps (trader, distributor, retailer) more transformations are possible as well as important packaging issues addressed. Packaging may also play an important role for increasing the shelf life of aliments (modified pressure, CO_2 and O_2 permeability etc.) and reducing food loss.

Each step of the transformation process (from genomic studies to the consumer plate) will be modelled by one or several agents. These agents will model the knowledge (rule based systems) required at each transformation step. Ontologies dedicated to the specific domain of transformation can be employed [24].

In the bottom part of the Fig. 1 the agents from the various steps are connected via communication/cooperation links. These links along with the agents will form the multi agent system used to model the agro-food chain.

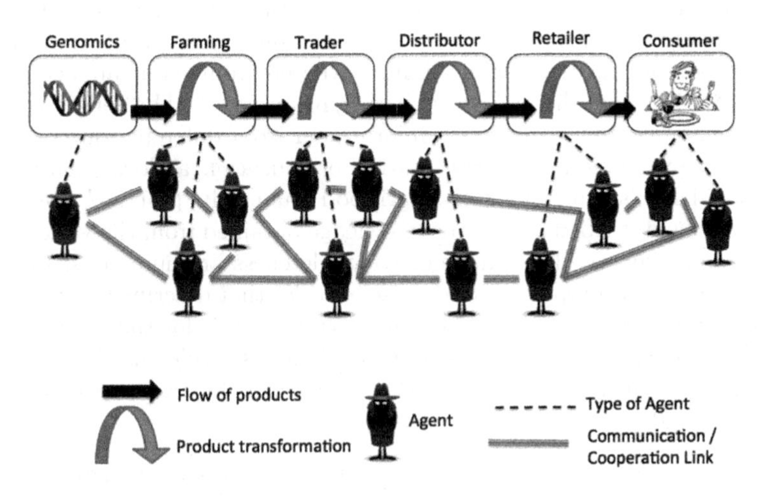

Fig. 1. Multi agent system modelisation of agri-food chains

As mentioned before, in AFC one or several agents will represent one unitary transformation. Please note that for each unitary transformation, within each individual agent, several knowledge representation challenges are to be addressed. First, the information to be represented at each step of the transformation is incomplete, imprecise and highly expressive. There are several ways of obtaining such information. For instance sensors can provide numerical information about the plant. Such information might be unreliable due to measurements errors. The numerical information has to be put in the context of symbolic information. Such symbolic information (transformation rules, ontological data) need to be represented in a logical language that allows for reasoning and for reuse. Linked Open Data can be employed for re-usability reasons. Expressive representation and reasoning languages will provide the possibility of deriving implicit information from explicitly represented knowledge.

In the next section we detail the next problem, the problem of agent to agent communication. We will focus on argumentation and negotiation. In Sect. 4 the multi agent interaction is studied.

3.1 Argumentation and Negotiation

The notion of one to one interaction among self-interested agents has been centred around argumentation and negotiation. Two conditions have to be fulfilled and namely bounded rationality and incomplete information. Let us start by addressing the last point and namely incomplete information. We will come back to bounded rationality at the end of this paper in Sect. 5.

Let us consider, as an illustrating example, the platform developed in the French Institute for Research in Agronomy (INRA) to link agronomy insights with socio-economic developments and behaviour of various stakeholders involved (farmers, consumers, biologists, industrial partners etc.). It aims at identifying ways and solutions to maintain the quality of production and satisfy the needs of the users, while limiting the environmental impact (see e.g. the MEANS initiative http://www6.inra.fr/meanseng/). The long-term ambition is to homogeneously integrate information from different sources, namely the regional production practices, market organization at local, national and international levels, and along the agri-food chains. In practical applications such as the one described above, the knowledge obtained from the various actors involved is incomplete. The causes of incompleteness are numerous. First, it is difficult to obtain a complete ontology (set of rules that describe the world) from domain experts. AGROVOC [21,30] can provide a basis for the ontology developer but the elicitation process is difficult. The basic rules used for reasoning might seem obvious for the domain expert. This calls for two important aspects to be considered:

- First, the representation language needs to be expressive enough in order for implicit knowledge to be derived from explicit knowledge. Existential rules, that allow for existential variables in the head of the rules are especially useful. The existential variables allow to represent variables that are unknown (same mechanism as value invention in tuple generating dependencies in databases) [8].
- Second, the incompleteness can be used as a way to help experts focus on the parts of the ontology that need expanding. One can use explanation facilities of query answering in presence of incompleteness [3–6]. The experts, faced with the system explanation, can choose to enrich the knowledge base if the explanation (or the results) are not conform to their expectation.

When putting together the knowledge from several incomplete sources one needs to perform alignment in order to integrate the sources. Such alignment can be obtained using various methods from the literature. For instance, key discovery on the two datasets and the use of such keys as alignment candidate generators have been proven to significantly improve the state of the art [7,26]. Reasoning can be performed on the union of the sources that share the same vocabulary. In most cases, the union of several sources is inconsistent. As false implies anything, the inconsistent knowledge bases cannot be used as such for reasoning (as any conclusion could be derived). Different inconsistency methods have been devised in order to reason with such knowledge [22]. It is important at this step to make several observations.

In this paper we accept the idea that full specifications cannot be established in agrifood chains (thus we need to address incomplete information). On the other hand, several complementary points of view - possibly contradictory - can be expressed (nutritional, environmental, taste, etc.). We then need to assess their compatibility (or incompatibility) and identify solutions satisfying a maximum set of viewpoints. Several logical frameworks based on argumentation have been proposed in the literature where argumentation was used as a logical tool able to reason in presence of inconsistency. The reasoning process was either done using forward chaining reasoning or backwards chaining. In forward chaining reasoning all arguments and attacks were computed and extensions used in order to represent maximal consistent point of views over the argument and attack set. In backwards chaining an argument was investigated to be accepted or rejected based on the other arguments attacking it and their respective status (accepted or rejected). The two approaches come down to same semantic results, of course, but differ from a computational point of view as well as methodological [13,14,18,32,33].

Please note that argumentation theory can be used not only to deal with inconsistency but also to explain the decision made by the system to a user (as already explained above as a method to remove incompleteness). Argumentation gives the possibility of defining formal protocols of interaction between agents. This is particularly interesting when one of the agents in question is a human agent. We can design formal protocols that underpin the basis of human agent interaction. The notion of an explanatory dialogue as proposed by [3] is a way to offer an interactive explanation that takes place between the system and the user. Explanatory dialogues allow (including and not limited to) the user to ask follows-up questions, clarification questions, elaborate on previous explanations.

Social Attitudes and Affects. When reasoning about knowledge (using classical methods or using inconsistency tolerant reasoning mechanisms) different pieces of knowledge can be of different importance for a decision maker. Existing argumentation-based systems for inconsistent ontology need to take this aspect into account and deal with such preferences on data sources (where more important knowledge is considered to be preferred to less important knowledge). Many approaches exists in literature for dealing with preferences and attacks. The state of the art considers two roles of preferences. Either preferences can inhibit attacks [1] or preferences can be used in a latter stage as a way of filtering out extensions. The preferences relation on the arguments can be lifted to a preference relation on sets of arguments (extensions). The latter approach has been used in agronomy and sucesfully validated with domain experts [17].

Another way of handling preferences is to use mental states in order to model dominant agents. As explained before, a multi-agent system is composed of multiple autonomous agents, each capable of reacting to changes in the environment. The internal workings of an agent cannot be discerned by an external observer, and agents are thus treated as black boxes by other agents. One common approach to agent design involves ascribing agents with mental states based on folk psychology. Thus, for example, the family of BDI techniques [28] to agent design ascribe an agent with a set of beliefs, a set of desires, and a set of intentions which are derived from these beliefs and desires. An agent would then act in such a way so that it will attempt to fulfil its intentions. Approaches such as [2] could be used in order to refine human to human or human to agent interactions in multi agent systems.

4 Strategic Behavior Modelling

In this section we investigate the modelling of the strategic structure of the multi agent system (how can agents accomplish their intentions either alone or in cooperation). Cooperation means following a common protocol that serves the group objective. As already mentioned we distinguish between unconstrained and constrained cooperation problems.

Unconstrained cooperation is, for example, an alignment between two agents with the purpose of speaking the same logical language. We already discussed alignment issue in the previous section as such aspects are fundamental to ensure communication throughout several agents.

When discussing constrained cooperation we will focus on two methods. First, the strategic structure of a system can be logically represented using coalition logic in Sect. 4.1. Next, we explain how normative systems can impose a certain group behaviour in Sect. 4.2.

4.1 Multi Agent Ressource Allocation and Coalitions

The issue of flexible allocation of tasks to multiple problem solvers received attention from the early days of Artificial Intelligence. The tasks that need to be performed are announced from a central node and other nodes subsequently place bids on the tasks they can perform. The central node collects the bid for the task and awards the task to the best bidder. This works as an abstraction of a marked-based centralized distributed system for the determination of adequate allocations of heterogenous indivisible resources. In a Multi Agent Resource Allocation (MARA) system [12], there is central node a (let's call it the auctioneer) and a set of n nodes, $I = 1, ..., n$ (the bidders) which concurrently demand bundles of resources from a common set of available resources, $R = r_1, ..., r_m$, held by the auctioneer. The auctioneer broadcasts R to all n bidders, asking them to submit in a specified common language, the bidding language, their R-valuations over bundles of resources. Bidders i R-valuation, v_i, is a non-negative real function on $\mathcal{P}(R)$, expressing for each bundle the individual interest of bidder i in obtaining S. No bidder i knows the valuation of any other n 1 bidders, but all the participants in the system agreed on the outcome: based on bidders R-valuations, the auctioneer will determine a resources allocation specifying for each bidder i her obtained bundle O_i (its outcome). The task of the auctioneer finding a maximum value allocation for a given set of bidder valuations is called the Winner Determination Problem (WDP). This is a NP-hard problem, being equivalent to weighted setpacking.

An instance of the MARA problem is the problem of coalition formation (that models teamwork explicitly). A particular strength of multi-agent systems is the ability of agents to form coalitions that may achieve goals more efficiently than when agents act as individuals. Possible applications of coalition formation techniques in multi-agent systems include rescue coordination, supply chain management, e-commerce, etc. MARA in general and coalition formation in particular can be used in agrifood chains to model more general behaviour for chain organisation. Many approaches used for agrifood chains are myopic (they only refer to one transformation or one actor on the chain). Having a global view will allow certain optimisations that go beyond the individual transformations. Also a global view of the system will pave the way for seamless reverse engineering techniques where final specifications are used in order to derive (using backwards chaining) initial conditions needed for such specifications.

Three main issues studied in the context of MARA in general (that also apply in the context of MARA for agrifood chains) are [19,29]:

1. Optimization of a coalition value. In agrifood chains this could refer to the minimising of cost of products (cost in the broad sense - depending on the resources needed). This relies however on having full knowledge on partial costs which is infeasible in certain practical cases due to incomplete knowledge (discussed in the previous section).
2. Division of a coalition value between agents (e.g. the concepts of core or Shapley value). Studying such concepts could help highlight the steps in the agrifood chain transformation with most utility (or, the inverse, steps that could be avoided).
3. Generating the optimal division of agents into exhaustive and disjoint coalitions. Such divisions are called coalition structures and the this problem is called an optimal coalition structure generation problem (CSG). Lastly, the CSG problem could be used in order to optimise the agrifood chain in its totality.

4.2 Normative Reasoning

Another way of organising a multi agent system is by installing a set of norms that need to be behaved by all agents. Norm aware agents make use of concepts such as obligations, permissions, and prohibitions, to represent and reason about socially imposed goals and capabilities. Such agents are able to decide whether to act in a manner consistent with norms, or whether to ignore them. Typically, norms are imposed on a set of agents in order to increase the overall utility of the system (often at the cost of individual utility), or reduce computational or communication overhead [11].

Norms, such as obligations, permissions and prohibitions, place soft constraints upon an agent. Typically, ignoring an obligation (i.e. violating it) means that a sanction is applied to the agent, but the agent may still choose to ignore a norm in some situations. An agent is said to be norm-aware if it is able to reason about the norms that apply to it. A multi-agent system containing norm-aware agents has a number of advantages over simpler multi-agent systems. Norms allow agents to assume, by default, that other agents will behave in a certain way, reducing the complexity of their reasoning. Norms are typically declarative, and have a great deal of explanatory power. Norms thus form a good programming and understanding metaphor for both creating agents, and understanding their actions in specific situations.

A norm may be defined in terms of five components. First, a norm has a type, for example, an obligation, or a permission. Second, a norm has an activation condition, identifying the situations in which the norm affects some agents. Third, a norm imposes some normative condition on the affected agent; if this normative condition does not hold, the norm is not being followed. Fourth, norms have a termination, or expiration condition, identifying the situations after which the norm no longer affects the agent. Finally, the norm must identify the agents which it affects. These agents are referred to as the norm targets.

During its lifecycle, an abstract norm becomes instantiated. While instantiated, its normative condition may evaluate to true or false at different times. Finally, the norms expiration condition evaluates to true, after which the instantiated norm is deleted. It is possible to construct this condition as a query to the knowledge base, and from this, determine whether the norm is violated or not.

A normative environment is used to keep track of the abstract (generic norms) and instantiated norms (norms applying to one agent during a given time lapse) within the system. Since norms may be instantiated and expire as time passes, the normative environment must, at each time point, identify which norms exist in the system.

In [15] the authors proposed a rich model for tracking and determining the status norms may be represented graphically via a logical language represented as a graph. The framework presented is intended to capture the evolution of a norm over time, allowing for its instantiation and expiration, as well as recording the time periods during which a norm was complied with or violated. Since the internal structure of such a norm is somewhat complex, some technique for explaining why a norm is in a certain state is required, and a visual model for explaining norm status useful for human agent interaction.

In both cases (modelling the constrained cooperation as a multi agent resource allocation problem or as a normative system) we could impose the global view on the system that was lacking in the state of the art that only considered the localised optimisation within a transformation. Furthermore, we can also model important ethical aspects which are very important to consider but difficult to take into account in existing systems.

The main difference between the two approaches is the elicitation aspects. While in MARA one needs to rely on a complete knowledge of the utility of the various coalitions in the normative approach such requirement is no longer imposed. One can state the different norms that the system should respect and then each individual agent will comply or not to the respective norm.

5 Discussion

In this paper we provided a proposal for modelling AFC problems as MAS problems. The contribution of the paper lays in the unifying framework that such modelling could being into KRR problems in AFC. While certain approaches in KRR for AFC already employ multi agent systems techniques (such as argumentation) an unified framework where the chain can be studied in its globality could prove to be beneficial. Such claim is supported by the advantages of modelling supply chain management as multi agent system modelling proved by the state of the art. While supply chain management is a particular case of agrifood chain modelling, in general in agrifood chain modelling the incompleteness and uncertainty of the knowledge makes the problem much harder. This is yet another reason to benefit from the uncertainty reasoning in multi agent systems.

As mentioned before the agents we consider here are rational agents. It could be (especially in an argumentation setting) that we do not want to consider solely rational agents. Indeed, the cognitive biases should be taken into account. Detecting and highlighting such biases (which could be common in domain experts due to the narrowness of their expertise) might be able to prevent decision errors in chain management.

We conclude this paper by a quick remark about implementation aspects. As already explained the aim of the paper is to be a position paper about the benefits of modelling AFC as MAS. Of course such modelisation should be followed in practice by judicial implementations. One of the most important aspects to consider is the flexibility of the system that should be extendible in time. Another important aspects is the seamless integration with LOD ontologies (in terms of compatibility with Web Services, SPARQL endpoints and expressivity).

Acknowledgments. The first author acknowledges the support of ANR grants ASPIQ (ANR-12-BS02-0003), QUALINCA (ANR-12-0012) and DURDUR (ANR-13-ALID-0002). The work of the first author was carried out part of the research delegation at INRA MISTEA and INRA IATE CEPIA Axe 5, Montpellier.

References

1. Amgoud, L., Vesic, S.: Two roles of preferences in argumentation frameworks. In: Liu, W. (ed.) ECSQARU 2011. LNCS, vol. 6717, pp. 86–97. Springer, Heidelberg (2011)
2. Anderson, K., et al.: The TARDIS framework: intelligent virtual agents for social coaching in job interviews. In: Reidsma, D., Katayose, H., Nijholt, A. (eds.) ACE 2013. LNCS, vol. 8253, pp. 476–491. Springer, Heidelberg (2013)
3. Arioua, A., Croitoru, M.: Formalizing explanatory dialogues. In: Beierle, C., Dekhtyar, A. (eds.) SUM 2015. LNCS, vol. 9310, pp. 282–297. Springer, Heidelberg (2015)
4. Arioua, A., Tamani, N., Croitoru, M.: On conceptual graphs and explanation of query answering under inconsistency. In: Hernandez, N., Jäschke, R., Croitoru, M. (eds.) ICCS 2014. LNCS, vol. 8577, pp. 51–64. Springer, Heidelberg (2014)

5. Arioua, A., Tamani, N., Croitoru, M.: Query answering explanation in inconsistent datalog +/− knowledge bases. In: Chen, Q., Hameurlain, A., Toumani, F., Wagner, R., Decker, H. (eds.) DEXA 2015. LNCS, vol. 9261, pp. 203–219. Springer, Heidelberg (2015)
6. Arioua, A., Tamani, N., Croitoru, M., Buche, P.: Query failure explanation in inconsistent knowledge bases: a dialogical approach. In: Bramer, M., Petridis, M. (eds.) Research and Development in Intelligent Systems XXXI, pp. 119–133. Springer, Switzerland (2014)
7. Atencia, M., Chein, M., Croitoru, M., David, J., Leclère, M., Pernelle, N., Saïs, F., Scharffe, F., Symeonidou, D.: Defining key semantics for the RDF datasets: experiments and evaluations. In: Hernandez, N., Jäschke, R., Croitoru, M. (eds.) ICCS 2014. LNCS, vol. 8577, pp. 65–78. Springer, Heidelberg (2014)
8. Baget, J.-F., Mugnier, M.-L., Rudolph, S., Thomazo, M.: Walking the complexity lines for generalized guarded existential rules. In: IJCAI Proceedings-International Joint Conference on Artificial Intelligence, vol. 22, p. 712 (2011)
9. Beneventano, D., Bergamaschi, S., Sorrentino, S., Vincini, M., Benedetti, F.: Semantic annotation of the cerealab database by the agrovoc linked dataset. Ecol. Inform. **26**, 119–126 (2015)
10. Bizer, C., Heath, T., Berners-Lee, T.: Linked data-the story so far. In: Semantic Services, Interoperability and Web Applications: Emerging Concepts, pp. 205–227 (2009)
11. Briggs, W., Cook, D.: Flexible social laws. In: International Joint Conference on Artificial Intelligence, vol. 14, pp. 688–693. Citeseer (1995)
12. Chevaleyre, Y., Dunne, P.E., Endriss, U., Lang, J., Maudet, N., RodrÍGuez-Aguilar, J.A.: Multiagent resource allocation. Knowl. Eng. Rev. **20**(02), 143–149 (2005)
13. Croitoru, M., Fortin, J., Mosse, P., Buche, P., Guillard, V., Guillaume, C.: Biosourced and biodegradable packaging design using argumentation to aggregate stakeholder preferences. In: EFFoST 2012 Annual Meeting, p. 1 (2012)
14. Croitoru, M., Fortin, J., Nir, O.: Arguing with preferences in ecobiocap. In: COMMA 2012: Computational Models of Argument, vol. 245, pp. 51–58. IOS Press (2012)
15. Croitoru, M., Oren, N., Miles, S., Luck, M.: Graphical norms via conceptual graphs. Knowl.-Based Syst. **29**, 31–43 (2012)
16. Croitoru, M., Thomopoulos, R., Tamani, N.: A practical application of argumentation in French agrifood chains. In: Laurent, A., Strauss, O., Bouchon-Meunier, B., Yager, R.R. (eds.) IPMU 2014, Part I. CCIS, vol. 442, pp. 56–66. Springer, Heidelberg (2014)
17. Croitoru, M., Thomopoulos, R., Vesic, S.: Introducing preference-based argumentation to inconsistentontological knowledge bases. In: Chen, Q., et al. (eds.) PRIMA 2015. LNCS, vol. 9387, pp. 594–602. Springer, Heidelberg (2015)
18. Guillard, V., Buche, P., Destercke, S., Tamani, N., Croitoru, M., Menut, L., Guillaume, C., Gontard, N.: A decision support system to design modified atmosphere packaging for fresh produce based on a bipolar flexible querying approach. Comput. Electron. Agric. **111**, 131–139 (2015)
19. Ieong, S., Shoham, Y.: Marginal contribution nets: a compact representation scheme for coalitional games. In: Proceedings of the 6th ACM Conference on Electronic Commerce, pp. 193–202. ACM (2005)
20. Jennings, N.R., Sycara, K., Wooldridge, M.: A roadmap of agent research and development. Auton. Agents Multi-agent Syst. **1**(1), 7–38 (1998)

21. Lauser, B., Sini, M., Liang, A., Keizer, J., Katz, S.: From agrovoc to the agricultural ontology service/concept server. an owl model for creating ontologies in the agricultural domain. In: Dublin Core Conference Proceedings. Dublin Core DCMI (2006)
22. Lembo, D., Lenzerini, M., Rosati, R., Ruzzi, M., Savo, D.F.: Inconsistency-tolerant semantics for description logics. In: Hitzler, P., Lukasiewicz, T. (eds.) RR 2010. LNCS, vol. 6333, pp. 103–117. Springer, Heidelberg (2010)
23. Lukose, D.: World-wide semantic web of agriculture knowledge. J. Integr. Agricu. **11**(5), 769–774 (2012)
24. Muljarto, A.-R., Salmon, J.-M., Neveu, P., Charnomordic, B., Buche, P.: Ontology-based model for food transformation processes - application to winemaking. In: Closs, S., Studer, R., Garoufallou, E., Sicilia, M.-A. (eds.) MTSR 2014. CCIS, vol. 478, pp. 329–343. Springer, Heidelberg (2014)
25. Olfati-Saber, R., Fax, A., Murray, R.M.: Consensus and cooperation in networked multi-agent systems. Proc. IEEE **95**(1), 215–233 (2007)
26. Pernelle, N., Saïs, F., Symeonidou, D.: An automatic key discovery approach for data linking. Web Semant. Sci. Serv. Agents World Wide Web **23**, 16–30 (2013)
27. Pokharel, S., Sherif, M.A., Lehmann, J.: Ontology based data access and integration for improving the effectiveness of farming in nepal. In: Proceedings of the 2014 IEEE/WIC/ACM International Joint Conferences on Web Intelligence (WI) and Intelligent Agent Technologies (IAT), vol. 02, pp. 319–326. IEEE Computer Society (2014)
28. Rao, A.S., Georgeff, M.P., et al.: Bdi agents: From theory to practice. In: ICMAS 1995, pp. 312–319 (1995)
29. Shehory, O., Kraus, S.: Formation of overlapping coalitions for precedence-ordered task-execution among autonomous agents. In: Proceedings of ICMAS 1996, pp. 330–337 (1996)
30. Soergel, D., Lauser, B., Liang, A., Fisseha, F., Keizer, J., Katz, S.: Reengineering thesauri for new applications: the agrovoc example. J. Digital Inform. **4**(4), 1–23 (2006)
31. Tamani, N., Croitoru, M., Buche, P.: Conflicting viewpoint relational database querying: an argumentation approach. In: Proceedings of the 2014 International Conference on Autonomous Agents and Multi-agent Systems, pp. 1553–1554. International Foundation for Autonomous Agents and Multiagent Systems (2014)
32. Tamani, N., Mosse, P., Croitoru, M., Buche, P., Guillard, V.: A food packaging use case for argumentation. In: Closs, S., Studer, R., Garoufallou, E., Sicilia, M.-A. (eds.) MTSR 2014. CCIS, vol. 478, pp. 344–358. Springer, Heidelberg (2014)
33. Tamani, N., Mosse, P., Croitoru, M., Buche, P., Guillard, V., Guillaume, C., Gontard, N.: Eco-efficient packaging material selection for fresh produce: industrial session. In: Hernandez, N., Jäschke, R., Croitoru, M. (eds.) ICCS 2014. LNCS, vol. 8577, pp. 305–310. Springer, Heidelberg (2014)
34. Tamani, N., Mosse, P., Croitoru, M., Buche, P., Guillard, V., Guillaume, C., Gontard, N.: An argumentation system for eco-efficient packaging material selection. Comput. Electron. Agric. **113**, 174–192 (2015)
35. Thomopoulos, R., Croitoru, M., Tamani, N.: Decision support for agri-food chains: A reverse engineering argumentation-based approach. Ecol. Inform. **26**, 182–191 (2015)
36. Van Der Hoek, W., Wooldridge, M.: On the logic of cooperation and propositional control. Artif. Intell. **164**(1), 81–119 (2005)

Predictive Model Based on the Evidence Theory for Assessing Critical Micelle Concentration Property

Ahmed Samet[1(✉)], Théophile Gaudin[2], Huiling Lu[2,3], Anne Wadouachi[3], Gwladys Pourceau[3], Elisabeth Van Hecke[2], Isabelle Pezron[2], Karim El Kirat[1], and Tien-Tuan Dao[1]

[1] Sorbonne University, Université de technologie de Compiègne, CNRS, UMR 7338 Biomechanics and Bioengineering, Compiègne, France
ahmed.samet@utc.fr

[2] Sorbonne University, Université de technologie de Compiègne, EA 4297 Transformations Intégrées de la Matière Renouvelable, Compiègne, France

[3] Université de Picardie Jules Verne, CNRS, FRE 3517 Laboratoire de Glycochimie, des Antimicrobiens et des Agroressources, Amiens, France

Abstract. In this paper, we introduce an uncertain data mining driven model for knowledge discovery in chemical database. We aim at discovering relationship between molecule characteristics and properties using uncertain data mining tools. In fact, we intend to predict the Critical Micelle Concentration (CMC) property based on a molecule characteristics. To do so, we develop a likelihood-based belief function modelling approach to construct evidential database. Then, a mining process is developed to discover valid association rules. The prediction is performed using association rule fusion technique. Experiments were conducted using a real-world chemical databases. Performance analysis showed a better prediction outcome for our proposed approach in comparison with several literature-based methods.

Keywords: Evidential data mining · Chemical database · Association rule · Associative classifier

1 Introduction

Data mining is generally held to be generically a discipline of the field of Knowledge Discovery, or Knowledge Discovery in Databases (KDD). It is usually defined as the process of identifying valid, novel, potentially useful, and ultimately understandable patterns from large collections of data. Then, causal rules are derived from those patterns. Frequent patterns and valid rules can be used to test hypotheses (or verification goals) or to autonomously find entirely new patterns (discovery goals) [1]. Discovery goals could be predictive (requiring predictions to be made using the data in the database) [2]. On the other hand, there

J.P. Carvalho et al. (Eds.): IPMU 2016, Part I, CCIS 610, pp. 510–522, 2016.
DOI: 10.1007/978-3-319-40596-4_43

has been an explosion in the availability of publicly accessible chemical information, including chemical structures of small molecules, structure-derived properties and associated biological activities in a variety of assays [3,4]. These data sources provide a significant opportunity to develop and apply computational tools to extract and understand the underlying structure-activity relationships. These techniques remain sensitive to the presence of imperfect data [5]. Recent years, we have noticed the emergence of uncertain data mining tools [6–8] that contribute to seek hidden pertinent information under the presence of uncertainty and imprecision. However, to the best of our knowledge, uncertain data mining tools have not yet been used to discover pertinent knowledges neither to predict in chemical databases.

In this work, we are interested in evidential data mining in chemical databases. The latter provides a generalizing framework for probabilistic and binary data mining disciplines [9]. Recently, mining over evidential databases has flourished by several contributions and the introduction of new support and confidence measures [10,11]. In addition, it has been applied on several fields such as healthcare [12], cheminformatics [13], etc.

From methodological point of view, we intend to apply an uncertain data mining-driven approach to analyze a chemical database. The chemical database contains records of amphiphilic molecules[1]. We aim at predicting physico-chemical properties of a new molecule from their structural characteristics. The imprecision within the data is modelled using evidence theory. Methodologically, we transform the chemical database into an evidential database with a likelihood-based approach. Once the imprecision examined, valid association rules are selected and used for the prediction of Critical Micelle Concentration (CMC) property.

This paper is organized as follows: in Sect. 2, the state-of-the-art works of evidential data mining are briefly recalled. In Sect. 3, we introduce our uncertain data mining driven approach. A new likelihood-based model for imprecision consideration is presented. The performance of our proposed approach was studied on a real-world chemical database in Sect. 4. Finally, we conclude and sketch potential issues for the future work.

2 Preliminaries

2.1 Evidential Database

An evidential database stores either uncertain and imprecise data [14] via the evidence theory. An evidential database, denoted by $\mathcal{EDB}$, with n columns and d rows where each column i ($1 \leq i \leq n$) has a domain Θ_i of discrete values. Each cell of a row j and a column i contains a normalized Belief Basic Assignment (BBA) $m_{ij} : 2^{\theta_i} \rightarrow [0, 1]$ as follows:

[1] An amphiphilic molecule is chemical compound possessing both hydrophilic (water-loving, polar) and lipophilic (fat-loving) properties.

$$\begin{cases} m_{ij}(\emptyset) = 0 \\ \sum_{A \subseteq \theta_i} m_{ij}(A) = 1. \end{cases} \quad (1)$$

An *item* corresponds to a focal element[2]. Two different itemsets (a.k.a patterns) can be related via either the inclusion or the intersection operator. Indeed, the inclusion operator for evidential itemsets [11] is defined as follows, where X and Y are two evidential itemsets:

$$X \subseteq Y \iff \forall x_i \in X, x_i \subseteq y_j \quad (2)$$

x_i and y_j are respectively the ith and the jth element of X and Y. For the same evidential itemsets X and Y, the intersection operator is defined as follows:

$$X \cap Y = Z \iff \forall z_k \in Z, z_k \subseteq x_i \text{ and } z_k \subseteq y_j. \quad (3)$$

An *evidential association rule* R is a causal relationship between two itemsets that can be written in the following form $R : X \rightarrow Y$ such that $X \cap Y = \emptyset$.

Example 1. We aim at developing a predictive model for a chemical database. The evidential database records information about several molecules. Table 1 shows an example of an evidential database.

Table 1. Evidential database $\mathcal{EDB}$

Molecule	Head Family (HF)?	Carbon Number (N_c)?	CMC?
M1	$m_{11}(Glucose) = 1.0$	$m_{21}(7) = 0.9$	$m_{31}(12) = 0.8$
		$m_{21}(7 \cup 8) = 0.1$	$m_{31}(12 \cup 50) = 0.2$
M2	$m_{12}(Glucosamine) = 1.0$	$m_{22}(8) = 0.8$	$m_{32}(12) = 0.7$
		$m_{22}(8 \cup 10) = 0.2$	$m_{32}(0.2 \cup 12) = 0.3$

The first transaction means that the molecule M1 is a Glucose head family type of the frame of discernment $\Theta_{HF} = \{Glucose, Glucosamine\}$. The second attribute reflects the Critical Micelle Concentration (CMC) discretized in the following frame of discernment $\Theta_{CMC} = \{0.2,\ 12,\ 50\}$. M1 has a CMC close to 12 mM and could be some doubt if it belongs to around 50 millimolar (mM) CMC class. In Table 1, $\{HF = Glucose\}$ is an item and $\{HF = Glucose\} \times \{CMC = 12 \cup 50\}$ is an itemset such that $\{HF = Glucose\} \subset \{HF = Glucose\} \times \{CMC = 12 \cup 50\}$ and $\{HF = Glucosamine\} \cap \{HF = Glucosamine\} \times \{CMC = 0.2 \cup 12\} = \{HF = Glucosamine\}$. $\{N_c = 8\} \rightarrow \{CMC = 12\}$ is an association rule.

In the following subsection, we recall the definition of *belief-based*, *precise-based* support and confidence measures that estimate the pertinence of patterns and association rules.

[2] Each subset A of 2^{Θ}, fulfilling $m(A) > 0$, is called a focal element.

2.2 Support and Confidence Measures

As is the case for probabilistic data mining [8], the support within the evidential context is based on expectation. Two support family approaches were proposed. The first support measure was proposed by [11] and called the belief-based support measure. It is considered as the lower bound for the support. It is written as follows:

$$Sup_{T_j}^{Bel}(X) = \prod_{i \in [1 \ldots n]} Sup_{T_j}^{Bel}(x_i) = \prod_{i \in [1 \ldots n]} Bel(x_i). \tag{4}$$

Thus, the belief-based support in the entire database is computed as follows:

$$Sup_{\mathcal{EDB}}^{Bel}(X) = \frac{1}{d} \sum_{j=1}^{d} Sup_{T_j}^{Bel}(X). \tag{5}$$

Since the belief-based support is a lower estimation of the support, it is obvious that in some cases that an itemset I could have a higher support value. Another measure was introduced by Samet et al. [15] that provides a medium estimation. The precise measure Pr is defined by:

$$Pr(x_i) = \sum_{x \subseteq \Theta_i} \frac{|x_i \cap x|}{|x|} \times m_{ij}(x) \qquad \forall x_i \in 2^{\Theta_i}. \tag{6}$$

The evidential support of an itemset $X = \prod_{i \in [1 \ldots n]} x_i$ in the transaction T_j (i.e., Pr_{T_j}) is then computed as follows:

$$Pr_{T_j}(X) = \prod_{x_i \in \Theta_i, i \in [1 \ldots n]} Pr(x_i). \tag{7}$$

Thus, the support $Sup_{\mathcal{EDB}}$ of the itemset X becomes:

$$Sup_{\mathcal{EDB}}(X) = \frac{1}{d} \sum_{j=1}^{d} Pr_{T_j}(X). \tag{8}$$

A new metric for confidence computing based on the precise-based support measure is introduced in [10]. For an association rule $R : R_a \rightarrow R_c$, the confidence is computed as follows:

$$Conf(R) = \frac{\sum_{j=1}^{d} Sup_{T_j}(R_a) \times Sup_{T_j}(R_c)}{\sum_{j=1}^{d} Sup_{T_j}(R_a)}. \tag{9}$$

Example 2. We consider the same problem described in Example 1. The precise support of the itemset $\{HF = Glucose\} \times \{N_c = 7\}$ in the evidential database is equal to $\frac{1\times0.95+0}{2} = 0.475$. It is superior to the one computed with the belief-based support which is equal to $\frac{1\times0.9+0}{2} = 0.45$. Finally, the association rule $\{N_c = 8\} \rightarrow \{CMC = 12\}$ has $\frac{0.05\times0.9+0.9\times0.85}{0.05+0.9} = 0.85$ as confidence in the evidential database.

3 Uncertain Data Mining Approach for CMC Prediction

In the following, we introduce our uncertain data mining approach for amphiphilic molecule's CMC prediction. The provided approach, shown in Fig. 1 consists in three stages. Imprecision within the raw database is processed when evidential database is constructed upon the use of likelihood modelling approach. Then, frequent patterns and valid association rules are retrieved with EDMA mining algorithm [10]. The selected valid association rules are then used to compute the CMC of an amphiphilic molecule.

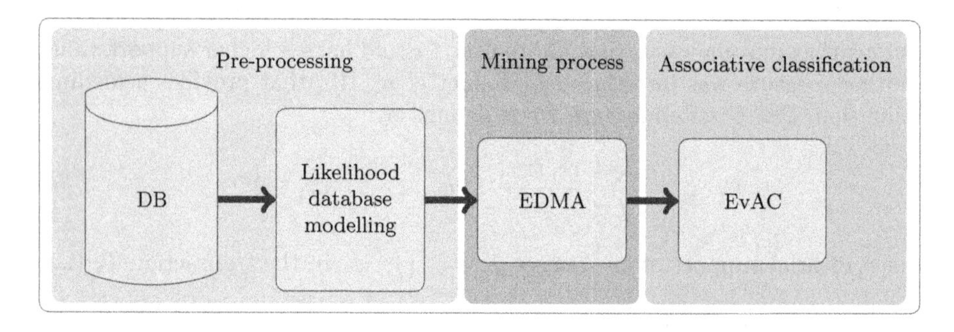

Fig. 1. Evidential data mining based model for the prediction of CMC property

3.1 Likelihood Modelling for Input Data

Let us assume the class-conditional probability densities $f(x|\omega_i)$ to be known. Having observed x, the likelihood function is a function from Θ to $[0, +\infty)$ defined as $L(\omega_i|x) = f(x|\omega_i)$, for all $k \in \{1, \ldots, K\}$. Shafer [16] proposed to derive from L a belief function on Θ defined by its plausibility function. Starting from axiomatic requirements, Appriou [17] proposed another method based on the construction of I belief functions $m_i(.)$. The idea consists in taking into account separately each class and evaluating the degree of belief given to each of them. In this case, the focal elements of each BBA m_i are the singleton $\{\omega_i\}$, its complement $\overline{\omega_i}$, and Θ. This model, hereafter referred to as the Separable Likelihood-based (SLB) method, has the following expression:

$$\begin{cases} m_i(\{\omega_i\}) = 0 \\ m_i(\overline{\omega_i}) = \alpha_i \cdot \{1 - R \cdot L(\omega_i|x)\} \\ m_i(\Theta) = 1 - \alpha_i \cdot \{1 - R \cdot L(\omega_i|x)\}. \end{cases} \tag{10}$$

where α_i is a coefficient that can be used to model external information such as reliability, and R is a normalizing constant that can take any value in the range $(0, (max_i(L(\omega_i|x)))^{-1}]$. A second model satisfying the axiomatic requirements [17] can be written as follows:

$$\begin{cases} m_i(\{\omega_i\}) = \frac{\alpha_i \cdot R \cdot L(\omega_i|x)}{1+R \cdot L(\omega_i|x)} \\ m_i(\overline{\omega_i}) = \frac{\alpha_i}{1+R \cdot L(\omega_i|x)} \\ m_i(\Theta) = 1 - \alpha_i \end{cases} \tag{11}$$

Both BBA models could be used to consider imprecision within the data. However, we retain the second BBA model since it is more informative.

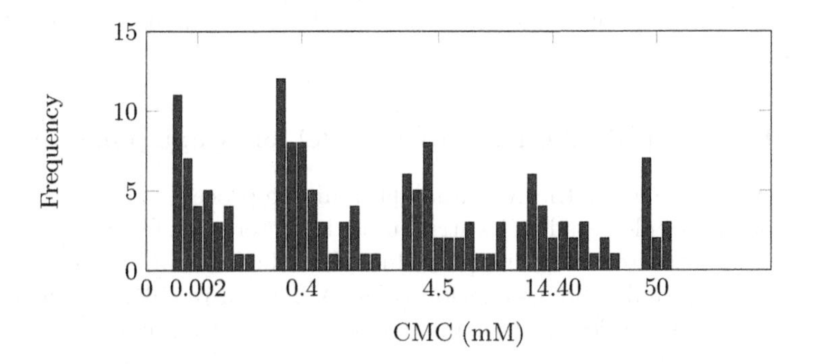

Fig. 2. Frequency of appearance histogram of amphiphilic molecule's CMC

Now, we intend to model each molecule x with a BBA that expresses its membership to the CMC classes. To do so, we distinguish between two types of data (i.e., proprieties and descriptors of molecule) in the database: categoric and numeric data. A categoric data, such as the head family in Table 1, are represented by *certain* BBA. A BBA is called a certain BBA when it has one focal element, which is a singleton. It is representative of perfect knowledge and the absolute certainty. The numeric data are transformed into a BBA using the likelihood model. In fact, from the histogram of frequency in Fig. 2, we construct a probability density function (pdf). A pdf is constructed on each CMC peak. The computed pdf corresponds to the class-conditional probability density. Five class-conditional probability density functions are constructed from Fig. 2. A class-conditional pdf is computed for the following CMC classes: CMC around 0.002, 0.4, 4.5, 14.40 and 50 mM. Once the class-conditional probability functions are constructed, we compute the BBA that expresses the membership of x to each CMC class following the model detailed in Eq. (10). The computed BBAs are then combined to obtain the final BBA that expresses the membership of x to all CMC classes:

$$m = \oplus_{i \in I} m_i \tag{12}$$

where $\oplus$ is the Dempster's rule of combination between two BBAs which is defined as:

$$(m_1 \oplus m_2)(A) = m_{\oplus}(A) = \frac{1}{1 - m(\emptyset)} \sum_{B \cap C = A} m_1(B) \times m_2(C) \qquad \forall A \subseteq \Theta, A \neq \emptyset \tag{13}$$

where $m(\emptyset)$ represents the conflict mass between m_1 and m_2, is defined as:

$$m(\emptyset) = \sum_{B \cap C = \emptyset} m_1(B) \times m_2(C). \tag{14}$$

Once the evidential database is constructed, we mine frequent patterns and valid association rules. The methodology for patterns and association rules mining over evidential amphiphilic molecules database is detailed in the following subsection.

3.2 Data Mining Predictive-Based Model for Amphiphile Molecules

In the following, we detail the evidential data mining process to predict the CMC property of a molecule based on its structural characteristics. Once the evidential database is constructed with the procedure described in Subsect. 3.1, we mine frequent patterns and valid association rules. A pattern is called frequent (resp. valid for association rules) if its computed support (resp. confidence) is higher than or equal to a fixed threshold *minsup* (resp. *minconf*). In our model, we use EDMA algorithm [10] to retrieve frequent patterns and valid association rules from the evidential database. EDMA generates patterns having a precise support higher than *minsup* in a level-wise manner. The valid association rules are then deduced from frequent patterns. Each pattern of size k gives $2^k - 2$ different association rules. The retrieved association rules are of help for predicting the CMC value of a molecule. To do so, from the set of all valid association rules $\mathcal{R}$, we retain only those that have a CMC item within the conclusion part such that:

$$\mathcal{RI} = \{R : R_a \rightarrow R_c \in \mathcal{R} | \exists y \in \Theta_{CMC},\ y \in R_C\}. \tag{15}$$

The set $\mathcal{RI}$ represents the set of all prediction rules. They are the input of the EvAC algorithm (see Algorithm 1). EvAC is an associative classifier algorithm that fuses interesting association rules for prediction purposes. Indeed, EvAC algorithm classifies the data with fusion techniques and $FILTRATE\_LARGE\_PREMISE(.)$ function (line 1) allows to filtrate the rules and to retain only those with the largest premise, having intersection with the under classification instance X. In fact, the set of the largest premise rules $\mathcal{R}_{large}$ are more precise than those with the shortest premise. Once found, they are considered as independent sources and are combined (line 2 to 4). The fusion is operated on association rules modelled into BBAs with Dempster's rule of combination (see Eq. (13)). The function *argmax* in line 5 allows the retention of

Algorithm 1. Evidential Associative Classification (EvAC) algorithm

Require: $\mathcal{R}, X, \Theta_C$
Ensure: *Class*

1: $\mathcal{R}_{large} \leftarrow FILTRATE\_LARGE\_PREMISE(\mathcal{R}, X, \Theta_C)$
2: **for all** $r \in \mathcal{R}_{large}$ **do**
3: $\quad m \leftarrow \begin{cases} m(\{r.conclusion\}) = conf(r) \\ m(\Theta_C) = 1 - conf(r) \end{cases}$
4: $\quad m_{\oplus} \leftarrow m_{\oplus} \oplus m$
5: $Class \leftarrow argmax_{H_k \in \Theta_C} BetP(H_k)$
6: **function** FILTRATE_LARGE_PREMISE($\mathcal{R}, X, \Theta_C$)
7: $\quad max \leftarrow 0$
8: $\quad$ **for all** $r \in \mathcal{R}$ **do**
9: $\quad\quad$ **if** $r.conclusion \in \Theta_C$ & $X \cap r.premise \neq \emptyset$ **then**
10: $\quad\quad\quad$ **if** $size(r.premise) > max$ **then**
11: $\quad\quad\quad\quad \mathcal{R}_{large} \leftarrow \{r\}$
12: $\quad\quad\quad\quad max \leftarrow size(r.premise)$
13: $\quad\quad\quad$ **else**
14: $\quad\quad\quad\quad$ **if** $size(r.premise) = max$ **then**
15: $\quad\quad\quad\quad\quad \mathcal{R}_{large} \leftarrow \mathcal{R}_{large} \cup \{r\}$
16: $\quad$ **return** $\mathcal{R}_{large}$

the hypothesis that maximizes the pignistic probability which is computed as follows:

$$BetP(H_n) = \sum_{A \subseteq \Theta} \frac{|H_n \cap A|}{|A|\,(1 - m(\emptyset))} \times m(A) \qquad \forall H_n \in \Theta. \tag{16}$$

Example 3. Let us assume a new molecule Mx, depicted in Table 2, we intend to predict its CMC. Table 3 is a numerical example of evidential rules' fusion using EvAC. The extracted classification association rules are modelled as BBAs. The decision with pignistic probability gives the $\{CMC = 12\}$ class which is naturally the case. The result is interpreted as the molecule Mx belongs to the $\{CMC = 12\}$ class and its CMC is highly possible centred around 12.

Table 2. The evidential transaction X under classification

Molecule	Head family?	Carbon Number?	CMC?
Mx	$m_{11}(Glucose) = 1.0$	$m_{21}(7) = 0.8$ $m_{21}(7 \cup 8) = 0.2$	?

Table 3. Numerical example of rule's fusion

Rule	Confidence	${}^{w}m_{R_l}^{\Theta_C}$	BetP
$R_1 : \{HF = Glucose\}, \{CN = 8\} \rightarrow \{CMC = 12\}$	0.9	$m_{R_1}^{\Theta_C}(\{12\}) = 0.9$	
		$m_{R_1}^{\Theta_C}(\Theta_C) = 0.1$	
$R_2 : \{HF = Glucose\}, \{CN = 7\} \rightarrow \{CMC = 12\}$	0.9	$m_{R_2}^{\Theta_C}(\{12\}) = 0.9$	$BetP(\{12\}) = 0.993$
		$m_{R_2}^{\Theta_C}(\Theta_C) = 0.1$	
$R_3 : \{HF = Glucose\}, \{CN = 7 \cup 8\} \rightarrow \{CMC = 50\}$	0.1	$m_{R_3}^{\Theta_C}(\{50\}) = 0.1$	$BetP(\{50\}) = 0.007$
		$m_{R_3}^{\Theta_C}(\Theta_C) = 0.9$	
$R_4 : \{HF = Glucose\}, \{CN = 7 \cup 8\} \rightarrow \{CMC = 12 \cup 50\}$	1	$m_{R_4}^{\Theta_C}(12 \cup 50) = 1$	

4 Empirical Results

Knowledge extraction from a chemical database is of great interest to the identification of useful molecules for a specific purpose. In this real case study, we aimed at predicting the relationships between structural characteristics and the physico-chemical properties of the amphiphile molecules. In particular, we focused on the prediction of the Critical Micelle Concentration (CMC) of each molecule by using its structural properties. The database is established from the domain literature using a systematic review process. Each retrieved paper is scanned and reviewed by two domain experts. Relevant information of structural characteristics and related physico-chemical properties are extracted and stored into a raw database for further processing. A transformation process, as the one described in Subsect. 3.1, is performed to establish an evidential database from raw data. The database after transformation and processing contains 199 amphiphile molecules (i.e., rows) detailed in 24 attributes (structural characteristics and related physico-chemical properties) (i.e., columns). The amphiphile molecule evidential database contains over 10^9 items (i.e., focal elements) after transformation.

Table 4. Classification accuracy

Method	EvAC$^{\text{Like-Pr}}$	EvAC$^{\text{ECM-Pr}}$	EvAC$^{\text{Like-Bel}}$	EvAC$^{\text{ECM-Bel}}$	CMAR [18]	N. Net	KNN	SVM
%	65.83	63.83	49.20	49.20	58.29 %	38.66	43.21	34.84

Figure 3 shows the number of extracted frequent patterns for two measures: the belief and the precise support measures. Those measure were evaluated for both likelihood-based and ECM-based imprecision modelling [19] for evidential database construction. The results show that precise-based support associated to an Evidential C-Means (ECM) for imprecision modelling provides the highest number of frequent patterns with a peak of 87423 comparatively to likelihood-based that has a peak of 50415. It is important to highlight that the belief-based

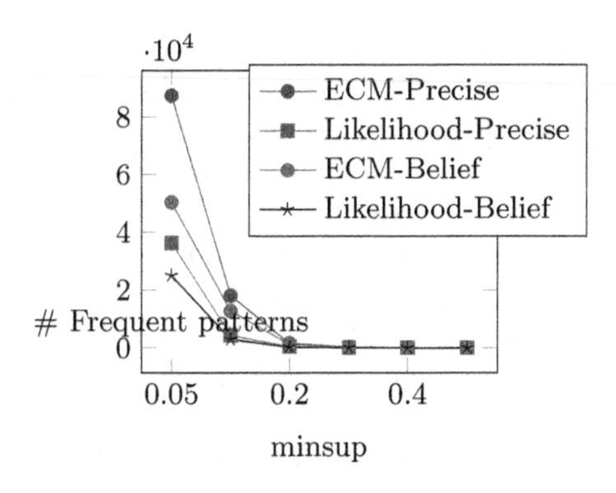

Fig. 3. Number of retrieved frequent patterns from the database.

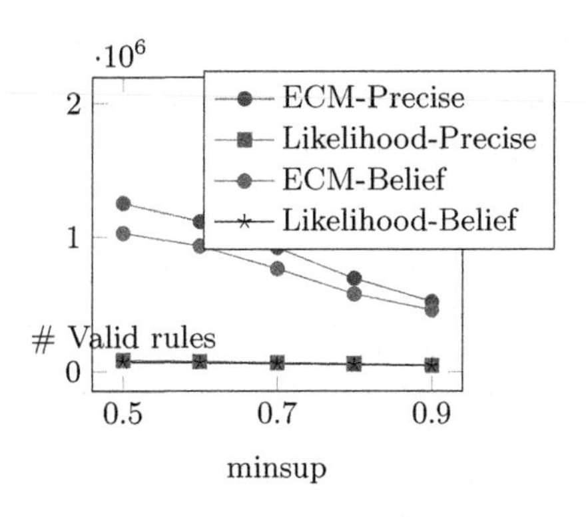

Fig. 4. Number of retrieved valid association rules from the database.

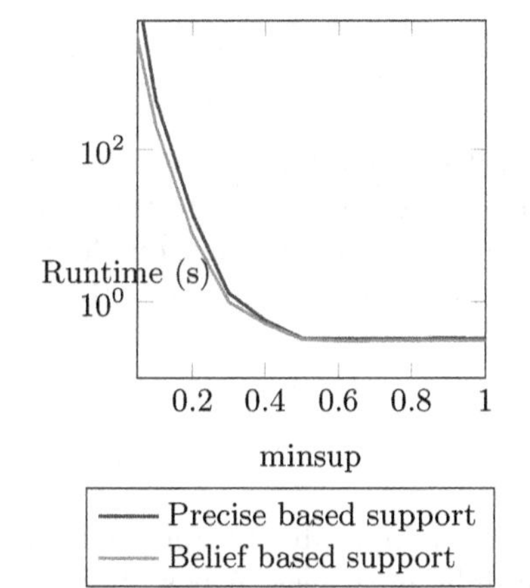

Fig. 5. Runtime relatively to minsup value

Θ_{CMC}	{0.004}	{0.4}	{4.5}	{14.40}	{50}
Recall	0.84	0.76	0.69	0.52	0.63
Precision	0.72	0.72	0.69	0.81	0.58
F-measure	0.77	0.74	0.69	0.63	0.60

Fig. 6. Recall, Precision and F-measure for EvAC on the CMC classification

support measure always provides a lower number of frequent patterns than the precise one for both likelihood and ECM database construction. This confirms that the belief-based support is a pessimistic measure.

Figure 4 highlights the number of association rules that will be used for classification. It is important to emphasize that the number of association rules depends on the number of the retrieved frequent patterns. The runtime for an extraction algorithm with the precise support is slightly higher than the belief-based one in Fig. 5. The belief measure provides better runtime thanks to the mathematical simplicity of computing the belief function. To evaluate the accuracy of our algorithm, we perform a cross validation classification. The accuracy of EvAC Algorithm is given in Table 4. The classification with a likelihood imprecision modelling approach is as efficient as those provided with ECM. Technically, the reduction of the association rules, number-wise, when it is done correctly, helps to improve results. Indeed, the classification process as demonstrated in Algorithm 1 relies on rules merging using the Dempster's rule of combination. An important number of association rules in addition to the characteristics of Dempster's combination rule behaviour misleads the fusion process to errors. In addition, the classification accuracy depends on the quality of the discretization. The use of the likelihood approach for evidential database construction provides a better handling to the uncertainty with BBAs. In contrast, the result of the k-NN, the Neural Networks and SVM using the Weka software, in Table 4, are obtained after going through a PKIDiscretization. The comparison shows that our proposed framework performs more efficiently.

In Fig. 6, we scrutinize the performance of the classification process under a likelihood database construction and the precise support with the *Recall*, *Precision* and *F-measure* relatively to each CMC class. We report the F_1 score which is the harmonic mean between precision and recall. Specifically, the F_1 score is:

$$F_1 = \frac{2 \times Prec \times Rec}{Prec + Rec}, \quad Prec = \frac{tp}{tp + fp}, \quad Rec = \frac{tp}{tp + fn}$$

with tp, fp, fn denoting true positives, false positives, and false negatives. Several recall values are low such as the CMC={14.16} comparatively to the other classes. This results can be explained by the proximity of centroid clusters found by ECM. In fact, CMC={14.16} and CMC={16.09} could be merged into one representative class for a better detection.

5 Conclusion

In this paper, we introduced new uncertain data mining driven approach for physico-chemical property prediction of amphiphilic molecules. A new imprecision modelling approach based on likelihood is provided. The likelihood modelling approach is used to construct evidential database. As illustrated in the experiment section, the proposed approach provided an interesting performance on a real-world chemical database. In future work, we will be interested in confronting our results the to other uncertain data mining approaches such as probabilistic and fuzzy databases. Furthermore, the performance of mining algorithm could be improved by adding specific heuristics to reduce focal elements through evidential database construction process.

Acknowledgement. This work was performed, in partnership with the SAS PIVERT, within the frame of the French Institute for the Energy Transition (Institut pour la Transition Energétique (ITE) P.I.V.E.R.T. (www.institut-pivert.com) selected as an Investment for the Future ("Investissements d'Avenir"). This work was supported, as part of the Investments for the Future, by the French Government under the reference ANR-001-01.

References

1. Seeja, K., Zareapoor, M.: Fraudminer: A novel credit card fraud detection model based on frequent itemset mining. Sci. World J. 2014 (2014). http://dx.doi.org/10.1155/2014/252797
2. Chen, Z., Chen, G.: Building an associative classifier based on fuzzy association rules. Int. J. Comput. Intell. Syst. **1**(3), 262–273 (2008)
3. Dehaspe, L., Toivonen, H., King, R.D.: Finding frequent substructures in chemical compounds. In: Proceeding of the Fourth International Conference on Knowledge Discovery and Data Mining (KDD 1998), New York City, New York, USA, pp. 30–36 (1998)
4. King, R.D., Srinivasan, A., Dehaspe, L.: Warmr: a data mining tool for chemical data. J. Comput. Aided Mol. Des. **15**(2), 173–181 (2001)
5. Sarfraz Iqbal, M., Golsteijn, L., Öberg, T., Sahlin, U., Papa, E., Kovarich, S., Huijbregts, M.A.: Understanding quantitative structure-property relationships uncertainty in environmental fate modeling. Environ. Toxicol. Chem. **32**(5), 1069–1076 (2013)
6. Weng, C.H., Chen, Y.L.: Mining fuzzy association rules from uncertain data. Knowl. Inf. Syst. **23**(2), 129–152 (2010)
7. Leung, C.S., MacKinnon, R., Tanbeer, S.: Fast algorithms for frequent itemset mining from uncertain data. In: Proceeding of IEEE International Conference on Data Mining (ICDM), Shenzhen, China, pp. 893–898, December 2014
8. Tong, Y., Chen, L., Cheng, Y., Yu, P.S.: Mining frequent itemsets over uncertain databases. In: Proceedings of the VLDB Endowment, vol. 5(11), pp. 1650–1661 (2012)
9. Samet, A., Lefevre, E., Ben Yahia, S.: Evidential database: a new generalization of databases? In: Proceedings of 3rd International Conference on Belief Functions, Belief 2014, Oxford, UK, pp. 105–114 (2014)
10. Samet, A., Lefevre, E., Ben Yahia, S.: Classification with evidential associative rules. In: Proceedings of 15th International Conference on Information Processing and Management of Uncertainty in Knowledge-Based Systems, Montpellier, France, pp. 25–35 (2014)
11. Hewawasam, K.R., Premaratne, K., Shyu, M.L.: Rule mining and classification in a situation assessment application: A belief-theoretic approach for handling data imperfections. Trans. Sys. Man Cyber. Part B **37**(6), 1446–1459 (2007)
12. Samet, A., Dao, T.T.: Mining over a reliable evidential database: Application on amphiphilic chemical database. To appear in Proceeding of 14th International Conference on Machine Learning and Applications, IEEE ICMLA 2015, Miami, Florida (2015)
13. Nouaouri, I., Samet, A., Allaoui, H.: Evidential data mining for length of stay (LOS) prediction problem. In: Proceeding of 11th IEEE International Conference on Automation Science and Engineering, CASE 2015, Gothenburg, Sweden, 2015, pp. 1415–1420 (2015)
14. Lee, S.: Imprecise and uncertain information in databases: an evidential approach. In: Proceedings of Eighth International Conference on Data Engineering, Tempe, AZ, pp. 614–621 (1992)
15. Samet, A., Lefevre, E., Ben Yahia, S.: Mining frequent itemsets in evidential database. In: Proceedings of the Fifth International Conference on Knowledge and Systems Engeneering, Hanoi, Vietnam, pp. 377–388 (2013)

16. Shafer, G.: A Mathematical Theory of Evidence. Princeton University Press, Princeton (1976)
17. Appriou, A.: Multisensor signal processing in the framework of the theory of evidence. In: Application of Mathematical Signal Processing Techniques to Mission Systems, pp. 5–1 (1999)
18. Li, W., Han, J., Pei, J.: Cmar: accurate and efficient classification based on multiple class-association rules. In: Proceeding of IEEE International Conference on Data Mining (ICDM), San Jose, California, USA, pp. 369–376 (2001)
19. Samet, A., Lefèvre, E., Ben Yahia, S.: Evidential data mining: precise support and confidence. J. Intell. Inf. Syst. 1–29 (2016). http://dx.doi.org/10.1007/s10844-016-0396-5

Fuzzy Methods in Data Mining and Knowledge Discovery

An Incremental Fuzzy Approach to Finding Event Sequences

Trevor P. Martin[1,2](✉) and Ben Azvine[2]

[1] Machine Intelligence and Uncertainty Group, University of Bristol, Bristol BS8 1UB, UK
trevor.martin@bristol.ac.uk
[2] BT TSO Security Futures, Adastral Park, Ipswich IP5 3RE, UK

Abstract. Recent years have seen increasing volumes of data generated by online systems, such as internet logs, physical access logs, transaction records, email and phone records. These contain multiple overlapping sequences of events related to different individuals and entities. Information that can be mined from these event sequences is an important resource in understanding current behaviour, predicting future behaviour and identifying non-standard patterns and possible security breaches. Statistical machine learning approaches have had some success but do not allow human insight to be included easily. We have recently presented a framework for representing sequences of related events, with scope for assistance from human experts. This paper describes the framework and presents a new algorithm which (i) allows the addition of new event sequences as they are identified from data or postulated by a human analyst, and (ii) allows subtraction/removal of sequences that are no longer relevant. Examination of the sequences can be used to further refine and modify general patterns of events.

Keywords: Event sequences · Incremental algorithm · Fuzzy · X-mu

1 Introduction

Collaborative intelligence aims to combine the power of machines with the interpretive skills, insight and lateral thinking provided by human analysts. The role of the computer in this partnership is to gather data, transform it algorithmically, and provide visualisation. The role of the human is to provide creativity and insight in analysing and understanding the data, and to extract "knowledge" - which may be in the form of predictive rules, normal and unusual patterns in the data, further insight into underlying mechanisms, etc.

In order to implement a successful collaborative intelligent system, it is necessary to exchange knowledge between the components - in particular between humans and machines, although in a multi-agent system we may also need to consider human-human and machine-machine exchange. Typically, machine processing is centred on well-defined entities and relations which may range from the

J.P. Carvalho et al. (Eds.): IPMU 2016, Part I, CCIS 610, pp. 525–536, 2016.
DOI: 10.1007/978-3-319-40596-4_44

flat table structures of database systems through graph-based representations and up to ontological approaches involving formal logics. Human language and communication, on the other hand, is based on a degree of vagueness and ambiguity enabling efficient transmission of information between humans without the need for precise definition of every term used. There is a fundamental mis-match between the representations used by machine and human components in a collaborative intelligent system. Even quantities that can be measured precisely (such as height of a person or building, volume of a sound, amount of rainfall, colour of an object, etc.) are usually described in human language using non-precise terms such as *tall, loud, quite heavy, dark green*, etc. More abstract properties such as *beautiful landscape, delicious food, pleasant weather, clear documentation, corporate social responsibility,* are fundamentally ill-defined, whether they are based on a holistic assessment or reduced to a combination of lower-level, measurable quantities. However, we are generally able to judge the degree to which a particular instance possesses such a property. Zadeh's initial formulation of fuzzy sets [1] was inspired primarily by the flexibility of definitions in natural language.

Large volumes of data are generated by monitoring and recording systems, such as internet logs, phone records, GPS monitors and physical access logs (e.g. to buildings), financial transactions, etc. Linking records together into sequences (whether within a single data source or across multiple sources) is a complex task which is ideally suited to the notion of collaborative intelligence. Specific problems in extracting sequences of related events include determination of what makes events "related', how to find groups of "similar" sequences, identification of typical sequences, and detection of sequences that deviate from expected patterns, where the notion of "expected" can either be derived from previous observations or from human analysis.

The ability to incorporate human knowledge and expertise is an area which distinguishes collaborative intelligence from (widely used) statistical machine learning approaches. In cases where insufficient data is available, or where the data lead to incorrect conclusions, machine learning is not reliable and human insight is required. For example, in the emergence of a previously unseen malware threat, a human analyst could use knowledge (e.g. of a zero-day exploit) to identify the likely behaviour before a statistically significant body of data has been gathered to train a machine learning system. An example of incorrect conclusions could come from records of card-based entry/exit barriers where "tailgating" occurs, i.e. an individual follows someone else through the barrier without swiping their card. Such data will give misleading behaviour patterns.

The issues involved in discerning event sequences are strongly linked to the concept of information granulation introduced by Zadeh [2] to formalise the process of dividing a group of objects into sub-groups (granules) based on "indistinguishability, similarity, proximity or functionality". In this view, a granule is a fuzzy set whose members are (to a degree) equivalent. In a similar manner, humans are good at dividing events into related groups, both from the temporal perspective (event A occurred *a few minutes before* event B but involves the same entities) and from the perspective of non- temporal properties (event C is *very similar* to event D because both involve *similar entities/activities*).

However, at the same time it is necessary to recognise that most machine-based algorithms require crisp, well-defined boundaries when processing data. In this work, we use the $X - \mu$ method to translate consistently between the (generally fuzzy) human knowledge representation and the (generally crisp) data required by machines. We describe a compact and expandable *sequence pattern* representation, which allows the addition of new event sequences as they are identified, and subtraction of sequences that are no longer relevant. The main contribution of the paper is the presentation of the incremental algorithms to add and remove sequences. Although the algorithm for sequence addition was presented briefly in [3], this paper contains a more detailed explanation. The algorithm for sequence subtraction has not been previously published (other than patent [4]).

2 Background

2.1 X-Mu Approach - Conversion between Fuzzy and Crisp

Human intelligence includes the ability to identify a group of related entities (e.g. physical objects, events, abstract ideas) and to subdivide them into smaller subgroups at an appropriate level of granularity for the task at hand. Such groups are rarely specified by "necessary and sufficient" conditions, but are better modelled by membership functions, where we can compare different entities and judge whether one belongs more strongly to the set than another. In the classical fuzzy approach, for any predicate on a universe U, we introduce a membership function

$\mu : U \rightarrow [0,\ 1]$

representing the degree to which each value in U satisfies the predicate. Within a universal set, the absolute value of the membership function for an element is generally less important than the relative value, compared to other elements. Whilst the end points 0 and 1 obviously correspond to classical non-membership and full membership in the set, other values are most useful in comparing the strength of membership (e.g. Bill Gates belongs more strongly than Larry Ellison to the set of *rich people*). In this interpretation of fuzzy sets, there is an underlying assumption that membership values are *commensurable*, i.e. that membership of (say) 0.8 in the set of *rich people* can be interpreted in the same way as membership of 0.8 in the set of *tall people* or membership 0.8 for a temperature value in the set of temperatures *near-freezing*. Such commensurability is routinely assumed in fuzzy control applications (for example, an inverted pendulum where a membership in a set of cart velocities might be combined with membership in a set of angular accelerations). We adopt the commensurability assumption in this paper. The interested reader is referred to [5,6].

Fuzzy approaches typically require modification of crisp algorithms to allow set-valued variables. This is most apparent in fuzzifications of arithmetic, where a single value is replaced by a fuzzy *interval.* For example, calculating the average age of four employees known to be 20, 30, 50 and 63 is inherently simpler than when the ages are given as *young, quite young, middle-aged* and *approaching retirement.* In the latter case, we must handle interval arithmetic AND membership grades.

In a similar fashion, querying a database to find employees who are aged over 60 is simpler than finding employees *approaching retirement age.*

The $X - \mu$ method [6] recasts the fuzzy approach as a mapping from membership to universe, allowing us to represent a set, interval or single value that varies with membership, e.g. the mid-point of an interval. This natural idea is difficult to represent in standard fuzzy theory, even though it arises frequently e.g. the cardinality of a discrete fuzzy set or the number of answers returned in response to a fuzzy query.

Since there is generally a set of values which satisfy the predicate to some degree, we must modify algorithms to handle sets of values rather than single values. These sets represent equivalent values - that is, values which cannot be distinguished from each other. In this work, we are dealing with events that are equivalent because their attributes are indiscernible - however, these sets of events may vary according to membership, interpreted as the degree to which elements can be distinguished from each other. The approach described in the next section assumes we have crisp equivalence classes. We allow fuzziness in the definition of sets used by human experts, and use the $X - \mu$ method to ensure we have crisp sets, by working at a specific membership level. The $X - \mu$ method also allows us to work with *intensional* definitions of equivalence classes (parameterised by membership), but we do not cover this aspect here.

3 Directed Acyclic Sequence Graphs (DASG): Graph Representation of Event Sequences

For any sequence of events, we create a directed graph representation in which each edge represents a set of indiscernible events. Clearly for reasons of storage and searching efficiency it is desirable to combine event sequences with common sub-sequences, as far as possible, whilst only storing event sequences that have been observed. This problem is equivalent to dictionary storage, where we are dealing with single letters rather than sets of events, and we can utilise efficient solutions that have been developed to store dictionaries. In particular, we adopt the notion of a DAWG (directed acyclic word graph) [7]. Words with common letters (or events) at the start and/or end are identified and the common paths are merged to give a minimal graph, in the sense that it has the smallest number of nodes for a DAWG representing the set of words (event sequences). Several algorithms for creating minimal DAWGs have been proposed. In the main, these have been applied to creation of dictionaries and word checking, efficient storage structure for lookup of key-value pairs and in DNA sequencing (viewed as a variant of dictionary storage). Most methods (e.g. [8,9]) assume that all words (letter sequences) are available and can be presented to the algorithm in a specific order. Sgarbas [7] developed an incremental algorithm which allowed additional data to be added to a DAWG structure, preserving the minimality criterion (i.e. assuming the initial DAWG represented the data in the most compact way, then the extended DAWG is also in the most compact form).

We assume that data is presented in a standard object-attribute-value table format (e.g. CSV), with additional attributes calculated as necessary. We assume that the data arrives in a sequential manner, either row by row or in larger groups which can be processed row-by-row. Each row represents an event; there may be several unrelated event sequences within the data stream but we assume events in a single sequence arrive in time order. It is not necessary to store the data once it has been processed, unless required for later analysis.

Table 1. Sample data from the VAST 2009 MC1 dataset

eventID	Date	Time	Emp	Entrance	Direction
1	jan-2	7:30	10	b	in
2	jan-2	13:30	10	b	in
3	jan-2	14:10	10	c	in
4	jan-2	14:40	10	c	out
5	jan-2	9:30	11	b	in
6	jan-2	10:20	11	c	in
7	jan-2	13:20	11	c	out
8	jan-2	14:10	11	c	in
9	jan-2	14:30	11	c	out
10	jan-3	9:20	10	b	in
11	jan-3	10:40	10	c	in
12	jan-3	14:00	10	c	out
13	jan-3	14:40	10	c	in
14	jan-3	16:50	10	c	out
15	jan-3	9:00	12	b	in
16	jan-3	10:20	12	c	in
17	jan-3	12:30	12	c	out
18	jan-3	14:30	12	c	in
19	jan-3	15:00	12	c	out

An example, used throughout the rest of the paper, is shown in Table 1. This is a small subset of benchmark data taken from mini-challenge 1 of the VAST 2009 dataset[1], which gives swipecard data showing employee movement into a building and in and out of a classified area within the building. No data is provided on exiting the building.

The DASG representation assumes that we can subdivide the attributes into the following categories:

- Event identifier - a key value which uniquely identifies a row of the table. In the example, *eventID* takes this role.

[1] http://hcil2.cs.umd.edu/newvarepository/benchmarks.php.

- Event Sequencer - one or more attributes with an associated total order, used to determine whether one event precedes or succeeds another. In the example, *Date* or *Time* or both could take this role.
- Event Linkage - one or more attributes with an equivalence relation that determines whether two events are linked (part of the same sequence). For example, in Table 1 we define a sequence of events involving the same employee, where there is no more than 8 h between contiguous events.
- Event Categorisation - one or more attributes with an equivalence relation that determines whether two events (in different sequences) can be considered as examples of the same event category. For example, we might group together events that happen at approximately the same time, and/or involving the same swipecard actions (building+in, classified+in or classified+out). The definition of *approximately the same time* can be fuzzy, but must be made crisp (via $X - \mu$) to define the equivalence relation.
- Recorded Data - for subsequent analysis, we can record one or more of the attributes associated with an event. This may be as simple as counting the number of instances, or may involve more sophisticated processing such as association rules between events.

There is no restriction on the number of attributes. We have selected three employees for illustration purposes; rows in the initial table were ordered by date/time, but have been additionally sorted by employee here to make the sequences obvious. In this data,

Emp = set of employee ids = {10, 11, 12}
Date, Time= date / time of event
Entry points = {B - building, C - classified section}
Access direction = {in, out}

We first define the linkage relations, to detect candidate sequences. Here, for a candidate sequence of n events:

$$S_1 = (o_{11}, o_{12}, o_{13}, \ \ldots \ , o_{1n})$$

we define the following computed quantities:

$$\begin{aligned} ElapsedTime \quad \Delta T_{ij} &= Time\,(o_{ij}) - Time\,(o_{ij-1}) \\ with \quad \Delta T_{i1} &= Time\,(o_{i1}) \end{aligned}$$

and restrictions (for $j > 1$):

$$\begin{aligned} Date\,(o_{ij}) &= Date\,(o_{ij-1}) \\ 0 < Time\,(o_{ij}) - Time\,(o_{ij-1}) &\le T_{thresh} \\ Emp\,(o_{ij}) &= Emp\,(o_{ij-1}) \\ (Action\,(o_{ij-1})\,, Action\,(o_{ij})) &\in AllowedActions \\ \text{where} \quad Action\,(o_{ij}) &= (Entrance\,(o_{ij})\,,\ Direction\,(o_{ij})) \end{aligned}$$

where the relation *AllowedActions* is specified in Table 2. These constraints can be summarised as

Table 2. Allowed actions (row = first action, column = next action)

	b,in	c,in	c,out
b,in	x	x	
c.in			x
c,out	x	x	

- events in a single sequence refer to the same employee
- successive events in a single sequence conform to allowed transitions between locations and are on the same day, within a time (T_{thresh}) of each other.

We choose a suitable threshold e.g. $T_{thresh} = 8$, ensuring anything more than 8 h after the last event is a new sequence. We identify candidate sequences by applying the linkage relations. Any sequence has either been seen before or is a new sequence. In Table 1, candidate sequences are made up of the events:
$1 - 2 - 3 - 4$,
$5 - 6 - 7 - 8 - 9$,
$10 - 11 - 12 - 13 - 14$,
$15 - 16 - 17 - 18 - 19$
We also define the *EventCategorisation* equivalence classes used to compare events in different sequences. Here,

$$EquivalentAction = I_{Action}$$
$$\text{For direction } In, \quad EquivalentEventTime = \{[7], [8], \ldots\}$$
$$\text{For direction } Out, \quad EquivalentElapsedTime = \{[0], [1], [2], \ldots\}$$

where I is the identity relation and the notation [7] represents the set of start times from 7:00–7:59. As mentioned in Sect. 2.1, fuzzy equivalence classes are converted to crisp sets at a specific membership or to intensional definitions, parameterised by membership. We represent each identified sequence as a path labelled by its event categorisations (Fig. 1). The algorithms presented in the next section allow us to incrementally add unseen sequences into a minimal DASG which represents exactly the set of sequences seen so far (Fig. 2). Nodes are labelled by unique numbering; since the graph is deterministic, each outgoing edge is unique. An edge can be specified by its start node and event categorisation, or by its event categorisation if there is no ambiguity about its start node.

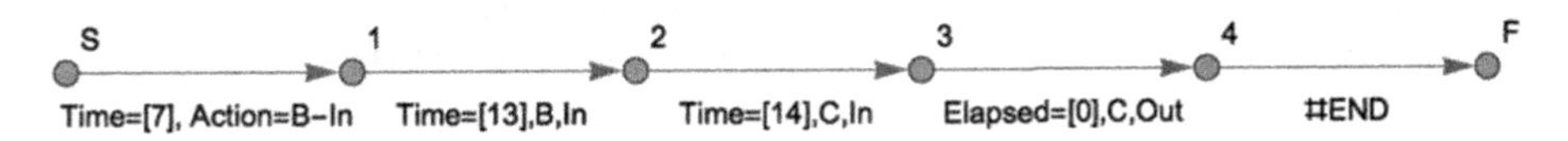

Fig. 1. DASG representation of the first sequence (events 1-2-3-4) from Table 1. The labels show the categorisation attributes, namely the equivalence class of the event time and the entrance and direction.

```
Algorithm ExtendGraph
Input :Graph G (minimal current sequence), start node S, end node F
CandidateSequence Q[0 - NQ] representing the candidate sequence;
        each element is an event identifier. The sequence is not
         already present in the graph and is terminated by #END
Output : updated minimal graph, incorporating the new sequence
Local variables :  Node startNode, newNode, endNode, matchNode
        Edge currentEdge, matchEdge
        Categorisation currentCategorisation
        integer seqCounter;

startNode = S
seqCounter = 0
WHILE  EventCategorisation(S[seqCounter]) ∈ OutgoingEdges(StartNode)
     currentEdge = (startNode, EventCategorisation(Q[seqCounter] )
     endNode = End (currentEdge)
     IF InDegree (endNode) > 1
     THEN
          newNode =CreateNewNode({currentEdge}, OutgoingEdges(endNode))
          IncomingEdges(endNode) = IncomingEdges (endNode) -
     currentEdge
          startNode = newNode
     ELSE
          startNode = endNode
     seqCounter++
ENDWHILE

WHILE seqCounter < NQ                          // create new path
     currentEdge = (startNode, EventCategorisation (S[seqCounter]) )
     startNode = CreateNewNode({currentEdge}, { })
     seqCounter++
ENDWHILE

currentCategorisation = #END
currentEdge = (startNode, #END )      // last edge, labelled by #END

IncomingEdges(F) = IncomingEdges (F) + currentEdge
endNode = F
nextEdgeSet = {currentEdge}

WHILE  nextEdgeSet contains exactly one element (i.e currentEdge)
                AND ExistsSimilarEdge(currentEdge, endNode)

     matchNode = StartOfSimilarEdge(currentEdge, endNode)
     startNode = Start (currentEdge)
     IncomingEdges(endNode) = IncomingEdges (endNode) - {currentEdge}
     nextEdgeSet = IncomingEdges (startNode)
     IncomingEdges (matchNode)= nextEdgeSet∪IncomingEdges (matchNode)
     endNode = matchNode
     currentEdge ∈ edgeSet          // choose any element,
END WHILE                           //  "while" loop terminates if >1
```

Fig. 2. Algorithm to extend a minimal graph by incremental addition of a sequence of edges

Standard definitions are used for *InDegree, OutDegree, IncomingEdges*and *OutgoingEdges* of a node, giving respectively the number of incoming and outgoing edges, the set of incoming edges and the set of outgoing edges. We also apply functions *Start* and *End* to an edge, to find or set its start and end nodes respectively and *EdgeCategorisation* to find its categorisation class.
Finally, let the function $ExistsSimilarEdge(edge, endnode)$ return true when:
edge has end node *endnode*, event categorisation *L*and start node *S1*
AND
a second, distinct, edge has the same end node and event categorisation L but a different start node $S2$
AND
$S1$ and $S2$ have $OutgoingEdges(S1) == OutgoingEdges(S2)$
If such an edge exists, its start node is returned by the function
$StartOfSimilarEdge(edge, endnode)$

The function $MergeNodes(Node1, Node2)$ deletes $Node2$ and merges its incoming and outgoing edges with those of $Node1$.

The function $CreateNewNode(Incoming, Outgoing)$ creates a new node with the specified sets of incoming and outgoing edges.

The algorithm proceeds in three distinct phases (corresponding to the three *while* loops in Fig. 2. In the first and second parts, we move step-by-step through the new event sequence and the graph, beginning at the start node S. If an event categorisation matches an outgoing edge, we follow that edge to the next node and move on to the next event in the sequence. If the new node has more than one incoming edge, we must copy[2] it; the copy takes the incoming edge that was just followed, and the original node retains all other incoming edges. Both copies have the same set of output edges. This part of the algorithm finds other sequences with one or more common starting events. If at some point, we reach a node where there is no outgoing edge matching the next event's categorisation, we create new edges and nodes for the remainder of the sequence, eventually connecting to the end node F. Note that as the sequence is new, we must reach a point at which no outgoing edge matches the next event's categorisation; if this happens at the start node S then the first stage is (effectively) omitted. Finally, in the third stage, we search for sequences with one or more common ending events. Where possible, the paths are merged.

The advantage of this algorithm is that it allows *incremental* modification, so that new sequences can be added at any time. Although the example shows the sequence patterns derived from data, it is also possible for a human expert to specify and add a sequence pattern without it having been seen in the data. Hence (for example) a previously unseen cyber-attack sequence could be added to the DASG and the matching event sequence would be detected as soon as it occurred. In contrast, a purely data-driven method would first flag the new sequence as unknown (not matching any pattern in the graph), and would only recognise subsequent occurrences after graph updating (Fig. 3).

[2] The copy and merge operations are also used when removing sequences.

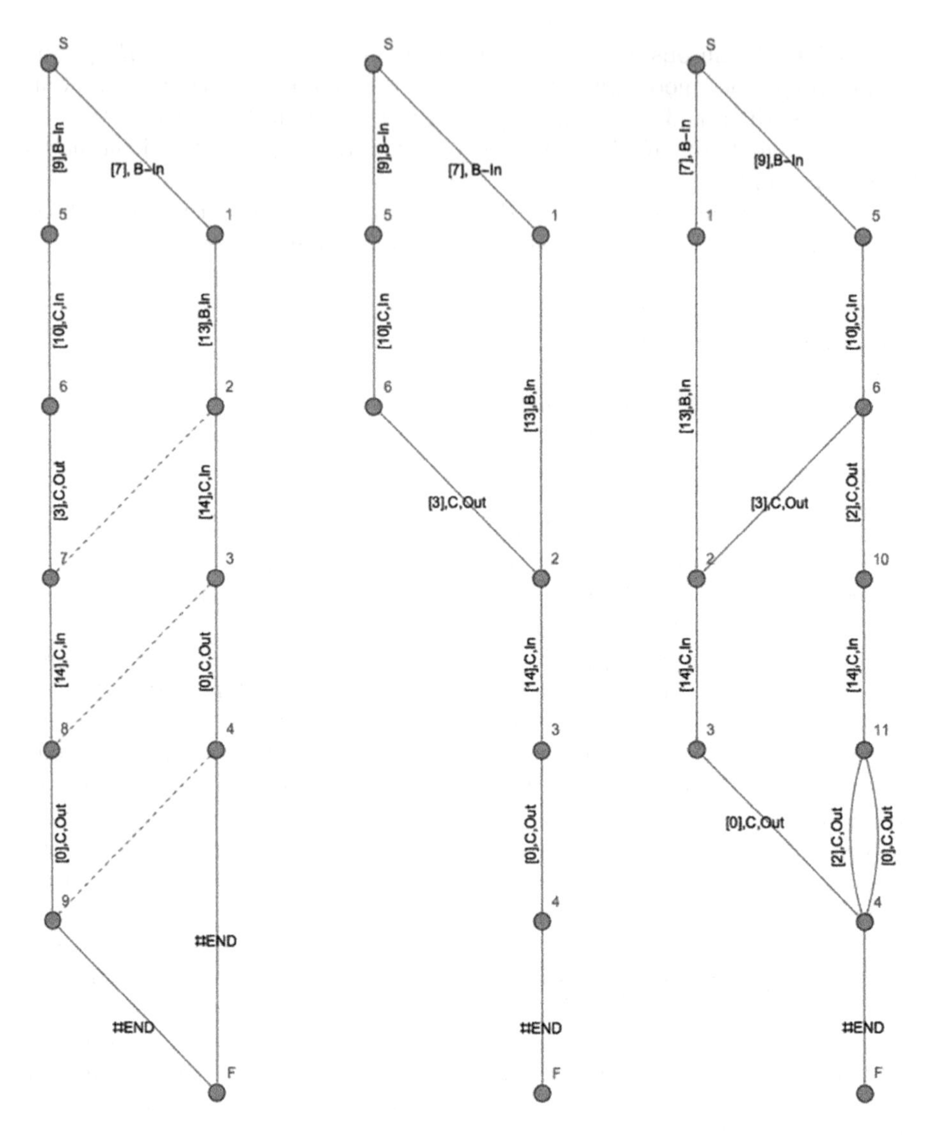

Fig. 3. Adding further sequences: (left) graph after phase 1 of adding the second sequence, dotted lines indicate nodes which are identical in phase 3; (centre) final graph (after identical nodes have been merged) representing sequences 1 and 2; (right) graph representing all four sequences

The representation also allows straightforward removal of sequences. For example, if it is known that a sequence will never be seen again because it is screened out by a different process, or because external changes make it impossible, then processing and storage efficiency are improved by removing the sequence pattern from the graph. However, we must take care not to remove

```
Algorithm ReduceGraph
Input :Graph G, start node S, end node F, the current DASG (minimal)
       Sequence C[0 - N0]  representing the sequence of event categories
       to be removed. Each element is an event categorisation i.e. a
       start-node and C[i] uniquely specifies an edge.
       NB the sequence must be present in the graph and there must be at
       least one sequence in the graph after removal.

Output : updated minimal graph, excluding the removed sequence

Local variables :  Node startNode, endNode, potentialDuplicates (LIST)
       Edge currentEdge
       integer seqCounter;

startNode = S
endNode =  F
seqCounter = 0
currentEdge = (startNode, C[0])

WHILE  endNode > startNode   AND  seqCounter < N0

    IF  OutDegree(currentEdge.END) > 1 AND InDegree(currentEdge.END)==1
    THEN startNode =  currentEdge.END
    ELSE
        IF OutDegree(currentEdge.END)==1 AND InDegree(currentEdge.END) > 1
        THEN endNode =  currentEdge.END
    ELSE
        IF OutDegree(currentEdge.END) >1 AND InDegree(currentEdge.END) >1
    THEN
        ADD currentEdge.END to potentialDuplicates
    ENDIF
    currentEdge = (currentEdge.END, C[seqCounter])
    seqCounter++
ENDWHILE

IF (endNode > startNode)
THEN
    Duplicate each node in potentialDuplicates
    Delete each edge in Sequence[] from startNode to endNode
    Apply 3rd phase of ExtendGraph (merge final edges)
ENDIF
```

Fig. 4. Algorithm to reduce a minimal graph by incremental removal of an existing sequence of edges

any edge which is part of another pattern. Figure 4 shows the algorithm. The process is straightforward - we consider nodes where the indegree ≥ 1, and the outdegree ≥ 1 (four possibilities).

- If a node has indegree one and outdegree greater than one, then the path up to and including this node must be retained, because it contributes to paths other than the path to be deleted. In this case, any potential duplicates up to this point can be discarded as the nodes are not altered.
- If a node has indegree greater than one and outdegree equal to one, then the path from this node to the end must be retained, because it contributes to paths other than the path to be deleted.

– If a node has indegree one and outdegree one, it (and its incoming and outgoing edges) can potentially be removed, depending on the path.
– Finally a node with both in- and out- degree greater than one might require modification and is marked as a potential duplicate.

4 Summary

The DASG representation allows us to store event sequence patterns in a compact directed graph format, with efficient incremental algorithms to add a previously unseen pattern to the graph, and to remove a pattern from the graph. Sequence patterns can be generated from data or by a human expert. The DASG representation allows fuzzy specification of categories and equivalence relations, which are converted to crisp relations using the $X - \mu$ approach. An efficient implementation of the DASG is possible by compiling the graph into a set of instructions for a virtual machine (described in [4,10]).

References

1. Zadeh, L.A.: Fuzzy sets. Inf. Control **8**(3), 338–353 (1965)
2. Zadeh, L.A.: Toward a theory of fuzzy information granulation and its centrality in human reasoning and fuzzy logic. Fuzzy Sets Syst. **90**, 111–127 (1997)
3. Martin, T.P., Azvine, B.: Representation and identification of approximately similar event sequences. In: Andreasen, T., et al. (eds.) FQAS 2015. AISC, vol. 400, pp. 87–99. Springer International Publishing, Switzerland (2016)
4. Martin, T.P., Azvine, B.: Patent pct/gb2014/000378: Sequence identification (2014)
5. Dubois, D., Prade, H.: Gradualness, uncertainty and bipolarity: making sense of fuzzy sets. Fuzzy Sets Syst. **192**, 3–24 (2012)
6. Martin, T.P.: The x-mu representation of fuzzy sets. Soft Comput. **19**, 1497–1509 (2015)
7. Sgarbas, K.N., Fakotakis, N.D., Kokkinakis, G.K.: Optimal insertion in deterministic dawgs. Theor. Comput. Sci. **301**(1–3), 103–117 (2003)
8. Revuz, D.: Minimization of acyclic deterministic automata in linear time. Theor. Comput. Sci. **92**, 181–189 (1992)
9. Hopcroft, J.E., Ullman, J.D.: Introduction to Automata Theory, Languages, and Computation. Addison-Wesley, Reading (1979)
10. Martin, T.P., Azvine, B.: A virtual machine for event sequence identification using fuzzy tolerance. In: IEEE International Conference on Fuzzy Systems (FUZZ-IEEE). IEEE Press (to appear)

Scenario Query Based on Association Rules (SQAR)

Carlos Molina[1(✉)], Belen Prados-Suárez[2], and Daniel Sanchez[3]

[1] Department of Computer Sciences, University of Jaen, Jaen, Spain
carlosmo@ujaen.es
[2] Department of Software Engineering, University of Granada, Granada, Spain
belenps@ugr.es
[3] Department of Computer Science and Artificial Intelligence, University of Granada, Granada, Spain
daniel@decsai.ugr.es

Abstract. In the last years association rules are being applied to support decision making. However, the main concern is in the precision and not in the interpretability of their results, so they produce large sets of rules difficult to understand for the user. A comprehensible system should work according to the human decision making process, which is quite based on the case study and the scenario projection. Here we propose an *association rule based system for scenario query* (SQAR), where the user can perform *"what if...?"* queries, and get as response what usually happens under similar scenarios. Even more we enrich our proposal with a hierarchical structure that allows the definition of scenarios with different detail levels, to comply with the needs of the user.

Keywords: OLAP · What if queries · Association rules

1 Introduction

Association Rules (AR) have been widely used with different purposes [18], mainly for problem description but also for tasks like classification, as in the case of the predictive association rules or associative classifiers, the multi-class classification methods, or the class association rules algorithms (CAR).

Lots of these methods are applied to support decision making [21], however, as Huysmans et al. states in [6], they are mainly focused on improving the accuracy of the classification; which results in systems that offer to the user a set of rules, usually very large, that is difficultly understandable. It is leading to a growing concern about improving the interpretability and comprehensibility of the data mining results [10], which is bringing proposals combining the association rules with other techniques like decision trees [8].

However, to create an understandable system for the user, it seems logic to study how the humans make their decisions, and develop a system according to it. In this sense, research points the study of cases and the analysis and projection of scenarios

J.P. Carvalho et al. (Eds.): IPMU 2016, Part I, CCIS 610, pp. 537–548, 2016.
DOI: 10.1007/978-3-319-40596-4_45

as the most widely spread practices in decision making environment, due to the impact they have in the quality and the biases of the decisions taken [12,20]. This way it is possible to find proposals using scenarios to support decision making in medicine [11], education [5] or even sports [19]. Even more, recent proposals point the methods based on scenarios as better than probabilistic models [4] or those based on predictions [14].

In OLAP *what if* queries [15] are a type of analysis where the user indicate an hypothetical scenario indicating some values for the datacube (or changes over them) and the system builds a new datacube under these assumptions. The user can then query this new datacube to analyse the results of this hypothesis [3]. This kind of analysis is time consuming due to the calculation needed to build the datacube. There are proposals to improve the efficiency of the process [7,16]. In all the cases, if the user changes the hypothetical query the system has to build a new datacube for this scenario.

This is why we propose here a comprehensible association rule based system, oriented to support decision making based on scenarios: *Scenario Query based on Association Rules* (SQAR). The main idea consists on creating a set of association rules, and build on it a knowledge base for an inference system (one for all scenarios). This system allows the user to query about a given scenario, through queries of the type *"what if..?"*, to which the user gets as response the elements or situations that usually take place under similar situations.

Even more we propose the use a multidimensional model as starting point, which allows taking advantage of the hierarchies defined on it to compose scenarios with different detail levels, and hence to use concepts nearer to the user.

It is explained in Sect. 2 where the underlying fuzzy multidimensional model and the association rules extraction method are presented. Next, in Sect. 3 the scenario oriented query system is described. In Sect. 4 we pose an simple example of use to illustrate the operation of the system; and finally, in Sect. 5, we show our conclusions.

2 Fuzzy Multidimensional Model

The base for the system structure is a multidimensional model that stores the data and allows querying on it. In this section we briefly present its basics, and a more detailed explanation of the structure and operations can be found in [13].

2.1 Multidimensional Structure

The structure of the fuzzy multidimensional model, starts with next definitions:

Definition 1. *A dimension is a tuple* $d = (l, \leq_d, l_\perp, l_\top)$ *where* $l = l_i, i = 1, ..., n$ *such that each* l_i *is a set of values* $l_i = \{c_{i1}, ..., c_{in}\}$ *and* $l_i \cap l_j = \emptyset$ *if* $i \neq j$*, and* $\leq_d$ *is a partial order relation between the elements of* l *so that* $l_i \leq_d l_k$ *if* $\forall c_{ij} \in l_i \Rightarrow \exists c_{kp} \in l_k / c_{ij} \subseteq c_{kp}$*.* $l_\perp$ *and* $l_\top$ *are two elements of* l *such that* $\forall l_i \in l \; l_\perp \leq_d l_i \leq_d l_\top$*.*

We denote level to each element l_i. To identify the level l of the dimension d we will use $d.l$. The two special levels $l_\perp$ and $l_\top$ will be called *base level* and *top level* respectively. The partial order relation in a dimension is what gives the hierarchical relation between levels.

Definition 2. *For each pair of levels l_i and l_j such that $l_j \in H_i$, we have the relation $\mu_{ij} : l_i \times l_j \to [0,1]$ and we call this the* **kinship relation**.

If we use only the values 0 and 1 and we only allow an element to be included with degree 1 by an unique element of its parent levels, this relation represents a crisp hierarchy. But if we relax these conditions and allow the use of values in the interval [0,1] with no other limitation, we have a fuzzy hierarchical relation.

Definition 3. *We say that any pair (h, α) is a* **fact** *when h is an m-tuple on the attributes domain we want to analyze, and $\alpha \in [0,1]$.*

The value α controls the influence of the fact in the analysis. The imprecision of the data is managed by assigning an α value representing this imprecision. Now we can define the structure of a fuzzy DataCube.

Definition 4. *A DataCube is a tuple $C = (D, l_b, F, A, H)$ such that $D = (d_1, ..., d_n)$ is a set of dimensions, $l_b = (l_{1b}, ..., l_{nb})$ is a set of levels such that l_{ib} belongs to d_i, $F = R \cup \emptyset$ where R is the set of facts and $\emptyset$ is a special symbol, H is an object of type history, and A is an application defined as $A : l_{1b} \times ... \times l_{nb} \to F$, giving the relation between the dimensions and the facts defined.*

2.2 Operations

Once we have the structure of the multidimensional model, we need the operations to analyze the data in the datacube. In this section we present the elements needed to apply the normal operations (roll-up, drill-down, pivot and slice).

Definition 5. *An aggregation operator G is a function $G(B)$ where $B = (h, \alpha)/(h, \alpha) \in F$ and the result is a tuple (h', α').*

The parameter required by the operator can be seen as a fuzzy bag, a construction which may have a group of elements that can be duplicated, and each one has a degree of membership.

Definition 6. *For each value a belonging to d_i we have the set*

$$F_a = \begin{cases} \bigcup\limits_{l_i \in H_{l_i}} F_b / b \in l_j \wedge \mu_{ij}(a,b) > 0 & \text{if } l_i \neq l_b \\ \{h / h \in H \wedge \exists a_1, ..., a_n A(a_1, ..., a_n) = h\} & \text{if } l_i = l_b \end{cases} \quad (1)$$

The set F_a represents all the facts that are related to the value a.

With this structure, the basic operations over datacubes are defined (definition and properties in [13]).

2.3 Association Rule Extraction: COGARE

Our proposal is based on an association rule extraction algorithm that works over datacubes called *Complexity Guided algorithm for Association Rule Extraction* COGARE [9]. The method extracts rules over datacubes and diminishes the complexity by using the concepts defined in the hierarchy over each dimension to reduce the number of rules, as indicated in Fig. 1.

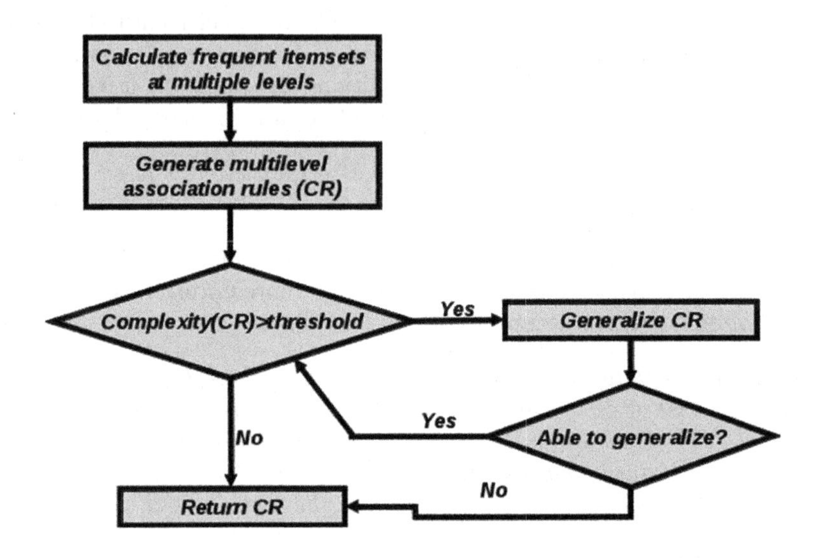

Fig. 1. COGARE algorithm

There are three main stages in the method:

- *Itemsets generation:* the algorithm uses the *Apriori* algorithm [1] adapted to the multidimensional model. It takes into account the hierarchy of the dimension when an itemset is not frequent, and looks for a generalization that may be frequent (bottom-up approach). In this generalization the support threshold is adapted such that the more general the items are, the higher the threshold is (see [9] for details).
- *Rules generation:* the algorithm extracts rules with an *Apriori* like algorithm.
- *Rules generalization:* the rule set obtained in the previous stage is generalized to reduce the complexity of the result. In this step, the algorithm tries to generalize the elements in the association rules by defining the items at a higher level (more abstract). If the generalized rules include another ones (i.e. they are defined over more concrete values but represent the same knowledge) those are deleted. On each step, the quality of the rule set is controlled so if it decreases down to an established threshold, the operation is not applied.

The associations rules obtained with this process are pruned according to a *certainty factor (CF)* [17] instead of *confidence* to avoid some of the well known problems of this quality measure. The method obtains an average complexity reduction of 39 % [9] and this is why we have chosen it as starting point for our proposal.

3 Scenario Query Based on Association Rules System

Using as base the fuzzy multidimensional system presented above, we add now the elements needed for the SQAR system. Figure 2 shows the system structure, where two modules are required:

- *Inference system:* the association rules will be used as knowledge base for an inference system. The user will give some data for the needed scenario and the inference system will use the rules to get other elements that normally appear in that case.
- *User interface:* the interface allows the user to describe the scenario and is also used to show the related inferred elements. We have implemented an interactive interface in the sense that user can refine the scenario after the inference process, adding or deleting elements.

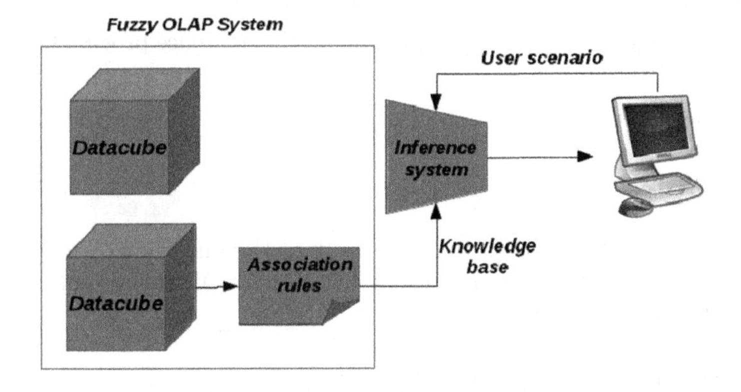

Fig. 2. SQAR System structure

3.1 Inference System

One of the main problems to use association rules in an inference system, is that it is not possible to propagate the values of the quality measures (support and confidence) when applying the rules; as states Balcazar et al. [2] "the confidence of the inferred items can not be propagated from the known items and their confidences".

However, this issue can be solved using the *CF*. This measure was first proposed for an expert system on Medicine called MYCIN [17]. This system used

CF to enable inexact reasoning, so the authors proposed an inference system to deal with CF and propagate the values. This is why the algorithm described in Sect. 2.3 to extract association rules uses the CF: since or rules have been obtained with this factor, they can directly be used in an inference system.

What we propose here is to incorporate the rules obtained by COGARE process as knowledge base for MYCIN inference system in a interactive interface. With this approach we hide the rules to the user so he/she does not have to interpret them. The user only has to establish the known values that define the scenario of interest (i.e. the elements in the dimensions of the datacube), and the system will apply the obtained rules to infer other values that are related to the ones introduced.

3.2 User Interaction

We propose an interactive process to solve the user's scenario queries:

1. The user chooses the values that define the scenario for one or several dimensions of the datacube.
2. The inference process is applied and the elements related to the ones selected are shown to the user.
3. User may interact with the system adding new values or deleting one of the previously selected.
4. After each change, the system applies again the inference process showing the new results to the user.

In the next section we present an example of use in the medical field.

4 Example

This example has three parts: first we introduce the underlying fuzzy datacube of the medical case, then we show the operation of the OLAM system, and finally the scenario query is exemplified.

4.1 Medical DataCube: $C_{Medical}$

This schema is defined over data collected for non-postponed surgeries which were carried out in hospitals in Granada between 2002 and 2004. For the facts, we only consider the data when the patients are from Granada. There are 50185 facts with one variable (*amount*) and 6 dimensions:

- *Patient:* the dimension that models patient data. The most detailed levels consider the different combinations of sex and age of each patient (the base level therefore has 2 sexes for 101 possible ages, totalling 202 values). As shown in Fig. 3, over this level we group the patients according to their sex (level *sex*) and age (level *age*). Over this last one, we define (level *group*) what we can consider to be *young*, *adult* and *old* patients using linguistic terms over the concrete values, grouping this way the values more naturally for user. The last level groups all the values so we have called it *all* with a single value (*all*).

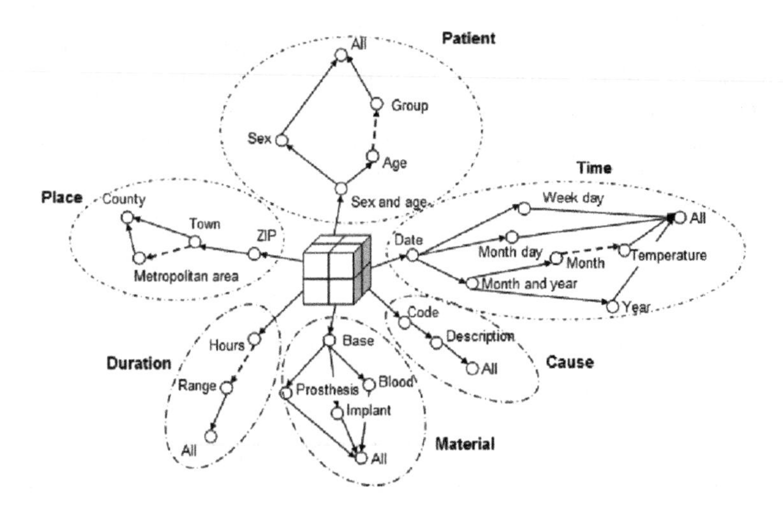

Fig. 3. Multidimensional schema over medical data

- *Time:* in this dimension we consider the date when the surgeries took place. Over this level, we have defined a normal hierarchy over dates: week day, month day, month, month and year, and year. The level *Temperature* represents information about the average temperature of each month in Granada using the labels *cold*, *warm* and *hot* to group the values. The relationships between the month and the temperature are not crisp because the user normally considers these concepts with imprecision. The definition of the relationships are shown in Fig. 4.
- *Place:* this dimension stores information about where the patients live. Since the definition of the metropolitan area of Granada is not clear, we have used a fuzzy relation to establish the relationship between this level and the towns.
- *Duration:* we also consider the amount of time that each operation took. The level *Range* groups this information according to three categories: *normal*, *long* and *very long* duration. These groups have been defined imprecisely as shown in Fig. 5.

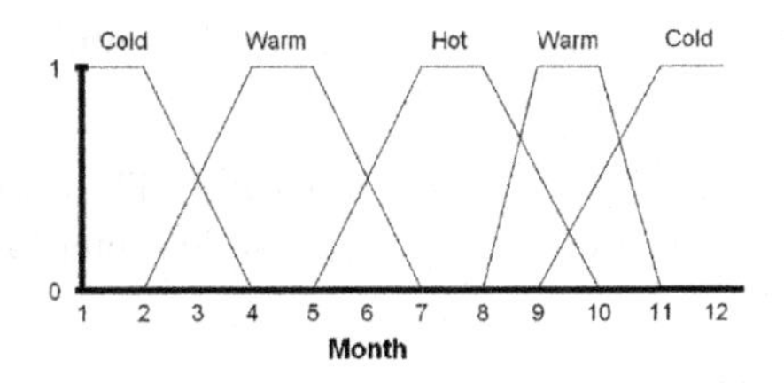

Fig. 4. Definition of level *Temperature* in dimension *Time* for $C_{Medical}$

- *Material:* we want to analyze whether any materials were required for the operations, i.e. blood, prothesis, implants. The dimension *Material* models this information.
- *Cause:* in this dimension we model the causes of the surgery according to the codes established by the WHO. We consider the 9 main categories as the base level and the description on them.

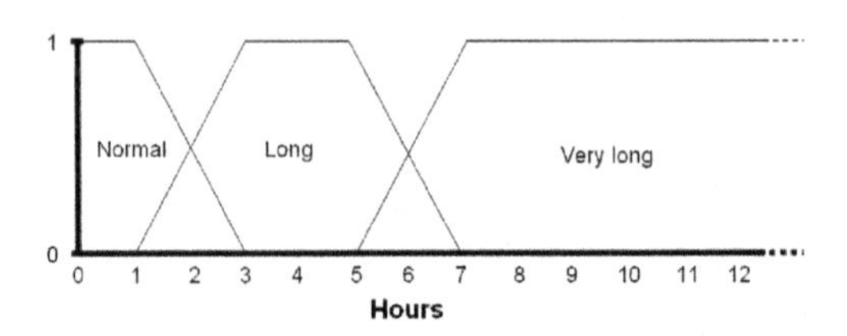

Fig. 5. Definition of level *Range* in dimension *Duration* for $C_{Medical}$

Measures: the only measure we consider is the number of surgeries performed with exactly the same values for all the dimensions we have built. This measure has been called the *amount*.

4.2 OLAM Process

To extract the rules we use COGARE algorithm over the datacube $C_{Medical}$. The main parameters used are:

- *Support:* 0.01.
- *CF:* 0.4 (minimum value of CF to consider a rule).
- *Abstraction function:* We use the *Generality* as abstraction function defined as the number of elements grouped by the item compared to the total number of elements at most detailed level of the dimension (the base level) (see [9] for more details).
- *Complexity due to number of rules:* in this case we use the function N defined as a relation between the number of rules and the number of possible items in the datacube (see [9] for details).

After the process we get 164 association rules. These rules will be used in the next step to enable the user to ask for possible scenarios.

4.3 *What-if...* Process

Figure 6 shows the initial screen of the system.

Let us suppose that a medical doctor is interested on the surgeries related to *infectious diseases*. In the interface the user chooses this kind of disease

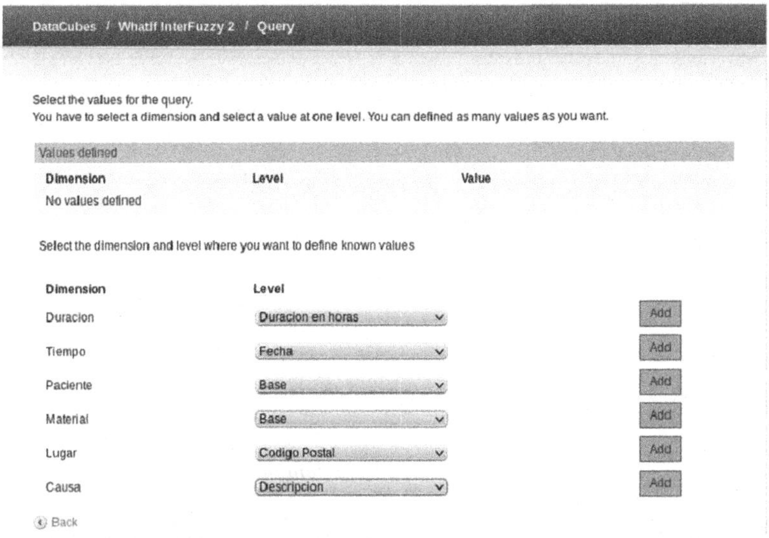

DataCubes / WhatIf InterFuzzy 2 / Query

Select the values for the query.
You have to select a dimension and select a value at one level. You can defined as many values as you want.

Values defined

Dimension	Level	Value
No values defined		

Select the dimension and level where you want to define known values

Dimension	Level	
Duracion	Duracion en horas	Add
Tiempo	Fecha	Add
Paciente	Base	Add
Material	Base	Add
Lugar	Codigo Postal	Add
Causa	Descripcion	Add

Back

Fig. 6. Initial screen

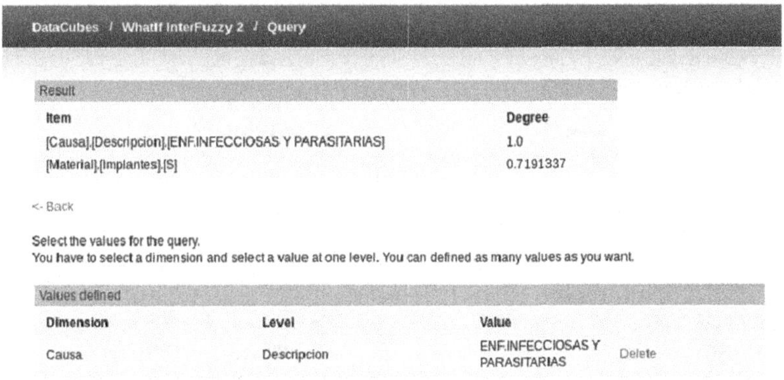

DataCubes / WhatIf InterFuzzy 2 / Query

Result

Item	Degree
[Causa].[Descripcion].[ENF.INFECCIOSAS Y PARASITARIAS]	1.0
[Material].[Implantes].[S]	0.7191337

<- Back

Select the values for the query.
You have to select a dimension and select a value at one level. You can defined as many values as you want.

Values defined

Dimension	Level	Value	
Causa	Descripcion	ENF.INFECCIOSAS Y PARASITARIAS	Delete

Fig. 7. Added *Infectious diseases*

from dimension *Cause*, level *Description*. The system then applies the inference process, and answers (Fig. 7) that the need of *implants* in these surgeries is normal ($CF = 0.719$).

Now the user wants to refine the query adding some information about the sex of the patient, *Female*. The system infers again with this new information (Fig. 8), giving as result that the CF of the use of implant during the operations increases (from 0.719 to 0.923).

Now the doctor decides to check for other relation: deletes the sex of the patient and adds the group of age, choosing *elder*. With this information, the system gets a new relation (Fig. 9): in almost all scenarios the patient lives in

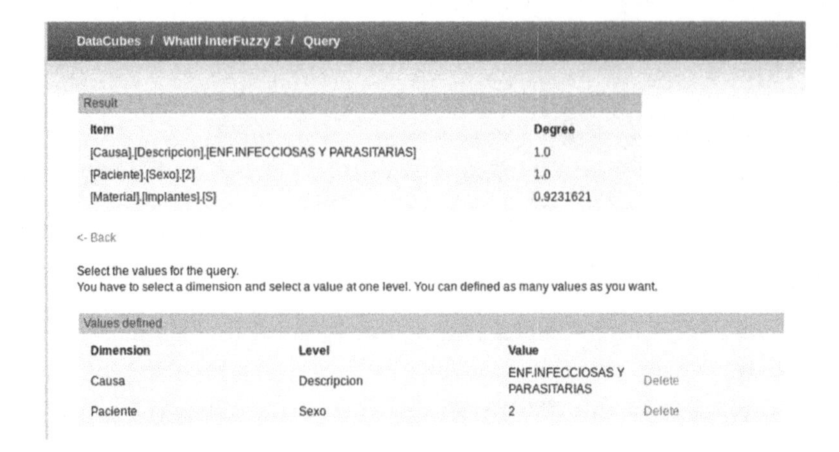

Fig. 8. Added *sex female*

DataCubes / WhatIf InterFuzzy 2 / Query

Result

Item	Degree
[Causa].[Descripcion].[ENF.INFECCIOSAS Y PARASITARIAS]	1.0
[Paciente].[Grupo].[3. Mayor]	1.0
[Lugar].[Provincia].[Granada]	0.9989109
[Material].[Sangre].[N]	0.9946451
[Material].[Implantes].[S]	0.97625244
[Duracion].[Duracion].[1. Normal]	0.41191706

<- Back

Select the values for the query.
You have to select a dimension and select a value at one level. You can defined as many values as you want.

Values defined

Dimension	Level	Value	
Causa	Descripcion	ENF.INFECCIOSAS Y PARASITARIAS	Delete
Paciente	Grupo	3. Mayor	Delete

Fig. 9. Deleted *sex female*, and added *age elder*

the capital *Granada* ($CF = 0.999$), during the operation *implants* are needed ($CF = 0.976$), but *no blood* ($CF = 0.995$) is required. The last information given is that the duration of the surgeries in this scenario is usually *normal* ($CF = 0.412$).

5 Conclusions

As can be seen in this paper we have proposed an intuitive query system based on association rules where the user asks for different scenarios and obtains an easily comprehensible response indicating what usually happens in similar scenarios, according to the knowledge base and the inference process.

The process is enriched with the use of hierarchies, that enables the operation with a terminology near to the final user.

This hiding of the rule system complexity with a comprehensible response; the use of scenarios, a decision making process habitual for the humans; and the close to the user terminology enhances the possibility of bring to the rule based decision support systems closer non-expert users.

Acknowledgements. This research is partially supported by projects Spanish Ministry of Economy and Competitiveness and the European Regional Development Fund (Fondo Europeo de Desarrollo Regional - FEDER) under project TIN2014-58227-P *Descripción lingüística de información visual mediante técnicas de minería de datos y computación flexible* and TIC1582 *Mejora de la Accesibilidad a la información mediante el uso de contextos e interpretaciones adaptadas al usuario* of the Consejeria de Economia, Innovacion, Ciencia y Empleo from Junta de Andalucia (Spain).

References

1. Agrawal, R., Sritkant, R.: Fast algorithms for mining association rules in large databases. In: Proceedings of 20th International Conference on Very Large Data Bases, pp. 478–499 (1994)
2. Balcázar, J.L.: Redundancy, deduction schemes, and minimum-size bases for association rules. Logical Methods Comput. Sci. 6(2) (2010). http://arxiv.org/abs/1002.4286
3. Balmin, A., Papadimitriou, T., Papakonstantinou, Y.: Hypothetical queries in an olap environment. In: VLDB, vol. 220, p. 231 (2000)
4. Conrado, C., de Oude, P.: Scenario-based reasoning and probabilistic models fordecision support. In: 2014 17th International Conference on Information Fusion (FUSION), pp. 1–9, July 2014
5. Georgopoulos, V., Chouliara, S., Stylios, C.: Fuzzy cognitive mapscenario-based medical decision support systems for education. In: 2014 36th Annual International Conference of the IEEE Engineering in Medicine and Biology Society (EMBC), pp. 1813–1816, August 2014
6. Huysmans, J., Dejaeger, K., Mues, C., Vanthienen, J., Baesens, B.: An empirical evaluation of the comprehensibility of decision table, tree and rule based predictive models. Decis. Support Syst. **51**(1), 141–154 (2011). http://www.sciencedirect.com/science/article/pii/S0167923610002368
7. Lakshmanan, L.V., Russakovsky, A., Sashikanth, V.: What-if olap queries with changing dimensions. In: IEEE 24th International Conference on Data Engineering, ICDE 2008, pp. 1334–1336. IEEE (2008)
8. Liao, S.H., Chu, P.H., Hsiao, P.Y.: Data mining techniques and applications: Adecade review from 2000 to 2011. Expert Syst. Appli. **39**(12), 11303–11311 (2012). http://www.sciencedirect.com/science/article/pii/S0957417412003077
9. Marín, N., Molina, C., Serrano, J.M., Vila, A.: A complexity guided algorithm for association rule extraction on fuzzy datacubes. IEEE Trans. Fuzzy Syst. **16**, 693–714 (2008)
10. Martin-Barragan, B., Lillo, R., Romo, J.: Interpretable support vector machines for functional data. Eur. J. Oper. Res. **232**(1), 146–155 (2014). http://www.sciencedirect.com/science/article/pii/S0377221712006406

11. Medeiros, S., Epelbaum, V., Moreira de Souza, J., Pimentel Esteves, M.: Fuzzy prospective scenarios in strategic planning in large-group decision. In: 2013 IEEE 17th International Conference on Computer Supported Cooperative Work in Design (CSCWD), pp. 43–48, June 2013
12. Meissner, P., Wulf, T.: Cognitive benefits of scenario planning: Its impact on biases and decision quality. Technol. Forecast. Soc. Chang. **80**(4), 801–814 (2013). http://www.sciencedirect.com/science/article/pii/S0040162512002375 scenario Method: Current developments in theory and practice
13. Molina, C., Rodriguez-Ariza, L., Sanchez, D., Vila, A.: A new fuzzy multidimensional model. IEEE Trans. Fuzzy Syst. **14**(6), 897–912 (2006)
14. Pang, J., Liu, L., Li, S.: A comparative study between "prediction-response" and "scenario-response" in unconventional emergency decision-making management. In: 2011 Fourth International Joint Conference on Computational Sciences and Optimization (CSO), pp. 649–652, April 2011
15. Pendse, N., Creeth, R.: The Olap Report. Business Intelligence, London (1995)
16. Saxena, G., Narula, R., Mishra, M.: New dimension value introduction for in-memory what-if analysis. arXiv preprint arXiv:1302.0351 (2013)
17. Shortliffe, E., Buchanan, B.: A model of inexact reasoning in medicine. Math. Biosci. **23**(3), 351–379 (1975)
18. Solanki, S., Patel, J.: A survey on association rule mining. In: 2015 Fifth International Conference on Advanced Computing Communication Technologies (ACCT), pp. 212–216, February 2015
19. Stella, N., Peacock, J., Chuan, T.: Evaluating decisions made in common basketball game scenarios. In: 2012 Southeast Asian Network of Ergonomics Societies Conference (SEANES), pp. 1–5, July 2012
20. Wang, Z., Yu, Z., Jiang, Y., Hong, X.: Research on scenarios management fortesting decision-making behavior in decision simulation environment. In: 2012 International Conference on Systems and Informatics (ICSAI), pp. 1213–1217, May 2012
21. Zhengmeng, C., Haoxiang, J.: A brief review on decision support systems and it's applications. In: 2011 International Symposium on IT in Medicine and Education (ITME), vol. 2, pp. 401–405 (2011)

POSGRAMI: Possibilistic Frequent Subgraph Mining in a Single Large Graph

Mohamed Moussaoui[1,2(✉)], Montaceur Zaghdoud[3], and Jalel Akaichi[4]

[1] BESTMOD, Institut Superieur de Gestion, University of Tunis, Tunis, Tunisia
mohamed.moussaoui.com@gmail.com
[2] Central Polytechnic College, Tunis, Tunisia
[3] Information System Department, Prince Sattam bin Abdulaziz University, Al Kharj, Saudi Arabia
zaghdoud@psau.edu.sa
[4] Information System Department, King Khalid University, Abha, Saudi Arabia
jalel.akaichi@kku.edu.sa

Abstract. The frequent subgraph mining has widespread applications in many different domains such as social network analysis and bioinformatics. Generally, the frequent subgraph mining refers to graph matching. Many research works dealt with structural graph matching, but a little attention is paid to semantic matching when graph vertices and/or edges are attributed. Therefore, the discovered frequent subgraphs should become more pruned by applying a new semantic filter instead of using only structural similarity in the graph matching process. In this paper, we present POSGRAMI, a new hybrid approach for frequent subgraph mining based principally on approximate graph matching. To this end, POSGRAMI first uses an approximate structural similarity function based on graph edit distance function. POSGRAMI then uses a semantic vertices similarity function based on possibilistic information affinity function. In fact, our proposed approach is a new possibilistic version of existing approach in literature named GRAMI. This paper had shown the effectiveness of POSGRAMI on some real datasets. In particular, it achieved a better performance than GRAMI in terms of processing time, number and quality of discovered subgraphs.

Keywords: Graph mining · Frequent subgraph mining · Approximate graph matching · Possibility theory · Possibilistic similarity

1 Introduction

The main issue of the graphs is related to their famous characteristics to model complex structures such as protein structures in biology that represent natural products of living cells, chemical compound analysis and social networks which can exhibit both

The author is very thankful to the executive directors and the organizers for their kind invitation to participate in the conference.

J.P. Carvalho et al. (Eds.): IPMU 2016, Part I, CCIS 610, pp. 549–561, 2016.
DOI: 10.1007/978-3-319-40596-4_46

community structures of political or cultural organization and computer networks. Indeed, graphs demonstrate today a powerful mathematic tool when representing objects and their relationship in many different domains. One of the ways used to present a large or multi-graphs is to extract all of its frequent subgraphs that occur frequently in a set of graphs. The frequent subgraphs are widely used in several applications, including classification [1] and modelling of user profiles [2]. In the literature, several algorithms have been proposed to allow the extraction of frequent subgraphs in large graph in an efficient way such as GRAMI [3]. Yet, these techniques are usually based on the structural aspect and don't integrate the semantic aspect of the graph. In other words, these techniques can generally efficiently extract a large set of frequent subgraphs, and provide basic statistical information such as support, number of nodes and number of edges of each subgraph. In addition, the data are generally imperfect and imprecise due to noise, incompleteness and inaccuracies in real world graph. Due to this issue, the similarity between subgraphs is often inexact and isomorphism between graphs is not complete in real applications. In this paper, we propose a novel framework, POSGRAMI, of possibilistic frequent subgraph mining in a single large graph. Our approach based on approximate technique for solving the frequent subgraph mining problem. To solve the problem of inexact graph matching, POSGRAMI uses hybrid graph matching process which combines approximate structural graph matching and semantic graph matching based upon possibilistic similarity. To summarize, we make the following contributions:

- We propose POSGRAMI, a possibilistic approach for frequent subgraph mining in a single large graph. POSGRAMI go beyond discovering only frequent subgraphs with structural similarity approach to prunes them by applying semantic filter.
- We introduce two similarity measures to detect an approximate structural and semantic graph matching using respectively the graph edit distance function and the information affinity.

Several applications for this possibilistic approach can be suggested such as: social network analysis and community detection.

2 Related Works

Generally, the challenge in frequent subgraph mining comes from the costly graph isomorphism that is known to be NP-complete in their generalization. This task is also known as the graph matching problem. The concept of graph matching consists of finding an exact or inexact matching between the nodes of two graphs. Several methods of exact graph matching have been proposed. The most popular algorithm is that of tree-based search techniques. The problem of exact graph matching is closely related to that of graph isomorphism which consists of finding an exact mapping among nodes and edges of two graphs. Several performance enhancements were proposed such as ranging from CSP based techniques [4] and search order optimization [5]. However, the exact graph matching cannot be very useful in real world where inexact correspondence may exist. This second category of graph matching defines approximate or inexact matching methods, where a strict correspondence between the two graphs being compared does not need to be found. In a real application, exact matching methods are often inapplicable due to the distortions or errors in the underlying data. Many works have been proposed for frequent subgraph mining based on approximate graph matching. In GRAMI [3], the authors formulate the frequent subgraph mining as a constraint satisfaction problem. In addition, GRAMI proposed two extensions to its core algorithm namely CGRAMI and AGRAMI. The first extension is a version that supports structural and semantic constraints in order to prune undesirable matches and limit the search space. The second extension is an approximate version, which approximates subgraph frequencies. Yet, GRAMI cannot handle the uncertainty inherent in real applications. Consequently, interesting frequent subgraph may be lost. In most existing approaches presented above are warranted for find the optimal solution, but needs exponential time and space due to the NP-completeness of the inexact graph matching problem.

3 Theoretical Framework

In this section, we present the fundamental definitions of frequent subgraph mining, possibility theory and possibilistic similarity.

Graph Edit Distance Function. This distance measure is defined as the length of the shortest sequence of edit operations required to transform one graph G1 into the other G2.

$$GED(G1,\ G2) = 1 - \frac{|mcs(G1,\ G2)|}{\max(|G1|,\ |G2|)}$$

With |mcs (G1, G2)| is the maximum number of nodes of common subgraphs of two graphs G1 and G2. |G1| and |G2| are respectively number of nodes of graph G1 and graph G2.

Approximate Structural Similarity. We introduced a new structural similarity function based upon Graph edit distance function. Given two graphs G1 and G2 and the distance edit function between these two graphs Distance GED (G1, G2). So, to get a similarity value less than 1, structural similarity STS (G1, G2) should be defined by:

$$\mathrm{STS}(\mathrm{G1},\mathrm{G2}) = 1 - \mathrm{GED}(\mathrm{G1},\mathrm{G2})$$

$$\mathrm{STS}(\mathrm{G1},\mathrm{G2}) = \frac{|mcs(G1, G2|}{\max(|G1|, |G2|)}$$

3.1 Possibility Theory: An Overview

Possibility theory [9] is a modern and simple uncertainty theory devoted to handle some types of uncertainty. It represents the state of incomplete knowledge by the notion of possibility distribution which corresponds to a mapping from the universe of discourse $\Omega = \{\omega_1, \omega_2, \omega_n\}$ to the scale L = [0, 1] encoding our knowledge on the real world states. From a possibility distribution, two dual measures can be derived:

Possibility and Necessity measures. Given a possibility distribution π on the universe of discourse Ω, the corresponding possibility and its dual necessity measures of any event A ⊆ Ω are, respectively, determined by the formulas:

$$\Pi(A) = \max_{\omega A} \pi(\omega) \; and \; N(A) = \min_{\omega A}(1 - \Pi(\omega)) = 1 - \Pi(A)$$

Π(A) evaluates at which level A is consistent with our knowledge represented by while N(A) evaluates at which level A is certainly implied by our knowledge represented by π.

3.2 Possibilistic Similarity Measure

The mathematical concept of similarity measures is used fundamentally to compare two objects and it reflects the degree of closeness between them by possibility distributions on the same universe of discourse. Several similarity measures have been proposed such as information closeness [7] and information affinity [8, 10]. In fact, we adopt the latter because the information affinity satisfies interesting properties such as non-negativity and non-degeneracy and permutation. The information affinity, denoted by *InfoAff* takes into account a classical informative distance, namely, the Manhattan distance along with the inconsistency measure. The information affinity between two possibility distributions π1 and π2 is defined as follows:

$$InfoAff(\pi 1, \pi 2) = 1 - \frac{\mathrm{d}(\pi 1, \pi 2) + \mathrm{Inc}(\pi 1 \wedge \pi 2)}{2}$$

$$\mathrm{d}\;(\pi_1,\; \pi_2) = \frac{1}{n}\sum\nolimits_{i=1}^{n} |\pi 1(\omega i) - \pi 2(\omega i)|$$

d (π1, π2) represents the Manhattan distance between π1 and π2 and Inc(π1 ∧ π2) tells us about the degree of conflict between the two distributions. In this paper, we use this affinity function as graph semantic similarity function SMS between two graphs G1 and G2 from possibilistic *InfoAff* defined as possibilistic distance between two possibility distributions π1 of G1 and π2 of G2 as follows:

$$\mathrm{SMS}\;(\mathrm{G1},\; \mathrm{G2}) = 1 - InfoAff(\mathrm{G1},\; \mathrm{G2})$$

3.3 Hybrid Structural-Semantic Similarity

Given two graphs G1 and G2 and based upon both structural distance GED and semantic similarity measure SMS defined above, we introduce here a new concept of hybrid structural semantic graph similarity (HGS) between two graphs G1 and G2 by using linear combination of two similarity quantities: structural similarity (STS) and semantic similarity (SMS) as follows:

$$\mathrm{HGS}\ (\mathrm{G1},\ \mathrm{G2}) = \alpha * \mathrm{STS}\ (\mathrm{G1},\ \mathrm{G2}) + (1-\alpha)\ \mathrm{SMS}\ (\mathrm{G1},\ \mathrm{G2})$$

$$\mathrm{HGS}(\mathrm{G1},\mathrm{G2}) = \alpha\frac{|mcs(G1,G2)|}{\max(|G1|,|G2|)} + (1-\alpha)(1-\frac{d(\pi 1,\pi 2)+Inc(\pi 1 \wedge \pi 2)}{2})$$

4 POSGRAMI: Possibilistic Frequent Subgraph Mining

GRAMI [3] proposed a frequent subgraph mining algorithm as a constraint satisfaction problem solving approach which is illustrated by Algorithm 1. It is one of algorithms which are based on an approximate graph matching in an efficient way.

Algorithm 1. Approximate GRAMI (G, min_sup, α)

Input: Graph G, minimum support min_sup, approximation parameter α
Output: All subgraphs S of G
Begin
1. Ω ← Ø, S ← Ø, X ← Ø
2. Let E the set of all frequent edge of G
3. **Foreach** e ∈ E **do**
4. S ← Structural_ semantic_CSP(G, CSP, E) // subgraph generation
5. **End**
6. Let consider the subgraph S to subgraph S_CSP
7. X← ISFREQUENTCSP(S_CSP, τ, α) // subgraph isomorphism test
8. Ω ← Ω ∪ X
9. Return Ω

End

Subgraph similarity search is usually not exact due to inaccuracies in real world graph applications because it's difficult to meet exactly the same subgraph as a part of many different graphs. Consequently, we proposed a new algorithm named POSGRAMI which tried to solve this problem by using an approximate graph or subgraph matching based upon possibility similarity of vertices and structure. It uses an approximate matching process that considers two different subgraphs with small differences in nodes or vertices structure as similar subgraphs. To enhance the GRAMI algorithm, POSGRAMI algorithm uses possibility theory for frequent subgraph mining. Which allowed us to achieve search space reductions by applying a possibility distribution for each node. Therefore, the possibilistic graph matching approach addresses the frequent subgraph mining problem without exhaustively enumerating all isomorphism in the graph. In other words, possibilistic graph mining approach searches not only frequent subgraphs with exact similarity, but by extends it to both inexact structural similarity search and semantic graph similarity search. POSGRAMI approach has three distinct steps to mine the possibilistic frequent subgraphs. In the first step, possibilistic graph candidate generation consists principally of traversing the graph and explores frequent subgraph search space. For each candidate we can check whether it is frequent by computing its support. The graph candidate generation process explored in a depth first search but doing so requires computing the possibilistic code to avoid duplicates. Therefore, this step avoids the generation of some insignificant subgraphs for further knowledge discovery conveniences. The second step, possibilistic graph matching allows to detect the approximate graph matching between candidate subgraphs generated. This step is based on hybrid structural-semantic similarity computation (HGS). This measure looks like a semantic and structural filter which allows only some candidates to be definitely considered by applying restriction using given tolerance threshold, Hyb-sim, to similarity between graph candidates generated by the first step. To calculate the structural and semantic similarity between the candidate subgraphs, we use two similarity measures using respectively the graph edit distance function and the information affinity. The third step, subgraph support computation consists of determining the frequency for each grown subgraph. A subgraph g is frequent if its occurrence count is greater than a user-defined minimum support. In the main POSGRAMI algorithm Algorithm 2 exposed bellow, a graph G is considered as input where vertices and edges are labeled in a specific domain that can be null.

Algorithm 2.: POSGRAMI (G, *min_sup*, *Hyb-sim*)

Input: A directed acyclic graph G, the minimum support *min_sup*, the tolerance threshold *Hyb-sim*
Output: The set of frequent subgraphs S
Begin
1. $\Omega \leftarrow \emptyset$
2. Let E the set of all frequent edge of G
3. **Foreach** e $\in$ E **do**
4. X $\leftarrow$ Subgraph_Generation(G, *Hyb-sim*, *min_sup*, E)
5. $\Omega \leftarrow \Omega \cup X$
6. Remove e from G and E
7. **End**
8. Return Ω

End

Given vertices' label possibilistic matrix, the new possibilistic algorithm for mining approximate subgraphs starts by finding a set E that covers all frequent edges in the graph (i.e. with support greater or equal to minimum support *min_sup*). Then, *Subgraph_Generation* method is recursively called for each frequent edge to grow the graphs and find all their frequent descendants. This procedure takes as input a subgraph S and tries to extend it with the frequent edges of E. In this work, we adopt the DFScode canonical form as in gSpan [11] to reject already generated extensions. The DFScode strategy is a technique for traversing or searching tree or graph data structures. It starts at the root and explores as far as possible along each branch before backtracking. *Subgraph_Generation* procedure eliminates the members of Cs that do not satisfy the support threshold *min_sup*, their extensions are also infrequent. The relative importance of nodes and edges in a graph is obtained through the numerical component affected to each node of a possibilistic graph G. Following, we introduce a new concept of possibilistic graph matching, *Possibilistic_GM*. This concept allows to detect the possibilistic similarity between subgraphs and to measure the frequency of these subgraphs. This algorithm focuses on the structural and/or semantic similarity between subgraphs. Indeed, in POSGRAMI algorithm, similarity between a pair of isomorphic subgraphs depends on how similar their possibility distributions.

4.1 Structural-Semantic Subgraph Generation

The exploration space of the frequent subgraphs forms a partial order. It can be searched in a depth first search or breadth first search order but doing so requires computing the possibilistic code to avoid duplicates. We include background knowledge to generate candidate subgraphs based on the numerical component affected to each node. Let assume that the corresponding variables in each node contain the possibility of being important and the possibility of being unimportant of each node according to the context of the application as presented in Algorithm 3.

Algorithm 3. Subgraph_Generation (G, t, *min_sup*, E)

Input: A directed acyclic graph G, the minimum support *min_sup*, the tolerance threshold *Hyb-sim*, the set of frequent edges E
Output: All frequent subgraphs that extend S
Begin
1. Ω ← S
2. Cs ← Ø //The candidate set
3. **Foreach** edge e ∈ E and V ∈ S
4. **If** e can be used to extend V **then**
5. Xt ← The extension of S with e
6. **End**
7. **Else**
8. Cs ← Cs ∪ Xt
9. **End**
10. **End**
11. **Foreach** Ci ∈ Cs **do**
12. Ps←Possibilistic_GM(*min_sup*, t, Cs)
13. Ω ← Ω ∪ Ps
14. **End**
15. Return Ω

End

Hence, each node can have one or more extension by one edge. Yet, we find that some extensions are useless for the context of the application. We find also some that subgraphs can be structurally frequent but without semantic meaning. This allows for take advantage of the respective pruning of smaller subgraphs to prune insignificant assignments. In various cases, a subgraph can be rejected without search. This strategy avoids the expensive search procedure. Indeed, POSGRAMI stores all embedding to generate only refinements that actually appear and to achieve fast approximate isomorphism testing in the next step.

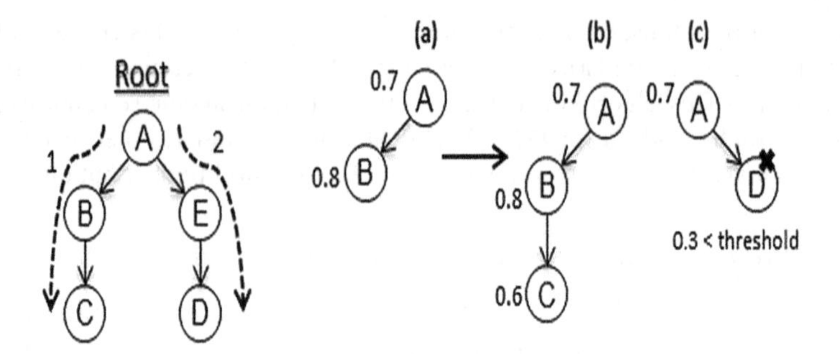

Fig. 1. Example of structural-semantic subgraph generation.

Table 1. Vertices possibility distribution

Possibility distribution	Π(A)	Π(B)	Π(C)	Π(D)	Π(E)
Important information	0.7	0.8	0.6	0.3	0.5
Unimportant information	0.1	0.55	0.22	0.85	1

Figure 1 illustrates a part of a subgraph generation tree consisting of subgraph (a) which is extended to (b). Let assume that the corresponding variables in each node respectively contain the possibility of being important and the possibility of being unimportant of each node according to the context of the application as shown in Table 1. Each node can have one or more extension by one edge. Yet, we find that some extensions are useless for the context of the application. We find also some subgraphs can be structurally frequent but without semantic meaning. For example, let consider the node A as the root in the graph. In this case, A may be extended by B and D. The substructure A-B can be structurally frequent and significant but the substructure A-C is not significant. This allows for take advantage of the respective pruning of smaller subgraphs to prune insignificant assignments. Consequently, this strategy avoids the expensive search procedure. We try, in the second step, to detect not only frequent subgraphs with exact similarity, but by extending it to non-exact similarity search using possibilistic similarity measure, i.e. information affinity [10].

4.2 Possibilistic Frequent Subgraph Mining Algorithm

Given a graph G proposed as input graph to POSGRAMI algorithm. After computing its three essential steps, we obtain a set of frequent subgraphs that occur frequently in this input graph using possibilistic similarity. To solve the problem of approximate or inexact graph matching, POSGRAMI uses two similarity measures: an approximate structural similarity function based on graph edit distance function and also a semantic vertice's similarity function based on possibilistic information affinity function. The procedure *Possibilistic_GM* of Algorithm 4 bellow describes the possibilistic frequent subgraph mining approach.

Algorithm 4. Possibilistic_GM (Cs, *min_sup*, *Hyb-sim*)

Input: The set of candidate subgraphs Cs, the minimum support *min_sup*, the tolerance threshold *Hyb-sim*
Output: All frequent subgraphs S
Begin

```
1.  Foreach g_i ∈ Cs do
2.      g.count ← 0 // graph frequency counter
3.      Πv_i ← Vertice_possibility distribution of g_i
4.  Foreach C_j ∈ Cs do
5.  // Extraction of Maximum Common Subgraph and calculus of Graph
    Edit Distance//
6.      Max-Common-SG(gi, gj_MCSG, GED);
7.  // Calculus of Semantic Similarity and Semantic Distance//
8.      Πv_j ← Vertice_possibility distribution of g_j
9.      Sim_vert ← Similarity (Πv_i, Πv_j)
10.     STD ← 1-Sim_vert;
11. // Calculs of Hybrid Structural-Semantic Similarity //
12.     Hybrid_G_Similarity(GED,STD);
13. // Hybrid Structural-Semantic Similarity Filtering //
14. If (Sim_vert ≥ Hyb-sim) then
15.     g.count ← g.count + 1
16. End
17.     If g.count ≥ min_sup then
18.       Ω ← Ω ∪ g_i
19.     End
20.   End
21. End
22. Return Ω
```

End

The possibilistic graph matching allows comparing two objects described by possibility distributions on the same universe of discourse. The approximate graph matching concept is very useful for real applications due to the fact that similarity between subgraphs is often not complete in real world. This concept helps to reduce the exponential number of frequent subgraphs. In the final step of our new frequent graph mining approach, an embedding list is used to determine the frequency (support) of each candidate subgraph. The support of a subgraph S is simply the number of occurrences of S in the graph G. The subgraph S is considered as frequent if its support is greater or equal to minimum support. The advantage of this innovative approach compared to the old one is that the newer is able to extract some possibilistic frequent subgraphs, particularly for datasets where traditional enumerative methods fail completely. The possibility theory provides some principal justifications over the usual mathematical modeling of probabilities in handling uncertainty which occur in many real-life problems.

5 Experimental Study

Generally, in a frequent subgraph mining approach two aspects are emphasized, namely the number of extracted frequent subgraphs and their significance. For our experimental comparison, we reused a source code obtained from of GRAMI [3].

Table 2 summarizes the main characteristics of the datasets. The first dataset models the social news of Twitter. Each node represents a Twitter user and each edge represents an interaction between two users. The original graph does not have possibility distribution, so we randomly added distribution to the nodes. This dataset was previously used in many studies such as [3] and is publicly available via the following link (socialcomputing.asu.edu/datasets/Twitter). The second dataset models the Microsoft co-authorship information. Nodes represent authors and are labeled with the author's field of interest. Edges represent interaction between two authors and are labeled with the number of co-authored papers. The third dataset models the search engine and digital library for scientific and academic papers CiteSeer. Each node represents a publication and each edge represents citations between them.

Table 2. Benchmark datasets

Dataset	|N|	|E|	Density
Twitter	11.316.811	85.331.846	Dense
MiCo	100.000	1.080.298	Dense
CiteSeer	3.312	4,732	Medium

5.1 Empirical Results

The experiments are measured using the UNIX time command. All experiments are conducted using Java on a Linux machine with 8 cores running at 3.20 GHz with 32 GB RAM and 1 TB disk. To evaluate the quality of the discovered frequent subgraphs using the information gain which is one of the most popular interestingness measures in data mining. Given a set of training examples Ω and a possibility distribution *Pos*. The information gain of *Pos* is computed using the following formulas:

$$InformationGain(\Omega, Pos) = Entropy(\Omega) - Entropy(\Omega|Pos)$$

$$Entropy(\Omega) = -\sum_{i=1}^{\Omega} p(xi) log p(xi)$$

Where $p\ (xi)$ is the probability of getting the *xi* value when randomly selecting an example from the set. The information gain is measured separately for each subgraph in order to measure how each subgraph is informative for the considered task. The average value of information gain is computed for all the frequent subgraph extracted by POSGRAMI and those discovered by the approximate version of GRAMI with different minimum support, *min_sup*. The retained hybrid similarity threshold is 65 %. Table 3 shows that POSGRAMI is able to extract a set of subgraphs that are more

Table 3. Comparison of information gain of the possibilistic subgraphs with those discovered by AGRAMI

Min supp	Twitter		Mico		CiteSeer	
	POSGRAMI	GRAMI	POSGRAMI	GRAMI	POSGRAMI	GRAMI
1000	0.326	0.285	0.463	0.336	0.520	0.423
2000	0.388	0.312	0.336	0.245	0.498	0.264
3000	0.224	0.115	0.288	0.325	0.292	0.341
4000	0.513	0.478	0.382	0.329	0.376	0.338

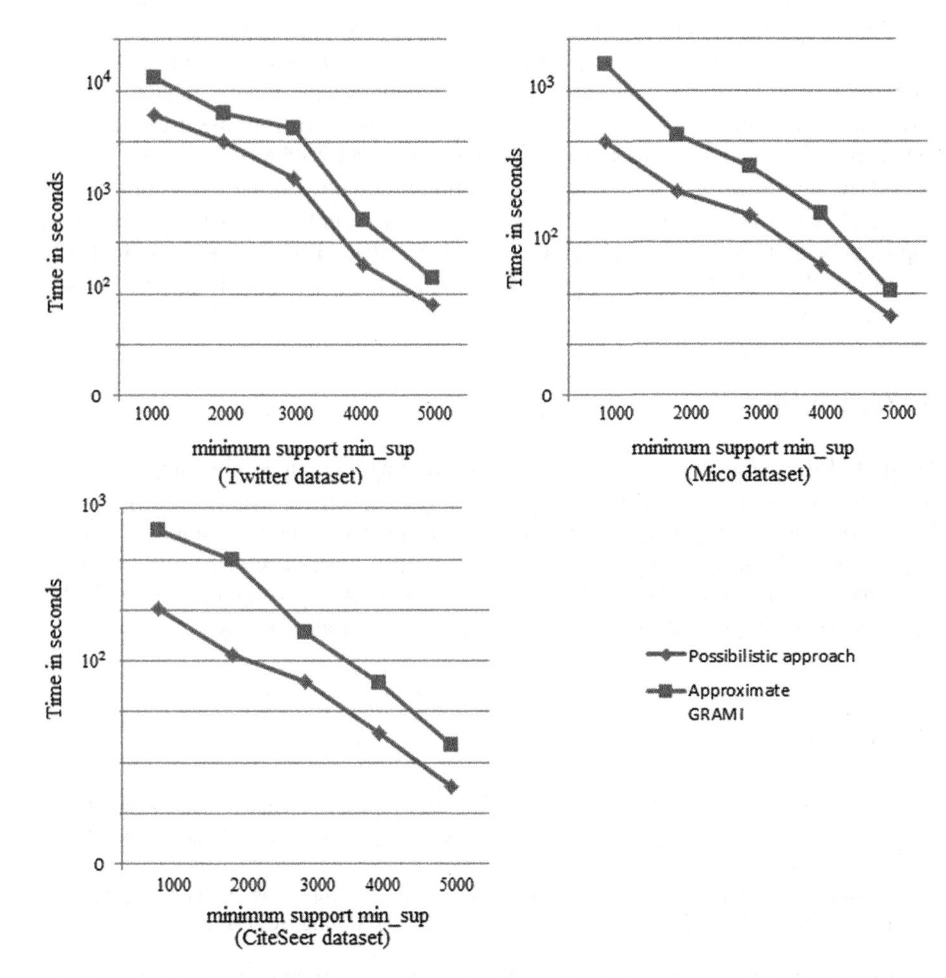

Fig. 2. Runtime of frequent subgraph mining for possibilistic approach and GRAMI among different minimum support with hybrid similarity threshold = 65 %.

informative than those extracted by the approximate GRAMI. Whereas, the quality of the sets of frequent subgraphs discovered by GRAMI did not even reach the information gain value of the whole set of frequent subgraphs. This proves the reliability of POSGRAMI and shows that using the uncertainty theory. It enables possibility theory to better detects approximate similarity between subgraphs and thus to mine a set of subgraph that are most informative.

5.2 Runtime Analysis

In this section, we evaluate the runtime of POSGRAMI compared to that of the approximate GRAMI. Firstly, we tried different minimum frequency threshold in order to obtain a reasonable number of frequent subgraphs from each dataset. Secondly, we have to fix the value of the approximation parameter *Hyb-sim* between 0 and 1 ($0 < Hyb\text{-}sim \leq 1$). If the value of the hybrid similarity equal to 1, that means no approximation matching. Figure 2 illustrates the evolution of runtime using different values of *min_sup* (minimum support) ranging from 1000 to 5000 with a step-size of 1000. The figure shows a difference in execution time between the two approaches.

Figure 2 shows that possibilistic approach is scalable and more robust in real-world applications that usually deal with huge amounts of data. The empirical study clearly shows that POSGRAMI is much more time efficient than GRAMI under different sets of parameters.

6 Conclusion

POSGRAMI is a new possibilistic frequent subgraph graph mining approach that integrates both structural and semantic aspects. The semantic aspect is a precious help in the discovery of knowledge. The success of this theory is due to its ability to handle uncertainty and imprecision in simple way and to offer a classified semantics to natural language statements. In short, the comparison experiment verifies that POSGRAMI performs better in general than GRAMI in terms of time and quality.

References

1. Deshpande, M., Kuramochi, M., Wale, N., Karypis, G.: Frequent substructure-based approaches for classifying chemical compounds. IEEE Trans. Knowl. Data Eng. **17**(8), 1036–1050 (2005)
2. Guralnik, V., Karypis, G.: A scalable algorithm for clustering sequential data. In: Proceedings of IEEE International Conference on Data Mining, ICDM 2001, pp. 179–186 (2001)
3. Elseidy, M., Abdelhamid, E., Skiadopoulos, S., Kalnis, P.: GRAMI: frequent subgraph and pattern mining in a single large graph. PVLDB **7**(7), 517–528 (2014)
4. McGregor, J.: Relational consistency algorithms and their application in finding subgraph and graph isomorphisms. Inf. Sci. **19**(3), 229–250 (1979)

5. He, H., Singh, A.K.: Graphs-at-a-time: query language and access methods for graph databases, pp. 405–418. ACM (2008)
6. Chaoji, V., Al Hasan, M., Salem, S., Besson, J., Origami, M.J.Z.: A novel and effective approach for mining representative orthogonal graph patterns. Stat. Anal. Data Min. **1**(2), 67–84 (2008)
7. Yan, X.: Closegraph: mining closed frequent graph patterns. In: Proceedings of the Ninth ACM SIGKDD International Conference on Knowledge Discovery and Data Mining, Series. KDD 2003, pp. 286–295. ACM, New York (2003)
8. Benferhat, S., Dubois, D., Kaci, S., Prade, H.: Modeling positive and negative information in possibility theory. Int. J. Inf. Syst. (IJIS) **23**(10), 1094–1118 (2008)
9. Zadeh, L.A.: Fuzzy sets as a basis for a theory of possibility. FuzzySets Syst. **100**, 9–34 (1999)
10. Jenhani, I., Ben Amor, N., Elouedi, Z., Benferhat, S., Mellouli, K.: Information affinity: a new similarity measure for possibilistic uncertain information. In: Mellouli, K. (ed.) ECSQARU 2007. LNCS (LNAI), vol. 4724, pp. 840–852. Springer, Heidelberg (2007)
11. Yan, X.: gSpan: Graph-based substructure pattern mining. In: Proceedings of the 2002 IEEE International Conference on Data Mining, Series. ICDM 2002. IEEE Computer Society, Washington, DC (2002)

Mining Consumer Characteristics from Smart Metering Data through Fuzzy Modelling

Joaquim L. Viegas(✉), Susana M. Vieira, and João M.C. Sousa

IDMEC, Instituto Superior Técnico, Universidade de Lisboa,
Lisboa, Portugal
{joaquim.viegas,susana.vieira,jmsousa}@tecnico.ulisboa.pt

Abstract. The electricity market has been significantly changing in the last decade. The deployment of smart meters is enabling the logging of huge amounts of data relating to the operations of utilities with the potential of being translated into knowledge on consumers and enable personalized energy efficiency programs. This paper proposes an approach for mining characteristics of a residential consumers (income, education and having children) from high-resolution smart meter data using transparent fuzzy models. The system consists in: (1) extraction of comprehensive consumption features from smart meter data, (2) use of fuzzy models in order to estimate the characteristics of consumers, and (3) knowledge extraction from the fuzzy models rules. Accurate estimates of consumer income and education level were not achieved (60 % accuracy), for the presence of children accuracies of over 70 % were achieved. Performance is comparable to the state of the art with the addition of model interpretability and transparency.

Keywords: Fuzzy modelling · Smart metering · Classification · Residential electricity consumers · Smart grid

1 Introduction

The energy sector has started to accept demand side management (DSM) as a necessity to achieve higher efficiency and reduce costs through management of peak load. DSM programs can target consumers with a wide array of goals and behaviours. Consumers who are inefficient in their consumption, using outdated and inefficient appliances or having poor insulation in their homes, can be the target of education programs or be offered incentives in exchange of buying more efficient appliances. Consumers who appear to have a more flexible demand for energy can be offered special tariffs in order to motivate them to reduce or shift their consumption at hours of peak system demand.

Technological advancement in the fields of metering, communications and computation are enabling utilities to monitor and save huge amounts of data related to their operation. The deployment of electricity meters with two-way communication capabilities is enabling the logging of the consumption of users

J.P. Carvalho et al. (Eds.): IPMU 2016, Part I, CCIS 610, pp. 562–573, 2016.
DOI: 10.1007/978-3-319-40596-4_47

with high resolution. The number of smart meters has surpassed the number of traditional one-way communication meters in the United States [1]. Close to 45 million smart meters are already installed in three Member States (Finland, Italy and Sweden) of the European Union (EU), representing 23 percent of the envisaged installation in the EU by 2020 [2].

The high resolution consumption data communicated by smart meters has the potential to give utilities the knowledge and insights needed to engage their consumers with the right energy efficiency programs, personalizing their services to the specific characteristics and needs of each customer. Through knowledge discovery and data mining, using computational intelligence techniques, the characteristics of a household can be extracted from their consumption patterns and used by decision makers and marketers in order to design the right programs and engage with the right customers. This topic also as strong implications for consumer privacy, indicating that households need to be aware and engage with those who capture this type of data, urging them of the importance of privacy protection.

In this context, this paper proposes and evaluates a transparent system to predict consumer characteristics from smart metering data. The designed system enables: (1) extraction of consumption features from high frequency smart meter data, (2) derivation of fuzzy models to infer consumer characteristics and (3) knowledge extraction from the fuzzy models rules.

The aim of the system is to derive transparent and interpretable models relating the consumption behaviour and characteristics of electricity customers, providing decision makers and marketers in the utility industry with valuable insights for the design and management of DSM programs. The proposed system is evaluated using real data from a smart metering trial run in Ireland [3].

Similarly to other works published in this topic, accurate estimates of consumer income and education level were not achieved, while for the inference of presence of children accuracies of over 70 % were achieved. The performance results are in line with the ones presented in other works which give no focus to model interpretability.

This paper is structured as follows: Sect. 2 presents the related work. Section 3 presents the system design. Section 4 presents the evaluation of the proposed system through an use case and Sect. 5 presents the main conclusions of this paper.

2 Related Work

The use of static data related to household characteristics, e.g., income, number of inhabitants, education, construction year and appliances in relation to static or dynamic energy consumption data is being studied in order to find the main drivers of residential energy consumption. In [4–6] factor analysis and linear regression are used to find the main determinants of energy consumption in residential settings, such as weather data, household characteristics and demographics. In [7] demographic data and psychological and belief related data is studied in comparison to energy consumption. In [8–11] consumptions profiles obtained via clustering are correlated to household characteristics. In [12] a methodology is presented for the characterization of medium voltage (MV) electricity customers through clustering and posterior modelling for which the classification of new customers is stated as a possible application.

In line with the proposed system, [13–15] present studies on the prediction of household information based on smart meter data. In Fusco et al. (2012) [13], the authors attempt to classify the presence of kids, specific appliances, employment status and education levels of residents through a wide range of features and classification methods, finding difficulties in achieving accurate predictions. In [15], occupancy states are inferred from consumption time series using hidden Markov models, achieving good prediction accuracy. In [14], a system to automatically estimate specific characteristics from a household is proposed, such as socio-economic status, its dwelling and its appliance stock, achieving promising results in the prediction of some of the characteristics.

Contrasting the work published in the topic, the proposed system has a strong focus on model interpretability and transparency, not only enabling the inference of household characteristics but also the possibility to understand the relation between the consumption dynamics and the specific targeted characteristics. In comparison with the approach proposed in [11], the proposed system deals with the inverse problem, deriving household characteristics from smart metering data in contrast to the inference of customer consumption patterns from the same characteristics.

3 System Design

This section presents the design of the proposed system for the mining of consumer characteristics from smart meter data. The smart metering data is first used for the extraction of relevant features that represent the consumption dynamics. The fuzzy model then infers the consumer characteristics from the extracted features. Figure 1 depicts the proposed system.

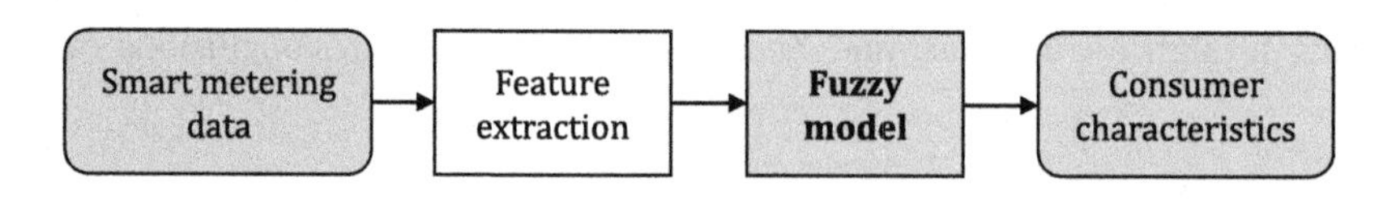

Fig. 1. Proposed system.

3.1 Feature Extraction and Transformation

The smart meter data is high frequency consumption data of households. Two different types of features were extracted in order to derive the models: (1) load indices which characterize the daily electricity consumption dynamics, and (2) average (A), maximum (M) and minimum (m) absolute daily consumption.

The LI are shape indices derived from the smart meter data, these are proposed in [16] and used for the characterization of medium-voltage customers in [12]. LI are used in this paper with the intention of obtaining models of easy interpretation and to explain what consumption characteristics are the most relevant in the characterization of consumers. The indices are presented in Table 1. i_1 is the load factor, i_2 is the off-peak factor, i_3 is the night impact coefficient, i_4 is the lunch impact coefficient and i_5 is the modulation coefficient at off-peak hours. P_{max}, P_{min}, P_{av} are, respectively, the maximum (M), minimum (m) and average A consumption of the corresponding periods. In order to characterize periods which are larger than one day (e.g. a month or season) the mean of the index is used.

Table 1. Normalized indices to characterize electricity customers' behaviour

Parameter	Definition	Periods
Daily P_{av}/P_{max}	$i_1 = P_{av,day}/P_{max,day}$	1 day
Daily $P_{min,day}/P_{max,day}$	$i_2 = P_{min,day}/P_{max,day}$	1 day
Night impact	$i_3 = 1/3P_{av,night}/P_{av,day}$	1 day and 8 h night (from 23 h to 06 h)
Lunch impact	$i_4 = 1/8P_{av,lunch}/P_{av,day}$	1 day and 3 h lunch from (12 h to 15 h)
Daily P_{min}/P_{av}	$i_5 = P_{min,day}/P_{av,day}$	1 day

3.2 Modelling

The proposed system make use of Takagi-Sugeno fuzzy models (FM) [17] to infer consumer characteristics from the consumption features extracted. Support Vector Machines (SVM) [18] are also used in the evaluation section for comparison purposes and verify the suitability of the FM.

Takagi-Sugeno Fuzzy Inference System. Fuzzy models are "grey box" and transparent models that allow the approximation of non-linear systems with no previous knowledge of the system to be modelled. Fuzzy inference systems have the advantage, in comparison to other non-linear modelling techniques, to not only provide transparency but also linguistic interpretation in the form of rules.

In this work, FM are derived from the data. These consist in fuzzy rules where each rule describes a local input-output relation. With FM, each discriminant function consists, for the binary classification case, in rules of the type

$$\begin{aligned} R_i \ : \ & \text{If } x_1 \text{is } A_{i1} \text{ and ... and } x_M \text{ is } A_{iM} \\ & \text{then } d_i(\mathbf{x}) = f_i(\mathbf{x}), i = 1, 2, ..., K \end{aligned} \quad (1)$$

where f_i is the consequent function of rule R_i. The output of the discriminant function $d_i(\mathbf{x})$ can be interpreted as a score (or evidence) for the positive example given the input feature vector $\mathbf{x}$. The degree of activation of the ith rule is given by $\beta_i = \prod_{j=1}^{M} \mu_{A_{ij}}(\mathbf{x})$, where $\mu_{A_{ij}}(\mathbf{x}) : \mathbb{R} \rightarrow [0,1]$. The discriminant output is computed by aggregating the individual rules contributions: $d(\mathbf{x}) = \frac{\sum_{i=1}^{K} \beta_i f_i(\mathbf{x})}{\sum_{i=1}^{K} \beta_i}$. A sample $\mathbf{x}$ is considered positive if the score is higher than a certain γ threshold $d_i(\mathbf{x}) > \gamma$.

The number of rules K and the antecedent fuzzy sets A_{ij} are determined by fuzzy clustering in the product space of the input variables. FCM is used to determine the cluster centres and the number of clusters was determined through cross-validation. The consequent functions $f_i(\mathbf{x})$ are linear functions determined by ordinary-least squares (OLS) in the space of the input and output variables.

Support Vector Machines. SVM [18] are a popular machine learning method for classification. Given non separable training vectors in two classes Support Vector Classification (SVC) finds the hyper plane that maximizes the margin between the training points of classes 0 and 1, allowing some points to be inside the margin. The classifier finds linear boundaries in the input feature space or can make use of the kernel trick in order to work in a transformed non-linear feature space.

4 Experimental Results

This section presents the experimental evaluation of the proposed system using real smart meter data. The data used and its processing are first described, followed by the performance achieved by the modelling approaches and the interpretation of the rules (membership and consequent functions).

4.1 CER Dataset

The proposed system is tested using data from 4232 Irish households monitored for one and a half year. The dataset consists of electricity consumption data logged at 30 min intervals and surveys responded before the start of the trial. This dataset resulted from an electricity customer behaviour trial by the Irish Commission for Energy Regulation (CER). The data is stored and maintained by the Irish Social Science Data Archive (ISSDA) [3].

The mean hourly consumption for the four seasons is pictured in Fig. 2. Consumption follows the typical residential dynamic with a small peak in the morning and lunch time, a larger one at the end of the afternoon and low consumption during the night. As expected, the mean consumption in winter presents the highest values due to the heating needs.

The binary consumer characteristics used as target of inference are:

- **Income:** Positive class indicates a high income ($> 50,000$ euros)
- **Education:** Positive class indicates the survey respondent has superior education (university)
- **Children:** Positive class indicates there are children in the consumers household

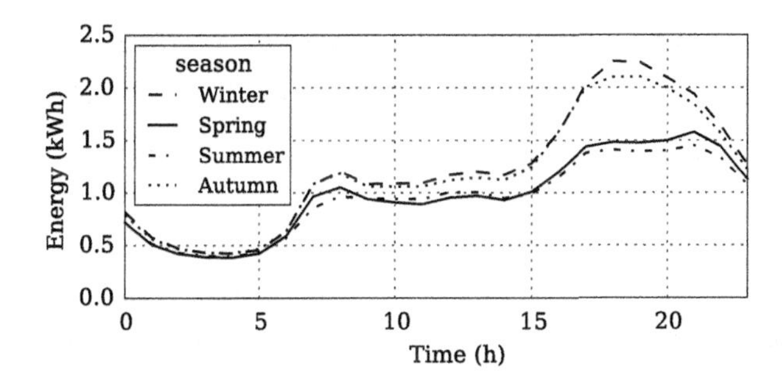

Fig. 2. Hourly aggregated mean seasonal consumption of all consumers.

Only consumers who responded to these three questions were considered in the evaluation, resulting in a total of 1287 households.

4.2 Model Performance

Feature extraction was done only for working days, excluding weekends and holidays. This is done in order maximize the probability of extracting features from days in which consumers follow a routine. The data was also seasonally separated to reduce the effect of evolving dynamics along the year.

The inference of the consumer characteristics was done using the two of models presented in the system design section and the results are obtained using 10-fold cross validation. The FM is derived using 2 and 3 clusters and sigmoid membership function. The parameters of the SVM are optimized in each fold through grid search (C={0.01, 0.1, 1, 10, 100}, γ={0.01, 0.045, 02, 1, 10})

In a binary classification task the true positive (TP) and false positive (FP) are the number of consumers correctly and incorrectly identified to segment "1" and the true negative (TN) and false negative (FN) are the consumers correctly and incorrectly identified to segment "0". The accuracy is obtained following Eq. 2. The area under the curve (AUC) is equal to the area under the receiver operating (ROC) curve of the classifiers [19].

$$Accuracy = \frac{TP + TN}{TP + FP + TN + FN} \tag{2}$$

The results of the evaluation are summarized in Fig. 3, which pictures the mean AUC and accuracy for the tested modelling approaches for each one of the target characteristics. The designed system was not able to achieve acceptable performance in the prediction of income and education but the presence of children was possible with around 70 % accuracy. Overall, the FM with 2 clusters resulted in the best performance both in accuracy and AUC.

Detailed results are presented in Table 2. It shows that the performance throughout the different seasons and modelling techniques was similar, with the exception of the Autumn data resulting in significantly poorer results in income and education inference.

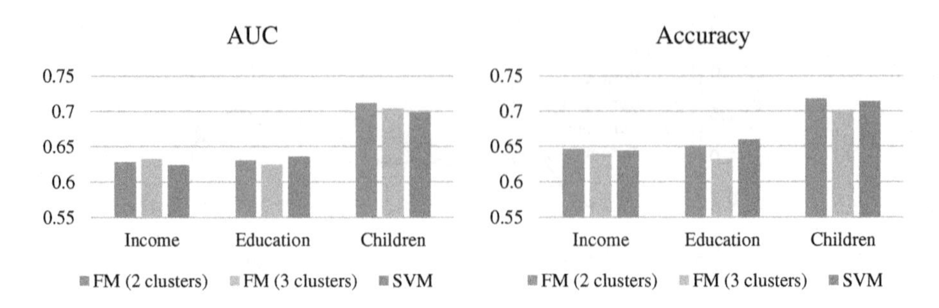

Fig. 3. Hourly aggregated mean seasonal consumption of all consumers. (Color figure online)

Table 2. Results on the prediction of characteristics from smart meter data

Fuzzy Model (2 clusters)					
Variable	Measure	Winter	Spring	Summer	Autumn
Income	AUC	0.619	0.628	0.607	0.595
	Accuracy	0.627	0.646	0.627	0.618
Education	AUC	0.618	0.616	0.630	0.598
	Accuracy	0.627	0.621	0.651	0.622
Children	AUC	0.700	0.694	0.674	0.712
	Accuracy	0.693	0.684	0.683	0.718
Fuzzy Model (3 clusters)					
Income	AUC	0.612	0.633	0.614	0.598
	Accuracy	0.622	0.639	0.629	0.630
Education	AUC	0.610	0.614	0.624	0.587
	Accuracy	0.625	0.631	0.632	0.604
Children	AUC	0.685	0.688	0.673	0.705
	Accuracy	0.687	0.685	0.689	0.702
SVM					
Variable	Measure	Winter	Spring	Summer	Autumn
Income	AUC	0.623	0.624	0.620	0.591
	Accuracy	0.636	0.644	0.630	0.615
Education	AUC	0.624	0.610	0.635	0.587
	Accuracy	0.630	0.616	0.660	0.612
Children	AUC	0.700	0.696	0.676	0.708
	Accuracy	0.700	0.682	0.677	0.714

The resulting performance is very similar to the performance achieved in Beckel et al. (2014) [14], where accuracies under 60 % for high income estimation and around 70 % for children were obtained.

4.3 Model Interpretation

Through the use of FM for the modelling of the target characteristics it is possible to extract the membership functions and consequent functions (rules) used for inference. The following paragraphs attempt to interpret the models and extract insights which could be of value for decision makers or marketers in the electricity utility industry.

Figure 4 pictures the best separated membership functions for the inference of income. Equations 3 and 4 presents the rules extracted considering the pictured variables, separated in what can be interpreted as high and low load factor and consumption groups. It is interesting to note that for the high load factor and consumption rule the load factor as a positive relationship with high income and the off-peak factor a negative relationship, while for the low load factor and consumption rule is the opposite.

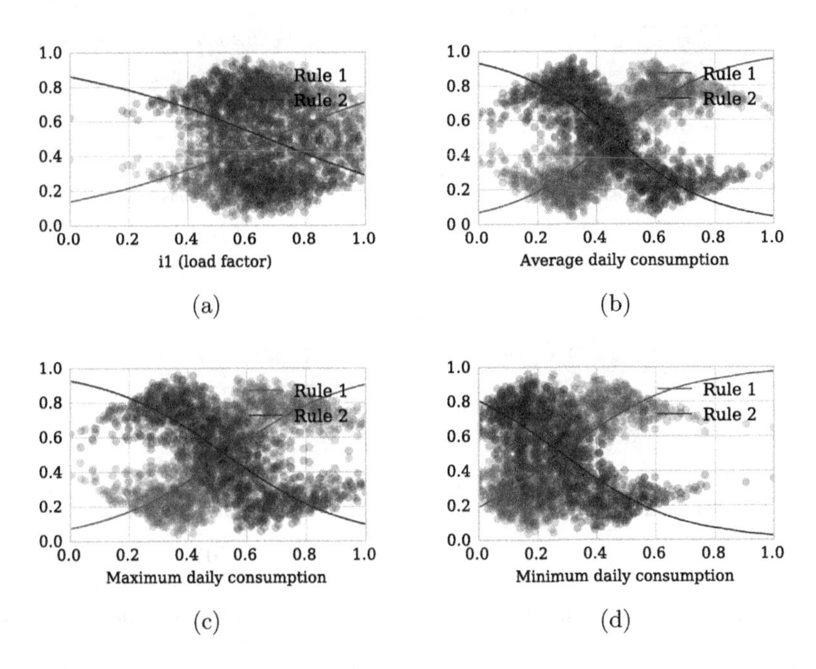

Fig. 4. Input membership function for Income inference using Spring data. (Color figure online)

Rules extracted from the FM for income inference using Spring data:

$$\begin{aligned} R_1: &\quad \text{If } i_1 \text{ is high and } A \text{ is high and } M \text{ is high and } m \text{ is high and ...} \\ \text{then } d_1(\mathbf{x}) &= 0.84 - 0.67i1 + 0.33i2 + 0.15i3 - 0.91i4 + 1.40i5 - 0.09A + 1.74m - 0.20M \end{aligned} \tag{3}$$

$$\begin{aligned} R_2: &\quad \text{If } i_1 \text{ is low and } A \text{ is low and } M \text{ is low and } m \text{ is low and ...} \\ \text{then } d_2(\mathbf{x}) &= 0.11 + 0.25i1 - 0.65i2 + 0.18i3 - 0.42i4 + 2.34i5 - 0.59A + 1.74m + 0.61M \end{aligned} \tag{4}$$

Figure 5 pictures the best separated membership functions for the inference of education. Equations 5 and 6 presents the rules extracted considering the pictured variables, separated in what can be interpreted as high and low off-peak factor and consumption groups.

It is interesting to note that for the high off-peak factor and consumption rule the load factor seemingly has a much lower influence on the result in comparison with the low off-peak factor and consumption rule, for which it is related in a positive fashion (for consumers with lower consumption and lower off-peak consumption, a more linear and constant consumption pattern is related with and higher level of education). For the high off-peak factor and consumption rule and high consumption during the night is related with a lower education.

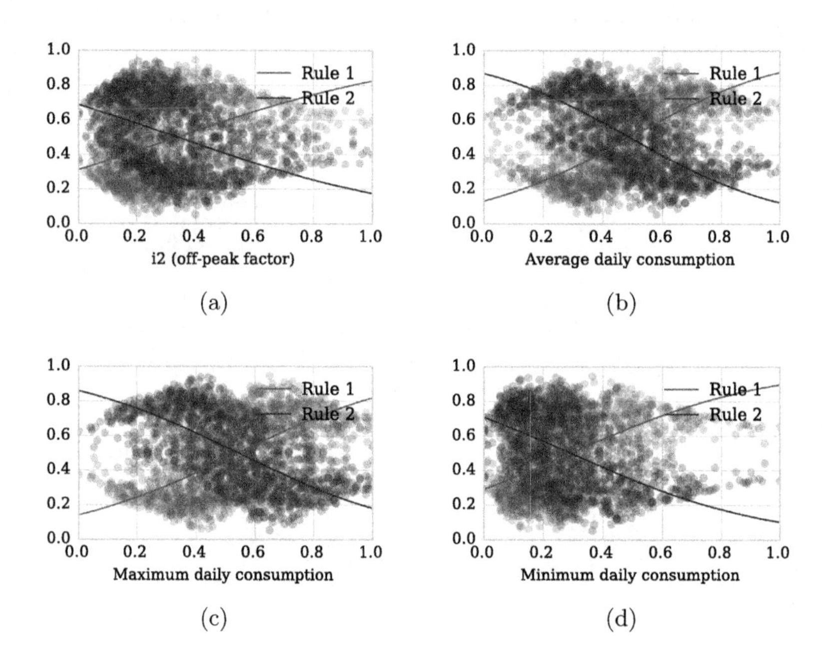

Fig. 5. Input membership function for education inference using Summer data. (Color figure online)

Rules extracted from the FM for education inference using Summer data:

$$R_1: \text{ If } i_2 \text{ is high and } A \text{ is high and } M \text{ is high and } m \text{ is high and ...}$$
$$\text{then } d_1(\mathbf{x}) = 0.50 - 0.19i1 - 0.22i2 - 1.28i3 + 0.12i4 - 0.49i5 + 0.66A - 0.31m - 0.53M \tag{5}$$

$$R_2: \text{ If } i_2 \text{ is low and } A \text{ is low and } M \text{ is low and } m \text{ is low and ...}$$
$$\text{then } d_2(\mathbf{x}) = 0.40 + 1.23i1 - 0.53i2 + 0.19i3 + 0.14i4 + 0.73i5 - 0.46A + 0.59m + 0.18M \tag{6}$$

Figure 6 pictures the best separated membership functions for the inference of the presence of children in the household. Equations 7 and 8 presents the rules extracted considering the pictured variables, separated in what can be interpreted as high and low load factor and consumption groups. It is interesting to note that for high load factor and consumption rule the night impact factor is related to having children (if a consumer has high and linear consumption, the amount of consumption at night has a relationship with having children) while for the low consumption and load factor rule the opposite is verified.

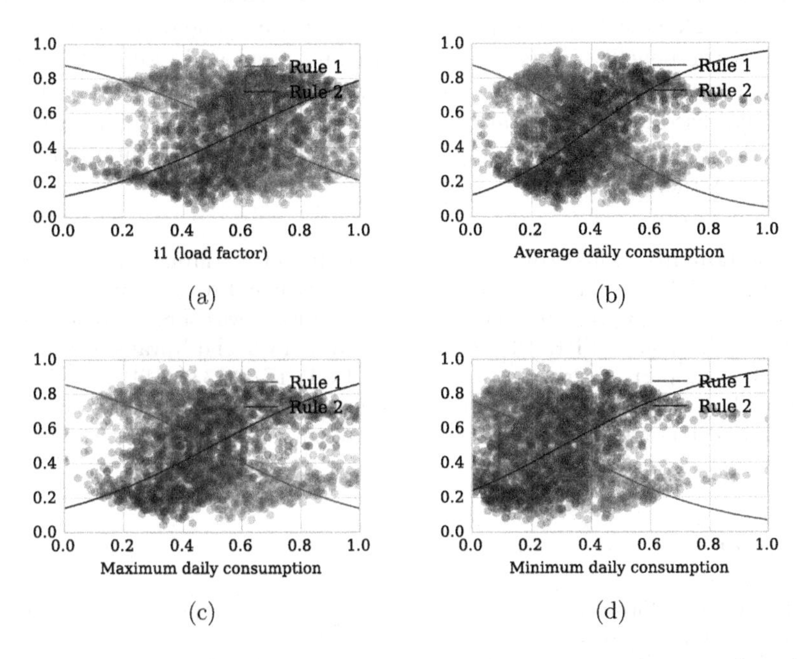

Fig. 6. Input membership function for Children inference using Autumn data. (Color figure online)

Rules extracted from the FM for children inference using Autumn data:

$$\begin{aligned} R_1: &\quad \text{If } i_1 \text{ is low and } A \text{ is low and } M \text{ is low and } m \text{ is low and } \ldots \\ \text{then } d_1(\mathbf{x}) &= 0.41 - 0.27i1 - 0.28i2 - 0.51i3 + 0.01i4 - 0.67i5 - 0.83A + 0.26m + 0.45M \end{aligned} \tag{7}$$

$$\begin{aligned} R_2: &\quad \text{If } i_1 \text{ is high and } A \text{ is high and } M \text{ is high and } m \text{ is high and } \ldots \\ \text{then } d_2(\mathbf{x}) &= 0.89 + 0.9i1 - 0.51i2 + 1.74i3 - 0.63i4 + 0.01i5 - 0.87A + 0.69m - 0.10M \end{aligned} \tag{8}$$

5 Conclusions

This paper proposes a system design for the estimation of consumer characteristics based on the mining of smart meter data using fuzzy modelling. Extracted rules are transparent and interpretable, enabling the extraction of valuable knowledge for marketers and decision makers in the electricity sector.

Evaluation with real data from smart trials run in Ireland show the promise of the approach, achieving reasonably accurate estimate of the presence of children in a household but not achieving acceptable accuracy for the estimation of high income and high education consumers. The results are in line with similar studies but present the benefit of the extraction of transparent and interpretable rules.

In the future the system needs to be more thoroughly tested, using different datasets and used for the estimation of other characteristics. Deeper research on consumption extracted features should also be sought after, potentially resulting in higher accuracies.

Acknowledgments. This work was supported by FCT through IDMEC, under Project SusCity: Urban data driven models for creative and resourceful urban transitions, MITP-TB/CS/0026/2013. The work of J.L. Viegas was supported by the PhD in Industry Scholarship SFRH/BDE/95414/2013 from FCT and Novabase. S.M. Vieira acknowledges support by Program Investigador FCT (IF/00833/2014) from FCT, co-funded by the European Social Fund (ESF) through the Operational Program Human Potential (POPH). Acknowledgement to FCT, through IDMEC, under LAETA, project UID/EMS/50022/2013.

References

1. U.S. Energy Information Administration. Annual electric power industry report. Technical report (2013)
2. Commission Européenne. Benchmarking smart metering deployment in the EU-27 with a focus on electricity (2014)
3. ISSDA. Data from the Commission for Energy Regulation. www.ucd.ie/issda
4. Sanquist, T.F., Orr, H., Shui, B., Bittner, A.C.: Lifestyle factors in U.S. residential electricity consumption. Energy Policy **42**, 354–364 (2012)
5. Kavousian, A., Rajagopal, R., Fischer, M.: Determinants of residential electricity consumption: using smart meter data to examine the effect of climate, building characteristics, appliance stock, and occupants' behavior. Energy **55**, 184–194 (2013)
6. Bedir, M., Hasselaar, E., Itard, L.: Determinants of electricity consumption in Dutch dwellings. Energy Build. **58**, 194–207 (2013)
7. Sütterlin, B., Brunner, T.A., Siegrist, M.: Who puts the most energy into energy conservation? A segmentation of energy consumers based on energy-related behavioral characteristics. Energy Policy **39**(12), 8137–8152 (2011)
8. Wijaya, T., Ganu, T., Chakraborty, D.: Consumer segmentation and knowledge extraction from smart meter and survey data. In: Proceedings of the 2014 SIAM International Conference on Data Mining (SDM14), pp. 226–234 (2014)

9. Rhodes, J.D., Cole, W.J., Upshaw, C.R., Edgar, T.F., Webber, M.E.: Clustering analysis of residential electricity demand profiles. Appl. Energy **135**, 461–471 (2014)
10. Viegas, J.L., Vieira, S.M., Melício, R., Mendes, V.M.F., Sousa, J.M.C.: Electricity demand profile prediction based on household characteristics. In: Proceedings of the 12th International Conference on the European Energy Market (2015)
11. Viegas, J.L., Vieira, S.M., Sousa, J.M.C.: Fuzzy clustering and prediction of electricity demand based on household characteristics. In: Proceedings of the16th World Congress of the International Fuzzy Systems Association (IFSA) and the 9th Conference of the European Society for Fuzzy Logic and Technology (EUSFLAT) (2015)
12. Ramos, S., Duarte, J.M., Duarte, F.J., Vale, Z.: A data-mining-based methodology to support MV electricity customers characterization. Energy Build. **91**, 16–25 (2015)
13. Fusco, F., Wurst, M., Yoon, J.W.: Mining residential household information from low-resolution smart meter data. In: Proceedings of the 21st International Conference on Pattern Recognition, pp. 3545–3548 (2012)
14. Beckel, C., Sadamori, L., Staake, T., Santini, S.: Revealing household characteristics from smart meter data. Energy **78**, 397–410 (2014)
15. Albert, A., Rajagopal, R.: Smart meter driven segmentation: what your consumption says about you. IEEE Trans. Power Syst. **28**(4), 4019–4030 (2013)
16. Chicco, G., Napoli, R., Postolache, P., Scutariu, M., Toader, C.: Customer characterization options for improving the tariff offer. IEEE Trans. Power Syst. **18**(1), 381–387 (2003)
17. Takagi, T., Sugeno, M.: Fuzzy Identification of Systems and Its Application to Modeling and Control (1985)
18. Cortes, C., Vapnik, V.: Support-vector networks. Mach. Learn. **297**, 273–297 (1995)
19. Hanley, J.A., McNeil, B.J.: The meaning and use of the area under a receiver operating characteristic (ROC) curve. Radiology **143**(4), 29–36 (1982)

Soft Computing for Image Processing

Approximate Pattern Matching Algorithm

Petr Hurtik(✉), Petra Hodáková, and Irina Perfilieva

Centre of Excellence IT4Innovations, Institute for Research
and Applications of Fuzzy Modeling, University of Ostrava, 30. dubna 22,
701 03 Ostrava, Czech Republic
{petr.hurtik,petra.hodakova,irina.perfilieva}@osu.cz
http://irafm.osu.cz/

Abstract. We propose a fast algorithm of image pattern (instance) matching which is based on an efficient encoding of the pattern and database images. For each image, the encoding produces a matrix of the F-transform components. The matching is then realized by comparing the F-transform components of the pattern and the database images. The optimal setting of the algorithm parameters is discussed, the success rate and the run time are exhibited.

Keywords: Pattern matching · Searching algorithm · Image searching · F-transform

1 Introduction

Exact pattern matching involves finding all occurrences of a pattern (instance) in a string, where the length of the pattern is shorter, or equal than that of the string. The exact pattern matching is still widely used in a variety of text searches engines.

The pattern matching was formed by string searching. Let us recall the most known string searching algorithms: Rabin-Karp [1], Knuth-Morris-Pratt [2] and Boyer-Moore [3]. Although these algorithms were proposed for strings, some of them may be extended for multidimensional data.

Generally, in pattern matching (especially for higher dimensions) the focus can be shifted to *inexact or approximate matching*. The latter allows small deviations from the zero distances to make the comparison still relevant.

The aim of this contribution is to design a fast algorithm of approximate pattern matching consisting in finding all occurrences of a given instance in a given database. To present possibilities of our algorithm in a real situation, we focus on two-dimensional data, i.e., images. The main idea of our approach is to work with approximations (codes) of images instead of the original ones. We use the technique of F-transform [4] to encode the given data. We remind that the F-transform produces dimensional reduced approximations of the original data. The advantage of F-transform is that we can choose how much the original data are reduced and by that still ensure a good quality of the data approximation.

J.P. Carvalho et al. (Eds.): IPMU 2016, Part I, CCIS 610, pp. 577–587, 2016.
DOI: 10.1007/978-3-319-40596-4_48

This property was used in an image reduction [5] where we showed that the F-transform can reduce an image more precisely than other tested algorithms (Bilinear, Bicubic and Lanczos resampling).

We follow up our previous work [6] where the F-transform based pattern matching was proposed for one dimensional data. In this contribution, we deal with an approximate pattern matching for r-dimensional data. We focus on the optimal setting of the F-transform parameters in order to obtain an efficient algorithm with respect to a success rate and a computational time.

The structure of the paper is as follows: the problem is formally defined and the technique of F-transform is recalled in Sect. 2 - Preliminaries. Section 3 analyzes a proper setting of a fuzzy partition and decision thresholds and then the algorithm steps are formulated. Experiments are demonstrated in Sect. 4.

2 Preliminaries

In this section, we formulate the problem we are focused on and briefly recall the main concepts of the F-transform which is the technique used to encode data in our approximate pattern matching algorithm.

2.1 Problem Formulation

We are given a database $\mathbf{I_{Dat}} = \{f_1, f_2, \ldots\}$ of multidimensional objects: strings, images, etc. Each object f_i is represented by a function of r variables:

$$f_i : D_i \to \mathbb{R},$$

where $D_i \subset \mathbb{R}^r$, $r \geq 1$, is the domain of f_i.

Additionally, we are given an *instance (pattern)* f_p of one or several objects from the database $\mathbf{I_{Dat}}$ where $f_p : D_p \to \mathbb{R}$ and $D_p \subset \mathbb{R}^r$.

The goal is to find all occurrences of the instance f_p in the database $\mathbf{I_{Dat}}$.

We distinguish between a full coincidence and a proper inclusion of f_p in one or several objects from $\mathbf{I_{Dat}}$.

The approximate matching will be realized by computing distances between the F-transform components of the instance and every object from the database. The minimal distance that is greater than a predefined threshold indicates that f_p matches the corresponding object from $\mathbf{I_{Dat}}$. The details are in Sect. 3.

2.2 F-Transform for Functions with r Variables

The idea of our method is to work with encoding data instead of the original data. We propose to encode the data by using the F-transform. It was mentioned in the previous section that data are generally represented by a function of r variables. Let us recall the main concepts of the F-transform for functions with r variables and a fuzzy partition of r-dimensional domain, see [4,7] for more definitions and properties.

We denote the common domain of all functions with r variables by $D^r = [a_1, b_1] \times \cdots \times [a_r, b_r]$ and vectors $\mathbf{x} = (x^1, \ldots, x^r)$ elements of D^r. Let us first introduce the notion of fuzzy partition for $D^1 = [a, b]$. Then we will extend it for the domain D^r.

Definition 1. *Let $c_0 = c_1 < \ldots < c_n = c_{n+1}$ be fixed nodes within $[a, b]$ such that $c_1 = a$, $c_n = b$ and $n > 2$. We say that fuzzy sets $A_1, \ldots, A_n : [a, b] \to [0, 1]$ identified with their membership functions defined on $[a, b]$ form a fuzzy partition of $[a, b]$ if the following conditions hold true for each $k = 1, \ldots, n$:*

1. *$A_k(c_k) = 1$;*
2. *$A_k(\mathbf{x}) = 0$ if $\mathbf{x} \in [a, b] \setminus (c_{k-1}, c_{k+1})$;*
3. *$A_k(\mathbf{x})$ is continuous on $[c_{k-1}, c_{k+1}]$;*
4. *$A_k(\mathbf{x})$ strictly increases on $[c_{k-1}, c_k]$ and strictly decreases on $[c_k, c_{k+1}]$.*

The membership functions $A_1, \ldots, A_n$ are called basic functions. *A point $\mathbf{x} \in [a, b]$ is* covered *by the basic function A_k if $A_k(\mathbf{x}) > 0$.*

If the nodes $c_1, \ldots, c_n$ are *h-equidistant*, i.e., for all $k = 2, \ldots, n$, $c_k = c_{k-1} + h$, where

$$h = (b - a)/(n - 1) \tag{1}$$

and two additional properties hold for $k = 2, \ldots, n - 1$:

5. $A_k(c_k - \mathbf{x}) = A_k(c_k + \mathbf{x})$ for all $\mathbf{x} \in [0, h]$;
6. $A_k(\mathbf{x}) = A_{k-1}(\mathbf{x} - h)$ and $A_{k+1}(\mathbf{x}) = A_k(\mathbf{x} - h)$ for all $\mathbf{x} \in [c_k, c_{k+1}]$;

then the fuzzy partition $A_1, \ldots, A_n$ is *h-uniform.*

Moreover, the fuzzy partition is called *Ruspini partition* if for all $\mathbf{x} \in [a, b]$ holds the *Ruspini condition*

$$\sum_{k=1}^{n} A_k(\mathbf{x}) = 1.$$

The following definition extends the concept of fuzzy partition of D^1 to D^r.

Definition 2. *Let a fuzzy partition of an interval $[a_j, b_j]$ be given by basic functions $A_1^j, \ldots, A_{n_j}^j$, $n > 2$, for $j = 1, \ldots, r$. Then the fuzzy partition of D^r is given by the fuzzy Cartesian product $\{A_1^1, \ldots, A_{n_1}^1\} \times_\odot \{A_1^2, \ldots, A_{n_2}^2\} \times_\odot \cdots \times_\odot \{A_1^r, \ldots, A_{n_r}^r\}$ with respect to the product t-norm of these r fuzzy partitions.*

Let us now recall the definition of the direct F-transform for function with r variables defined at discrete points.

Definition 3. *Let $\{A_1^1, \ldots, A_{n_1}^1\} \times_\odot \{A_1^2, \ldots, A_{n_2}^2\} \times_\odot \cdots \times_\odot \{A_1^r, \ldots, A_{n_r}^r\}$ be a fuzzy partition of D^r and let a function $f : D^r \to \mathbb{R}$ be known at points $(p_1^1, \ldots, p_1^r), \ldots, (p_N^1, \ldots, p_N^r)$ such that for each $(k_1, \ldots, k_r)$ where $k_j = 1, \ldots, n_j$ and $j = 1, \ldots, r$, there exists $i = 1, \ldots, N : A_{k_1}^1(p_i^1) \cdots A_{k_r}^r(p_i^r) > 0$. We say that* *a ν-tuple* $\mathbf{F}_{\mathbf{n_1 n_2 \ldots n_r}}[f] = [F_{k_1 \ldots k_r}]$ *of real numbers where* $\nu = (n_1 \cdot n_2 \ldots n_r)$ *is the discrete direct F-transform of* f *with respect to the given fuzzy partition if*

$$F_{k_1 \ldots k_r} = \frac{\sum_{i=1}^{N} f(p_i^1, \ldots, p_i^r) A_{k_1}^1(p_i^1) \cdots A_{k_r}^r(p_i^r)}{\sum_{i=1}^{N} A_{k_1}^1(p_i^1) \cdots A_{k_r}^r(p_i^r)} \tag{2}$$

for each r-tuple $k_1 \ldots k_r$.

Let us remark that the F-transform can approximate the original function with an arbitrary precision. The quality of the approximation depends on the parameter h of the chosen fuzzy partition. It holds, the smaller h, the better approximation by the F-transform components, see [4] for more details.

3 Algorithm

The purpose of this contribution is to propose a method for matching and identifying place of the pattern f_p with the particular $f_i \in \mathbf{I_{Dat}}$. Before we define particular steps of the algorithm, let us introduce the main ideas:

1. Application of the F-transform to f_p and $f_i \in \mathbf{I_{Dat}}$, $i = 1, \ldots, d$, w.r.t. the same h-uniform fuzzy partition. By this part, we encode the original data into the reduced representation in the form of components.
2. Comparison of the components of f_p with the components of all $f_i \in \mathbf{I_{Dat}}$, $i = 1, \ldots, d$, i.e., sliding comparison. This part is realized by computing corresponding distances.
3. Matching of f_p with a particular $f_i \in \mathbf{I_{Dat}}$. The final match is realized with respect to the minimal distance between the components and considering a predefined decision threshold.

The ideas of the algorithm are described in a general way. Now we focus on two following details which will be analyzed more deeply:

Q1: The F-transform is dependent on a chosen fuzzy partition. How to determine the particular fuzzy partition, i.e., what is the optimal value of the parameter h?

Q2: The final matching is realized by the threshold of decision making. Is it possible to derive the threshold automatically or it has to be specified by a user?

Fuzzy partition specification: It was mention in Sect. 2.2, the smaller h (the more F-transform components) the better approximation of the function. On the other hand, the bigger h (the less F-transform components) the less comparisons of the components and therefore, the faster processing. The question is what is the optimal value for h?

It is clear that when we compute a distance between two components (one from a pattern and one from a database image) which correspond to the same part of those images, the distance is zero. What happen when the corresponding compared parts of the image are shifted by δ? How much is the distance of the corresponding components influenced by the δ and by the parameter h? We describe this dependency by an "error" which denotes a ratio between the distance of components (between the pattern and the database) with respect to the δ and the standard deviation of all components of the pattern. The dependency is illustrated on a graph in Fig. 1. We observe that the error is increasing for $\delta \in [0, h/2]$ and decreasing for $\delta \in [h/2, h]$ (because of the sliding comparison). A next graph in Fig. 2 illustrates the influence of the chosen h on the average error with respect to the shift δ. We observe that there is no influence of the chosen h on the average error. Therefore, we can choose the parameter h as big as possible (with respect to the size of the pattern) to make the processing faster.

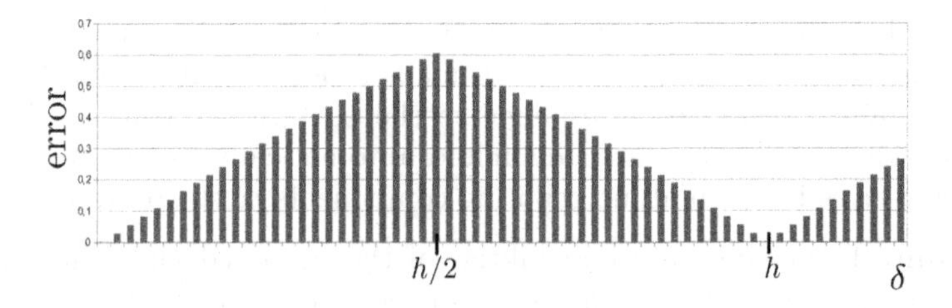

Fig. 1. The error denotes a ratio between the distance of components (with respect to the δ) and the standard deviation of all components of the pattern.

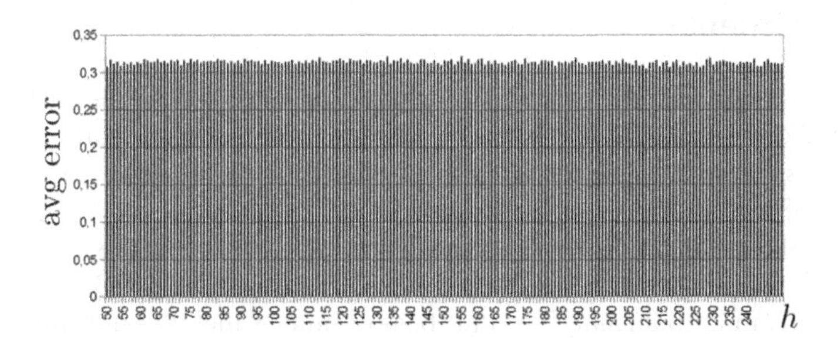

Fig. 2. Dependency of the average error on the chosen h with respect to the shift δ.

Once we specify the parameter h, we need to establish h-uniform fuzzy partition with this parameter. As we mentioned above, the error is increasing with respect the shift δ and maximum for $\delta = h/2$. In order to avoid this increasing error and make it as much constant as possible, we propose to use two h-uniform fuzzy partitions shifted in their position by $h/2$. This solution gives us the error almost constant for arbitrary δ, see a graph in Fig. 3.

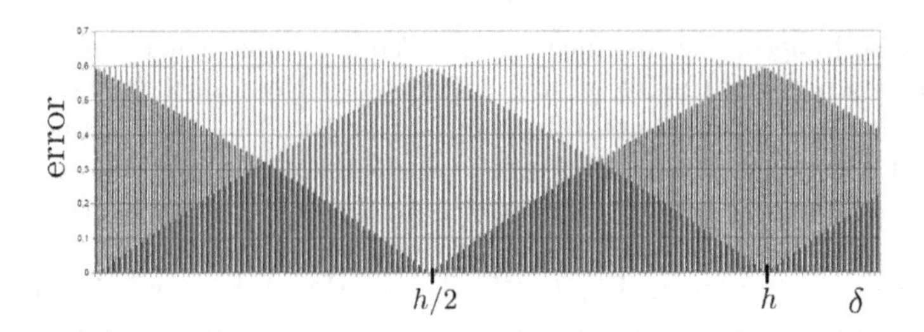

Fig. 3. Dependency of the error on $\delta \in [0, h/2]$. The errors of two h-uniform fuzzy partitions shifted by $h/2$ are displayed in blue and red, the "final" error is displayed in yellow. (Color figure online)

Decision Making Threshold: The algorithm marks one (or more) image(s) in the database where the pattern can be included. Each marked image has assigned the corresponding distance. To make a decision, if the marked image with the minimal distance corresponds to the pattern, we compare it with a threshold. The threshold value is delimited automatically as the average value of the differences between all neighboring components of the pattern.

Algorithm. To demonstrate possibilities of the proposed algorithm in a real situation, we present the particular steps of the algorithm over two-dimensional data, i.e., images. Let us remark that generally an image is defined as a discrete function of two variables $f : [1, N_1] \times [1, N_2] \to \mathbb{R}$ defined on an array of pixels $\{(p_i, p_j) | i = 1, \ldots, N_1; j = 1, \ldots, N_2\}$. The image pattern matching algorithm is described in details in Fig. 4.

4 Experiments

In this section, we present experiments tested on real-life images with different scenarios. We created a large database of images and tested several patterns. We present the success rate and the computational time of the tests. The particular conditions of the presented experiments are as follows.

The database consists of 1292 images with high-resolutions (8 Mpx, 12 Mpx). The physical size of the database stored in a hard-disk is 4076 MB in jpeg format and 32558 MB in bmp format. The database of images includes in total $1.11 \cdot 10^{10}$ pixels. Let us remark that the computational time of all experiments was measured on a low-powered notebook without any parallelism.

The pre-processing of the algorithm can be realized only once and its computational time is not included in the total computational time. In this step, the F-transform components w.r.t. several values of h are computed for all images from the database. The components are stored in .txt file without any compression. The size of the original database is reduced as follows:

$$h = 20: \ 413 \text{ MB}, \quad h = 40: \ 101 \text{ MB}, \quad h = 60: \ 44 \text{ MB},$$
$$h = 80: \ 28 \text{ MB}, \quad h = 100: \ 15 \text{ MB}, \quad h = 800: \ 0.2 \text{ MB}.$$

Procedure: *FTransMatch*(I_{Dat}, f_p, H)
inputs:
$I_{Dat} = \{f_1, \ldots, f_d\}$: *a database of d images*
f_p: *a pattern image*
$H = \{20, 40, \ldots\}$: *a set of values of a parameter h*
outputs: *a set of tuples* $< i, x, y, Dist_i >$:
i: *an index of a database image*
x, y: *coordinates in the i-th database image where the pattern* f_p *was matched*
$Dist_i$: *a distance between the pattern and the i-th database image*
pre-processing:
1. **for each** $h \in H$
2. **for each** $f_i \in I_{Dat}, i = 1, \ldots, d$
3. $\mathbf{F}_{\mathbf{n_{1_i} n_{2_i}}}[f_i] = [F_{k_{1_i} k_{2_i}}]_{k_{1_i}=1,\ldots,n_{1_i};\ k_{2_i}=1,\ldots,n_{2_i}};$
$$F_{k_{1_i} k_{2_i}} = \frac{\sum_{j_1=1}^{N_{1_i}} \sum_{j_2=1}^{N_{2_i}} f_i(p_{j_1}^1, p_{j_2}^2) A_{k_{1_i}}^1(p_{j_1}^1) A_{k_{2_i}}^2(p_{j_2}^2)}{\sum_{j_1=1}^{N_{1_i}} \sum_{j_2=1}^{N_{2_i}} A_{k_{1_i}}^1(p_{j_1}^1) A_{k_{2_i}}^2(p_{j_2}^2)}$$
w.r.t. *h*-uniform fuzzy partition of $[1, N_{1_i}] \times [1, N_{2_i}]$
4. $\mathbf{F}_{\mathbf{n_{1_i}-1\, n_{2_i}-1}}[f_i] = [F_{k_{1_i} k_{2_i}}]_{k_{1_i}=1,\ldots,n_{1_i}-1;\ k_{2_i}=1,\ldots,n_{2_i}-1};$
$$F_{k_{1_i} k_{2_i}} = \frac{\sum_{j_1=h/2}^{N_{1_i}} \sum_{j_2=h/2}^{N_{2_i}} f_i(p_{j_1}^1, p_{j_2}^2) A_{k_{1_i}}^1(p_{j_1}^1) A_{k_{2_i}}^2(p_{j_2}^2)}{\sum_{j_1=h/2}^{N_{1_i}} \sum_{j_2=h/2}^{N_{2_i}} A_{k_{1_i}}^1(p_{j_1}^1) A_{k_{2_i}}^2(p_{j_2}^2)}$$
w.r.t. *h*-uniform fuzzy partition of $[h/2, N_{1_i}] \times [h/2, N_{2_i}]$
processing:
1. $h_p = \sqrt{N_{1_p} N_{2_p} / 15}$
2. *choose* $h_j \in H$ *such that* $|h_j - h_p| = \min_{h \in H} |h - h_p|$
3. $\mathbf{F}_{\mathbf{n_{1_p} n_{2_p}}}[f_p] = [F_{k_{1_p} k_{2_p}}]_{k_{1_p}=1,\ldots,n_{1_p};\ k_{2_p}=1,\ldots,n_{2_p}};$
$$F_{k_{1_p} k_{2_p}} = \frac{\sum_{j_1=1}^{N_{1_p}} \sum_{j_2=1}^{N_{2_p}} f_p(p_{j_1}^1, p_{j_2}^2) A_{k_{1_p}}^1(p_{j_1}^1) A_{k_{2_p}}^2(p_{j_2}^2)}{\sum_{j_1=1}^{N_{1_p}} \sum_{j_2=1}^{N_{2_p}} A_{k_{1_p}}^1(p_{j_1}^1) A_{k_{2_p}}^2(p_{j_2}^2)}$$
w.r.t. h_j-uniform fuzzy partition of $[1, N_{1_p}] \times [1, N_{2_p}]$
4. $\theta = \frac{\sum_{s=1}^{n_{1_p}-1} \sum_{t=1}^{n_{2_p}-1} |F_{s_p,t_p} - F_{s+1_p,t_p}| + |F_{s_p,t_p} - F_{s_p,t+1_p}|}{2(n_{1_p}-1)(n_{2_p}-1)}$
5. **for each** $i = 1, \ldots, d$
6. **for each** $x = 1, \ldots, n_{1_i} - n_{1_p}$
7. **for each** $y = 1, \ldots, n_{2_i} - n_{2_p}$
8. $T_i^{xy} \subset \mathbf{F}_{\mathbf{n_{1_i} n_{2_i}}}[f_i]$ *such that* $T_i^{xy} = [F_{k_{1_i} k_{2_i}}]_{k_{1_i}=x,\ldots,x+n_{1_p}-1;\ k_{2_i}=y,\ldots,y+n_{2_p}-1}$
9. $T_{i,h/2}^{xy} \subset \mathbf{F}_{\mathbf{n_{1_i}-1\, n_{2_i}-1}}[f_i]$ *such that* $T_{i,h/2}^{xy} = [F_{k_{1_i} k_{2_i}}]_{k_{1_i}=x,\ldots,x+n_{1_p}-1;\ k_{2_i}=y,\ldots,y+n_{2_p}-1}$
10. $Dist_i(\mathbf{F}_{\mathbf{n_{1_p} n_{2_p}}}[f_p], T_i^{xy}) = \sum_{k_{1_i}=1}^{n_{1_p}} \sum_{k_{2_i}=1}^{n_{2_p}} |F_{k_{1_p} k_{2_p}} - F_{k_{1_i} k_{2_i}}|$
11. $Dist_{i,h/2}(\mathbf{F}_{\mathbf{n_{1_p} n_{2_p}}}[f_p], T_{i,h/2}^{xy}) = \sum_{k_{1_i}=1}^{n_{1_p}} \sum_{k_{2_i}=1}^{n_{2_p}} |F_{k_{1_p} k_{2_p}} - F_{k_{1_i} k_{2_i}}|$
12. **if** $Dist_i + Dist_{i,h/2} < \theta$ **then** store $< i, x \cdot h, y \cdot h, Dist_i + Dist_{i,h/2} >$
13. **end;**

Fig. 4. F-transform based pattern matching algorithm

Let us emphasize that the extreme reduction is obtained for $h = 800$ where the database size is reduced to 0.2 MB. If we use rar compression, we obtain the file of the size 0.04 MB, i.e., 100000x smaller than the original database in jpeg format.

The first experiment tested 40 image patterns selected by a human. The patterns differ in size and content. All patterns include some texture, e.g., face, tree, car etc. The result of this test was 100 % success rate. The mean computational time was 200 ms of matching one pattern in the whole database (1292 images). An example is shown in Fig. 5 (*Top*).

Fig. 5. Three examples of the pattern matching - patterns (*left*), matched images (*right*). *Top:* The pattern was a part of an image. It was searched and matched successfully. *Middle:* The pattern was a part of an image and it was modified by adding a black square. It was searched and matched successfully. *Bottom:* The pattern was a homogeneous part of an image (only 255 values; the pattern is magnified to be more visible in the images.). It was searched and matched incorrectly to the image on the left. The original image is on the right. Remark: the pattern was matched to both the images but the incorrect one had the minimal distance.

The second experiment tested 500 image patterns created by a computer via a script. The script selects a random-sized pattern from a random position in a random image from the database. The result of this test was 90 % success rate. The mean computational time was 300 ms per pattern. The reason why we did not achieve 100 % success rate is the homogeneity of patterns. Some patterns contained only homogeneous areas (for instance sky) and the algorithm identified those patterns with more images from the database where the correct one did not achieve the minimal distance. The example of an incorrect matching is presented in Fig. 5 (*Bottom*). In such cases, the algorithm can be modified to return more than one "best match" result. For example, when the algorithm is set to return 3 % of the best matched images from the database, the achieved success rate is 99 %.

Let us remark that we also tried to search for patterns which are little modified with respect to the original database images they are created from. We experimented with manually created modifications, see an example in Fig. 5 (*Middle*). For automatized test we modified patterns by Gaussian convolution with the mask (5 × 5) pixels. Even in such cases, we achieved the same success rate and computational time as in the previous experiments.

5 Discussion

In this section, we discuss some problems which arose while the algorithm was designed.

"Miss-match" Problem. The problem called "miss-match" is a situation when the pattern is matched with the incorrect image (see Fig. 5 (*Bottom*)). This problem can be formally described as follows:

$$f_p \subset f_i \text{ and } f_p \not\subset f_j; \quad f_i, f_j \in \mathbf{I_{Dat}}$$

$$\text{but} \quad Dist_i(\mathbf{F_{n_{1_p} n_{2_p}}}[f_p], T_i^{xy}) > Dist_j(\mathbf{F_{n_{1_p} n_{2_p}}}[f_p], T_j^{xy}).$$

The problem of miss-match can arise when the pattern is a part of some image and the components of the pattern and of the corresponding part in the image are shifted by some unknown δ. This situation is displayed in Fig. 6.

To avoid the miss-match problem, we use the two h-uniform fuzzy partitions shifted by $h/2$ for each image comparison (described in Sect. 3).

Higher Degree F-transform. In the proposed algorithm, we use the ordinary F-transform to encode the original data. Let us remark that there exists also a higher degree F-transform and generally, the higher degree F-transform gives the better approximation of the function. Therefore, we investigated the usage of the first degree F-transform [8] and the standard arithmetic mean.

Similarly to the Sect. 3 where we discussed the specification of the fuzzy partition and the parameter h, we investigated the dependency of the error on the chosen encoding method. A graph in Fig. 7 illustrates the dependency of the error on the shift δ with respect to the ordinary F-transform (F^0-transform), the F^1-transform and the standard arithmetic mean (over non-overlapping sub-areas of the function). This graphical demonstration was obtained by a numerical experiment with the fixed parameter $h = 50$ and with the shift $\delta \in \{0, 1, \ldots, 200\}$.

The graph in Fig. 7 confirms that for the shift $\delta \geq 2h$ (the case when the two compared components are not overlapped) the error is constant for all three

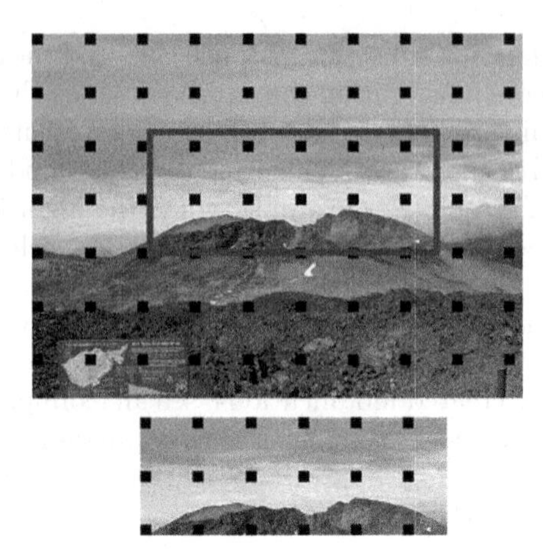

Fig. 6. Illustration of different positions of the pattern components (*bottom*) with respect to the components of the original image (*top*).

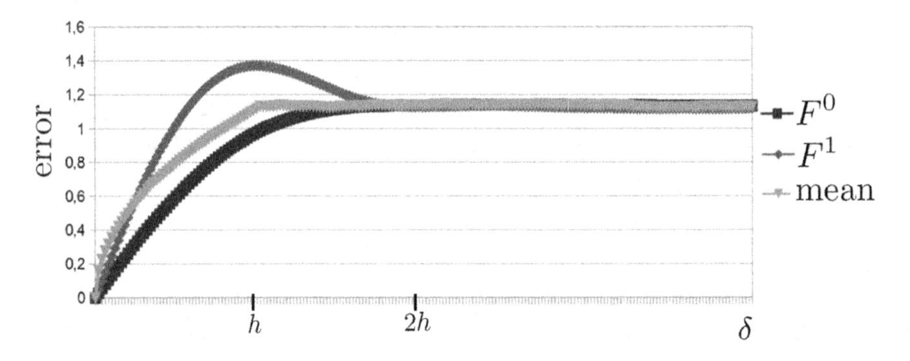

Fig. 7. Dependency of the error on shift δ with respect to the ordinary F-transform, F^1-transform and arithmetic mean (Colour figure online)

methods. We also observe that in the case of the F^1-transform, the error for the shift $\delta \in \{h/2, \ldots, 2h\}$ (two compared components overlap) is bigger than in the case of non-overlapped components. This leads to the miss-match problem and therefore the F^1-transform is not suitable for this comparison. We conclude that the most suitable method to avoid the miss-match problem is the ordinary F-transform (F^0-transform).

6 Conclusion

In this contribution, we proposed the pattern matching algorithm. The algorithm was presented for two dimensional data (images) but the method is universal and can be easily extended to process r-dimensional data in general. The possibility of modification of the algorithm into arbitrary dimension is one of the most important benefit of the proposed algorithm.

We focused on the algorithm design in details. The core idea lies in encoding the original data by the F-transform in order to reduce the dimension and obtain a simplified representation. We investigated particular steps of the algorithm and discussed settings of all parameters. The algorithm is suitable for the exact pattern matching as well as for the approximate matching. The algorithm can return either one best result identified with the pattern or the list of best results.

The proposed methodology was tested on several experiments with the result of 90 % success rate (at least). In the tested 4 GB database with more than thousand big-resolution images, the algorithm excelled with the computational speed of 200–300 ms per pattern.

The general usage of the algorithm is searching in the real data when we face to a problem of big sized data. The data are too big to be loaded into a fast memory (RAM) and then they have to be stored in a slower memory (hard disk) and swapped. The proposed algorithm uses the reduced representation of data and then they can be loaded to the fast memory directly.

Acknowledgment. This research was partially supported by the NPU II project LQ1602 "IT4Innovations excellence in science" provided by MŠMT.

References

1. Karp, R.M., Rabin, M.O.: Efficient randomized pattern-matching algorithms. IBM J. Res. Dev. **31**(2), 249–260 (1987)
2. Knuth, D.E., Morris Jr., J.H., Pratt, V.R.: Fast pattern matching in strings. SIAM J. Comput. **6**(2), 323–350 (1977)
3. Boyer, R.S., Moore, J.S.: A fast string searching algorithm. Commun. ACM **20**(10), 762–772 (1977)
4. Perlieva, I.: Fuzzy transforms: Theory and applications. Fuzzy Sets Syst. **157**(8), 993–1023 (2006)
5. Perfilieva, I., Hurtik, P., Di Martino, F., Sessa, S.: Image reduction method based on the F-transform. Soft. Comput. (2015). doi:10.1007/s00500-015-1885-0
6. Hurtik, P., Hodáková, P., Perlieva, I.: Fast string searching mechanism. In: Proceedings of the 2015 Conference of the International Fuzzy Systems Association and the European Society for Fuzzy Logic and Technology. AISR, vol. 89, pp. 412–418. Atlantis Press (2015)
7. Štěpnička, M.: Fuzzy Transform and its Applications to Problems in Engineering Practice (Thesis). University of Ostrava, Ostrava (2008). http://irafm.osu.cz/f/PhD_theses/Stepnicka.pdf
8. Perlieva, I., Daňková, M., Bede, B.: Towards a higher degree f-transform. Fuzzy Sets Syst. **180**(1), 3–19 (2011)

Image Reconstruction by the Patch Based Inpainting

Pavel Vlašánek(✉) and Irina Perfilieva

Institute for Research and Applications of Fuzzy Modeling,
University of Ostrava, Ostrava, Czech Republic
{pavel.vlasanek,irina.perfilieva}@osu.cz
http://irafm.osu.cz/

Abstract. The paper is focused on demonstration of image inpainting technique using the F-transform theory. Side by side with many algorithms for the image reconstruction we developed a new method of patch-based filling of an unknown (damaged) image area. The unknown area is proposed to be recursively filled by those known patches that have non-empty overlaps with the unknown area and are the closest ones among others from a database. We propose to use the closeness measure on the basis of the F^1-transform.

Keywords: Image processing · F-transform · Inpainting

1 Introduction

Inpainting belongs to a group of sophisticated methods/algorithms to replace the lost or corrupted parts of partially given images. The technique can be used to fill places after an object was cut away or small defects were removed. The inpainting can be approximately divided into two groups: a *patch-based substitution* [1–4] and a *recursive filling, which is usually based on partial differential equations* [5–8]. This classification is not mutually exclusive, so there are methods that use both approaches. In this paper, we propose a new patch-based inpainting technique using the sparse image representation and F^1-transform [9].

The F^k-transform theory [10] performs a transformation of the original universe of functions (images) into a universe of their vectors or matrices of components for which further computations are easier. In this respect, the F-transform is useful in applications such as image compression, fusion, reconstruction, etc. The F-transform degree k states for a polynomial degree of the components. The F^1-transform components are polynomials of the first degree, whereas F^0-transform components are scalars.

The structure of this contribution is as follows. The preliminaries and state of the art are provided in Sect. 2. The theory of F-transform is described in Sect. 3. Section 4 describes a concrete application of the F-transform to the image inpainting problem, and Sect. 5 contains some examples. The conclusion is provided in Sect. 6.

J.P. Carvalho et al. (Eds.): IPMU 2016, Part I, CCIS 610, pp. 588–598, 2016.
DOI: 10.1007/978-3-319-40596-4_49

2 Preliminaries and State of the Art

Let us fix the following notation to use throughout the paper. Image I is a 2D function such as $I : [0, M] \times [0, N] \rightarrow [0, 255]$, where $M + 1$ denotes the image width, $N + 1$ denotes the image height, and $[0, 255]$ denotes the pixel intensity. We denote $[0, M]_{\mathbb{Z}} = \{0, 1, 2, \ldots, M\}$, $[0, N]_{\mathbb{Z}} = \{0, 1, 2, \ldots, N\}$ and $[0, 255]_{\mathbb{Z}} = \{0, 1, 2, \ldots, 255\}$. Image I is assumed to be partially defined: it is defined (known) on the area Φ and undefined (unknown, damaged) on the area Ω. The border between these areas is denoted by $\delta\Omega$ and assumed to be unknown. The notation is illustrated in Fig. 1.

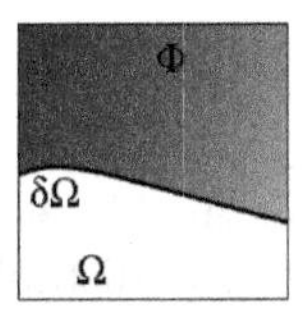

Fig. 1. Two areas where image I is defined (Φ) and undefined (Ω).

In the recursive inpainting process, Ω should be recursively filled so that every pixel $\omega \in \delta\Omega \cup \Omega$ should be replaced by some pixel $\phi \in \Phi$. In patched-based inpainting, a rectangular patch $\Psi \in \Phi$ is the most favorable choice. We denote Ψ_ϕ for a rectangular patch centered at pixel ϕ. If pixel ω is unknown (belongs to $\delta\Omega \cup \Omega$), then the patched-based inpainting consists of finding a patch from the known area, e.g., Ψ_ϕ, which is the most similar to Ψ_ω, so that

$$\Psi_\phi = \arg\min_{\phi'} d(\Psi_\omega, \Psi_{\phi'}). \tag{1}$$

Function d is usually selected as the Euclidian distance. However, the pioneers Efros and Leung [4] used $d(\Psi_\omega, \Psi_\phi)$ as the sum of the square differences between Ψ_ω and Ψ_ϕ, which is defined as follows

$$d(\Psi_\omega, \Psi_\phi) = \sum_{x=0}^{\Psi_w} \sum_{y=0}^{\Psi_h} (\Psi_\omega(x, y) - \Psi_\phi(x, y))^2,$$

where $\Psi_w + 1$ and $\Psi_h + 1$ are the patch width and height, respectively.

A priority of unknown pixel selection is very important. A number of the known pixels in Ψ_ω is taken into the consideration. The higher number means higher priority. In general, the highest number of known pixels have patches Ψ_ω centered at the border $\delta\Omega$. This however, requires to determine such pixels (remember that they are considered as unknown). In our contribution, we propose to determine the border pixels using the operation of erosion (mathematical morphology operation) described later in the paper. Inside the border, pixels are selected according to their indexation. This way of processing is known as *onion-peel*.

3 F-Transform

We recall the definition of the F-transform [10] and the notion of a fuzzy partition[1] at the beginning. Fuzzy sets (*basic functions*) $A_0, \dots, A_m, 1 < m < M$, which are identified with their membership functions $A_0, \dots, A_m : [0, M] \to [0, 1]$, establish a *fuzzy partition* of $[0, M]$ with nodes $0 = x_0 < x_1 < \dots < x_m = M$, if the following conditions are fulfilled:

(1) $A_k : [0, M] \to [0, 1], \ A_k(x_k) = 1$;
(2) $A_k(x) = 0$ if $x \notin (x_{k-1}, x_{k+1}), \ k = 0, \dots, m$;
(3) $A_k(x)$ is continuous on $[0, M]$;
(4) $A_k(x)$ strictly increases on $[x_{k-1}, x_k]$ and strictly decreases on $[x_k, x_{k+1}]$, where $k = 1, \dots, m$;
(5) $\sum_{k=0}^{m} A_k(x) = 1, \ x \in [0, M]$.

We say that the fuzzy partition $A_0, \dots, A_m$ is an *h-uniform fuzzy partition* if nodes $x_k = hk, \ k = 0, \dots, m$ are equidistant, $h = M/m$, and two additional properties are met:

(6) $A_k(x_k - x) = A_k(x_k + x), \ x \in [0, h], \ k = 0, \dots, m$;
(7) $A_k(x) = A_{k-1}(x - h), \ k = 1, \dots, m, \ x \in [x_{k-1}, x_{k+1}]$.

Parameter h will be referred to as a *radius* of a partition. Similarly, a partition of $[0, N]$ by $B_1, \dots B_n$ can be defined.

3.1 $\mathbf{F^0}$-Transform

The F^0-transform of an image is given by a corresponding matrix of components. Let the fuzzy partition of $[0, M]$ and $[0, N]$ be given by basic functions $A_0, \dots, A_m : [0, M] \to [0, 1]$ and $B_0, \dots, B_n : [0, N] \to [0, 1]$, respectively. We remark that the set of pixels $P = \{(i, j) \in [0, M]_{\mathbb{Z}} \times [0, N]_{\mathbb{Z}}\}$ with integer coordinates is *sufficiently dense with respect to the chosen partitions.* Thus, $(\forall k)(\exists x \in [0, M]_{\mathbb{Z}}) \ A_k(x) > 0$, and $(\forall l)(\exists y \in [0, N]_{\mathbb{Z}}) \ B_l(y) > 0$, which follows from condition 5) above.

We call the $m \times n$ matrix of real numbers $\mathbf{F^0_{mn}}[I] = (F^0_{kl})$ *the (discrete) F^0-transform* of image I with respect to $\{A_0, \dots, A_m\}$ and $\{B_0, \dots, B_n\}$, if for all $k = 0, \dots, m; l = 0, \dots, n$,

$$F^0_{kl} = \frac{\sum_{y=0}^{N} \sum_{x=0}^{M} I(x, y) A_k(x) B_l(y)}{\sum_{y=0}^{N} \sum_{x=0}^{M} A_k(x) B_l(y)}. \tag{2}$$

The elements F^0_{kl} are called the *components of the F^0-transform.* In the terminology of kernels or masks, the expression (2) can be rewritten as a convolution with *sliding window* [11]. In image processing, the F^0-transform components determine an average intensity value over the appropriate area. The illustration is in Fig. 2.

[1] For the sake of simplicity, we consider this notion for a one-dimensional universe.

Fig. 2. F^0-transform components of four 5×5 image areas.

3.2 F^1-Transform

In this section, we recall the (direct) F^1-transform as it was presented in [12]. Let $\{A_k \times B_l \mid k = 0, \ldots, m, l = 0, \ldots, n\}$ be a fuzzy partition of $[0, M] \times [0, N]$. Let $L_2^1(A_k) \subseteq L_2(A_k)$ $(L_2^1(B_l) \subseteq L_2(B_l))$[2] be spanned by the two orthogonal polynomials

$$P_k^0(x) = 1, \quad P_k^1(x) = x - x_k,$$
$$(Q_l^0(y) = 1, \quad Q_l^1(y) = y - y_l),$$

where 1 denotes the respective constant function.

Analogously, let $L_2^1(A_k \times B_l) \subseteq L_2(A_k \times B_l)$ be spanned by the three orthogonal polynomials

$$S_{kl}^{00}(x, y) = 1, \quad S_{kl}^{10}(x, y) = x - x_k, \quad S_{kl}^{01}(x, y) = y - y_l.$$

Let $I \in L_2([0, M] \times [0, N])$, and F_{kl}^1 be the orthogonal projection of $I|_{[x_{k-1}, x_{k+1}] \times [y_{l-1}, y_{l+1}]}$ on subspace $L_2^1(A_k \times B_l)$, $k = 0, \ldots, m$, $l = 0, \ldots, n$.

We say that matrix $\mathbf{F}_{\mathbf{mn}}^{\mathbf{1}}[I] = (F_{kl}^1)$, $k = 0, \ldots, m$, $l = 0, \ldots, n$ is the F^1-transform of I with respect to $\{A_k \times B_l \mid k = 0, \ldots, m, l = 0, \ldots, n\}$, and F_{kl}^1 is the corresponding F^1-transform component.

The F^1-transform components of I are linear polynomials in the form of

$$F_{kl}^1(x, y) = c_{kl}^{00} + c_{kl}^{10}(x - x_k) + c_{kl}^{01}(y - y_l),$$

[2] $L_2(A_k)$ is a Hilbert space of square-integrable functions $f : [x_{k-1}, x_{k+1}] \to \mathbb{R}$, whose *weighted inner product* $\langle f, g \rangle_k$ is given by

$$\langle f, g \rangle_k = \int_{x_{k-1}}^{x_{k+1}} f(x)g(x)A_k(x)dx,$$

where the *weight function* is equal to A_k.

where the coefficients are

$$
\begin{aligned}
c_{kl}^{00} &= \frac{\sum_{y=0}^{N}\sum_{x=0}^{M} I(x,y)A_k(x)B_l(y)}{\sum_{y=0}^{N}\sum_{x=0}^{M} A_k(x)B_l(y)},\\
c_{kl}^{10} &= \frac{\sum_{y=0}^{N}\sum_{x=0}^{M} I(x,y)(x-x_k)A_k(x)B_l(y)}{\sum_{y=0}^{N}\sum_{x=0}^{M}(x-x_k)^2 A_k(x)B_l(y)},\\
c_{kl}^{01} &= \frac{\sum_{y=0}^{N}\sum_{x=0}^{M} I(x,y)(y-y_l)A_k(x)B_l(y)}{\sum_{y=0}^{N}\sum_{x=0}^{M}(y-y_l)^2 A_k(x)B_l(y)}.
\end{aligned} \tag{3}
$$

In image processing, the F^1-transform components determine an average intensity value and an average gradient of the appropriate area, as illustrated in Fig. 3.

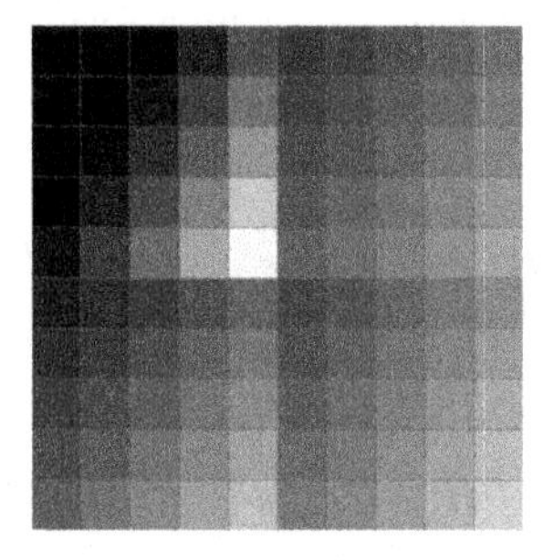

Fig. 3. F^1-transform components of four 5×5 image areas. The areas are identical to those in Fig. 2.

4 Application to Inpainting Process

As previously mentioned, image inpainting aims at filling the unknown area. Image I with a designated unknown area Ω is the input of this problem. Ω must be replaced (filled) using patches from the known area Φ, where $\Phi = [0, M] \times [0, N] \backslash \Omega$. The following method is our main contribution to the topic. It consists of four sequential steps that are characterized below.

4.1 Creation of a Database of Patches

The known area Φ of image I is partitioned into partially overlapped patches. The patch has a rectangular shape with a predefined size. The number of patches q is determined by their size. The patches Ψ_ϕ^i where $i = 1, \ldots, q$ are stored in the database D_Ψ, which will be used later for the reconstruction.

Database Optimization. Many database patches are frequently notably similar to one another. To avoid duplicity, we propose to compare distances between already included patches and a new one Ψ^n. The following distance is used:

$$d(\Psi^n, \Psi^d) = \sum_{k=0}^{m} \sum_{l=0}^{n} |\Psi^n(k,l) - \Psi^d(k,l)|; \forall \Psi^d \in D_\Psi. \quad (4)$$

If there is at least one Ψ^d such that the value (4) is less than a certain threshold, the patch Ψ^n is not included into D_Ψ. The appropriate choice of the threshold is a subject of our future research. Currently, we use a value from the interval [0..5].

4.2 Border Extraction

The unknown area Ω is surrounded by the border $\delta\Omega$. We consider square-shape patches Ψ centered at unknown pixels. For the sake of reconstruction, it is desirable to have as many known pixels in Ψ as possible. This condition is fulfilled if the corresponding centers are on the border $\delta\Omega$, which implies that we must determine the border of Ω, process its pixels, consequently update the border, etc.

For the purpose of border extraction, we propose to use the operator of mathematical morphology [13] - *erosion*. The erosion is defined as follows

$$\Phi \ominus T = \{z \in [0, M] \times [0, N] | T_z \subseteq \Phi\},$$

where T is a structuring element and z is a translation vector. For the purpose of border determination, we consider square-shaped structuring elements of the size 3×3. The border is determined as follows

$$\delta\Omega = \Phi - (\Phi \ominus T).$$

The illustration is in Fig. 4.

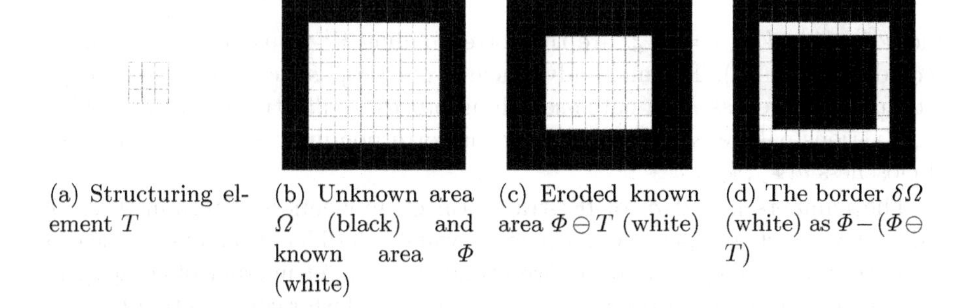

(a) Structuring element T (b) Unknown area Ω (black) and known area Φ (white) (c) Eroded known area $\Phi \ominus T$ (white) (d) The border $\delta\Omega$ (white) as $\Phi - (\Phi \ominus T)$

Fig. 4. Illustration of the border extraction using mathematic morphology erosion. For better illustration, the colors are inverted.

The type of reconstruction based on the primary processing of border pixels is commonly known as onion peel.

4.3 Patch Searching

We take a pixel ω from the border $\delta\Omega$ and create the patch Ψ_ω, whose center is at this pixel. Let s be the mask of the damaged (unknown) part in Ψ_ω as shown in Fig. 5. We apply s to each patch in the database D_Ψ and create a new database D_Ψ^s, where each patch Ψ_ϕ^s is partially damaged by s. The goal is to find the closest patch Ψ_ϕ^s from D_Ψ^s to the patch Ψ_ω.

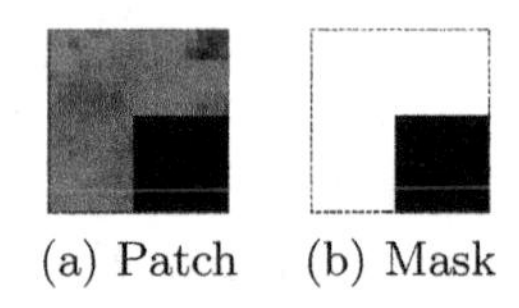

(a) Patch (b) Mask

Fig. 5. Patch Ψ_ω (left) with unknown area determined by the mask s.

We propose to measure closeness between patches by the mean average of distances between coefficients of their F^1-transform components, i.e.

$$d(\Psi_1,\Psi_2) = \frac{d^{00}(\Psi_1,\Psi_2) + d^{01}(\Psi_1,\Psi_2) + d^{10}(\Psi_1,\Psi_2)}{3},$$

where

$$d^{00}(\Psi_1,\Psi_2) = \frac{\sum_{k=0}^{m}\sum_{l=0}^{n} |\Psi_1^{00}(k,l) - \Psi_2^{00}(k,l)|}{mn},$$
$$d^{01}(\Psi_1,\Psi_2) = \frac{\sum_{k=0}^{m}\sum_{l=0}^{n} |\Psi_1^{01}(k,l) - \Psi_2^{01}(k,l)|}{mn},$$
$$d^{10}(\Psi_1,\Psi_2) = \frac{\sum_{k=0}^{m}\sum_{l=0}^{n} |\Psi_1^{10}(k,l) - \Psi_2^{10}(k,l)|}{mn},$$

and Ψ_i^{00}, Ψ_i^{01}, Ψ_i^{10}, $i = 1, 2$, are the corresponding F^1-transform coefficients in accordance with (3). The main advantage of our proposal consists in its computational effectiveness. The patch-searching process is illustrated in Fig. 6 where show various patches Ψ_ϕ^s together with the corresponding values of the measure of closeness $d(\Psi_\omega, \Psi_\phi^s)$.

The numbers n and m of fF^1-transform components is determined by the radius h of the chosen partition. The exact value of h differs from one application to another. A smaller value of h corresponds to a larger number of components and greater computation time. A larger h leads to a faster computation but with a higher risk of a wrong assignment of Ψ_ϕ^s to Ψ_ω.

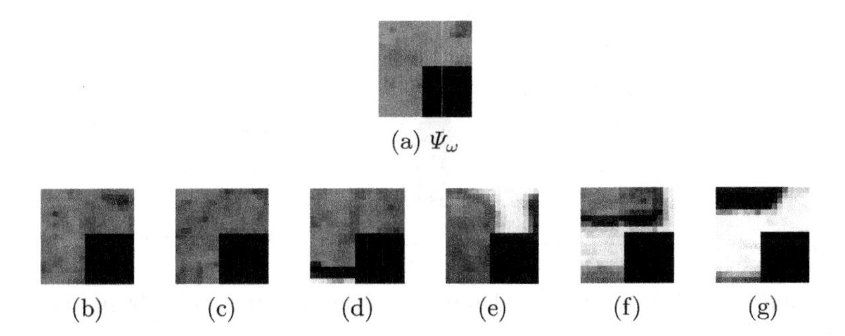

Fig. 6. (a) Original patch Ψ_ω with an unknown area (colored in black) whose mask is s; some patches from D_Ψ^s with the corresponding measures of closeness $d(\Psi_\omega, \Psi_\phi^s)$ as follows: (b) 7.79; (c) 8.94; (d) 14.19; (e) 20.02; (f) 29.44; (g) 46.02.

4.4 Patch Reconstruction

After the closest patch Ψ_ϕ^s is selected, we replace (reconstruct) the partially known Ψ_ω by the fully known $\Psi_\phi \in D_\Psi$, which corresponds to Ψ_ϕ^s. The illustration is in Fig. 7.

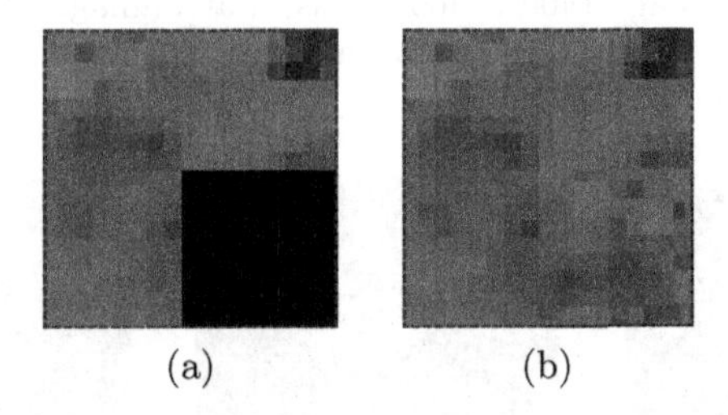

Fig. 7. (a) Patch Ψ_ω centered around the pixel $\omega \in \delta\Omega$; (b) The closest patch Ψ_ϕ^s from the database D_Ψ.

5 Examples and Comparison

There are no established benchmarks to verify a quality of inpainting methods. Therefore, in order to compare the proposed method with other ones we choose a visual quality estimation. We start with the example where inpainting is used for erasing of unwanted objects. In Fig. 8, we demonstrate the ability of our method to delete a relatively large object and reconstruct the erased part by the corresponding texture.

In the following two examples, we artificially created damaged (unknown) areas of various geometrical shapes and then reconstructed them by the proposed method. We used two measures RMSE and SSIM and estimated the quality of the

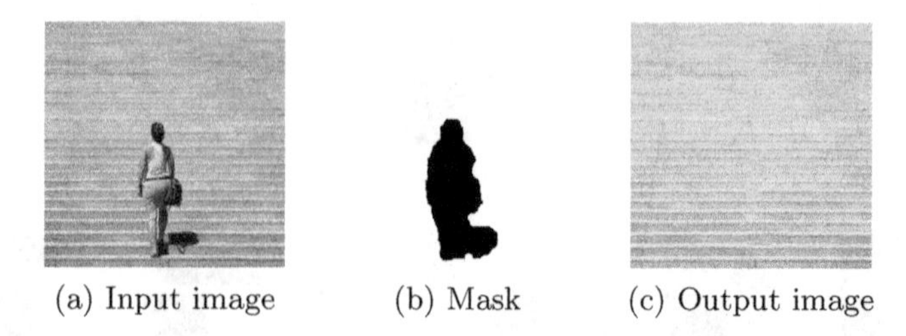

(a) Input image (b) Mask (c) Output image

Fig. 8. Application of the proposed inpainting method to the problem "erase and reconstruct".

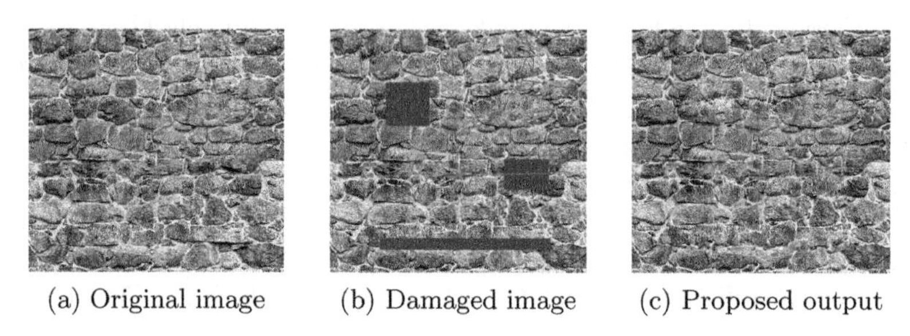

(a) Original image (b) Damaged image (c) Proposed output

Fig. 9. Illustration of the proposed inpainting technique to the problem "erase and reconstruct". Original image (400 × 400 pixels) was damaged by erasing highlightd areas and then reconstructed. The original and reconstructed images were compared by RMSE (22.0426), and SSIM (0.916244).

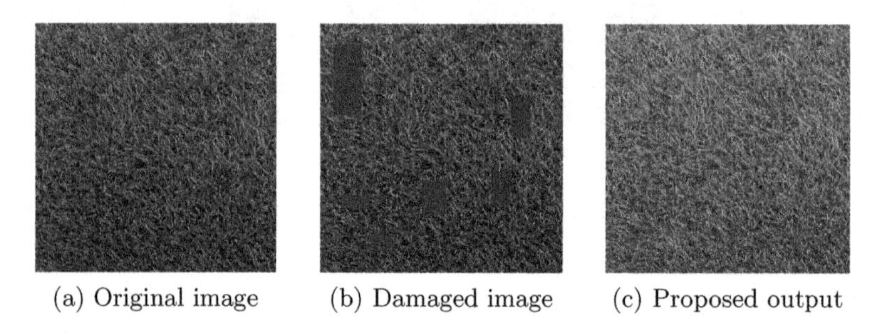

(a) Original image (b) Damaged image (c) Proposed output

Fig. 10. Illustration of the proposed inpainting technique to the problem "erase and reconstruct". Original image (300 × 300 pixels) was damaged by erasing highlightd areas and then reconstructed. The original and reconstructed images were compared by RMSE (13.7273), and SSIM (0.91897).

obtained reconstruction. In Figs. 9 and 10, the illustration of the reconstruction together with the estimation of the quality is given.

We made a visual comparison with *exemplar-based inpainting* technique published in [14]. Images and details are in Figs. 11 and 12.

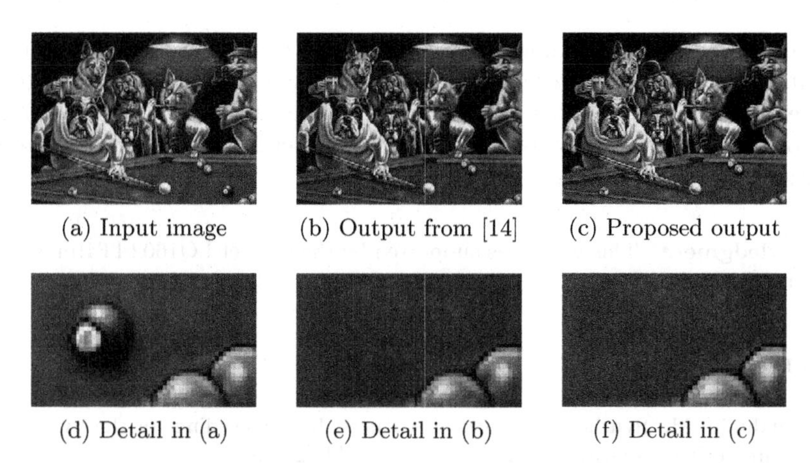

(a) Input image (b) Output from [14] (c) Proposed output

(d) Detail in (a) (e) Detail in (b) (f) Detail in (c)

Fig. 11. Comparison between two inpainting methods: the proposed one and published in [14]. The black billiard ball was erased. Reconstruction of the erased area by the proposed method has less visible artifacts.

(a) Input image (b) Output from [14] (c) Proposed output

(d) Detail in (a) (e) Detail in (b) (f) Detail in (c)

Fig. 12. Comparison between two inpainting methods: the proposed one and published in [14]. The figure of a man was erased. Reconstruction of the erased area by the proposed method has less visible artifacts.

6 Conclusion

The paper describes an ongoing study in patch-based inpainting, which is inspired by the F^1-transform technique. We propose to create feature vectors that consist of the F^1-transform coefficients and use them in computation of closeness. Every partially known patch is replaced by the closest patch from the available database. Moreover, we propose a priority in the selection of patches.

We decomposed the entire reconstruction process into the sequence of steps: creation of a patch database, border extraction, patch searching and patch reconstruction. All steps are illustrated.

Acknowledgment. This work was supported by the project LQ1602 IT4Innovations excellence in science.

References

1. Ashikhmin, M.: Synthesizing natural textures. In: Proceedings of the Symposium on Interactive 3D Graphics, pp. 217–226. ACM (2001)
2. De Bonet, J.S.: Multiresolution sampling procedure for analysis and synthesis of texture images. In: Proceedings of the 24th Annual Conference on Computer Graphics and Interactive Techniques, pp. 361–368. ACM Press/Addison-Wesley Publishing Co. (1997)
3. Efros, A.A., Freeman, W.T.: Image quilting for texture synthesis and transfer. In: Proceedings of the 28th Annual Conference on Computer Graphics and Interactive Techniques, pp. 341–346. ACM (2001)
4. Efros, A.A., Leung, T.K.: Texture synthesis by non-parametric sampling. In: Proceedings of the Seventh IEEE International Conference on Computer Vision, vol. 2, pp. 1033–1038. IEEE (1999)
5. Ballester, C., Caselles, V., Verdera, J., Bertalmio, M., Sapiro, G.: A variational model for filling-in gray level and color images. In: Proceedings of the Eighth IEEE International Conference on Computer Vision, ICCV, vol. 1, pp. 10–16. IEEE (2001)
6. Bertalmio, M., Bertozzi, A.L., Sapiro, G.: Navier-stokes, fluid dynamics, and image and video inpainting. In: Proceedings of the IEEE Computer Society Conference on Computer Vision and Pattern Recognition, CVpPR, vol. 1, pp. I–355. IEEE (2001)
7. Bertalmio, M., Sapiro, G., Caselles, V., Ballester, C.: Image inpainting. In: Proceedings of the 27th Annual Conference on Computer Graphics and Interactive Techniques, pp. 417–424. ACM Press/Addison-Wesley Publishing Co. (2000)
8. Chan, T.F., Shen, J.: Nontexture inpainting by curvature-driven diffusions. J. Vis. Commun. Image Represent. **12**(4), 436–449 (2001)
9. Perfilieva, I., Daňková, M., Bede, B.: Towards a higher degree F-transform. Fuzzy Sets Syst. **180**(1), 3–19 (2011)
10. Perfilieva, I.: Fuzzy transforms: Theory and applications. Fuzzy Sets Syst. **157**(8), 993–1023 (2006)
11. Vlašánek, P., Perfilieva, I.: F-transform and discrete convolution. In: European Society for Fuzzy Logic and Technology. Atlantis Press (2015)
12. Perfilieva, I., Hodáková, P., Hurtík, P.: Differentiation by the F-transform and application to edge detection. Fuzzy Sets Syst. (2014)
13. Serra, J.: Image Analysis and Mathematical Morphology. Academic Press Inc., London (1983)
14. Object removal. http://white.stanford.edu/teach/index.php/Object_Removal

Similarity Measures for Radial Data

Carlos Lopez-Molina[1,2(✉)], Cedric Marco-Detchart[1], Javier Fernandez[1], Juan Cerron[1], Mikel Galar[1], and Humberto Bustince[1,3]

[1] Dpto. Automatica y Computacion, Universidad Publica de Navarra, Pamplona, Spain
carlos.lopez@unavarra.es
[2] Department of Mathematical Modeling, Statistics and Bioinformatics, Ghent University, Ghent, Belgium
[3] Institute of Smart Cities, Universidad Publica de Navarra, Pamplona, Spain

Abstract. Template-based methods for image processing hold a list of advantages over other families of methods, e.g. simplicity and ability to mimic human behaviour. However, they also demand a careful design of the pattern representatives as well as that of the operators in charge of measuring/detecting their presence in the data. This work presents a method for fingerprint analysis, specifically for singular point detection, based on template matching. The matching process sparks the need for similarity measures able to cope with radial data. As a result, we introduce the concepts of Restricted Radial Equivalence Function (RREF) and Radial Similarity Measure (RSM), further used to evaluate the perceptual closeness of scalar and vectorial pieces of radial data, respectively. Our method, which goes by the name of Template-based Singular Point Detection method (TSPD), has qualitative advantages over other alternatives, and proves to be competitive with state-of-the art methods in quantitative terms.

Keywords: Fingerprint analysis · Singular point detection · Radial data · Restricted equivalence function · Similarity measure

1 Introduction

Among biometrical identity authentication systems, those based on fingerprint analysis are the most popular alternative for a large range of applications, e.g. card-less payments or access control. Fingerprint-based authentication systems carry out the individual authentication analyzing, mostly, ridge patterns in the fingertips. A simple division of the context of fingerprint authentication can be made by discriminating *identification* and *verification* tasks. Identification refers to the localization of an individual in a database, provided one or more of its fingerprints; verification refers to the confirmation of a claimed identity in a database. For both tasks, having an accurate way to perform one-to-one fingerprint comparisons (usually known as *fingerprint matching*) is critical.

Fingerprint matching is not trivial, and often involves significant computational effort. For fingerprint verification this is not a major problem, since the

J.P. Carvalho et al. (Eds.): IPMU 2016, Part I, CCIS 610, pp. 599–611, 2016.
DOI: 10.1007/978-3-319-40596-4_50

input fingerprint is only compared with those corresponding to the claimed identity. However, in the context of identification, the time and resources spent at each matching become of paramount importance. In fact, several strategies have been developed to minimize the number of such comparisons in an identification process. The most popular of such strategies is *classification* [8,9], which consists of comparing the input image only to those belonging to the same so-called class. Such class is assigned according to the global structure of the ridges in the fingertip. Although several classification schemes and strategies have been proposed, most of the applications use the five major classes in the Henry system [18]: *arch*, *tented arch*, *left loop*, *right loop* and *whorl*. These classes can also be depicted in terms of the number and relative position of the so-called Singular Points (SPs). SPs are locations of the fingerprint in which abnormal ridge patterns occur, and can be categorized in two types: *cores* (where ridges are bent) and *deltas* (where the ridge flow diverges). Considering that SPs are defined as local ridge patterns, we decided to study a SP detection method which could be based on template matching. As a final goal, this SP detection shall be used as support information to decide which class does a fingerprint belong to.

Template matching is a recurrent solution in digital image processing and roots in very early cognitive abilities in human vision system [20]. A priori, the only information needed in a template matching system is an expression of the items to be identified (the templates) and a comparison measure able to detect occurrences of the template in the input data. Examples of template-based methods for image processing range from low-level feature detectors [5,11,22], to composite object detectors (e.g. the eye detector in [16]). In the SP detection scenario, defining a template is straightforward, since the SPs themselves are defined by their visual properties. However, the design of a comparison measure able to model the fit of such templates at each location of a fingerprint involves major challenges. Since a human fingerprint is composed of lines (ridges), we demand the templates to be represented as a set of lines (equivalently, orientations). As a consequence, the comparison measures used to quantify the matching of each template at each position of the fingerprint must consider the radial nature of the data. Put to more strict terms, the similarity measure must be able to cope with radial data.

Different mathematical theories have been used to model, leading to what Zadeh referred to as *a vast armamentarium* of techniques for comparison [28]. Axiomatic representations of metric and non-metric comparison frameworks have appeared in the literature (e.g. [25] for set-based similarity, or [6] for T-indistinguishability). Despite the variety of measures and inspirations, none of such frameworks is prepared for radial data. In this work we present the concepts of Restricted Equivalence Function (REF) and Similarity Measure (SM) for radial data. Our study is motivated by our specific problem, but also fuelled by the increasing relevance of radial data in applied research. Tasks in highly researched topics, e.g. computer vision, often involve handling radial data in different flavours (e.g., angular, vectorial or tensorial data [29]) and consequently demand well-defined operators for different tasks, including data comparison.

The proposed operators fulfil these demands and, moreover, become a key the completion of our template-based SP detection method.

The remainder of the work is as follows. In Sect. 2 we review some standard notation on radial data. Section 3 is devoted to introduce the concepts of RREF and RSM, while in Sect. 4 we present our proposal for SP detection in fingerprints. Section 5 includes an experimental study in which we illustrate the performance of our SP detection method, compared to other well-known methods in the literature. Finally, Sect. 6 gathers some conclusions and a brief discussion on potential future evolutions of our method.

2 Preliminaries

Among the areas in which fuzzy set theory has played a relevant role, data similarity modelling is one of the most prominent ones. The reason is that notions like similarity, closeness or likeliness are inherently bounded to human perception or interpretation. Hence, different proposals have appeared to effectively model the comparison of information. Among these, we find fuzzy metric spaces [15], with interesting advantages over classical metric spaces in terms of interpretability [10] or equivalence and similarity measures [3], which we take as inspiration to develop measures that can handle radial data. In this work we make use of three well-known concepts, namely Restricted Equivalence Function (REFs, [3]), Similarity Measures (SMs, [27]), and Aggregation Operators (AgOps, [2]). Due to space constraints, we do not review these concepts explicitly, but we refer to the referenced works for further information.

In this work we define an equivalence relation R to relate angles in with identical orientation and direction. Let $a, b \in \mathbb{R}$, we have aRb if and only if $a = b + 2k\pi$, where $k \in \mathbb{Z}$. In this way, the equivalence class $[a] = \{b \mid bRa\}$ is the set containing all the data associated with the same angle. Note also that any semiopen interval whose width is 2π (i.e., an interval of the form $[\omega, \omega+2\pi[$) is the quotient set (a set which contains one element and only one of each equivalence class). In particular, the most common quotient sets for radial data are $[0, 2\pi[$ and $[-\pi, \pi[$. In this work, we consider the quotient set $\Omega = [0, 2\pi[$, on which we define the classical operations sum ($a \oplus b = [a+b]$) and difference ($a \ominus b = [a-b]$), where $[t]$ denotes the only element $z \in \Omega$ such that zRt. In this context, we refer to *mirroring* as the mapping: $m : \Omega \longrightarrow \Omega$ such that $m(a) = 0 \ominus a$.

3 Comparison of Radial Data

Radial data has been a subject of analysis since mid-18th century [7]. Most of the literature on radial data is based on adapting distributions to circular set-ups, in order to abilitate statistical analysis of radial data. One of the problems receiving meager attention in radial data is data comparison. In fact, to the best of our knowledge, no comprehensive analysis of the quantification of similarity between two angles has been performed in the literature. There have been proposals for radial data comparison using the *sample median direction* [7] or the *sample modal* *direction* [19], as well as some metrics on $[0, 2\pi[$. For example, the angular metric, given by $\delta^*(a,b) = \min(|b-a|, 2\pi-|b-a|)$, which represents the amplitude of the shortest arc encompassing two angles. However, no development has been made on interpretable measures able to adapt to human perception or evaluation.

Radial data introduces very interesting novelties for similarity modelling, specifically the fact that, for radial data, increasing the farness of two elements will eventually lead to increasing their closeness. This apparent contradiction, which is only such contradiction in linear data, roots on the very nature of radial data. This section is devoted to develop functions that are able to measure the perceived similarity between scalar and vector angular data.

3.1 Restricted Radial Equivalence Functions

Definition 1. *A mapping $r_\theta : \Omega^2 \to [0,1]$ is called a Restricted Radial Equivalence Function (RREF) associated with the metric δ if it satisfies the following:*

(RR1) $r_\theta(a,b) = r_\theta(b,a)$ *for all* $a, b \in \Omega$*;*
(RR2) $r_\theta(a,b) = 1$ *if and only if* $\delta(a,b) = 0$*;*
(RR3) $r_\theta(a,b) = 0$ *if and only if* $\delta(a,b)$ *is maximum;*
(RR4) $r_\theta(a,b) = r_\theta(m(a), m(b))$ *for all* $a, b \in \Omega$*;*
(RR5) For all $a, b, c, d \in \Omega$*, if* $\delta(b,c) \leq \delta(a,d)$*, then* $r_\theta(b,c) \geq r_\theta(a,d)$*.*

Definition 1 is not a direct extension of that of REF to radial data. Differences arise from the absence of monotonicity in radial data, hampering the interpretation of closeness, farness and relative sorting, and are mostly reflected in (RR5). Nevertheless, the spirit and semantics of RREFs are those of REFs. In this work, we only consider RREFs associated with the angular metric $\delta^*(a,b) = \min(|b-a|, 2\pi - |b-a|)$.

Proposition 1. *Let r_θ be a RREF associated with the metric δ^*. For all $a_1, b_1, a_2, b_2 \in \Omega$, if $\delta^*(a_1,b_1) = \delta^*(a_2,b_2)$ then $r_\theta(a_1,b_1) = r_\theta(a_2,b_2)$.*

Proof. *Let $a_1, b_1, a_2, b_2 \in \Omega$ such that $\delta^*(a_1,b_1) = \delta^*(a_2,b_2)$. According to (RR5), $\delta^*(a_1,b_1) \leq \delta^*(a_2,b_2)$ implies $r_\theta(a_1,b_1) \geq r_\theta(a_2,b_2)$. Analogously, $\delta^*(a_2,b_2) \leq \delta^*(a_1,b_1)$ implies $r_\theta(a_2,b_2) \geq r_\theta(a_1,b_1)$ so the equality holds.*

Proposition 1 implies that, if $h : \Omega^2 \to [0,1]$ satisfies (RR5) with respect to the metric δ^*, then it also satisfies (RR4), since $\delta(a,b) = \delta(m(a), m(b))$.

Proposition 2. *Let φ and ψ be automorphisms of the intervals $[0,1]$ and $[0,\pi]$, respectively. The mapping $t : \Omega^2 \to [0,1]$ given by*

$$t(a,b) = \varphi^{-1}\left(1 - \left(\frac{1}{\pi}\psi\left(\delta^*(a,b)\right)\right)\right) \tag{1}$$

is a RREF.

Proof. *Direct by the properties of the metric δ^*.*

Some examples of RREFs constructed as in Proposition 2 are included in Fig. 1.

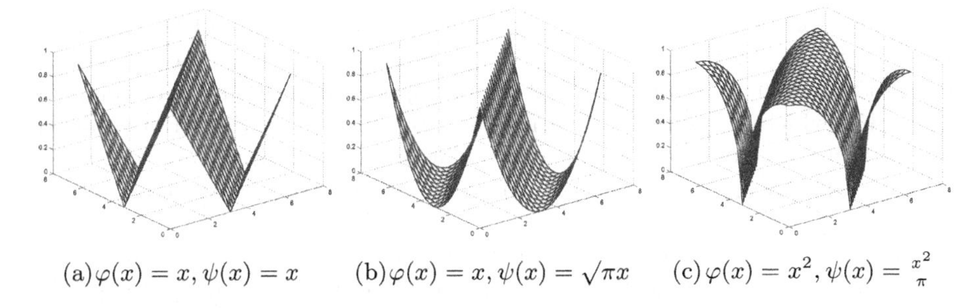

(a) $\varphi(x) = x, \psi(x) = x$ (b) $\varphi(x) = x, \psi(x) = \sqrt{\pi x}$ (c) $\varphi(x) = x^2, \psi(x) = \frac{x^2}{\pi}$

Fig. 1. Restricted radial equivalence functions generated as in Proposition 2 from automorphisms in the unit interval (namely φ) and in $[0, \pi]$ (namely ψ).

3.2 Radial Similarity Measures

Definition 2. *A mapping* $s_\theta : \Omega^k \times \Omega^k \to \mathbb{R}^+$ *is said to be a k-ary Radial Similarity Measure (RSM) associated with the metric* δ^* *if it satisfies the following:*

(SR1) $s_\theta(\boldsymbol{a}, \boldsymbol{b}) = s_\theta(\boldsymbol{b}, \boldsymbol{a})$ *for all* $\boldsymbol{a}, \boldsymbol{b} \in \Omega^k$*;*
(SR2) $s_\theta(\boldsymbol{a}, \boldsymbol{b}) = 0$ *if and only if* $d^*(a_i, b_i) = \pi$ *for all* $i \in \{1, \ldots, k\}$*;*
(SR3) $s_\theta(\boldsymbol{c}, \boldsymbol{c}) = Max_{\boldsymbol{a},\boldsymbol{b} \in \Omega^k} s_\theta(\boldsymbol{a}, \boldsymbol{b})$ *for all* $c \in \Omega^k$*;*
(SR4) For all $\boldsymbol{a}, \boldsymbol{b}, \boldsymbol{c}, \boldsymbol{d} \in \Omega^k$, *if* $\delta^*(\boldsymbol{a}, \boldsymbol{d}) \geq \delta^*(\boldsymbol{b}, \boldsymbol{c})$ *then* $s_\theta(\boldsymbol{a}, \boldsymbol{d}) \leq s_\theta(\boldsymbol{b}, \boldsymbol{c})$, *where* $\delta^*(\boldsymbol{a}, \boldsymbol{d}) \geq \delta^*(\boldsymbol{b}, \boldsymbol{c})$ *implies that* $\delta^*(a_i, d_i) \geq \delta^*(b_i, c_i)$ *for all* $i \in \{1, \ldots, k\}$.

Proposition 3. *Let* r_θ *be a RREF and let* f *be a k-ary aggregation function such that* $f(\boldsymbol{x}) = 0$ *if and only if* $x_i = 0$ *for all* $i \in \{1, \ldots, k\}$ *and* $f(\boldsymbol{x}) = 1$ *if and only if* $x_i = 1$ *for all* $i \in \{1, \ldots, k\}$. *The function* $s_{\theta_{[f, r_\theta]}} : \Omega^k \times \Omega^k$, *given by*

$$s_{\theta_{[f,r_\theta]}}(\boldsymbol{a}, \boldsymbol{b}) = f(r_\theta(a_1, b_1), \ldots, r_\theta(a_k, b_k)) \tag{2}$$

is a k-ary radial similarity measure that satisfies

- $s_{\theta_{[f,r_\theta]}}(\boldsymbol{a}, \boldsymbol{b}) = s_{\theta_{[f,r_\theta]}}(\boldsymbol{b}, \boldsymbol{a})$ *for all* $\boldsymbol{a}, \boldsymbol{b} \in \Omega^k$*;*
- $s_{\theta_{[f,r_\theta]}}(\boldsymbol{a}, \boldsymbol{b}) = 0$ *if and only if* $d^*(a_i, b_i) = \pi$ *for all* $i \in \{1, \ldots, k\}$*;*
- $s_{\theta_{[f,r_\theta]}}(\boldsymbol{a}, \boldsymbol{b}) = 1$ *if and only if* $a_i = b_i$ *for all* $i \in \{1, \ldots, k\}$*;*
- *For all* $\boldsymbol{a}, \boldsymbol{b}, \boldsymbol{c}, \boldsymbol{d} \in \Omega^k$, *if* $\delta^*(\boldsymbol{a}, \boldsymbol{d}) \geq \delta^*(\boldsymbol{b}, \boldsymbol{c})$ *then* $s_{\theta_{[f,r_\theta]}}(\boldsymbol{a}, \boldsymbol{d}) \leq s_{\theta_{[f,r_\theta]}}(\boldsymbol{b}, \boldsymbol{c})$, *where* $\delta^*(\boldsymbol{a}, \boldsymbol{d}) \geq \delta^*(\boldsymbol{b}, \boldsymbol{c})$ *implies that* $\delta^*(a_i, d_i) \geq \delta^*(b_i, c_i)$ *for all* $i \in \{1, \ldots, k\}$.
- $s_{\theta_{[f,r_\theta]}}(\boldsymbol{a}, \boldsymbol{b}) = s_{\theta_{[f,r_\theta]}}(m(\boldsymbol{a}), m(\boldsymbol{b}))$ *for all* $\boldsymbol{a}, \boldsymbol{b} \in \Omega^k$ *where* $m(\boldsymbol{a}) = (m(a_1), \ldots, m(a_k))$.

4 A Novel Proposal for Singular Point Detection

In this section we present a framework for SP detection based on templates, which is referred to as *Template-based SP Detection* method (TSPD method).

To the best of our knowledge, no author has proposed the use of templates to represent SPs, probably due to the lack of reliable comparison methods that can handle the matching score. The most similar approach is the usage of complex filters [17], which are convolved with the complex representation of orientation maps[1]. From our point of view, template-matching is a natural strategy for SP detection, mostly because the very definition of SP is vague and based on human perception.

Any template matching-based framework for image processing is based on a three-fold nucleus: (a) a fit representation of the input data, (b) templates describing the patterns to be searched, and (c) a reliable tool to quantify the similarity between both representations. Since our framework is based on mimicking human perception, our aim is to maintain all three components as faithful as possible to the human comprehension of the problem. Consequently, we elaborate on the ridge-like representation of fingerprints (a) and templates (b), while employing RSMs for (c).

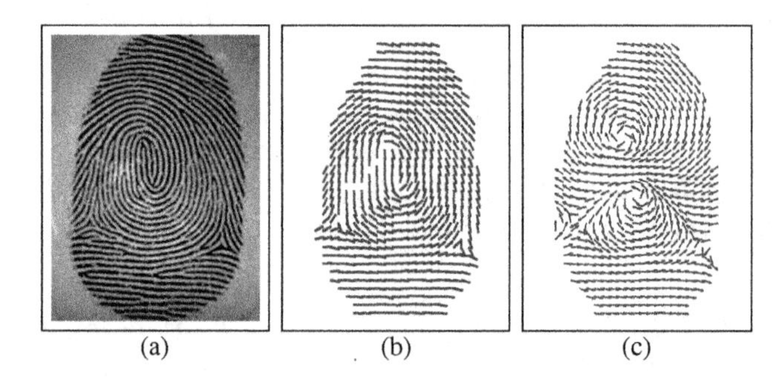

Fig. 2. Whorl image generated with SFinGe (a), together with its Orientation Map (OM) (b) and Squared Orientation Map (SqOM) (c). Blocks are composed of 12×12 pixels.

(a) *Fingerprint representation using Orientation Maps (OMs).* How to represent the ridges in a fingerprint has been often studied, but most of the authors agree on using OMs [24]. OMs are constructed from the division of fingerprint image into disjoint blocks; each of those blocks is assigned a unique orientation given by the majority ridge orientation of its pixels. The best-known procedure for OM generation is the gradient method [1], in which the orientation of the ridges is computed pixel-wise as the perpendicular to the gradient direction. Gradients are usually computed with Sobel masks, although many other options are elegible [23]. Figure 2 includes a OM of a whorl-typed fingerprint. In Fig. 2(b) cores take oriented cup-like patterns, which are dependent on the specific orientation of each SP;

[1] The method in [17] is indeed used as baseline contender for the experiments in Sect. 5.

deltas, however, produce triangular orientation patterns. Some authors have explored an alternative representation of ridges that is better fitted than OMs to our goals, as mentioned earlier. This representation, namely Squared OM (SqOM) [13], is created from an OM by multiplying each of its values (orientations) by 2 (consequently, orientations are translated to directions). In our proposal we use SqOMs. As seen in Fig. 2(c), this produces interesting improvements in the representation of SPs w.r.t. that in OMs. Specifically, we aim at exploiting the fact that cores, in SqOMs, produce either clockwise or anticlockwise vector flows. In both cases, they become rotation-invariant. Regarding deltas, the improvement is not as decisive, since their appearance does not become rotation-invariant. Still, using SqOMs simplifies the design of the templates, and is kept as standard representation of fingerprints in the TSPD method.

(b) *Templates for SP representation.* The templates in our framework must be a minimal set capturing the way in which SPs appear (or are perceived) in a SqOM. Cores manifest as either clockwise or anticlockwise vector flows. As a consequence, there is only need for two templates. Moreover, these templates can be functionally represented in a very simple manner.
Let the origin $(0, 0)$ represent the center of a template T of size $(2n + 1) \times (2n + 1)$. The orientation at a position $(x, y) \in [-n, n]^2$ of a core template is given by

$$T(x,y) = \begin{cases} \text{atan2}(y,x) & \text{if it is a clockwise core, and} \\ \text{atan2}(-y,-x) & \text{if it is an anticlockwise core,} \end{cases} \tag{3}$$

where $\text{atan2}(y, x)$ is the well-known sign-sensitive version of the arctangent of $\frac{y}{x}$, *i.e.* the anticlockwise angle of the vector (x, y) with respect to the positive x-axis. Note that the center of the template has no value, and hence contains no information for the matching process.
Regarding deltas, template design becomes more intricate. In a general manner, a delta is represented as a triangular pattern in the OM, and becomes a symmetric pattern with vectors opposing each other in two orthonormal directions in the SqOM (see Fig. 2). None of those representations is rotation-invariant, and consequently an orientation-dependent template must be created to represent delta SPs. The orientation at a position $(x, y) \in [-n, n]^2$ of a delta SP template with orientation $\alpha \in [0, \pi]$ is given by

$$T_\alpha(x,y) = \text{atan2}(-(\cos(\alpha)y - \sin(\alpha)x), \sin(\alpha)y + \cos(\alpha)x). \tag{4}$$

Figure 3 displays the delta SP template for different values of α. In such templates we can observe how the delta pattern is composed of two orthonormal axis, one resembling an *attractor* to the origin, the other one being a *repeller* from it.
In the template definitions above there are two decisions to be taken in the instantiation of templates. The first decision is on the number of delta SP templates to be used; that is, how many different values of α are used to

produce templates. Theoretically, a greater number of templates will lead to more accurate detections, although presumably coupled to a better fitting of abnormal ridge occurrences that do not correspond to SPs. Also, more templates induce a higher computational effort.The second decision relates to the size of the templates. Indeed the size of the templates must be dependent upon the size of the blocks in the SqOM, as well as upon the expected granularity of the fingerprint capturing process. These parameters are further discussed in Sect. 5.2.

(c) *Comparison of SqOMs and templates.* The comparison of SqOMs and templates is done in the simplest possible manner. For each template we produce a similarity map with the same dimensions as the SqOM. Each position of such similarity maps corresponds to the value yielded by the RSM between the template and the neighbourhood of the block. Finally, all the similarity maps corresponding to the same type of SP are fused using the max operator. In this way, we obtain a graded representations of the presence of each type of SP.

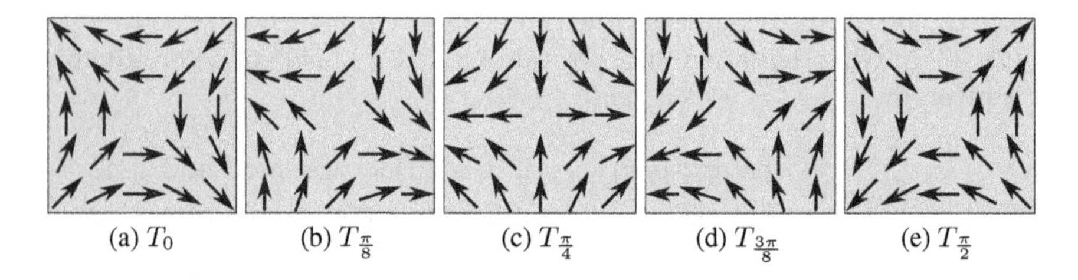

(a) T_0 (b) $T_{\frac{\pi}{8}}$ (c) $T_{\frac{\pi}{4}}$ (d) $T_{\frac{3\pi}{8}}$ (e) $T_{\frac{\pi}{2}}$

Fig. 3. Examples of delta SP templates T_α generated as in Eq. (4) with different values of α.

5 Experiments

The TSPD method has interesting qualitative advantages compared to other methods, e.g. the simplicity of the process and its high interpretability. In this section we compare the performance of our method to that of the most relevant SP detection methods in the literature, in quantitative terms. The results of the TSPD method have been compared with those of the Poincarè method [14], as well as to those of the one proposed by Liu [17]. The former method has been selected because it is the most used SP detection method in literature, whereas the latter method is included because it holds strong similarities to ours.

5.1 Quantification of the Results

This experiment uses the most common dataset for fingerprint analysis, as it is the National Institute of Standards and Technology Special Database 4

(NIST-4) [26]. This database, published by the FBI (Federal Bureau of Investigation), contains 4000 rolled-ink fingerprints (of 512×480 pixels) from 2000 fingertips. It has been, historically, the most popular dataset for the validation of fingerprint classification. We have manually labelled the first 1000 fingerprints from NIST-4 database[2]. Labelling has been carried out according to the specifications given in the specialized literature and has been thoroughly revised by multiple reviewers.

In this experiment we have quantified the performance of each procedure in correctly and accurately detecting SPs.For each dataset we have created a confusion matrix which accounts for the success and fallout in SP detection. After extracting the SPs for a fingerprint, we first compute the best-possible matching between the cores in the automatic solution to those in the ground truth, forcing a one-to-one correspondence. Each matched core in the automatic solution accounts for as True Positive (TP). Then, each unmatched core in the automatic solution and in the ground truth are tagged as False Positive (FP) and False Negative (FN), respectively. Finally, in case both the automatic solution and the ground truth contain less than two cores, the missing SPs are taken as correct predictions, and consequently are accounted for as True Negatives (TN). The process is analogous for the deltas, whose results are stored in an independent confusion matrix. Note that each fingerprint can generate more than two hits in the confusion matrix, if SPs are both missed (FNs) and misdetected (FPs). Note also that in the matching we consider some tolerance in the correspondence of SPs tagged by the automatic method to those in the ground truth. For the present experiment, this spatial tolerance is set to 10 % of the length of the image diagonal.

The results generated with the above-mentioned procedure lead to two confusion matrices for each dataset, one for cores and one for deltas. From such matrices, we have generated different scalar interpretations of the quality of the results. More specifically, we consider precision (PREC) and recall (REC), which are further combined to produce the so-called $F_{0.5}$ measure. Moreover, we also measure the percentage of fingerprints in which all the SPs (cores and deltas) have been correctly detected.

5.2 Experimental Procedure

In this experiment, the techniques used for OM/SqOM computation, smoothing and segmentation are identical for each of the three SP detection methods. Firstly, the image is divided into non-overlapping blocks of 10×10 pixels. Secondly, the image is segmented to avoid false SP detections in the ridge abnormalities occurring at the fingertip boundaries. Thirdly, to compute the gradients for the OM we use the well-known Sobel operators [23]. The resulting matrix is the OM, which is further regularized using a flat mask of 5×5 blocks [13]. Once the OM is generated, each of the methods needs to be customized:

[2] The labels for this experiment can be downloaded from [4].

- *TSPD method.* In order to preserve the fairness of the comparison, we have considered a very basic configuration, which can be seen as the baseline configuration of the method. This configuration involves only 4 templates (two for each type of SP), all of them having 5×5 blocks. In the case of the delta SP templates, we take $\alpha \in \{0, \frac{\pi}{2}\}$. This is, objectively, the minimum set of templates to be used. However, the usage of 5 RSM is expected to shed light on the impact of the RSMs on the final results, as well as to illustrate the flexibility of the TSPD method.
  In this experiment, RREFs are constructed as in Proposition 2, *i.e.* from pairs of automorphisms (φ, ψ); The automorphism φ is given by $\varphi(x) = x^{0.5}$, while the automorphism ψ is given by $\psi(x) = \frac{x^{e_i}}{\pi^{e_i - 1}}$, with $e_i \in (0.5, 0.75, 1, 1.5, 2)$. In the remainder of this paper, each of such combinations of automorphisms is referred to as R_i, with i representing the index in e_i. In this way, R_4 involves the use of $e_4 = 1.5$ in the construction of ψ. The thresholds used in combination with each of the five R_i are 0.7, 0.8, 0.85, 0.9 and 0.95, respectively[3].
- *Poincarè method.* This method consists of computing the difference between each orientation in a 3×3 neighbourhood and its clockwise successor. Those differences are further summed up to produce the Poincarè index in each block [18]. This index takes value 0, $\frac{1}{2}$ or $-\frac{1}{2}$, indicating the absence of a SP, the presence of a core or the presence of a delta, respectively. Although other authors have used other configurations of the neighbourhood [12,21], specially regarding its size, we maintain the frequently used 3×3 size.
- *Liu's method.* In this method SqOMs are filtered with first order complex filters at different scales. More specifically, the large scale filters are used to discriminate the real SPs from spurious responses, while the fine scale ones determine their precise location. The threshold used for discrimination of SPs is set to 0.7. Regarding the scales we consider, as in [17], filters of $s \times s$ blocks, with $s \in \{3, 5, 7, 9\}$.

5.3 Results

The performance obtained by each method is listed in Fig. 4, including the average $F_{0.5}$ for each type of SP and the percentage of fingerprints for which the method achieved a perfect detection (which we refer to as Perfect Detection Percentage, PDP). That is, the rate of fingerprints for which each method gathered the exact number of SPs.

In Fig. 4 TSPD-R_5 is the best performer overall, obtaining the greatest PDP (69.90 %), as well as the best average $F_{0.5}$ measure (0.861). Although the other competing methods stay relatively close to those values, especially when it comes to the PDP, TSPD-R_5 shows the most consistent results, not plunging in one single aspect of the evaluation.

The relevance of the RSMs in the TSPD is put to the spotlight by the results in Fig. 4. Interestingly, TSPD-R_3, with φ and ψ being identity functions, is

[3] The threshold for each method, including Liu's and Poincaré, has been manually set to optimize the results.

	Quant.	Template-based SP detection TSPD-R_1	TSPD-R_2	TSPD-R_3	TSPD-R_4	TSPD-R_5	Poincaré	Liu
Cores	TP	897	869	875	900	868	759	**922**
	FP	273	148	147	194	132	**40**	329
	FN	93	121	115	90	122	231	**68**
	TN	776	888	889	844	905	**982**	721
	PREC	.767	.854	.856	.823	.868	**.950**	.737
	REC	.906	.878	.884	.909	.877	.767	**.931**
	$F_{0.5}$	.831	.866	.870	.864	**.872**	.849	.823
	Avg. dist.	14.4	13.9	**13.7**	13.9	13.8	13.8	16.4
Deltas	TP	**871**	841	844	864	839	728	767
	FP	340	169	176	236	159	53	**29**
	FN	**102**	132	129	109	134	245	206
	TN	731	891	884	827	898	1006	**1011**
	PREC	.719	.833	.827	.785	.841	.932	**.964**
	REC	**.895**	.864	.867	.888	.862	.748	.788
	$F_{0.5}$	.798	.848	.847	.834	.851	.830	**.867**
	Avg. dist.	11.2	10.7	10.4	10.6	**10.3**	13.5	11.5
	Avg. $F_{0.5}$	.815	.857	.859	.849	**.861**	.839	.845
	PDP (%)	56.70	67.80	68.10	66.50	**69.90**	66.40	54.50

Fig. 4. Results gathered by each SP detection method on the NIST dataset (1000 fingerprints).

not the best performer. This fact indicates that a sensible choice of automorphisms in the construction of the RREFs can optimize the result of the TSPD method. In Fig. 4 we also observe that Liu's method obtains a PDP similar to our worst configurations (54.50 %). Despite being the best method detecting cores (922), Liu's method also produces 329 false cores detections, significantly more than those by the Poincarè method (40) and any of the configurations of the TSPD. The behaviour is reversed in the analysis of the deltas, having that the precision of Liu's method is very high (29 FPs and 1011 TNs), but comes together to a relatively low recall (induced by 767 TPs). This evidence is consistent with those by Galar *et al.* [9]. As for the Poincarè method, we observe a high PDP; still, the average $F_{0.5}$ is rather mean because of its difficulties in detecting cores (231 FNs) and deltas (245 FNs). From this results, we conclude that TSPD-R_5 obtains the best results, showing the best trade-off between successes (TPs, TNs) and failures (FPs, FNs).

From the results in this experiment we conclude that the TSPD method is, at least, competitive with the contending methods. Although it involves a certain parameter setting (that of the RSMs and thresholds), this is not radically different from what happens in other SP detection methods (including Liu's method). Interestingly, the RREF leading to the best results in the TSPD method is not that constructed with the pair of automorphisms R_3, indicating that non-linear modelling of dissimilarity can play a role in real applications. Specifically, the best-performing version is that using the pair of automorphisms R_5, since it outperforms all of the other versions of the TSPD method in terms of Combined F and PDP.

It is worth noting that the TSPD method has advantages over its counterparts other than pure performance. For example, it holds interesting visualization properties when it comes to error correction, partly derived from the simplicity of the method. Indeed, we have not exploited the potential use of multi-scale templates yet as Liu's method does. Finally, the model is flexible and configurable, and results can be adapted to each application due to the usage of parametrizable RSMs.

6 Conclusions

This work has two main contributions. First, we have adapted the concepts of Restricted Equivalence Function (REF) and Similarity Measure (SM) to radial environments. The resulting operators, namely Restricted Radial Equivalence Function (RREF) and Radial Similarity Measure (RSM), capture the expected behaviour and semantics of the original operators, but at the same time embrace the cyclic nature of radial data. Second, we have illustrated the validity of the operators for an image processing task, such as fingerprint analysis. In order to do so, we have presented a framework for Singular Point (SP) detection based on templates, which requires the use of RSMs at the template matching stage. This framework, namely Template-based Singular Point Detection (TSPD) method, shows promising results and portrays the usefulness of RSMs for the comparison of radial data in scenarios in which imprecision and ambiguities occur.

Acknowledgements. The authors gratefully acknowledge the financial support of the Spanish Ministry of Science (project TIN2013-40765-P), as well as the financial support of the Research Foundation Flanders (FWO project 3G.0838.12.N).

References

1. Bazen, A.M., Gerez, S.H.: Systematic methods for the computation of the directional fields and singular points of fingerprints. IEEE Trans. Pattern Anal. Mach. Intell. **24**(7), 905–919 (2002)
2. Beliakov, G., Pradera, A., Calvo, T.: Aggregation Functions: A Guide for Practitioners, Studies in Fuzziness and Soft Computing, vol. 221. Springer, Heidelberg (2007)
3. Bustince, H., Barrenechea, E., Pagola, M.: Restricted equivalence functions. Fuzzy Sets Syst. **157**(17), 2333–2346 (2006)
4. Cerron, J., Marco-Detchart, C., Lopez-Molina, C., Bustince, H., Galar, M.: Singular point location for NIST-4 database (2015). https://giara.unavarra.es/datasets/solNIST4.zip
5. Chaudhuri, S., Chatterjee, S., Katz, N., Nelson, M., Goldbaum, M.: Detection of blood vessels in retinal images using two-dimensional matched filters. IEEE Trans. Med. Imaging **8**(3), 263–269 (1989)

6. De Baets, B., De Meyer, H., Naessens, H.: A class of rational cardinality-based similarity measures. J. Comput. Appl. Math. **132**(1), 51–69 (2001)
7. Fisher, N.I.: Statistical Analysis of Circular Data. Cambridge University Press, Cambridge (1993)
8. Galar, M., Derrac, J., Peralta, D., Triguero, I., Paternain, D., Lopez-Molina, C., García, S., Benítez, J.M., Pagola, M., Barrenechea, E., Bustince, H., Herrera, F.: A survey of fingerprint classification part I: taxonomies on feature extraction methods and learning models. Knowl. Based Syst. **81**, 76–97 (2015)
9. Galar, M., Derrac, J., Peralta, D., Triguero, I., Paternain, D., Lopez-Molina, C., García, S., Benítez, J.M., Pagola, M., Barrenechea, E., Bustince, H., Herrera, F.: A survey of fingerprint classification part II: experimental analysis and ensemble proposal. Knowl. Based Syst. **81**, 98–116 (2015)
10. Gregori, V., Morillas, S., Sapena, A.: Examples of fuzzy metrics and applications. Fuzzy Sets Syst. **170**(1), 95–111 (2011)
11. Hueckel, M.H.: An operator which locates edges in digitized pictures. J. ACM **18**(1), 113–125 (1971)
12. Karu, K., Jain, A.K.: Fingerprint classification. Pattern Recogn. **29**(3), 389–404 (1996)
13. Kass, M., Witkin, A.: Analyzing oriented patterns. Comput. Vis. Graph. Image Process. **37**(3), 362–385 (1987)
14. Kawagoe, M., Tojo, A.: Fingerprint pattern classification. Pattern Recogn. **17**(3), 295–303 (1984)
15. Kramosil, I., Michálek, J.: Fuzzy metrics and statistical metric spaces. Kybernetika **11**(5), 336–344 (1975)
16. Li, Y., Qi, X., Wang, Y.: Eye detection by using fuzzy template matching and feature-parameter-based judgement. Pattern Recogn. Lett. **22**(10), 1111–1124 (2001)
17. Liu, M.: Fingerprint classification based on Adaboost learning from singularity features. Pattern Recogn. **43**, 1062–1070 (2010)
18. Maltoni, D., Maio, D., Jain, A.K., Prabhakar, S.: Handbook of Fingerprint Recognition. Springer, Heidelberg (2009)
19. Mardia, K.V., Jupp, P.E.: Directional Statistics. Wiley, New York (2000)
20. Marr, D.: Vision. MIT Press, Massachusetts (1982)
21. Nyongesa, H.O., Al-Khayatt, S., Mohamed, S.M., Mahmoud, M.: Fast robust fingerprint feature extraction and classification. J. Intell. Rob. Syst. **40**(1), 103–112 (2004)
22. Poli, R., Valli, G.: An algorithm for real-time vessel enhancement and detection. Comput. Meth. Programs Biomed. **52**(1), 1–22 (1997)
23. Prewitt, J.M.S.: Object enhancement and extraction. In: Lipkin, B., Rosenfeld, A. (eds.) Picture Processing and Psychopictorics, pp. 75–149. Academic Press, New York (1970)
24. Turroni, F., Maltoni, D., Cappelli, R., Maio, D.: Improving fingerprint orientation extraction. IEEE Trans. Inf. Forensics Secur. **6**(3), 1002–1013 (2011)
25. Tversky, A.: Features of similarity. Psychol. Rev. **84**(4), 327–352 (1977)
26. Watson, C.I., Wilson, C.L.: NIST Special Database 4, Fingerprint Database. Technical report, U.S. National Institute of Standards and Technology (1992)
27. Xuecheng, L.: Entropy, distance measure and similarity measure of fuzzy sets and their relations. Fuzzy Sets Syst. **52**(3), 305–318 (1992)
28. Zadeh, L.A.: Similarity relations and fuzzy orderings. Inf. Sci. **3**(2), 177–200 (1971)
29. Zhang, F., Hancock, E.R.: New Riemannian techniques for directional and tensorial image data. Pattern Recogn. **43**(4), 1590–1606 (2010)

Application of a Mamdani-Type Fuzzy Rule-Based System to Segment Periventricular Cerebral Veins in Susceptibility-Weighted Images

Francesc Xavier Aymerich[1,2(✉)], Pilar Sobrevilla[3], Eduard Montseny[2], and Alex Rovira[1]

[1] MR Unit, Department of Radiology (IDI), Vall d'Hebron University Hospital, Vall d'Hebron Research Institute (VHIR), Autonomous University of Barcelona, 08035 Barcelona, Spain
{xavier.aymerich, alex.rovira}@idi.gencat.cat
[2] ESAII Department, Universitat Politècnica de Catalunya, 08003 Barcelona, Spain
{xavier.aymerich, eduard.montseny}@upc.edu
[3] MAII Department, Universitat Politècnica de Catalunya, 08034 Barcelona, Spain
pilar.sobrevilla@upc.edu

Abstract. This paper presents an algorithm designed to segment veins in the periventricular region of the brain in susceptibility-weighted magnetic resonance images. The proposed algorithm is based on a Mamdani-type fuzzy rule-based system that enables enhancement of veins within periventricular regions of interest as the first step. Segmentation is achieved after determining the cut-off value providing the best trade-off between sensitivity and specificity to establish the suitability of each pixel to belong to a cerebral vein. Performance of the algorithm in susceptibility-weighted images acquired in healthy volunteers showed very good segmentation, with a small number of false positives. The results were not affected by small changes in the size and location of the regions of interest. The algorithm also enabled detection of differences in the visibility of periventricular veins between healthy subjects and multiple sclerosis patients.

Keywords: Brain · Fuzzy rule-based systems · Image segmentation · Magnetic resonance imaging

1 Introduction

Susceptibility-weighted imaging (SWI) is a noninvasive magnetic resonance imaging (MRI) technique that takes advantage of the magnetic susceptibility effects of paramagnetic deoxygenated hemoglobin [1]. Because of this capability, SWI can be used to visualize venous structures in the brain, providing valuable complementary information for the diagnosis and treatment of patients with neurological disorders such as multiple sclerosis (MS) [2]. A quantitative method to determine the number of veins detected on

J.P. Carvalho et al. (Eds.): IPMU 2016, Part I, CCIS 610, pp. 612–623, 2016.
DOI: 10.1007/978-3-319-40596-4_51

SWI would be of value for monitoring MS severity, progression, and the response to therapy [3].

Segmentation of venous structures over the entire brain with SWI is extremely complex. This is partly because certain regions, including the periventricular white matter, contain numerous small veins, most of them very thin and difficult to differentiate from their surroundings in these sequences. Many of the available methods used for segmenting veins in SW images are adaptations of techniques designed and used for segmenting bright arteries from a dark background [4]. Several methods for cerebrovascular segmentation have been proposed, and a detailed review is provided by Lesage et al. [5]. Focusing only on SWI, two main approaches have been used for segmentation of brain venous structures [6]. The first is based on the use of a statistical local thresholding algorithm [3]. The second approach involves application of scale-space analysis based on vesselness filters, which can be used to directly visualize venous structures [7]. Segmentation is then done using thresholding [8] or an active contour model [9]. Some examples of vesselness filters [10–13] are based on Frangi's [7] and Sato's vesselness filter [8].

Detection of venous blood pixels in SW images addressed to segmenting venous structures in the brain is subject to several factors that imply inherent uncertainty. Most cerebral veins are tiny, thin structures existing in an environment where noise, non-homogeneity, artifacts, and partial volume effects introduce varying degrees of vagueness that affect their detection and the definition of their paths. Fuzzy rule-based systems (FRBSs) are important areas in which fuzzy logic and fuzzy set theory are applied. In contrast to classical rule-based systems, FRBS deal with fuzzy rules instead of classical logical ones, and their success resides in their approximation to human perception and reasoning, and their intuitive handling and simplicity [14].

Some examples of FRBS use in MRI have been reported [15, 16], and two studies, conducted by Forkert et al. [17, 18], have described FRBS application to solve the problem of segmenting vasculature in MR images. However, Forkert's work focuses on 3D time-of-flight (TOF) magnetic resonance angiography (MRA) rather than magnetic resonance venography with SWI. To our knowledge, the approach presented here is the first application of a FRBS to segment veins in SW images.

2 Materials and Methods

2.1 Image Datasets

Image datasets for training and test purposes were obtained from 13 healthy individuals (10 women and 3 men), with a mean age of 36.7 years (range, 28–50 years). Images of one of these subjects were used for training and the images of the other 12 were used to test the algorithm. To evaluate the capability of the algorithm to detect differences in vein visibility between patients and healthy subjects, we also studied 13 relapsing-remitting multiple sclerosis (RRMS) patients (9 women and 4 men) with a mean age of 37.1 years (range, 28–46 years), mean disease duration of 10.3 years (range, 0.83–22.0 years), and an average Expanded Disability Status Scale (EDSS) score of 3.1. All images were acquired on a Siemens Magnetom Trio 3.0 T scanner

(Siemens, Erlangen, Germany) with a 12-channel array head coil using a 3D fast-low angle single shot sequence (repetition time [TR]/echo time [TE], 32 ms/24.6 ms; flip angle, 15°; matrix, 320 × 320; voxel size, 0.78 × 0.78 × 3.0 mm^3; iPAT factor, 2). Fifty-two parallel contiguous axial slices covering the whole brain were acquired using this sequence. The study was approved by the Clinical Research Ethics Committee of Hospital Universitari Vall d'Hebron in Barcelona (Spain).

2.2 Proposed Method

The proposed algorithm for segmenting veins on SWI has 5 main steps: image selection, preprocessing, definition of regions of interest, enhancement of veins, and segmentation of veins.

The first step is selection of 4 contiguous SW slices from the brain MRI examination of a subject, I^n $(1 \leq n \leq 4)$, where the regions of interest can be visualized. Then, structures outside the intracranial region have to be removed in I^n $(1 \leq n \leq 4)$ by applying to I^n a procedure based on the brain surface extraction (BSE) algorithm [19] included in the Medical Image, Processing, Analysis, and Visualization (MIPAV) software package, version 4.4.1 (Center for Information Technology, NIH, Bethesda, Maryland, USA). Thus, the I_B^n $(1 \leq n \leq 4)$ images are obtained. These images are then normalized, so that the normalized value for each pixel p located at position (i,j) in I_B^n is given by:

$$I_N^n(i,j) = \begin{cases} \text{int}\left(1000\frac{I_B^n(i,j)}{h_{\max}}\right) & \text{if } I_B^n(i,j) \leq 4h_{\max} \\ 4000 & \text{otherwise} \end{cases} \quad (1)$$

where h_{max} is the gray-level value greater than 150 associated with the maximum frequency in the smoothed histogram $H_S(I_B^n)$ obtained by averaging the frequency values whose distances in $H(I_B^n)$ were less than or equal than 2.

In the next step, pixels belonging to veins are enhanced to best differentiate them from their surroundings within $ROI(I_N^n)$ by applying a Mamdani-type FRBS to the images I_N^n $(1 \leq n \leq 4)$ to obtain the associated Adequacy images, I_A^n $(1 \leq n \leq 4)$. The inputs of this FRBS are perceptual features of vein pixels within ROIs evaluated with low-level operators, and the output is the adequacy of these pixels to belong to veins.

Finally, a cut-off value is applied to the *Adequacy images* to obtain the segmented images, I_{SV}^n $(1 \leq n \leq 4)$, whose value for a pixel p located at position (i, j) equals one if it belongs to $ROI(I_N^n)$ and $I_A^n(i,j) > c$, and is zero otherwise. ROC curve analysis was used to select the best cut-off value.

2.3 Implementation of the Method

This section describes how the algorithm steps were implemented.

Selection of Images, Preprocessing, and Definition of ROIs. From the 52 axial slices acquired in the examination of one healthy volunteer, we selected 4 contiguous axial slices, I_t^n $(1 \leq n \leq 4)$, where the ROIs were visualized. The training images I_t^n were then normalized following the preprocessing procedure previously described, obtaining I_{Nt}^n. In each hemisphere of I_{Nt}^n we then manually defined a 9.38 mm by 42.97 mm rectangular ROI in the periventricular region mainly occupied by white matter from the corona radiata, $ROI(I_{Nt}^n)$. The pixels within the 8 rectangular ROIs obtained are the only ones included in the next steps.

Enhancement of veins within the ROIs

Selection of features. Alook inside the white rectangles in Fig. 1(a) shows that veins in SW images are visualized as mainly linear structures showing a darker gray level than their immediate surroundings. Comparison of the linear venous structures in the ROIs and the dark structures within the ellipses of the magnified image in Fig. 1(b) shows that the veins are lighter than the wider vessels in other locations, and some of them show short discontinuities. In addition, certain thin, dark structures are seen outside the ROIs, such as areas of cortex with a high iron content (Fig. 1(b), white arrows), whose characteristics may cause them to be mistaken for veins.

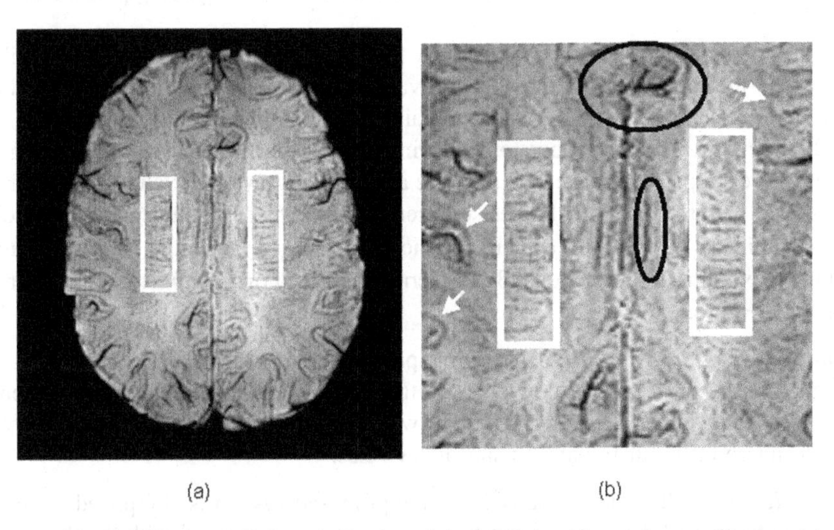

Fig. 1. Example of SW image after application of the BSE algorithm and normalization process. (a) Location of ROIs (white rectangles). (b) Magnification of the area containing ROIs in which black ellipses enclose blood vessels showing better differentiation than most of those within the ROIs, and white arrows point to some cortex locations with high iron content, which, because of their perceptual features, could be mistaken for vein vessels.

Taking into consideration these factors, 3 features are essential to detect vein pixels within periventricular ROIs: Gray-level, Thinness, and Linearity.

Selection of operators to evaluate the features. Several low-level operators were analyzed, seeking those that best characterized *Gray-level*, *Thinness*, and *Linearity*.

Selection of the most appropriate operator to evaluate *Gray-level* involved analysis of 9 operators: *gray-level of the central pixel*, and *maximum*, *minimum*, *mean* and *median gray-level* values of the pixels covered by 3 × 3 and 5 × 5 raster windows. The *standard deviations of the gray-level* values within these windows and the *gray-level differences* between the central pixel and its neighbors were the 26 operators analyzed for *Thinness*. Lastly, to select the operator to evaluate *Linearity*, we used kernels that enabled detection of horizontal, vertical, or oblique (+45 and –45 degrees) single-pixel-wide lines. Application of these kernels required inverting the gray scale of the images in order to detect dark linear structures.

To obtain the operators, pixels corresponding to veins located within the ROIs in the training set images I^n_{Nt} $(1 \leq n \leq 4)$, $ROI(I^n_{Nt})$, were manually labeled using MRIcro software [20]. These were the reference images I^n_{REFVt} $(1 \leq n \leq 4)$, in which the value assigned to a pixel in the *(i, j)* position was 1 if it belonged to a vein and 0 otherwise. Moreover, since our analysis was focused on finding the best operators to evaluate the features of vein pixels, pixels within $I^n_{REFV_t}$ were divided into 3 reference sets: S_1, comprising pixels labeled as veins and belonging to very thin veins; S_2, including pixels labeled as veins without thinness restrictions; and S_3, comprising pixels that were not labeled.

The operator selected to evaluate each feature had to maximize the separability, defined by the ratio between the absolute value of the mean value difference and the maximum of standard deviation values obtained for the reference sets S_r and S_s $(r, s \in \{1, 2, 3\}, r \neq s)$. Then, the operators showing the best performance for evaluating the features of each pixel *p* located at *(i,j)* were as follows: gray-level value, *gl(i,j)*, for the *Gray-level* feature; the third highest difference between the central and surrounding pixels within a 5 × 5 window centered on the pixel, $dif^3_{5x5}(i,j)$, for the *Thinness* feature; and the maximum of 4 directional 3 × 3 kernels centered on *(i,j)*, *maxK3x3(i,j)*, for the *Linearity* feature.

Proposed Mamdani-type FRBS. The principal elements of the FRBS were defined and designed based on *a priori* knowledge, in this case, expert judgment and experience. The underlying knowledge of the system was then explicitly translated into a set of easy to interpret linguistic labels using fuzzy rules.

Knowledge Base: The data base (DB) of the proposed system is comprised of sets of linguistic terms and the membership function partitions associated with the three input variables introduced in previous section (*Gray-level*, *Thinness*, and *Linearity*), and the output variable (*Adequacy* of a pixel to belong to a venous blood vessel). The sets of linguistic terms considered for the input and output variables are given as follows:

$L_{Gray\text{-}level}$ = {Dark, Medium-Dark, Light} = $\{D, MD, LG\}$
$L_{Thinness}$ = {Low, Medium, High} = $\{L_{thin}, M_{thin}, H_{thin}\}$
$L_{Linearity}$ = {Low, Medium, High} = $\{L_{lin}, M_{lin}, H_{lin}\}$
$L_{Adequacy}$ = {Very Poor, Poor, Fair, Good, Excellent} = $\{VP, P, F, G, E\}$

To define the semantics of these linguistic labels, we adopted trapezoidal-shaped membership functions (MFs). The MFs were defined by the quadruple (*sl*, *cl*, *cu*, *su*),

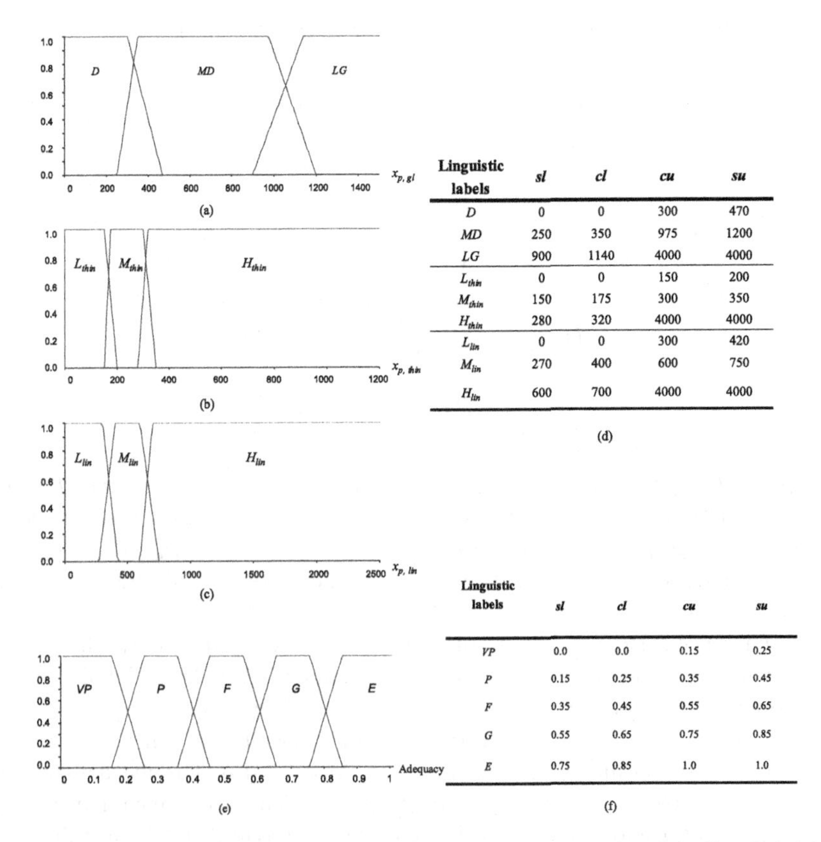

Linguistic labels	*sl*	*cl*	*cu*	*su*
D	0	0	300	470
MD	250	350	975	1200
LG	900	1140	4000	4000
L_{thin}	0	0	150	200
M_{thin}	150	175	300	350
H_{thin}	280	320	4000	4000
L_{lin}	0	0	300	420
M_{lin}	270	400	600	750
H_{lin}	600	700	4000	4000

(d)

Linguistic labels	*sl*	*cl*	*cu*	*su*
VP	0.0	0.0	0.15	0.25
P	0.15	0.25	0.35	0.45
F	0.35	0.45	0.55	0.65
G	0.55	0.65	0.75	0.85
E	0.75	0.85	1.0	1.0

(f)

Fig. 2. Trapezoidal-shaped membership functions defining the semantics of the linguistic labels. Input variables associated with (a) *Gray-level*, (b) *Thinness*, and (c) *Linearity* features. (d) Output variable associated with *Adequacy*. Values defining the shape of each membership function are given in (d) for input variables, and in (f) for the output variable.

where sl and su, and cl and cu are the lower and upper bounds of the support and the core, respectively. The shapes of the MFs can be seen in Fig. 2, which show the fuzzy partitions defined on the domains of each variable.

MFs associated with the input variables were obtained by analyzing the values of the features within and around $ROI(I^n_{Nt})$ defined in I^n_{Nt} $(1 \leq n \leq 4)$, whereas the uniform partition corresponding to the output variable involved 5 levels of adequacy.

The rule base (RB) is represented in compact format by the decision table shown in Table 1. To achieve good performance, the 27 rules conforming the RB were obtained taking into account expert human knowledge and attending to a trade-off between the number of pixels associated with true and false positives. The pixel values for the features fire each rule for the eight ROIs defined in the images of $ROI(I^n_{Nt})$.

Inference Engine: The fuzzification interface establishes a mapping from crisp input values to fuzzy sets defined in the universe of discourse of this input, $U = U_{gl}$ x U_{thin} x U_{lin}. To do so, for each pixel p belonging to an ROI we obtained the

Table 1. Decision table describing the rule base of the proposed FRBS

		Linearity		
Gray-level	*Thinness*	L_{lin}	M_{lin}	H_{lin}
D	L_{thin}	*VP*	*VP*	*VP*
	M_{thin}	*VP*	*VP*	*P*
	H_{thin}	*VP*	*P*	*F*
MD	L_{thin}	*F*	*G*	*F*
	M_{thin}	*F*	*E*	*E*
	H_{thin}	*P*	*F*	*G*
LG	L_{thin}	*P*	*F*	*P*
	M_{thin}	*P*	*F*	*F*
	H_{thin}	*VP*	*P*	*P*

corresponding input vector: $x_p = (x_{gl}, x_{thin}, x_{lin}) = (gl(i,j), dif^3_{5x5}(i,j), maxK3x3(i,j))$. Then, we sought the degree to which it belonged to each of the fuzzy sets defined in U; that is, $\mu_D(x_{gl})$, $\mu_{MD}(x_{gl})$, $\mu_{LG}(x_{gl})$, $\mu_{L,thin}(x_{thin})$, $\mu_{M,thin}(x_{thin})$, $\mu_{H,thin}(x_{thin})$, $\mu_{L,lin}(x_{lin})$, $\mu_{M,lin}(x_{lin})$, $\mu_{H,lin}(x_{lin})$, which were obtained via the corresponding trapezoidal membership functions.

The inference system and defuzzification interface selected to derive the fuzzy and crisp outputs were the classical ones used by Mamdani [21]. These involve the minimum t-norm for both the conjunctive operator, T, which derives the rules firing according to the decision table, and the implication operator, I, which determines the output of the compositional rule of inference. Mode A-FATI is the defuzzification interface, where the aggregation operator, G, is modeled by the maximum t-conorm, while the defuzzification method, D, is the center of gravity (COG), such that:

$$\mu_{A_k}(x_p) = Min(\mu_{A_{k,gl}}(x_{gl}),\ \mu_{A_{k,thin}}(x_{thin}),\ \mu_{A_{k,lin}}(x_{lin})),\quad k = 1, \ldots, 27 \tag{2}$$

$$\mu_{B'_k}(y) = Min(\mu_{A_k}(x_p),\ \mu_{B_k}(y)),\quad k = 1, \ldots, 27 \tag{3}$$

$$\mu_{B'}(y) = \underset{1 \le k \le 27}{Max}\left(\mu_{B'_k}(y)\right) \tag{4}$$

$$y_p = COG(\mu_{B'}(y)) = \frac{\int_Y y\mu_{B'}(y)dy}{\int_Y \mu_{B'}(y)dy} \approx \frac{\sum_{i=1}^{N} y_i \mu_{B'}(y_i)}{\sum_{i=1}^{N} \mu_{B'}(y_i)} \tag{5}$$

where y_i was the i-value in the uniform partition of the adequacy interval [0, 1] in N values, where the value of N was 100. The values y_p so obtained provided the adequacy of each pixel p in $ROI(I^n_{Nt})$ to be a vein pixel.

Segmentation of Veins. After applying the FRBS to the training images, I^n_{Nt} $(1 \le n \le 4)$, we obtained the associated *Adequacy images*, I^n_{At} $(1 \le n \le 4)$: For each pixel p, the value of $I^n_{At}(p)$ was the value obtained if the pixel belonged to the ROI. If not, the value was set at zero.

We then sought the cut-off value, "c", providing the best segmentation results. We considered 71 cut-off values, c_i $(1 \leq i \leq 71)$, in the interval [0.2, 0.9] and applied them to the *Adequacy images*, I^n_{At} $(1 \leq n \leq 4)$. For each c_i, four segmented images were obtained, I^n_{SVt} $(1 \leq n \leq 4)$, whose value for a pixel p located at position (i, j) equaled one if it belonged to $ROI(I^n_{Nt})$ and I^n_{At} $(i,j) > c_i$, and was zero otherwise. To know the number of true positives (TP), false positives (FP), true negatives (TN), and false negatives (FN), a pixel-by-pixel comparison was performed between I^n_{SVt} and the reference image, I^n_{REFVt}, manually labeled by an expert to determine the correctness of the classification within $ROI(I^n_{Nt})$.

To select the best cut-off value, we used an ROC plot, representing the trade-off between sensitivity (S), that is the true positive rate (TPR), and the false positive rate (FPR), for each c_i value. The true positive and false positive rates are given by S = TPR = TP/(TP + FN), and FPR = FP/(TN + FP). Applying this process, we obtained 71 points (FPRci, TPRci), depicted in the ROC space. The best trade-off value is defined by the closest point to the upper left corner, located at (0.0316, 0.9394), which in our case was c = 0.58.

3 Results

To evaluate the performance of the algorithm, we selected quality indices that enabled pixel-by-pixel comparison within the ROIs of the reference and segmented images. The indices were also defined based on the number of TP, FP, TN, and FN. Then, in addition to the true and false positive rates we also considered specificity (SPC), accuracy (ACC), and the Dice coefficient (DC). These quality indices are defined by SPC = TN/(TN + FP), ACC = (TP + TN)/(TP + FP + TN + FN), and DC = 2TP/(2TP + FP + FN).

3.1 Evaluation of the Algorithm on Training and Test Images

Training Images. The results for the training images are presented in the third row of Table 2, which shows the mean (m) and standard deviation (σ) values for the indices S, SPC, ACC, and DC in the comparison between the segmented and reference images. As was expected, and as can be inferred from the values greater than 0.9 obtained for sensitivity and Dice coefficient (third and sixth columns in Table 2), the vast majority of pixels labeled as veins within the reference images were detected by the algorithm. Moreover, the specificity and accuracy values (fourth and fifth columns of Table 2) point to a low number of false positives. Finally, the low standard deviation values with regard to the mean values obtained for all the indices studied indicate that there were no large differences in terms of detection quality between the regions analyzed.

Test Images. To help in the interpretation of the results for the 96 regions of interest, $ROI(I^n_{NT})$, selected in the test images, the fourth row of Table 2 shows the mean (m) and

Table 2. Quality index values obtained with the proposed algorithm and other approaches

		S	SPC	ACC	DC
Algorithm	Group	*m ()*	*m ()*	*m ()*	*m ()*
Proposed FRBS	Training	0.938 (0.017)	0.968 (0.001)	0.962 (0.005)	0.916 (0.022)
	Test	0.914 (0.044)	0.969 (0.023)	0.957 (0.017)	0.910 (0.031)
Frangi's	Test	0.776 (0.054)	0.886 (0.040)	0.847 (0.028)	0.707 (0.031)
Sato's	Test	0.796 (0.054)	0.866 (0.045)	0.851 (0.029)	0.720 (0.029)

standard deviation (σ) for S, SPC, ACC, and DC. As can be seen, the mean values for the test and training images are similar for all the indices except sensitivity, which has a slightly lower mean value for the test images. Furthermore, the σ values for the test images are somewhat higher than those for the training images, but this is because of a greater data heterogeneity from the test set images. Therefore, despite the higher σ values, the results indicate a similar performance for the test and training images.

Figure 3 illustrates the performance of the algorithm. Column (a) is a magnified depiction of I^n_{NT} around $ROI(I^n_{NT})$, the overlays in column (b) correspond to the manual marking of vein pixels by an expert, I^n_{REFVT}, and the overlays in column (c) are the results of vein segmentation, I^n_{SVT}. The vast majority of pixels labeled as veins in

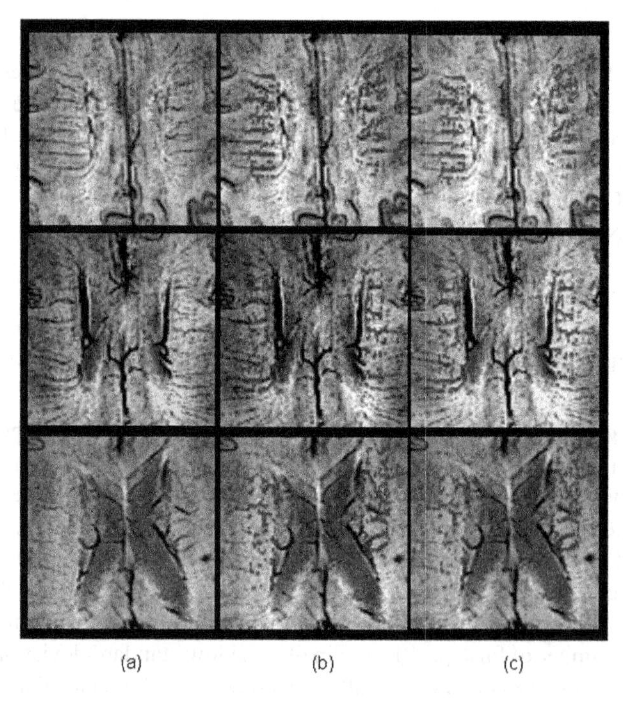

Fig. 3. Examples of the results obtained by processing three test slices corresponding to different anatomical locations. (a) Magnification of the original image area including ROIs. Columns (b) and (c) show overlays of the reference delineated and the segmentation results, respectively

the reference images were included in the segmentation. Only a very small number of false positives and false negatives occurred (see quality indices in Table 2). The false positives were mainly due to factors such as movement during image acquisition, small differences in the width of some vessels between the reference and segmented images, and visualization of other thin structures rich in iron. The false negatives were mainly due to partial volumes in some locations within the ROIs.

3.2 Dependence on the Window Properties

Location and size of the window selected for ROI delineation must be taken into account to avoid false positives. Structures surrounding periventricular white matter may have characteristics that make them difficult to differentiate from veins when they are partially included in ROIs and an automatic segmentation is applied.

The previous analysis was done using ROIs defined by 9.38 mm × 42.97 mm rectangular windows (W_1). As an initial approach to evaluate the influence of the size and position of the window, we tested a smaller (8.59 mm × 39.06 mm) window (W_2). Although the size of W_2 was not greatly different from W_1, the use of a smaller window helped to select positions with a lower possibility of presenting FP.

To evaluate the effect of using one or another window in the test images, we compared the quality indices obtained when ROIs were derived using W_1 or W_2 (Table 3). Means and standard deviations for S, SPC, ACC, and DC were very similar using the two windows, and p-values for the differences (bottom row) were not significant.

Table 3. Quality index values obtained for test images using windows: W_1 and W_2.

	S	SPC	ACC	DC
Window	*m ()*	*m ()*	*m ()*	*m ()*
W_1	0.914 (0.044)	0.961 (0.023)	0.957 (0.017)	0.910 (0.030)
W_2	0.916 (0.044)	0.970 (0.023)	0.958 (0.018)	0.913 (0.031)
p-value	0.827	0.839	0.822	0.643

3.3 Comparison with Other Methods

To assess the gain of the proposed approach, we compared it with two publicly available methods developed for problems similar to the one proposed here. These two approaches were based on implementation of vesselness enhancement filters (VEF), developed by Frangi [7] and Sato [8], and included in the VMTK module [22] of the 3Dslicer software [23]. First, the optimum values for the parameters of the approaches based on these filters were selected by evaluation of their segmentation performance on $ROI(I^n_{Nt})$. We then used these approaches to analyze $ROI(I^n_{NT})$.

The results obtained using the filters on $ROI(I^n_{NT})$ are shown in Table 2. Although the approaches based on the VEF yielded high values for the four quality indices, the proposed approach showed better performance: a significant improvement ($p < 0.001$, ANOVA test) of around 10 % to 15 % in S, SPC, and ACC, and around 26 % in DC.

3.4 Evaluation of the Ability to Detect Changes in the Visibility of Veins

To evaluate the applicability of the proposed Mamdani-type FRBS algorithm for clinical purposes, we considered an approach similar to that reported by Ge et al. [3]. The authors demonstrated a significant reduction in the visibility of veins in the periventricular white matter in RRMS patients compared to controls.

In this evaluation, we included images from 13 RRMS patients and 13 healthy controls. The proposed FRBS was applied to the ROIs defined in these images, and the segmented images were obtained. Following segmentation, the mean number of venous blood voxels within the ROIs was 108.60 ($\sigma = 14.70$) in RRMS patients and 126.94 ($\sigma = 18.70$) in healthy controls, and there exists a significant difference ($p = 0.01$, Student's *t*-test) between these two groups, as was reported by Ge et al. [3].

4 Conclusion

The algorithm proposed here is based on a fuzzy rule-based system that allows segmenting periventricular venous vasculature in SW MR images. The algorithm entails initial enhancement of cerebral veins in defined ROIs using an FRBS to determine the adequacy of each pixel to belong to a vein, and applies a cut-off value for final segmentation of these structures. The results obtained from healthy volunteers showed very good segmentation with a very small number of false positives. The method also proved to be robust and applicable to the study of periventricular veins in MS.

References

1. Haacke, E.M., Xu, Y.B., Cheng, Y.C.N., Reichenbach, J.R.: Susceptibility weighted imaging (SWI). MRM **52**, 1–40 (2004)
2. Mittal, S., Wu, Z., Neelavalli, J., Haacke, E.M.: Susceptibility-weighted imaging: technical aspects and clinical applications, Part 2. Am. J. Neuroradiol. **30**, 232–252 (2009)
3. Ge, Y., Zohrabian, V.M., Osa, E., et al.: Diminished visibility of cerebral venous vasculature in MS by susceptibility-weighted imaging at 3.0 Tesla. JMRI **29**, 1190–1194 (2009)
4. Suri, J.S., Liu, K.C., Reden, L., et al.: A review on MR vascular image processing: skeleton versus nonskeleton approaches: part II. IEEE Trans. IT Biomed. **6**, 338–350 (2002)
5. Lesage, D., Angelini, E.D., Bloch, I., Funka-Lea, G.: A review of 3D vessel lumen segmentation techniques: models, features and extraction schemes. MIA **13**, 819–845 (2009)
6. Haacke, E.M., Reichenbach, J.R.: Susceptibility Weighted Imaging in MRI: Basic Concepts and Clinical Applications. Willey, Hoboken (2011)
7. Frangi, A.F., Niessen, W.J., Hoogeveen, R.M., et al.: Model-based quantitation of 3-D magnetic resonance angiographic images. IEEE Trans. Med. Imaging **18**, 946–956 (1999)
8. Sato, Y., Nakajima, S., Shiraga, N., et al.: 3D multi-scale line filter for segmentation and visualization of curvilinear structures in medical images. MIA **2**, 143–168 (1998)
9. Lorenz, C., Carlsen, I.C., Buzug, T.M., et al.: A multi-scale line filter with automatic scale selection based on the Hessian matrix for medical image segmentation. In: ter Haar, R.B., Florack, L., Koenderink, J., et al. (eds.) Scale-Space Theory in Computer Vision, pp. 152–163 (1997)

10. Grabner, G., Dal-Bianco, A., Hametner, S., et al.: Group specific vein-atlasing: an application for analyzing the venous system under normal and MS conditions. JMRI **40**, 655–661 (2014)
11. Manniesing, R., Viergever, M.A., Niessen, W.J.: Vessel enhancing diffusion - a scale space representation of vessel structures. MIA **10**, 815–825 (2006)
12. Koopmans, P.J., Manniesing, R., Niessen, W.J., et al.: MR venography of the human brain using susceptibility weighted imaging at very high field strength. MAGMA **21**, 149–158 (2008)
13. Zivadinov, R., Poloni, G.U., Marr, K., et al.: Decreased brain venous vasculature visibility on susceptibility-weighted imaging venography in patients with multiple sclerosis is related to chronic cerebrospinal venous insufficiency. BMC Neurol. **11**, 128 (2011). http://www.biomedcentral.com/1471-2377/11/128
14. Nauck, D., Kruse, R.: Obtaining interpretable fuzzy classification rules from medical data. AIM **16**, 149–169 (1999)
15. Mehta, S.B., Chaudhury, S., Bhattacharyya, A., Jena, A.: Soft-computing based diagnostic tool for analyzing demyelination in magnetic resonance images. ASOC **10**, 529–538 (2010)
16. Zarandi, M.H.F., Zarinbal, M., Izadi, M.: Systematic image processing for diagnosing brain tumors: a type-II fuzzy expert system approach. ASOC **11**, 285–294 (2011)
17. Forkert, N.D., Schmidt-Richberg, A., Fiehler, J., et al.: Fuzzy-based vascular structure enhancement in Time-of-Flight MRA images for improved segmentation. Methods Inf. Med. **50**, 74–83 (2011)
18. Forkert, N.D., Schmidt-Richberg, A., Fiehler, J., et al.: 3D cerebrovascular segmentation combining fuzzy vessel enhancement and level-sets with anisotropic energy weights. MRI **31**, 262–271 (2013)
19. Shattuck, D.W., Sandor-Leahy, S.R., Schaper, K.A., et al.: Magnetic resonance image tissue classification using a partial volume model. Neuroimage **13**, 856–876 (2001)
20. Rorden, C., Brett, M.: Stereotaxic display of brain lesions. Behav. Neurol. **12**, 191–200 (2000)
21. Mamdani, E.H.: Application of fuzzy algorithms for control of simple dynamic plant. Proc. Inst. Electr. Eng. London **121**, 1585–1588 (1974)
22. Haehn, D.: The Vascular Modeling Toolkit in 3D Slicer (Modules: VMTKVessel Enhancement) (2010). http://www.slicer.org/slicerWiki/index.php/Modules:_VMTKVessel Enhancement. Accessed 27 March 2015
23. Pieper, S., Lorensen, B, Schroeder, W., Kikinis, R.: The NA-MIC kit: ITK, VTK, pipelines, grids and 3D slicer as an open platform for the medical image computing community. In: 3rd IEEE International Symposium on Biomedical Imaging: Macro to Nano, vol. 1–3, pp. 698–701 (2006)

On the Use of Lattice OWA Operators in Image Reduction and the Importance of the Orness Measure

Daniel Paternain[1,2(✉)], Gustavo Ochoa[3], Inmaculada Lizasoain[3,4], Edurne Barrenechea[1,2], Humberto Bustince[1,2], and Radko Mesiar[5]

[1] Departamento de Automatica Y Computacion, Universidad Publica de Navarra, Campus Arrosadia Sn, 31006 Pamplona, Spain
{daniel.paternain,edurne.barrenechea,bustince}@unavarra.es
[2] Institute of Smart Cities, Universidad Publica de Navarra, Campus Arrosadia Sn, 31006 Pamplona, Spain
[3] Departamento de Matemáticas, Universidad Publica de Navarra, Campus Arrosadia Sn, 31006 Pamplona, Spain
{ochoa,ilizasoain}@unavarra.es
[4] Institute for Advanced Materials, Universidad Publica de Navarra, Campus Arrosadia Sn, 31006 Pamplona, Spain
[5] Department of Mathematics and Descriptive Geometry, Faculty of Civil Engineering, Slovak University of Technology, Radlinskho 11, 813 68 Bratislava, Slovakia
mesiar@math.sk

Abstract. In this work we investigate the use of OWA operators in color image reduction. Since the RGB color scheme can be seen as a Cartesian product of lattices, we use the generalization of OWA operators to any complete lattice. However, the behavior of lattice OWA operators in image processing is not easy to predict. Therefore, we propose an orness measure that generalizes the orness measure given by Yager for usual OWA operators. With the aid of this new measure, we are able to classify each OWA operator and to analyze how its properties affect the results of applying OWA operators in an algorithm for reducing color images.

Keywords: Image reduction · Lattice OWA operators · Orness

1 Introduction

Aggregation functions, in general, and OWA operators, in particular, are very important tools in real-world applications, where the need to fuse or aggregate several inputs into a single representative output frequently arises. If we focus in image processing, aggregation functions are key tools in fields such as filtering and denoising, stereo vision, reconstruction or reduction [2,5,10,11]. In this work we focus on the latter.

When we deal with grayscale images, usual version of aggregation functions (defined on $[0,1]$) can be used. In particular, in [12,13] some examples can be

J.P. Carvalho et al. (Eds.): IPMU 2016, Part I, CCIS 610, pp. 624–634, 2016.
DOI: 10.1007/978-3-319-40596-4_52

seen, as well as in [11] where a specific family of OWA operators, called centered OWA operators, have been used. However, when we deal with color images (for example, those based on the RGB color scheme) aggregation functions are not trivially extended. One possible solution was given in [1], where aggregations on a cartesian product of lattices (product lattice) were used.

From another point of view, in [9] the generalization of OWA operators to any complete lattice was proposed. Since it is easy to prove that the cartesian product of lattices is, in fact, a lattice, it is clear that lattice OWA operators can be also used in color image processing.

In this work we study the use of lattice OWA operators defined in [9] in color image reduction, where images are defined on the RGB color scheme. Since it is possible to define many lattice OWA operators (just by taking many weighting vectors), our idea is to establish a classification of OWA operators using a generalization of the orness measure given by Yager [15].

Taking into account the proposed orness measure, we apply several OWA operators in color image reduction and we study whether the orness of each OWA operator determines its goodness in reduction algorithms.

The structure of the paper is as follows: in Sect. 2 we recall several preliminary definitions related to lattices and in Sect. 3 we recall the generalization of OWA operators to any complete lattice. In Sect. 4 we define our proposed orness measure and in Sect. 5 we study the effect of OWA operators in image reduction. We finish, in Sect. 6 with conclusions and future research.

2 Preliminaries

Throughout this paper $(L, \leq_L)$ will denote a complete lattice, i.e., a partially ordered set in which all subsets have both a supremum and an infimum. 0_L and 1_L will respectively stand for the least and the greatest elements of L. A lattice L is said to be complemented if for each $a \in L$ there exists some $b \in L$ such that $a \wedge b = 0_L$ and $a \vee b = 1_L$ [6]. A subset M of L is called a sublattice of $(L, \leq_L)$ if whenever a, $b \in M$, then both $a \wedge b$ and $a \vee b$ belong to M.

Definition 1 (see [3]). *A map $T : L \times L \to L$ ($S : L \times L \to L$) is said to be a t-norm [resp. t-conorm] on $(L, \leq_L)$ if it is commutative, associative, increasing in each component and has a neutral element 1_L [resp. 0_L].*

Remark 1. For any $n > 2$, $T(a_1, \ldots, a_n)$ will denote

$$T(\ldots(T(T(a_1, a_2), a_3), \ldots a_{n-1}), a_n).$$

Note that, for any permutation σ of the elements $1, \ldots, n$,

$$T(a_1, \ldots, a_n) = T(a_{\sigma(1)}, \ldots, a_{\sigma(n)}).$$

Throughout this paper $(L, \leq_L, T, S)$ will denote a complete lattice endowed with a t-norm T and a t-conorm S. As usual, L^n will denote the cartesian product $L \times \cdots \times L$ and L^{I_n} will stand for the set of all the n-ary lattice intervals $[a_1, \ldots, a_n]$ with $a_1 \leq_L \cdots \leq_L a_n$ contained in L.

Recall that an n-ary aggregation function (see [8]) is a function $M : L^n \to L$ such that:

1. $M(a_1, \ldots, a_n) \leq_L M(a'_1, \ldots, a'_n)$ whenever $a_i \leq_L a'_i$ for $1 \leq i \leq n$.
2. $M(0_L, \ldots, 0_L) = 0_L$ and $M(1_L, \ldots, 1_L) = 1_L$.

It is said to be *idempotent* if $M(a, \ldots, a) = a$ for every $a \in L$ and it is called *symmetric* if, for every permutation σ of the set $\{1, \ldots, n\}$, $M(a_1, \ldots, a_n) = M(a_{\sigma(1)}, \ldots, a_{\sigma(n)})$.

A wide family of both symmetric and idempotent aggregation functions was introduced by Yager in [14] on the lattice $L = [0, 1]$, the real unit interval:

Definition 2 (Yager [14]). *Let $\alpha = (\alpha_1, \cdots, \alpha_n) \in [0, 1]^n$ be a weighting vector with $\alpha_1 + \cdots + \alpha_n = 1$. An n-ary* ordered weighted average operator *or* OWA operator *is a map $F_\alpha : [0, 1]^n \to [0, 1]$ given by*

$$F_\alpha(a_1, \cdots, a_n) = \alpha_1 b_1 + \cdots + \alpha_n b_n,$$

where $(b_1, \ldots, b_n)$ is a rearrangement of $(a_1, \cdots, a_n)$ satisfying that $b_1 \geq \cdots \geq b_n$.

With the purpose of classifying these operators, Yager introduced in [15] (see also [4]) an orness measure for each OWA operator F_α, which depends only on the weighting vector $\alpha = (\alpha_1, \ldots, \alpha_n)$, in the following way:

$$orness(F_\alpha) = \frac{1}{n-1} \sum_{i=1}^{n} (n-i)\alpha_i = F_\alpha \left(1, \frac{n-2}{n-1}, \cdots, \frac{1}{n-1}, 0\right). \quad (1)$$

3 Lattice OWA Operators

In this section we recall the generalization of OWA operators to any complete lattice endowed with a t-norm T and a t-conorm S [9]. In order to generalize these operators, we demand certain conditions about the weighting vector.

Definition 3 ([9]). *Consider any complete lattice $(L, \leq_L, T, S)$. A lattice vector $\alpha = (\alpha_1, \ldots, \alpha_n) \in L^n$ is said to be a* weighting vector in $(L, \leq_L, T, S)$ *if $S(\alpha_1, \ldots, \alpha_n) = 1_L$ and it is called a* distributive weighting vector in $(L, \leq_L, T, S)$ *if it also satisfies that for any $a \in L$,*

$$S\left(T(a, \alpha_1), \ldots, T(a, \alpha_n)\right) = T\left(a, S(\alpha_1, \ldots, \alpha_n)\right)$$

and consequently equal to a.

Evidently, OWA operators need an arrangement of inputs. Therefore, the main difficulty arises in converting a vector $(a_1, \ldots, a_n) \in L^n$ into another vector $(b_1, \ldots, b_n) \in L^n$ such that $b_1 \geq_L \cdots \geq_L b_n$.

Definition 4 ([9]). *Let* $(L, \leq_L, T, S)$ *be a complete lattice. For any vector* $(a_1, \ldots, a_n) \in L^n$*, an n-dimensional lattice interval* $[b_n, \ldots, b_1]$ *is defined by*

- $b_1 = a_1 \vee \cdots \vee a_n \in L.$
- $b_2 = [(a_1 \wedge a_2) \vee \cdots \vee (a_1 \wedge a_n)] \vee [(a_2 \wedge a_3) \vee \cdots \vee (a_2 \wedge a_n)] \vee \cdots \vee [a_{n-1} \wedge a_n] \in L.$

$\vdots$

- $b_k = \bigvee \{a_{j_1} \wedge \cdots \wedge a_{j_k} \mid j_1 < \cdots < j_k \in \{1, \ldots, n\}\} \in L.$

$\vdots$

- $b_n = a_1 \wedge \cdots \wedge a_n \in L.$

Now, we recall the definition of lattice OWA operator.

Definition 5 ([9]). *Let* $(L, \leq_L, T, S)$ *be a complete lattice. For each distributive weighting vector* $\alpha = (\alpha_1, \ldots, \alpha_n) \in L^n$*, the function* $F_\alpha : L^n \to L$ *given by*

$$F_\alpha(a_1, \ldots, a_n) = S\left(T(\alpha_1, b_1), \cdots, T(\alpha_n, b_n)\right) \quad (a_1, \ldots, a_n) \in L^n$$

*is called an n-*ary OWA operator*, where* $[b_n, \ldots, b_1]$ *is the n-dimensional lattice interval obtained from* $(a_1, \ldots, a_n)$ *using Definition 4.*

4 Orness Measure for Lattice OWA Operators

In this section we study a methodology to analyze the behavior of an OWA operator by means of an orness measure. Concretely, we focus on a wide class of complete lattices that includes all the finite lattices:

(MFC) *For any* $a, b \in L$ *with* $a \leq_L b$*, there exists some maximal chain with a finite length, named* l*, between* a *and* b*,*

$$a = a_0 <_L a_1 <_L \cdots <_L a_l = b.$$

The maximality means that, for any $0 \leq i \leq l - 1$ there is no $c \in L$ with $a_i <_L c <_L a_{i+1}$.

Definition 6. *Let* $(L, \leq_L, T, S)$ *be a complete lattice satisfying condition (MFC). For any* $a, b \in L$*, the distance from* a *to* b *is the length of the shortest maximal chain between* a *and* b*, denoted by* $d(a, b)$*.*

In [15], Yager also defined, for any weighting vector on $[0, 1]$, the concept of quantifier. Given a weighting vector $\alpha = (\alpha_1, \ldots, \alpha_n) \in [0, 1]^n$, the quantifier $Q_\alpha : \{0, 1, \ldots, n\} \to [0, 1]$ is given by

$$Q_\alpha(k) = \begin{cases} 0 & \text{if } k = 0 \\ \alpha_1 + \cdots + \alpha_k & \text{otherwise} \end{cases} \quad (2)$$

The same concept can be extended to a distributive weighting vector on an arbitrary lattice L.

Definition 7. *Let $(L, \leq_L, T, S)$ be a complete lattice satisfying condition (MFC). For any distributive weighting vector, $\alpha = (\alpha_1, \ldots, \alpha_n) \in L^n$, define the qualitative quantifier $Q_\alpha : \{0, 1, \ldots, n\} \to L$ by means of*

$$Q_\alpha(0) = 0_L,$$
$$Q_\alpha(k) = S(\alpha_1, \ldots, \alpha_k) \text{ for } k = 1, \ldots, n$$

Now, we are ready to give the orness measure of any lattice OWA operator.

Definition 8. *Let $(L, \leq_L, T, S)$ be a complete lattice satisfying condition (MFC). For any distributive weighting vector in $(L, \leq_L, T, S)$, $\alpha = (\alpha_1, \ldots, \alpha_n) \in L^n$, consider the qualitative quantifier $Q_\alpha : \{0, 1, \ldots, n\} \to L$ defined in Definition 7. For each $k = 1, \ldots, n$, call $m(k) = d\,(Q_\alpha(k-1), Q_\alpha(k))$. If $m = m(1) + \cdots + m(n)$, then define*

$$orness(F_\alpha) = \frac{1}{n-1} \sum_{i=1}^{n} (n-i) \frac{m(i)}{m}. \quad (3)$$

Some interesting properties of the proposed orness measure (see [7] for an axiomatic definition of orness) are proven in the following results.

Theorem 1. *Let $(L, \leq_L, T, S)$ be a complete lattice satisfying condition (MFC). If $\alpha = (\alpha_1, \ldots, \alpha_n) \in L^n$ and $\beta = (\beta_1, \ldots, \beta_n) \in L^n$ are distributive weighting vectors in $(L, \leq_L, T, S)$ with $F_\alpha = F_\beta$, then $orness(F_\alpha) = orness(F_\beta)$.*

Proposition 1. *Let $(L, \leq_L, S, T)$ be a complete lattice satisfying condition (MFC) and $F_\alpha : L^n \to L$ an arbitrary OWA operator.*

1. *If $\alpha = (1_L, 0_L, \ldots, 0_L) \in L^n$, then $orness(F_\alpha) = 1$.*
2. *$orness(F_\alpha) = 0$ if and only if F_α is the AND-operator.*
3. *If $\alpha = (0_L, \ldots 0_L, \overset{(k)}{1_L}, 0_L, \ldots, 0_L) \in L^n$ with $1 < k < n$, then $orness(F_\alpha) = \frac{n-k}{n-1}$, as in the case of standard orness.*

5 Image Reduction Using Lattice OWA Operators

In this work we consider images in the RGB color scheme. Given an image of M rows and N columns, if we define $L = \{0, 1, \ldots, 255\}$, then the color image can be seen as a mapping $\{1, \ldots, M\} \times \{1, \ldots, N\} \to L \times L \times L$. Since L is a chain, the product $L \times L \times L$ is, in fact a lattice [1].

Image reduction consists in diminishing or compacting the spatial resolution of an image preserving its original properties to some extent. That is, an image reduction algorithm receives an image and returns a new image representing the same scene but with a lower number of pixels.

Suppose that, starting from an image of $M \times N$ pixels, we want to obtain a new image of $M' \times N'$ with $M' < M$ and $N' < N$. One possibility consists in finding numbers $p, s \in \mathcal{N}$ such that $M = M'p$ and $N = N's$ (if M' or N' are not multiple of M or N, respectively, we take the largest $p, s \in \mathcal{N}$ such that $M < M'p$ and $N < N's$), so that we can establish a mapping between a set of adjacent $p \times s$ pixels of the original image and one pixel of the reduced image. Following this procedure, the objective of image reduction consists in aggregating (or fusing) $p \times q$ pixels into a single representative pixel. Since we know that each pixel of an RGB color image is an element of the lattice $L \times L \times L$, we can reduce image by defining OWA operators

$$F_\alpha : (L \times L \times L)^{p \times s} \to L \times L \times L.$$

Next, we show a reduction algorithm based on lattice OWA operators:

Reduction algorithm
Require: Image Q of $M \times N$ pixels; weighting vector α.
Ensure: Image Q' of $M' \times N'$ pixels with $M' < M$ and $N' < N$.
1: Find numbers $p, s \in \mathcal{N}$ such that $M = M'p$ and $N = N's$
2: Divide the image Q into non-overlapping blocks of $p \times s$ pixels
3: **for** each block in Q **do**
4: Take the $p \times s$ pixels of the block, namely $q_1, \ldots, q_{p \times s} \in L \times L \times L$
5: Apply OWA operator

$$F_\alpha(q_1, \ldots, q_{p \times s})$$

6: Take the result of F_α as a pixel of the reduced image Q'
7: **end for**

Observe that in the proposed algorithm, the same OWA operator is applied to every block within the image. Although it would be possible to apply different OWA operators, we are interested in analyzing the behavior of each operator in the whole image. Therefore, if we take t different weighting vectors, we will obtain t different reduced images by applying t times the reduction algorithm, each time with a different weighting vector.

Consider the following set of weighting vectors:

$$\begin{aligned}
\alpha_1 = &((255,192,128),(192,255,192),(128,192,255),\\
&(64,128,192),(0,64,128),(0,0,64),\\
&(0,0,0),(0,0,0),(0,0,0))\\
\alpha_2 = &((192,128,64),(255,192,128),(192,255,192),\\
&(128,192,255),(64,128,192),(0,64,128),\\
&(0,0,64),(0,0,0),(0,0,0))\\
\alpha_3 = &((128,64,0),(192,128,64),(255,192,128),\\
&(192,255,192),(128,192,255),(64,128,192),\\
&(0,64,128),(0,0,64),(0,0,0))\\
\alpha_4 = &((64,0,0),(128,64,0),(192,128,64),\\
&(255,192,128),(192,255,192),(128,192,255),\\
&(64,128,192),(0,64,128),(0,0,64))\\
\alpha_5 = &((0,0,0),(64,0,0),(128,64,0),\\
&(192,128,64),(255,192,128),(192,255,192),\\
&(128,192,255),(64,128,192),(0,64,128))\\
\alpha_6 = &((0,0,0),(0,0,0),(64,0,0),\\
&(128,64,0),(192,128,64),(255,192,128),\\
&(192,255,192),(128,192,255),(64,128,192))\\
\alpha_7 = &((0,0,0),(0,0,0),(0,0,0),\\
&(64,0,0),(128,64,0),(192,128,64),\\
&(255,192,128),(192,255,192),(128,192,255))\\
\alpha_8 = &((0,0,0),(0,0,0),(0,0,0),\\
&(0,0,0),(64,0,0),(128,64,0),\\
&(192,128,64),(255,192,128),(192,255,255))
\end{aligned}$$

It is difficult to predict the effect of each weighting vector just by analyzing the weights. However, we can use our proposed orness measure in order to classify each weighting vector. Recall that, as it has been explained in [11], better reduced images are obtained when the operator is far from pure OR and AND operators. That is, better results will be obtained by those weighting vectors with central orness value. The orness of each weighting vector can be seen in Fig. 1.

We have applied our reduction algorithm to the original image in Fig. 1 of 510×510 pixels, obtaining 8 reduced images of 170×170 pixels (Fig. 2). Observe that, as long as the orness of the weighting vector decreases, the brightness of the color intensities of the images also decreases.

In order to analyze the quality of each image and following the ideas in [1,11], we have reconstructed the images into its original size and we have compared them with the original image. In this sense, the closer the reconstructed with the original image, the better the OWA operator used to reduce the image. We have

Table 1. Orness measure of OWA operators constructed by weighting vectors α_1 to α_8.

W. vector	Orness
α_1	0.9735
α_2	0.9202
α_3	0.8400
α_4	0.7329
α_5	0.5632
α_6	0.4382
α_7	0.3132
α_8	0.1985

Fig. 1. Original Mandrill image (510 × 510 pixels)

used the well-known method of bilinear interpolation (implemented in Matlab using *imresize*) in order to enlarge the image and the measure of MSE, PSNR and SSIM [16] in order to compare the images. Recall that lower values of MSE indicate that the compared images are very close. On the contrary, lower values of SSIM and PSNR indicate that the images are very different. Table 2 shows the quality measure for each OWA operator considered. Observe that the best results in terms of MSE, PSNR and SSIM are obtained by the same weighting vector (α_5) and similar results are obtained with α_6. However, when the orness of the OWA operator either increases or decreases, the quality of the reduced images always decreases. In this sense, we can conclude that the weighting vectors associated to central orness measure are better for image reduction than others.

Another interesting effect of OWA operators in image reduction is the fact that we can use them in order to highlight some specific zones of the image attending at the color properties. For example, if we want to highlight bright

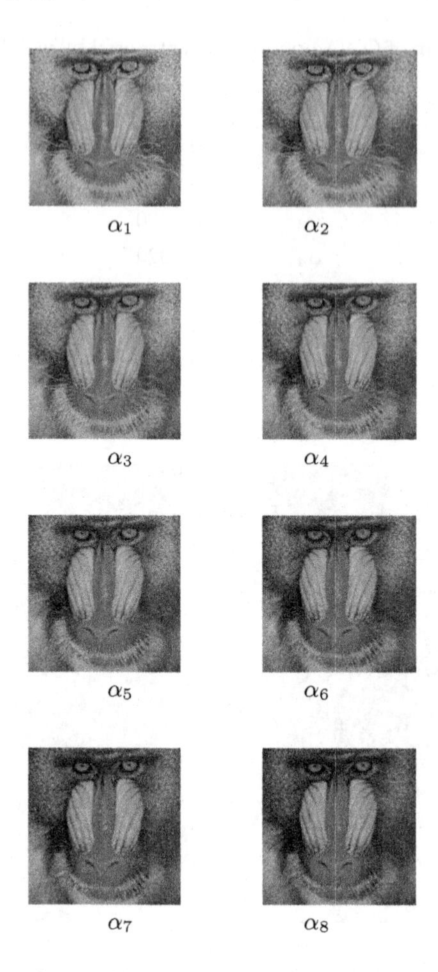

Fig. 2. Reduced images obtained by applying reduction algorithm taking $\alpha_1, \ldots, \alpha_8$ as weighting vectors.

Table 2. Quality measures (MSE, PSNR and SSIM) of enlarged images obtained from Fig. 2.

W. vector	Orness	MSE	PSNR	SSIM
α_1	0.9735	1442	16.54	0.6562
α_2	0.9292	1156	17.50	0.6659
α_3	0.8400	842.0	18.88	0.6792
α_4	0.7329	626.2	20.16	0.6908
α_5	0.5632	528.9	20.90	0.6942
α_6	0.4382	533.8	20.86	0.6890
α_7	0.3132	633.9	20.11	0.6756
α_8	0.1985	828.1	18.95	0.6550

(dark) zones. In this case, we are not interested in obtaining a reduced image which is similar to the original and, therefore, we do not consider weighting vectors with central orness measures. If we want to highlight bright (dark) zones, it is better to consider OWA operators with high (low) orness measure. This effect can be seen in Fig. 3, where two different OWA operators have been used (those corresponding to weighting vectors α_1 and α_8). Observe that in the image obtained using α_1, the bright areas have been highlighted and enlarged, while the opposite effect is obtained by means of α_8.

Fig. 3. Original image (row 1) and two reduced images (row 2) using α_1 (left) and α_8 (right).

6 Conclusions

In this paper we have investigated the use of lattice OWA operators in an application to image reduction. In order to analyze the behavior of each operator in the reduced image, we have proposed a new orness measure that determines the proximity of each lattice OWA operator with respect to the pure OR operator.

After applying several OWA operators to our reduction algorithm, we have seen that when we are interested in keeping the intensity properties of the original image we have to use lattice OWA operators with an orness measure close to 0.5. However, when we want to highlight certain areas attending at the intensities of that area, we have to move to specific lattice OWA operators. In these cases, the proposed orness measure allows to choose the appropriate operator.

Acknowledgments. D. Paternain and H. Bustince have been partially supported by Spanish project TIN2013-40765-P. R. Mesiar has been partially supported by grant APVV-14-0013.

References

1. Beliakov, G., Bustince, H., Paternain, D.: Image reduction using means on discrete product lattice. IEEE Trans. Image Process. **21**, 1070–1083 (2012)
2. Beliakov, G., Bustince, H., Calvo, T.: A Practical Guide to Averaging Functions. Springer, Heidelberg (2016)
3. De Baets, B., Mesiar, R.: Triangular norms on product lattices. Fuzzy Sets Syst. **104**, 61–75 (1999)
4. Dujmović, J.J.: A generalization of some functions in continuous mathematical logic - evaluation functions and its applications (In Serbo-Croatian). In: Proceedings of the Informatica Conference, Bled, paper d27, Yugoslavia (1973)
5. Galar, M., Fernandez, J., Beliakov, G., Busince, R.: Interval-valued fuzzy sets applied to stereo matching of color images. IEEE Trans. Image Process. **20**, 1949–1961 (2011)
6. Grätzer, G.: General Lattice Theory. Birkhäuser Verlag, Basel (1978)
7. Kishor, A., Singh, A.K., Pal, N.R.: Orness measure of OWA operators: a new approach. IEEE Trans. Fuzzy Syst. **22**, 1039–1045 (2014)
8. Komorníková, M., Mesiar, R.: Aggregation functions on bounded partially ordered sets and their classification. Fuzzy Sets Syst. **175**, 48–56 (2011)
9. Lizasoain, I., Moreno, C.: OWA operators defined on complete lattices. Fuzzy Sets Syst. **224**, 36–52 (2013)
10. González-Hidalgo, M., Massanet, S.: A fuzzy mathematical morphology based on discrete t-norms: fundamentals and applications to image processing. Soft Comput. **18**, 2297–2311 (2014)
11. Paternain, D., Fernandez, J., Bustince, H., Mesiar, R., Beliakov, G.: Construction of image reduction operators using averaging aggregation functions. Fuzzy Sets Syst. **261**, 87–111 (2015)
12. Perfilieva, I.: Fuzzy transforms: theory and applications. Fuzzy Sets Syst. **157**, 993–1023 (2006)
13. Perfilieva, I., Kreinovich, V.: F-transform in view of aggregation functions. In: Bustince, H., Fernandez, J., Mesiar, R., Calvo, T. (eds.) Aggregation Functions in Theory and in Practise. AISC, vol. 228, pp. 393–400. Springer, Heidelberg (2013)
14. Yager, R.R.: On ordered weighting averaging aggregation operators in multicriteria decision-making. IEEE Trans. Syst. Man Cybern. **18**, 183–190 (1988)
15. Yager, R.R.: Families of OWA operators. Fuzzy Sets Syst. **59**, 125–148 (1993)
16. Zhou, W., Boviz, A.C., Sheikh, H.R., Simoncelli, E.P.: Image quality assessment: from error visibility to structural similarity. IEEE Trans. Image Process. **13**, 600–612 (2004)

A Methodology for Hierarchical Image Segmentation Evaluation

J. Tinguaro Rodríguez[1(✉)], Carely Guada[1], Daniel Gómez[2], Javier Yáñez[1], and Javier Montero[1]

[1] Faculty of Mathematics, Complutense University of Madrid, 28040 Madrid, Spain
{jtrodrig, cguada, jayage, javier_montero}@ucm.es
[2] Faculty of Statistics, Complutense University of Madrid, 28040 Madrid, Spain
dagomez@estad.ucm.es

Abstract. This paper proposes a method to evaluate hierarchical image segmentation procedures, in order to enable comparisons between different hierarchical algorithms and of these with other (non-hierarchical) segmentation techniques (as well as with edge detectors) to be made. The proposed method builds up on the edge-based segmentation evaluation approach by considering a set of reference human segmentations as a sample drawn from the population of different levels of detail that may be used in segmenting an image. Our main point is that, since a hierarchical sequence of segmentations approximates such population, those segmentations in the sequence that best capture each human segmentation level of detail should provide the basis for the evaluation of the hierarchical sequence as a whole. A small computational experiment is carried out to show the feasibility of our approach.

Keywords: Image segmentation · Hierarchical network clustering · Edge-based image segmentation evaluation

1 Introduction

Image segmentation, understood as the identification of connected and homogenous regions of an image, is an essential tool in today applications of image processing. There is a wide variety of approaches to segmentation as well as of techniques to perform it [2], in the same way that also different approaches exist regarding how to evaluate the performance of segmentation procedures (see for instance [3, 6, 9, 10, 12, 13]).

A technique based on segmentation is that of hierarchical image segmentation [1], whose aim is to produce a consistent sequence of segmentations identifying the objects in an image with different levels of detail. Hierarchical segmentation is a relevant extension of segmentation since various applications (see for instance [5]) require different detail levels to be simultaneously available, in such a way that object identification is consistent through the different detail levels. However, contrarily to non-hierarchical segmentation, there are relatively few hierarchical segmentation procedures, and these usually find the problem that there are not clearly accepted approaches on how to evaluate a hierarchical segmentation algorithm.

J.P. Carvalho et al. (Eds.): IPMU 2016, Part I, CCIS 610, pp. 635–647, 2016.
DOI: 10.1007/978-3-319-40596-4_53

The aim of this paper is to propose a method to enable evaluation of hierarchical segmentation procedures by an edge-based segmentation evaluation approach (see [3, 10]), possibly today's most widely accepted and extended segmentation evaluation methodology. The main argument behind the proposed method is that a hierarchical sequence of segmentations captures or approximates the different possible levels of detail humans may use or that may be needed in an application. Thus, given a set of reference human segmentations, which constitute a sample of such different possible detail levels, the hierarchical sequence should be evaluated as a whole through the segmentations in the sequence that best approximate each reference's level of detail.

A detailed description of the proposed method is given in Sect. 3, and a small computational experiment is carried out in Sect. 4 to illustrate its feasibility. This experiment evaluates the Divide-and-Link [7, 8], so the basics of this hierarchical segmentation algorithm are described in Sect. 2.3 upon the basis of formal definitions of the segmentation (Sect. 2.1) and hierarchical segmentation (Sect. 2.2) problems. Finally, some conclusions are shed in Sect. 5.

2 Preliminaries

In this section, we remind the standard definitions of image segmentation and hierarchical image segmentation, providing the formal basis from which our proposal will be developed in next section. We also describe here an algorithm that performs hierarchical image segmentation, and that will be used in Sect. 4 to illustrate the proposed evaluation methodology.

Thus, let us start by recalling that a digital image I is often regarded as a graph whose nodes are the pixels, and whose edges represent the neighborhood relationships of such pixels in the image. Formally, in this approach an $(r \times s)$-digital image is decomposed in its set of pixels $V = \{p_{ij} = (i,j) | 1 \leq i \leq r, 1 \leq j \leq s\}$ and a set of edges $E = \{e = \{p_{ij}, p_{i'j'}\} | p_{ij}, p_{i'j'} \in V$ are neighbor pixels$\}$, in such a way that two pixels $p, p' \in V$ are considered to be adjacent to each other in the image if and only if $e = \{p, p'\} \in E$. Therefore, $G = (V, E)$ is a graph constituted by the image's pixels as well as by their neighbor relationships. The graph G can be assumed to be connected.

Neighboring relationships between pixels are usually introduced through proximity topologies specifying who the neighbors of any given pixel are. Possibly, the simplest such topology is that specifying that the neighbors of a given a pixel p_{ij} are the four horizontally and vertically adjacent pixels $p_{i-1,j}$, $p_{i+1,j}$, $p_{i,j-1}$ and $p_{i,j+1}$ (see Fig. 1). Different topologies are possible specifying both a greater number (8, 12, etc.) of neighbors and more complex patterns of proximity relationships. For simplicity, however, in this paper we assume the just introduced 4-connectivity.

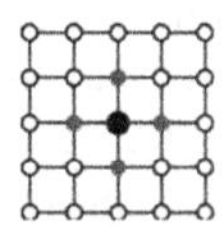

Fig. 1. Proximity topology associating 4 neighbors (in red) to a given pixel (black dot). (Color figure online)

In a (network-based) segmentation context, the previous graph $G = (V, E)$ is usually complemented by a dissimilarity measure d that informs on the spectral differences between (unordered) pairs of neighbor pixels. Notice that such dissimilarity measure d is defined over the edges in E. That is, given an edge $e = \{p, p'\} \in E$, the value $d_e \geq 0$ quantifies how different the pixels $p, p' \in V$ are in terms of its spectral information, in such a way that the greater d_e is, the more dissimilar p and p' are. Different alternatives are possible in order to construct such dissimilarity measure from the spectral information of the pixels, depending on the nature of such spectral information (white intensities, RGB, multispectral data, etc.) as well as on the specific problem being faced and the context it belongs to. Once a measure d verifying desirable properties is chosen, we denote by $D = \{d_e | e \in E\}$ the set of all dissimilarities between adjacent nodes. The available information of a digital image I is therefore summarized by the image network $N(I) = \{G, D\}$.

2.1 Image Segmentation

Image segmentation is commonly understood as an image processing technique dealing with the task of identifying the different objects that appear in a digital image. Typically, objects in an image occupy regions formed by adjacent and connected pixels. This idea explains why a graph-based approach to digital images is useful for segmentation. Moreover, it provides a basis for formalizing the notion of image segmentation itself. Here we recall two equivalent definitions of image segmentation: a first, more natural one, based on the elements to be separated (the nodes of the graph); the second one is instead based on the edges which link those elements.

Definition 1. Given an image network $N(I) = \{G = (V, E), D\}$, a family $S = \{R_1, \ldots, R_t\}$ with $R_j \subset V$ for all $j \in \{1, \ldots, t\}$ constitutes a segmentation of image I if and only if the following conditions holds:

(a) Non overlapping regions, i.e., for all $i \neq j$, $R_i \cap R_j = \emptyset$.
(b) Covering: $\bigcup_{j=1}^{t} R_j = V$.
(c) Connectivity of all regions: for all $j \in \{1, \ldots, t\}$, the subgraph $(R_j, E_{|Rj})$ is a connected graph.

In the previous definition, different objects in the image are associated to the different regions or connected sets of pixels R_j in S. However, in a network context it is possible to characterize image segmentation through an alternative but equivalent approach, instead based on the edge set $B \subset E$ that identifies the boundaries between the different regions composing the segmentation of the image network $N(I)$. Particularly, the segmentation can be univocally characterized through the minimal set of edges which separate the regions of the segmentation:

Definition 2. Given an image network $N(I) = \{G = (V, E), D\}$, a subset $B \subset E$ characterizes a segmentation of I if and only if the number of connected components of the partial graph $G(E-B) = (V, E-B)$ decreases when any edge of B is deleted.

In this way, given a boundary edges set B verifying Definition 2, the family $S = \{R_1, \ldots, R_t\}$ of connected components of the partial graph $G(E-B)$ constitutes a segmentation in the sense of Definition 1. The reciprocal is also true, that is, given a family $S = \{R_1, \ldots, R_t\}$ of connected pixels regions verifying the conditions of Definition 1, the set B of edges whose endpoint pixels lie in different regions $R_i \neq R_j$ of S is a boundary edges set in the sense of Definition 2. More generally, it is possible to state the following result establishing the equivalence of the two previous definitions of image segmentation (a proof can be found for instance in [7]).

Theorem 1. Given an image network $N(I) = \{G = (V, E), D\}$, let $\mathcal{S}^n(N(I))$ be the set of all *node* segmentations (in the sense of Definition 1) and $\mathcal{S}^e(N(I))$ be the set of all *edge* segmentations (in the sense of Definition 2). Then, there exists a natural bijection $\phi : \mathcal{S}^n(N(I)) \rightarrow \mathcal{S}^e(N(I))$ that assigns to each node segmentation S an edge segmentation B given by the boundaries of the regions in S.

Particularly, if a subset of edges $B \subset E$ is an edge segmentation of the image network $N(I)$, then any of its edges links two different regions of the corresponding node segmentation. The following proposition provides necessary and sufficient conditions for this property to hold for a set of edges B. Notice that this lemma (a proof can be found in [7]) introduces the structure of *forest* (a partial graph without cycles) in an image network, which will be essential in the algorithm for hierarchical image segmentation described in Sect. 2.3.

Proposition 1. Given an image network $N(I) = \{G = (V, E), D\}$, a subset of edges $B \subset E$ is an (edge) segmentation if and only if there exist a spanning forest $G(F) = (V, F)$ and a subset $F' \subset F$ verifying the following conditions:

(a) $F' \subset B$;
(b) $F - F' \not\subset B$;
(c) For all $e = \{p, q\} \in E - F, e \in B$ if and only if p and q belong to different connected components of the partial graph $G(F - F')$.

2.2 Hierarchical Image Segmentation

The notion of hierarchical image segmentation addresses the problem of providing different levels of segmentation detail of the objects in an image. For instance, having different levels of segmentation detail may be useful for data mining or knowledge discovery tasks [4]. Basically, a hierarchical set of image segmentations is a *consistent* sequence of segmentations of an image at different levels of detail. By consistent we mean that the segmentations at coarser levels of detail can be obtained by merging regions from segmentations at finer levels of detail.

The previous definitions of image segmentation can be easily extended to the hierarchical case by just recalling that if $S, S' \mathcal{S}^n(N(I))$ are two segmentations, then S is *finer* than S' if for all $R \in S$ there exist a region $R' \in S'$ such that $R \subset R'$.

Definition 3. Given an image network $N(I) = \{G = (V, E), D\}$, a family $\mathcal{S} = \{S^1, \ldots, S^k\}$ of node segmentations of $N(I)$ is a hierarchical segmentation of $N(I)$ when the following holds:

(a) $S^i \in \mathcal{S}^n(N(I))$ for all $i \in \{1, \ldots, k\}$ (i.e. each S^i is a node-based segmentation of $N(I)$);
(b) S^{i+1} is finer than S^i.

A similar definition can be established in terms of edge-based segmentations, just taking into account that given two edge segmentations $B, B' \in \mathcal{S}^e(N(I))$, B is finer than B' if $B' \subset B$.

Definition 4. Given an image network $N(I) = \{G = (V, E), D\}$, a family $\mathfrak{B} = \{\mathfrak{B}, \ldots, \mathfrak{B}^{\mathfrak{k}}\}$ of edge segmentations of $N(I)$ is a hierarchical segmentation of $N(I)$ when the following holds:

(a) $B^i \in \mathcal{S}^e(N(I))$ for all $i \in \{1, \ldots, k\}$ (i.e. each S^i is a edge-based segmentation of $N(I)$);
(b) B^{i+1} is finer than B^i.

2.3 Divide and Link: An Algorithm for Hierarchical Image Segmentation

In this section, we briefly recall the basics of the Divide and Link (D&L) algorithm, introduced in [8] in the context of social network analysis, and adapted to perform hierarchical image segmentation in [7]. This algorithm will be used later in Sect. 4 to illustrate the proposed evaluation methodology.

Given an image network $N(I) = \{G = (V, E), D\}$, the D&L algorithm is based on a divisive process that begins with the trivial segmentation formed by the whole image V, and ends after a predefined number of divisive stages K are performed. In each stage h of the algorithm $(1 \leq h \leq K)$ a segmentation S^h of the image is obtained, which is finer than that of the previous stage. These successive divisive stages are ruled by the dissimilarity measures d_e of the edges $e \in E$, as well as by a set of K predefined thresholds $\alpha_1 > \alpha_2 > \ldots > \alpha_K$, in such a way two adjacent pixels p and q belonging to the same region at a given stage h will be separated into different regions in the next stage $h + 1$ of the hierarchical process if and only if $d_{\{p,q\}} \geq \alpha_{h+1}$. To this aim, the algorithm selects in each stage h a set of divisive edges (those with $d_e \geq \alpha_h$) and a set of linking edges (with $d_e < \alpha_h$), and arranges them in a single ordered list used to build a spanning forest. This structure allows avoiding potential inconsistencies (as it has no cycles), in such a way that, after deleting the divisive edges, the regions of this stage's segmentation are obtained as the connected components of the forest. The pseudocode of the algorithm can be found in [7]. Figure 2 illustrates the different segmentations obtained for the well-known *Church* digital color image through a divisive process with $K = 5$ stages. The dissimilarity measure d was computed as the L_1 spectral distance between adjacent pixels (assuming a 4-connectivity topology as in Fig. 1).

Fig. 2. Original *Church* image and the segmentations obtained in 5 steps of the D&L algorithm through the corresponding thresholds α_h.

3 Edge-Based Evaluation for Hierarchical Segmentation

In this section we introduce a methodology for evaluating hierarchical segmentation techniques as the just exposed Divide-and-Link algorithm. This proposal belongs to the category of supervised segmentation evaluation methods, and more particularly to the subcategory of edge-based segmentation evaluation (see [3, 10, 13]). As such, it departs from a reference image dataset, depicting objects that are delimited by borders or boundaries specified by humans. These human-segmented images provide the ground truth against which machine-segmentations are compared.

Notice that, even although the methods to be compared produce segmentations (i.e. pixel regions occupied by an object) and not boundary maps (i.e. just the borders of such objects), the edge-based approach exploits the close relationships between the segmentation and edge-detection problems to allow evaluating a segmentation in terms of the boundaries between segments or regions (region-based evaluation [9] is also possible, but it is way less extended as only a few small reference datasets are available). At this respect, borders between two regions of a segmented image are usually assigned to the boundary pixels of the larger segment, but any other approach that generates a boundary map from a segmented image will be as valid in terms of the method proposed here.

A first problem that arises when trying to develop a supervised evaluation approach in the context of hierarchical segmentation is the lack of hierarchically human-segmented reference images. That is, reference datasets provide a single ground truth boundary map for each human that segments an image. And although different humans may segment the image with a different level of detail, normally these different human segmentations of a given reference image does not form a consistent hierarchy (in the sense above discussed of providing a sequence of finer segmentations). Thus, it is not possible to apply the edge-based reference datasets directly to evaluate any hierarchical segmentation technique.

However, although a priori unknown, the different detail levels of different human segmentations of an image may be soundly captured or approximated through the sequence of machine segmentations of a hierarchical procedure. That is, given a human segmentation and a hierarchical sequence of machine segmentations, there always exists an element of the sequence that best approximates the detail level of the human one. Thus, the whole hierarchical sequence can be taken as an approximation to the different levels of detail that a collection of different-sensitivity humans performing segmentation may use. And our point is that, in order to evaluate the performance of a hierarchical segmentation procedure, we should focus precisely on those segmentations of the sequence that best approximate each human reference's level of detail.

Thus, let us formalize these ideas by specifying the process to evaluate a hierarchical segmentation algorithm on a reference image. Suppose that for this image there are available L independent human segmentations GT^l $(1 \leq l \leq L)$ providing ground truth references for the evaluation. Suppose also that the hierarchical algorithm provides a set of K machine segmentations S^k $(1 \leq k \leq K)$ structured hierarchically, all with the same size as the reference image. Then, as discussed above, given a human segmentation GT^l, it is successively matched against each machine segmentation $S^k (k = 1, \ldots, K)$, providing the usual counts of true positives, false positives and false negatives. These are respectively denoted by TP_{kl}, FP_{kl} and FN_{kl}, and are typically used to compute precision and recall scores as follows:

$$P_{kl} = \frac{TP_{kl}}{TP_{kl} + FP_{kl}}, \quad R_{kl} = \frac{TP_{kl}}{TP_{kl} + FN_{kl}}$$

These scores are in turn used to compute an F-measure F_{kl} for each combination of human-machine segmentations. The F-measure is widely used as a criterion to evaluate the performance of diverse techniques, particularly segmentation algorithms, and is obtained as the harmonic mean of precision and recall:

$$F_{kl} = 2 \cdot \frac{P_{kl} \cdot R_{kl}}{P_{kl} + R_{kl}}$$

In this way, it is possible to report a single measure evaluating how good each machine boundary map of the hierarchical sequence approximates the boundaries (and thus also the level of detail) of a given human map.

Following the discussion above, we then select the map of the sequence $S^1, \ldots, S^K$ with the best F_{kl}-measure for each human ground truth GT_l, that is identified by the index

$$k_l^* = arg \max_{1 \leq k \leq K} F_{kl}$$

Thus, the machine map $S^{k_l^*}$ is the element of the hierarchical sequence that provides a best approximation of the segments (i.e. the boundaries) provided by human l. As several human segmentations are usually available for a given image, we then average the precision $P_{k_l^* l}$ and recall $R_{k_l^* l}$ scores of the machine maps $S^{k_l^*}$ selected for each of the different available human maps $GL_l (l = 1, \ldots, L)$ in order to obtain global precision P and recall R scores for the considered image:

$$P = \frac{1}{L}\sum_{l=1}^{L} P_{k_l^* l} \quad R = \frac{1}{L}\sum_{l=1}^{L} R_{k_l^* l}$$

These scores P and R can then be combined as usual in a global F-measure measuring the overall performance goodness of the hierarchical segmentation procedure being evaluated on a given reference image. As discussed, such global F-measure has to be understood as a measure of the extent up to which a given hierarchical segmentation is able to capture the different detail levels that a set of different human segmentations of such an image may contain.

Remark 1. Notice that the previously referred counts of true positives, false positives and false negatives are usually dependent on a distance threshold δ that specifies the tolerance level to small boundary localization errors. Machine boundary pixels that lie closer than a distance δ from a human boundary pixel and that can be matched in a one-to-one correspondence to human boundary pixels (through solving a sparse minimum cost bipartite assignment problem) are counted as true positives. Only those machine pixels that cannot be matched counts as false positives. Unmatched human boundary pixels count as false negatives. In this sense, just let us remark that our method can be used taking into account any tolerance to localization error.

Remark 2. Another important remark concerns the usage of precision-recall (or PR) curves, relatively similar to ROC curves but replacing fallout by precision in the x-axis. PR curves are commonly used as rich descriptors of performance for (non-hierarchical) segmentation techniques that rely on a set of parameters. Each particular configuration of the parameters produces a pair of precision-recall scores, and through a search of the parameter space a PR curve is obtained that approximates the Pareto frontier of the algorithm (for a given human segmentation) in terms of the precision-recall criteria. Then, as the F-measure curve is typically unimodal, the maximum F attained at the points of the PR curve is reported as the overall measure of performance [10].

At this point, it is important to remark that a hierarchical procedure as the above exposed D&L algorithm depends on a set of K parameters $\alpha_1 < \alpha_2 < \ldots < \alpha_K$ for each possible number of detail levels K, i.e. for each sequence of K hierarchically-related segmentations. By varying these parameters, a different PR curve may be obtained for each value of K following the method proposed above. That is, we claim that each hierarchical sequence $\{S^1, \ldots, S^K\}$ (associated to parameters $\alpha_1, \ldots, \alpha_K$) of machine segmentations has to be evaluated *globally* with respect to the whole set of L human segmentations (possibly providing different reference levels of detail) through the global P and R scores (and the corresponding F-measure) obtained as above. Thus, different specifications of the parameters $\alpha_1, \ldots, \alpha_K$ do produce different (P, R) pairs, which in turn may be used to build the PR curve of the hierarchical procedure for a given K. Then, two different approaches may be devised to choose a particular value of K:

1. Consider a predefined value for K, for instance on the basis of a priori knowledge about the different levels of detail actually present on the human reference

segmentations. Or simply take $K = L$, assuming that possibly each one of the L human reference segmentations is performed at a different level of detail.
2. Search for the optimal K i.e. that providing the best PR curve, in the sense of either reaching the greatest maximum F-measure or the greatest AUC (area-under-the-curve).

Notice that, although the second, search-based approach would allow uncovering the best configuration of the parameters of the hierarchical procedure, it may be unpractical since the number of parameters to be adjusted in order to compute adequate PR curves for each K grows linearly with K. For this reason, we find the first approach to be more practical and particularly taking $K = L$ is encouraged unless reliable a priori knowledge about the present levels of detail is available.

To sum up, the proposed evaluation methodology for hierarchical segmentation can be seen as an extension of the state-of-the-art edge-based segmentation evaluation methodology [1, 10], in the sense that the former builds up upon the latter by just enabling hierarchical segmentation techniques to be introduced in comparisons. Particularly, the proposed method allows a sequence of segmentations to be evaluated *as a whole*, viewing it as an approximation to the *whole* set of available human references. As commented above, the lack of reference datasets and the consequent difficulties for developing a proper evaluation methodology is possibly one of the main problems found in the development of hierarchical segmentation techniques. Thus, we hope that the proposed methodology may contribute to alleviate this bottleneck problem of hierarchical segmentation evaluation.

4 Computational Example

Now we describe a small computational experiment to illustrate the proposed evaluation methodology. This experiment compares the performance of the exposed Divide-and-Link (D&L) hierarchical segmentation algorithm with that of a small set of well-known edge-detection techniques on just a single reference image and its associated human segmentations. The same *Church* image as in Fig. 2 together with a set of $L = 5$ human segmentations (shown in the first column of Fig. 3) will provide the ground truth for the evaluation. Besides D&L, the Canny [4], Prewitt [11] and Sobel (see [2]) edge-detectors (as implemented in MATLAB R2010b) will constitute the set of procedures to be compared. For simplicity, we do no compute the PR curves of the detectors, and just the precision-recall scores obtained through the detectors' MATLAB default parameters will be reported. The D&L algorithm is set up asking for a hierarchy of $K = L = 5$ segmentations, to be obtained with arbitrary thresholds $\alpha_1,\ldots,\alpha_5$ (those used in Fig. 2) and by using the 4-connectivity topology together with the L_1 dissimilarity measure (i.e. d_e is given by the sum of the absolute differences in RGB intensities of adjacent pixels p,q such that $e = \{p, q\}$). A boundary map is derived from each of the 5 D&L segmentations by declaring as boundary just the pixels in the larger region from each pair of adjacent pixels belonging to different regions. Counts of true positives, false negatives and false positives are obtained through pixel-to-pixel matching between the human and the machine boundary maps, i.e. localization error is not tolerated (so $\delta = 0$). Finally, precision and recall scores as well as the respective

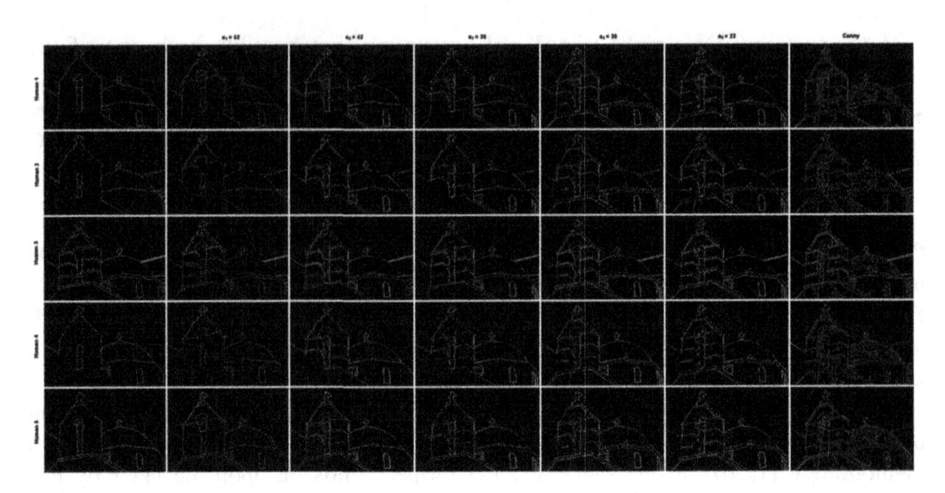

Fig. 3. Human segmentations ground truth (first column) and matchings of these with the machine boundary maps generated through different D&L thresholds and the Canny detector. Green pixels are true positives, yellow ones are false positives and magenta false negatives. (Color figure online)

F-measures are computed for each pair of human and machine segmentations as described in Sect. 3. Let us remark that this experiment is neither intended to constitute a rigorous evaluation of the D&L algorithm nor a detailed comparison of this procedure with state-of-the-art techniques, but just to provide an illustrative example intended to show the feasibility of the proposed evaluation methodology and hopefully contributing to a better understanding of this paper.

Results for each human-D&L segmentation matching are shown in Table 1 below. Highest *F*-measures and precision-recall scores for each human are bolded. Notice that the two D&L segmentations with $\alpha_2 = 42$ and $\alpha_3 = 36$ consistently obtain the best *F*-measures of the D&L hierarchical sequence. They are the most precise detectors and

Table 1. *F*-measures and pairs of scores (precision,recall) for each matching of human-D&L segmentations.

	D&L $\alpha_1 = 52$	D&L $\alpha_2 = 42$	D&L $\alpha_3 = 36$	D&L $\alpha_4 = 30$	D&L $\alpha_5 = 22$
Human 1	F = 0.07 (0.15,0.05)	**F = 0.27** (**0.21**,0.37)	F = 0.27 (0.20,0.40)	F = 0.21 (0.14,0.42)	F = 0.21 (0.14,**0.45**)
Human 2	F = 0.05 (0.10,0.03)	F = 0.17 (0.13,0.26)	**F = 0.18** (**0.13**,0.30)	F = 0.14 (0.09,0.32)	F = 0.14 (0.09,**0.34**)
Human 3	F = 0.04 (0.16,0.03)	F = 0.23 (0.25,0.21)	**F = 0.25** (**0.25**,0.25)	F = 0.22 (0.18,0.27)	F = 0.22 (0.18,**0.30**)
Human 4	F = 0.08 (0.13,0.05)	**F = 0.26** (**0.19**,0.43)	F = 0.25 (0.17,0.46)	F = 0.19 (0.12,0.47)	F = 0.18 (0.11,**0.50**)
Human 5	F = 0.03 (0.06,0.02)	F = 0.24 (**0.20**,0.30)	**F = 0.25** (0.19,0.33)	F = 0.19 (0.13,0.34)	F = 0.19 (0.13,**0.36**)

obtain the best precision-recall balance. As expected, the lower the threshold α_k is, the greater recall is obtained, as a finer level of detail is introduced. In this sense, these illustrative results suggest that two different (but relatively similar) detail levels are being used along the 5 human segmentations. Particularly, humans 1 and 4 seem to provide segments with a somehow lower detail level than humans 2, 3 and 5 (humans 2 and 3 signal the electric lines at the bottom-right, while humans 2 and 5 provide more details of the bell tower base and top).

Thus, following the discussion in last section, the evaluation of the D&L algorithm should focus on those segmentations of the hierarchical sequence that best approximate the detail level of each human segmentation. This is accomplished in Table 2 below, combining the best D&L results (in terms of *F*) for each human with the results obtained by the Canny, Sobel and Prewitt detectors. Again, best scores for each human are bolded. The D&L best-matches consistently obtain a significantly higher precision than the other detectors, at the cost of a small difference with Canny's recall. Overall, after averaging the precision and recall scores along all humans, D&L obtains the highest *F*-measure, clearly outperforming the Sobel and Prewitt detectors and obtaining a better precision-recall balance than Canny (of course, just for the MATLAB default parameters of the three detectors other than D&L). Anyway, besides the particular results obtained, this experiment shows the feasibility of the proposed evaluation methodology, as just intended.

Table 2. Comparison of the different detectors performance along the various human segmentations. D&L results come from selecting in Table 1 the partitions with maximum F-measure for each human. Overall results are obtained by averaging the (precision,recall) pairs.

	D&L	Canny	Sobel	Prewitt
Human 1	**F = 0.27** (0.**21**,0.37)	F = 0.15 (0.09,**0.41**)	F = 0.13 (0.10,0.22)	F = 0.13 (0.09,0.21)
Human 2	**F = 0.18** (**0.13**,0.30)	F = 0.11 (0.06,**0.33**)	F = 0.09 (0.06,0.16)	F = 0.09 (0.06,0.16)
Human 3	**F = 0.25** (**0.25**,0.25)	F = 0.17 (0.12,**0.27**)	F = 0.14 (0.13,0.14)	F = 0.13 (0.12,0.14)
Human 4	**F = 0.26** (**0.19**,0.43)	F = 0.13 (0.08,**0.46**)	F = 0.13 (0.08,0.25)	F = 0.12 (0.08,0.25)
Human 5	**F = 0.25** (**0.19**,0.33)	F = 0.15 (0.09,**0.35**)	F = 0.14 (0.10,0.20)	F = 0.13 (0.10,0.20)
Overall	**F = 0.25** (**0.19**,0.33)	F = 0.14 (0.09,**0.36**)	F = 0.13 (0.10,0.20)	F = 0.12 (0.09,0.19)

5 Conclusions

A method to enable the evaluation of hierarchical segmentation algorithms through a supervised edge-based approach has been introduced in this paper. This makes possible comparing different hierarchical procedures between them and against other non-hierarchical segmentation techniques (as well as with edge detectors). The main point we made is that hierarchical segmentation can be understood as an approximation to the (a priori unknown) different detail levels with which an image may be segmented. Thus, given a sample of these different possible levels of detail as that provided by a set of reference human segmentations, a hierarchical sequence of segmentations should be evaluated by selecting the set of segmentations of the hierarchical sequence that best approximate each human segmentation level of detail. The proposed methodology is based on the state-of-the-art edge-based segmentation evaluation framework, just extending it to allow hierarchical segmentation procedures to be evaluated and compared. Hopefully this may contribute to solve or at least alleviate the bottleneck problem of hierarchical segmentation evaluation.

Possibly the main advantage of a hierarchical approach to segmentation as that of the D&L algorithm is that it allows working at different levels of detail in a consistent way. Although other (non-hierarchical) segmentation procedures may provide different detail levels by varying the parameters configuration, typically the so-obtained segmentations are not hierarchical, i.e. they do not consistently identify objects in the sense that successive partitions enable a higher detail levels without modifying the objects already detected in previous segmentations. The same may apply for the boundaries respectively resulting from a hierarchical segmentation technique (applying a consistent method to decide the boundaries of the detected regions) and a usual edge-detector.

At this respect, with this paper we would also like to emphasize that image processing algorithms should always depart from precise formal definition of the particular problems they address (as we have done defining segmentation and hierarchical segmentation). Although related, segmentation, edge detection and image classification (to say just three) indeed constitute different image processing problems. In our opinion, a precise and accepted formal definition of the different problems being addressed should be a main objective of research for the image processing community, since it would allow for a better categorization and relative comparison of the different existing techniques, as well as for the development a formal study of the relationships holding between the different problems, and therefore also between the various techniques that address them.

As a particular future research objective derived from this work, we are working in a rigorous and updated evaluation and comparison of several hierarchical and non-hierarchical segmentation procedures, which will also allow a further validation of the proposed methodology.

Acknowledgment. This research has been supported by the Government of Spain, grant TIN2012-32482, and by the Government of Madrid, Grant S2013/ICE-2845.

References

1. Arbelaez, P., Maire, M., Fowlkes, C., Malik, J.: Contour detection and hierarchical image segmentation. IEEE Trans. Pattern Anal. Mach. Intell. **33**(5), 898–916 (2011)
2. Basavaprasad, B., Ravindra, S.H.: A survey on traditional and graph theoretical techniques for image segmentation, Int. J. Comput. Appl. **1**, 38–46 (2014)

3. Bowyer, K., Kranenburg, C., Dougherty, S.: Edge detector evaluation using empirical ROC curves. Comput. Vis. Image Underst. **84**(1), 77–103 (2001)
4. Canny, J.: A computational approach to edge detection. IEEE Trans. Pattern Anal. Mach. Intell. **8**, 679–698 (1986)
5. Castillo-Ortega, R., Chamorro-Martínez, J., Marín, N., Sánchez, D., Soto-Hidalgo, J.M.: Describing images via linguistic features and hierarchical segmentation. In: Proceedings of the WCCI 2010 IEEE World Congress on Computational Intelligence, pp. 1104–1111 (2010)
6. Correia, P., Pereira, F.: Objective evaluation of video segmentation quality. IEEE Trans. Image Process. **12**(2), 186–200 (2003)
7. Gómez, D., Yáñez, J., Guada, C., Rodríguez, J.T., Montero, J., Zarrazola, E.: Fuzzy image segmentation based upon hierarchical clustering. Knowl. Based Syst. **87**, 26–37 (2015)
8. Gómez, D., Zarrazola, E., Yáñez, J., Montero, J.: A divide-and-link algorithm for hierarchical clustering in networks. Inf. Sci. **316**, 308–328 (2015)
9. Lee, S., Chung, S., Park, R.: A comparative performance study of several global thresholding techniques for segmentation. Comput. Vis. Graphs Image Process. **52**, 171–190 (1990)
10. Martin, D., Fowlkes, C., Malik, J.: Learning to detect natural image boundaries using local brightness, color and texture cues. IEEE Trans. Pattern Anal. Mach. Intell. **26**(5), 530–549 (2004)
11. Prewitt, J.M.S.: Object enhancement and extraction. In: Lipkin, B.S., Rosenfeld, A. (eds.) Picture Processing and Psychopictorics. Academic Press, New York (1970)
12. Shin, M.C., Goldgof, D.B., Bowyer, K.W.: Comparison of edge detector performance through use in an object recognition task. Comput. Vis. Image Underst. **84**(1), 160–178 (2001)
13. Zhang, H., Fritts, J.E., Goldman, S.A.: Image segmentation evaluation: a survey of unsupervised methods. Comput. Vis. Image Underst. **110**, 260–280 (2008)

Higher Degree F-transforms Based on B-splines of Two Variables

Martins Kokainis[1](✉) and Svetlana Asmuss[1,2]

[1] Department of Mathematics, University of Latvia, Zellu 25, Riga 1002, Latvia
{martins.kokainis,svetlana.asmuss}@lu.lv
[2] Institute of Mathematics and Computer Science, University of Latvia, Raina Bulvaris 29, Riga 1459, Latvia

Abstract. The paper deals with the higher degree fuzzy transforms (F-transforms with polynomial components) for functions of two variables in the case when two-dimensional generalized fuzzy partition is given by B-splines of two variables. We investigate properties of the direct and inverse F-transform in this case and prove that using B-splines as basic functions of fuzzy partition allows us to improve the quality of approximation.

Keywords: F-transform · Two-dimensional fuzzy partition · Approximation error · B-splines

1 Introduction

The concept of fuzzy transform (F-transform or F^0-transform) was introduced in 2001 [11] (see also the key paper [9]) and generalized to the case of higher degree (F^m-transform, $m \geq 1$) in 2011 [12] by I. Perfilieva with co-authors. Initially, the technique of F-transform was described for functions of one variable. The extension for the two-dimensional case (i.e. for functions of two variables) has been introduced in [17] and developed in [7,8].

There is a number of papers dealing with fuzzy transforms with respect to a fuzzy partition with specially designed basic functions including fuzzy partitions based on splines (see, e.g., [1,5]). In both mentioned cases the ordinary F-transforms (i.e. with the classical components-numbers) for functions of one variable have been considered. The construction of higher degree fuzzy transforms with respect to a fuzzy partition given by central, odd degree B-splines of one variable was presented by the authors at the previous FUZZ-IEEE conference (IEEE International Conference on Fuzzy Systems) in Istanbul in 2015 (see [6]). At the current stage our main focus area refers to the case of two variables. Let us note that in last years exactly two-dimensional F-transforms have been widely applied in image processing for solving problems of image compression, reconstruction, denoising, fusion, edge detection (see, e.g., [2,3,13,14]). Our motivation to develop and investigate the technique of spline-based F-transform

J.P. Carvalho et al. (Eds.): IPMU 2016, Part I, CCIS 610, pp. 648–659, 2016.
DOI: 10.1007/978-3-319-40596-4_54

for the case of several variables has been caused by the rapid development of such applications.

Our research focuses on the F^m-transform with respect to a generalized uniform fuzzy partition given by B-splines of two variables of degree $2k_1 - 1$ and $2k_2 - 1$ respectively. We generalize for the two-dimensional case the main result of [6] and obtain that in this case the inverse F^m-transform is precise for polynomials of two variables of degree $r_1 \leq 2k_1 - 1$ and $r_2 \leq 2k_2 - 1$ when $r_1 + r_2 \leq 2m + 1$. On the basis of this result we obtain error estimations for approximation by the inverse F^m-transform and prove that using B-splines may improve the approximation properties of the technique of higher degree two-dimensional F-transform.

The paper is structured as follows. Section 2 contains preliminaries on fuzzy partitions, F^m-transforms (direct and inverse) and B-splines. Section 3 describes a special design of fuzzy partition based on B-splines of two variables and the corresponding F-transform constructions. Section 4 identifies the degree of polynomials for which the inverse two-dimensional F^m-transform based on B-splines is precise. Section 5 is devoted to approximation properties of the inverse F^m-transform described above. Finally, Sect. 6 concludes the results.

2 Preliminaries

We will use the following notation:

- $[n..m]$ (for integers n, m with $n \leq m$) – the set $\{n, n+1, \ldots, m\}$;
- $\mathbb{P}_l$ – the set of univariate polynomials of degree at most l;
- $\mathbb{P}_{l_1,l_2} = \text{span}\left\{x^i y^j \mid i \in [0\,..\,l_1], j \in [0\,..\,l_2]\right\}$ – the set of bivariate polynomials s.t. the degree of the first variable is at most l_1 and the degree of the second variable is at most l_2.

2.1 Fuzzy Partition

Fuzzy partition of a rectangle $[a, b] \times [c, d] \subset \mathbb{R}^2$ is commonly defined via fuzzy partitions of the intervals $[a, b]$ and $[c, d]$.

Generalized 1D Fuzzy Partition (see, e.g., [3,10]). Suppose that $[a, b] \subset \mathbb{R}$, $a < b$, and $N \in \mathbb{N}$ is chosen. Let $h > 0$ and $h' > h/2$. Let $t_0, \ldots, t_N$ be fixed nodes s.t. $a < t_0 < \ldots < t_N < b$ and the following requirements are satisfied:

1. $t_{i+1} = t_i + h$ for all $i \in [0\,..\,N-1]$;
2. $\bar{E}_i \subset [a, b]$ for all $i \in [0\,..\,N]$, where $E_i := (t_i - h', t_i + h')$ and $\bar{E}_i$ is the closure of E_i (the constraint $h' > h/2$ ensures that the adjacent sets E_i and E_{i+1} intersect and the whole interval (a, b) is covered);
3. $\bigcup_{j=0}^{N} \bar{E}_j = [a, b]$.

The sets E_i depend on the parameter h'; for the sake of simplicity, this parameter is not included in the notation of E_i.

Definition 1. *Fuzzy sets* $A_0, \dots, A_N : [a,b] \to [0,1]$ *are said to constitute a generalized* (h,h')*-uniform fuzzy partition of* $[a,b]$ *if the following conditions are satisfied:*

- $A_i(t) > 0$ *if* $t \in E_i$, *and* $A_i(t) = 0$ *if* $t \in [a,b] \setminus E_i$, $i \in [0\,..\,N]$;
- A_i *is continuous on* $\bar{E}_i$, $i \in [0\,..\,N]$;
- $\sum_{j=0}^{N} A_j(t) > 0$ *for all* $t \in (a,b)$, $i \in [0\,..\,N]$.
- $A_i(t_i - t) = A_i(t_i + t)$ *for all* $t \in [0,h']$, $i \in [0\,..\,N]$;
- $A_i(t) = A_{i+1}(t+h)$ *for all* $t \in \bar{E}_i$ *and* $i \in [0\,..\,N-1]$.

It is easy to see that then there is an even function $A : [-H,H] \to \mathbb{R}$, where $H = h'/h$, such that for all $i \in [0\,..\,N]$ and $t \in \bar{E}_i$, $A_i(t) = A\left(\frac{t-t_i}{h}\right)$. The function A is said to be the generating function of the partition $A_0, \dots, A_N$.

Definition 2. *Suppose that an interval* $I \subset [a,b]$ *is such that*

$$\sum_{i=0}^{N} A_i(t) = 1 \quad \textit{for all } t \in I.$$

Then the fuzzy partition $A_0,\dots,A_N$ *is said to fulfill the Ruspini condition on* I.

2D Fuzzy Partition (see, e.g., [3,10]). Suppose that the fuzzy sets $A_0,\dots,A_{N_1}$ establish a fuzzy partition of $[a,b]$ and $B_0,\dots,B_{N_2}$ establish a fuzzy partition of $[c,d]$. Then the fuzzy sets $A_i \times B_j$, $i \in [0\,..\,N_1]$, $j \in [0\,..\,N_2]$, are said to constitute a fuzzy partition of the Cartesian product $[a,b] \times [c,d]$. The membership function of $A_i \times B_j$ is equal to the product $A_i(x)B_j(y)$ of the corresponding membership functions A_i, B_j.

Moreover, if $A_0,\dots,A_{N_1}$ fulfill the Ruspini condition on $I \subset [a,b]$ and B_0, $\dots$, B_{N_2} fulfill the Ruspini condition on $J \subset [c,d]$, then $A_i \times B_j$, $i \in [0\,..\,N_1]$, $j \in [0\,..\,N_2]$, satisfy the Ruspini condition on $I \times J$ [4], in the sense that

$$\sum_{i=0}^{N_1} \sum_{j=0}^{N_2} A_i(x)B_j(y) = 1 \quad \text{for all } (x,y) \in I \times J.$$

2.2 Higher Degree F-transforms

Suppose that fuzzy sets $A_0,\dots,A_{N_1}$ establish a generalized (h_1,h_1')-uniform fuzzy partition of $[a,b]$, and $B_0,\dots,B_{N_2}$ establish a generalized (h_2,h_2')-uniform fuzzy partition of $[c,d]$. Denote the fixed nodes of $[a,b]$ by $t_i^{(1)}$, $i \in [0\,..\,N_1]$; similarly, the fixed nodes of $[c,d]$ are denoted by $t_j^{(2)}$, $j \in [0\,..\,N_2]$. Let

$$E_i^{(1)} := (t_i^{(1)} - h_1', t_i^{(1)} + h_1'), \quad E_j^{(2)} := (t_j^{(2)} - h_2', t_j^{(2)} + h_2'),$$

and $\bar{E}_i^{(1)}$, $\bar{E}_j^{(2)}$ stand for the closure of $E_i^{(1)}$, $E_j^{(2)}$, respectively.

For any $i \in [0\,..\,N_1]$, $j \in [0\,..\,N_2]$, let $L_2(A_i \times B_j)$ be the Hilbert space of all square-integrable functions $f : \bar{E}_i^{(1)} \times \bar{E}_j^{(2)} \to \mathbb{R}$, equipped with the inner product $\langle \cdot, \cdot \rangle_{i,j}$, defined as

$$\langle f, g \rangle_{i,j} = \int\limits_{\bar{E}_i^{(1)} \times \bar{E}_i^{(2)}} f(t_1, t_2) g(t_1, t_2) A_i(t_1) B_j(t_2)\, \mathrm{d}t_1\, \mathrm{d}t_2,$$

and the associated norm $\|\cdot\|_{i,j}$[1]. By $L_2^m(A_i \times B_j)$, $m \in \mathbb{N} \cup \{0\}$, we denote the subspace of $L_2(A_i \times B_j)$, spanned by the bivariate polynomials of degree at most m. Finally, by $L_2(A, B)$ we shall denote the set of all $f : [a, b] \times [c, d] \to \mathbb{R}$ s.t. its restriction on any rectangle $\bar{E}_i^{(1)} \times \bar{E}_j^{(2)}$ belongs to $L_2(A_i \times B_j)$.

Fix any $f \in L_2(A, B)$. Its direct F^m-transform is defined as the $(N_1 + 1) \times (N_2 + 1)$ matrix whose (i, j)-th component is the orthogonal projection of f on $L_2^m(A_i \times B_j)$.

Let $F_{i,j}^m(f, x, y)$ be the value of the (i, j)-th component of the direct F^m-transform of f at $(x, y) \in [a, b] \times [c, d]$. Then the inverse F^m-transform of f (evaluated at (x, y)), which we denote by $\mathcal{F}^m(f, x, y)$, is defined as

$$\mathcal{F}^m(f, x, y) = \frac{\sum_{i=0}^{N_1} \sum_{j=0}^{N_2} F_{i,j}^m(f, x, y) A_i(x) B_j(y)}{\sum_{i=0}^{N_1} \sum_{j=0}^{N_2} A_i(x) B_j(y)}.$$

If $(x, y) \in I \times J$ and the chosen fuzzy partition satisfies the Ruspini condition on $I \times J$, this expression is simplified to

$$\mathcal{F}^m(f, x, y) = \sum_{i=0}^{N_1} \sum_{j=0}^{N_2} F_{i,j}^m(f, x, y) A_i(x) B_j(y).$$

Suppose that P_i^l, Q_j^l, $l \in [0\,..\,m]$, are univariate polynomials s.t. $\deg P_i^l = \deg Q_j^l = l$ and

$$\int\limits_{E_i^{(1)}} P_i^{l_1}(x) P_i^{l_2}(x) A_i(x)\, \mathrm{d}x = \int\limits_{E_j^{(2)}} Q_j^{l_1}(y) Q_j^{l_2}(y) B_j(y)\, \mathrm{d}y = \begin{cases} 1, & l_1 = l_2, \\ 0, & l_1 \neq l_2. \end{cases}$$

Then the bivariate polynomials $P_i^{l_1} \times Q_j^{l_2}$ satisfy

$$\left\langle P_i^{l_1} \times Q_j^{l_2}, P_i^{l_3} \times Q_j^{l_4} \right\rangle_{ij} = \begin{cases} 1, & l_1 = l_3 \text{ and } l_2 = l_4, \\ 0, & \text{otherwise}, \end{cases}$$

[1] In fact, this way a semi-norm $\|\cdot\|_{i,j}$ in $L_2(A_i \times B_j)$ is defined. The identification of functions f and g, which are almost everywhere equal in the sense $\|f - g\|_{i,j} = 0$, turns $\|\cdot\|_{i,j}$ from a semi-norm into a norm and the semi-definite sesquilinear form $\langle \cdot, \cdot \rangle_{i,j}$ into an inner product. Further on, this technical detail will be ignored.

and the direct F^m-transform components can be computed as

$$F^m_{i,j}(f,x,y) = \sum_{\substack{l_1+l_2 \le m,\\ l_1,l_2 \ge 0}} c^{l_1,l_2}_{i,j}(f) P^{l_1}_i(x) Q^{l_2}_j(y),$$

where

$$c^{l_1,l_2}_{i,j}(f) = \frac{\left\langle f, P^{l_1}_i \times Q^{l_2}_j \right\rangle_{i,j}}{\left\langle P^{l_1}_i \times Q^{l_2}_j, P^{l_1}_i \times Q^{l_2}_j \right\rangle_{i,j}}. \tag{1}$$

When there is no ambiguity, notation $c^{l_1,l_2}_{i,j}(f)$ will be simplified to $c^{l_1,l_2}_{i,j}$.

2.3 B-splines

Univariate B-splines of degree n are [15,16] piecewise (w.r.t. a given domain partition) polynomial (of degree n) functions in $C^{n-1}(\mathbb{R})$, characterized by having minimal support.

Central B-splines [16] are even B-splines that have 1-equidistant knots. For each integer $n \ge 0$ the central B-spline is unique (up to a constant factor). The properties of B-splines and construction of a fuzzy partition using central B-spline as the generating function are described in more details in [6].

3 Two Dimensional Fuzzy Partition Generated by B-splines

Fix $N_1, N_2, k_1, k_2 \in \mathbb{N}$. By A and B we denote the central B-splines of degree $2k_1 - 1$ and $2k_2 - 1$, respectively.

Let a rectangle $[a,b] \times [c,d] \subset \mathbb{R}^2$ be fixed. Denote $h_1 = (b-a)/(N_1 + 2k_1)$ and $h_2 = (d-c)/(N_2 + 2k_2)$. By $t^{(1)}_i$ we denote the h_1-equidistant nodes: $t^{(1)}_i = a + h_1\,(i + k_1)$, $i \in [-k_1 \,..\, N_1 + k_1]$; then

$$a = t^{(1)}_{-k_1} < t^{(1)}_0 < t^{(1)}_{N_1} < t^{(1)}_{N_1+k_1} = b.$$

Similarly, by $t^{(2)}_j$ we denote the h_2-equidistant nodes $t^{(2)}_j = c + h_2\,(j + k_2)$, $j \in [-k_2 \,..\, N_2 + k_2]$. By $A_i(x)$ and $B_j(y)$ we denote the functions $A\left(\frac{x - t^{(1)}_i}{h_1}\right)$ and $B\left(\frac{y - t^{(2)}_j}{h_2}\right)$, respectively, $i \in [0 \,..\, N_1]$, $j \in [0 \,..\, N_2]$.

We consider the generalized fuzzy partition of $[a,b] \times [c,d]$ formed by the functions $A_i(x) \cdot B_j(y)$, where $(i,j) \in [0 \,..\, N_1] \times [0 \,..\, N_2]$.

An illustration of the fuzzy partition of $[1,3] \times [2,4]$, generated by B-splines of degree 3 in each variable, is shown in Fig. 1.

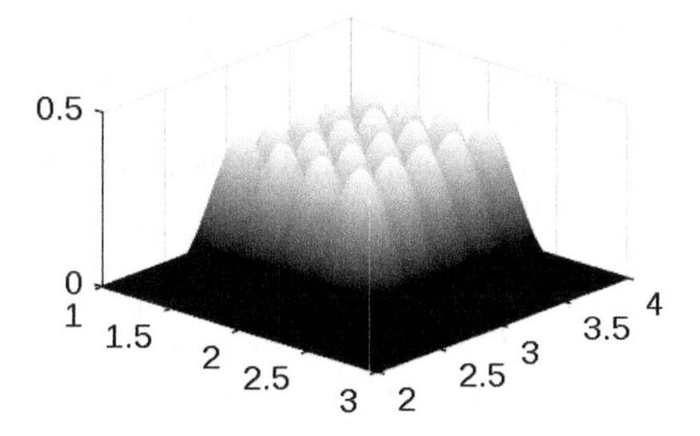

Fig. 1. Fuzzy partition of $[1,3]\times[2,4]$ when $k_1=k_2=2$, $N_1=3$, $N_2=4$

3.1 Spaces $L_2(A\times B)$ and $L_2(A_i\times B_j)$

Let $L_2(A\times B)$ stand for the Hilbert space, which consists of square integrable functions $f:[-k_1,k_1]\times[-k_2,k_2]\to\mathbb{R}$, with the inner product $\langle\cdot,\cdot\rangle$ given by

$$\langle f,g\rangle=\int_{-k_1}^{k_1}\int_{-k_2}^{k_2} f(x,y)g(x,y)A(x)B(y)\,\mathrm{d}x\,\mathrm{d}y$$

and the induced norm $\|f\|=\langle f,f\rangle^{0.5}$. Space $L_2^m(A\times B)$ is defined as the closed linear subspace of $L_2(A\times B)$, spanned by the set of linearly independent system of polynomials $\{x^iy^j\;|0\le i,j\le m,\;i+j\le m\}$ (restricted to $[-k_1,k_1]\times[-k_2,k_2]$).

Define the analogous one-dimensional concepts: $L_2(A)$ is the Hilbert space of square integrable functions $f:[-k_1,k_1]\to\mathbb{R}$ with the inner product $\langle\cdot,\cdot\rangle_A$ given by $\langle f,g\rangle_A=\int_{-k_1}^{k_1} f(t)g(t)A(t)\,\mathrm{d}t$ (and the induced norm $\|\cdot\|_A$); space $L_2^m(A)$ is defined as the closed linear subspace of $L_2(A)$, spanned by the closed set of linearly independent system of polynomials $\{1,t,t^2,\dots,t^m\}$ (restricted to $[-k_1,k_1]$). Spaces $L_2(B)$ and $L_2^m(B)$ are defined similarly.

Orthogonal (w.r.t. $\langle\cdot,\cdot\rangle_A$) monic polynomials in $L_2(A)$ are constructed as follows:

$$P^0(t)\equiv 1,\qquad P^{l+1}(t)=\chi^l(t)-\sum_{j=0}^{l}\lambda_j^lP^j(t),\quad l\ge 0,$$

where $\lambda_j^l=\langle\chi^l,P^j\rangle_A\left\|P^j\right\|_A^{-2}$, $\chi^l(t)=t^{l+1}$. By inductive arguments (and the fact that A is an even function, but $\langle\cdot,\cdot\rangle_A$ is defined as an integral over a symmetric interval), it can be seen that P^l is an odd function iff l is odd and P^l is an even function iff l is even. Another obvious but important property is that P^l is orthogonal (w.r.t. $\langle\cdot,\cdot\rangle_A$) to every polynomial of degree less than l. Similarly define orthogonal (w.r.t $\langle\cdot,\cdot\rangle_B$) monic polynomials Q^l with $\deg Q^l=l$.

Then polynomials $P^{l_1}(x)\cdot Q^{l_2}(y)$ and $P^{s_1}(x)\cdot Q^{s_2}(y)$ are orthogonal (for $(l_1,l_2)\ne(s_1,s_2)$, w.r.t. $\langle\cdot,\cdot\rangle$); moreover, $L_2^m(A\times B)$ is spanned by the polynomials $P^{l_1}(x)\cdot Q^{l_2}(y)$ with $0\le l_1+l_2\le m$.

Fix $i\in[0\,..\,N_1]$, $j\in[0\,..\,N_2]$ and consider the space $L_2(A_i\times B_j)$. By changing variables in the corresponding integral, the inner product $\langle\cdot,\cdot\rangle_{ij}$ can be obtained from $\langle \cdot, \cdot \rangle$ as follows: $\langle f, g \rangle_{i,j} = h_1 h_2 \left\langle \tilde{f}, \tilde{g} \right\rangle$, where by $\tilde{p}(x, y)$ the function $p\left(xh_1 + t_i^{(1)}, yh_2 + t_j^{(2)}\right)$ is denoted.

Hence the basis of $L_2^m(A_i \times B_j)$ are formed by the polynomials $P_i^{l_1}(t_1)Q_j^{l_2}(t_2)$, $l_1, l_2 \geq 0$, $l_1 + l_2 \leq m$, where

$$P_i^l(t_1) = P^l\left(\frac{t_1 - t_i^{(1)}}{h_1}\right) \text{ and } Q_j^l(t_2) = Q^l\left(\frac{t_2 - t_j^{(2)}}{h_2}\right).$$

Moreover, $\left\|P_i^{l_1} Q_j^{l_2}\right\|_{i,j}^2 = h_1 h_2 \left\|P^{l_1} Q^{l_2}\right\|^2$.

3.2 Ruspini Condition

Let

$$[\hat{a}, \hat{b}] = [t_{k_1-1}^{(1)}, t_{N_1-k_1+1}^{(1)}] \quad \text{and} \quad [\hat{c}, \hat{d}] = [t_{k_2-1}^{(2)}, t_{N_2-k_2+1}^{(2)}]. \tag{2}$$

The fuzzy partition $A_0, \ldots, A_{N_1}$ of $[a, b]$ fulfills [6, Eq. 13] the Ruspini condition on the interval $[\hat{a}, \hat{b}]$; similarly, the fuzzy partition $B_0, \ldots, B_{N_2}$ fulfills the Ruspini condition on $[\hat{c}, \hat{d}]$. Consequently, $A_i \times B_j$, $(i, j) \in [0 \mathinner{..} N_1] \times [0 \mathinner{..} N_2]$, (the fuzzy partition of $[a, b] \times [c, d]$) fulfills the Ruspini condition on $[\hat{a}, \hat{b}] \times [\hat{c}, \hat{d}]$.

It follows that for all $(x, y) \in [\hat{a}, \hat{b}] \times [\hat{c}, \hat{d}]$ the inverse F^m-transform of a function f can be computed as

$$\mathcal{F}^m(f, x, y) = \sum_{\substack{i \in [0 \mathinner{..} N_1] \\ j \in [0 \mathinner{..} N_2]}} \sum_{\substack{l_1 + l_2 \leq m \\ l_1, l_2 \geq 0}} c_{i,j}^{l_1,l_2} P_i^{l_1}(x) Q_j^{l_2}(y) A_i(x) B_i(y),$$

where $c_{i,j}^{l_1,l_2}$ is defined as in (1); moreover, when f is a polynomial of degree at most m, we have $\mathcal{F}^m(f, x, y) = f(x, y)$.

For example, in Fig. 1, the Ruspini condition is fulfilled on the interval $[13/7, 15/7] \times [11/4, 13/4]$.

4 Approximation of Polynomials

From now on we will assume that the fuzzy partition and polynomials P^{l_1}, Q^{l_2}, $P_i^{l_1}$, $Q_j^{l_1}$ are defined as in the previous section. First we state the following result; its proof is technical and therefore omitted here.

Lemma 1. *Suppose that* $r_1, r_2 \in \mathbb{N} \cup \{0\}$ *and a polynomial* f *is defined as*

$$f(x, y) = \sum_{l_1=0}^{r_1} \sum_{l_2=0}^{r_2} a^{l_1,l_2} P^{l_1}(x) Q^{l_2}(y).$$

Let $c_{i,j}^{l_1,l_2}$ be defined as in (1) for all $l_1 \in [0\mathinner{..}r_1]$, $l_2 \in [0\mathinner{..}r_2]$, $i \in [0\mathinner{..}N_1]$ and $j \in [0\mathinner{..}N_2]$. Then for all $l_1 \in [0\mathinner{..}N_1]$, $l_2 \in [0\mathinner{..}N_2]$ there exists a polynomial $p_{l_1,l_2} \in \mathbb{P}_{r_1-l_1,r_2-l_2}$ s.t.

$$c_{i,j}^{l_1,l_2} = p_{l_1,l_2}(i,j) \quad \textit{for all } (i,j) \in [0\mathinner{..}N_1] \times [0\mathinner{..}N_2].$$

Our main result is that for the described fuzzy partition the inverse F^m-transform coincides with p on the Ruspini rectangle for all polynomials p of degree at most $2m+1$ (provided that their degree in x or y variable does not exceed $2k_1 - 1$ or $2k_2 - 1$, respectively).

Theorem 1. *Suppose that $f \in \mathbb{P}_{r_1,r_2}$, $r_1 \leq 2k_1 - 1$, $r_2 \leq 2k_2 - 1$, $r_1 + r_2 \leq 2m+1$. Then*

$$\mathcal{F}^m(f,x,y) = f(x,y) \quad \textit{for all } (x,y) \in [\hat{a},\hat{b}] \times [\hat{c},\hat{d}] \quad (\textit{defined by } (2)).$$

Proof. Suppose that $r_1 + r_2 \geq m+1$, otherwise the claim is trivially satisfied. For all $(i,j) \in [0\mathinner{..}N_1] \times [0\mathinner{..}N_2]$ we can express

$$\sum_{l_1=0}^{r_1} \sum_{l_2=0}^{r_2} c_{i,j}^{l_1,l_2} P_i^{l_1}(x) Q_j^{l_2}(y) = f(x,y), \tag{3}$$

where $c_{i,j}^{l_1,l_2}$ is defined as in (1).

Fix nonnegative integers l_1, l_2 so that $l_1 + l_2 > m$ and $l_1 \leq 2k_1 - 1$, $l_2 \leq 2k_2 - 1$. We will show that

$$\sum_{\substack{i \in [0\mathinner{..}N_1] \\ j \in [0\mathinner{..}N_2]}} c_{i,j}^{l_1,l_2} P_i^{l_1}(x) Q_j^{l_2}(y) A_i(x) B_j(y) = 0. \tag{4}$$

By Lemma 1, there exists a polynomial $\bar{p} \in \mathbb{P}_{r_1-l_1,r_2-l_2}$ such that $c_{i,j}^{l_1,l_2} = \bar{p}(i,j)$ for all $(i,j) \in [0\mathinner{..}N_1] \times [0\mathinner{..}N_2]$. It means that there are such reals a_{s_1,s_2} that

$$c_{i,j}^{l_1,l_2} = \sum_{s_1=0}^{r_1-l_1} \sum_{s_2=0}^{r_2-l_2} a_{s_1,s_2} i^{s_1} j^{s_2} \text{ for all } (i,j) \in [0\mathinner{..}N_1] \times [0\mathinner{..}N_2].$$

By linearity, to prove (4), it suffices to show that

$$\sum_{\substack{i \in [0\mathinner{..}N_1] \\ j \in [0\mathinner{..}N_2]}} i^{s_1} j^{s_2} P_i^{l_1}(x) Q_j^{l_2}(y) A_i(x) B_j(y) = 0 \tag{5}$$

for all $s_1 \in [0\mathinner{..}r_1 - l_1]$ and $s_2 \in [0\mathinner{..}r_2 - l_2]$. Notice that (5) can be factored as

$$\left(\sum_{i=0}^{N_1} i^{s_1} P_i^{l_1}(x) A_i(x) \right) \left(\sum_{j=0}^{N_2} j^{s_2} Q_j^{l_2}(y) B_j(y) \right) = 0.$$

Recall that [6, Claim 3]

$$\sum_{i=1-k_1}^{k_1} p(i)P^l(\tau - i)A(\tau - i) = 0 \tag{6}$$

for all $\tau \in [0, 1]$ and univariate polynomials p whose degree does not exceed $\min\{l - k_1, 2k_1 - 1 - l\}$. A similar statement holds for Q^l and B.

Let $p(x) = x^{s_1}$ and $q(y) = y^{s_2}$. Notice that $s_1 \le r_1 - l_1$ and $s_2 \le r_2 - l_2$; it immediately follows that $s_1 + l_1 \le r_1 \le 2k_1 - 1$ and $s_2 + l_2 \le r_2 \le 2k_2 - 1$. Moreover,

$$s_1 + s_2 \le (r_1 + r_2) - (l_1 + l_2) \le (2m + 1) - (m + 1) = m \le l_1 + l_2 - 1.$$

It follows that at least one of the inequalities $s_1 \le l_1 - 1$, $s_2 \le l_2 - 1$ holds. Suppose that $s_1 \le l_1 - 1$ (the other case is similar), then $\deg p \le \min\{l_1 - 1, 2k_1 - l_1\}$, thus by (6) we have

$$\sum_{i=1-k_1}^{k_1} i^{s_1} P^{l_1}(\tau - i)\, A(\tau - i) = 0 \quad \text{for all } \tau \in [0, 1]. \tag{7}$$

There is such $i_0 \in [k_1 - 1 \mathinner{..} N_1 - k_1]$ that $x \in [t_{i_0}^{(1)}, t_{i_0+1}^{(1)}]$ (remark: it is possible that there are several such indices i_0, i.e., $x = t_{i'}^{(1)}$ for some $i' \in [k_1 \mathinner{..} N_1 - k_1 - 1]$; then any of these indices can be chosen, say, $i_0 = i'$).

Then $\operatorname{supp} A_i \subset [i_0 - k_1 + 1 \mathinner{..} i_0 + k_1]$; letting $\tau = (x - t_{i_0}^{(1)})/h_1 \in [0, 1]$ we have $(x - t_{i_0+i}^{(1)})/h_1 = \tau - i$, and $P_i^{l_1}(x)A_i(x) = P^{l_1}(\tau - i)\, A(\tau - i)$. Hence

$$\sum_{i=0}^{N_1} i^{s_1} P_i^{l_1}(x)A_i(x) = \sum_{i=1-k_1}^{k_1} i^{s_1} P^{l_1}(\tau - i)\, A(\tau - i) = 0,$$

where the last equality is due to (7). But this implies that (4) holds and we conclude that

$$\mathcal{F}^m(f, x, y) = \sum_{\substack{i \in [0 \mathinner{..} N_1] \\ j \in [0 \mathinner{..} N_2]}} \sum_{\substack{l_1 \in [0 \mathinner{..} r_1] \\ l_2 \in [0 \mathinner{..} r_2]}} c_{i,j}^{l_1, l_2} P_i^{l_1}(x) Q_j^{l_2}(y) A_i(x) B_j(y).$$

Now from (3) and the Ruspini condition it follows that

$$\mathcal{F}^m(f, x, y) = f(x, y) \sum_{\substack{i \in [0 \mathinner{..} N_1] \\ j \in [0 \mathinner{..} N_2]}} A_i(x) B_j(y) = f(x, y).$$

5 Approximation of Smooth Functions

Theorem 2. *Suppose that $k_1, k_2 \in \mathbb{N}$ and non-negative integers r, m satisfy $r \le \min\{2k_1 - 1, 2k_2 - 1, 2m+1\}$. Let f be a function from $C^{r+1}([a,b] \times [c,d])$ and $\mathcal{F}^m(f,\cdot)$ be its inverse F^m-transform with respect to the generalized fuzzy partition based on two-variable B-splines of degree $2k_1 - 1$ and $2k_2 - 1$, respectively.*

Then for all (x,y) from $[\hat{a},\hat{b}] \times [\hat{c},\hat{d}]$ (defined by (2)) the following estimation

$$f(x,y) - \mathcal{F}^m(f,x,y) = O(h^{r+1})$$

holds, where $h = \max\{h_1, h_2\}$.

Proof. We apply for function f the Taylor formula with the Lagrange form of the remainder:

$$f(x,y) = p_r(f,x,y) + \frac{1}{(r+1)!}\,\mathrm{d}^{r+1} f\left(x_0 + \theta_1(x - x_0), y + \theta_2(y - y_0)\right),$$

where $p_r(f,\cdot,\cdot)$ is the Taylor polynomial for f, $\deg p_r \le r$, (x_0, y_0) is the point of expansion, $\theta_1, \theta_2 \in [0,1]$.

Then

$$f(x,y) - \mathcal{F}^m(f,x,y) = R_r(x,y) - \mathcal{F}^m(R_r,x,y),$$

where $R_r(x,y) = f(x,y) - p_r(x,y)$. Taking into account that all partial derivatives

$$\frac{\partial^{r_1+r_2} f}{\partial x^{r_1} \partial y^{r_2}} \quad \text{for } r_1, r_2, \text{ s.t. } r_1 + r_2 \le r+1,$$

are continuous on $[a,b]\times[c,d]$, we obtain that $R_r(x,y) = O(h^{r+1})$ holds whenever $x - x_0 = O(h)$ and $y - y_0 = O(h)$.

To estimate the approximation error $f(x,y) - \mathcal{F}^m(f,x,y)$ we assume that $x \in [t_{i_0}^{(1)}, t_{i_0+1}^{(1)}]$, $y \in [t_{j_0}^{(2)}, t_{j_0+1}^{(2)}]$ and take $x_0 = t_{i_0}^{(1)}$ and $y_0 = t_{j_0}^{(2)}$. Clearly, this choice ensures that $R_r(x,y) = O(h^{r+1})$. To estimate $\mathcal{F}^m(R_r,x,y)$, we use the formula

$$\mathcal{F}^m(R_r,x,y) = \sum_{i=i_0-k_1+1}^{i_0+k_1} \sum_{j=j_0-k_2+1}^{j_0+k_2} \sum_{\substack{l_1+l_2 \le m \\ l_1,l_2 \ge 0}} c_{i,j}^{l_1,l_2}(R_r) P_i^{l_1}(x) Q_j^{l_2}(y) A_i(x) B_j(y),$$

where $c_{i,j}^{l_1,l_2}(R_r)$ is defined as in (1). Let us show that $c_{i,j}^{l_1,l_2}(R_r) = O(h^{r+1})$, $P_i^{l_1}(x) = O(1)$, $Q_j^{l_2}(y) = O(1)$, then (since clearly $A_i(x)B_j(y) = O(1)$) the estimation $\mathcal{F}^m(R_r,x,y) = O(h^{r+1})$ will immediately follow.

Let $(i,j) \in [i_0 - k_1 + 1 \,..\, i_0 + k_1] \times [j_0 - k_2 + 1 \,..\, j_0 + k_2]$ and $l_1 \in [0 \,..\, r_1]$, $l_2 \in [0 \,..\, r_2]$ be fixed. By Cauchy inequality,

$$\left| \left\langle R_r, P_i^{l_1} Q_j^{l_2} \right\rangle_{i,j} \right| \le \|R_r\|_{i,j} \cdot \left\| P_i^{l_1} Q_j^{l_2} \right\|_{i,j} = \|R_r\|_{i,j} \left\| P^{l_1} Q^{l_2} \right\| \sqrt{h_1 h_2}.$$

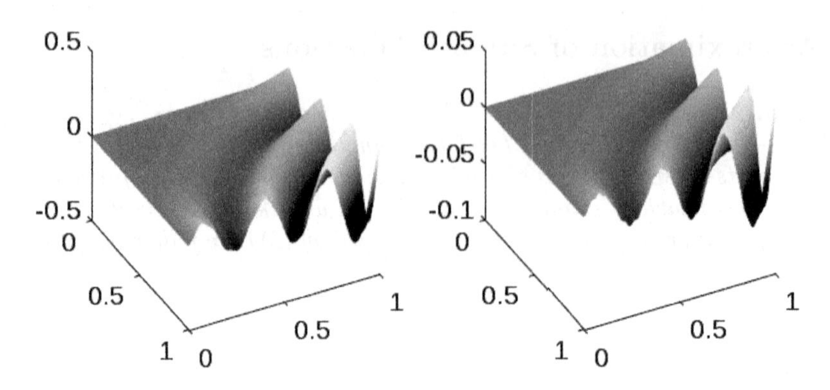

Fig. 2. Difference between $f(x,y)=\sin(2\pi^2xy)$ and its inverse F^1-transform

Therefore $\left|c_{i,j}^{l_1,l_2}(R_r)\right| \leq \frac{\|R_r\|_{i,j}}{\left\|P^{l_1}Q^{l_2}\right\|\sqrt{h_1h_2}}$. On the other hand, from the mean value theorem we have

$$\|R_r\|_{i,j}^2 = 4k_1k_2h_1h_2R_r^2(t_1,t_2)A_i(t_1)B_j(t_2)$$

for some $(t_1,t_2) \in E_i^{(1)} \times E_i^{(2)}$. Since $R_r(t_1,t_2) = O(h^{r+1})$, we obtain that $c_{i,j}^{l_1,l_2}(R_r) = O(h^{r+1})$.

Since $|i-i_0| \leq k_1+1$ and $P_i^{l_1}(x) = P^{l_1}\left(\frac{x-t_i^{(1)}}{h_1}\right)$, we have $x-t_i^{(1)} = O(h)$ and $\frac{x-t_i^{(1)}}{h_1} = O(1)$, thus also $P_i^{l_1}(x) = O(1)$. Similarly, $Q_j^{l_2}(y) = O(1)$. We conclude that $\mathcal{F}^m(R_r,x,y) = O(h^{r+1})$ and, as claimed,

$$f(x,y) - \mathcal{F}^m(f,x,y) = R_r(x,y) - \mathcal{F}^m(R_r,x,y) = O(h^{r+1}).$$

Example. In Fig. 2 we show the difference between the original function $f(x,y) = \sin(2\pi^2xy)$ and its inverse F^1-transform in two cases: with less basic functions (the left figure: $N_1 = N_2 = 17$) and with more basic functions (the right figure: $N_1 = N_2 = 27$) (for the sake of clarity $[a,b]$ and $[c,d]$ are chosen so that in both cases the Ruspini condition is fulfilled on $[0,1]\times[0,1]$). In both cases the fuzzy partitions are generated by cubic B-splines.

6 Conclusion

The paper is devoted to the continuous version of F^m-transforms based on B-splines of two variables. It is proved that it is possible to improve the approximation properties of two-dimensional fuzzy transforms by using B-splines as a tool for generating a fuzzy partition. Taking into account that a large part of applications (including applications in image processing) of the technique of F-transforms corresponds to the discrete case, we consider the extension of our approach to this case as one of main directions for our future research. We see that the obtained results will allow us to receive good approximative properties also for two-dimensional discrete F^m-transforms based on B-splines of two variables.

References

1. Bede, B., Rudas, I.J.: Approximation properties of fuzzy transforms. Fuzzy Sets Syst. **180**(1), 20–40 (2011)
2. Daňková, M., Hodáková, P., Perfilieva, I., Vajgl, M.: Edge detection using F-transform. In: Proceedings of ISDA 2011, pp. 672–677, Spain (2011)
3. Hodáková, P.: Fuzzy (F-)transform of functions of two variables and its applications in image processing. Ph.D. thesis, University of Ostrava, Ostrava, Czech Republic (2014)
4. Hodáková, P., Perfilieva, I.: F^1-transform of functions of two variables. In: Montero, J., Pasi, G., Ciucci, D. (eds.) Proceedings of EUSFLAT 2013. Advances in Intelligent Systems Research, Milan, Italy, vol. 32, pp. 547–553. Atlantis Press (2013)
5. Kodorane, I., Asmuss, S.: On approximation properties of spline based F-transform with respect to fuzzy m-partition. In: Montero, J., Pasi, G., Ciucci, D. (eds.) Proceedings of EUSFLAT 2013. Advances in Intelligent Systems Research, Milan, Italy, vol. 32, pp. 772–779. Atlantis Press (2013)
6. Kokainis, M., Asmuss, S.: Approximation properties of higher degree F-transforms based on B-splines. In: IEEE International Conference on Fuzzy Systems FUZZ-IEEE 2015, Istanbul, Turkey (2015)
7. Perfilieva, I.: Fuzzy approach to solution of differential equations with imprecise data: application to reef growth problem. In: Demicco, R., Klir, G. (eds.) Fuzzy Logic in Geology, pp. 275–300. Academic Press, Cambridge (2003)
8. Perfilieva, I.: Fuzzy transforms. In: Peters, J.F., Skowron, A., Dubois, D., Grzymała-Busse, J.W., Inuiguchi, M., Polkowski, L. (eds.) Transactions on Rough Sets II. LNCS, vol. 3135, pp. 63–81. Springer, Heidelberg (2004)
9. Perfilieva, I.: Fuzzy transforms: theory and applications. Fuzzy Sets Syst. **157**(8), 993–1023 (2006)
10. Perfilieva, I.: F-transform. In: Kacprzyk, J., Pedrycz, W. (eds.) Handbook of Computational Intelligence, pp. 113–130. Springer, Heidelberg (2015)
11. Perfilieva, I., Chaldeeva, E.: Fuzzy transformation and its applications. In: Proceedings of the 4th Czech-Japan Seminar on Data Analysis and Decision Making under Uncertainty, pp. 116–124. Czech Republic (2001)
12. Perfilieva, I., Danková, M., Bede, B.: Towards a higher degree F-transform. Fuzzy Sets Syst. **180**(1), 3–19 (2011)
13. Perfilieva, I., Daňková, M., Vajgl, M., Hodáková, P.: F-transform based image fusion. In: Ukimura, O. (ed.) Image Fusion, pp. 3–22. INTECH, Rijeka (2011)
14. Perfilieva, I., Hodáková, P., Hurtik, P.: Differentiation by the F-transform and application to edge detection. Fuzzy Sets Syst. **288**, 96–114 (2016)
15. Schoenberg, I.J.: On interpolation by spline functions and its minimal properties. In: On Approximation Theory / Über Approximationstheorie. ISNM, vol. 5, pp. 109–129. Birkhäuser (1964)
16. Schoenberg, I.J.: Cardinal Spline Interpolation. CBMS, vol. 12. SIAM, Philadelphia (1973)
17. Štěpnička, M., Valášek, R.: Fuzzy transforms for function with two variables. In: Proceedings of the 6th Czech-Japan Seminar on Data Analysis and Decision Making under Uncertainity, pp. 100–107. Czech Republic (2003)

Gaussian Noise Reduction Using Fuzzy Morphological Amoebas

Manuel González-Hidalgo, Sebastia Massanet(✉), Arnau Mir, and Daniel Ruiz-Aguilera

Department of Mathematics and Computer Science,
University of the Balearic Islands, Cra. de Valldemossa, km 7,5, 07122 Palma, Spain
{manuel.gonzalez,s.massanet,arnau.mir,daniel.ruiz}@uib.es

Abstract. Many image processing and computer vision applications require a preprocessing of the image to remove or reduce noise. Gaussian noise is a challenging type of noise whose removal has led to the proposal of several noise filters. In this paper we present a novel version of the morphological filters based on amoebas with the aim to incorporate fuzzy logic into them to achieve a better treatment of the uncertainty. The experimental results show that the proposed algorithm outperforms the classical amoeba-based filters both from the visual point of view and the quantitative performance values for images corrupted with Gaussian noise with standard deviation from 10 to 30.

Keywords: Noise reduction · Fuzzy mathematical morphology · Morphological amoebas · t-norms · Fuzzy implications

1 Introduction

One of the most important problems in image processing is the image noise reduction. In many cases, due to various reasons, the image acquiring process is affected by the introduction of unwanted information. There are mainly three types of noise models: the impulsive noise, in which only some of the pixels of the image are affected; the additive noise, where the value of a random variable with a certain distribution is added to all the pixels of the image; and the multiplicative noise, which depends on the intensity level (an example would be the speckle noise). This article deals with the additive noise introduced by a Gaussian distribution (called Gaussian noise).

Different techniques and methods have been proposed to reduce Gaussian noise, see [20] for a recent review. One of these approaches is the spatial domain, which includes local or non-local filters, that attempt to take advantage of the correlations which exist in most natural images [20]. Another well-known filters are Wiener, mean, Gaussian, anisotropic, median or bilateral. Another category is the transform domain filtering methods, where the images are represented by orthonormal basis with a series of coefficients. Examples of this technique are the wavelets based methods [20]. Finally, we can cite also learning-based filtering methods [6,23] and mathematical morphology based filters [1,17].

J.P. Carvalho et al. (Eds.): IPMU 2016, Part I, CCIS 610, pp. 660–671, 2016.
DOI: 10.1007/978-3-319-40596-4_55

Although all these methods deal well enough with the reduction of the Gaussian noise, in many cases this noise reduction implies an alteration of relevant information of the image such as contours or different textures. These alterations can cause serious problems with further processing of the image. Trying to solve this last problem, a new proposal was made by Lerallut et al. [16], where the concept of morphological amoebas was introduced. The main idea of this approach is to use dynamic structuring elements in the application of the morphological operators, in opposition to what is usually done in classic filtering, where there is only one kernel or structuring element, which does not depend on the pixel where is being applied.

From another side, the fuzzy mathematical morphology generalizes the binary morphology [19] using concepts and techniques of the theory of fuzzy sets [3,13]. This theory allows a better processing and a representation with higher flexibility of the uncertainly and the ambiguity present in each level in an image. The four basic morphological operations are dilation, erosion, closing and opening and because the grey level images can be viewed as fuzzy sets (see [18]), morphological fuzzy operators can be defined using fuzzy tools. Therefore, conjunctions (continuous t-norms and uninorms, see [12]) and their residuals implications have been used. Fuzzy mathematical morphology has been applied in different studies to remove salt-and-pepper noise [9,10], obtaining competitive results. The idea of this work is to bring the morphological amoebas into the fuzzy mathematical morphology, in order to apply them to reduce Gaussian noise. To accomplish this goal, we propose an algorithm that based on the ideas behind some non-local filters such as [4] assigns grey-level values to the pixels of the amoeba according to a similarity measure.

The structure of the communication is as follows. In the next section, the Gaussian noise model and the definitions and properties of the fuzzy morphological operators are recalled. In Sect. 3 the morphological amoebas are studied and defined from the fuzzy mathematical morphology point of view, and a new algorithm based on them is proposed. Then, in Sect. 4 the numerical results of the performed experiments are analysed. In the last section some conclusions and future work are discussed.

2 Preliminaries

2.1 Noise Models

Noisy sensors or transmission channels are the main sources of noise in digital images. One type of noise is the Gaussian noise, that is created during the acquisition phase due to poor illumination of sensors, high temperature, transmission, etc. It has a Gaussian distribution being one of the toughest types of noise to reduce. Its probability density function is given by $p(x) = \frac{1}{\sigma\sqrt{2\pi}}e^{-\frac{(x-\mu)^2}{2\sigma^2}}$, where μ is its mean and σ, its standard deviation.

Let F be the original image and $\overline{F}$ be the image corrupted with Gaussian noise of mean μ and standard deviation σ. If x is a pixel and $F(x)$ and $\overline{F}(x)$ are the grey level values of images F and $\overline{F}$, respectively, we have $\overline{F}(x) = F(x)+G(x)$ where $G(x)$ is a random value of a Gaussian distribution of mean μ and standard deviation σ.

2.2 Fuzzy Mathematical Morphology

Fuzzy morphological operators are defined using fuzzy operators such as fuzzy conjunctions, like t-norms, and fuzzy implications. More details on these logical connectives can be found in [14] and [2], respectively. Let us recall the definitions of t-norms and fuzzy implications.

Definition 1 [14]. *A* t-norm *is a commutative, associative, non-decreasing function* $T : [0,1]^2 \to [0,1]$ *with neutral element* 1*, i.e.,* $T(1,x) = x$ *for all* $x \in [0,1]$.

Definition 2 [2]. *A binary operator* $I : [0,1]^2 \to [0,1]$ *is a* fuzzy implication *if it is non-increasing in the first variable, non-decreasing in the second one and it satisfies* $I(0,0) = I(1,1) = 1$ *and* $I(1,0) = 0$.

A well-known way to obtain fuzzy implications is the residuation method. Given a t-norm T, the binary operator $I_T(x,y) = \sup\{z \in [0,1] \mid T(x,z) \leq y\}$ is a fuzzy implication called the *residual implication* or the *R-implication* of T.

The different t-norms and the different residual implications that will be considered in Sect. 4 are displayed in Table 1.

Table 1. Considered t-norms with their residual implications.

Name	t-norm	Residual implication
Minimum	$T_{\mathrm{M}}(x,y) = \min(x,y)$	$I_{\mathrm{GD}}(x,y) = \begin{cases} 1 & \text{if } x \leq y \\ y & \text{if } x > y \end{cases}$
Nilpotent Minimum	$T_{\mathrm{nM}}(x,y) = \begin{cases} 0 & \text{if } x+y \leq 1 \\ \min(x,y) & \text{otherwise} \end{cases}$	$I_{\mathrm{FD}}(x,y) = \begin{cases} 1 & \text{if } x \leq y \\ \max(1-x,y) & \text{if } x > y \end{cases}$
Łukasiewicz	$T_{\mathrm{LK}}(x,y) = \max(x+y-1,0)$	$I_{\mathrm{LK}}(x,y) = \min(1, 1-x+y)$

Using the previous operators, we can define the basic fuzzy morphological operators such as dilation and erosion. We will use the following notation: T denotes a t-norm, I a fuzzy implication, F a grey-level image, and B a grey-level structuring element (see [13] for formal definitions), d_F denotes the set of points where F is defined and $T_v(F)$ is the translation of a fuzzy set F by $v \in \mathbb{R}^n$ defined by $T_v(F)(x) = F(x-v)$.

Definition 3 [18]. *The* fuzzy dilation $D_T(F,B)$ *and the* fuzzy erosion $E_I(F,B)$ *of* F *by* B *are the grey-level images defined by*

$$D_T(F,B)(y) = \sup_{x \in d_F \cap T_y(d_B)} T(B(x-y), F(x)),$$
$$E_I(F,B)(y) = \inf_{x \in d_F \cap T_y(d_B)} I(B(x-y), F(x)).$$

From the fuzzy erosion and the fuzzy dilation, the fuzzy opening and the fuzzy closing of a grey-level image F by a structuring element B can be defined.

Definition 4 [7]. *The* fuzzy closing $C_{T,I}(F, B)$ *and the* fuzzy opening $O_{T,I}(F, B)$ *of* F *by* B *are the grey-level images defined by*

$$C_{T,I}(F, B)(y) = E_I(D_T(F, B), -B)(y), \quad O_{T,I}(F, B)(y) = D_T(E_I(F, B), -B)(y).$$

A more detailed account on these operators, their properties and applications can be found in [7,8,11,18]. In particular, when I is the R-implication of T, most of the desirable algebraical properties of a mathematical morphology hold.

3 Introducing Fuzziness in Morphological Amoebas

In this section, we review the literature concerning morphological amoebas, the morphological operators derived from them and some of the applications where these operators have been used. In particular, the definition of a morphological amoeba and the idea and motivation behind its introduction is recalled. After that, we introduce the definition of a fuzzy morphological amoeba and the basic morphological operators (erosion, dilation, closing and opening) derived from them. From these operators, a filtering algorithm for Gaussian noise reduction based on these operators is proposed.

3.1 Morphological Amoebas

As we have already commented, mathematical morphology uses an structuring element to process the image. Traditionally, this structuring element is usually a fixed, space-invariant image in all the pixels of the image. However, this approach has an important drawback. Since the local features of an image change in general all over the image, a fixed structuring element may not be the best option to process all the pixels of the image. In particular, fixed structuring elements may remove thin elements or displace contours leading to a bad performance of the concrete algorithm applied to the image. This effect is particularly highlighted in noise reduction applications. In this field, the performance of a filter is usually evaluated in terms of a balance between the noise reduction or removal and the preservation of the features (edges, texture) of the image. This trade-off is problematic because if one wants to design a morphological filter to remove high percentages of noise, a large structuring element must be chosen. However, in this case, the small details of the image are removed hindering the restoration of the image.

To overcome this problem, adaptive structuring elements have been proposed. These structuring elements are able to adapt themselves to the content of the image in each pixel keeping the details of the image. One of the most important and successful approaches in this topic is the one based on morphological amoebas introduced in [16]. An amoeba is an organism which is capable of altering its shape by extending and retracting its pseudopods, temporary extensions of its cell. The behaviour and ability of this organism induced the proposal of some morphological operators based on adaptive structuring elements, called *morphological amoebas* which try to reproduce this behaviour. Thus, for each pixel of the image, the shape of the amoeba is computed through the concept of amoeba distance.

Definition 1 [16]. *The* amoeba distance d_λ *with parameter* $\lambda > 0$, $\lambda \in \mathbb{R}$, *between two pixels* x, y *of an image* F *is given by:*

$$d_\lambda(x,y) = \begin{cases} 0 & \text{if } x = y, \\ \min\limits_{\sigma(x,y)} L(\sigma(x,y)) & \text{if } x \neq y, \end{cases}$$

where $\sigma(x,y) = \{x = x_0, x_1, \ldots, x_n = y\}$ *is a path between these two pixels and* $L(\sigma(x,y))$ *denotes the length of the path* $\sigma(x,y)$ *and it is given by*

$$L(\sigma(x,y)) = \sum_{i=0}^{n} (1 + \lambda \cdot |F(x_i) - F(x_{i+1})|).$$

It is necessary to bound the extension of the amoeba. In practice, fixed a pixel x, we will only compute the amoeba distance of this pixel x to all the pixels in a concrete window W_x centred in x. Thus, the amoeba is limited by this window and it is also limited to those pixels with an amoeba distance lower than a maximum amoeba distance $d_{\max}$. Specifically, $A(x) = \{y \mid y \in W_x, d_\lambda(x,y) \leq d_{\max}\}$.

Once the construction of the amoeba is established, in [16], the morphological operators based on morphological amoebas are defined.

Definition 2 [16]. *The* morphological dilation $D(F)$ *and the* morphological erosion $E(F)$ *based on morphological amoebas of an image* F *are defined as*

$$D(F)(x) = \max_{z \in A(x)} F(z), \quad E(F)(x) = \min_{z \in A(x)} F(z),$$

where $A(x)$ *is the amoeba of pixel* x.

Remark 1. The morphological dilation and morphological erosion given in Definition 2 are equivalent to the t-morphological operators for grey-level images (see [18] for more details) where now, the maximum and the minimum are not computed using a fixed structuring element, but on the pixels of the amoeba.

From these two basic morphological operators, the morphological closing and the morphological opening based on morphological amoebas can be defined.

Definition 3 [16]. *The* morphological closing $C(F)$ *and the* morphological opening $O(F)$ *based on morphological amoebas of an image* F *are defined as* $C(F) = E(D(F))$ *and* $O(F) = D(E(F))$, *respectively.*

Note that when the closing or the opening based on morphological amoebas are computed, every pixel is affected by two amoebas, the one computed from the original image when the dilation is applied and then, the one computed from the dilated image when the erosion is applied. These two morphological operators are the basis of the so-called *alternate filters* (see [19,21]), given by combinations of openings and closings. Although in some morphological frameworks, in particular for fixed structuring elements, the filters $C(O(F))$, $O(C(F))$, $O(O(F))$ and $C(C(F))$ are increasing and idempotent, this is not true in the amoeba framework due to the change of the amoeba in each step of the filter. When this occurs, the so-called *Alternate Sequential Filters* ASF (see [19,21]), which are combinations of the alternate filters, can be defined. These operators based on morphological amoebas are defined in [16] and their performance is visually assessed. Nevertheless, the authors state that the basic filters that apply the median or the mean over the amoebas are computationally less expensive and provide similar results than the ASF filters. We will denote by mean(F) and median(F) these two amoeba-based filters. These two last filters will be considered in Sect. 4 and compared with the filter presented in this paper.

3.2 Fuzzy Morphological Amoebas and Derived Alternate Filters

Once we have recalled the motivation and the definitions of the classical morphological amoebas, we want to introduce fuzziness into their definition and to propose fuzzy morphological operators based on these fuzzy morphological amoebas. The main problem that we have to overcome to achieve this goal is related to the grey-level values of the pixels of the amoeba. While in classical grey-level t-morphological operators, the structuring element has no grey-level values and it is only used to select those pixels from which the maximum or the minimum is computed, in fuzzy morphological operators the grey-level pixels of the structuring element play a key role (see Definition 3). Therefore, the fuzzy morphological amoeba FA at a pixel x of the image F is computed through the following steps:

1. The classical morphological amoeba $A(x)$ at that pixel is computed using Definition 1 within a window centred at x of size Lws.
2. For each pixel y of the amoeba, we consider a window centred at y of size $Sws < Lws$ and a window of the same size centred at x and we construct the subimages F_y and F_x of F given by these two windows.
3. A degree of similarity between F_y and F_x is computed using the following similarity measure:

$$EQ_{\sigma DI}(F_y, F_x) = 1 - \frac{1}{(Sws)^2} \sum_{z \in d(F_y) \cap T_{x-y}(d_{F_x})} |F_y(z) - F_x(z - x + y, z - x + y)|.$$

   This measure has interesting properties [5]: $0 \leq EQ_{\sigma DI} \leq 1$ and larger values of $EQ_{\sigma DI}$ are indicators of a greater similarity between F_x and F_y.
4. We assign a value of $sgl \in [0,1]$ to the grey-level of the pixels y of the amoeba with a lowest value $EQ_{\sigma DI}(F_y, F_x)$. We assign 1 to the grey-level of the pixels y of the amoeba such that $EQ_{\sigma DI}(F_y, F_x) = 1$. The grey-levels of the remaining pixels of the amoeba are assigned proportionally, that is

$$FA_x(y) = 1 - (1 - EQ_{\sigma DI}(F_y, F_x)) \cdot \frac{1 - sgl}{1 - \min_{z \in A(x)} EQ_{\sigma DI}(F_z, F_x)}.$$

We will denote by $FA_x^{sgl,Lws,Sws}$ the fuzzy morphological amoeba at a pixel x computed using the previous steps, to make explicit the dependence of the amoeba to the parameters sgl, Lws and Sws. The idea behind this algorithm is analogous to the one presented in [4]. There, it is stated that every small window in a natural image has many similar windows in the same image. As greater is the similarity between the small window centred at a pixel x and the one centred at a pixel y, greater should be the weight of y into the denoising of x. Thus, as greater is the similarity, greater is the grey-level value of a pixel of the amoeba. Taking into account the previous discussion, the fuzzy morphological operators given in Definition 3 can be rewritten using these fuzzy morphological amoebas.

Definition 5. *The* fuzzy morphological dilation $D_T(F)$, erosion $E_I(F)$, closing $C_{T,I}(F)$ *and* opening $O_{T,I}(F)$ *of an image* F *using the fuzzy morphological amoebas* $FA_x^{sgl,Lws,Sws}$ *with* $x \in d_F$*, denoted by* FA_x *for short, are respectively the grey-level images defined by for all* $y \in d_F$

$$\begin{aligned} D_T(F)(y) &= \sup_{x \in d_F \cap d_{FA_y}} T(FA_y(x), F(x)), \\ E_I(F)(y) &= \inf_{x \in d_F \cap d_{FA_y}} I(FA_y(x), F(x)), \\ C_{T,I}(F)(y) &= E_I(D_T(F))(y), \quad O_{T,I}(F)(y) = D_T(E_I(F))(y). \end{aligned}$$

Since it can be checked that $FA_x^{sgl,Lws,Sws}(x) = 1$ for all $x \in d_F$, the following proposition gives some inclusions between the fuzzy morphological operators based on fuzzy morphological amoebas.

Proposition 6. *Let* F *be an image. The following inclusions hold:*

$$E_I(F) \subseteq F \subseteq D_T(F), \qquad E_I(F) \subseteq O_{T,I}(F), \qquad C_{T,I}(F) \subseteq D_T(F).$$

From the fuzzy morphological closing and opening, we can define the alternate filters $C_{T,I}(O_{T,I}(F))$, $O_{T,I}(C_{T,I}(F))$, $O_{T,I}(O_{T,I}(F))$ and $C_{T,I}(C_{T,I}(F))$. These operations can be useful to reduce Gaussian noise as it will be shown in Sect. 4. In Fig. 1, the images obtained after applying the fuzzy morphological operators based on fuzzy morphological amoebas to an image corrupted with Gaussian noise are depicted. In addition, the result obtained by the filter $O_{T,I}(C_{T,I}(F))$ is also included. The Gaussian noise has been notably reduced while the details and texture of the image have been preserved.

4 Experiments

In this section we develop some experimental results to show the potential of the alternate filters defined from fuzzy morphological operators based on fuzzy morphological amoebas. Specifically, we have compared the performances of the alternate filter $O_{T,I}(C_{T,I})$ using T_{M}, T_{nM} and T_{LK} and their corresponding R-implications with the well-known methods of classic amoebas using mean and median filters (see previous section for details) and the Kuwahara filter [15].

(a) Original image (b) Gaussian $\sigma = 15$ (c) Fuzzy dilation (d) Fuzzy erosion

(e) Fuzzy closing (f) Fuzzy opening (g) Alt. filter OC

Fig. 1. Images obtained applying some fuzzy morphological operators based on fuzzy morphological amoebas with $T = T_{\mathrm{M}}$ and its residual implication.

4.1 Framework

For the experiments, 15 images of the miscellaneous volume of the USC-SIPI image database of the University of South Carolina[1] have been selected and corrupted with Gaussian noise with $\mu = 0$ and $\sigma \in \{10, 15, 20, 25, 30\}$. The six filters have been applied to these images and their results have been compared.

In addition to the visual comparison of the filtered images obtained by the filters, the restoration performance will be quantitatively measured by the widely used performance objective measure SSIM. Let F_1 and F_2 be two images of dimensions $M \times N$. In the following, we suppose that F_1 is the original noise-free image and F_2 is the restored image for which some filter has been applied. The structural similarity index measure (SSIM) was introduced in [22] under the assumption that human visual perception is highly adapted for extracting structural information from a scene. The measure is defined as follows:

$$\mathrm{SSIM}(F_2, F_1) = \frac{(2\mu_1\mu_2 + C_1)}{(\mu_1^2 + \mu_2^2 + C_1)} \cdot \frac{(2\sigma_{12} + C_2)}{(\sigma_1^2 + \sigma_2^2 + C_2)},$$

where μ_k, $k = 1, 2$ are the means of the images F_1 and F_2 respectively, σ_k^2 is the variance of each image, σ_{12} is the covariance between the two images, $C_1 = (0.01 \cdot 255)^2$ and $C_2 = (0.03 \cdot 255)^2$. Larger values of SSIM ($0 \leq SSIM \leq 1$) are indicators of better capabilities for noise reduction and image recovery.

[1] This image database can be downloaded from http://sipi.usc.edu/database/misc.tar.gz.

Table 2. Parameters of the alternate filter $O_{T,I}(C_{T,I})$ for the three considered t-norms and their corresponding R-implications. For all the configurations, it has been used $d_{\max} = 0.196$.

t-norm	σ	LwS	SwS	λ	sgl	σ	LwS	SwS	λ	sgl	σ	LwS	SwS	λ	sgl
T_{M}		7	3	7.0	0.118		7	3	3.2	0.118		9	3	2.1	0.118
T_{nM}	10	7	3	4.5	0.039	15	7	3	3.0	0.039	20	9	3	2.1	0.039
T_{LK}		9	3	4.3	0.004		9	3	3.0	0.004		9	3	2.15	0.004
T_{M}		9	3	1.8	0.0118		9	3	1.55	0.118					
T_{nM}	25	9	3	1.75	0.039	30	9	3	1.55	0.039					
T_{LK}		9	3	1.8	0.004		9	3	1.5	0.004					

Table 3. Mean ($\overline{x}$) and standard deviation (s) of the SSIM values obtained by the considered filters and number of images in which every configuration $O_{T,I}(C_{T,I})$ outperforms the other three methods.

	$\sigma = 10$			$\sigma = 15$		
Method	$\overline{x}$	s	# Images	$\overline{x}$	s	# Images
$O_{T,I}(C_{T,I})$ (T_{nM})	0.9105	0.0527		0.8596	0.0695	
$O_{T,I}(C_{T,I})$ (T_{M})	0.8891	0.0678		0.8545	0.0719	
$O_{T,I}(C_{T,I})$ (T_{LK})	0.9135	0.0525		0.8615	0.0716	
Classic amoebas (mean)	0.8901	0.8090	(12, 12, 11)	0.8085	0.1109	(12, 12, 12)
Classic amoebas (median)	0.8947	0.0688	(12, 12, 12)	0.8201	0.0976	(12, 12, 12)
Kuwahara	0.8774	0.0572	(13, 7, 13)	0.8372	0.0652	(13, 12, 13)
	$\sigma = 20$			$\sigma = 25$		
$O_{T,I}(C_{T,I})$ (T_{nM})	0.8089	0.0815		0.7619	0.0893	
$O_{T,I}(C_{T,I})$ (T_{M})	0.8081	0.0816		0.7605	0.0903	
$O_{T,I}(C_{T,I})$ (T_{LK})	0.8103	0.0818		0.7618	0.0899	
Classic amoebas (mean)	0.7418	0.1273	(12, 12, 12)	0.6727	0.1404	(12, 13, 13)
Classic amoebas (median)	0.7560	0.1148	(12, 12, 12)	0.6891	0.1283	(12, 13, 13)
Kuwahara	0.7937	0.0731	(13, 12, 12)	0.7494	0.0799	(11, 10, 12)
	$\sigma = 30$					
$O_{T,I}(C_{T,I})$ (T_{nM})	0.6992	0.0984				
$O_{T,I}(C_{T,I})$ (T_{M})	0.7143	0.0946				
$O_{T,I}(C_{T,I})$ (T_{LK})	0.7173	0.0922				
Classic amoebas (mean)	0.6117	0.1460	(14, 13, 13)			
Classic amoebas (median)	0.6282	0.1364	(14, 13, 11)			
Kuwahara	0.7073	0.0838	(7, 11, 11)			

4.2 Experimental Results

First of all, in Table 2, for every σ, the best parameters values for the filter $O_{T,I}(C_{T,I})$ with the three different t-norms and residual implications have been established according to the SSIM values. These parameters values have been used for the comparison with the other considered filters.

In Table 3, we show the mean, standard deviation and the number of images in which every configuration of $O_{T,I}(C_{T,I})$ outperforms the other three filters.

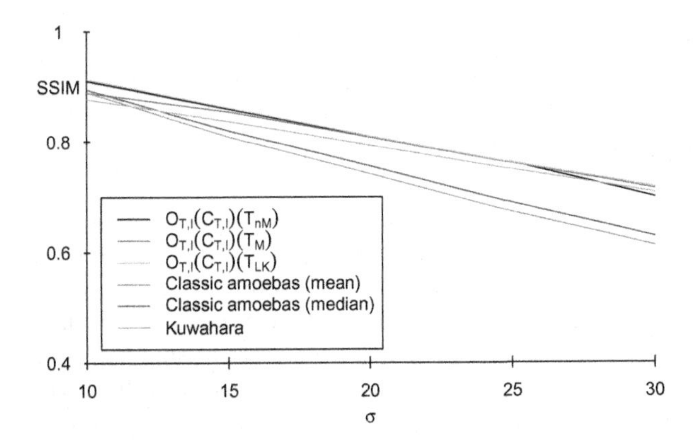

Fig. 2. Plot of the mean of the SSIM measures of the 15 images.

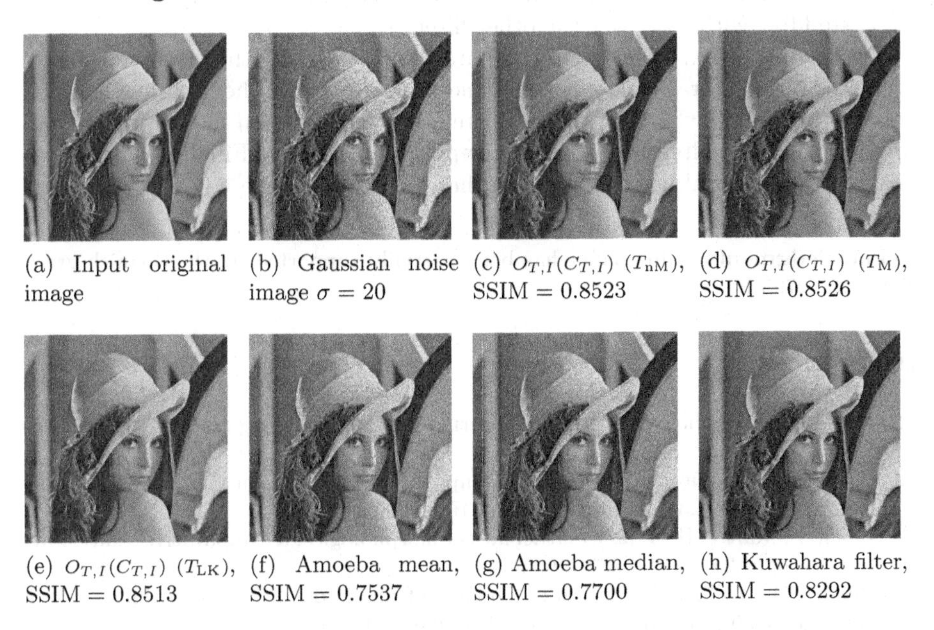

(a) Input original image (b) Gaussian noise image $\sigma = 20$ (c) $O_{T,I}(C_{T,I})$ (T_{nM}), SSIM = 0.8523 (d) $O_{T,I}(C_{T,I})$ (T_{M}), SSIM = 0.8526

(e) $O_{T,I}(C_{T,I})$ (T_{LK}), SSIM = 0.8513 (f) Amoeba mean, SSIM = 0.7537 (g) Amoeba median, SSIM = 0.7700 (h) Kuwahara filter, SSIM = 0.8292

Fig. 3. Results obtained by the different filters for the corrupted image in (b).

For instance, with $\sigma = 10$, $O_{T,I}(C_{T,I})$ with T_{nM} or T_{M} outperform in 12 of the 15 images the classic amoeba-based mean filter, while with T_{LK} outperforms in 11 of the images. As it can be seen, the $O_{T,I}(C_{T,I})$ with T_{LK} and its residual implication has the highest mean for all the different values of σ (see Fig. 2). Moreover, the Wilcoxon statistical test concludes that the differences in average with the second filter are statistically significant for all σ values except for $\sigma = 30$. In addition to a higher mean value, the alternate filters $O_{T,I}(C_{T,I})$ and Kuwahara have the lowest standard deviation of the SSIM measure. This fact ensures the robustness of this approach. In Fig. 3, we display the results obtained by the different filters for the "Lenna" image corrupted with Gaussian noise $\sigma = 20$.

5 Conclusions and Future Work

In this paper, we have introduced the concept of fuzzy morphological amoebas which are morphological amoebas that incorporate fuzziness in their expressions. These amoebas are used to define fuzzy morphological operators. From these operators, an alternate filter based on this theory has been proposed to remove Gaussian noise in images. The filter has been tested using the t-norms T_{M}, T_{nM} and T_{LK} and it has been shown that it outperforms, both from the visual point of view and the quantitative performance measure, other well-known Gaussian filters as the classical amoebas filters based on the mean and the median and the Kuwahara filter. The Wilcoxon statistical test ensures the superiority of the alternate filter with T_{LK} over the other filters.

As a future work, we want to compare this filter with other Gaussian filters presented in the literature, to use some method to estimate the σ of the Gaussian noise to determine which set of parameters has to be chosen in each image and to consider other alternate filters to improve the results. Finally, we want to analyse the potential use of fuzzy morphological amoebas in edge detection.

Acknowledgments. This paper has been partially supported by the Spanish grant TIN2013-42795-P.

References

1. Angulo, J.: Morphological bilateral filtering. SIAM J. Imaging Sci. **6**(3), 1790–1822 (2013)
2. Baczyński, M., Jayaram, B.: Fuzzy Implications. Studies in Fuzziness and Soft Computing, vol. 231. Springer, Berlin, Heidelberg (2008)
3. Bloch, I., Maître, H.: Fuzzy mathematical morphologies: a comparative study. Pattern Recogn. **28**, 1341–1387 (1995)
4. Buades, A., Coll, B., Morel, J.: A review of image denoising algorithms, with a new one. Multiscale Model. Simul. **4**(2), 490–530 (2005)
5. Bustince, H., Pagola, M., Barrenechea, E.: Construction of fuzzy indices from fuzzy DI-subsethood measures: application to the global comparison of images. Inf. Sci. **177**, 906–929 (2007)
6. Chatterjee, P., Milanfar, P.: Clustering-based denoising with locally learned dictionaries. IEEE Trans. Image Process. **18**(7), 1438–1451 (2009)
7. De Baets, B.: Fuzzy morphology: a logical approach. In: Ayyub, B.M., Gupta, M.M. (eds.) Uncertainty Analysis in Engineering and Science: Fuzzy Logic. Statistics, and Neural Network Approach, pp. 53–68. Kluwer Academic Publishers, Norwell (1997)
8. González-Hidalgo, M., Massanet, S.: Closing and opening based on discrete t-norms. Applications to natural image analysis. In: Galichet, S., Montero, J., Mauris, G. (eds.) Proceedings of EUSFLAT-LFA-2011 Advances in Intelligent Systems Research, vol. 17, pp. 358–365. Atlantis Press (2011)

9. González-Hidalgo, M., Massanet, S., Mir, A., Ruiz-Aguilera, D.: A fuzzy filter for high-density salt and pepper noise removal. In: Bielza, C., et al. (eds.) Advances in Artificial Intelligence. LNCS, vol. 8109, pp. 70–79. Springer, Berlin, Heidelberg (2013)
10. González-Hidalgo, M., Massanet, S., Mir, A., Ruiz-Aguilera, D.: High-density impulse noise removal using fuzzy mathematical morphology. In: Pasi, G., Montero, J., Ciucci, D. (eds.) Proceedings of the 8th conference of the European Society of Fuzzy Logic and Technology Conference (EUSFLAT 2013), pp. 728–735. Atlantis Press, Milano, Italy (2013)
11. González-Hidalgo, M., Mir-Torres, A., Ruiz-Aguilera, D., Torrens, J.: Applications of morphological operators based on uninorms. In: ESTYLF 2008, pp. 203–210. European Centre for Soft Computing, Asturias (2008)
12. González-Hidalgo, M., Mir-Torres, A., Ruiz-Aguilera, D., Torrens, J.: Image analysis applications of morphological operators based on uninorms. In: Proceedings of the IFSA-EUSFLAT 2009 Conference, Lisbon, Portugal, pp. 630–635 (2009)
13. Kerre, E., Nachtegael, M.: Fuzzy Techniques in Image Processing. Studies in Fuzziness and Soft Computing, vol. 52. Springer, New York (2000)
14. Klement, E., Mesiar, R., Pap, E.: Triangular Norms. Kluwer Academic Publishers, London (2000)
15. Kuwahara, M., Hachimura, K., Eiho, S., Kinoshita, M.: Processing of ri-angiocardiographic images. In: Preston Jr., J., Onoe, M. (eds.) Digital Processing of Biomedical Images, pp. 187–202. Springer, New York (1976)
16. Lerallut, R., Decencire, E., Meyer, F.: Image filtering using morphological amoebas. Image Vis. Comput. **25**(4), 395–404 (2007)
17. Mendiola-Santibañez, J.D., Terol-Villalobos, I.R.: Filtering of mixed gaussian and impulsive noise using morphological contrast detectors. Image Process. IET **8**(3), 131–141 (2014)
18. Nachtegael, M., Kerre, E.: Classical and fuzzy approaches towards mathematical morphology. In: Kerre, E.E., Nachtegael, M. (eds.) Fuzzy techniques in image processing, Chapter 1. Studies in Fuzziness and Soft Computing, vol. 52, pp. 3–57. Physica-Verlag, New York (2000)
19. Serra, J.: Image Analysis and Mathematical Morphology, vols. 1, 2. Academic Press, London (1982, 1988)
20. Shao, L., Yan, R., Li, X., Liu, Y.: From heuristic optimization to dictionary learning: A review and comprehensive comparison of image denoising algorithms. IEEE Trans. Cybern. **44**(7), 1001–1013 (2014)
21. Soille, P.: Morphological Image Analysis. Springer, Berlin, Heidelberg (1999)
22. Wang, Z., Bovik, A.C., Sheikh, H.R., Simoncelli, E.P.: Image quality assessment: from error visibility to structural similarity. IEEE Trans. Image Process. **13**(4), 600–612 (2004)
23. Yan, R., Shao, L., Liu, Y.: Nonlocal hierarchical dictionary learning using wavelets for image denoising. IEEE Trans. Image Process. **22**(12), 4689–4698 (2013)

Clustering

Proximal Optimization for Fuzzy Subspace Clustering

Arthur Guillon[1(✉)], Marie-Jeanne Lesot[1], Christophe Marsala[1], and Nikhil R. Pal[2]

[1] Sorbonne Universités, UPMC Univ Paris 06, CNRS, LIP6 UMR 7606, 4 place Jussieu, 75005 Paris, France
{Arthur.Guillon,Marie-Jeanne.Lesot,Christophe.Marsala}@lip6.fr
[2] Indian Statistical Institute, Calcutta 700 108, West Bengal, India
nikhil@isical.ac.in

Abstract. This paper proposes a fuzzy partitioning subspace clustering algorithm that minimizes a variant of the FCM cost function with a weighted Euclidean distance and a penalty term. To this aim it considers the framework of proximal optimization. It establishes the expression of the proximal operator for the considered cost function and derives PFSCM, an algorithm combining proximal descent and alternate optimization. Experiments show the relevance of the proposed approach.

Keywords: Fuzzy partitioning clustering · Subspace clustering · Proximal descent

1 Introduction

Subspace clustering [1] is an unsupervised machine learning task that aims at partitioning data into groups with strong internal similarity and external dissimilarity (just as clustering) while also discovering the best subspaces to represent these clusters. The identified subspaces are required to be minimal, yet sufficient to describe the clusters they contain.

The definition of subspace clustering requires the identification of the clusters and of their subspaces to be simultaneous: indeed, if either clusters or their subspaces are known beforehand, the problem reduces to finding the subspaces or correct description of the clusters, respectively. In addition, as opposed to feature selection, different clusters are most of the time discovered in different subspaces.

As briefly sketched in Sect. 2, there exist several families of techniques and algorithms to solve the subspace clustering problem, as well as various representations of the subspaces, depending on the intended application of the subspace clustering.

This paper places itself in the partitioning paradigm in a fuzzy setting and produces clusters identified by a center. Moreover, it discovers axis-parallel subspaces, which are thus identified by weights on the original data features. An original cost function formalises these concepts and adds, to a FCM cost function [3]

J.P. Carvalho et al. (Eds.): IPMU 2016, Part I, CCIS 610, pp. 675–686, 2016.
DOI: 10.1007/978-3-319-40596-4_56

with weighted Euclidean distance, a penalty term expressing constraints to identify the relevant subspaces.

As this penalty term is not differentiable, standard optimization techniques such as alternate optimization are not available. This paper introduces a novel optimization scheme, exploiting tools from the proximal descent theory [8]. The utilisation of such techniques is still relatively new in machine learning and in clustering in particular [9].

This paper proposes an innovative implementation of this theoretical paradigm in the fuzzy subspace clustering framework. It establishes a theorem giving the expression of the proximal operator allowing the optimization of the considered cost function. Finally, it proposes an algorithm, called PFSCM, standing for Proximal Fuzzy Subspace C-Means, using this result to solve the subspace clustering problem through the combination of proximal descent and alternate optimization.

This paper is structured as follows: in Sect. 2, related works and the scientific context of subspace clustering are summed up. A new cost function is presented and studied in Sect. 3. In Sect. 4, the implementation of proximal descent is studied to optimize the proposed function, leading to the update equation from which the PFSCM algorithm is derived. This algorithm is then experimentally validated in Sect. 5.

2 Related Works

Subspace clustering [1] can be seen as a combination of clustering and feature selection tasks, the latter being performed locally for each cluster. It aims at identifying both a data decomposition into homogeneous and distinct subgroups and the subspaces in which these clusters are defined. Figure 1 gives an example of such clusters, contained in axis-parallel subspaces: although the data are 3-dimensional, cluster c_1 actually lives in the plane $z = 0$ and cluster c_2 in the plane $y = 0$.

A large number of approaches to the subspace clustering problem have been explored in machine learning as well as in data mining or computer vision. A list can be found in [12]. This paper focuses on iterative partitioning techniques. The k-subspace algorithm [13] generalises the k-means approach, alternating between

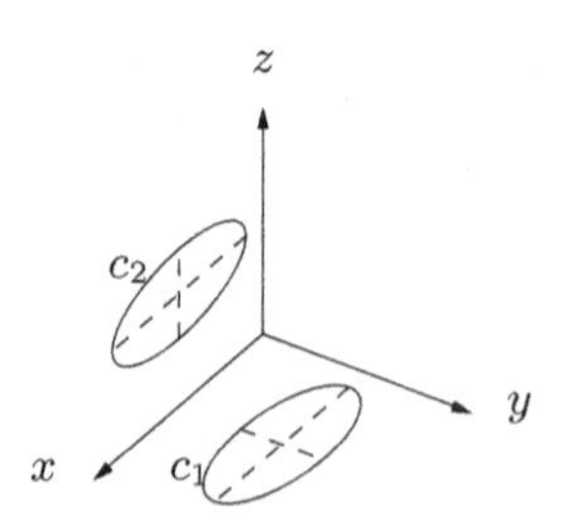

Fig. 1. Two clusters, contained in two different planes: c_1 in (x, y) and c_2 in (x, z).

the assignation of points to the clusters and the estimation of subspaces to fit these clusters. Witten and Tibshirani [14] propose a reformulation of the k-means minimization problem into a maximization problem with a weighted distance. An ℓ_1-based constraint is added in order to produce sparse weight vectors and identify the subspaces. Qiu et al. [10] adapt this framework to fuzzy clustering and compare it to some usual subspace clustering algorithms. Both the crisp and fuzzy variants of these algorithms heavily modify the original k-means function to formulate a maximization problem with a ℓ_1-regularization term, in order to identify minimal subspaces.

Closer to the original k-means paradigm, the fuzzy c-means clustering algorithm [3] has been adapted to the context of subspace clustering. Keller and Klawonn [6] adapt the FCM cost function to use a weighted Euclidean distance. Denoting $(x_i)_{i=1}^n \in \mathbb{R}^d$ the datapoints of dimension d, c the number of clusters, $(u_{ri}) \in [0,1]$ for $i \in \{1,\ldots,n\}$ and $r \in \{1,\ldots,c\}$ the fuzzy membership degree of x_i to cluster C_r, $\mu_r \in \mathbb{R}^d$ the center of cluster C_r and $(w_{rj}) \in [0,1]$ the weight of dimension j for cluster C_r, they study the following cost function:

$$J_{K\&K}(C,U,W) = \sum_{r=1}^{c}\sum_{i=1}^{n} u_{ri}^m \sum_{j=1}^{d} w_{rj}^v (x_{ij} - \mu_{rj})^2 \tag{1}$$

where $m, v \in \mathbb{R}$ are fuzzifiers which can be tuned by the user to specify the level of fuzziness of the corresponding parameters and C, U, W are respectively the matrices containing the centers (μ_r), the memberships (u_{ri}) and the weights (w_{rj}). The function is minimized under the following constraints:

- (C1) $\forall i \in \{1,\ldots,n\}, \sum_{r=1}^{c} u_{ri} = 1$ and (C2) $\forall r \in \{1,\ldots,c\}, \sum_{i=1}^{n} u_{ri} > 0$;
- (C3) $\forall r \in \{1,\ldots,c\}, \sum_{j=1}^{d} w_{rj} = a \neq 0$.

The first two constraints $(C1)$ and $(C2)$ are similar to the FCM ones. Constraint $(C3)$ on the weights (w_{rj}), where a is a user-defined parameter, is specific to the subspace clustering problem and prevents the trivial solution such that $\forall r, \forall j, w_{rj} = 0$. The minimization of Eq. (1) under these constraints produces a solution to the fuzzy subspace clustering problem. The computed weights (w_{rj}) indicate how close the points assigned to C_r are in dimension j. Figure 1 illustrates the relation between Eq. (1) and subspace clustering: cluster c_1 lies in the (x, y) plane. In the z dimension, its points are very close to its center; therefore, minimizing $J_{K\&K}$ amounts to maximizing w_{1z} rather than w_{1x} and w_{1y}.

Borgelt [4] generalises Keller and Klawonn's work and proposes to slightly change the weights, so that the algorithm completely selects dimensions by attributing a null weight to others. He introduces the following cost function, where the terms u_{ri}^m and w_{rj}^v are replaced with general fuzzification functions g and h [7], which are supposed to be convex and differentiable on the $[0,1]$ interval:

$$J_B(C,U,W) = \sum_{r=1}^{c}\sum_{i=1}^{n} h(u_{ri}) \sum_{j=1}^{d} g(w_{rj})(x_{ij} - \mu_{rj})^2 \tag{2}$$

This function is optimized under the same constraints as $J_{K\&K}$. Experimental results on artificial data show that Borgelt's algorithm better selects subspaces.

Both Keller & Klawonn and Borgelt functions are differentiable in each parameter on the considered domains, which allows to retain the technical framework of fuzzy c-means alternate optimization. They derive their algorithms from the corresponding cost function through the usual Lagrangian technique and obtain three update equations for parameters C, U and W.

3 Proposed Cost Function for Fuzzy Subspace Clustering

In this section, a new cost function is introduced to model the subspace clustering problem and a study of its properties is conducted.

3.1 A Weighted Fuzzy c-Means Function

Using the same notations as in Sect. 2, we propose the following cost function:

$$J(C,U,W) = \sum_{r=1}^{c}\sum_{i=1}^{n} u_{ri}^{m} \sum_{j=1}^{d} w_{rj}^{2}(x_{ij}-\mu_{rj})^{2} + \gamma\sum_{r=1}^{c}|\sum_{j=1}^{d}(w_{rj}) - \alpha| \quad (3)$$

under the classic FCM constraints (C1) and (C2).

The first term is the same as Keller & Klawonn's cost function, except for the weight fuzzifier v which is set to 2, in order to simplify further mathematical analysis of the function: it corresponds to a FCM cost function with a locally weighted Euclidean distance.

The second term adds a cost to the function which prevents the sum of the weights of each cluster C_r from being too far from the user-defined parameter α which plays the same role as a in Keller & Klawonn's (C3) constraint: for $\alpha \neq 0$, it prevents the trivial solution $W = 0$. The user-defined parameter $\gamma \in \mathbb{R}$ is used to balance out the two terms: it only needs to be large enough to penalize trivial solutions. This term can also be interpreted as an "inlined" constraint that incorporates constraint (C3), which does not need to be optimized through the particular Lagrangian method, but rather allows the use of new optimization techniques.

The cost function J in Eq. (3) thus conveys the idea of finding a solution to the subspace clustering problem with a relaxed constraint, inspired by ℓ_1-regularization [11].

3.2 Minimization of the Cost Function

Using the cost function J, solving the subspace clustering problem amounts to finding the parameters (C^*, U^*, W^*) that minimize J. This function can be decomposed as follows:

$$J(C, U, W) = F(C, U, W) + \gamma G(W)$$

$$\text{where } F(C, U, W) = \sum_{r=1}^{c} \sum_{i=1}^{n} u_{ri}^{m} \sum_{j=1}^{d} w_{rj}^{2} (x_{ij} - \mu_{rj})^2$$

$$G(W) = \sum_{r=1}^{c} G_r(W_r) = \sum_{r=1}^{c} |\sum_{j=1}^{d} (w_{rj}) - \alpha|$$

The function J verifies several properties of interest, which motivate and validate the technique used in the next section. First, J is a convex function of W, as it is the sum of convex functions.

F is differentiable in all three parameters and it can be shown that its gradient is Lipschitz-continuous for fixed C and U. These properties guarantee good performances of well-known optimization algorithms, such as gradient descent.

For fixed W, minimizing J under constraints is equivalent to minimizing F. As for fuzzy c-means, this can be done through alternate optimization. From the above function and constraints, the two classic update equations for membership degree and cluster centers are derived:

$$u_{ri} = \frac{d_{ri}^{\frac{2}{1-m}}}{\sum_{s=1}^{c} d_{si}^{\frac{2}{1-m}}} \quad \text{where} \quad d_{ri}^2 = \sum_{j=1}^{d} w_{rj}^2 (x_{ij} - \mu_{rj})^2 \tag{4}$$

$$\text{and } \mu_{rj} = \frac{\sum_{i=1}^{n} u_{ri}^{m} \cdot x_{ij}}{\sum_{i=1}^{n} u_{ri}^{m}} \tag{5}$$

These two equations are used in the PFSCM algorithm described in Sect. 4 to update the terms u_{ri} and μ_r in order to find the minimum of J.

Function G is convex but not differentiable in the variable W, which prevents the derivation of an update term for weight optimization and motivates the use of proximal descent, proposed in the next section.

4 Proximal Descent for Weight Optimization

As it is not differentiable everywhere, the function J previously defined cannot be optimized by classic alternate optimization. This section proposes a new algorithm, PFSCM (which stands for Proximal Fuzzy Subspace C-Means), based on an advanced technique of convex optimization: proximal descent [9].

In this section, the parameter of interest is the matrix of weights W, while C and U are fixed. Therefore, $J(C, U, W)$ is noted $J(W)$ for the sake of simplicity.

4.1 Proximal Descent

The cost function has the form $J(W) = F(W) + \gamma G(W)$, where both functions are convex but only F is differentiable and classic optimization techniques thus cannot be applied. As this general form of function has gained interest in the machine learning community (for example, when the second function G is a regularization term), proximal descent has been studied as an alternative to these techniques [2].

When $\gamma = 0$, usual optimization techniques would suggest to seek for the minimum of F (0 in the particular case of Eq. (3)) by iterating some update equation. For example, gradient descent considers a general equation of the form $W^{t+1} = W^t - \eta \cdot \nabla F(W^t)$, where t is the iteration index and η is a descent step size. This simple optimization scheme provides an iterative algorithm in order to minimize any convex function F, starting from any W^0 and iterating until convergence.

As the function G is not differentiable, its gradient ∇G does not exist for each W^t. Proximal descent enriches gradient descent in the following way:

$$W^{t+1} = \text{prox}_{\frac{\gamma}{L}G}\left(W^t - \frac{1}{L}\nabla F(W^t)\right) \tag{6}$$

$$\text{where } \text{prox}_{\frac{\gamma}{L}G}(W') = \underset{W}{\text{argmin}}\left\{\frac{1}{2}\|W - W'\|^2 + \frac{\gamma}{L}G(W)\right\} \tag{7}$$

where $L > 0$ is a descent step size, similar to η. That is, in order to solve a global minimization problem, proximal descent solves a minimization problem as defined by Eq. (7) at each step of the iteration.

Proximal descent can be understood as a technique of separating the descent in two phases: first for the function F, then for G. Such a descent scheme is also known as the "forward-backward" algorithm. In order to solve Eq. (7), proximal descent approximates F around the current point of the iterative descent, W^t:

$$\begin{aligned} &\underset{W}{\text{argmin}}\left\{F(W^t) + \langle\nabla F(W^t), W - W^t\rangle + \gamma G(W) + \frac{L}{2}\|W - W^t\|^2\right\} \\ &= \underset{W}{\text{argmin}}\left\{\frac{1}{2}\|W - (W^t - \frac{1}{L}\nabla F(W^t))\|^2 + \frac{\gamma}{L}G(W)\right\} \end{aligned}$$

Here again, if $\gamma = 0$, this problem has a simple solution: gradient descent scheme $W^{t+1} = W^t - \frac{1}{L}\nabla F(W^t)$, hence the scheme given in Eq. (6).

The key ingredient to efficiently implement the descent scheme defined by Eq. (6) is the notion of proximal operator: it provides a closed-form expression to the optimization problem defined by Eq. (7), which is often counter-intuitive, yet simple to implement.

4.2 Efficient Weight Optimization with Proximal Operators

We establish in the following theorem a proximal operator for the penalty term $G(W) = \gamma \sum_{r=1}^{c} |\sum_{j=1}^{d}(w_{rj}) - \alpha|$. Let K be the vector $(1,\ 1, \ldots 1) \in \mathbb{R}^{1\times d}$, such that $K \cdot K^{\intercal} = d$.

Theorem 1. *Let* $G_r(W_r) = |\sum_{j=1}^{d}(w_{rj}) - \alpha|$ *and* $L \in \mathbb{R}$.

$$\mathrm{prox}_{\frac{\gamma}{L}G_r}(W_r) = W_r + \frac{1}{d}K^{\mathsf{T}} \cdot (\alpha + \mathrm{prox}_{\frac{\gamma d}{L}|\cdot|}(K \cdot W_r - \alpha) - K \cdot W_r) \tag{8}$$

where $\mathrm{prox}_{\lambda|\cdot|}(x) = \mathrm{sign}(x)\max(|x| - \lambda, 0)$.

Moreover, $prox_{\frac{\gamma}{L}G}(W) = \left(prox_{\frac{\gamma}{L}G_r}(W_r)\right)_{r=1\ldots c} \in \mathbb{R}^{d\times c}$.

Proof. The proof uses results from [5,9]. First, $G_r(W_r) = \phi(K \cdot W_r)$ where $\phi(x) = |x - \alpha|$. Using the translation and semi-orthogonal linear transform properties [5]:

$$\begin{aligned}\mathrm{prox}_{G_r}(W_r) &= W_r + \frac{1}{d}K^{\mathsf{T}} \cdot \left(\mathrm{prox}_{\phi}(K \cdot W_r) - K \cdot W_r\right)\\ &= W_r + \frac{1}{d}K^{\mathsf{T}} \cdot \left(\alpha + \mathrm{prox}_{d|\cdot|}(K \cdot W_r - \alpha) - K \cdot W_r\right)\end{aligned}$$

Hence the expression of $\mathrm{prox}_{\frac{\gamma}{L}G_r}$ by the postcomposition property [9]. Finally, $\mathrm{prox}_{\frac{\gamma}{L}G}$ is computed using the separable sum property of proximal operators [9]. □

Equation (8) gives the expression of a proximal operator for the G function which can be used to efficiently implement the scheme defined in Eq. (6) to update the current estimation of W.

As for gradient descent, the choice of L matters for the actual convergence of the descent, as well as for its speed. We observe that setting $L = \mathrm{trace}(H^{-1})$ yields good results, where H is the Hessian matrix of F (as a function of W). As F is simple enough, H is a diagonal matrix and does not depend on W.

4.3 A Fuzzy Subspace Algorithm: PFSCM

Using the previous mathematical results, we propose the PFSCM algorithm for fuzzy subspace clustering (see Algorithm 1). PFSCM combines alternate optimization of k-means-style algorithms for differentiable parameters with proximal descent for the optimization of the weights.

Initialization is a typical issue of k-means-like algorithms. In this paper, initial centers are randomly chosen and each cluster receives uniform weights for all dimension. As most partitioning algorithms, the number c of clusters to identify must be set by the user, as well as constants $\gamma > 0$ and $\alpha > 0$.

The algorithm then iterates the update of all three parameters U, μ and W, much like alternate optimization in k-means algorithm. It consists of two alternate inner loops: the regular parameters μ and U are optimized separately from W, which requires the special optimization procedure described in the previous subsection. Parameters μ and U are optimized one last time at the end of the algorithm, in order to guarantee that the result takes the final computed weights into account.

The convergence criteria are defined as the distance between the current and the previous values of the parameters being optimized. In particular, convergence for (μ, U) is defined as $\|\mu_t - \mu_{t+1}\|_2 < \varepsilon \vee \|U_t - U_{t+1}\|_2 < \varepsilon$.

Data: X: data matrix
Parameters: c,γ,α: numbers;
Variables: μ, U, W: arrays;
W$_{last}$: array
Initialization: $W_r \leftarrow (1,\ 1, \ldots\ 1)$ for each C_r;
$\mu \leftarrow$ random centers
Output: μ, U, W
repeat
  repeat
    Update U according to Equation (4);
    Update μ according to Equation (5)
  until convergence(μ, U);
  repeat
    Update W according to Equation (7)
  until convergence(W);
  W$_{last}$ $\longleftarrow$ W
until convergence(W$_{last}$);
Update U and μ one last time.

Algorithm 1: The proximal fuzzy subspace clustering PFSCM algorithm

PFSCM outputs U, C and W. In order to exploit the result of the algorithm, it may be of interest to extract the dimension associated to each cluster. To that aim we propose to post-process the matrix W using an additional parameter *cut* to cut out the irrelevant dimensions in a simple fashion: a dimension j for a cluster C_r is considered relevant if $w_{rj} > cut$.

5 Experimental Study

The proposed PFSCM algorithm has been tested on artificial data in order to study its ability to correctly identify centers of non-circular clusters, as well as the dimensions that are relevant to describe the clusters. The results show the effectiveness of PFSCM in detecting the clusters and their subspaces. Moreover, PFSCM is compared to Keller & Klawonn's algorithm [6] and shows to provide a better estimation of the dimensionality of the subspaces.

5.1 Illustrative Example

This subsection presents an illustrative experiment in $d = 2$ dimensions, similar to the example given in Keller and Klawonn [6] and graphically represented in Fig. 2: four clusters are generated, one of them (the top red one in Fig. 2) being circular while the others have a very low variance in one dimension. PFSCM is run with $c = 4$, $m = 2$, $\alpha = 1$ and $\gamma = 1000$.

In Fig. 2 the points are colored according to the cluster C_r for which u_{ri} is maximum and Table 1 presents the weights computed for each dimension and cluster. It can be observed that PFSCM correctly identifies the desired clusters

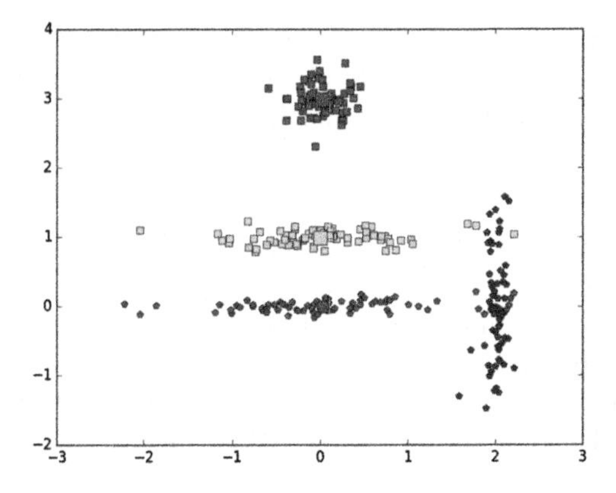

Fig. 2. Clustering example in two dimensions (Color figure online)

Table 1. Computed weights for the example given in Fig. 2. Column w_1 (resp. w_2) denotes the weight associated to the x-axis (resp. y-axis).

	Red cluster		Yellow cluster		Green cluster		Blue cluster	
	w_1	w_2	w_1	w_2	w_1	w_2	w_1	w_2
Weights	0.528	0.472	0.063	0.937	0.027	0.973	0.964	0.036

and their dimensions: the two weights (w_1, w_2) found for the circular cluster are similar, whereas the horizontal (respectively vertical) clusters verify $w_2 \gg w_1$ (respectively $w_1 \gg w_2$).

It is worth noting that, for this specific instance, some points close to the blue cluster are assigned to one of the horizontal clusters, as it minimizes the cost function. This kind of inliers is frequent in subspace clustering problems, and naturally leads to the use of fuzzy membership values (u_{ri}). In a similar fashion, moving the vertical cluster towards the center of the whole figure leads to the "stealing" of some points of the red cluster by the blue one. However, it can be observed that the identified dimensions for the circular cluster stay relatively stable (both $w_{rj} > cut = 0.2$), failing to recognize a non-flat cluster only 4 times out of 100 in the specific situation where all generated centers are vertically aligned.

5.2 Experimental Protocol

Considered Data. In order to validate PFSCM, the previous experiment is generalized to higher dimensions, more precisely to artificial data of dimension $d \in \{5, 7, 9, 11, 13, 15\}$. For each experiment, $k = 4$ centers $c_1, \ldots, c_4$ are generated randomly in the hypercube $[-3, 3]^d$ with a minimum (Euclidean) distance of 0.3 between the centers. Then, d_r dimensions $j_1, \ldots j_{d_r}$ are randomly picked, with d_r randomly chosen between 1 and $d-3$. Dimensions $j_1, \cdots, j_{d_r}$ are thereafter called the "relevant dimensions" for cluster C_r.

For each cluster, 100 points are generated according to a Gaussian distribution, with variance $v < 0.1$ for dimensions $j_1, \ldots, j_{d_r}$ and $v \in [0.5, 0.9]$ for other dimensions. The generated points in cluster r in dimension j thus follow $X_r \sim \mathcal{N}(c_r, v_j)$.

Algorithm Parameters. Keller & Klawonn's algorithm is initialized with FCM centers and uses $m = v = 2$, $a = 1$ and $c = 4$. PFSCM is ran with $m = 2$, $\alpha = 1$, $\gamma = 1000$ and $c = 4$. Both algorithms use the same convergence criterion, with $\varepsilon = 10^{-4}$.

The parameter *cut* is set to $\frac{1}{2d}$, which is a simple rule of thumb to identify the dimensions selected as relevant by the algorithms in each considered dimension d.

Quality Criteria. Both algorithms are evaluated on three metrics in order to qualify their results and their ability to discover the desired clusters and subspaces, and their dimensions.

First, let $\delta = \sum_{r=1}^{4} \|\mu_r - c_r\|_2$ be the sum of the Euclidean distances between the generated centers and the computed ones (μ_r): this metric is a standard quality criterion for evaluating the produced clusters. A low value means that the computed centers are close to the original ones.

We also consider θ defined as the percentage of clusters for which all relevant dimensions are correctly identified by the algorithm: the relevant dimensions are correctly identified if $w_{rj} > cut \Leftrightarrow j \in \{j_1, \cdots j_{d_r}\}$.

Finally, for the clusters for which the relevant dimensions have been correctly identified, let the weight ratio $\phi = \frac{\omega_1}{\omega_{j_{d_r}}}$ where ω_1 is the largest computed weight and $\omega_{j_{d_r}}$ the smallest computed weight for the relevant dimensions. This metric computes the distortion of the cluster between the relevant dimensions, as estimated by both algorithms.

5.3 Experimental Results

The results of the experiment are presented in Table 2 in the form of the means and standard deviations of the three criteria, computed over 100 runs of each algorithm. Both algorithms sometimes produce bad results, identifying centers too far from the generated ones, which distort the means and standard deviations of the previous metrics. Such outliers (less than 2 % of the runs) have been cleaned out of the results.

It can be observed that PFSCM correctly identifies the generated centers (as shown by δ), and produces stable results in each dimension, as shown by the low standard deviation. Moreover the algorithm, along with the proposed cut ratio $cut = \frac{1}{2d}$, performs well in selecting the relevant dimensions of the subspaces (as shown by θ). Finally, the weights ratio ϕ is relatively stable when the number of dimensions increases.

Keller & Klawonn's algorithm correctly identifies the centers (c_r) and the difference with PFSCM is not meaningful. However it appears to miss out some relevant dimensions of the generated subspaces. This is a general feature of the

Table 2. Comparison between PFSCM and Keller & Klawonn's algorithm

	d	δ		ϕ		θ
		Mean	SD	Mean	SD	%
PFSCM	5	0.90	0.67	2.51	1.41	76
	7	0.98	0.81	3.08	1.72	79
	9	0.90	0.50	3.96	2.09	80
	11	0.88	0.33	4.35	2.01	83
	13	0.97	0.40	4.78	1.99	83
	15	0.90	0.10	5.22	1.84	91
K&K	5	1.27	1.03	2.61	1.78	43
	7	1.55	1.38	3.12	2.29	39
	9	1.39	1.18	4.05	3.01	31
	11	1.26	0.90	4.50	3.48	28
	13	1.42	1.29	4.68	3.60	25
	15	1.21	1.06	8.05	3.27	10

algorithm, which can be seen in [6] as well: although the most relevant dimension is almost always identified, Keller & Klawonn's algorithm gives a much smaller weight to the other relevant dimensions, which is also shown by the larger mean for ϕ. This feature can be modulated by tuning the value of the fuzzifier v, but then this modification affects the weights of each dimension, including the most relevant one.

In summary, PFSCM identifies the same clusters as Keller & Klawonn's algorithm, but produces a better estimation of the dimensions of the subspaces. It is also more regular when the dimension d increases.

6 Conclusion and Future Works

This paper introduces a new approach to solve the fuzzy subspace clustering problem with a cost function involving non-differentiable terms. Advanced optimization techniques are explored, which replace the standard update equations of fuzzy c-means-like algorithms.

Experiments on synthetic data show the relevance of the proposed approach, that appears to correctly identify all the relevant dimensions and not more, whereas Keller & Klawonn's algorithm tends to underestimate the number of relevant dimensions. This provides more information about the importance of each dimension for the subspaces and clusters.

Future works will aim at generalizing this approach around the same key ideas: a differentiable function matching the specification of the problem and one or several penalty functions, expressing constraints on the shape of the solution. The introduction of regularization terms for parameters other than W will also be studied. Finally, more efficient descent schemes will be considered, in order to speed up the descent.

References

1. Agrawal, R., Gehrke, J., Gunopulos, D., Raghavan, P.: Automatic subspace clustering of high dimensional data for data mining applications. In: Proceedings of the ACM SIGMOD International Conference on Management of Data, pp. 94–105. ACM (1998)
2. Bach, F., Jenatton, R., Mairal, J., Obozinski, G., et al.: Convex optimization with sparsity-inducing norms. In: Optimization for Machine Learning, pp. 19–53 (2012)
3. Bezdek, J.C.: Pattern Recognition with Fuzzy Objective Function Algorithms. Kluwer Academic Publishers, Norwell (1981)
4. Borgelt, C.: Fuzzy subspace clustering. In: Fink, A., Lausen, B., Seidel, W., Ultsch, A. (eds.) Advances in Data Analysis, Data Handling and Business Intelligence. Studies in Classification, Data Analysis, and Knowledge Organization, pp. 93–103. Springer, Heidelberg (2010)
5. Combettes, P.L., Pesquet, J.C.: Proximal splitting methods in signal processing. In: Bauschke, H.H., Burachik, R.S., Combettes, P.L., Elser, V., Luke, D.R., Wolkowicz, H. (eds.) Fixed-Point Algorithms for Inverse Problems in Science and Engineering. Springer Optimization and Its Applications, vol. 49, pp. 185–212. Springer, New York (2011)
6. Keller, A., Klawonn, F.: Fuzzy clustering with weighting of data variables. Int. J. Uncertain. Fuzziness Knowl. Based Syst. **8**(06), 735–746 (2000)
7. Klawonn, F., Höppner, F.: What is fuzzy about fuzzy clustering? Understanding and improving the concept of the fuzzifier. In: Berthold, M., Lenz, H.-J., Bradley, E., Kruse, R., Borgelt, C. (eds.) IDA 2003. LNCS, vol. 2810, pp. 254–264. Springer, Heidelberg (2003)
8. Moreau, J.J.: Fonctions convexes duales et points proximaux dans un espace hilbertien. CR Acad. Sci. Paris Sér. A Math. **255**, 2897–2899 (1962)
9. Parikh, N., Boyd, S.: Proximal algorithms. Found. Trends Optim. **1**(3), 123–231 (2013)
10. Qiu, X., Qiu, Y., Feng, G., Li, P.: A sparse fuzzy c-means algorithm based on sparse clustering framework. Neurocomputing **157**, 290–295 (2015)
11. Tibshirani, R.: Regression shrinkage and selection via the lasso. J. R. Stat. Soc. Series B (Methodological) **58**, 267–288 (1996)
12. Vidal, R.: A tutorial on subspace clustering. IEEE Signal Proces. Mag. **28**(2), 52–68 (2010)
13. Wang, D., Ding, C., Li, T.: K-subspace clustering. In: Buntine, W., Grobelnik, M., Mladenić, D., Shawe-Taylor, J. (eds.) ECML PKDD 2009, Part II. LNCS, vol. 5782, pp. 506–521. Springer, Heidelberg (2009)
14. Witten, D.M., Tibshirani, R.: A framework for feature selection in clustering. J. Am. Stat. Assoc. **105**, 713–726 (2010)

Participatory Learning Fuzzy Clustering for Interval-Valued Data

Leandro Maciel[1](✉), Rosangela Ballini[1], Fernando Gomide[2], and Ronald R. Yager[3]

[1] Department of Economics Theory, Institute of Economics, University of Campinas, Campinas, São Paulo, Brazil
{maciel,gomide}@dca.fee.unicamp.br, ballini@unicamp.br

[2] Department of Electrical Engineering and Industrial Automation, School of Electrical and Computer Engineering, University of Campinas, Campinas, São Paulo, Brazil

[3] Machine Intelligence Institute, Iona College, New Rochelle, NY, USA
yager@panix.com

Abstract. This paper suggests an interval participatory learning fuzzy clustering (iPL) method for partitioning interval-valued data. Participatory learning provides a paradigm for learning that emphasizes the pervasive role of what is already known or believed in the learning process. iPL clustering method uses interval arithmetic, and the Hausdorff distance to compute the (dis)similarity between intervals. Computational experiments are reported using synthetic interval data sets with linearly non-separable clusters of different shapes and sizes. Comparisons include traditional hard and fuzzy clustering techniques for interval-valued data as benchmarks in terms of corrected Rand (CR) index for comparing two partitions. The results suggest that the interval participatory learning fuzzy clustering algorithm is highly effective to cluster interval-valued data and has comparable performance than alternative hard and fuzzy interval-based approaches.

Keywords: Fuzzy clustering · Participatory learning · Interval data

1 Introduction

Data clustering plays an important role in several engineering and data processing domains. As a method of unsupervised learning, clustering can be applied in a wide range of fields such as data mining, pattern recognition, bioinformatics, computer vision, image processing, information retrieval, etc. Clustering has recently become a subject of great interest, mainly due the explosive growth in the use of databases and the huge volume of data stored in them [2].

Clustering methods aim to partition a set of items into clusters or groups within a given cluster have a high degree of similarity, whereas items belonging to different clusters have a high degree of dissimilarity [1]. Clustering techniques can broadly divided into hierarchical and partitioning methods. In hierarchical methods, a complete hierarchy is built, i.e., a nested sequence of partitions of

J.P. Carvalho et al. (Eds.): IPMU 2016, Part I, CCIS 610, pp. 687–698, 2016.
DOI: 10.1007/978-3-319-40596-4_57

the input data is constructed. On the other hand, in partitioning techniques the objective is to obtain a single partition of the input data into a fixed number of clusters by optimizing and objective function, in general.

Hard and fuzzy methods are the most popular partitioning clustering techniques. Partitioning hard clustering methods construct disjointed clusters, in which each object of the data set must be assigned to precisely one cluster, whereas fuzzy partitioning clustering considers the idea of partial membership of each patter in a given cluster, providing a fuzzy partition of the data. One of the first fuzzy clustering methods was provided by [3]. The objective function in this case was based on an adequacy criterion defined by Euclidian distances. This method was further generalized by [4] which proposes the well-known fuzzy C-means algorithm. [5] suggested the first adaptive fuzzy clustering algorithm, which uses a quadratic distance defined by a full covariance matrix estimated for each cluster[1]. Recently, in [7] fuzzy K-means clustering algorithms based on adaptive quadratic distances are suggested. The distances are computed using diagonal fuzzy covariance matrices estimated globally or defined by diagonal fuzzy covariance matrices estimated locally for each cluster.

All these methods consider objects described by real-valued variables, i.e., the patterns to be grouped are usually represented as a vector of quantitative or qualitative measurements where each pattern takes a single value for each variable. When handling real world complex data, these models are very restrictive, since they do not take into account variability and/or uncertainty inherent to the data. Instead of real numbers, variable must assume sets of categories or intervals, possibly even with frequencies or weights, which comprise a new domain of multivariate analysis, pattern recognition and artificial intelligence: Symbolic Data Analysis (SDA) [1]. SDA includes methods such as clustering, factorial techniques, decision trees, etc., for managing aggregated data described by multi-valued variables represented by categories, intervals, sets, or probability distributions [8].

There are several clustering methods for symbolic data, which differ in the type of the considered symbolic data, in their structures and/or in the considered clustering criteria. Considering the problem of partitioning clustering methods for symbolic data, [9] extended the classical K-means method in order to manage data characterized by numerical and categorical variables[2]. A dynamic clustering algorithm for symbolic data was suggested by [10]. The method considers context-independent proximity functions, where the cluster representatives are probability distributions vectors.

[11] addressed several clustering algorithms for symbolic data described by intervals variables by the generalization of classical data analysis approaches. Considering adaptive distances, [12] proposed partitioning clustering methods for interval data based on city-block distances. Further, [7] suggested an algorithm using an adequacy criterion based on adaptive Hausdorff distances.

---

[1] [6] also used adaptive distances for quantitive data partition, however, the method concerns hard partition.

[2] A brief literature review of hierarchical clustering methods for symbolic data can be found in [1].

In terms of fuzzy clustering, [1] extends the fuzzy C-means clustering method for symbolic interval data. The author presented adaptive and non-adaptive fuzzy C-means clustering methods for interval data using an adequacy criterion based on squared Euclidian distances between vectors of intervals. Similarly, [13] introduced partitioning fuzzy K-means clustering models for interval-valued data based on adaptive quadratic distances that change at each algorithm iteration and can be either the same for all clusters and their representatives.

In [2], fuzzy Kohonen clustering networks for partitioning interval data are suggested. More recently, [14] proposed an interval-valued possibilistic fuzzy C-means clustering algorithm. The model use both fuzzy memberships and possibilistic typicalities to model the uncertainty implied in the data sets, and develop solutions to overcome the difficulties caused by type-2 fuzzy sets, such as the construction of footprint of uncertainty, type-reduction and defuzzification.

The aim of this paper is to suggest a fuzzy clustering approach for symbolic interval-valued data based on the Participatory Learning (PL) paradigm introduced by [15]. Participatory learning provides a paradigm for learning that emphasizes the pervasive role of what is already known or believed in the learning process. Central to this framework is the idea that for new information to contribute to learning, it must display some compatibility or consistency with what is already believed [16]. Therefore, the focus of this paper is to extend the participatory learning clustering algorithm, proposed by [17], to the situation in which the observations are interval-valued data. It suggests the interval participatory learning fuzzy clustering (iPL) approach as a new framework to cluster interval data. iPL clustering method uses interval arithmetic [18,19], and the Hausdorff distance to compute the (dis)similarity between intervals. Computational experiments are reported using synthetic interval data sets with linearly non-separable clusters of different shapes and sizes. The results of iPL are compared against traditional hard and fuzzy clustering techniques for interval-valued data, as suggested in [1], in terms of corrected Rand (CR) index [20] for comparing two partitions.

This paper proceeds as follows. After this introduction, Sect. 2 gives a brief reminder of the interval arithmetic adopted in this work. Section 3 details the participatory learning fuzzy algorithm for interval-valued data based on the Hausdorff distance for intervals. Computational experiments using synthetic data are shown in Sect. 4 in order to show the usefulness of the suggested approach against traditional hard and fuzzy clustering methods for interval data. Finally, Sect. 5 concludes the paper and suggests issues for further investigation.

2 Interval Arithmetic

In this paper an interval x is a closed bounded set of real numbers:

$$x = [x^L, x^U] \in \Im, \tag{1}$$

where $\Im = \{[x^L, x^U] : x^L, x^U \in \Re, x^L \leq x^U\}$ is the set of closed intervals of the real line $\Re$, x^L the lower bound, and x^U the upper bound of the interval. An m-dimensional interval vector $\mathbf{x}$ is an ordered m-tuple of intervals $\mathbf{x} = [x_1, x_2, \ldots, x_m]^T$, where $x_j = [x_j^L, x_j^U] \in \Im$, $j = 1, \ldots, m$.

Interval arithmetic extends traditional arithmetic to operate on intervals. This paper uses the arithmetic operations introduced by Moore [19]:

$$\begin{aligned} x + y &= [x^L + y^L, x^U + y^U], \\ x - y &= [x^L - y^U, x^U - y^L], \\ xy &= \left[\min\{x^L y^L, x^L y^U, x^U y^L, x^U y^U\}, \max\{x^L y^L, x^L y^U, x^U y^L, x^U y^U\}\right], \\ x/y &= x\,(1/y)\,, \text{ with } 1/y = [1/y^U, 1/y^L]. \end{aligned} \tag{2}$$

Interval arithmetic subsumes classic arithmetic. This means that if an operation of interval arithmetic takes real numbers as operands, considering them as intervals of length zero, then we obtain the same result as if the operation were performed using traditional arithmetic.

The interval participatory learning fuzzy model requires a metric to measure distances and (dis)similarities between intervals. This paper uses the Hausdorff distance instead of the Euclidean distance as commonly used by the literature [7]. If $\mathbf{x}$ and $\mathbf{y}$ are vectors of intervals, then the Hausdorff distance between $\mathbf{x}$ and $\mathbf{y}$, denoted by $dH(\mathbf{x}, \mathbf{y})$, is

$$dH(\mathbf{x}, \mathbf{y}) = \sum_{j=1}^{m} \left(\max\left\{|x_j^L - y_j^L|, |x_j^U - y_j^U|\right\}\right). \tag{3}$$

The next section addresses iPL, the interval participatory learning fuzzy clustering method for interval-valued data.

3 Interval Participatory Learning Fuzzy Clustering Method

This section provides the interval participatory learning fuzzy clustering method (iPL). One of the main characteristics of a clustering algorithm is to naturally partition a data set $\mathbf{X} = \{\mathbf{x}_1, \ldots, \mathbf{x}_N\}$, intervals $\mathbf{x}_k$ in this paper, $k = 1, 2, \ldots, N$, in c, $2 \leq c \leq N$, fuzzy subsets of $\mathbf{X}$, where c is the number of clusters and N the number of observations. The main idea is to extend the participatory learning fuzzy clustering method suggested by [17], a form of unsupervised fuzzy clustering algorithm, to deal with interval-valued data. Next, iPL clustering method is formulated. Further, the corresponding algorithm is described as well.

3.1 Interval Participatory Learning Fuzzy Clustering

Participatory learning (PL) assumes that model learning depends on what the system already knows about the model. Therefore, the current model is part of the process itself and influences the way in which new observations are used for self-organization. An essential property of PL is that the impact of new data in inducing self-organization or model revision depends on its compatibility with the current rule base structure or, equivalently, its compatibility with the current cluster structure [17].

In participatory learning clustering, the object of learning is cluster structure. Cluster structures are defined by cluster centers (or prototypes). Formally, let $\mathbf{V} = \{\mathbf{v}^1, \dots, \mathbf{v}^c\}$, $\mathbf{v}^i \in [0,1]^m$, $i = 1, \dots, c$, be a variable that encodes the belief of a system, i.e., the clusters centers of an initial cluster structure. The aim is to learn the value of this variable. It is assumed that the knowledge about the value of the variable comes in a sequence of observations $\mathbf{x}_k \in [0,1]^m$, where $\mathbf{x}_k$ is the k-th observation of the system[3].

The aim of the participatory mechanism is to learn the value of $\mathbf{v}^i$, using data $\mathbf{x}_k$. In other words, each $\mathbf{x}_k$, $k = 1, 2, \dots, N$, is used as a vehicle to learn about $\mathbf{v}^i$, where N stands for the number of observations. We say that the learning process is participatory if the contribution of each data $\mathbf{x}_k$ to the learning process depends upon its acceptance by the current estimate of $\mathbf{v}^i$ being valid. Implicit in this idea is that, to be useful and to contribute to the learning of $\mathbf{v}^i$, observations $\mathbf{x}_k$ must somehow be compatible with current estimates of $\mathbf{v}^i$.

Let $\mathbf{v}^i_k$ be the estimate of $\mathbf{v}^i$ after k observation. According to participatory learning principles, to be relevant for the learning process, $\mathbf{x}_k$ must be close to $\mathbf{v}^i_k$. It means that the system is willing to learn from information that is not too different from the current beliefs. In order to update the estimate or belief of $\mathbf{v}^i$, a smoothing like algorithm is applied [17]:

$$\mathbf{v}^i_{k+1} = \mathbf{v}^i_k + G^i_k(\mathbf{x}_k - \mathbf{v}^i_k), \tag{4}$$

where

$$G^i_k = \alpha \rho^i_k \tag{5}$$

with $\alpha \in [0,1]$ as the learning rate and ρ^i_k the compatibility degree between $\mathbf{x}_k$ and $\mathbf{v}^i_k$.

This formulation allows for different perceptions of similarity for different components of the vectors[4]. In [17], the compatibility measure ρ^i_k is defined as:

$$\rho^i_k = 1 - d^i_k, \tag{6}$$

where $d^i_k = ||\mathbf{x}_k - \mathbf{v}^i_k||$ is the Euclidian distance and $||\cdot||$ is a norm. Since in this paper we are considering interval-valued data, the Euclidian distance is replaced by the Hausdorff distance for intervals as in Eq. (3). Therefore, the compatibility measure in iPL is calculated as follows:

$$\rho^i_k = 1 - dH^i_k = 1 - \sum_{j=1}^{m} \left(\max \left\{ |x^L_{j,k} - v^{i,L}_{j,k}|, |x^U_{j,k} - v^{i,U}_{j,k}| \right\} \right), \tag{7}$$

where m is the dimension of the data.

[3] One must note that in iPL, both data $\mathbf{x}$ and cluster centers $\mathbf{v}$ are intervals, as defined in Sect. 2.

[4] Notice that the formulation in (4) uses the conventional vector quantization rule to update the cluster centers as in the fuzzy self-organizing neural network [21], which considers the distance from a new data point to the cluster surfaces.

In order to account the situation when a stream of conflicting observations arises during a certain period of time iPL clustering considers an arousal mechanism that monitors the compatibility of the current beliefs with the observations. This information is translated into an arousal index used to influence the learning process. The higher the arousal rate, the less confident is the system with the current belief, and conflicting observations become important o update the beliefs [17].

Formally, let $a_k^i \in [0, 1]$ be the arousal index of cluster i at k. The arousal is updated as follows:

$$a_{k+1}^i = a_k^i + \beta(1 - \rho_{k+1}^i - a_k^i), \tag{8}$$

The value of $\beta \in [0, 1]$ controls the rate of change of arousal: the closer β is to one, the faster the system is to sense compatibility variations. The arousal accounts for observations that are declared incompatible with the current system beliefs. In data clustering this means that if a data point $\mathbf{x}_k$ is far enough from all cluster centers, then there is enough motivation to create a new cluster. If the arousal index is greater than a threshold value $\tau \in [0, 1]$, a new cluster is created:

$$\mathbf{v}_k^{i+1} = \mathbf{x}_k. \tag{9}$$

Otherwise, the most compatible cluster with $\mathbf{x}_k$ is updated using (4).

The way iPL clustering takes into account the arousal mechanism is to incorporate the arousal index (8) into (5); that is, we assume

$$G_k^i = \alpha(\rho_k^i)^{1-a_k^i}. \tag{10}$$

When $a_k^i = 0$, $G_k^i = \alpha\rho_k^i$, which is the iPL procedure with no arousal. If the arousal index increases, the similarity measure has a reduced effect. The arousal index can be interpreted as the complement of the confidence we have in the truth of the current belief, the rule base structure. The arousal mechanism monitors the performance of the system by observing the compatibility of the current model with the observations. Therefore learning is dynamic in the sense that (4) can be viewed as a belief revision strategy whose effective learning rate (10) depends on the compatibility among new data, the current cluster structure and on model confidence as well.

Note that the learning rate is modulated by compatibility. Conventional learning models, have no participatory considerations and the learning rate is usually set small to avoid undesirable oscillations due to spurious values of data far from cluster centers. While protecting against the influence of noisy data, low learning rate slow down learning. Participatory learning allows higher values of the learning rate and the compatibility index lowers the effective learning rate when large deviations occur. On the other hand, high compatibility increases the effective rate, speeding up the learning process.

Whenever a cluster center is updated or a new cluster added, the iPL fuzzy clustering procedure should verify whether redundant clusters are created. Updating a cluster center using (4) can push a given center closer to another one and a redundant cluster may be formed. Thus, a mechanism to exclude redundancy is needed. One such mechanism is to verify if similar outputs due to distinct rules are produced. In iPL clustering, a cluster center is declared redundant whenever its similarity with another center is greater than or equal to a threshold value $\lambda \in [0, 1]$. If this is the case, we can either maintain the original cluster center or replace it by the average between the new data and the current cluster center. As in (7), the compatibility index among cluster centers, i and j, is computed as:

$$\rho_k^{i,j} = 1 - dH_k^{i,j}. \tag{11}$$

Therefore, if $\rho_k^{i,j} \geq \lambda$, the cluster i is declared redundant.

3.2 iPL Algorithm

The interval participatory learning fuzzy clustering (iPL) algorithm is summarized in this section. Initialization of the method includes the setting of initial values for the cluster centers $\mathbf{V}^0$ (step 1). Particularly, in this paper two random points of $\mathbf{X}$ are chosen to set $\mathbf{V}^0$. The corresponding fuzzy partition matrix is also calculated. The fuzzy partition is represented by a membership matrix $\mathbf{U} \in \Re^{(N \times c)}$ whose element $u_{i,k} \in [0, 1]$, $i = 1, 2, \ldots, c$, is the membership degree of the k-th data point $\mathbf{x}_k$ to the i-th cluster, the one with center $\mathbf{v}^i$. The membership degrees, are computed using:

$$u_{i,k} = \left(\sum_{j=1}^{c} \left(\frac{dH_k^i}{dH_k^j} \right)^{2/\eta - 1} \right)^{-1}, \tag{12}$$

where η is the fuzzification parameter, whose default value is $\eta = 2$.

Initial values for α, β, τ and λ must be set by the user (step 3). Next, for each observation, the compatibility (step 7) and arousal (step 10) indexes are computed for all clusters. If the arousal index of a current observation $\mathbf{x}_k$ is greater than the threshold τ, the data $\mathbf{x}_k$ is declared as the center of a new cluster (steps 12–13). Otherwise, the closest center to $\mathbf{x}_k$ is updated (step2 15–16). Further, the compatibility index among all clusters are calculated (step2 18–22). If the compatibility index of two clusters is greater than the threshold λ, redundant clusters are excluded (steps 23–25).

Two stop criteria are considered in iPL. The algorithm ends when either the maximum number of iterations ($t_{\max}$) is reached or there is no significant variation of the cluster centers from an iteration to the next. After stopping, the fuzzy partition matrix is updated considering the cluster centers produced during the last iteration. The detailed steps of the iPL algorithm are given next.

Interval participatory learning fuzzy clustering algorithm

1. set $c = 2$ and choose $\mathbf{V}^0$ randomly
2. compute $\mathbf{U}^0$ from $\mathbf{V}^0$
3. select values for α, β, τ and λ
4. set $a_0^i = 0$, $i = 1, \ldots, c$, and $t = 1$
5. **for** $k = 1, \ldots, N$ **do**
6.   **for** $i = 1, \ldots c$ **do**
7.   compute compatibility $\rho_k^i = 1 - dH_k^i$
8.   **end for**
9.   **for** $i = 1, \ldots c$ **do**
10.   compute arousal $a_k^i = a_{k-1}^i + \beta(1 - \rho_k^i - a_{k-1}^i)$
11.   **end for**
12.   **if** $a_k^i \geq \tau$, $\forall\, i \in \{1, \ldots, c\}$ **then** create a new cluster center
13. $\mathbf{v}_k^{c+1} = \mathbf{x}_k$
14.   **else** update the most compatible cluster center $\mathbf{v}_k^s$
15. $\mathbf{v}_{k+1}^s = \mathbf{v}_k^s + \alpha(\rho_k^s)^{1-a_k^s}(\mathbf{x}_k - \mathbf{v}_k^s)$
16. $s = \arg\max_i \{\rho_k^i\}$, $i = 1, \ldots, c$
17.   **end if**
18.   **for** $i = 1, \ldots c - 1$ **do**
19.   **for** $j = i + 1, \ldots, c$ **do**
20.   compute compatibility among all clusters $\rho_k^{i,j} = 1 - dH_k^{i,j}$
21.   **end for**
22.   **end for**
23.   **if** $\exists\, i \mid \rho_k^{i,j} \geq \lambda$
24.   exclude $\mathbf{v}_k^i$
25.   **end if**
26.   update $\mathbf{U}$
27. **end for**
28. compute $error = \max_i |dH^i(\mathbf{v}_t^i - \mathbf{v}_{t-1}^i)|$, $i = 1, \ldots, c$
29. **if** $error > \epsilon$ or $t \leq t_{\max}$ **then** $t = t + 1$ **and** return to step 5
30. **else** stop
31. **end if**
32. update fuzzy partition matrix $\mathbf{U}$

4 Computational Experiments

The interval IPL approach introduced in this paper is an unsupervised learning clustering method for interval-valued data, useful for a variety of problems in data mining, pattern recognition, image processing, etc. This section illustrates interval data clustering with iPL using synthetic interval data sets with linearly non-separable clusters of different shapes and sizes.

4.1 Data

This work considers the same data point configuration presented in [1,12]. Synthetic data comprise two data sets of 350 points in $\Re^2$, in which each point is

Table 1. Mean and variances of the bi-variate normal distributions of the clusters in data sets 1 and 2.

	Data set 1			Data set 2		
Parameter	Cluster 1	Cluster 2	Cluster 3	Cluster 1	Cluster 2	Cluster 3
μ_1	28	60	45	45	60	52
μ_2	22	30	38	22	30	38
σ_1^2	100	9	9	100	9	9
σ_2^2	9	144	9	9	144	9

represented by the vector $\mathbf{z} = (z_1, z_2)$, $z_j \in \Re$, $j = 1, 2$. The 350 points in each data set are drawn from three bi-variate normal distributions of independent components. From each data set three clusters of unequal sizes and shapes are constructed: two clusters with an ellipsoidal shape and size 150 and one cluster with a spherical shape and size 50. Therefore, the two data sets have three clusters each. Table 1 shows the configuration parameters of the two data sets, where μ stands for the mean and σ^2 the variance. Further, it is supposed that there is no correlation between the two components in the bi-variate distributions.

Figure 1 illustrates data sets 1 and 2. Data set 1 contains well-separated clusters, whereas data set 2 shows overlapping clusters. Each point (z_1, z_2) of the two data sets is considered as the 'seed' of rectangle to build interval data sets as in [1]. Each rectangle is a vector of two intervals defined by: $([(z_1 - \gamma_1)/2, (z_1 + \gamma_1)/2], [(z_2 - \gamma_2)/2, (z_2 + \gamma_2)/2])$, where γ_1 and γ_2 are the width and height of the rectangle, respectively. In Fig. 2 the two synthetic interval data sets built from data set 1 and 2 are shown, respectively, for he case when γ_1 and γ_2 are drawn randomly from $[1, 8]$.

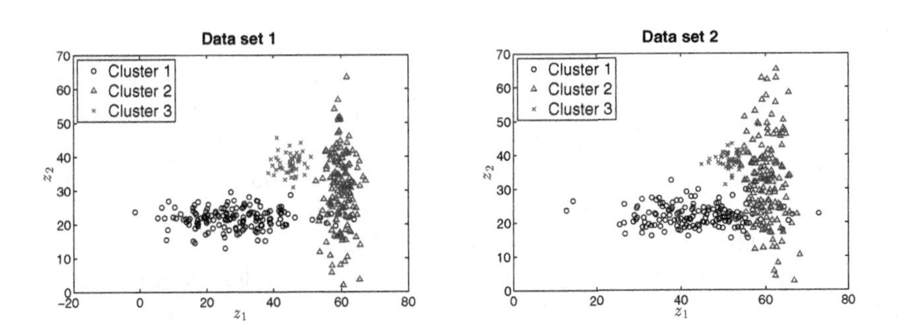

Fig. 1. Data sets 1 and 2 with well-separated and overlapping clusters, respectively.

4.2 Performance Measurement

As alternative, this paper compares the results of iPL with two hard clustering methods and two fuzzy clustering approaches for interval-valued data. The inter-

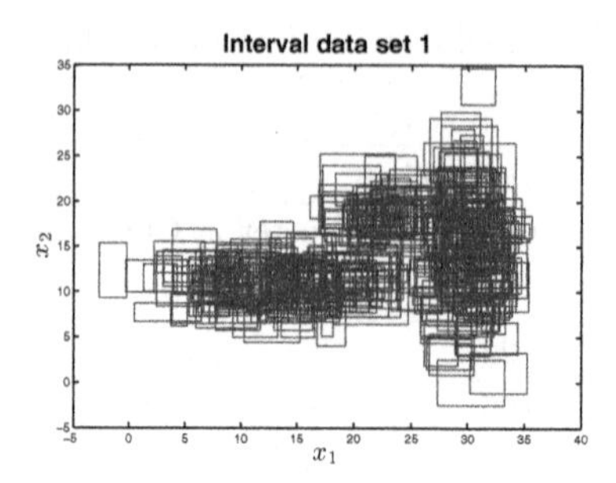

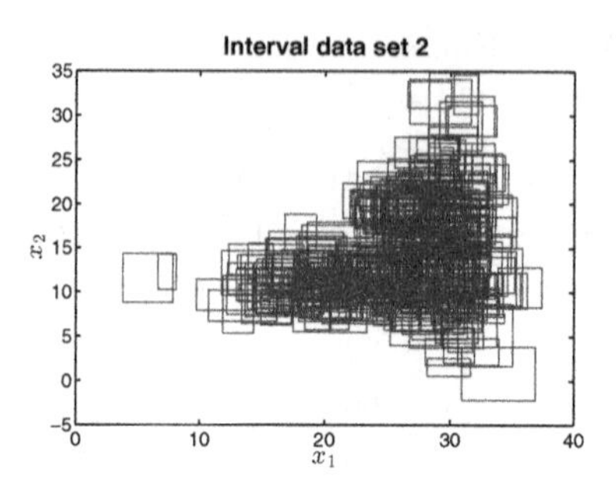

Fig. 2. Interval data sets 1 and 2 with well-separated and overlapping clusters, respectively.

val hard clustering methods differ in terms of the distance measures: using city-block distance (Hard$_{CB}$) [12] and Hausdorff distance (Hard$_H$) [7]. Otherwise, the fuzzy methods comprise fuzzy c-means approaches extended for interval-valued data as proposed by [1]. They also differ in terms of distance measures: using quadratic Euclidian distance (FCM$_E$) and Hausdorff distance (FCM$_H$).

The results are compared according to an external validity index. The key idea is to compare the a priori partition (known) with the partition obtained from the clustering algorithm. As in [1,7,13], this work uses the corrected Rand (CR) index [20] for comparing two partitions. It measures the similarity between a *a priori* partition and a partition furnished by a partitioning clustering algorithm. CR takes values on the interval $[-1, 1]$, where the value 1 indicates perfect agreement between partitions, whereas values near 0 (or negatives) correspond to clusters agreement [1].

4.3 Results

As an experiment, the construction on the interval data was processed for different values of γ_1 and γ_2 as follows: $[1, 8]$, $[1, 16]$, $[1, 24]$, $[1, 32]$ and $[1, 40]$. For each interval, we run the algorithms 100 times, therefore, the average corrected Rand CR index is calculated. Simulations were also conducted to find the best performance in terms of CR index suggested the following control parameters for iPL: $\alpha = 0.08$, $\beta = 0.2$, $\tau = 0.17$ and $\lambda = 0.19$.

Table 2 shows the values of the average of the CR index from the clustering methods for interval data sets 1 and 2 for the different intervals of parameters γ_1 and γ_2. Best results are in bold. Regarding the data configurations presenting well-separated clusters (data set 1), i.e., interval data set 1 (Table 2), the average CR indices are better for iPL and the models that use the Hausdorff distance (Hard$_H$ and FCM$_H$). iPL and FCM$_H$ are the methods with better performance but very similar regardless the range of the predefined intervals in Table 2.

Considering data configurations presenting overlapping clusters (data set 2), the fuzzy clustering approaches FCM$_H$ and iPL clearly outperforms the other methods (Table 2). The remaining models present similar values of CR index

Table 2. Methods average CR index for interval data sets 1 and 2

Interval data set 1					
Intervals	Hard_{CB}	Hard_H	FCM_E	FCM_H	iPL
$[1, 8]$	0.935	0.923	0.837	**0.964**	0.958
$[1, 16]$	0.931	0.931	0.844	0.979	**0.981**
$[1, 24]$	0.892	0.909	0.822	0.957	**0.963**
$[1, 32]$	0.773	0.912	0.761	**0.919**	0.911
$[1, 40]$	0.701	0.886	0.739	**0.868**	0.857
Interval data set 2					
Intervals	Hard_{CB}	Hard_H	FCM_E	FCM_H	iPL
$[1, 8]$	0.470	0.448	0.430	**0.832**	0.780
$[1, 16]$	0.432	0.434	0.443	**0.812**	0.773
$[1, 24]$	0.406	0.418	0.425	**0.739**	0.679
$[1, 32]$	0.389	0.412	0.409	**0.630**	0.591
$[1, 40]$	0.373	0.393	0.390	**0.527**	0.486

average. In this case, FCM_H shows better results than iPL, which is due to the adaptive property of FCM_H, i.e., the method uses adaptive parameters in the Hausdorff distances for each cluster.

5 Conclusion

This paper suggested an interval participatory learning fuzzy clustering (iPL) method for partitioning interval-valued data. iPL clustering method uses interval arithmetic, and the Hausdorff distance to compute the (dis)similarity between intervals. Using synthetic interval data sets, computational experiments compared iPL against hard and fuzzy clustering methods in terms of corrected Rand (CR) index for comparing two partitions. Results indicate the high potential of iPL approach to cluster interval-valued data mainly for data configurations presenting overlapping clusters. Further studies shall consider the application of iPL for real data, as well as its use in fuzzy inference models.

Acknowledgments. The authors thank the Brazilian Ministry of Education (CAPES), the Brazilian National Research Council (CNPq), and the Research of Foundation of the State of São Paulo (FAPESP) for financial support.

References

1. Carvalho, F.A.T.: Fuzzy c-means clustering methods for symbolic interval data. Pattern Recogn. Lett. **28**, 423–437 (2007)

2. Almeida, C.W.D., Souza, R.M.C.R., Candeias, A.L.B.: Fuzzy Kohonen clustering networks for interval data. Neurocomputing **99**, 65–75 (2013)
3. Dunn, J.C.: A fuzzy relative to the ISODATA process and its use in detecting compact, well-separated clusters. J. Cybern. **3**, 32–57 (1974)
4. Bezdek, J.C.: Pattern Recognition with Fuzzy Objective Function Algorithms. Plenum Press, New York (1981)
5. Gustafson, D.E., Kessel, W.C.: Fuzzy clustering with a fuzzy covariance matrix. In: Proceedings of the IEEE Conference on Decision and Control, pp. 761–766, San Diego, CA (1979)
6. Diday, E., Govaert, G.: Classification automatic avec distances adaptatives. R.A.I.R.O. Informatique Comput. Sci. **11**(4), 329–349 (1977)
7. Carvalho, F.A.T., Tenório, C.P., Cavalcanti Jr., N.L.: Partitional fuzzy clustering methods based on adaptive quadratic distances. Fuzzy Sets Syst. **157**, 2833–2857 (2006)
8. Billard, L., Diday, E.: From the statistics of data to the statistics of knowledge: symbolic data analysis. J. Am. Stat. Assoc. **98**, 470–487 (2003)
9. Ralambondrainy, H.: A conceptal version of the k-means algorithm. Pattern Recogn. Lett. **16**, 1147–1157 (1995)
10. Verde, R., Carvalho, F.A.T., Lechevallier, Y.: A dynamical clustering algorithm for symbolic data. In: Tutorial on Symbolic Data Analysis held during the 25th Annual Conference on the Gesellschaft fur Klassifikation. University of Munich, 13 March 2001
11. Bock, H.H.: Clustering algorithms and Kohonen maps for symbolic data. J. Jpn. Soc. Comput. Stat. **15**, 1–13 (2002)
12. Souza, R.M.C.R., Carvalho, F.A.T.: Clustering of interval data based on city-block distances. Pattern Recogn. Lett. **25**(3), 353–365 (2004)
13. Carvalho, F.A.T., Tenório, C.P.: Fuzzy K-means clustering algorithms for interval-valued data based on adaptive quadratic distances. Fuzzy Sets Syst. **161**, 2978–2999 (2010)
14. Ji, Z., Xia, Y., Sun, Q., Cao, G.: Interval-valued possibilistic fuzzy C-means clustering algorithm. Fuzzy Sets Syst. **253**, 138–156 (2014)
15. Yager, R.: A model of participatory learning. IEEE Trans. Syst. Man Cybern. **20**(5), 1229–1234 (1990)
16. Yager, R.: Participatory learning with granular observations. IEEE Trans. Fuzzy Syst. **17**(1), 1–12 (2009)
17. Silva, L., Gomide, F., Yager, R.: Participatory learning in fuzzy clustering. In: Proceedings of the 14th IEEE International Conference on Fuzzy Systems, pp. 857–861, Reno, Nevada, USA (2005)
18. Hickey, T., Ju, Q., Emden, M.: Interval arithmetic: from principles to implementation. J. ACM **48**(5), 1038–1068 (2001)
19. Moore, R.E., Kearfott, R.B., Cloud, M.J.: Introduction to Interval Analysis. SIAM Press, Philadelphia (2009)
20. Hubert, L., Arabie, P.: Comparing partitions. J. Classif. **2**, 193–218 (1985)
21. Leng, G., McGinnity, T.M., Prasad, G.: An approach for on-line extraction of fuzzy rules using a self-organizing fuzzy neural network. Fuzzy Sets Syst. **150**(2), 211–243 (2005)

Fuzzy c-Means Clustering of Incomplete Data Using Dimension-Wise Fuzzy Variances of Clusters

Ludmila Himmelspach(✉) and Stefan Conrad

Institute of Computer Science, Heinrich-Heine-Universität Düsseldorf, 40225 Düsseldorf, Germany
{himmelspach,conrad}@cs.uni-duesseldorf.de

Abstract. Clustering is an important technique for identifying groups of similar data objects within a data set. Since problems during the data collection and data preprocessing steps often lead to missing values in the data sets, there is a need for clustering methods that can deal with such imperfect data. Approaches proposed in the literature for adapting the fuzzy c-means algorithm to incomplete data work well on data sets with equally sized and shaped clusters. In this paper we present an approach for adapting the fuzzy c-means algorithm to incomplete data that uses the dimension-wise fuzzy variances of clusters for imputation of missing values. In experiments on incomplete real and synthetic data sets with differently sized and shaped clusters, we demonstrate the benefit over the basic approach in terms of the assignment of data objects to clusters and the cluster prototype computation.

Keywords: Clustering · Fuzzy c-means (FCM) · Incomplete data · Missing values

1 Introduction

Clustering is one of the important and primarily used techniques for the automatic knowledge extraction from large amounts of data. Its aim is to identify groups, so-called clusters, of similar objects within a data set. Data clustering is used in many fields, including database marketing, image processing, bioinformatics, text mining, and many others. The quality of the data plays an important role in the clustering process and might affect the clustering results. However, problems during the data collection and data preprocessing are often inevitable and might lead to uncertain, erroneous, or missing values in the data sets. Since the completion or correction of data is often expensive or even impossible, there is a need for clustering methods that can deal with such imperfect data.

In the literature, several approaches for adapting the fuzzy c-means (FCM) algorithm to incomplete data have been proposed [1–3]. Like the basic FCM algorithm, these methods assume that the clusters are equally sized and shaped. Even if using the Euclidean distance function as a similarity criterion, in practice,

J.P. Carvalho et al. (Eds.): IPMU 2016, Part I, CCIS 610, pp. 699–710, 2016.
DOI: 10.1007/978-3-319-40596-4_58

FCM often reliably identifies differently shaped and sized clusters in complete data sets. Therefore, the FCM algorithm is widely used as a stand-alone clustering method as well as in hybrid hierarchical clustering methods [4]. However, the experiments conducted in [5] showed that FCM adapted to incomplete data produces less accurate clustering results on incomplete data with differently sized and scattered clusters than in the case of clusters with equal sizes and volumes. In [6], we proposed an approach for adapting FCM to incomplete data that takes the cluster scatters into account during the imputation of missing values. Since this method computes the cluster dispersions on the basis of completely available features, it cannot be applied on data sets where missing values occur in all features. Moreover, the experiments showed that this approach produces less accurate cluster prototypes than the basic approach. We assume the reason in the calculation and the use of the cluster dispersions. The cluster dispersions are computed on the basis of completely available features, but they are used for the imputation of missing values in the other features. Therefore, if clusters have different extents in different dimensions, the way of using the cluster dispersions for the imputation of missing values as it is used in our previous approach does more harm than good. In this paper, we present an approach for adapting the fuzzy c-means algorithm to incomplete data that uses the dimension-wise fuzzy variances of clusters for imputation of missing values. Our approach involves the cluster shapes and volumes for missing value imputation in a simple and computationally inexpensive way. In experiments on real and synthetic data sets with differently sized and shaped convex clusters, we demonstrate the capabilities of our new approach and show the benefit over the basic approach in terms of the assignment of data objects to clusters and the cluster prototype computation.

The remainder of the paper is organized as follows. In the next section we give a short overview of the basic fuzzy c-means algorithm and the methods for adapting FCM to incomplete data. In Sect. 3 we describe our idea for missing values imputation using the dimension-wise fuzzy variance of clusters and present the modified FCM algorithm. The evaluation results of our method and the comparison with the basic approach are presented in Sect. 4. In Sect. 5 we close the paper with a short summary and the discussion of future research.

2 Approaches for Fuzzy Clustering of Incomplete Data

2.1 Fuzzy c-Means Algorithm (FCM)

The fuzzy c-means algorithm (FCM) is a well known clustering algorithm that partitions a given data set $X = \{x_1, ..., x_n\}$ in a d-dimensional metric data space into c clusters that are represented by their cluster prototypes $V = \{v_1, ..., v_c\}$. Unlike the k-means algorithm [7], which assigns each data object to exactly one cluster, fuzzy c-means algorithm assigns data items to clusters with membership degrees [8]. The membership degree $u_{ik} \in [0, 1]$ expresses the relative degree to which the data point x_k with $1 \leq k \leq n$ belongs to the cluster C_i, $1 \leq i \leq c$.

The objective function of the fuzzy c-means algorithm is defined as follows:

$$J_m(X,U,V) = \sum_{i=1}^{c}\sum_{k=1}^{n} u_{ik}^m \cdot d^2(v_i, x_k) \,. \tag{1}$$

The similarity between the data items and the cluster prototypes is expressed by the squared distance function. The parameter m, $m > 1$, is the *fuzzification parameter* which determines the vagueness of the resulting partitioning. The objective function of the fuzzy c-means algorithm is minimized using an *alternating optimization (AO)* scheme [8]. The objective function is alternately optimized over the membership degrees and the cluster prototypes in an iterative process.

The algorithm begins with the initialization of the cluster prototypes v_i which can be either the first c data items of the data set or c randomly chosen data items or c randomly chosen points in the data space. Alternatively, the membership degrees can be initialized. In the first iteration step the membership degrees of each data item to each cluster are updated according to Formula (2).

$$u_{ik} = \begin{cases} \dfrac{(d^2(v_i,x_k))^{\frac{1}{1-m}}}{\sum\limits_{l=1}^{c}(d^2(v_l,x_k))^{\frac{1}{1-m}}} & \text{if } I_{x_k} = \emptyset, \\ \lambda, \lambda \in [0,1] \text{ with } \sum_{v_i \in I_{x_k}} u_{ik} = 1 & \text{if } I_{x_k} \neq \emptyset, v_i \in I_{x_k}, \\ 0 & \text{if } I_{x_k} \neq \emptyset, v_i \notin I_{x_k}, \end{cases} \tag{2}$$

where $I_{x_k} = \{v_i \mid d^2(v_i,x_k) = 0\}$. In the second iteration step the new cluster prototypes are calculated based on all data items depending on their membership degrees to the cluster (see Formula (3)).

$$v_i = \frac{\sum\limits_{k=1}^{n}(u_{ik})^m x_k}{\sum\limits_{k=1}^{n}(u_{ik})^m}, \qquad 1 \le i \le c \,. \tag{3}$$

The iterative process continues as long as the cluster prototypes change up to a value ϵ. Although the fuzzy c-means algorithm is known as a stable and robust clustering algorithm that does not often get stuck in a local optimum [9,10], it is sensible to evaluate the algorithm for different initializations to achieve the optimal partitioning results.

2.2 Different Approaches for Fuzzy Clustering of Incomplete Data

In the literature, several approaches for adapting fuzzy clustering algorithms to incomplete data have been proposed. Some of them such as the *whole-data strategy FCM (WDSFCM)* and the *partial distance strategy FCM (PDSFCM)* [1,2] carry out the analysis only on the basis of available values. Other methods like the *optimal completion strategy FCM (OCSFCM)* [1,2], the *nearest prototype strategy FCM (NPSFCM)* [1,2], and the *distance estimation strategy FCM (DESFCM)* [3] impute missing feature values or distances in an additional iteration step of the fuzzy c-means algorithm. In experiments described in [1,5], the lowest missclassification errors have been obtained by PDSFCM, OCSFCM, and NPSFCM. OCSFCM and NPSFCM tend to strengthen the clustering structure of incomplete data which turned out to be beneficial e.g. for determining the optimal number of clusters using cluster validity indexes (CVIs). In the following we focus on the description of OCSFCM because it provides the basis for our approach.

Optimal Completion Strategy FCM (OCSFCM). The idea of the *optimal completion strategy (OCS)* is to iteratively compute the missing values as the additional variables over which the objective function is minimized [1,2]. The fuzzy c-means algorithm is modified by adding an additional iteration step where the missing values are updated according to Formula (4).

$$x_{kj} = \frac{\sum_{i=1}^{c}(u_{ik})^m v_{ij}}{\sum\limits_{i=1}^{c}(u_{ik})^m}, \qquad 1 \leq k \leq n \text{ and } 1 \leq j \leq d\ . \tag{4}$$

In this way, the missing values are imputed by the weighted means of all cluster centers in each iteration step.

The advantage of this approach is that missing values are imputed during the clustering process. However, the drawback of the OCSFCM is that the calculation of the cluster prototypes and the imputation of the missing values influence each other because the algorithm does not distinguish between the available and imputed feature values. In [11] the author proposed to diminish the influence of imputed values to the calculation of cluster prototypes by reducing the membership degrees of incomplete data items depending on the number of missing values. The resulting algorithm loses the property of a probabilistic fuzzy clustering algorithm, though. In our approach we tackle this problem by computing the cluster prototypes only on the basis of available feature values.

3 FCM Clustering of Incomplete Data Using Dimension-Wise Fuzzy Variances of Clusters

In [6], we have proposed an approach for adapting FCM to incomplete data which can be regarded as an extension of OCSFCM that takes the dispersions of clusters into account for the imputation of missing values. The experimental results have shown the benefits of this approach over the basic OCSFCM algorithm. However, since the cluster dispersions are computed on the basis of completely available features but used for the imputation of missing values in the other features, this approach is restricted to data sets with equally shaped clusters, i.e. clusters that have the same extents in different dimensions. Therefore, in our new approach we impute missing values taking the dimension-wise fuzzy variances of clusters into account. The idea of our approach is imputing missing feature values of incomplete data items depending on both the distances between the cluster prototypes and the incomplete data items and the extents of clusters in the corresponding dimensions.

3.1 A New Membership Degree Using Dimension-Wise Fuzzy Variances of Clusters

For each cluster C_i, $1 \leq i \leq c$, we calculate its fuzzy variance w_{ij} in each dimension j, $1 \leq j \leq d$, as an averaged squared distance of data items to their cluster prototypes according to Formula (5).

$$w_{ij} = \frac{\sum_{k=1}^{n} u_{ik}^{m} i_{kj} (x_{kj} - v_{ij})^2}{\sum_{k=1}^{n} u_{ik}^{m} i_{kj}}, \tag{5}$$

where

$$i_{kj} = \begin{cases} 1, & \text{if } x_{kj} \text{ is available} \\ 0 & \text{else} \end{cases} \quad \text{for } 1 \leq j \leq d,\ 1 \leq k \leq n. \tag{6}$$

We calculate the dimension-wise fuzzy variances of clusters using only available feature items. In this way, we avoid the influence of the imputed feature values on the calculation of cluster variances. That makes sense because missing values are imputed by values close to the corresponding feature values of the cluster prototypes. Therefore, taking both the available and the imputed feature values into account for the calculation of cluster variances would reduce the real variances of clusters. Furthermore, using all available feature items for the calculation of the dimension-wise fuzzy variances of clusters makes our approach applicable on incomplete data sets where missing values occur in all features and data items. Calculating the fuzzy variances of clusters for each feature prevents the influence of the distorted estimation of cluster variances in single dimensions to the whole cluster variances in the case of a large amount of missing values in those features.

We integrate the fuzzy variances of clusters in the new membership degree update formula for the imputation of missing values as follows:

$$u_{(ik)j}^{w} = \frac{\left(w_{ij}^{-1} (v_{ij} - x_{kj})^2\right)^{\frac{1}{1-m}}}{\sum_{l=1}^{c} \left(w_{lj}^{-1} (v_{lj} - x_{kj})^2\right)^{\frac{1}{1-m}}} \quad \forall x_{kj} \text{ with } i_{kj} = 0, \tag{7}$$

$1 \leq k \leq n$, $1 \leq i \leq c$, and $1 \leq j \leq d$. Note that we also compute the membership degrees for the imputation of missing values for each dimension. That implies that a missing value of an incomplete data item in a particular feature is updated depending on the distances between the corresponding values of the cluster prototypes and the last imputed value and the extents of clusters in that feature. The larger the variance of a cluster and the smaller the distance between the imputed data item and the cluster center in that particular dimension, the higher the new membership degree is. If all clusters are equally sized and shaped, then the membership degree $u^w_{(ik)j}$ depends only on the distances between the data items and the cluster prototypes. As in the basic OCS approach the imputation of missing values in our approach depends on the imputed values in the previous iteration.

3.2 FCM for Incomplete Data Using Dimension-Wise Fuzzy Variances of Clusters (FCMDFVC)

We integrate the new membership degree for imputation of missing values in the basic OCSFCM algorithm and refer the resulting algorithm to as *Fuzzy c-Means Algorithm for Incomplete Data using Dimension-wise Fuzzy Variances of Clusters (FCMDFVC)*. The working principle of FCMDFVC is depicted in Algorithm 1 and is basically the same as of OCSFCM.

The algorithm begins with the initialization of cluster prototypes and missing values. The membership degrees are updated in the first iteration step in the same way as in the basic FCM and OCSFCM. The available and the imputed feature values are not distinguished. Since we want to avoid the influence of the imputed values to the calculation of the cluster prototypes, we compute the cluster prototypes only on the basis of available feature values as in PDSFCM according to Formula (8).

$$v'_{ij} = \frac{\sum_{k=1}^{n} (u_{ik})^m i_{kj} x_{kj}}{\sum_{k=1}^{n} (u_{ik})^m i_{kj}} \qquad \text{for } 1 \le i \le c,\ 1 \le j \le d. \tag{8}$$

If the termination condition is not reached, in the third iteration step, missing values are imputed depending on the distances to the cluster prototypes and the fuzzy variances of clusters according to Formula (9).

$$x_{kj} = \frac{\sum_{i=1}^{c} (u^w_{(ik)j})^m v'_{ij}}{\sum_{i=1}^{c} (u^w_{(ik)j})^m}, \qquad 1 \le k \le n \text{ and } 1 \le j \le d\ . \tag{9}$$

Basically, the new membership degree for the imputation of missing values can also be integrated in NPSFCM. In this case, the computation of cluster prototypes using only available feature values is essential because the computation of cluster prototypes and the imputation of missing values influence each other even more than in OCSFCM.

Algorithm 1. FCMDFVC(X, c, m, ϵ)

Require: X is a d-dimensional incomplete data set with n data items, $2 \leq c \leq n$ is a number of clusters, $m > 1$ is a fuzzification parameter, $\epsilon > 0$ is a termination accuracy

1: Initialize the set of data centers $v' = \{v'_1, ..., v'_c\}$
2: Initialize all missing values x_{kj} in X with random values in the data space
3: $v = \{\}$
4: **repeat**
5: $\quad v = v'$
6: $\quad$ Calculate the membership degrees u_{ik} of each data item x_k to each cluster C_i according to Formula (2) // *Step 1*
7: $\quad$ Calculate the set of new cluster prototypes $v' = \{v'_1, ..., v'_c\}$ according to Formula (8) // *Step 2*
8: $\quad$ **if** $\|v - v'\| > \epsilon$ **then**
9: $\quad\quad$ Impute the missing values x_{kj} according to Formula (9) // *Step 3*
10: $\quad$ **end if**
11: **until** $\|v - v'\| < \epsilon$
12: **return** v'

4 Data Experiments

In this section we compare our approach with the basic OCSFCM algorithm in terms of the assignment of data objects to clusters and the cluster prototypes computation. In order to assess the impact of single modifications of the basic method, in our experiments, we also tested the OCSFCM algorithm that computes the cluster prototypes only using the available feature values and the FCMDFVC approach that computes cluster prototypes using available and imputed feature values.

4.1 Test Data and Experimental Setup

We tested the four above-mentioned approaches on different real and synthetic data sets. For the sake of brevity, here we only report the results obtained on the *wine* data set from the UCI Machine Learning Repository [12] and three synthetic data sets with different properties. The *wine* data set consists of 178 data items, each with 13 features representing the results of a chemical analysis of three wine types. Corresponding to the wine types, the data items are distributed in three classes with 59, 71, and 48 instances. The synthetic data sets are depicted in Fig. 1. Each of the data sets consist of 2000 data items generated by the compositions of three and five 3-dimensional Gaussian distributions respectively. The data items are distributed in three clusters with 400, 700, and 900 instances in the data set *3D-3* and in five clusters with 200, 350, 450, 700, and 300 instances in the data sets *3D-5* and *3D-5-h*. All clusters have different magnitudes and partly overlap. The data set *3D-5-h* was generated from the data set *3D-5* by moving the clusters so that two groups of two and three differently sized clusters build a hierarchical structure in the resulting data set. Using the data set *3D-5-h*, we wanted to find out if the ordering of clusters in the data sets affects the performance of the clustering algorithms adapted to incomplete data.

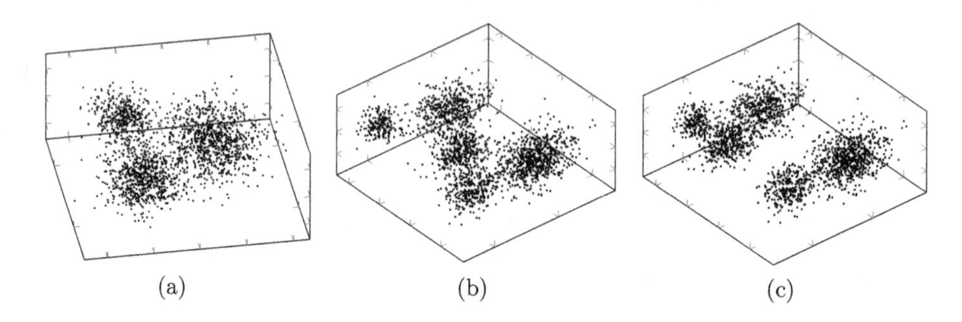

Fig. 1. Test data: (a) 3D-3, (b) 3D-5, (c) 3D-5-h.

In our experiments we first clustered the complete data sets with the basic FCM algorithm and used the resulting clusterings as a baseline for the comparison. Then we generated incomplete data sets by removing values in all features with different probabilities according to the general missing-data pattern [13]. The percentage of missing values was calculated in relation to all feature values in the data sets. In the resulting incomplete data sets, missing values were equally distributed in all features. The general missing-data pattern is common but challenging for clustering methods because missing values occur in many data items. Since we did not adapted the clustering algorithms to incomplete data with conditionally missing values, we deleted the values from the test data sets according to the common missing-data mechanisms MCAR [13]. We clustered the incomplete data sets with the four above-mentioned clustering algorithms. We initialized the cluster prototypes with random values in the data space at the beginning of the algorithms to create the test conditions as real as possible. We used the Frobenius norm distance for the stopping criterion $\|V - V'\| < \epsilon$ defined in Formula (10).

$$\|V - V'\|_F = \sqrt{\sum_{i=1}^{c}\sum_{j=1}^{d} |v_{ij} - v'_{ij}|^2} \quad \text{for } 1 \le i \le c,\ 1 \le j \le d. \tag{10}$$

In all our experiments we set the value ϵ to 0.0001.

4.2 Experimental Results

The performance results of the fuzzy c-means clustering algorithms adapted to incomplete data are shown in Figs. 2, 3, 4, and 5. For the sake of uniformity, we marked the clustering algorithms that only use the available feature items for the computation of cluster prototypes with an asterisk in the legend. That means that our new approach described in Sect. 3 is listed as FCMDFVC* in the legend.

We evaluated the partitioning results produced by the clustering approaches using different crisp and fuzzy indexes. Here, we only report the results for the subset similarity index [14] for comparing the fuzzy partitions produced by the clustering algorithms. Although in previous publications the partitioning results produced by the fuzzy c-means algorithms for incomplete data were evaluated using crisp similarity indexes, in our opinion, it makes more sense to use fuzzy indexes to compare the resulting membership degrees. We used the Frobenius norm distance between the terminal prototypes produced by FCM on complete data and the terminal cluster prototypes produced by the four fuzzy clustering approaches on incomplete data. Since there were significant variations in the results from trial to trial, the figures below show the averaged results obtained over 100 trials. For the sake of clarity, we omitted the standard deviations in the diagrams.

Figure 2 presents the performance results for the basic OCSFCM and our approach produced on the incomplete *wine* data set. According to these results the computation of cluster prototypes only using available feature values seems to be the only factor that improved the partitioning results on incomplete data. Our proposal to impute missing values using the dimension-wise fuzzy variance of clusters seems to worsen the partitioning results. Indeed, in almost all experiments this approach produced the least accurate partitioning results and the highest prototype error among all approaches. As we mentioned above, the reason is that the imputation of missing values depends on the imputation in the previous iteration step and influences the computation of the cluster prototypes. Therefore, it is essential to compute the cluster prototypes only on the basis of available feature items in our approach that produced similarly good results as OCSFCM using available feature values for the computation of cluster prototypes.

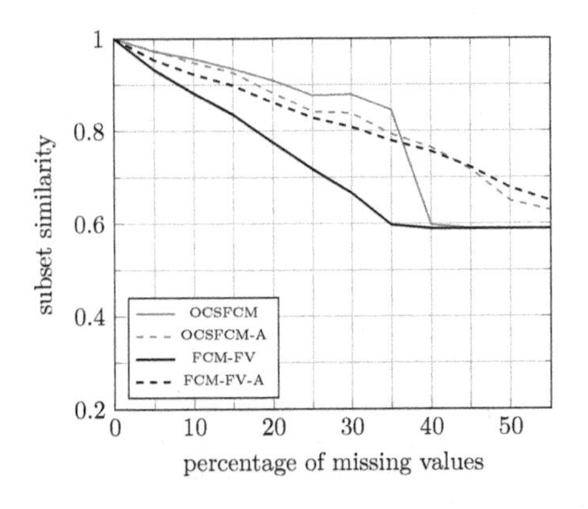

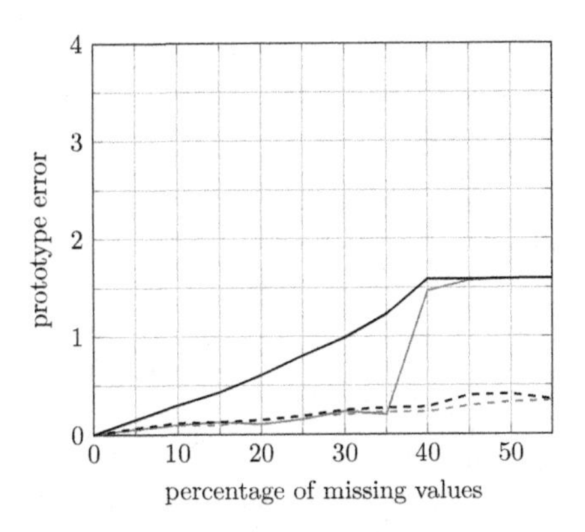

Fig. 2. Averaged results of 100 trials using incomplete *wine* data set.

Since the *wine* data set has 13 attributes, we cannot visualize it and cannot comment on the obtained results. Figures 3 and 4 present the performance results produced by the four approaches on the 3-dimensional data sets with three and five clusters. As the diagrams show, our new approach produced slightly more accurate partitioning results and much more accurate terminal cluster prototypes. While the performance results of OCSFCM and OCSFCM* were similar on the data set *3D-3*, the OCSFCM approach using available feature values for the computation of cluster prototypes produced on average much inaccurate partitioning results on the data set *3D-5*. This is due to the fact that unlike other methods this approach produced comparably good clustering results only in few of 100 trials and performed poorly on average.

Figure 5 shows the performance results for the four approaches produced on the incomplete *3D-5-h* data set where clusters were moved together building two groups. All clustering approaches produced unstable results on this data set. The basic FCM algorithm produced comparably accurate results on the complete *3D-5-h* data set, we obtained 0.9949 for the crisp subset similarity averaged over

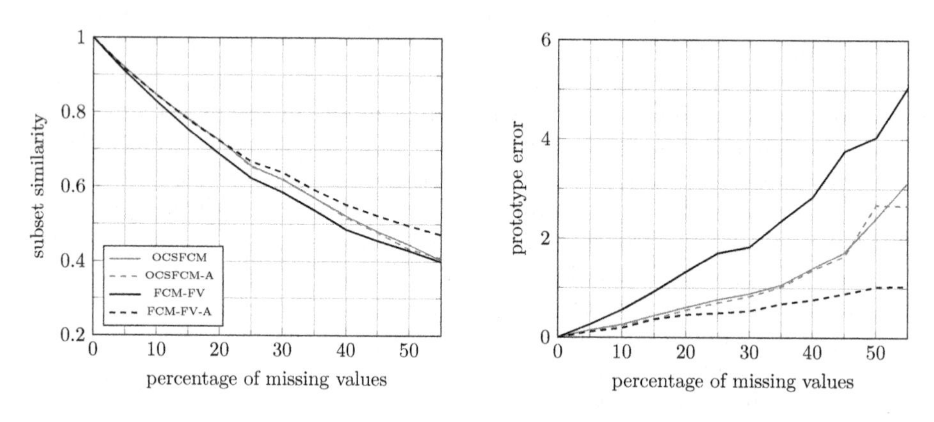

Fig. 3. Averaged results of 100 trials using incomplete *3D-3* data set.

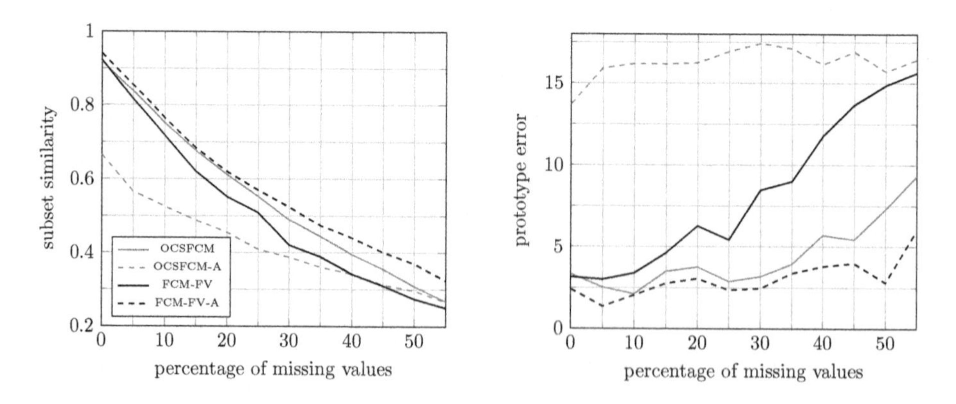

Fig. 4. Averaged results of 100 trials using incomplete *3D-5* data set.

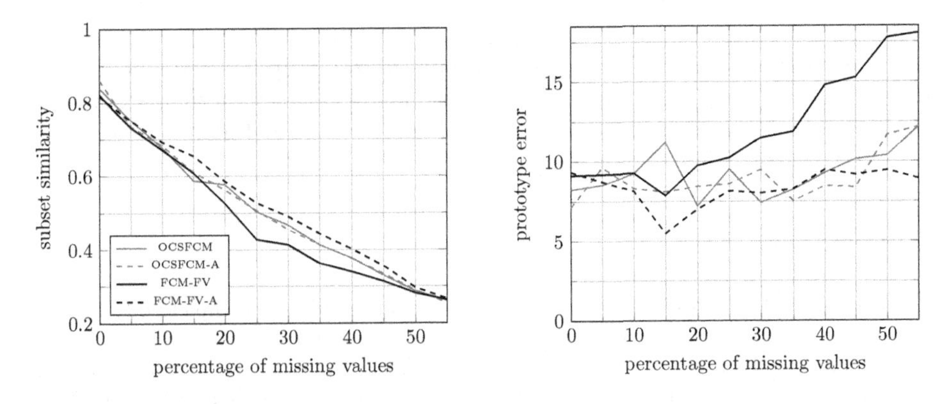

Fig. 5. Averaged results of 100 trials using incomplete *3D-5-h* data set.

100 trials. That shows that the spatial arrangement of clusters in the data set has a strong influence on the clustering results produced on incomplete data.

5 Conclusions and Future Work

Approaches proposed in the literature for adapting the fuzzy c-means (FCM) algorithm to incomplete data assume that the clusters are equally sized and shaped. They produce less accurate clustering results on incomplete data with differently sized and scattered clusters than on the data sets with clusters of equal sizes and volumes. Our previous approach presented in [6] for adapting FCM to incomplete data that takes the cluster scatters into account during the imputation of missing values computes the cluster dispersions on the basis of completely available features, but uses them for the imputation of missing values in the other features. Therefore, if clusters have different extents in different dimensions, it fails to work. In this paper, we presented a new approach for adapting the fuzzy c-means algorithm to incomplete data that uses the dimension-wise fuzzy variances of clusters for imputation of missing values. Our approach involves the cluster shapes and volumes for missing value imputation and computes cluster prototypes only using the available feature values. In experiments on real and synthetic data sets with differently sized and shaped clusters, we demonstrated that our new approach produces slightly more accurate partitioning results and much more accurate terminal cluster prototypes than the basic OCSFCM method.

In all our experiments we used incomplete data with missing values MCAR because we did not adapted our approach to incomplete data with a conditional absence of values. In the future we plan to apply the idea of using the class specific probabilities for imputation of missing values as in the approach presented in [15] in order to improve the performance of our approach on data sets with missing values MAR and NMAR. Furthermore, in our experiments we assumed the optimal number of clusters to be known. Determining the optimal number of clusters on incomplete data is still a challenging problem. Therefore, we want to analyze whether the partitioning results produced by our approach on incomplete data with differently sized and scattered clusters are better than the partitioning produced by the basic OCSFCM for determining the optimal number of clusters using the cluster validity indexes.

References

1. Hathaway, R.J., Bezdek, J.C.: Fuzzy *c*-means clustering of incomplete data. IEEE Trans. Syst. Man Cybern. Part B **31**(5), 735–744 (2001)
2. Timm, H., Döring, C., Kruse, R.: Fuzzy cluster analysis of partially missing datasets. In: Proceedings of the European Symposium on Intelligent Technologies, Hybid Systems and Their Implementation on Smart Adaptive Systems (EUNITE 2002), pp. 426–431 (2002)
3. Sarkar, M., Leong, T.-Y.: Fuzzy K-means clustering with missing values. In: Proceedings of the American Medical Informatics Association Annual Symposium, pp. 588–592 (2001)
4. van der Laan, M.Y., Pollard, K.S.: A new algorithm for hybrid hierarchical clustering with visualization and the bootstrap. J. Stat. Plann. Infer. **117**(2), 275–303 (2003)
5. Himmelspach, L., Conrad, S.: Clustering approaches for data with missing values: comparison and evaluation. In: Proceedings of the Fifth IEEE International Conference on Digital Information Management (ICDIM 2010), pp. 19–28 (2010)
6. Himmelspach, L., Conrad, S.: Fuzzy clustering of incomplete data based on cluster dispersion. In: Hüllermeier, E., Kruse, R., Hoffmann, F. (eds.) IPMU 2010. LNCS, vol. 6178, pp. 59–68. Springer, Heidelberg (2010)
7. MacQueen, J.: Some methods for classification and analysis of multivariate observations. In: Proceedings of the Fifth Berkeley Symposium on Mathematical Statistics and Probability, Volume 1: Statistics, pp. 281–297 (1967)
8. Bezdek, J.C.: Pattern Recognition with Fuzzy Objective Function Algorithms. Kluwer Academic Publishers, Norwell (1981)
9. Kruse, R., Döring, C., Lesot, M.-J.: Fundamentals of fuzzy clustering. In: Advances in Fuzzy Clustering and Its Applications, pp. 1–30 (2007)
10. Klawonn, F., Kruse, R., Winkler, R.: Fuzzy clustering: more than just fuzzification. Fuzzy Sets Syst. **281**, 272–279 (2015)
11. Timm, H.: Fuzzy-Clusteranalyse: Methoden zur Exploration von Daten mit fehlenden Werten sowie klassifizierten Daten. Ph.D. thesis, Germany (2002)
12. Asuncion, A., Newman, D.J.: UCI Machine Learning Repository (2007). http://www.ics.uci.edu/~mlearn/MLRepository.html
13. Little, R.J.A., Rubin, D.B.: Statistical Analysis with Missing Data, 2nd edn. Wiley, New York (2002)
14. Runkler, T.A.: Comparing partitions by subset similarities. In: Hüllermeier, E., Kruse, R., Hoffmann, F. (eds.) IPMU 2010. LNCS, vol. 6178, pp. 29–38. Springer, Heidelberg (2010)
15. Timm, H., Döring, C., Kruse, R.: Different approaches to fuzzy clustering of incomplete datasets. Int. J. Approximate Reasoning **35**(3), 239–249 (2004)

On a Generalized Objective Function for Possibilistic Fuzzy Clustering

József Mezei[1,3(✉)] and Peter Sarlin[2,3]

[1] Faculty of Social Sciences, Business and Economics, Åbo Akademi University, Turku, Finland
jmezei@abo.fi
[2] Department of Economics, Hanken School of Economics, Helsinki, Finland
peter@risklab.fi
[3] RiskLab Finland at Arcada University of Applied Sciences, Helsinki, Finland

Abstract. Possibilistic clustering methods have gained attention in both applied and theoretical research. In this paper, we formulate a general objective function for possibilistic clustering. The objective function can be used as the basis of a mixed clustering approach incorporating both fuzzy memberships and possibilistic typicality values to overcome various problems of previous clustering approaches. We use numerical experiments for a classification task to illustrate the usefulness of the proposal. Beyond a performance comparison with the three most widely used (mixed) possibilistic clustering methods, this also outlines the use of possibilistic clustering for descriptive classification via memberships to a variety of different class clusters. We find that possibilistic clustering using the general objective function outperforms traditional approaches in terms of various performance measures.

Keywords: Possibilistic clustering · Membership function · Typicality values · Classification

1 Introduction

Clustering is one of the central tasks in pattern recognition and machine learning, and aims at partitioning a set of data points into groups of "similar" observations. Fuzzy clustering methods rely on set-theoretical notions introduced by Zadeh [23], motivated by the imprecision present in many (if not all) real life phenomena. The essential idea behind fuzzy sets (i.e., degree of belonging to sets) naturally translates to clustering algorithms: elements can belong to several (overlapping) fuzzy clusters. In fuzzy clustering, the fuzzy c-means (FCM) clustering algorithm [2] is the best known and used method. Since the FCM memberships do not always explain the degrees of belonging for the data well, Krishnapuram and Keller [11] proposed a possibilistic approach to clustering. However, the performance of Krishnapuram and Keller's approach depends heavily on the parameter initialization. This has been pointed out several times in the literature [6], and resulted in different modifications of the original possibilistic fuzzy clustering algorithm.

J.P. Carvalho et al. (Eds.): IPMU 2016, Part I, CCIS 610, pp. 711–722, 2016.
DOI: 10.1007/978-3-319-40596-4_59

In this paper, we focus on a subset of fuzzy clustering methods (objective function-based algorithms) that include the most widely applied variants [8]. Objective function-based fuzzy clustering algorithms have been applied in various domains from economics [12], systems modeling [19] and finance [1] to health-related decision making problems [13] and image segmentation [5]. In the paper, we formulate a general objective function that can be used as the basis of a fuzzy clustering task, and which includes the most important previous models as special cases. The objective function synthesizes various terms appearing in different fuzzy-possibilistic clustering models in the literature [11,15,16,21]. The new model is proposed with the main task of solving a two-class classification task. Beyond a performance comparison with the three most widely used (mixed) possibilistic clustering methods, this also outlines the use of possibilistic clustering for descriptive classification via memberships to a variety of different class clusters. In contrast to previous contributions, we perform a clustering only on data belonging to one of the classes. To predict the class for a new observation, we assess whether any cluster has a distance below a given optimized threshold. In the numerical experiments presented in the paper, we show that possibilistic clustering using the general objective function outperforms previous approaches in terms of various predictive performance measures.

The rest of the paper is structured as follows. In Sect. 2, we discuss different types of partitions utilized as the basis of classification algorithms, whereas Sect. 3 specifies the general objective function to be used in possibilistic clustering. Section 4 presents the numerical experiments that illustrate the advantages of the proposed objective function. Finally, Sect. 5 concludes with a discussion of the results and potential future research directions.

2 From Fuzzy Partitions to Fuzzy Clustering

In a general setting, the input data for clustering consists of n observations: $\mathbf{x}_i = [x_{i1}, x_{i2} \dots x_{im}]$ for $i = 1, \dots, n$, with every observation described by m measurement variables. As the basis of clustering, a partition has to be determined to assign the observations to one (or more) clusters. There are two general approaches used in various clustering algorithms: partitions based on

- crisp sets: every object belongs to exactly one cluster;
- fuzzy sets: every object can belong to several clusters with different degrees.

If we denote the number of clusters with c, a crisp partition can be described by the matrix $\mathbf{U} = [\mu_{ij}]_{c \times n}$, where

$$\mu_{ij} \in \{0,1\}, 1 \leq i \leq c, 1 \leq j \leq n \tag{1}$$

$$\sum_{i=1}^{c} \mu_{ij} = 1, 1 \leq j \leq n \tag{2}$$

$$0 < \sum_{j=1}^{n} \mu_{ij} < n, 1 \leq i \leq c \tag{3}$$

Traditional clustering algorithms, such as the c-means, are based on crisp partitions. By modifying the required properties in various ways (preserving (3) to avoid trivial solutions), we can obtain more general classes of partitions.

The most widely used alternative to crisp clustering, usually termed as (probabilistic) fuzzy c-means clustering [2], defines a fuzzy partition (as discussed first in [3]) that can be described by extending the possible values for the cluster memberships, μ_{ij}, to the $[0, 1]$ interval. Formally, we modify (1) into

$$\mu_{ij} \in [0,1], 1 \le i \le c, 1 \le j \le n \tag{4}$$

while keeping the other two conditions unchanged. With this formulation, observations can belong to several clusters with different degrees specified by the *membership value* μ_{ij}. A typical problem in clustering which can be tackled more appropriately by employing fuzzy partitions is the case of observations on the boundaries between clusters.

While fuzzy clustering improves on traditional clustering from various perspectives, the assumption requiring the sum of membership values to be equal to 1 for every observation forces outlier points to belong to at least one cluster to a large degree. To handle this issue, possibilistic partitions as the basis of clustering can be defined as in [11], by keeping (4) and modifying (2) into

$$\exists \mu_{ij} > 0, 1 \le j \le n \tag{5}$$

This condition ensures that every observation belongs to at least one cluster to some degree, but does not need to belong to a high degree to clusters overall.

2.1 Objective Function-Based Possibilistic Clustering

Based on one (or the combination of several) of the described types of partitions, various clustering methodologies can be developed. There is a large number of different variations of fuzzy clustering methods that can be classified based on different properties. The most commonly used distinction divides the algorithms into two main groups: (i) methods that aim at finding a fuzzy partition using global criteria for optimality in the form of an objective function, and (ii) methods generalizing the previous approaches by allowing the user to choose among multiple update equations for the prototypes and membership degrees without considering a particular criterion function. Although there are several approaches belonging to the second group [17], objective function-based approaches dominate the literature. According to these approaches, one group of parameters (e.g., the membership degrees) are optimized holding the other group (e.g., the cluster prototypes or cluster centers) fixed and vice versa following an iterative updating scheme. This requires the utilization of a distance function (usually Euclidean distance) to determine the membership values. Depending on the structure of the objective function, various parameters restricting the size and shape of clusters can be included in the algorithm.

Since the proposal of Krishnapuram and Keller [11], the first approach that uses possibilistic partition as the basis of clustering, there have been many contributions in the literature to extend and improve this original version of possibilistic fuzzy clustering (PFC). In contrast to the membership values used in c-means clustering, the values reflecting the level of assignment of an observation to a cluster are termed as typicality values; in the following we will also employ this term when referring to possibilistic memberships. The proposed algorithm offers important advantages compared to traditional fuzzy c-means algorithm especially when dealing with outliers and noisy data, however, it can result in identical clusters unless a proper initialization of the typicality values (see [21] for a detailed discussion on this issue). For this reason, when this original possibilistic clustering algorithm is used, a typical approach is to initialize the values using fuzzy c-means algorithm. The first main improvement on PFC was proposed in [15], where a mixed possibilistic c-means clustering ($MPFC$) approach is introduced based on the idea of incorporating membership and typicality in the same objective function. As it was pointed out in [21], this model still has issues with regard to membership initialization and in many cases can result in identical clusters. For this reason, they extend the model by controlling the shape of resulting clusters with a repulsion term in the objective function. A different extension is proposed in [16] in the form of a possibilistic c-means fuzzy clustering model ($PCFC$) by applying a limiting term on the typicality values in the objective function. Additionally, there have been different approaches extending possibilistic clustering with interval-valued membership and typicality values. Min et al. [14] proposed the interval-valued possibilistic clustering algorithm by incorporating interval-valued fuzzy sets into the PFC model of Krishnapuram and Keller [11]. Recently, Jie et al. [9] proposed an interval-valued extension of the algorithm by Pal et al. [15].

3 A General Objective Function for Mixed Fuzzy-Possibilistic Clustering

In this section, we start by formulating a general objective function that synthesizes different terms appearing in various proposals to combining the improvements of the models. According to these points, a general objective function, which to the knowledge of the authors has not been formulated before, can be specified as:

$$J(\mathbf{X},\mathbf{U},\mathbf{B}) = \sum_{i=1}^{c}\sum_{j=1}^{n}\left(a\mu_{ij}^{m} + bt_{ij}^{\lambda}\right) d^2(x_i,c_j) + \sum_{i=1}^{c}\eta_i\sum_{j=1}^{n}(1-t_{ij})^{\lambda} + \sum_{i=1}^{c}\gamma_i\sum_{j=1,j\neq i}^{c}\frac{1}{\zeta d^2(c_i,c_j)} \tag{6}$$

where the following notations are used:

- μ_{ij} is the membership degree that is normalized for every observation, t_{ij} is the typicality value that corresponds to non-normalized (possibilistic) membership degrees, while c_i is the center of cluster i.
- a and b specify the weights of membership and typicality in the clustering algorithm.
- m and λ specify the fuzziness value for membership and typicality: the closer this value is to 1, the more crisp the clustering is. Without any parameter optimization, the generally recommended value to use is 2.
- η_i specifies the boundary of the fuzzy cluster in terms of the typicality value (i.e. when typicality becomes less than 0.5 indicating being distinct from the cluster).
- γ_i is a weighting factor to compensate for clusters being too close to each other.
- ζ specifies the minimal acceptable distance among clusters.

This general form of the objective function allows for the use of normalized (fuzzy) and possibilistic partitions in the form of membership and typicality values. The most widely used fuzzy clustering approaches are all based on different special cases of this general objective function:

- the fuzzy c-means [2] ($a = 1, b = 0, \eta_i = 0, \gamma_i = 0$);
- the possibilistic fuzzy clustering in [11] ($a = 0, b = 1, \gamma_i = 0$);
- the mixed c-means clustering model [15] ($a = 1, b = 1, \eta_i = 0, \gamma_i = 0$);
- the possibilistic c-means clustering model [16] ($\gamma_i = 0$);
- the extended possibilistic clustering model [21] ($a = 0$).

An important feature of most of the objective function-based clustering algorithms is that the terms in the function can be optimized individually resulting in the updating formula for membership and typicality values. As the specified general objective function still maintins the property of separability in terms of the variables, as in other possibilistic clustering algorithms, the optimal clustering can be found by sequentially updating memberships and typicalities and recalculating cluster centers. The update formulas can be specified based on the modification applied to [16,21] as:

$$\mu_{i,j} = \frac{1}{\sum_{k=1}^{c} \left(\frac{d(x_i,c_j)}{d(x_i,c_k)}\right)^{\frac{2}{m-1}}} \tag{7}$$

$$t_{ij} = \frac{1}{1 + \left(\frac{b}{\eta_i} d(x_i, c_j)^2\right)^{\frac{2}{\lambda_1 - 1}}} \tag{8}$$

$$c_i = \frac{\sum_{j=1}^{n} \left(a\mu_{ij}^{m} + bt_{ij}^{\lambda}\right) x_j - \gamma_i \sum_{j=1, j\neq i}^{c} \frac{1}{\zeta d^4(c_i, c_j)} c_j}{\sum_{j=1}^{n} \left(a\mu_{ij}^{m} + bt_{ij}^{\lambda}\right) - \gamma_i \sum_{j=1, j\neq i}^{c} \frac{1}{\zeta d^4(c_i, c_j)}} \tag{9}$$

In general, the centroid representing a cluster is calculated by aggregating the observations assigned to the cluster. In the crisp case, most frequently this aggregation is the simple arithmetic mean while in case of fuzzy clustering, the membership-weighted average is utilized to obtain the cluster center. In our case, the cluster centers are updated based on the distance from the weighted average of the observations, with the weights being the weighted averages of the membership and typicality values.

3.1 Clustering as the Basis of Binary Classification Tasks

As the last step of any fuzzy clustering algorithm, one can obtain a crisp clustering by reducing the fuzzy partition into a crisp partition. Based on the membership or typicality values, the typical approach to obtain a hard clustering is to assign the observation to the cluster with the highest corresponding membership. Our main purpose for utilizing the proposed approach is to apply it as a basis for tackling binary classification problems. That is, additionally to the variables used in the clustering algorithm, we suppose that the data set contains an additional class variable indicating to which of 2 possible classes an observation belongs.

The traditional approach to build a classification model based on the result of fuzzy clustering utilizes the obtained crisp clustering assignment and combines it with the class information. The class of every cluster is defined as the class with the highest number of observations assigned to the cluster. This class assignment then can be used to specify the class of new observations based on the cluster with minimal distance between the observation and corresponding cluster centers.

In contrast to this, we apply a different approach in case of binary classification problems, i.e. two output classes, from now on termed as positive and negative cases. The main idea is that the clustering is performed only on the observations belonging to one of the output classes, e.g. only on observations classified as positive. After the clusters describing different types of positive cases are obtained, both negative and positive cases can be mapped onto the clustering in terms of fuzzy membership and typicality values. As every cluster is classified as positive, it is a reasonable assumption that observations belonging to negative cases will result in higher distance values from each of the cluster centers compared to positive cases, as clusters represent the aggregated behaviour of observations in the positive class. According to this reasoning, for every observation, we use the minimum of the average of membership and typicality values for each cluster, as the basis of the classification process. If an observation has a low (with respect to the chosen threshold) minimum average membership/typicality, it implies that it is far from all the cluster centers, and will be classified as a negative case, while observations with high (with respect to the chosen threshold) minimum average value will be classified as positive.

Note 1. As an important reason for using this procedure we note, that in practice, there are many typical classification problems in which one output class, usually the less frequent one, is of crucial interest. Important examples include identifying patients with a disease, detecting potentially risky borrowers applying for bank loans or detecting machines with high probability of failure in the near future. By performing the clustering first only on this more interesting class, additionally to the classification model, we can use the clustering the traditional way by characterizing and understanding different groups within the positive cases. While we do not focus on this perspective, this would provide an additional benefit compared to traditional classification methods.

3.2 Summary of the Proposed Approach

The following steps summarize the most important components of performing possibilistic fuzzy clustering using a general objective function:

- Initialize the clustering process with the membership values generated by a run of fuzzy c-means clustering. Additionally, the following clustering parameters are defined: maximum number of iterations, range for the possible number of clusters to be tested, fuzziness and typicality parameters.
- Update the membership and typicality values based on the Formulas (7), (8) and sequentially recalculate the cluster centers by calculating the distance from the weighted average of the observations.
- When a specified stopping criterion is reached in terms of the number of iterations or change in the membership and typicality values, stop the updating and record the cluster centroids as the final cluster centers.
- Predict the class for new observations by comparing the minimum of the weighted average of membership and typicality values among all the clusters.

4 Numerical Experiments

In this section, we will perform a numerical comparison of the performance of the various possibilistic clustering methods discussed in the paper. In the experiments, the clustering methods are used as the basis for a classification task with two classes. The main idea behind the experiments is that the clustering is performed only on the observations belonging to one of the output classes (the positive cases). As a result of this clustering step, we obtain the cluster centers for the specified number of clusters.

In the next step, the membership and typicality degrees for all the data points with respect to each cluster can be calculated according to the formula specified in the previous section. According to the reasoning that in average observations belonging to the second output class (negative cases) will result in higher distance values from each of the cluster centers, we use the minimum of the average of membership and typicality values, as the basis of the classification process. If an observation has a low minimum average uncertainty (membership and typicality), it implies that it is far from all the cluster centers, and will be classified as a negative case, while observations with high minimum value will be classified as positive.

In the last step, an optimal threshold on this possibilistic likelihood represented by the minimum value is identified with the Usefulness measure (i.e., a preference weighted average of type I and II errors, see [18] for further information). For a given threshold, the calculation of this measure makes use of the confusion matrix that is specified through the following four values: (i) true positive (TP), i.e. positive cases classified correctly as positive; (ii) true negative (TN), i.e. negative cases classified correctly as negative; (iii) false positive (FP), i.e. negative cases classified incorrectly as positive; (iv) false negatives (FN), i.e. positive cases classified incorrectly negative. Based on these notations, one can define type I error as $T_1 = FN/(FN + TP)$ (the share of misclassified positive cases to the total number of positive cases), and type II errors $T_2 = FP/(TN + FP)$ (the share of misclassified negative cases to the total number of negative cases). By specifying the preference between making type I and type II errors as μ and using the notations P_1 and P_2 for the probabilities of positive and negative cases, respectively, the loss function can be defined as $L(\mu) = \mu T_1 P_1 + (1-\mu) T_2 P_2$. The absolute usefulness of a classification model can be defined by comparing the loss function to using the model of assigning every observation to the most frequent class: $U_a(\mu) = \min(\mu P_1, (1 - \mu)P_2) - L(\mu)$. Relative usefulness compares absolute usefulness with a perfect model (model with loss function value 0). The optimal threshold for the classification problem is chosen as the value which results in maximal relative usefulness.

4.1 Data

In the experiments, we utilize four datasets that are frequently used in the literature to assess classification performance of algorithms. We include small datasets with few attributes (Haberman and Transfusion), a dataset with large number of attributes and few observations (Ionosphere), and a moderately large dataset (Adult).

- **Haberman dataset** [7]: This dataset stores information on patients who have undergone surgery for treating breast cancer. This dataset has 306 entries with 3 features: age of the patient, the year of surgery, number of positive axillary nodes detected. There are two resulting classes: *the patient survived 5 years or longer* (225 observations) and *the patient died within 5 year* (81 observations).
- **Blood Transfusion Service Center dataset** [22]: The data descends from the Blood Transfusion Service Center of Hsin-Chu City in Taiwan. The dataset contains 749 entries described by 4 features (months since last donation; total number of donations; total blood donated; months since first donation. The data is classified into two different classes recording whether the donor has donated blood or not.
- **Ionosphere dataset** [20]: The data was collected by a radar system in Goose Bay, Labrador. The target of the measurement was free electrons in the ionosphere. There are 34 features in the dataset with 351 observations. The observations are classified as either *good* (showing evidence of structure in the ionosphere), or *bad* (no sign of structure).

– **Adult dataset** [10]: The dataset contains information extracted from the 1994 census data in the United States. The dataset contains 14 attributes describing social and demographic information about citizens registered in the census. The classification task is to predict whether a person's income exceeds 50,000 dollars a year or not. The dataset includes 32561 observations.

4.2 Classification Performance Measures

A basic, yet somewhat limited, measure of prediction performance is accuracy $(TP+TN/(TP+TN+FP+FN))$: the proportion of correctly classified cases. For problems with imbalanced classes, a popular performance measure is the area under the curve (AUC), which is based on the Receiver Operating Characteristic (ROC). The ROC curve [4] depicts the true positive and false positive rates based on the threshold chosen in case of a probabilistic classifier output to determine the output class, and AUC measures the area under the ROC curve. The maximum value of AUC is 1, and the closer the value is to 1, the higher the probability that the classifier assigns the right class to the data point. It is important to note that the value of AUC can be misleading in the case of a dataset with imbalanced classes, but still this is one of the most widely used evaluation measures.

4.3 Results

As a comparison to the results of the various possibilistic clustering algorithms, we utilized three of the most widely used classification algorithms (support vector machines (SVM), k-nearest neighbour (kNN), and classification and regression trees ($CART$)) to obtain a reasonable baseline for assessing classification performance. The results can be seen in Table 1. Although we focus only on the classification performance, we note here that additionally to this, and in contrast to many traditional classification methods, by identifying natural clusters in a dataset, we can obtain a descriptive representation of the various types of a given class present in the dataset. Consequently, if the classification performance of a clustering algorithm is as good as another classification algorithm, this additional benefit can justify its use in various contexts.

In the numerical experiments, we included four different clustering approaches: (i) PFC from [11], (ii) $MPFC$ from [15], (iii) $PCFC$ from [16],

Table 1. Results of the preliminary test of three classification algorithm

Dataset	Haberman		Transfusion		Ionosphere		Adult	
Method	AUC	Accuracy	AUC	Accuracy	AUC	Accuracy	AUC	Accuracy
SVM	0.62	71.2 %	0.69	77.7 %	0.68	74.7 %	0.74	76.7 %
kNN	0.55	65.9 %	0.60	72.5 %	0.61	72.5 %	0.67	71.7 %
$CART$	0.59	62.9 %	0.66	67.5 %	0.65	70.7 %	0.71	74.2 %

Table 2. Results for the Haberman dataset

	c	TN	TP	FP	FN	AUC	Acc	U_{rel}
PFC	15	130	50	95	31	0.61	0.59	0.12
$MPFC$	15	186	25	39	56	0.60	0.69	0.11
$PCFC$	14	164	36	61	45	0.60	0.65	0.12
$GPFC$	15	176	34	49	47	0.65	0.68	0.16

Table 3. Results for the transfusion dataset

	c	TN	TP	FP	FN	AUC	Acc	U_{rel}
PFC	20	498	49	72	129	0.63	0.73	0.11
$MPFC$	19	564	9	6	169	0.58	0.69	0.06
$PCFC$	16	523	28	47	150	0.58	0.74	0.08
$GPFC$	20	499	50	71	128	0.64	0.73	0.17

Table 4. Results for the ionosphere dataset

	c	TN	TP	FP	FN	AUC	Acc	U_{rel}
PFC	6	74	156	52	69	0.67	0.66	0.22
$MPFC$	15	45	200	81	25	0.65	0.70	0.23
$PCFC$	13	68	162	58	63	0.64	0.70	0.21
$GPFC$	17	79	166	47	59	0.65	0.70	0.24

and (iv) the approach based on the generalized objective function formulated in the paper, denoted as $GPFC$. In the experiments, we did not optimize parameter values, we utilized the most commonly used values: $a = b = 0.5, m = \lambda = 2, \eta_i = \gamma_i = 1$

As we can observe from the results of the analysis for the four datasets in Tables 2, 3, 4, and 5, clustering based on the formulated objective function outperforms the other approaches in terms of AUC and relative usefulness in all the cases and is never worse in terms of accuracy. With respect to individual elements of the confusion matrix, we cannot conclude any definite result as it shows better performance compared to other approaches (i) for negative cases in the analysis of the Haberman and Adult datasets, and (ii) for positive cases in the analysis of the transfusion and ionosphere datasets. When we compare the results to the performance of the traditional classification algorithms, we can state (at least for the considered datasets) that we can obtain close to similar performance in terms of AUC and accuracy. By accounting for other potential benefits that can result from analyzing clusters more thoroughly (visually or even with simple descriptive measures), it indicates that the proposed approach can be a potential alternative in similar tasks.

Table 5. Results for the adult dataset

	c	TN	TP	FP	FN	AUC	Acc	U_{rel}
PFC	20	18798	5678	5922	2163	0.72	0.75	0.24
$MPFC$	19	17678	6457	7042	1384	0.71	0.74	0.21
$PCFC$	19	18423	5987	6297	1854	0.73	0.75	0.22
$GPFC$	20	18947	6345	5773	1496	0.75	0.78	0.25

5 Conclusions

In this article, we formulated a generalized possibilistic clustering approach. The general objective function for mixed fuzzy-possibilistic clustering combines previous approaches as an extended and improved version of the original model by Krishnapuram and Keller [11]. The generalized approache is evaluated in terms of a binary classification problem, which implies a descriptive classifier via memberships to a variety of different class clusters. We have shown on four datasets that the use of the general objective function improves classification performance vis-à-vis previous possibilistic clustering algorithms.

There are several possible ways in which present work can be extended in the future. Firstly, in the numerical experiments we did not attempt to optimize various parameters of the objective function in order to identify the optimal initial values. For example, we used equal weights for the memberships and typicality values in the objective function. These could be optimized in various ways, such as based upon a cross-validation exercise. The approach needs also to be tested on large-volume datasets. Additionally, one could incorporate interval-valued fuzzy sets into the proposed clustering scheme by allowing for membership and typicality values to take the form of intervals in $[0, 1]$.

References

1. Alam, P., Booth, D., Lee, K., Thordarson, T.: The use of fuzzy clustering algorithm and self-organizing neural networks for identifying potentially failing banks: an experimental study. Expert Syst. Appl. **18**(3), 185–199 (2000)
2. Bezdek, J.C.: Pattern Recognition with Fuzzy Objective Function Algorithms. Kluwer Academic Publishers, Berlin (1981)
3. Bezdek, J.C., Harris, J.D.: Fuzzy partitions and relations; an axiomatic basis for clustering. Fuzzy Sets Syst. **1**(2), 111–127 (1978)
4. Bradley, A.P.: The use of the area under the roc curve in the evaluation of machine learning algorithms. Pattern Recogn. **30**(7), 1145–1159 (1997)
5. Chuang, K.S., Tzeng, H.L., Chen, S., Wu, J., Chen, T.J.: Fuzzy c-means clustering with spatial information for image segmentation. Comput. Med. Imaging Graph. **30**(1), 9–15 (2006)
6. Döring, C., Lesot, M.J., Kruse, R.: Data analysis with fuzzy clustering methods. Comput. Stat. Data Anal. **51**(1), 192–214 (2006)
7. Haberman, S.J.: Generalized residuals for log-linear models. In: Proceedings of the 9th international biometrics conference, pp. 104–122 (1976)

8. Höppner, F., Klawonn, F., Kruse, R., Runkler, T.: Fuzzy Cluster Analysis: Methods For Classification, Data Analysis and Image Recognition. Wiley, Hoboken (1999)
9. Ji, Z., Xia, Y., Sun, Q., Cao, G.: Interval-valued possibilistic fuzzy c-means clustering algorithm. Fuzzy Sets Syst. **253**, 138–156 (2014)
10. Kohavi, R.: Scaling up the accuracy of naive-bayes classifiers: a decision-tree hybrid. In: Proceedings of the Second International Conference on Knowledge Discovery and Data Mining, pp. 202–207 (1996)
11. Krishnapuram, R., Keller, J.M.: A possibilistic approach to clustering. IEEE Trans. Fuzzy Syst. **1**(2), 98–110 (1993)
12. Marghescu, D., Sarlin, P., Liu, S.: Early-warning analysis for currency crises in emerging markets: a revisit with fuzzy clustering. Intell. Syst. Account. Finance Manage. **17**(3–4), 143–165 (2010)
13. Masulli, F., Schenone, A.: A fuzzy clustering based segmentation system as support to diagnosis in medical imaging. Artif. Intell. Med. **16**(2), 129–147 (1999)
14. Min, J.H., Shim, E.A., Rhee, F.C.H.: An interval type-2 fuzzy pcm algorithm for pattern recognition. In: 2009 IEEE International Conference on Fuzzy Systems, pp. 480–483. IEEE (2009)
15. Pal, N.R., Pal, K., Bezdek, J.C.: A mixed c-means clustering model. In: Proceedings of the Sixth IEEE International Conference on Fuzzy Systems, vol. 1, pp. 11–21. IEEE (1997)
16. Pal, N.R., Pal, K., Keller, J.M., Bezdek, J.C.: A possibilistic fuzzy c-means clustering algorithm. IEEE Trans. Fuzzy Syst. **13**(4), 517–530 (2005)
17. Runkler, T.A., Bezdek, J.C.: Alternating cluster estimation: a new tool for clustering and function approximation. IEEE Trans. Fuzzy Syst. **7**(4), 377–393 (1999)
18. Sarlin, P.: On policymakers' loss functions and the evaluation of early warning systems. Econ. Lett. **119**(1), 1–7 (2013)
19. Setnes, M.: Supervised fuzzy clustering for rule extraction. IEEE Trans. Fuzzy Syst. **8**(4), 416–424 (2000)
20. Sigillito, V.G., Wing, S.P., Hutton, L.V., Baker, K.B.: Classification of radar returns from the ionosphere using neural networks. Johns Hopkins APL Techn. Dig. **10**, 262–266 (1989)
21. Timm, H., Borgelt, C., Döring, C., Kruse, R.: An extension to possibilistic fuzzy cluster analysis. Fuzzy Sets Syst. **147**(1), 3–16 (2004)
22. Yeh, I.C., Yang, K.J., Ting, T.M.: Knowledge discovery on rfm model using bernoulli sequence. Expert Syst. Appl. **36**(3), 5866–5871 (2009)
23. Zadeh, L.A.: Fuzzy sets. Inf. Control **8**(3), 338–353 (1965)

Seasonal Clustering of Residential Natural Gas Consumers

Marta P. Fernandes(✉), Joaquim L. Viegas, Susana M. Vieira, and João M.C. Sousa

IDMEC, Instituto Superior Técnico, Universidade de Lisboa, Lisbon, Portugal
{marta.fernandes,joaquim.viegas,susana.vieira,jmsousa}@tecnico.ulisboa.pt

Abstract. This paper proposes a methodology to define the seasonal load profiles of residential gas consumers using smart metering data. A detailed clustering analysis is performed using fuzzy c-means, k-means and hierarchical clustering algorithms with multiple clustering validity indices. The analysis is based on a sample of more than one thousand households over one year. The results provide evidence that crisp algorithms present the best clustering results overall. However, the fuzzy algorithm proves to be suited when the others generate clusters which are not representative of population groups. Compact and well defined seasonal clusters of gas consumers are obtained, where the representative profiles reflect the consumption patterns that vary according to the season of the year. The knowledge obtained with this methodology can assist decision makers in the energy utilities in developing demand side management programs, consumer engagement strategies, marketing, as well as in designing innovative tariff systems.

Keywords: Residential gas consumption · Clustering · Load profile · Smart metering

1 Introduction

The natural gas sector has lived through a significant change of framework, resulting from the ability of this energy source to respond to several challenges such as, the continuity of supply, flexibility, environmental preservation, economic efficiency and continued market liberalization.

In the case of gas consumption data analysis, there is still a lack of literature regarding clustering and consumer profiling. This data has the potential to give insights of great importance for utilities and policy makers. Significant insights can be derived by the knowledge of typical consumption curves of different consumers groups.

Computational intelligence methods have been applied in several smart grid applications. However, most of the efforts have been oriented to the electrical energy field, for which a considerable amount of literature has been published. Several papers use classical k-means [1–4], weighted fuzzy average k-means [4],

J.P. Carvalho et al. (Eds.): IPMU 2016, Part I, CCIS 610, pp. 723–734, 2016.
DOI: 10.1007/978-3-319-40596-4_60

fuzzy c-means [5–7], modified follow the leader and self organizing maps [4] and hierarchical clustering [3,4] when identifying and classifying electrical load profiles.

Another aspect of literature in this area investigates the impact of weather on energy consumption. Among the different sectors affected by weather risk, the gas sector is one of the most sensitive [8,9]. Gas supply costs usually increase with cold weather and decrease with warm weather. Furthermore, the gas usage typically varies with changes in heating season weather.

Clustering tools enable the discovery of representative patterns from a myriad of collected data. Therefore, this paper proposes to define the seasonal profiles of residential natural gas consumers, using three clustering algorithms namely K-means, Fuzzy C-means and Ward's hierarchical clustering. These algorithms were selected due to their characteristics, both K-means and Fuzzy C-means are classical partitioning algorithms, that have already been successfully used for the case of smart metering electricity data [2–7,10,11]. Euclidean distance was selected for both algorithms and the reason behind this is the interpretability of the results, since the Euclidean distance captures the (imposed) geometry between the clusters in a Euclidean space [12]. Ward's method was selected due to the fact that the clusters are formed in order to minimize the increase of the within-cluster sums of squares and because it has been applied to natural gas data [13].

Results obtained from the above algorithms are assessed and compared by means of suitable clustering validity indices. The consumption patterns are analyzed for the seasonal clusters obtained and conclusions concerning the specific characteristics of each cluster are drawn. High frequency gas consumption data from a smart metering trial conducted in Ireland [14] is analyzed for that purpose. The database is robust given that it has information of almost one thousand and a half households over one year and a half.

This is the paper outline. In Sect. 2, the data preprocessing methods are presented. In Sect. 3 the clustering techniques and the performance measures used are discussed. In Sect. 4 results are presented and discussed. Section 5 concludes the paper.

2 Pre-processing of Gas Consumption Data

Gas smart metering consumption data is composed of a large set of time stamped intervals with consumption values. In order to obtain consumers' profiles which can be easily interpreted, visualized and manipulated, there are a few pre-processing steps to be performed. The following paragraphs explain those steps.

Missing consumption data, resulting due to equipment malfunction or communication problems is ignored in the proposed methodology as the high volume of consumption data and aggregation reduce its impact on the overall data quality.

The context filtering process consists on selecting data which represents a specific context, defined, for example, by a temporal window (e.g. winter, summer, year), type of day (e.g. working day) and location.

Given the outliers potential to bias the results obtained from data, they are usually deleted or replaced by appropriate values. In this paper, a significant percentage of null consumption measurements was identified and the respective households were excluded from the study.

After the context filtering and outlier analysis, a dimensionality reduction is performed through data aggregation, in order to obtain a representative curve of the whole temporal window. The aggregation is characterized by the period used, e.g., hourly, daily, yearly and operator, e.g., sum, mean, median. Considering N samples and H input features, with $\mathbf{x} = [x_{i_1}, ..., x_{i_H}]$ where i=1:N, a matrix $X \in \mathbb{R}^{N \times H}$ is constructed for the selected period, which includes the available information in compact matrix format:

$$X = \begin{bmatrix} x_{1_1} & \cdots & x_{1_H} \\ \vdots & \ddots & \vdots \\ x_{N_1} & \cdots & x_{N_H} \end{bmatrix} \tag{1}$$

The final preprocessing consists of the normalization of the data for easier clustering and representation of the information. Normalization for numeric values may be performed, e.g., based on the mean, standard deviation, maximum value of the whole dataset. In this paper, the profiles are normalized with regards to all consumers' maximum hourly consumption, using the minimum-maximum normalization method.

3 Clustering

Clustering is an unsupervised learning task that aims at decomposing a given set of objects into subgroups or clusters based on similarity [15]. The goal is to divide the dataset in such a way that objects belonging to the same cluster are as similar as possible, whereas objects belonging to different clusters are as dissimilar as possible.

In the scope of this paper, clustering methods are used to find the groups of consumers which have similar consumption curves in some context, such as a season. After data pre-processing, various clustering configurations are tested using multiple clustering validity indices (CVIs).

3.1 K-Means

The objective of the k-means clustering algorithm is to partition the dataset $\mathbf{X}$ into a determined number of clusters (n_c). The set of clusters $\mathbf{C} = \{c_1, ..., c_{n_c}\}$, is required to be a partition of the dataset into non-empty pairwise disjoint subsets. The objective function, of the hard C-means can be written as follows:

$$\mathbf{J}_h(\mathbf{X}, \mathbf{U}_h, \mathbf{V}) = \sum_{i=1}^{n_c} \sum_{j=1}^{N} \mu_{ij} d_{ij}^2(\mathbf{x}_j, \overline{\mathbf{c}_i}) \tag{2}$$

where $\mathbf{V} = \{\overline{\mathbf{c}_1}, ..., \overline{\mathbf{c}_{n_c}}\}$ is the set of cluster centers and $\mathbf{U}_h$ is a $n_c \times N$ binary matrix called partition matrix. The distance measure d_{ij} in this paper is the Euclidean. The individual elements $\boldsymbol{\mu}_{ij} \in \{0, 1\}$ indicate the assignment of data to clusters, and it is required that:

$$\sum_{i=1}^{n_c} \boldsymbol{\mu}_{ij} = 1, \forall j \in \{1, ..., N\} \quad \text{and} \quad \sum_{j=1}^{N} \boldsymbol{\mu}_{ij} > 0, \forall i \in \{1, ..., n_c\}. \tag{3}$$

Cluster centers $\overline{\mathbf{c}}_i$ are computed as the mean of all data vectors assigned to them, using Eq. 4 for each cluster:

$$\overline{\mathbf{c}}_i = \frac{\sum_{j=1}^{N} \boldsymbol{\mu}_{ij} \mathbf{x}_j}{\sum_{j=1}^{N} \boldsymbol{\mu}_{ij}} \tag{4}$$

3.2 Fuzzy C-Means

Formally, a fuzzy cluster model of a given dataset $\mathbf{X}$ into n_c clusters is defined to be best when it minimizes the objective function:

$$\mathbf{J}_f(\mathbf{X}, \mathbf{U}_f, \mathbf{V}) = \sum_{i=1}^{n_c} \sum_{j=1}^{N} \boldsymbol{\mu}_{ij}^m d_{ij}^2 (\mathbf{x}_j, \overline{\mathbf{c}_i}) \tag{5}$$

under the constraints presented in Eq. 3 that have to be satisfied for membership degrees $\boldsymbol{\mu}_{ij} \in [0, 1]$ in $\mathbf{U}_f$. The parameter m, where $m > 1$, is called the fuzzifier or weighting exponent. Cluster centers $\overline{\mathbf{c}}_i$ are computed as the mean of all data vectors assigned to them, using Eq. 6 for each cluster:

$$\overline{\mathbf{c}}_i = \frac{\sum_{j=1}^{N} \boldsymbol{\mu}_{ij}^m \mathbf{x}_j}{\sum_{j=1}^{N} \boldsymbol{\mu}_{ij}^m} \tag{6}$$

3.3 Hierarchical Clustering

Hierarchical agglomerative clustering algorithms produce a sequence of data partitions of decreasing numbers of clusters at each step, where each clustering scheme results from the previous one by merging the two closest clusters into one.

The Ward's [16] minimum variance method is used in this paper. Ward's method represents a cluster by its center and in each iteration it finds the pair of clusters that leads to minimum increase in total within-cluster variance after merging. This increase is a weighted squared distance between cluster centers. The objective function is then the error sum of squares defined to be the squared Euclidean distance between clusters. The cluster centers are obtained using Eq. 4 for each cluster.

3.4 Clustering Validity Indices

Usually the number of (true) clusters in the given data is unknown in advance. However, it is required to specify n_c as an input parameter. The goal of all CVIs is to maximize intra-cluster similarity and minimize inter-cluster similarity. The following CVIs are used in the proposed methodology, to assess the best n_c:

- **Silhouette (Sil):** it reflects the compactness and separation of the clusters. A larger averaged Sil width indicates better quality of the clustering result.
- **Davies and Bouldin's index (DB):** estimates the cohesion based on the distance from the data points in a cluster to its centroid, and the separation based on the distance between centroids. The best n_c should minimize the value of the index.
- **Dunn's index(DI):** identifies clusters that are compact and well separated. For a given assignment of clusters, a higher DI indicates better clustering.
- **Weighted intra-inter cluster distance index (WI)**: it compares the homogeneity of the data to its separation. A WI equal to 1 describes a clustering where every pair of objects from different clusters has null similarity and at least one pair of objects from the same cluster has a non-zero similarity.
- **Xie and Beni's index (XB):** it aims to quantify the ratio of the total variation within clusters and the separation of clusters. The best n_c should minimize the value of the index.

4 Results and Discussion

4.1 Natural Gas Data Preprocessing

The natural gas consumption data used in this paper was provided by the Irish Social Science Data Archive (ISSDA) [17]. The data pre-processing was performed by considering the initial gas consumption data of 1493 households.

- **Missing data analysis:** In the database, 6 days were missing, which were ignored given that the high volume of consumption data and aggregation reduce its effect on the overall data quality.
- **Context filtering:** Only smart metering data from working days was used. The profiles were extracted seasonally, which means days were aggregated for each season of the year. Data from December of 2009 to 2010 was used, where the corresponding number of days to be aggregated was 243.
- **Outliers analysis:** From data analysis, a significant percentage of null consumption measurements was identified in the study period. All consumers with more than 90 % of null consumption measurements in a year were excluded. This percentage was considered suitable when analysing the data and with this criterion, a total of 63 consumers was excluded and 1430 were left.
- **Data aggregation:** The consumption data was sampled every half hour for each meter. It was aggregated hourly resulting in 24 features for each consumer in each season, where the operator used was the summation.

- **Data normalization:** The profiles were normalized with regards to all consumers' maximum hourly consumption for each season.

After data pre-processing, the authors were left with household gas consumption (kWh) for 1430 households over 243 working days and consumers' hourly aggregated data.

4.2 Seasonal Profiles

Two partitional algorithms, k-means (KM) and fuzzy c-means (FCM) and Ward's hierarchical clustering (HC), were used to obtain the seasonal profiles. For FCM, the fuzziness parameter m was varied between 1.25 and 2. The best result was obtained for $m = 1.25$ and only the results using this best value of m are presented in this section.

The number of clusters (n_c) was varied between 2 and 10 in order to obtain the best, or the most suitable, n_c for each algorithm and season. The authors consider a partition suitable for the application if it presents a uniform distribution, minimizes distances between curves of the same clusters and maximizes distances between different clusters. The selected clusters are considered well defined, compact and balanced, representing a significant number of consumers in the population.

Table 1 presents the results of the selected n_c based on the CVIs scores as well as expert decision, through visual analysis, for each algorithm and season. The best n_c was considered to be 3 for spring and summer, 4 for autumn and 5 for winter.

In Table 1, unlike the other CVIs, DI indicates 2 clusters as the best n_c consistently. Except for two events, Sil also indicates that n_c. For all seasons, the

Table 1. Selected n_c (with respective CVI score), for each algorithm and season.

Seasons	Algorithms	Sil	DB	DI	WI	XB
Spring	KM	2(0.31)	5(1.73)	2(0.84)	4(0.59)	2(0.82)
	HC	2(0.23)	6(1.99)	2(0.72)	3(0.74)	3(1.25)
	FCM	2(0.30)	4(1.82)	2(0.82)	5(0.58)	2(1.60)
Summer	KM	2(0.46)	5(1.40)	2(0.75)	5(0.76)	2(0.77)
	HC	4(0.30)	4(1.45)	2(0.62)	3(0.61)	4(1.69)
	FCM	2(0.45)	5(1.70)	2(0.73)	3(0.75)	2(0.91)
Autumn	KM	2(0.24)	2(1.80)	2(0.94)	3(0.62)	2(0.83)
	HC	2(0.17)	4(1.84)	2(0.78)	4(0.71)	4(1.24)
	FCM	2(0.24)	5(1.60)	2(0.93)	5(0.60)	2(0.88)
Winter	KM	2(0.23)	5(1.80)	2(0.88)	3(0.65)	5(1.10)
	HC	5(0.08)	5(2.05)	2(0.77)	5(0.63)	5(1.60)
	FCM	2(0.23)	5(1.63)	2(0.88)	5(0.67)	2(1.01)

two representative profiles obtained consisted of a separation between high and low consumption consumers. For a more interesting analysis of the consumers' consumption patterns, higher n_c has to be obtained, since with 2 clusters the majority of the consumers' profiles are significantly different from the representative ones, the clusters centers.

In the following figures, the consumers' profiles are represented by 24 features for each season, and the x-axis is the 24 h in a day.

Spring Profiles. The best n_c obtained for spring was 3 with the HC algorithm and clusters are presented in Fig. 1. Except for DB index, for all CVIs this n_c indicated better performance than a higher number.

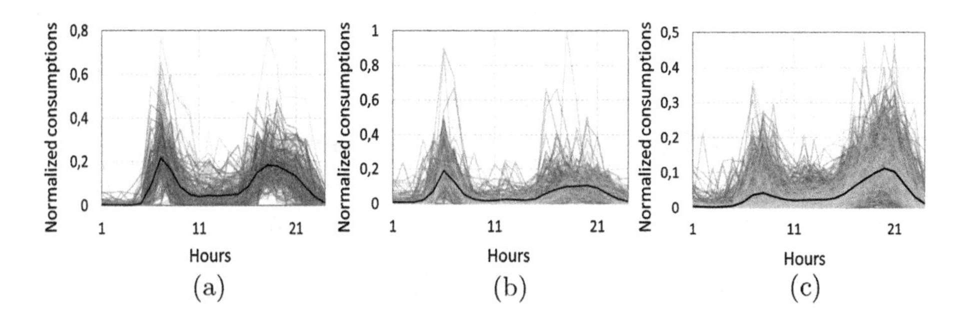

Fig. 1. Spring profiles for 3 clusters: (a) Cluster 1; (b) Cluster 2; (c) Cluster 3.

Summer Profiles. The best n_c obtained for summer was 3 with FCM algorithm and clusters are presented in Fig. 2. For a n_c of 3, better CVIs results were achieved with the other algorithms. However, for both the HC and KM algorithms one of the clusters had very few consumers, 17 and 74, respectively. Given that the aim is to obtain not only compact and well defined clusters, but also representative of the population, this number of consumers was not considered significantly representative and therefore these clusters were not selected.

Autumn Profiles. The best n_c obtained for autumn was 4 with the HC algorithm as presented in Fig. 3. In Table 1, the CVIs indicate a different n_c for other algorithms. Regarding the n_c indicated for the other algorithms, the HC presented better results than KM for a n_c of 3. In the case of FCM algorithm, CVIs indicated a n_c of 5, however the results obtained with 4 clusters produced by the HC presented compact and well defined clusters and consistent CVIs performance.

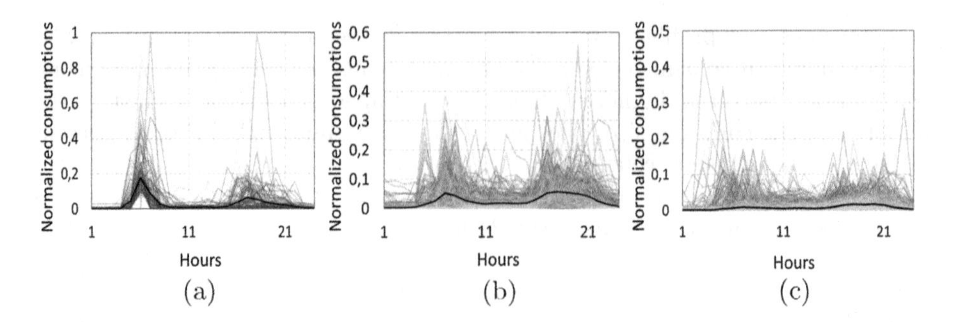

Fig. 2. Summer profiles for 3 clusters: (a) Cluster 1; (b) Cluster 2; (c) Cluster 3.

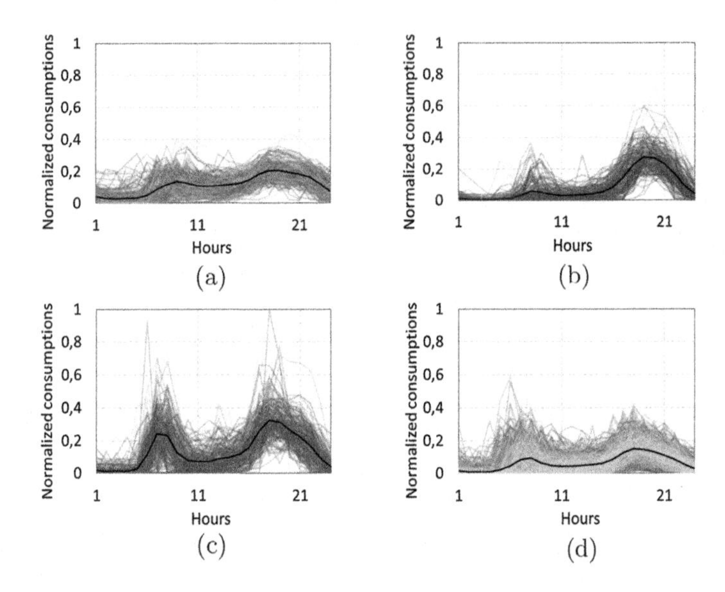

Fig. 3. Autumn profiles for 4 clusters: (a) Cluster 1; (b) Cluster 2; (c) Cluster 3; (d) Cluster 4.

Winter Profiles. The best n_c obtained for winter was 5 with KM algorithm as presented in Fig. 4. The higher number of profiles obtained may be related with a higher use of the energy source at this time of year. In Table 1, the CVIs indicated a n_c of 2, 3 and 5 as the best. For the case of a n_c equal to 3 clusters, only WI indicated this n_c as best for the KM algorithm and the clusters obtained were not so well defined as with 5 clusters. In Table 1, DB consistently indicated this n_c for all algorithms, moreover each of the remaining CVIs indicated it for at least one algorithm.

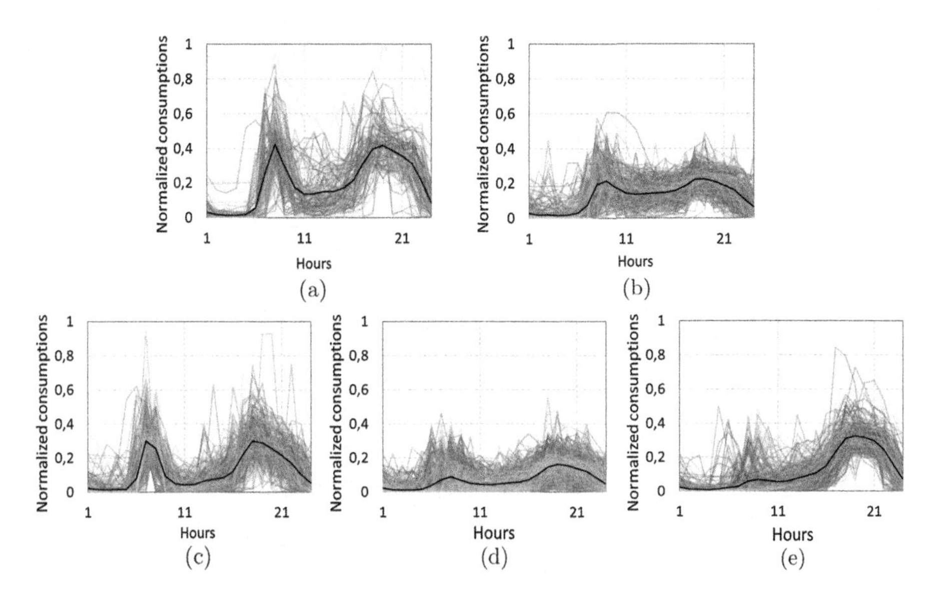

Fig. 4. Winter profiles for 5 clusters: (a) Cluster 1; (b) Cluster 2; (c) Cluster 3; (d) Cluster 4; (e) Cluster 5.

4.3 Representative Profiles

The consumers' representative profiles consist of their mean normalized consumptions, which are represented as load profiles (LP) in Fig. 5. All consumers' representative profiles are mainly characterized by:

- a morning and an evening peak consumption;
- the time at which the consumption starts to rise and to decline;
- the off-peak consumption.

All seasons have at least one LP with a marked difference between peaks. For the case of spring, LP 2 and LP 3, for summer LP 1, for autumn LP 2 and for winter LP 5. In spring and summer, the off-peak consumptions are lower than in the other two seasons, which are colder and characterized by a higher amount of gas use. All seasons have at least one LP with a peak consumption in a short period of time (approximately 3 h): LP 1 and LP 2 for spring, LP 1 for summer, LP 3 for autumn and LP 1 and LP 3 for winter. The larger peaks may take several hours of the day to rise and decline, which may be related with the heating systems programming.

In Table 2, clusters size is presented in terms of percentage of consumers for the clusters 1 to 5 (C1–C5), where each cluster is represented by the LP 1 to 5, respectively. The algorithms which generated those clusters are presented as well.

From Table 2, the consumption dynamics can be described for each season:

- **Spring**: the majority of consumers (66 %) is represented by LP 3, which corresponds to those who have the lowest morning peak consumption and a high one in the evening, as well as a low off-peak consumption during daytime and night;

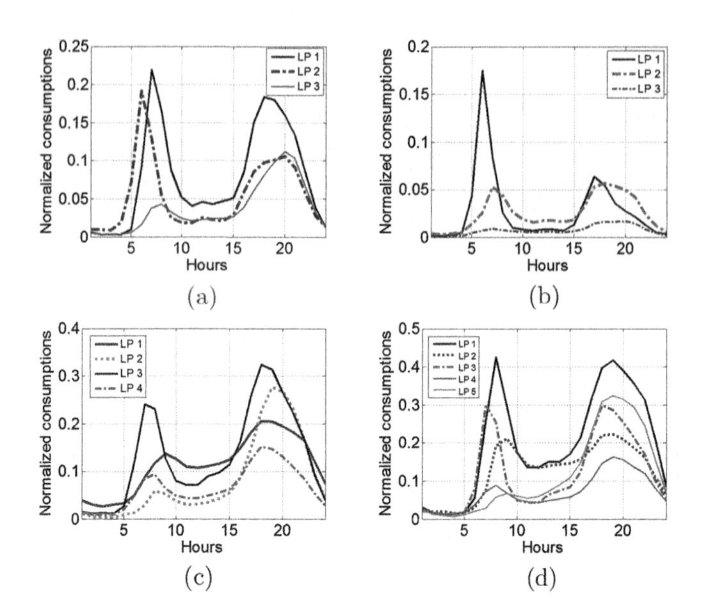

Fig. 5. Seasonal representative profiles for: (a) spring; (b) summer; (c) autumn; (d) winter.

Table 2. Percentage of consumers for the clusters (C1-C5) of each season and respective algorithms.

Season	C1	C2	C3	C4	C5	Algorithm
Spring	13 %	21 %	66 %	-	-	HC
Summer	9 %	22 %	69 %	-	-	FCM
Autumn	15 %	13 %	15 %	57 %	-	HC
Winter	8 %	20 %	19 %	36 %	17 %	KM

- **Summer**: the majority of consumers (69 %) is represented by LP 3, which corresponds to those who have approximately equal morning and evening peak consumption, as well as low off-peak consumption at night and daytime;
- **Autumn**: the majority of consumers (57 %) is represented by LP 4, which corresponds to those who have a low morning peak consumption and a slightly higher one in the evening. All LPs have a higher evening peak consumption than the morning one, as well as a low off-peak consumption during night, except for LP 1 which has the highest during daytime and night;
- **Winter**: the majority of consumers (36 %) is represented by LP 4, which corresponds to those who have a low morning peak consumption and the lowest in the evening, as well as the lowest daytime off-peak consumption. All LPs have a low off-peak consumption during the night until morning, compared to the consumption that they present throughout the day. This is the season which presents the more balanced clusters, in terms of size.

5 Conclusions

In this paper, residential natural gas consumers' representative profiles were defined, using high frequency gas consumption data from a smart metering trial conducted in Ireland. Consumption data was hourly aggregated by summation, for each season. Hence, each consumer had a 24 h seasonal profile. The final consumer representative seasonal profile, also referred as load profile (LP), consisted on the mean normalized consumption of the cluster to which the consumer belonged, and it was represented by the cluster center.

Different clustering algorithms were used, as well as CVIs to assess clustering performance. As result, 3 LPs were obtained using HC and FCM algorithms, for spring and summer, respectively, 4 LPs using HC for autumn, and 5 LPs using KM for winter, in a total of 15 LPs. While crisp algorithms were selected for three seasons, the FCM was selected only for summer. In this case, the other algorithms presented clusters with reduced size, which were not considered representative of a population group. Thus, a fuzzy algorithm, that allows the assignment of fuzzy membership degrees of the consumers to the clusters, may be suited for this type of situation.

The CVIs which best indicated a n_c corresponding to compact and well defined clusters, were DB, WI and XB. Other algorithms, CVIs and similarity distances can be applied to this data in order to achieve better results, with higher CVIs scores. Respectively, for all the cases and for most of them, DI and Sil consistently indicated 2 clusters. For all seasons, the two representative profiles obtained consisted of a separation between high and low consumption consumers, where the majority of the consumers' profiles were significantly different from the representative ones. Therefore, for a more interesting analysis of the consumers' consumption patterns, higher number of clusters had to be obtained.

The knowledge obtained with this methodology can assist decision makers in the energy utility industry in order to develop demand side management programs, consumer engagement strategies, marketing, demand forecasting tools as well as in designing more innovative and sophisticated tariff systems.

Acknowledgements. This work is supported by the Portuguese Government under the program SusCity, through FCT/MEC (under the Unit IDMEC - Pole IST, Research Group IDMEC/CSI), project MITP-TB/CS/0026/2013. The work of J. L. Viegas was supported by the PhD in Industry Scholarship SFRH/BDE/95414/2013 from FCT and Novabase. S. Vieira acknowledges support by Program Investigador FCT (IF/00833/2014) from FCT, co-funded by the European Social Fund (ESF) through the Operational Program Human Potential (POPH).

References

1. Azad, S., Wolfs, P., et al.: Identification of typical load profiles using k-means clustering algorithm. In: 2014 Asia-Pacific World Congress on Computer Science and Engineering (APWC on CSE), pp. 1–6. IEEE (2014)

2. Deepak Sharma, D., Singh, S.: Electrical load profile analysis and peak load assessment using clustering technique. In: PES General Meeting—Conference & Exposition, 2014 IEEE, pp. 1–5. IEEE (2014)
3. Kim, Y.I., Ko, J.M., Choi, S.H.: Methods for generating TLPs (typical load profiles) for smart grid-based energy programs. In: 2011 IEEE Symposium on Computational Intelligence Applications in Smart Grid (CIASG), pp. 1–6. IEEE (2011)
4. Bidoki, S., Mahmoudi-Kohan, N., Gerami, S.: Comparison of several clustering methods in the case of electrical load curves classification. In: 2011 16th Conference on Electrical Power Distribution Networks (EPDC), pp. 1–7. IEEE (2011)
5. Sathiracheewin, S., Surapatana, V.: Daily typical load clustering of residential customers. In: 2011 8th International Conference on Electrical Engineering/Electronics, Computer, Telecommunications and Information Technology (ECTI-CON), pp. 797–800. IEEE (2011)
6. Hossain, M.J., Kabir, A., Rahman, M.M., Kabir, B., Islam, M.R.: Determination of typical load profile of consumers using fuzzy c-means clustering algorithm. International Journal of Soft Computing and Engineering (IJSCE), ISSN-2231-2307(2011)
7. Lo, K., Zakaria, Z., Sohod, M.: Determination of consumers load profiles based on two-stage fuzzy c-means. In: Proceedings of the 5th WSEAS International Conference on Power Systems and Electromagnetic Compatibility, Corfu, Greece, pp. 212–217 (2005)
8. Zanotti, G., Gabbi, G., Laboratore, D.: Climate variables and weather derivatives: gas demand, temperature and seasonality effects in the italian case. Temperature and Seasonality Effects in the Italian Case (14/01/2003) (2003)
9. Dirks, J.A., Gorrissen, W.J., Hathaway, J.H., Skorski, D.C., Scott, M.J., Pulsipher, T.C., Huang, M., Liu, Y., Rice, J.S.: Impacts of climate change on energy consumption and peak demand in buildings: A detailed regional approach. Energy **79**, 20–32 (2015)
10. Yang, S.I., Shen, C., et al.: A review of electric load classification in smart grid environment. Renew. Sustain. Energy Rev. **24**, 103–110 (2013)
11. Viegas, J.L., Vieira, S.M., Sousa, J.M.C.: Fuzzy clustering and prediction of electricity demand based on household characteristics. In: Proceedings of the16th World Congress of the International Fuzzy Systems Association (IFSA) and the 9th Conference of the European Society for Fuzzy Logic and Technology (EUSFLAT) (2015)
12. Pkekalska, E., Duin, R.P.: The Dissimilarity Representation for Pattern Recognition: Foundations and Applications, vol. 64. World Scientific, River Edge (2005)
13. Brabec, M., Konár, O., Malỳ, M., Pelikán, E., Vondráček, J.: A statistical model for natural gas standardized load profiles. J. R. Stat. Soc. Ser. C (Appl. Stat.) **58**(1), 123–139 (2009)
14. CER: Smart metering information paper: gas customer behaviour trial findings report, Commission for Energy Regulation (CER). Technical report (2011)
15. de Oliveira, J.V., Pedrycz, W., et al.: Advances in Fuzzy Clustering and its Applications. Wiley, Chichester (2007)
16. Ward Jr., J.H.: Hierarchical grouping to optimize an objective function. J. Am. Stat. Assoc. **58**(301), 236–244 (1963)
17. ISSDA: Data from the Commission for Energy Regulation (CER). www.ucd.ie/issda

Author Index

Adam, Lukáš I-35
Agli, Hamza I-326
Aguiló, Isabel I-375
Akaichi, Jalel I-549
Alcalde, Cristina II-12
Almeida, Rui Jorge II-729, II-752
Almonnieay, Anas I-475
Amagasa, Michihiro II-213
Anderson, Derek T. I-78
Asiain, Maria Jose I-353
Asmuss, Svetlana I-648
Aycard, Olivier I-286
Aymerich, Francesc Xavier I-612
Azvine, Ben I-525
Azzoune, Hamid II-341

Bäck, Thomas II-187
Baczyński, Michał I-353, I-411
Bahri, Oumayma II-225
Baldi, Paolo II-136
Ballini, Rosangela I-687
Barrenechea, Edurne I-624, II-435
Bartoszuk, Maciej II-767, II-780
Baştürk, Nalan II-752
Batista, Fernando II-690
Baumard, Philippe I-338
Bedregal, Benjamín I-353, II-809
Beer, Frank II-393
Běhounek, Libor II-482
Beliakov, Gleb II-411, II-767, II-780
Belkasmi, Djamal II-341
Ben Amor, Nahla II-225
Boer, Arjen-Kars I-462
Bonnard, Philippe I-326
Bou, Félix II-123
Boudet, Laurence II-717
Boukezzoula, Reda I-262
Boukhris, Imen I-274, II-291
Boussarsar, Oumaima II-291
Bronselaer, Antoon II-305, II-317, II-367
Buche, Patrice I-498
Bühler, Ulrich II-393
Burusco, Ana II-12
Bustince, Humberto I-353, I-599, I-624, II-435

Cabrer, Leonardo II-108
Caldari, Letizia II-569
Cao, Nhung II-470
Cardin, Marta I-117
Cariou, Stéphane I-238
Carvalho, Joao Paulo II-690
Cena, Anna II-445
Cerron, Juan I-599
Cetintav, Bekir II-790
Charnomordic, Brigitte I-498
Chen, Jesse Xi II-667
Cinicioglu, Esma Nur I-313
Cintula, Petr II-95
Cobreros, Pablo II-161
Coletti, Giulianella II-569
Conrad, Stefan I-699
Coolen, Frank P.A. I-153, I-165
Coolen-Maturi, Tahani I-165
Coquin, Didier I-262
Cornejo, M. Eugenia II-69
Cornelis, Chris II-23
Couso, Inés II-279, II-595
Couturier, Pierre II-423
Crockett, Keeley II-656
Croitoru, Madalina I-498
Cruz, Carlos II-559
Cuzzocrea, Alfredo II-379

D'eer, Lynn II-23
Daňková, Martina II-482
Dao, Tien-Tuan I-510, II-253
De Miguel, Laura II-435
De Mol, Robin II-305, II-317
De Tré, Guy II-305, II-317, II-367
Demirel, Neslihan II-790
Denœux, Thierry I-253
Destercke, Sébastien II-619
Detyniecki, Marcin I-226

Dijkman, Remco M. II-729
Dimuro, Graçaliz I-353
Dubois, Didier I-363, II-279
Dvořák, Antonín II-495

Egré, Paul II-161
Eklund, Patrik I-437, II-61
El Kirat, Karim I-510
Elkano, Mikel II-435
Eļkins, Aleksandrs II-48
Elmi, Sayda II-199
Elouedi, Zied I-274, II-291, II-643
Esteva, Francesc II-123
Ezzati, Reza II-821

Fanlo, Jean-Louis I-238
Feng, Geng I-129
Fernandes, Marta P. I-723
Fernandez, Javier I-599
Fiorini, Nicolas I-238
Fortin, Jerome I-498
Franzoi, Laura II-3

Gagolewski, Marek II-445, II-767, II-780
Gajewski, Marek I-214
Galar, Mikel I-599
Garbay, Catherine I-286
Gasir, Fathi II-656
Gatt, Albert I-191
Gaudin, Théophile I-510
Geschke, Andrew II-411
Godo, Lluís II-123
Gómez, Daniel I-635
Gomide, Fernando I-687
Gonçalves, Rodrigo M.T. II-729
Gonzales, Christophe I-326, I-338
González-Hidalgo, Manuel I-660
Grabisch, Michel I-11
Grzegorzewski, Przemysław I-423
Guada, Carely I-635
Guillon, Arthur I-675
Gurler, Selma II-790

Hadjali, Allel II-199, II-253, II-341
Harispe, Sébastien I-238
Harmati, István Á. II-798
Havens, Timothy C. I-78
Held, Pascal II-678
Held, Susanne I-450
Himmelspach, Ludmila I-699
Ho, Chun-Hsing I-475
Hodáková, Petra I-577
Höhle, Ulrich I-437
Holčapek, Michal II-459, II-510, II-705
Honda, Aoi I-65
Hourbracq, Matthieu I-338
Hüllermeier, Eyke I-450
Hurtik, Petr I-577

Imoussaten, Abdelhak II-423

Jain, Akshay I-204
James, Simon II-411, II-767, II-780
Jiang, Tianqi I-204
Jones, Hazael I-498
Jurio, Aranzazu II-435

Kacprzyk, Janusz I-214
Kalina, Martin II-522
Kaymak, Uzay I-462
Keller, James M. I-78, I-204
Kóczy, László T. II-798
Kokainis, Martins I-648
Kortelainen, Jari I-437
Koshiyama, Adriano I-486
Kovanic, Pavel I-177
Kowalczyk, Wojtek II-175, II-187
Král, Pavol II-522
Kreinovich, Vladik II-459
Kroupa, Tomáš I-35
Kruse, Rudolf II-678

Labourey, Quentin I-286
Labreuche, Christophe I-23
Lai, Chieh-Ping I-475
Lefevre, Eric I-274, II-643
Lefort, Sébastien I-226
Legastelois, Bénédicte II-148
Lesot, Marie-Jeanne I-226, I-675, II-148
Leung, Carson K. II-379
Lian, Chunfeng I-253
Lizasoain, Inmaculada I-624
Lopez-Molina, Carlos I-599
Lu, Huiling I-510

Maciel, Leandro I-687
Mallek, Sabrine I-274

Marco-Detchart, Cedric I-599
Marín, Nicolás I-191
Marsala, Christophe I-675
Martin, Trevor P. I-525
Martínez-García, Miriam II-238
Mas, Margalida I-387
Massanet, Sebastia I-399, I-660, II-581
Mayag, Brice I-101
Medina, Jesús II-69
Medjkoune, Massissilia I-238
Mesiar, Radko I-3, I-58, I-624
Mezei, József I-711
Mir, Arnau I-660
Miranda, Enrique I-141
Miranda, Pedro I-11
Molina, Carlos I-537
Monserrat, Miquel I-387
Montero, Javier I-635
Montes, Ignacio I-141
Montmain, Jacky I-238, II-423
Montseny, Eduard I-612
Moreau, Aurélien II-329
Moreno, Antonio II-238
Moussaoui, Mohamed I-549
Murinová, Petra II-36

Nagata, Kiyoshi II-213
Nakata, Michinori II-355
Neiger, Reto I-450
Neveu, Pascal I-498
Nguyen, Linh II-705
Nguyen, Thanh-Long I-262
Nielandt, Joachim II-317
Niemyska, Wanda I-411
Nimmo, Dale II-411
Noguera, Carles II-95
Novák, Vilém II-36

Ochoa, Gustavo I-624
Okazaki, Yoshiaki I-65

Pagola, Miguel II-435
Pal, Nikhil R. I-675
Palmeira, Eduardo II-809
Paredes, Jorge I-486
Patelli, Edoardo I-129
Paternain, Daniel I-353, I-624
Peláez-Moreno, Carmen II-81
Pellerin, Denis I-286
Perfilieva, Irina I-577, I-588, II-459
Petrík, Milan II-532
Petturiti, Davide II-569
Pezron, Isabelle I-510
Pfannschmidt, Karlson I-450
Pivert, Olivier II-329
Poli, Jean-Philippe II-717
Portet, François I-191
Pourceau, Gwladys I-510
Prade, Henri I-363, II-605
Prados-Suárez, Belen I-537

Raddaoui, Badran II-253
Ramasso, Emmanuel I-299
Ramírez-Poussa, Eloísa II-69
Reformat, Marek Z. II-667
Revault d'Allonnes, Adrien II-148
Ribeiro, Bernardete II-633
Richard, Gilles II-605
Rico, Agnès I-363
Riera, Juan Vicente I-399, II-581
Ripley, David II-161
Rivieccio, Umberto II-108
Rodríguez, J. Tinguaro I-635
Rodriguez, Ricardo Oscar II-108, II-123
Rogger, Jonas II-95
Rombaut, Michèle I-286
Rosete-Suárez, Alejandro II-559
Rovira, Alex I-612
Ruan, Su I-253
Ruiz-Aguilera, Daniel I-387, I-660
Runkler, Thomas A. II-547

Sakai, Hiroshi II-355
Salgado, Cátia M. II-741
Sallak, Mohamed II-619
Samet, Ahmed I-510, II-253
Sánchez, Daniel I-191, I-537
Sanz, Jose II-435
Sarlin, Peter I-711
Saulnier, Hugo II-595
Scharnhorst, Volkher I-462
Schon, Walter II-619
Sedki, Karima II-265
Sgarro, Andrea II-3
Silva, Catarina II-633
Šipeky, Ladislav I-58
Šipošová, Alexandra I-58
Smits, Grégory II-329
Sobrevilla, Pilar I-612

Šostak, Alexander II-48
Sousa, João M.C. I-562, I-723, II-729, II-741
Sow, Diadie II-423
Štěpnička, Martin II-470, II-495
Strauss, Olivier II-595
Stupňanová, Andrea I-3
Suñer, Jaume I-375
Symeonidou, Danai I-498

Takáč, Zdenko I-353
Takahagi, Eiichiro I-91
Talbi, El-Ghazali II-225
Tanscheit, Ricardo I-486
Teheux, Bruno I-363
Thomopoulos, Rallou I-498
Tijus, Charles I-226
Timonin, Mikhail I-46
Tobji, Mohamed Anis Bach II-199
Tomlin, Leary I-78
Torrens, Joan I-375, I-387, I-399, II-581
Torres-Pérez, Isis II-559
Trabelsi, Asma II-643

Uljane, Ingrīda II-48
Ulutagay, Gozde II-790

Valls, Aida II-238
Valverde-Albacete, Francisco J. II-81
Van Hecke, Elisabeth I-510
van Loon, Saskia I-462
van Rooij, Robert II-161
van Stein, Bas II-175, II-187
Vantaggi, Barbara II-569
Vellasco, Marley I-486
Verdegay, José Luis II-559
Vetterlein, Thomas II-532
Vicente, Marco II-690
Viegas, Joaquim L. I-562, I-723
Vieira, Susana M. I-562, I-723, II-741
Vlašánek, Pavel I-588

Wadouachi, Anne I-510
Wagner, Christian I-78
Walter, Gero I-153
Wilbik, Anna I-462
Wuillemin, Pierre-Henri I-326, I-338

Yager, Ronald R. I-687, II-667
Yaghlane, Boutheina Ben II-199
Yahi, Safa II-265
Yáñez, Javier I-635
Yao, Yiyu II-23
Yenilmez, Taylan I-313
Yin, Yi-Chao I-165
Yu, Lanting II-619

Zadrożny, Sławomir I-214
Zaghdoud, Montaceur I-549
Ziari, Shokrollah II-821
Zibetti, Elisabetta I-226

Printed in the United States
By Bookmasters

Printed in the United States
By Bookmasters